PLANT TRANSCRIPTION FACTORS

PLANT TRANSCRIPTION FACTORS

EVOLUTIONARY, STRUCTURAL, AND FUNCTIONAL ASPECTS

Edited by

DANIEL H. GONZALEZ
Universidad Nacional del Litoral, Santa Fe, Argentina

AMSTERDAM • BOSTON • HEIDELBERG • LONDON
NEW YORK • OXFORD • PARIS • SAN DIEGO
SAN FRANCISCO • SINGAPORE • SYDNEY • TOKYO

Academic Press is an Imprint of Elsevier

Academic Press is an imprint of Elsevier
125, London Wall, EC2Y 5AS, UK
525 B Street, Suite 1800, San Diego, CA 92101-4495, USA
225 Wyman Street, Waltham, MA 02451, USA
The Boulevard, Langford Lane, Kidlington, Oxford OX5 1GB, UK

Copyright © 2016 Elsevier Inc. All rights reserved.

No part of this publication may be reproduced or transmitted in any form or by any means, electronic or mechanical, including photocopying, recording, or any information storage and retrieval system, without permission in writing from the publisher. Details on how to seek permission, further information about the Publisher's permissions policies and our arrangements with organizations such as the Copyright Clearance Center and the Copyright Licensing Agency, can be found at our website: www.elsevier.com/permissions.

This book and the individual contributions contained in it are protected under copyright by the Publisher (other than as may be noted herein).

Notices
Knowledge and best practice in this field are constantly changing. As new research and experience broaden our understanding, changes in research methods, professional practices, or medical treatment may become necessary.

Practitioners and researchers must always rely on their own experience and knowledge in evaluating and using any information, methods, compounds, or experiments described herein. In using such information or methods they should be mindful of their own safety and the safety of others, including parties for whom they have a professional responsibility.

To the fullest extent of the law, neither the Publisher nor the authors, contributors, or editors, assume any liability for any injury and/or damage to persons or property as a matter of products liability, negligence or otherwise, or from any use or operation of any methods, products, instructions, or ideas contained in the material herein.

British Library Cataloguing-in-Publication Data
A catalogue record for this book is available from the British Library

Library of Congress Cataloging-in-Publication Data
A catalog record for this book is available from the Library of Congress

ISBN: 978-0-12-800854-6

For information on all Academic Press publications
visit our website at http://store.elsevier.com/

Typeset by Thomson Digital
Printed and bound in the United States of America

Contents

List of Contributors	ix
Preface	xi

A

GENERAL ASPECTS OF PLANT TRANSCRIPTION FACTORS

1. Introduction to Transcription Factor Structure and Function
DANIEL H. GONZALEZ

1.1 Introduction: Transcription in Eukaryotes	3
1.2 Structure of Transcription Factors	4
1.3 DNA Recognition by Transcription Factors	4
1.4 DNA-Binding Domains	5
1.5 Protein–Protein Interactions	6
1.6 Regulation of Transcription Factor Action	7
1.7 Plant Transcription Factors	9
References	9

2. Methods to Study Transcription Factor Structure and Function
IVANA L. VIOLA, DANIEL H. GONZALEZ

2.1 Introduction	13
2.2 *In Vivo* Functional Studies	14
2.3 Methods for the Analysis of *In Vitro* Protein–DNA Interactions	18
2.4 Methods to Study Protein–DNA Interactions *In Vivo*	21
2.5 Analysis of Protein–Protein Interactions	25
References	29

3. General Aspects of Plant Transcription Factor Families
JONG CHAN HONG

3.1 Introduction	35
3.2 Overview of the Transcription Cycle in Eukaryotes	36
3.3 Components Involved in the Formation of the RNAPII Preinitiation Complex in Plants	39
3.4 Plant Transcription Factor Families	44
3.5 Major TF Families that are Conserved Across Eukaryotes	44
3.6 Plant-Specific TF Families	49
3.7 TFs without DBD but Interacting with DBD-Containing TFs	50
3.8 Conclusion	52
References	52

4. Structures, Functions, and Evolutionary Histories of DNA-Binding Domains of Plant-Specific Transcription Factors
KAZUHIKO YAMASAKI

4.1 Introduction	57
4.2 Description of Respective DBDs	59
4.3 Evolutionary History of Plant-Specific TFs	65
References	69

5. The Evolutionary Diversification of Genes that Encode Transcription Factor Proteins in Plants
TOSHIFUMI NAGATA, AENI HOSAKA-SASAKI, SHOSHI KIKUCHI

5.1 Introduction – Distinctive Features of TF Genes in Plants (*Arabidopsis* and Rice)	73
5.2 A Comparative Analysis of TF Genes between Plants and Animals	76
5.3 A Comparative Analysis of Transcription Factor Genes in 32 Diverse Organisms	76
5.4 The Appearance of New TF Gene Members During Evolution	92
5.5 The Different Evolutionary Methods of TF Genes in Animals and Plants	93
5.6 TF Gene Evolution and its Biological Function	94
5.7 Conclusion: the Regulatory Role of Individual Transcription Factors	96
References	96

B

EVOLUTION AND STRUCTURE OF DEFINED PLANT TRANSCRIPTION FACTOR FAMILIES

6. Structure and Evolution of Plant Homeobox Genes
IVANA L. VIOLA, DANIEL H. GONZALEZ

6.1 Introduction	101
6.2 Structure of the Homeodomain	102
6.3 Specific Contacts with DNA	102
6.4 Plant Homeodomain Families	104
6.5 The Evolution of Plant Homeobox Genes	108
References	110

7. Homeodomain–Leucine Zipper Transcription Factors: Structural Features of These Proteins, Unique to Plants

MATÍAS CAPELLA, PAMELA A. RIBONE, AGUSTÍN L. ARCE, RAQUEL L. CHAN

7.1 Homeoboxes and Homeodomains in Eukaryotic Kingdoms	114
7.2 Plant Homeoboxes	114
7.3 The Plant Homeodomain Superfamily	114
7.4 Different Domains Present in Homeodomain Transcription Factors	115
7.5 The HD-Zip Family	117
7.6 Target Sequences Recognized by the HD-Containing Transcription Factors	122
7.7 What do we Know about the Target Sequences of the HD-Zip Proteins?	123
7.8 Concluding Remarks	124
References	124

8. Structure and Evolution of Plant MADS Domain Transcription Factors

GÜNTER THEIßEN, LYDIA GRAMZOW

8.1 Introduction: Who Cares about MADS Domain Transcription Factors?	127
8.2 The Structure of MADS Domain Proteins	128
8.3 Evolution of MADS Domain Transcription Factors	132
8.4 Concluding Remarks	136
References	137

9. TCP Transcription Factors: Evolution, Structure, and Biochemical Function

EDUARDO GONZÁLEZ-GRANDÍO, PILAR CUBAS

9.1 Introduction	139
9.2 Evolution of TCP Proteins	139
9.3 The TCP Domain: Structure and Function	142
9.4 Activation and Repression Domains	147
9.5 TCP Factors as Intrinsically Disordered Proteins	148
9.6 Posttranslational Modifications of TCP	148
9.7 Concluding Remarks	149
References	149

10. Structure and Evolution of Plant GRAS Family Proteins

CORDELIA BOLLE

10.1 Presence of GRAS Proteins in Plants and other Organisms	153
10.2 Genomic Organization (Intron/Exon)	156
10.3 Structure of GRAS Proteins	157
10.4 Conclusion	160
References	160

11. Structure and Evolution of WRKY Transcription Factors

CHARLES I. RINERSON, ROEL C. RABARA, PRATEEK TRIPATHI, QINGXI J. SHEN, PAUL J. RUSHTON

11.1 Introduction	163
11.2 The Structure of the WRKY Domain	164
11.3 The Evolution of WRKY Genes	164
11.4 R Protein–WRKY Genes	173
11.5 Conclusion: a Reevaluation of WRKY Evolution	177
References	179

12. Structure, Function, and Evolution of the Dof Transcription Factor Family

SHUICHI YANAGISAWA

12.1 Discovery and Definition of the Dof Transcription Factor Family	183
12.2 Structure and Molecular Characteristics of Dof Transcription Factors	184
12.3 Molecular Evolution of the Dof Transcription Factor Family	188
12.4 Physiological Functions of Dof Transcription Factors	190
12.5 Perspective	194
References	194

13. NAC Transcription Factors: From Structure to Function in Stress-Associated Networks

DITTE H. WELNER, FARAH DEEBA, LEILA LO LEGGIO, KAREN SKRIVER

13.1 Introduction	199
13.2 NAC Structure	200
13.3 Evolution of NAC Proteins	202
13.4 NAC Proteins: from Structure to Interactions with DNA and other Proteins	204
13.5 NAC Networks in Abiotic Stress Responses	207
13.6 Conclusion	209
References	210

C

FUNCTIONAL ASPECTS OF PLANT TRANSCRIPTION FACTOR ACTION

14. Homeobox Transcription Factors and the Regulation of Meristem Development and Maintenance

KATSUTOSHI TSUDA, SARAH HAKE

14.1 Introduction	215
14.2 *KNOX* and *BELL*: TALE Superfamily Homeobox Genes	216
References	225

15. CUC Transcription Factors: To the Meristem and Beyond
AUDE MAUGARNY, BEATRIZ GONÇALVES, NICOLAS ARNAUD, PATRICK LAUFS

15.1 Introduction	230
15.2 Evolution and Structure of NAM/CUC3 Proteins	230
15.3 NAM/CUC3 Genes Define Boundaries in Meristems and Beyond	234
15.4 Multiple Regulatory Pathways Contribute to the Fine Regulation of NAM/CUC3 Genes	236
15.5 NAM/CUC3 Control Plant Development Via Modifications of the Cellular Behavior	242
15.6 Conclusion	243
References	244

16. The Role of TCP Transcription Factors in Shaping Flower Structure, Leaf Morphology, and Plant Architecture
MICHAEL NICOLAS, PILAR CUBAS

16.1 Introduction	250
16.2 TCP Genes and the Control of Leaf Development	250
16.3 TCP Genes and the Control of Shoot Branching	256
16.4 TCP Genes and the Control of Flower Shape	259
16.5 TCP Genes Affect Flowering Time	262
16.6 Concluding Remarks	262
References	263

17. Growth-Regulating Factors, A Transcription Factor Family Regulating More than Just Plant Growth
RAMIRO E. RODRIGUEZ, MARÍA FLORENCIA ERCOLI, JUAN MANUEL DEBERNARDI, JAVIER F. PALATNIK

17.1 Growth-Regulating Factors, a Plant-Specific Family of Transcription Factors	269
17.2 Control of GRF Activity	272
17.3 Role of GRFs in Organ Growth and other Developmental Processes	274
17.4 Conclusion and Perspectives	277
References	278

18. The Multifaceted Roles of miR156-targeted SPL Transcription Factors in Plant Developmental Transitions
JIA-WEI WANG

18.1 Introduction to Developmental Transitions	281
18.2 miR156 and its Targets	282
18.3 miR156-SPL Module in Timing Embryonic Development	283
18.4 miR156-SPL Module in Juvenile-to-Adult Phase Transition in Higher Plants	283
18.5 The miR156-SPL Module Regulates Flowering Time in Higher Plants	285
18.6 The miR156-SPL Module in Developmental Transitions in Moss	286
18.7 The miR156-SPL Module in other Developmental Processes	287
18.8 Perspectives	290
References	291

19. Functional Aspects of GRAS Family Proteins
CORDELIA BOLLE

19.1 The Role of GRAS Proteins in Development	296
19.2 The Role of GRAS Proteins in Signaling	301
19.3 General Principles of GRAS Function	303
19.4 Conclusion	307
References	307

20. DELLA Proteins, a Group of GRAS Transcription Regulators that Mediate Gibberellin Signaling
FRANCISCO VERA-SIRERA, MARIA DOLORES GOMEZ, MIGUEL A. PEREZ-AMADOR

20.1 About DELLAs and Gibberellins	313
20.2 GA Signaling through DELLAs	317
20.3 The Molecular Mechanism of DELLA Action: DELLA–Protein Interactions and Target Genes	319
20.4 Conclusion and Future Perspectives	324
References	324

21. bZIP and bHLH Family Members Integrate Transcriptional Responses to Light
MARÇAL GALLEMÍ, JAIME F. MARTÍNEZ-GARCÍA

21.1 The Role of Light in the Control of Plant Development: A Brief Introduction	329
21.2 PIFs: Factors that Link Light Perception, Changes in Gene Expression, and Plant Development	333
21.3 HFR1 and PAR1: Atypical bHLH Factors that Act as Transcriptional Cofactors	336
21.4 HY5: a Paradigm of a bZIP Member in Integrating Light Responses	337
21.5 Conclusions	339
References	339

22. What Do We Know about Homeodomain–Leucine Zipper I Transcription Factors? Functional and Biotechnological Considerations
PAMELA A. RIBONE, MATÍAS CAPELLA, AGUSTÍN L. ARCE, RAQUEL L. CHAN

22.1 HD-Zip Transcription Factors are Unique to Plants	344
22.2 Brief History of the Discovery of HD-Zip Transcription Factors	344
22.3 Expression Patterns of HD-Zip I Genes	344
22.4 Environmental Factors Regulate the Expression of HD-Zip I Encoding Genes	346
22.5 The Function of HD-Zip I TFs from Model Plants	346
22.6 HD-Zip I TFs from Nonmodel Species	349
22.7 Divergent HD-Zip I Proteins from Nonmodel Plants	351
22.8 Knowledge Acquired from Ectopic Expressors	352
22.9 HD-Zip I TFs in Biotechnology	352
22.10 Concluding Remarks	353
References	354

D

MODULATION OF PLANT TRANSCRIPTION FACTOR ACTION

23. Intercellular Movement of Plant Transcription Factors, Coregulators, and Their mRNAs
DAVID J. HANNAPEL

23.1 Introduction to Noncell-Autonomous Mobile Signals	359
23.2 Mobile Transcription Factors of the Shoot Apex in Protein Form	360
23.3 Mobile Root Transcription Factors	360
23.4 Transcription Factors and Coregulators that Move Long Distance through the Sieve Element System	361
23.5 Full-Length Mobile mRNAs and their Roles in Development	363
23.6 Conclusions	369
References	369

24. Redox-Regulated Plant Transcription Factors
YUAN LI, GARY J. LOAKE

24.1 Introduction	373
24.2 Concept of Redox Regulation	374
24.3 Redox Regulation of NPR1 During Plant Immunity	377
24.4 Redox Regulation of Basic Leucine Zipper Transcription Factors	378
24.5 Redox Regulation of MYB Transcription Factors	379
24.6 Redox Regulation of Homeodomain-Leucine Zipper Transcription Factors	380
24.7 Rap2.4a is Under Redox Regulation	381
24.8 Redox Regulation of Class I TCP Transcription Factors	381
24.9 Conclusion	382
References	382

25. Membrane-Bound Transcription Factors in Plants: Physiological Roles and Mechanisms of Action
YUJI IWATA, NOZOMU KOIZUMI

25.1 Introduction	385
25.2 bZIP Transcription Factors	386
25.3 NAC Transcription Factors	389
25.4 Conclusions and Future Perspectives	391
References	393

26. Ubiquitination of Plant Transcription Factors
SOPHIA L. STONE

26.1 The Ubiquitin Proteasome System	396
26.2 The Ubiquitin Proteasome System and Regulation of Transcription Factor Function	400
References	405

Index 411

List of Contributors

Agustín L. Arce   Instituto de Agrobiotecnología del Litoral, Universidad Nacional del Litoral, CONICET, Santa Fe, Argentina

Nicolas Arnaud   INRA, UMR1318, AgroParisTech, Institut Jean-Pierre Bourgin, Versailles, France

Cordelia Bolle   Department Biologie I, Lehrstuhl für Molekularbiologie der Pflanzen (Botanik), Biozentrum der LMU München, Planegg-Martinsried, Germany

Matías Capella   Instituto de Agrobiotecnología del Litoral, Universidad Nacional del Litoral, CONICET, Santa Fe, Argentina

Raquel L. Chan   Instituto de Agrobiotecnología del Litoral, Universidad Nacional del Litoral, CONICET, Santa Fe, Argentina

Pilar Cubas   Department of Plant Molecular Genetics, Centro Nacional de Biotecnología (CNB-CSIC), Madrid, Spain

Juan Manuel Debernardi   Instituto de Biología Molecular y Celular de Rosario (IBR), CONICET, Facultad de Ciencias Bioquímicas y Farmacéuticas, Universidad Nacional de Rosario, Rosario, Argentina

Farah Deeba   Department of Biology, University of Copenhagen, Copenhagen, Denmark; Department of Biochemistry, PMAS Arid Agriculture University Rawalpindi, Rawalpindi, Pakistan

María Florencia Ercoli   Instituto de Biología Molecular y Celular de Rosario (IBR), CONICET, Facultad de Ciencias Bioquímicas y Farmacéuticas, Universidad Nacional de Rosario, Rosario, Argentina

Marçal Gallemí   Centre for Research in Agricultural Genomics (CRAG), Consortium CSIC-IRTA-UAB-UB, Barcelona, Spain

Maria Dolores Gomez   Instituto de Biología Molecular y Celular de Plantas (IBMCP), Universidad Politécnica de Valencia-Consejo Superior de Investigaciones Científicas, Ciudad Politécnica de la Innovación (CPI), Valencia, Spain

Beatriz Gonçalves   INRA, UMR1318, AgroParisTech, Institut Jean-Pierre Bourgin, Versailles, France

Daniel H. Gonzalez   Instituto de Agrobiotecnología del Litoral (CONICET-UNL), Cátedra de Biología Celular y Molecular, Facultad de Bioquímica y Ciencias Biológicas, Universidad Nacional del Litoral, Santa Fe, Argentina

Eduardo González-Grandío   Department of Plant Molecular Genetics, Centro Nacional de Biotecnología (CNB-CSIC), Madrid, Spain

Lydia Gramzow   Department of Genetics, Friedrich Schiller University Jena, Jena, Germany

Sarah Hake   U.S. Department of Agriculture-Agricultural Research Service, Plant and Microbial Biology Department, Plant Gene Expression Center, University of California at Berkeley, Berkeley, CA, USA

David J. Hannapel   Plant Biology Major, Iowa State University, Ames, Iowa, USA

Jong Chan Hong   Division of Life Science, Applied Life Science (BK21 Plus Program), Plant Molecular Biology and Biotechnology Research Center, Gyeongsang National University, Jinju, Korea; Division of Plant Sciences, University of Missouri, Columbia, MO, USA

Aeni Hosaka-Sasaki   Plant Genome Research Unit, Agrogenomics Research Center, National Institute of Agrobiological Sciences (NIAS), Tsukuba, Ibaraki, Japan

Yuji Iwata   Graduate School of Life and Environmental Sciences, Osaka Prefecture University, Nakaku, Sakai, Osaka, Japan

Shoshi Kikuchi   Plant Genome Research Unit, Agrogenomics Research Center, National Institute of Agrobiological Sciences (NIAS), Tsukuba, Ibaraki, Japan

Nozomu Koizumi   Graduate School of Life and Environmental Sciences, Osaka Prefecture University, Nakaku, Sakai, Osaka, Japan

Patrick Laufs   INRA, UMR1318, AgroParisTech, Institut Jean-Pierre Bourgin, Versailles, France

Yuan Li   Institute of Molecular Plant Sciences, School of Biological Sciences, University of Edinburgh, King's Buildings, Edinburgh, UK

Gary J. Loake   Institute of Molecular Plant Sciences, School of Biological Sciences, University of Edinburgh, King's Buildings, Edinburgh, UK

Leila Lo Leggio   Department of Chemistry, University of Copenhagen, Copenhagen, Denmark

Jaime F. Martínez-García   Centre for Research in Agricultural Genomics (CRAG), Consortium CSIC-IRTA-UAB-UB; Institució Catalana de Recerca i Estudis Avançats (ICREA), Barcelona, Spain

Aude Maugarny   INRA, UMR1318, AgroParisTech, Institut Jean-Pierre Bourgin, Versailles, France

Toshifumi Nagata   Plant Genome Research Unit, Agrogenomics Research Center, National Institute of Agrobiological Sciences (NIAS), Tsukuba, Ibaraki, Japan

Michael Nicolas   Department of Plant Molecular Genetics, Centro Nacional de Biotecnología (CNB-CSIC), Madrid, Spain

Javier F. Palatnik   Instituto de Biología Molecular y Celular de Rosario (IBR), CONICET, Facultad de Ciencias Bioquímicas y Farmacéuticas, Universidad Nacional de Rosario, Rosario, Argentina

Miguel A. Perez-Amador Instituto de Biología Molecular y Celular de Plantas (IBMCP), Universidad Politécnica de Valencia-Consejo Superior de Investigaciones Científicas, Ciudad Politécnica de la Innovación (CPI), Valencia, Spain

Roel C. Rabara Texas A and M AgriLife Research and Extension Center, Dallas, TX, USA

Pamela A. Ribone Instituto de Agrobiotecnología del Litoral, Universidad Nacional del Litoral, CONICET, Santa Fe, Argentina

Charles I. Rinerson Texas A and M AgriLife Research and Extension Center, Dallas, TX, USA

Ramiro E. Rodriguez Instituto de Biología Molecular y Celular de Rosario (IBR), CONICET, Facultad de Ciencias Bioquímicas y Farmacéuticas, Universidad Nacional de Rosario, Rosario, Argentina

Paul J. Rushton Texas A and M AgriLife Research and Extension Center, Dallas, TX, USA

Qingxi J. Shen School of Life Sciences, University of Nevada, at Las Vegas, NV, USA

Karen Skriver Department of Biology, University of Copenhagen, Copenhagen, Denmark

Sophia L. Stone Department of Biology, Dalhousie University, Halifax NS, Canada

Günter Theißen Department of Genetics, Friedrich Schiller University Jena, Jena, Germany

Prateek Tripathi Molecular and Computational Biology Section, Dana and David Dornsife College of Letters, Arts and Sciences, University of Southern California, Los Angeles, CA, USA

Katsutoshi Tsuda U.S. Department of Agriculture-Agricultural Research Service, Plant and Microbial Biology Department, Plant Gene Expression Center, University of California at Berkeley, Berkeley, CA, USA

Francisco Vera-Sirera Instituto de Biología Molecular y Celular de Plantas (IBMCP), Universidad Politécnica de Valencia-Consejo Superior de Investigaciones Científicas, Ciudad Politécnica de la Innovación (CPI), Valencia, Spain

Ivana L. Viola Instituto de Agrobiotecnología del Litoral (CONICET-UNL), Cátedra de Biología Celular y Molecular, Facultad de Bioquímica y Ciencias Biológicas, Universidad Nacional del Litoral, Santa Fe, Argentina

Jia-Wei Wang National Key Laboratory of Plant Molecular Genetics, Institute of Plant Physiology and Ecology, Shanghai Institutes for Biological Sciences, Shanghai, China

Ditte H. Welner Joint BioEnergy Institute, Lawrence Berkeley National Laboratory, Berkeley, CA, USA

Kazuhiko Yamasaki Biomedical Research Institute, National Institute of Advanced Industrial Science and Technology (AIST), Tsukuba, Japan; RIKEN Quantitative Biology Center, Tsurumi-ku, Yokohama, Japan

Shuichi Yanagisawa Laboratory of Plant Functional Biotechnology, Biotechnology Research Center, The University of Tokyo, Yayoi 1-1-1, Bunkyo-ku, Tokyo, Japan

Preface

Transcription factors (TFs) are central regulators of gene expression and, as such, modulate essential aspects of organismal function, including cell differentiation, tissue and organ development, responses to hormones and environmental factors, metabolic networks, and disease resistance, among others. Intrinsic to TF action is their capacity to specifically interact with DNA sequences and with other proteins as part of transcriptional complexes involved in the regulation of gene expression. Accordingly, knowledge of the structure of TFs and of the molecular mechanisms involved in the establishment of these interactions is essential to understand TF action. Plants contain a vast number of TFs (about 10% of plant genes encode TFs) that were acquired at different stages of evolution and were adapted to perform specific functions.

This book is intended as a source of information for those interested in the study of plant TFs and the many processes they regulate. It contains basic information that can be useful to students and researchers entering the field as well as more specific chapters devoted to plant TF families. These specific chapters do not constitute a comprehensive list of what is known about the different TF families but are rather examples of how the study of the different aspects of a specific TF family can be useful to establish the molecular aspects of TF function. Section A deals with general aspects of plant TFs. It contains two introductory chapters that describe the basics of TF action and methods usually employed to study TFs, followed by chapters that discuss structural and evolutionary aspects of plant TF families and plant-specific TF DNA-binding domains. Sections B and C present information about the structure, evolution, and functional aspects of several plant TF families, with examples of families that arose at different stages of organismal evolution and were adapted to modulate specific aspects of plant developmental programs and responses. Finally, Section D contains chapters that discuss aspects of the posttranslational regulation of plant TF action by either intra- or intercellular movement, proteolytic processing, ubiquitination, or redox interconversions. The book is centered on TFs rather than on processes, understanding that there are excellent books that already describe plant regulatory networks and the TFs involved. These books, however, often describe interactions established by TFs in regulatory networks but do not deepen into the structural aspects of the TFs involved. I hope that through this book readers will acquire a general view of different aspects of plant TFs that eventually will help to fill the gap existing between the knowledge of the participation of a TF in a defined process and the establishment of the structural properties related to TF functions. In other words, to establish structural–functional relationships that explain in detail the molecular mechanisms involved in TF action.

I would like to thank all authors who generously contributed their chapters. As experts in the field, their contribution was essential for the assembly of this book. I would also like to thank Mary Preap, from Academic Press/Elsevier, for her valuable assistance, and people from my lab for their contributions along many years. Finally, I also acknowledge support from the Argentine Research Council (Consejo Nacional de Investigaciones Científicas y Técnicas, CONICET) and Agencia Nacional de Promoción Científica y Tecnológica (ANPCyT) for their support to my research activities.

Daniel H. Gonzalez

SECTION A

GENERAL ASPECTS OF PLANT TRANSCRIPTION FACTORS

1	*Introduction to transcription factor structure and function*	*3*
2	*Methods to study transcription factor structure and function*	*13*
3	*General aspects of plant transcription factor families*	*35*
4	*Structures, functions, and evolutionary histories of DNA-binding domains of plant-specific transcription factors*	*57*
5	*The evolutionary diversification of genes that encode transcription factor proteins in plants*	*73*

CHAPTER 1

Introduction to Transcription Factor Structure and Function

Daniel H. Gonzalez

Instituto de Agrobiotecnología del Litoral (CONICET-UNL), Cátedra de Biología Celular y Molecular, Facultad de Bioquímica y Ciencias Biológicas, Universidad Nacional del Litoral, Santa Fe, Argentina

OUTLINE

1.1	Introduction: Transcription in Eukaryotes	3	1.5 Protein–protein Interactions 6
1.2	Structure of Transcription Factors	4	1.6 Regulation of Transcription Factor Action 7
1.3	DNA Recognition by Transcription Factors	4	1.7 Plant Transcription Factors 9
1.4	DNA-binding Domains	5	References 9

1.1 INTRODUCTION: TRANSCRIPTION IN EUKARYOTES

In eukaryotes, various RNA polymerases are responsible for the transcription of nuclear genes (Roeder and Rutter, 1969). Particularly, RNA polymerase II is in charge of transcribing all protein-coding genes, in addition to several genes that encode noncoding RNAs (Kornberg, 2007; Cramer et al., 2008). RNA polymerase II promoters are composed of a big number of discrete DNA sequences (also named boxes or elements), usually located upstream of the transcription start site, but also within and downstream of transcribed regions (Lenhard et al., 2012). These sequences can be classified as basal promoter elements, upstream promoter elements, and enhancers, and are the binding sites of proteins named transcription factors, which influence the transcription of genes linked to them (Figure 1.1). Basal promoter elements are usually located near the transcription start site (Juven-Gershon et al., 2008) and are the binding site of general transcription factors that participate in the expression of most genes by promoting the binding of RNA polymerase II (Li et al., 1994; Orphanides et al., 1996; Roeder, 1996; Conaway and Conaway, 1997; Reese, 2003). The most common basal promoter element is the TATA box, recognized by TATA-box binding protein (TBP) (Peterson et al., 1990; Burley, 1996), a component of the general transcription factor II D (TFIID) (Horikoshi et al., 1990). Upstream promoter elements are very diverse. They are located further upstream of basal elements (up to several hundred base pairs of the transcription start site) and are recognized by specific transcription factors according to the type of elements present in each gene (Mitchell and Tjian, 1989; Ptashne and Gann, 1997; Lee and Young, 2000). In most genes, the presence of these elements is necessary for efficient transcription since the sole interaction of general transcription factors with the basal promoter is not enough to assemble a stable transcriptional complex (Gill, 1996; Stargell and Struhl, 1996; Struhl et al., 1998). In addition, many of these elements are required for the transcriptional regulation of gene expression under different circumstances, thus receiving the name of response elements. Enhancers are regions of the genome that affect the expression of particular genes linked to them (Stadhouders et al., 2012; Smallwood and Ren, 2013;

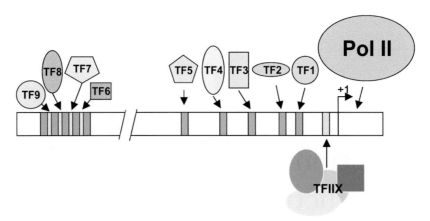

FIGURE 1.1 **The structure of eukaryotic promoters.** Eukaryotic gene promoters are composed of discrete binding sites for multiple transcription factors dispersed over long distances (usually several thousands of base pairs). General transcription factors (TFIIX) for RNA polymerase II (Pol II) interact with sequences located near the transcription start site (yellow). Specific transcription factors (TF1 to TF9) recognize particular sequences located in proximal promoter regions (blue; at hundreds of base pairs of the start site) or in enhancers (green; at thousands of base pairs of the start site). The transcriptional activity of a gene will be defined by the nature of the transcription factors bound in different regions of its promoter. The transcription start site is indicated by +1.

Levine et al., 2014). Usually, they are not classified as part of the promoter, although their action is required for the correct transcriptional activity of the corresponding genes. Enhancers contain groups of response elements and have the peculiarity of acting at long distances (up to several thousands of base pairs), through the formation of loops in DNA. The transcription characteristics of a gene will then be established by the nature of the different elements that compose its promoter region, including enhancers, and the interactions established by different proteins with these elements and among themselves. Apart from the obvious presence of the appropriate partners, the interaction of promoter elements with the corresponding binding proteins will be influenced by the structure of the chromatin in that particular region of the genome, which leads to an additional source of complexity (Li et al., 2007; Cairns, 2009; Venters and Pugh, 2009; Voss and Hager, 2014).

1.2 STRUCTURE OF TRANSCRIPTION FACTORS

Transcription factors are proteins that influence the transcription of genes by binding to defined regions of the genome (Latchman, 1997). Genes encoding transcription factors constitute 3–10% of all genes in eukaryotic genomes (Levine and Tjian, 2003; Harbison et al., 2004; Reece-Hoyes et al., 2005). A basic feature of transcription factors is that they contain DNA-binding domains that recognize specific sequences within the promoter regions of the genes they regulate (Figure 1.2A; Kummerfeld and Teichmann, 2006). By binding to these sequences they either increase or decrease the transcription of their target genes, thus acting as activators or repressors, respectively. Activation or repression is usually achieved through interaction with other components of the transcription apparatus (Figure 1.2B,C; Takagi and Kornberg, 2006), which makes protein–protein interactions another important feature of transcription factor action (Walhout, 2006). Activation or repression of gene expression can also be achieved through interaction with chromatin-modifying enzymes, which are then directed to modulate the accessibility of the transcription machinery to specific regions of the genome (Figure 1.2B,C). The capability of a transcription factor to act as a repressor or an activator is, in most cases, dependent on domains that are located outside, and act independently of, the DNA-binding domain (Figure 1.2A). This brings transcription factors a modular structure and the possibility of acquiring new properties by domain mixing or shuffling, a process used by evolution and researchers to generate new mechanisms of transcriptional regulation (Gossen and Bujard, 1992; Morgenstern and Atchley, 1999; Beerli et al., 2000; Ansari and Mapp, 2002; Traven et al., 2006; Liu et al., 2013).

1.3 DNA RECOGNITION BY TRANSCRIPTION FACTORS

The recognition of specific DNA sequences by transcription factors is achieved by interactions established between the side chains of amino acids of the DNA binding domain with nucleotides of the target site (Figure 1.3). For specific recognition, interactions must be established with the nucleotide bases that are located inside the DNA double-helical structure. For this reason, most transcription factors establish connections with DNA by binding to the major groove, although interactions through the minor groove have also been reported in several cases (Figure 1.3). Amino acid side chains of transcription factor DNA-binding domains can establish specific interactions with bases through hydrogen bonding and van der Waals contacts (Figure 1.3B,C; Shimoni and Glusker, 1995; Suzuki et al., 1995; Luscombe et al., 2001; Rohs et al., 2010). While these interactions determine the specificity, the strength, or affinity of the interaction is additionally determined by unspecific contacts established with the sugar phosphate backbone, including ionic interactions between DNA phosphates and positively charged residues of the DNA-binding domain. Another factor that influences the strength and specificity of the interaction is the topology of the DNA around the transcription factor-binding site (Pan et al., 2010). Curvatures in DNA are often required by transcription factors to bind their target genes efficiently (Rohs et al., 2009), and some transcription factors induce DNA bending upon binding (van der Vliet and Verrijzer, 1993), thus leading to changes that facilitate other processes, like DNA melting or the binding of additional proteins.

1.4 DNA-BINDING DOMAINS

DNA-binding domains adopt different structures, and the interaction of these domains with DNA can be established through alpha helices, beta sheets, or disordered regions (Figure 1.3; Pabo and Sauer, 1992). Usually, the DNA-binding domain forms a module that can be separated from the rest of the transcription factor structure without losing activity. This facilitates structural studies of the isolated DNA-binding domains or their complexes with DNA using techniques that require low molecular weight proteins, like crystallization or NMR. DNA-binding domains are named according to their structural characteristics, and most organisms contain several transcription factors that share the same type of DNA-binding domain. Accordingly, transcription factors are classified in families that usually receive the name of the respective DNA-binding domain (Table 1.1; Stegmaier et al., 2004; Vaquerizas et al., 2009; Charoensawan et al., 2010). Transcription factors that share the same type of DNA-binding domain (in other words, transcription factors from the same family) tend to have more similar DNA binding specificities than those that belong to different families. In any case, variations in DNA-binding specificity are often observed within the same family, and these are most often due to changes in specific residues of the DNA-binding domain (Berger et al., 2008; Noyes et al., 2008; Badis et al., 2009). Thus, changes in the amino acid residues of the DNA-binding domain are also used by evolution and researchers to create transcription factors with novel DNA-binding characteristics (Blancafort et al., 2004; Amoutzias et al., 2007; Joung and Sander, 2013).

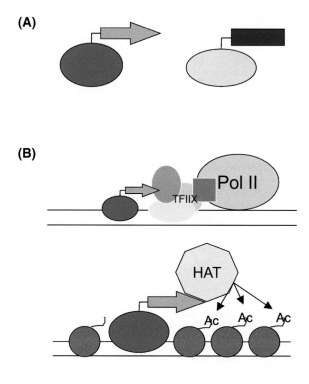

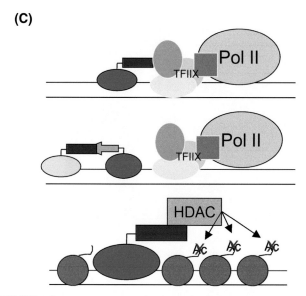

FIGURE 1.2 **The general structure of transcription factors.** (A) Most transcription factors have a modular structure. They contain a DNA binding domain involved in the recognition of specific DNA sequences (blue and yellow ovals). Usually, they also contain activation (green arrow) or repressor (red rectangle) domains that increase or decrease, respectively, the transcription of genes to which they bind. (B) Activation can be achieved through stabilizing protein–protein interactions with other components of the transcription machinery or the recruitment of chromatin modifying enzymes, like HAT, that relax nucleosome structure. (C) Repression can be achieved by interfering with activators or recruiting chromatin-modifying enzymes, like histone deacetylases (HDAC), that increase compaction of nucleosomes.

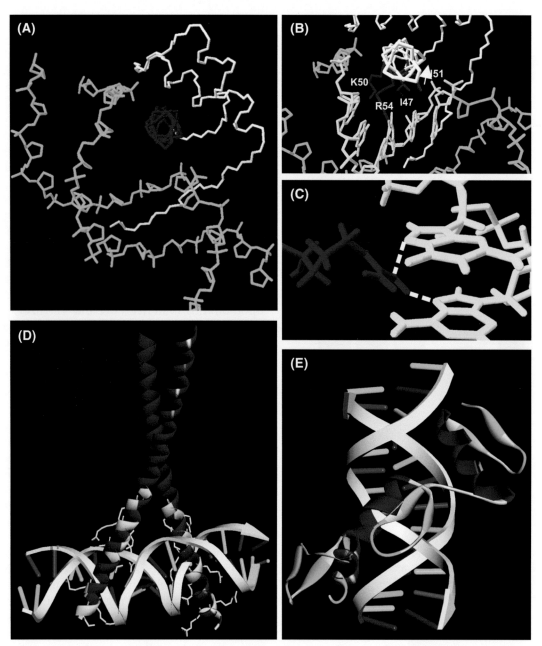

FIGURE 1.3 **DNA recognition by transcription factors.** (A) Specific interactions of the *Drosophila* bicoid homeodomain transcription factor with DNA (Baird-Titus et al., 2006) are established by an alpha helix (helix III, in red) that inserts into the major groove of DNA (green). Additional interactions are established by a disordered N-terminal arm (yellow) along the adjacent minor groove (only the path of backbone atoms is shown). (B) For specific recognition, the side arms of bicoid helix III amino acids Ile47, Lys50, Asn51, and Arg54 (red) establish specific contacts with bases of the DNA (yellow). (C) For example, bicoid Arg54 forms hydrogen bonds with G and A of the recognition sequence GGATTA. (D) A dimer of the yeast bZip transcription factor GCN4 (Keller et al., 1995, in red) interacts with DNA through alpha helices that run across two adjacent major grooves. Specific contacts are established mainly by basic amino acids (yellow). (E) Zinc finger transcription factors contain DNA-binding modules formed by adjacent alpha helices and beta hairpins. Each module binds DNA through the major groove and is connected to adjacent modules by a disordered region. The different modules wrap along the major groove of DNA (Kim and Berg, 1996).

1.5 PROTEIN–PROTEIN INTERACTIONS

Many transcription factors require the formation of dimers to bind DNA (Figure 1.4A,B). These dimers can form between two identical molecules (homodimers) or between different molecules (heterodimers), usually from the same transcription factor family (Amoutzias et al., 2008; Funnell and Crossley, 2012; Pogenberg et al., 2014). Typical transcription factors that require dimerization for binding are those of the bZip and bHLH families (Figure 1.4A,B), which use a basic region to interact with DNA and in which the dimerization domain

TABLE 1.1 Classification of Transcription Factors*

1. Superclass: Basic domains
  1.1 Class: Leucine zipper factors (bZIP)
  1.2 Class: Helix–loop–helix factors (bHLH)
  1.3 Class: Helix–loop–helix/leucine zipper factors (bHLH-ZIP)
  1.4 Class: NF-1
  1.5 Class: RF-X
  1.6 Class: bHSH

2. Superclass: Zinc-coordinating DNA-binding domains
  2.1 Class: Cys4 zinc finger of nuclear receptor type
  2.2 Class: Diverse Cys4 zinc fingers
  2.3 Class: Cys2His2 zinc-finger domain
  2.4 Class: Cys6 cysteine–zinc cluster
  2.5 Class: Zinc fingers of alternating composition

3. Superclass: Helix–turn–helix
  3.1 Class: Homeodomain
  3.2 Class: Paired box
  3.3 Class: Fork head/winged helix
  3.4 Class: Heat shock factors
  3.5 Class: Tryptophan clusters
  3.6 Class: TEA domain

4. Superclass: beta-Scaffold factors with minor groove contacts
  4.1 Class: Rel homology region (RHR)
  4.2 Class: STAT
  4.3 Class: p53
  4.4 Class: MADS box
  4.5 Class: beta-Barrel alpha helix transcription factors
  4.6 Class: TATA-binding proteins
  4.7 Class: HMG
  4.8 Class: Heteromeric CCAAT factors
  4.9 Class: Grainyhead
  4.10 Class: Cold-shock domain factors
  4.11 Class: Runt

0. Superclass: Other transcription factors
  0.1 Class: Copper fist proteins
  0.2 Class: HMGI(Y)
  0.3 Class: Pocket domain
  0.4 Class: E1A-like factors
  0.5 Class: AP2/EREBP-related factors

From http://www.gene-regulation.com/pub/databases/transfac/clSM.html

(either leucine zipper or helix–loop–helix motif) is intrinsic to the DNA-binding module (Massari and Murre, 2000; Amoutzias et al., 2007) (see Chapter 21). Other transcription factors that form dimers or higher order structures are able to bind DNA by themselves, but acquire increased specificity or affinity after complex formation (Goutte and Johnson, 1993; Mann and Chan, 1996; Zhong and Vershon, 1997). In an extreme case, a transcription factor will only be able to specifically bind DNA in the presence of a partner that induces a change to an active conformation (Goutte and Johnson, 1993).

Besides interactions that modify the DNA-binding properties of transcription factors, a multitude of other protein–protein interactions, established through regions located either within or outside the DNA-binding domain, are inherent to transcription factor action.

Transcriptional complexes are formed by a wealth of transcription factors bound to different regions of a gene (Burley and Kamada, 2002; Ogata et al., 2003). This implies that the protein–protein interactions of transcription factors, established either directly or through bridging proteins (Figure 1.4C; Takagi and Kornberg 2006), are essential to determining the stability of the transcriptional complex and, in this way, the transcription rate of a gene. Protein interaction domains that stabilize the transcriptional complex cause an increase in transcription and are then referred to as activation domains (Figure 1.3B; Blau et al., 1996). Activation domains can be classified according to the kind of amino acids that predominate in their structure, thus giving rise to acidic or hydrophobic activation domains (Hope et al., 1988; Regier et al., 1993; Drysdale et al., 1995). Some of these domains interact with general factors or coactivators (Vashee and Kodadek, 1995; Näär et al., 2001), which means that they function when associated with almost any transcription factor and in many different organisms. Other activation domains are more specific and will only function when the appropriate partner is present within the same cell. Due to the nature of eukaryotic transcriptional complexes, which require stabilization of the general transcription machinery through protein–protein interactions, the presence of activators is often required. In addition, transcription factors can act as repressors, usually interfering with the activity of activators by competing for the same target site in DNA or blocking their action through direct or indirect protein–protein interactions (Figure 1.3C; Gaston and Jayaraman, 2003; Reynolds et al., 2013). In addition, activators and repressors can influence transcription by changing the chromatin structure in the vicinity of their binding sites (Figure 1.3B,C). This is also achieved by protein–protein interactions with enzymes like histone acetylases (HAT) and deacetylases, which modify histones and change the strength of their interactions with DNA and, in this way, the accessibility of the transcription machinery (Wolffe et al., 1997; Deckert and Struhl, 2001).

1.6 REGULATION OF TRANSCRIPTION FACTOR ACTION

The transcriptional regulation of genes is usually brought about by regulation of the activity of transcription factors that influence their expression. The term "activity" in this context must be regarded in a broad sense, since often what is regulated is rather the presence of the transcription factor in the right place to effect transcription (that is, the nucleus), or its presence in the cell at all (in which case, the synthesis of the transcription factor is regulated; Figure 1.5). The location of transcription

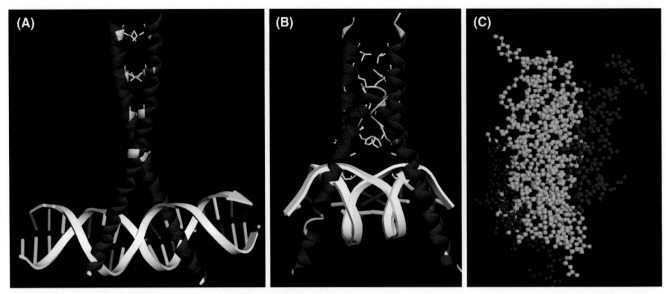

FIGURE 1.4 **Protein–protein interactions and transcription factor function.** bZip (A) and bHLH (B) transcription factors bind DNA only as dimers. Dimerization is facilitated by the interaction of regularly spaced hydrophobic amino acids (yellow) present at the dimer interface. The structures of yeast GCN4 (bZip) and mouse MyoD (bHLH) DNA-binding domains are shown (Ma et al., 1994; Keller et al., 1995). (C) The KIX domain of the coactivator CREB-binding protein (CBP) interacts with many transcription factors. The image shows KIX (green) bound to the activation domains of the transcription factors c-Myb (blue) and mixed lineage leukemia (MLL; red). The binding of each partner to KIX induces conformational changes that promote cooperative ternary complex formation (De Guzman et al., 2006).

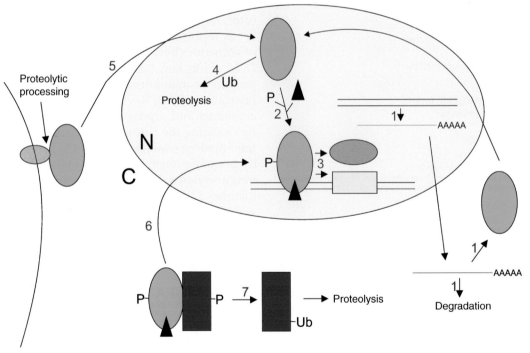

FIGURE 1.5 **Regulation of transcription factor action.** The "activity" of transcription factors (blue) can be modulated in several ways. (1) The synthesis of transcription factors can be modulated through regulation of their expression at the transcriptional, posttranscriptional, and translational levels. (2) Once synthesized, the DNA-binding activity of transcription factors can be regulated by ligand binding (black triangles) or posttranslational modifications (P for phosphorylation and many others). Not only the activity, but also the interaction with other components of the transcriptional machinery (3) or the proteolytic degradation of transcription factors (4) can be regulated. (5) Some transcription factors are retained outside the nucleus because they contain membrane-binding domains (gray). Regulated proteolytic processing allows migration to the nucleus of the soluble domain, containing the transcription factor activity. (6) Alternatively, transcription factors are retained outside the nucleus by protein partners (red) that inhibit nuclear import. Ligand binding or posttranslational modifications release the transcription factor from the inhibitory protein, which may also be targeted for degradation (7). N, nucleus; C, cytosol; Ub, ubiquitin.

factors within the cell can be regulated by the presence of proteins that either retain them in the cytosol or help them to translocate to the nucleus (Whiteside and Goodbourn, 1993; Kaffman and O'Shea, 1999). The translocation of proteins to the nucleus requires the presence of nuclear localization signals (NLS, usually stretches of positively charged residues exposed to the protein surface) in their structure, which are recognized by the nuclear import machinery (Stewart, 2007). While most transcription factors contain NLS in their structure, some of them require the formation of a complex with an NLS-containing partner for nuclear localization. Contrary to that, some transcription factors that contain NLS are retained in the cytosol by partners that block their NLS or their interaction with the import machinery. Interactions with ligands or posttranscriptional modifications that disrupt these complexes, sometimes leading to proteolysis of the inhibitory protein, allow migration to the nucleus and binding to the target gene(s). Retention of transcription factors outside the nucleus is sometimes also achieved by their interaction with membranes (Hoppe et al., 2001; Chapter 25). In this case, the activation of specific proteases leads to separation of the membrane-binding domain from the soluble domain that can migrate to the nucleus and interact with target genes. Interactions with ligands or posttranscriptional modifications can also be used to modulate the stability of transcription factors, which can then be targeted to degradation by the proteasome either in the presence or absence of these signaling events (Geng et al., 2012; Yao and Ndoja, 2012; Chapter 26). Posttranslational modifications can also directly modulate the activity of a transcription factor that is otherwise always present within the nucleus. Known modifications include phosphorylation, acetylation, glycosylation, ubiquitination, sumoylation, and redox-dependent changes, among others (Jackson and Tjian, 1988; Bohmann 1990; Bannister and Miska, 2000; Gill, 2003; Liu et al., 2005; Ndoja et al., 2014) In this case, regulation can be exerted on the DNA-binding capacity of the transcription factor or its ability to modulate transcription, acting on its capacity to interact with other proteins once bound to DNA.

1.7 PLANT TRANSCRIPTION FACTORS

The general characteristics of transcription in plants are similar to those of other eukaryotes. Accordingly, many plant transcription factor families are also present in fungi and animals, suggesting that they are ancient acquisitions. However, plants have also acquired transcription factors with novel characteristics (Chapter 3). Some of these factors, like those of the HD-Zip family (Ariel et al., 2007; Chapter 7), arose from the combination of known transcription factor elements also present in other domains of life (the homeodomain and leucine zipper, in this case), while others contain DNA-binding domains that are not present in transcription factors from other eukaryotes (Yamasaki et al., 2013; Chapter 4). Nevertheless, whether they are related to transcription factor families present in other organisms or not, the characteristics of the different plant transcription factor families, as well as their roles in transcriptional regulation of different processes, makes the study of plant transcription factors a necessary road to understand plant function at the molecular level.

Acknowledgments

I acknowledge support from Consejo Nacional de Investigaciones Científicas y Técnicas (CONICET, Argentina), Agencia Nacional de Promoción Científica y Tecnológica (ANPCyT, Argentina) and Universidad Nacional del Litoral.

References

Amoutzias, G.D., Robertson, D.L., Van de Peer, Y., Oliver, S.G., 2008. Choose your partners: dimerization in eukaryotic transcription factors. Trends Biochem. Sci. 33, 220–229.

Amoutzias, G.D., Veron, A.S., Weiner 3rd, J., Robinson-Rechavi, M., Bornberg-Bauer, E., Oliver, S.G., Robertson, D.L., 2007. One billion years of bZIP transcription factor evolution: conservation and change in dimerization and DNA-binding site specificity. Mol. Biol. Evol. 24, 827–835.

Ansari, A.Z., Mapp, A.K., 2002. Modular design of artificial transcription factors. Curr. Opin. Chem. Biol. 6, 765–772.

Ariel, F.D., Manavella, P.A., Dezar, C.A., Chan, R.L., 2007. The true story of the HD-Zip family. Trends Plant Sci. 12, 419–426.

Badis, G., Berger, M.F., Philippakis, A.A., Talukder, S., Gehrke, A.R., Jaeger, S.A., et al., 2009. Diversity and complexity in DNA recognition by transcription factors. Science 324, 1720–1723.

Baird-Titus, J.M., Clark-Baldwin, K., Dave, V., Caperelli, C.A., Ma, J., Rance, M., 2006. The solution structure of the native K50 bicoid homeodomain bound to the consensus TAATCC DNA-binding site. J. Mol. Biol. 356, 1137–1151.

Bannister, A.J., Miska, E.A., 2000. Regulation of gene expression by transcription factor acetylation. Cell. Mol. Life Sci. 57, 1184–1192.

Beerli, R.R., Dreier, B., Barbas 3rd, C.F., 2000. Positive and negative regulation of endogenous genes by designed transcription factors. Proc. Natl. Acad. Sci. USA 97, 1495–1500.

Berger, M.F., Badis, G., Gehrke, A.R., Talukder, S., Philippakis, A.A., Peña- Castillo, L., et al., 2008. Variation in homeodomain DNA binding revealed by high-resolution analysis of sequence preferences. Cell 133, 1266–1276.

Blancafort, P., Segal, D.J., Barbas 3rd, C.F., 2004. Designing transcription factor architectures for drug discovery. Mol. Pharmacol. 66, 1361 1371.

Blau, J., Xiao, H., McCracken, S., O'Hare, P., Greenblatt, J., Bentley, D., 1996. Three functional classes of transcriptional activation domain. Mol. Cell. Biol. 16, 2044–2055.

Bohmann, D., 1990. Transcription factor phosphorylation: a link between signal transduction and the regulation of gene expression. Cancer Cells 2, 337–344.

Burley, S.K., 1996. The TATA box binding protein. Curr. Opin. Struct. Biol. 6, 69–75.

Burley, S.K., Kamada, K., 2002. Transcription factor complexes. Curr. Opin. Struct. Biol. 12, 225–230.

Cairns, B.R., 2009. The logic of chromatin architecture and remodelling at promoters. Nature 461, 193–198.

Charoensawan, V., Wilson, D., Teichmann, S.A., 2010. Genomic repertoires of DNA-binding transcription factors across the tree of life. Nucleic Acids Res. 38, 7364–7377.

Conaway, R.C., Conaway, J.W., 1997. General transcription factors for RNA polymerase II. Prog. Nucleic Acid Res. Mol. Biol. 56, 327–346.

Cramer, P., Armache, K.J., Baumli, S., Benkert, S., Brueckner, F., Buchen, C., et al., 2008. Structure of eukaryotic RNA polymerases. Annu. Rev. Biophys. 37, 337–352.

Deckert, J., Struhl, K., 2001. Histone acetylation at promoters is differentially affected by specific activators and repressors. Mol. Cell. Biol. 21, 2726–2735.

De Guzman, R.N., Goto, N.K., Dyson, H.J., Wright, P.E., 2006. Structural basis for cooperative transcription factor binding to the CBP coactivator. J. Mol. Biol. 355, 1005–1013.

Drysdale, C.M., Dueñas, E., Jackson, B.M., Reusser, U., Braus, G.H., Hinnebusch, A.G., 1995. The transcriptional activator GCN4 contains multiple activation domains that are critically dependent on hydrophobic amino acids. Mol. Cell. Biol. 15, 1220–1233.

Funnell, A.P., Crossley, M., 2012. Homo- and heterodimerization in transcriptional regulation. Adv. Exp. Med. Biol. 747, 105–121.

Gaston, K., Jayaraman, P.S., 2003. Transcriptional repression in eukaryotes: repressors and repression mechanisms. Cell. Mol. Life Sci. 60, 721–741.

Geng, F., Wenzel, S., Tansey, W.P., 2012. Ubiquitin and proteasomes in transcription. Annu. Rev. Biochem. 81, 177–201.

Gill, G., 2001. Regulation of the initiation of eukaryotic transcription. Essays Biochem. 37, 33–43.

Gill, G., 2003. Post-translational modification by the small ubiquitin-related modifier SUMO has big effects on transcription factor activity. Curr. Opin. Genet. Dev. 13, 108–113.

Gossen, M., Bujard, H., 1992. Tight control of gene expression in mammalian cells by tetracycline-responsive promoters. Proc. Natl. Acad. Sci. USA 89, 5547–5551.

Goutte, C., Johnson, A.D., 1993. Yeast a1 and alpha 2 homeodomain proteins form a DNA-binding activity with properties distinct from those of either protein. J. Mol. Biol. 233, 359–371.

Harbison, C.T., Gordon, D.B., Lee, T.I., Rinaldi, N.J., Macisaac, K.D., Danford, T.W., et al., 2004. Transcriptional regulatory code of a eukaryotic genome. Nature 431, 99–104.

Hope, I.A., Mahadevan, S., Struhl, K., 1988. Structural and functional characterization of the short acidic transcriptional activation region of yeast GCN4 protein. Nature 333, 635–640.

Hoppe, T., Rape, M., Jentsch, S., 2001. Membrane-bound transcription factors: regulated release by RIP or RUP. Curr. Opin. Cell Biol. 13, 344–348.

Horikoshi, M., Yamamoto, T., Ohkuma, Y., Weil, P.A., Roeder, R.G., 1990. Analysis of structure-function relationships of yeast TATA box binding factor TFIID. Cell 61, 1171–1178.

Jackson, S.P., Tjian, R., 1988. O-glycosylation of eukaryotic transcription factors: implications for mechanisms of transcriptional regulation. Cell 55, 125–133.

Joung, J.K., Sander, J.D., 2013. TALENs: a widely applicable technology for targeted genome editing. Nat. Rev. Mol. Cell Biol. 14, 49–55.

Juven-Gershon, T., Hsu, J.Y., Theisen, J.W., Kadonaga, J.T., 2008. The RNA polymerase II core promoter – the gateway to transcription. Curr. Opin. Cell Biol. 20, 253–259.

Kaffman, A., O'Shea, E.K., 1999. Regulation of nuclear localization: a key to a door. Annu. Rev. Cell Dev. Biol. 15, 291–339.

Keller, W., König, P., Richmond, T.J., 1995. Crystal structure of a bZIP/DNA complex at 2.2 Å: determinants of DNA specific recognition. J. Mol. Biol. 254, 657–667.

Kim, C.A., Berg, J.M., 1996. A 2.2 Å resolution crystal structure of a designed zinc finger protein bound to DNA. Nat. Struct. Biol. 3, 940–945.

Kornberg, R.D., 2007. The molecular basis of eukaryotic transcription. Proc. Natl. Acad. Sci. USA 104, 12955–12961.

Kummerfeld, S.K., Teichmann, S.A., 2006. DBD: a transcription factor prediction database. Nucleic Acids Res. 34, D74–D81.

Latchman, D.S., 1997. Transcription factors: an overview. Int. J. Biochem. Cell Biol. 29, 1305–1312.

Lee, T.I., Young, R.A., 2000. Transcription of eukaryotic protein-coding genes. Annu. Rev. Genet. 34, 77–137.

Lenhard, B., Sandelin, A., Carninci, P., 2012. Metazoan promoters: emerging characteristics and insights into transcriptional regulation. Nat. Rev. Genet. 13, 233–245.

Levine, M., Cattoglio, C., Tjian, R., 2014. Looping back to leap forward: transcription enters a new era. Cell 157, 13–25.

Levine, M., Tjian, R., 2003. Transcription regulation and animal diversity. Nature 424, 147–151.

Li, B., Carey, M., Workman, J.L., 2007. The role of chromatin during transcription. Cell 128, 707–719.

Li, Y., Flanagan, P.M., Tschochner, H., Kornberg, R.D., 1994. RNA polymerase II initiation factor interactions and transcription start site selection. Science 263, 805–807.

Liu, H., Colavitti, R., Rovira, I.I., Finkel, T., 2005. Redox-dependent transcriptional regulation. Circulation Res. 97, 967–974.

Liu, W., Yuan, J.S., Stewart, Jr., C.N., 2013. Advanced genetic tools for plant biotechnology. Nat. Rev. Genet. 14, 781–793.

Luscombe, N.M., Laskowski, R.A., Thornton, J.M., 2001. Amino acid–base interactions: a three-dimensional analysis of protein–DNA interactions at an atomic level. Nucleic Acids Res. 29, 2860–2874.

Ma, P.C., Rould, M.A., Weintraub, H., Pabo, C.O., 1994. Crystal structure of MyoD bHLH domain-DNA complex: perspectives on DNA recognition and implications for transcriptional activation. Cell 77, 451–459.

Mann, R.S., Chan, S.-K., 1996. Extra specificity from extradenticle: the partnership between HOX and PBX/EXD homeodomain proteins. Trends Genet. 12, 258–262.

Massari, M.E., Murre, C., 2000. Helix-loop-helix proteins: regulators of transcription in eukaryotic organisms. Mol. Cell. Biol. 20, 429–440.

Mitchell, P.J., Tjian, R., 1989. Transcriptional regulation in mammalian cells by sequence-specific DNA binding proteins. Science 245, 371–378.

Morgenstern, B., Atchley, B.R., 1999. Evolution of bHLH transcription factors: modular evolution by domain shuffling? Mol. Biol. Evol. 16, 1654–1663.

Näär, A.M., Lemon, B.D., Tjian, R., 2001. Transcriptional coactivator complexes. Annu. Rev. Biochem. 70, 475–501.

Ndoja, A., Cohen, R.E., Yao, T., 2014. Ubiquitin signals proteolysis-independent stripping of transcription factors. Mol. Cell 53, 893–903.

Noyes, M.B., Christensen, R.G., Wakabayashi, A., Stormo, G.D., Brodsky, M.H., Wolfe, S.A., 2008. Analysis of homeodomain specificities allows the family-wide prediction of preferred recognition sites. Cell 133, 1277–1289.

Ogata, K., Sato, K., Tahirov, T.H., 2003. Eukaryotic transcriptional regulatory complexes: cooperativity from near and afar. Curr. Opin. Struct. Biol. 13, 40–48.

Orphanides, G., Lagrange, T., Reinberg, D., 1996. The general transcription factors of RNA polymerase II. Genes Dev. 10, 2657–2683.

Pabo, C.O., Sauer, R.T., 1992. Transcription factors: structural families and principles of DNA recognition. Annu. Rev. Biochem. 61, 1053–1095.

Pan, Y., Tsai, C.J., Ma, B., Nussinov, R., 2010. Mechanisms of transcription factor selectivity. Trends Genet. 26, 75–83.

Peterson, M.G., Tanese, N., Pugh, B.F., Tjian, R., 1990. Functional domains, and upstream activation properties of cloned human TATA binding protein. Science 248, 1625–1630.

Pogenberg, V., Consani Textor, L., Vanhille, L., Holton, S.J., Sieweke, M.H., Wilmanns, M., 2014. Design of a bZip transcription factor with homo/heterodimer-induced DNA-binding preference. Structure 22, 466–477.

Ptashne, M., Gann, A., 1997. Transcriptional activation by recruitment. Nature 386, 569–577.

Reece-Hoyes, J.S., Deplancke, B., Shingles, J., Grove, C.A., Hope, I.A., Walhout, A.J.M., 2005. A compendium of *C. elegans* regulatory transcription factors: A resource for mapping transcription regulatory networks. Genome Biol. 6, R110.

Reese, J.C., 2003. Basal transcription factors. Curr. Opin. Genet. Devel. 13, 114–118.

Regier, J.L., Shen, F., Triezenberg, S.J., 1993. Pattern of aromatic and hydrophobic amino acids critical for one of two subdomains of the VP16 transcriptional activator. Proc. Natl. Acad. Sci. USA 90, 883–887.

Reynolds, N., O'Shaughnessy, A., Hendrich, B., 2013. Transcriptional repressors: multifaceted regulators of gene expression. Development 140, 505–512.

Roeder, R.G., 1996. The role of general initiation factors in transcription by RNA polymerase II. Trends Biochem. Sci. 21, 327–335.

Roeder, R.G., Rutter, W.J., 1969. Multiple forms of DNA-dependent RNA polymerase in eukaryotic organisms. Nature 224, 234–237.

Rohs, R., Jin, X., West, S.M., Joshi, R., Honig, B., Mann, R.S., 2010. Origins of specificity in protein-DNA recognition. Annu. Rev. Biochem. 79, 233–269.

Rohs, R., West, S.M., Sosinsky, A., Liu, P., Mann, R.S., Honig, B., 2009. The role of DNA shape in protein-DNA recognition. Nature 461, 1248–1253.

Shimoni, L., Glusker, J.P., 1995. Hydrogen bonding motifs of protein side chains: Descriptions of binding of arginine and amide groups. Prot. Sci. 4, 65–74.

Smallwood, A., Ren, B., 2013. Genome organization and long-range regulation of gene expression by enhancers. Curr. Opin. Cell Biol. 25, 387–394.

Stadhouders, R., van den Heuvel, A., Kolovos, P., Jorna, R., Leslie, K., Grosveld, F., Soler, E., 2012. Transcription regulation by distal enhancers: who's in the loop? Transcription 3, 181–186.

Stargell, L.A., Struhl, K., 1996. Mechanisms of transcriptional activation in vivo: two steps forward. Trends Genet. 12, 311–315.

Stegmaier, P., Kel, A.E., Wingender, E., 2004. Systematic DNA-binding domain classification of transcription factors. Genome Inform. 15, 276–286.

Stewart, M., 2007. Molecular mechanism of the nuclear protein import cycle. Nat. Rev. Mol. Cell. Biol. 8, 195–208.

Struhl, K., Kadosh, D., Keaveney, M., Kuras, L., Moqtaderi, Z., 1998. Activation and repression mechanisms in yeast. Cold Spring Harb. Symp. Quant. Biol. 63, 413–421.

Suzuki, M., Brenner, S.E., Gerstein, M., Yagi, N., 1995. DNA recognition code of transcription factors. Protein Eng. 8, 319–328.

Takagi, Y., Kornberg, R.D., 2006. Mediator as a general transcription factor. J. Biol. Chem. 281, 80–89.

Traven, A., Jelicic, B., Sopta, M., 2006. Yeast Gal4: a transcriptional paradigm revisited. EMBO Rep. 7, 496–499.

van der Vliet, P.C., Verrijzer, C.P., 1993. Bending of DNA by transcription factors. BioEssays 15, 25–32.

Vaquerizas, J.M., Kummerfeld, S.K., Teichmann, S.A., Luscombe, N.M., 2009. A census of human transcription factors: function, expression and evolution. Nat. Rev. Genet. 10, 252–263.

Vashee, S., Kodadek, T., 1995. The activation domain of GAL4 protein mediates cooperative promoter binding with general transcription factors in vivo. Proc. Natl. Acad. Sci. USA 92, 10683–10687.

Venters, B.J., Pugh, F., 2009. A canonical promoter organization of the transcription machinery and its regulators in the Saccharomyces genome. Genome Res. 19, 360–371.

Voss, T.C., Hager, G.L., 2014. Dynamic regulation of transcriptional states by chromatin and transcription factors. Nat. Rev. Genet. 15, 69–81.

Walhout, A.J.M., 2006. Unraveling transcription regulatory networks by protein-DNA and protein–protein interaction mapping. Genome Res. 16, 1445–1454.

Whiteside, S.T., Goodbourn, S., 1993. Signal transduction and nuclear targeting: regulation of transcription factor activity by subcellular localisation. J. Cell Sci. 104, 949–955.

Wolffe, A.P., Wong, J., Pruss, D., 1997. Activators and repressors: making use of chromatin to regulate transcription. Genes Cells 2, 291–302.

Yamasaki, K., Kigawa, T., Seki, M., Shinozaki, K., Yokoyama, S., 2013. DNA-binding domains of plant-specific transcription factors: structure, function, and evolution. Trends Plant Sci. 18, 267–276.

Yao, T., Ndoja, A., 2012. Regulation of gene expression by the ubiquitin-proteasome system. Semin. Cell Dev. Biol. 23, 523–529.

Zhong, H., Vershon, A.K., 1997. The yeast homeodomain protein MATα2 shows extended DNA binding specificity in complex with Mcm1. J. Biol. Chem. 272, 8402–8409.

CHAPTER 2

Methods to Study Transcription Factor Structure and Function

Ivana L. Viola, Daniel H. Gonzalez

Instituto de Agrobiotecnología del Litoral (CONICET-UNL), Cátedra de Biología Celular y Molecular, Facultad de Bioquímica y Ciencias Biológicas, Universidad Nacional del Litoral, Santa Fe, Argentina

OUTLINE

2.1 Introduction	13
2.2 In Vivo Functional Studies	14
2.2.1 Overexpression or Ectopic Expression of Transcription Factors	14
2.2.2 Transactivation Systems for Expression of Transcription Factors	14
2.2.3 Expression of Dominant Negative Forms of Transcription Factors	16
2.2.4 Expression of Dysregulated Forms of Transcription Factors	16
2.2.5 Gain-of-Function Mutants	16
2.2.6 Expression of Fusions to Activating or Repressor Domains	17
2.2.7 Inducible Systems	17
2.3 Methods for the Analysis of In Vitro Protein–DNA Interactions	18
2.3.1 The EMSA Assay	18
2.3.2 Selex	18
2.3.3 Analysis of Transcription Factor–DNA Complexes by Footprinting Assays	20
2.3.4 Microarray-Based Identification of Transcription Factor Target Genes	21
2.4 Methods to Study Protein–DNA Interactions In Vivo	21
2.4.1 Chromatin Immunoprecipitation (ChIP) Assays	21
2.4.2 ChIP-chip and ChIP-Seq	22
2.4.3 DNA Adenine Methyltransferase Identification (DamID)	22
2.4.4 Identification of Transcription Factors Using the Yeast One-Hybrid Assay	24
2.4.5 Transient Assays to Analyze Protein–DNA Interactions In Vivo	24
2.5 Analysis of Protein–Protein Interactions	25
2.5.1 Methods for Protein–Protein Complex Identification	25
2.5.1.1 Yeast Two-Hybrid Assay (Y2H)	25
2.5.1.2 Tandem and One-step Tag-Based Affinity Purification	27
2.5.2 Methods for Verification of Protein–Protein Interactions	28
2.5.2.1 Coimmunoprecipitation	28
2.5.2.2 In Vivo Split Methods	28
2.5.2.3 Resonance Energy Transfer Methods	28
References	29

2.1 INTRODUCTION

The regulation of gene expression is a complex process controlled by a network of interactions between different regulatory proteins and *cis*-regulatory sequences present in the promoters of their target genes (Harbison et al., 2004). According to the basic combinatorial principles of gene regulation, a given DNA-binding protein contributes to the transcription of many genes with varying expression patterns by acting in conjunction with several other transcription factors, each possessing its own unique expression pattern. In addition, transcription factors can be regulated by posttranscriptional mechanisms, allowing them to be present in an active or inactive state or in different subcellular compartments. Chromatin structure and modification states play key roles in determining the competence of transcription, and most DNA-binding proteins can recognize a broad spectrum of DNA sequences with a wide range of affinities. Furthermore, most DNA-binding proteins are members of protein families, with each cell type containing several family members that recognize similar DNA sequences. Based on these considerations, it is likely that multiple proteins will be capable of binding a defined control element *in vitro*, including several members of a particular family of proteins, and perhaps members of another family that recognize a similar or overlapping sequence. The challenge is to determine which of these proteins is capable of performing the protein–protein and protein–DNA interactions that allow it to regulate endogenous genes. In conclusion, despite intensive work, deciphering the transcriptional code is proving to be more difficult than the genetic code (Harbison et al., 2004). The development of methodologies that allow the characterization of transcription factor DNA-binding specificities has been crucial to understanding transcriptional regulation. Transcription profiling can be applied to stable loss- and gain-of-function transcription factor mutants to identify the global expression changes that are associated with the mutant phenotype, thereby facilitating placement of the transcription factor in a developmental pathway or process. Combining DNA motif discovery and information on the up- or downregulation of target genes can lead to a deeper understanding of molecular modes of action and the specificity of particular transcription factors. In this chapter, the different strategies that can be used to elucidate transcription factor functions as well as a number of techniques commonly used to determine DNA-binding specificities *in vitro* and to identify their primary targets *in vivo* are described.

2.2 IN VIVO FUNCTIONAL STUDIES

To study transcription factor action *in vivo*, the same strategies usually used to study the function of genes, like the analysis of loss-of-function mutants and gene-silencing techniques, for example, can be used. Since many transcription factor families are composed of a large number of members with different degrees of functional overlap, sometimes the analysis of higher order mutants or alternative strategies must be used. Many of these strategies rely on the specific characteristics of transcription factor action and are discussed here.

2.2.1 Overexpression or Ectopic Expression of Transcription Factors

The simplest way to modify the action of a transcription factor in plants is to increase its expression by fusing its coding region to a strong promoter (Figure 2.1A). The most popular promoter for this purpose is the 35SCaMV promoter, which brings about high expression levels in most plant tissues. Thus, the use of this promoter and of most promoters used for a similar purpose not only brings about an increase in expression but also expression in cells where the transcription factor is not usually expressed, thus giving rise to ectopic expression too. The phenotypic or gene expression changes brought about by the overexpression/ectopic expression of the transcription factor can be used to infer what the real function of the transcription factor may be. The utility of this approach is mainly based on its simplicity. However, the results obtained are subject to artifacts and must be regarded with care. First, because many of the effects observed are probably due to the presence of the transcription factor in cells where it is normally absent, thus it is difficult to extrapolate how this refers to the "normal" function of the transcription factor. In addition, expression in unusually high levels is likely to distort many of the interactions involved in transcription factor action. Since the establishment of specific interactions with other proteins and DNA are essential for correct transcription factor action, any distortion of these interactions is likely to be the cause of many of the changes observed. Expression at high levels is likely to bring about unspecific protein–protein or protein–DNA interactions. In the case of transcription factor families, for example, a member of a family may be forced to interact with transcriptional components or response elements that are usually bound by a different member of the same family. In families that form heterodimers, overexpression of one member may disrupt the entire set of active molecules formed by the different interacting members of the family. Nevertheless, overexpression approaches are useful in many cases, provided that they are combined with other functional data. Overexpression or the ectopic expression of transcription factors is also useful for technological purposes to obtain plants with desired characteristics.

2.2.2 Transactivation Systems for Expression of Transcription Factors

Besides using expression over all parts of the plant to study transcription factor action, it may be useful to analyze the effect of expressing the transcription factor in defined groups of cells or developmental stages. For this purpose, the coding region of a transcription factor gene may be linked to promoter regions known to drive expression in defined places or stages (Figure 2.1A). In

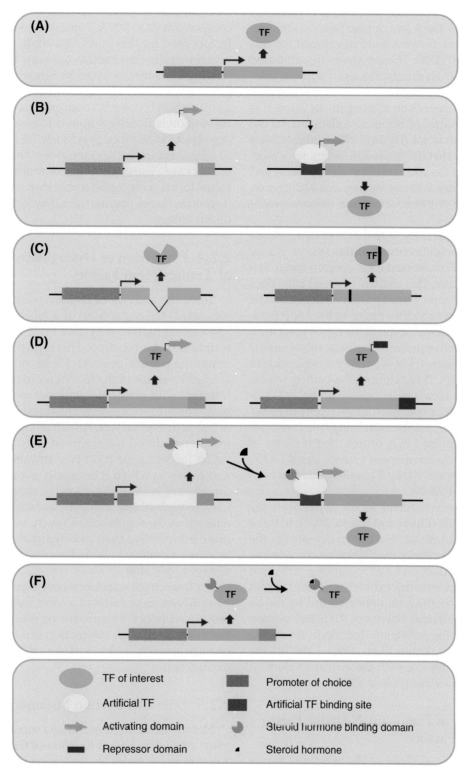

FIGURE 2.1 **Strategies to study transcription factor function.** (A) The expression level/pattern of a transcription factor of interest (blue) can be modified by expressing it from either a strong, constitutively active promoter (like the 35SCaMV promoter) or another type of promoter that brings about its ectopic expression (gray). (B) A two-component transactivation system can be used to modify the expression of a transcription factor. In this case, the promoter of interest (gray) drives the expression of an artificial transcriptional activator (yellow and green); this artificial protein binds to control elements (violet) located upstream of the coding region of the transcription factor under study, thus promoting its expression. (C) The expression of truncated or modified forms of transcription factors can lead to dominant negative effects by competing with the "normal" functions of endogenous transcription factors or generating constitutively active forms. (D) Fusions of transcription factors to strong activating (green arrow) or repressor (red rectangle) domains can be used to modify/enhance the action of transcription factors. (E) Inducible two-component transactivation systems can also be used. In this case, the artificial transcription factor of a system like the one described in (B) is fused to a steroid hormone-binding domain (orange) that retains the transcription factor in the cytosol. Addition of the steroid hormone (black section) activates the transcription factor that is then able to induce the gene of interest. (F) Inducible one-component systems are also available. In this case, the steroid hormone-binding domain (orange) is directly fused to the transcription factor under study, which becomes inducible by addition of the hormone (black section).

addition to the obvious strategy of cloning the selected promoters in front of the transcription factor coding region, two-component systems were developed for this purpose (Moore et al., 2006). These systems are composed of an activation construct that expresses an artificial transcription factor from a desired promoter and an effector construct in which expression of the gene of interest is located under the control of responsive elements for the artificial transcription factor (Figure 2.1B). Two requisites for these systems are that the responsive promoter is inactive in the absence of the artificial transcription factor and that this factor does not activate endogenous plant genes. Thus, nonplant transcription factors and responsive elements are usually used.

One of these systems is the pOp–LhG4 system (Moore et al., 1998). The effector construct of this system is a *lac* operator (from the *Escherichia coli lac* operon) fused to a plant minimal promoter. This construct, named pOp, does not direct gene expression when introduced in plants. The activation component encodes a fusion of the Lac repressor (LacI) to the activation domain of the GAL4 yeast transcription factor. When expressed in plants, this chimeric transcription factor, named LhG4, promotes expression of a gene linked to pOp. Thus, cloning the coding region of the transcription factor under study under the control of pOp allows its expression in different tissues/stages just by crossing plants containing this construct with plants that express LhG4 under the promoter of choice. A different two-component system, based on the *E. coli* LexA repressor and the *lexA* operator, has also been developed (Zuo et al., 2000).

Other two-component systems for use in plants are the GAL4/UASG systems (Elliott and Brand, 2008). In these systems, developed initially for use in *Drosophila*, the activating construct expresses the *Saccharomyces cerevisiae* GAL4 DNA-binding domain fused to a strong activation domain. The effector construct contains several copies of the upstream activating sequence bound by GAL4 (UASG) fused to a minimal promoter. An initial version of this system used the activation domain of the maize C1 transcription factor (Guyer et al., 1998). Later on, the system was adapted for use with the activation domain of the herpes simplex virus protein VP16.

2.2.3 Expression of Dominant Negative Forms of Transcription Factors

Sometimes, overexpression of the modified forms of transcription factors can be used to alter the function of endogenous processes related to their action (Figure 2.1C). The logic underlying this strategy is that if a modified transcription factor that is able to perform only part of its normal functions is expressed in plants, it will disrupt processes in which it is normally involved, simply because it will compete with related endogenous transcription factors for binding to other transcriptional components or to DNA. Typical versions of transcription factors used for this purpose include factors with active protein–protein interaction domains but inactive DNA-binding domains or vice versa. Since most of these strategies are based on successful competition of the modified transcription factor with endogenous components, overexpression is usually required to observe an effect. The drawbacks mentioned previously for the overexpression of native transcription factors are then also applicable in this case. The availability of modified transcription factor forms with considerably increased affinity for endogenous interacting partners may help to overcome this disadvantage.

2.2.4 Expression of Dysregulated Forms of Transcription Factors

Since the activity of transcription factors is highly regulated, the expression of a modified form that has lost this regulation may give hints about normal transcription factor function. This type of strategy is useful when the transcription factor becomes constitutively active as a result of the loss of regulation. Changes that interfere with posttranscriptional modifications or stability sometimes originate constitutively active versions of transcription factors. An advantage of this strategy is that the modified transcription factor can be expressed under its own gene promoter and then in the amounts and places in which it is usually expressed. If a knockout mutant is available, then the modified form can be used to replace the endogenous factor and analyze the effect of its dysregulation. As such, this strategy is much more informative than overexpression techniques, but requires a deeper knowledge of the system and is not always applicable.

For transcription factors whose expression is regulated by miRNAs, expression of a form that is resistant to the miRNA, obtained by introducing mutations that disrupt the mRNA–miRNA interaction but do not modify the coding capacity, can be used to analyze transcription factor function.

2.2.5 Gain-of-Function Mutants

Natural or induced mutations sometimes bring about gain rather than loss of function of transcription factors. Apart from strategies like activation tagging, designed for this purpose and others that provide similar information than overexpression approaches, activating mutations may be due to changes in promoter regions that bring about misexpression of the corresponding gene, mutations that modify the stability or the regulation of the activity of the transcription factor or mutations that bring about dominant negative forms.

2.2.6 Expression of Fusions to Activating or Repressor Domains

Another strategy that can be employed to study transcription factor function is fusion to strong activating or repressor domains (Figure 2.1D). The logic of this strategy is that binding of the modified transcription factor to its target genes will bring about defined changes (either increase or decrease, respectively) in their expression, and consequent phenotypic changes, that can be used to infer function. This strategy is particularly useful when members of families with redundant functions are under study. Conversion of the transcription factor into a strong repressor or activator usually overrides or counteracts the action of endogenous proteins with similar functions. An advantage of this strategy is that it does not usually require overexpression of the modified transcription factor to observe an effect. In this way, the transcription factor can be expressed under its own gene promoter. A disadvantage is that the normal function of the transcription factor is disrupted in any case, since the repressor or activating domains fused to it most likely establish different interactions with the transcriptional machinery than its native form. It can be postulated that this strategy provides information about the genes with which the transcription factor under study is normally able to interact within the plant, but not about the nature or the magnitude of the changes in expression brought about by this interaction.

A repressor motif that is usually used for this kind of strategy is derived from the EAR (ERF-associated amphiphilic repression) domain (Hiratsu et al., 2003). This is a plant-specific repression domain present in several transcription factors that brings about a decrease in gene expression by promoting changes in chromatin structure. This is due to interactions of the EAR domain with corepressors like TOPLESS and SAP18 (Kagale and Rozwadowski, 2011). A modified version of the EAR domain, called SRDX (LDL DLE LRL GFA), has been used to create chimeric repressor gene silencing technology (CRES-T; Mitsuda et al., 2011). SRDX has the advantage that it is a short amino acid sequence that can be fused to many different transcription factors to obtain a strong repressive form.

The most widely used activation domain is derived from the herpes simplex virus protein VP16 (Sadowski et al., 1988). This is an acidic activation domain that is functional in many organisms, including plants (Wilde et al., 1994).

2.2.7 Inducible Systems

Several inducible systems have been developed to express or activate transcription factors when desired. Induction or activation is usually attained by treatment of plants that contain the appropriate construct(s) with a chemical compound (e.g., a steroid hormone; Figure 2.1E). An advantage of this strategy is that it allows verification of the changes that take place soon after the transcription factor is expressed. In this way, changes not directly related with the action of the transcription factor are minimized. This is especially important when expression of the transcription factor brings about profound changes in plant development or growth or is even lethal to plants.

Most inducible systems are two-component systems that can be used to express any gene of interest in response to chemical treatment (Figure 2.1E). However, inducible one-component systems can be used to study plant transcription factors. These systems are based on the fusion of the transcription factor of interest to the hormone-binding domain of a steroid hormone Type I nuclear receptor (Figure 2.1F). These receptors are animal transcription factors that are activated by binding the corresponding hormone to the hormone-binding domain. In the absence of the hormone, they are retained in the cytosol and are then unable to affect gene expression. It has been shown that fusion of the hormone-binding domain to many different transcription factors grants them the same regulatory properties, which can then be used to regulate their action (Schena et al., 1991; Lloyd et al., 1994). Fusions to the glucocorticoid and estrogen receptors can be used to regulate transcription factor action by dexamethasone and 17-β-estradiol, respectively. Fusions to the insect ecdysone receptor, which is activated by methoxyfenozide, have also been reported (Martinez et al., 1999; Padidam et al., 2003). An advantage of these systems is that they do not require protein synthesis for activation, since the transcription factor is already present in the cell, though inactive, at the time of induction. In this way, activation in the presence of a protein synthesis inhibitor (usually cycloheximide) can be used to analyze genes that are directly modulated by the transcription factor, thus avoiding secondary effects.

Fusions to steroid hormone receptors have been used to assemble inducible two-component systems based on the two-component systems described earlier (Aoyama and Chua, 1997; Martinez et al., 1999; Zuo et al., 2000; Padidam et al., 2003; Craft et al., 2005). In these systems, the activation construct is made inducible by fusing the corresponding artificial transcription factor to a steroid hormone receptor (Moore et al., 2006).

In addition, other two-component systems have been used for inducible expression. One of them uses the tetracycline (Tet) repressor encoded in the bacterial transposable element Tn10, for tetracycline-induced expression of a derivative 35SCaMV promoter that contains the Tet repressor binding site (Gatz et al., 1992). Another system uses the ethanol-inducible *Aspergillus nidulans* ALCR transcription factor (ALCR) to activate a synthetic promoter composed of ALCR target sites (from the

A. nidulans alcA gene) fused to the 35SCaMV promoter (Caddick et al., 1998). This system has been widely used for inducible expression in plants.

2.3 METHODS FOR THE ANALYSIS OF *IN VITRO* PROTEIN–DNA INTERACTIONS

2.3.1 The EMSA Assay

To study DNA–protein interactions *in vitro*, a very simple, efficient, and widely used method, first described in 1981, is the electrophoretic mobility shift assay (EMSA, also known as gel mobility shift assay or gel retardation assay) (Fried and Crothers, 1981; Garner and Revzin, 1981). This assay is based on the fact that molecules of different size, shape or charge will have different electrophoretic mobilities in a nondenaturing gel. In the case of a DNA–protein complex, the interaction of the protein with DNA will generate a slower migrating species in relation to the free DNA (Figure 2.2A). EMSA can be carried out in simple steps that include labeling the DNA probe, preparation of the DNA–protein binding reaction and subsequent analysis on a native polyacrylamide gel. In general, oligonucleotides or DNA fragments ranking from 30 bp to 200 bp (base pairs) are used for optimal resolution of complexes versus free DNA, and the DNA can be labeled radioactively or with fluorescent dyes (Steiner and Pfannschmidt, 2009; Viola et al., 2012). In this way, proteins within a crude cell extract that specifically recognize a given *cis*-element can be analyzed by incubating a labeled DNA fragment with the extract to allow the formation of protein–DNA complexes. Extracts from different cell types or developmental stages can be analyzed to determine whether the binding activity is cell-specific or developmentally regulated. Through these gel retardation experiments it is possible to reveal a specific protein–DNA complex even when the protein is at low concentrations within the extract. To confirm that the detected protein–DNA complexes are the result of a specific interaction with a defined regulatory sequence present in the DNA fragment, a mutagenesis assay of this target sequence is necessary. In this case, the protein–DNA binding reactions will be achieved with the native and a mutant version of the labeled oligonucleotide (Figure 2.2A). Alternatively, competition assays can be established. In these assays, binding to a labeled DNA molecule is performed when there is an excess of unlabeled DNA carrying the same or different sequences. If the interaction is specific, then the binding to labeled DNA will be competed by the DNA carrying the same sequence but not by a different one. In any event, it is always advisable to include a "nonspecific" unlabeled DNA (like poly (dI/dC)) to avoid unspecific binding of proteins to labeled DNA.

Besides being useful for qualitative purposes, EMSA has the added advantage of being suitable for quantitative equilibrium and kinetic analyses (i.e., measurement of dissociation constants and association and dissociation rates; Gerstle and Fried, 1993). Furthermore, because of its very high sensitivity, EMSA makes it possible to resolve complexes of different protein or DNA stoichiometry (Fried and Daugherty, 1998) or even to detect conformational changes in proteins that might interact with the DNA of interest. A modification known as the electrophoretic supershift assay, in which inclusion of an antibody causes an additional retardment of the complex, can be used to identify the presence of a particular protein bound to DNA. A similar strategy can be used to analyze the formation of a complex between a given protein and another protein bound to DNA. Procedures that combine EMSA with western blotting or mass spectrometry have also been designed to identify DNA-binding proteins that recognize specific sequences (Hellman and Fried, 2007).

2.3.2 Selex

Transcription factors modulate gene expression through sequence–specific interactions with their DNA-binding sites (Dervan, 1986). Accordingly, the identification of the DNA sequences to which transcription factors bind is a first step in determining their functions in biological processes. Advances in the determination of transcription factor binding sites using *in vivo* and *in vitro* techniques has contributed to deducing transcriptional regulatory codes (Harbison et al., 2004; Badis et al., 2009). One of these approaches, SELEX (systematic evolution of ligands by exponential enrichment), provides an excellent tool for deciphering protein DNA-binding sequence specificities *in vitro*. Initially described by Oliphant et al. (1989) and Blackwell and Weintraub (1990), the SELEX strategy involves the progressive selection of the specific target sequence of a transcription factor from a large combinatorial double-stranded oligonucleotide library through successive steps of binding and amplification. In addition to serving as a technique to establish the *in vitro* DNA-binding specificity of a protein, SELEX is also a powerful tool in determining whether a particular protein binds DNA in a sequence-specific fashion or not (Chai et al., 2011).

In the SELEX assay, a very large oligonucleotide library containing all possible sequences of 20–30 bp flanked by nonrandom sequences of fixed length is used in a binding reaction with the transcription factor under study (Figure 2.2B). The fraction of DNA molecules bound to the protein is separated from the free DNA (i.e., by EMSA, nitrocellulose membrane filtration, using affinity surfaces and affinity tags, or crosslinking and antibody-based flow cytometry; Gopinath, 2007), and amplified by the polymerase chain reaction (PCR) using primers specific to the flanking sequences. Then, the population of amplified molecules is used in a new protein–DNA binding reaction.

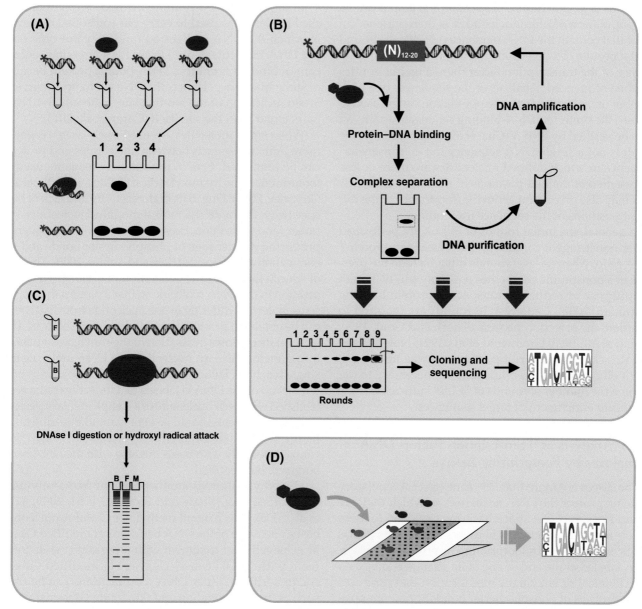

FIGURE 2.2 **Methods to study protein–DNA interactions in vitro.** (A) Electrophoretic mobility shift assay (EMSA). A protein extract or the recombinant protein of interest is incubated with a labeled DNA fragment and the protein–DNA complexes are resolved from free DNA by electrophoresis in a native polyacrylamide gel (lane 2 of gel image). To confirm the specificity of the interaction, a binding reaction with a mutated version of the DNA fragment is necessary (lane 4). Lanes 1 and 3 correspond to binding reactions without the addition of protein. (B) In the SELEX strategy, a labeled oligonucleotide library that contains a central random region between two constant arms is incubated with the purified recombinant transcription factor of interest (pink oval). Once the oligonucleotides selected by the protein are purified (e.g., by EMSA) the obtained population of DNA fragments is amplified by PCR and employed in a new round of selection. After the desired number of rounds of binding, selection, and PCR amplification, the DNA population is sequenced and the consensus sequence for the DNA-binding protein is obtained. (C) Footprinting assay. A DNA fragment labeled only in one strand is employed in a protein-binding reaction (tube B) and submitted to controlled digestion with DNAse I or attack with hydroxyl radical under conditions where only one cleavage per DNA molecule is produced. As a control, a reaction without protein is performed (tube F). The DNA regions protected by the bound protein can easily be identified by comparison of the patterns obtained after separating the bound (B) and free (F) DNA samples in a denaturing DNA sequencing gel. M, molecular marker. (D) Protein-binding microarray experiments use a high-density dsDNA microarray platform where every possible nucleotide sequence of a determinate length is represented at least once. The purified protein tagged with an epitope (e.g., GST) is allowed to bind those DNA probes containing their preferred sequences. After hybridization with a fluorophore-conjugated antibody specific to the epitope, the microarray is scanned and the spot intensity image is used to determine binding parameters.

By means of several rounds of selection and amplification, the population of oligonucleotides is then enriched in those that contain the DNA sequences specifically bound by the protein. To determine the consensus-binding sequence of the transcription factor these oligonucleotides are then sequenced. Analysis of the sequences of a high number of molecules from the population can be used to deduce the preferred DNA-binding sequence, which can then be verified by EMSA (Chai et al., 2011). Usually, the result is not a single DNA sequence but a "consensus" sequence in which defined nucleotides are more or less represented at defined positions of the target sequence. This may also give a hint about the importance of the different positions of the sequence for binding.

In general, the initial rounds of SELEX are performed under nonstringent conditions, such as a high protein/DNA ratio, whereas later rounds often include the presence of a nonspecific competitor (i.e., poly (dI/dC)) with the purpose of reducing nonspecific protein binding (Gopinath, 2007). Usually 4–18 rounds are required to complete the selection process (Tuerk and Gold, 1990; Huang et al., 1993; Grotewold et al., 1994; Nole-Wilson and Krizek, 2000) and competition assays with the different oligonucleotide populations are necessary to determine the number of rounds of SELEX sufficient for the significant enrichment of target sequences.

2.3.3 Analysis of Transcription Factor–DNA Complexes by Footprinting Assays

The discovery more than 35 years ago that regulatory proteins protect the DNA sequences to which they are bound from nuclease attack has been exploited to identify *cis*-regulatory elements in diverse organisms (Galas and Schmitz, 1978). In footprinting or protection assays, DNA labeled at the end of one of its strands is subjected to a chemical or enzymatic modification that produces the hydrolysis of phosphodiester bonds at random positions (Tullius, 1989; Figure 2.2C). The conditions of the modification reaction are adjusted such that only one cleavage per DNA molecule is obtained on average. As a consequence, a population of labeled molecules whose size extends from the labeled end to each cleavage site is obtained. The rationale of the assay is that if a bound protein protects certain regions of the DNA, molecules from the corresponding sizes will be absent or less represented in the population if the assay is applied to a protein–DNA complex. Comparing the patterns obtained after resolving samples of protein-bound and free DNA by denaturing polyacrylamide gel electrophoresis, the regions protected by the DNA-binding protein can be deduced (Figure 2.2C). The use of a chemical sequencing reaction of the same fragment or another molecular weight marker allows determination of the sequence of the protected region with single-nucleotide resolution.

In the DNase I footprinting method, deoxyribonuclease I (DNase I) is used to carry out controlled digestion in which the DNA is digested randomly but only once per DNA molecule. Since DNase I is a large molecule, it cannot bind adjacent to a DNA-bound protein because of steric hindrance. Hence, the DNase I footprint gives a broad indication of the binding site, generally 8–10 base pairs larger than the site itself (Carey et al., 2013).

When small molecules are used as cleavage reagents, more intimate contacts between nucleotides and proteins can be identified. One of the smallest reagents used in footprinting is the hydroxyl radical (Tullius, 1988; Viola and Gonzalez, 2011). Due to its high reactivity and lack of base specificity, the hydroxyl radical modifies nucleotides in a rather random fashion, irrespective of the DNA sequence, producing the cleavage of phosphodiester bonds and the elimination of nucleosides (Jain and Tullius, 2008). This lack of specificity allows the generation of cleavages at every position in the DNA molecule so that a pattern that represents molecules differing in one nucleotide from each other can be obtained, producing results of high resolution. This method then allows evaluation of the contacts established by a protein with each nucleotide of a DNA molecule in a single reaction (Tullius and Dombroski, 1986). Since only one strand of the DNA is labeled at a time, the contacts established with each nucleotide of a complementary pair can be distinguished. It should be kept in mind that, due to the nature of the modification (Balasubramanian et al., 1998), contacts not only with bases but also with the DNA backbone will be identified.

Dimethyl sulfate is another reagent frequently used for footprinting (Shaw and Stewart, 1994, 2009; Tron et al., 2005). This reagent methylates G and A nucleotide bases. Cleavage of the DNA backbone at modified Gs can be achieved after treatment with piperidine, while treatment with NaOH causes cleavage at modified Gs and As. In addition, other DNA modifications can be used such as carbethoxylation and ethylation (Wissmann and Hillen, 1991; Manfield and Stockley, 2009).

An alternative footprinting analysis involves modification of the target DNA before protein binding. In this case, DNA molecules with modifications that affect protein binding will not be incorporated in the protein–DNA complex. This will cause an enrichment of these molecules in the population of free DNA, while molecules with modifications that do not affect protein binding will be enriched in the population of bound DNA. Analysis of both populations in a denaturing polyacrylamide gel allows the identification of positions important for protein binding. This type of approach is called missing nucleoside analysis when the hydroxyl radical is used to modify DNA and methylation interference when dimethyl sulfate is used (Viola and Gonzalez, 2011). In the case of dimethyl sulfate, since specific positions of bases are modified (G N^7, facing the major groove, and A N^3, facing the minor groove), specific protein–DNA interactions can be monitored.

2.3.4 Microarray-Based Identification of Transcription Factor Target Genes

For many years, biochemical assays like those just described were used to characterize DNA–protein interactions. However, such approaches are generally laborious, not highly scalable and, although *in vitro* selections have permitted the sampling of a large number of potential DNA-binding sequences, the resulting sites provide only a partial view of the DNA-binding specificity of a transcription factor, as typically only the highest affinity binding sites are obtained. It is possible that lower affinity DNA sites that are functionally significant in transcriptional regulation of gene expression may not be detected (Walter et al., 1994; Amendt et al., 1999). New technologies have been developed that permit the analysis of DNA-binding sites at much higher resolution and in a more unbiased manner. Protein-binding microarrays (PBMs), based on DNA microarray technology, have been used for high-throughput characterization of the *in vitro* DNA-binding sequence specificity of transcription factors and other DNA-binding proteins. With this technology, it has been possible to identify sites that match the known DNA-binding motifs of transcription factors as well as new candidate regulatory sites (Bulyk et al., 1999, 2001; Mukherjee et al., 2004). Additionally, DNA-binding site data obtained from PBMs, combined with gene annotation data, comparative sequence analysis, and gene expression profiling, can be used to predict what genes are regulated by a given transcription factor. To date several groups have developed PBMs for high-throughput determination of the DNA-binding specificities of transcription factors (Mukherjee et al., 2004; Berger et al., 2006; Warren et al., 2006; Kim et al., 2009; Godoy et al., 2011).

In PBM experiments, a DNA-binding protein of interest is expressed with an epitope tag that serves a dual purpose: to isolate the protein by affinity purification and to achieve its detection by means of an epitope-specific reporter, such as an antibody. Alternatively, directly labeled proteins can be used in the assay. The protein is then incubated on a double-stranded DNA microarray and, after a wash to avoid nonspecific binding, the signal intensities obtained at the different array positions are measured (Figure 2.2D). Three types of DNA molecules can be used to construct the dsDNA array: short double-stranded oligonucleotides created by primer extension, short double-stranded DNAs created using self-hairpinning oligonucleotides, or longer double-stranded DNAs resulting from PCR amplification of genomic regions (Doi et al., 2002; Bulyk et al., 2001; Mukherjee et al., 2004). Microarrays made of long PCR products have the advantage that they cover more sequence space with relatively few microarray features (Ren et al., 2000; Lee et al., 2002). However, the probability of calling a binding event correctly is less for a single binding site that is embedded in a long rather than a short sequence. Moreover, depending on the number and types of binding sites present within a single region, interaction may occur once or several times by one or several proteins at various degrees of affinity. For accurate information about strength and specificity, arrays that consist of short synthetic double-stranded oligonucleotides exhibit superior performance (Warren et al., 2006; Berger and Bulyk, 2009).

PBM technology has several advantages over high-throughput *in vitro* selection methodologies. First, PBM data are more quantitative, since the signal within each spot on the microarray corresponds to numerous DNA–protein binding events. In addition, nonbinding sequences can be identified. Finally, PBMs can provide binding preference data for each DNA sequence variant present in the array (Bulyk, 2007).

2.4 METHODS TO STUDY PROTEIN–DNA INTERACTIONS *IN VIVO*

Key to understanding transcriptional regulation by transcription factors is the identification of their direct target genes. In general, the analysis of loss- and gain-of-function mutants that present altered phenotypes often provides the first clues to transcription factor function. However, additional approaches are required to elucidate the signal transduction cascades modulated by the transcription factor. For example, induction of expression or nuclear localization of a transcription factor and measurement of gene expression, either at short timepoints after induction and/or in the presence of cycloheximide, have been used to identify putative direct target genes (Ueki et al., 2009; Abel and Theologis, 1994; Yoo et al., 2007 Menges and Murray, 2002). Several methods used to obtain information about the specific DNA fragments bound by transcription factors *in vivo*, including chromatin immunoprecipitation (ChIP), Dam methylase identification (DamID), and the yeast one-hybrid assay, are described in the chapter.

2.4.1 Chromatin Immunoprecipitation (ChIP) Assays

Even though *in vitro* protein–DNA interaction assays, as EMSA or footprinting, permit the study of the DNA-binding specificities of transcription factors, these interactions may not reflect the situation in cells. Chromatin immunoprecipitation (ChIP) represents a valuable alternative to probing such interactions *in vivo* under physiological conditions and to estimating the density of protein at specific sites (Kuo and Allis, 1999). The first ChIP protocol was developed by Gilmour and Lis (1984, 1985) to monitor RNA polymerase/DNA association in *E. coli* and *Drosophila*. Basically, cells in the ChIP assay are treated with a crosslinking agent to covalently bind any DNA-binding protein to the chromatin. Then, the cells are lysed and the genomic DNA is isolated and sonicated to produce sheared chromatin. An antibody specific to the protein of interest is added to the sonicated material and used to isolate the protein with all attached DNA via immunoprecipitation (Figure 2.3A). The DNA is released by reversing the crosslinking and subsequently purified. Finally, to evaluate whether a suspected target gene is associated with a particular DNA-binding protein, a test for enrichment of the target fragment in the immuneprecipitate compared with controls is performed employing semiquantitative or quantitative PCR.

Formaldehyde is most commonly used to crosslink DNA-associated proteins to DNA in ChIP experiments because it offers several advantages: it easily enters cells and, unlike UV irradiation, crosslinks proteins to DNA and proteins to proteins so that it is possible to isolate DNA fragments indirectly associated to the protein of interest. In addition, the crosslinks are reversible by heat. Because formaldehyde rapidly inactivates enzymes, it provides a "snapshot" of the interactions occurring within the cell. However, formaldehyde can disrupt the epitopes necessary for immunoprecipitation, either by modifying the antigen (Das et al., 2004) or simply by denaturing the protein. For the ChIP assay, a highly specific antibody is necessary and in most cases epitope tags are engineered into the protein. A key determinant of a successful ChIP assay is the quality of the antibody, as some antibodies work poorly or not at all for ChIP. In the plant research community, commercial antibodies against HA, GFP, and FLAG tags are commonly used (Yamaguchi et al., 2014).

Depending on the approach, different controls may be used. If antibodies against an endogenous protein are employed, the use of a null mutant line as control is recommended (Wu et al., 2012), while a nontransformed parental line can be used in the case of an epitope-tagged protein (Yamaguchi et al., 2013). For genomic control, a locus unlikely to be bound by the factor of interest should be tested for each experimental and control ChIP reaction. Because of anatomical differences between animal and plant cells, such as rigid cell walls, high levels of cellulose and lignin, and large vacuoles in plant cells, several modifications are needed to establish efficient ChIP protocols for plant systems (Gendrel et al., 2002; Saleh et al., 2008; Kaufmann et al., 2010). Vacuum infiltration is applied to allow the formaldehyde to penetrate plant cells and nuclei. In addition, use of young plant tissues increases the yield of nuclei per gram of fresh weight (Yamaguchi et al., 2014).

ChIP experiments allow the ability of candidate proteins to bind the control region of interest *in vivo* to be examined. However, it is important to keep in mind that ChIP is not a functional assay (Wang et al., 2009). ChIP is further limited by its dependence on high-quality antibodies for the protein of interest and on protein–DNA interactions that are amenable to ChIP analysis. Some proteins are difficult to study using ChIP assays for a variety of reasons, including transient binding, inefficient crosslinking to DNA or epitope masking.

2.4.2 ChIP-chip and ChIP-Seq

ChIP assays can also be used to globally map the target sites of a protein in an entire genome. In this case, the population of DNA fragments obtained after immunoprecipitation is analyzed using genome microarrays or high-throughput DNA sequencing (Figure 2.3A). These techniques are called ChIP-chip and ChIP-Seq, respectively (Wu et al., 2006). In ChIP-chip, the inmunoprecipitated DNA fragments are labeled in an amplification reaction and subsequently hybridized to DNA microarrays to identify the protein-bound fragments. These microarrays can contain PCR-amplified genomic regions or an array of oligonucleotides designed to tile a portion of the genome ("tiling arrays"), allowing the definition of the genomic binding sites of a given protein (Winter et al., 2011; Yamaguchi et al., 2013). ChIP-Seq is a more recent alternative that uses next-generation DNA sequencing for identification of genomic sites enriched in the population of bound DNA fragments (Mardis, 2007).

2.4.3 DNA Adenine Methyltransferase Identification (DamID)

Another recently developed method, alternative to ChIP, utilizes the DNA adenine methyltransferase (Dam) from *E. coli* to allow the identification of *in vivo* binding sites of transcription factors (van Steensel and Henikoff, 2000). In the DamID assay, the target protein is expressed as a fusion molecule to the Dam that methylates adenine at the A N^6 position within the GATC DNA sequence (Brooks et al., 1983). Therefore, when the protein under study binds its native genomic binding sites, the Dam portion of the fusion will methylate GATC sites that are located in the vicinity of the bound region (Figure 2.3B). After genomic DNA extraction, the methylated sites are digested by means of the methyl-specific restriction enzyme *Dpn*I, which cuts only at methylated GATC sites. The smaller *Dpn*I digestion fragments are specifically amplified using a methylation-specific PCR protocol, labeled, and hybridized in a microarray in the same way as for a ChIP-chip assay (Vogel et al., 2007; Figure 2.3B). This method takes advantage of the absence of adenine methylation in eukaryotes and of a suitable GATC motif distribution in eukaryotic genomes (one site every 200–400 bp on average). This approach has been

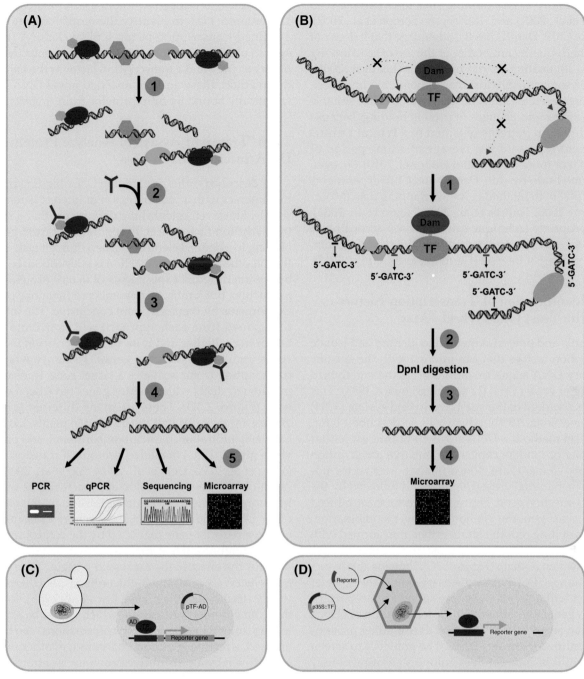

FIGURE 2.3 **Methods to study protein–DNA interactions *in vivo*.** (A) Chromatin immunoprecipitation (ChIP). Plant cells expressing the transcription factor of interest are treated with a crosslinking agent that covalently links DNA-binding proteins to DNA. Once purified, the genomic DNA is sonicated to produce fragments of 100–300 bp (1). The protein–DNA complexes corresponding to the protein of interest are specifically purified through immunoprecipitation using an antibody against the protein (2, 3). Finally, the crosslinks are reversed and the genomic fragments purified (4). Purified DNA can be analyzed by PCR or qPCR to confirm the hypothetical targets of the protein or by microarray (ChIP-chip) or sequencing (ChIP-Seq) for a genome-wide analysis (5). (B) DNA adenine methyltransferase identification (DamID). The DNA-binding protein of interest (TF, orange oval) is expressed and linked to the Dam methyltransferase from *E. coli* (violet oval); the Dam enzyme methylates the adenine of GATC sites present in the neighborhood of the target genes of the transcription factor (1). However, more distant sites are not modified. Upon isolation of genomic DNA (2), digestion using DpnI, which only cuts methylated GATCs, is performed (3). Then, adapters are ligated to the ends of cleaved GATCs for PCR amplification, labeling and hybridization in a microarray (4). (C) Yeast one-hybrid assay. The transcription factor under study or a protein from a cDNA library (pink oval) is expressed in yeast fused to the activation domain of GAL4 (AD, green oval) from plasmid pTF-AD. The yeast genome contains a fragment of a promoter of interest or specific DNA regulatory sequences (blue bar) fused to a minimal promoter (light blue bar). If the transcription factor binds to the promoter of interest, it activates transcription of a reporter gene (orange bar), which brings about a detectable signal or acts as a selectable marker. (D) Analysis of protein–DNA interactions in plant cells. A construct that overexpresses the transcription factor of interest (pink) is transiently introduced in plants along with a plasmid that contains the promoter of the putative target gene in front of a reporter gene. Moreover, a regulatory region of interest fused to a minimal promoter can be assayed. The expression levels of the reporter gene are indicative of the existence of protein–DNA interactions *in vivo*.

used to identify *in vivo* binding sites in *Drosophila* (van Steensel et al., 2001) and *Arabidopsis* (Tompa et al., 2002).

Unlike ChIP, DamID has the advantage that it does not require high-quality antibodies or the use of crosslinking reagents, eliminating the risk of artifacts (Lee et al., 2006). However, some proteins lose their genomic binding specificity when fused to Dam, and this method is less suitable for detecting rapid changes in protein binding, because the methylation patterns obtained in a typical DamID experiment represent the average of a time period of about 24 h or more (Helwa and Hoheisel, 2010). So, even if performed side-by-side, DamID and ChIP experiments suggest that both methods give overlapping data (Moorman et al., 2006; Tolhuis et al., 2006; Negre et al., 2006). The appropriate technique can be chosen according to specific advantages and constraints, the model system, and the protein of interest (Germann and Gaudin, 2011).

2.4.4 Identification of Transcription Factors Using the Yeast One-Hybrid Assay

A simple and powerful method to identify and isolate transcription factors that can interact with the specific regulatory DNA sequence of interest is the one-hybrid screening of yeast (Y1H) (Li and Herskowitz, 1993). This system is a variant of the yeast two-hybrid system (Y2H) (Fields and Song, 1989), which will be described later. In the Y1H method, a DNA segment of interest (either a promoter or tandem copies of a putative transcription regulatory element) are cloned in a reporter vector upstream of a minimal yeast promoter and followed by the reporter/selection gene. The reporter vector is introduced into the yeast genome via homologous recombination, and the resulting reporter strain is then transformed with a cDNA library that expresses fusions to a constitutive yeast activation domain (Figure 2.3C). Proteins able to recognize the specific DNA elements present in the reporter construct will activate the expression of the reporter/selection gene, thus allowing the identification of cDNA clones that encode the respective DNA-binding proteins to be identified (Sieweke, 2000). The potential to screen several million independent colonies simultaneously makes the Y1H system quick and extremely sensitive, allowing the cloning and identification of very low abundance transcription factors.

In the Y1H system, known *cis*-elements as well as noncharacterized DNA fragments of promoters can be used to search for interacting DNA-binding proteins present in the cDNA library (Ouwerkerk and Meijer, 2001). In addition, there is a high level of conservation of *cis*-elements across species, and *cis*-elements from well-characterized plants such as *Arabidopsis* and rice can often be used successfully to isolate transcription factors from other species for which neither genomic nor EST data are yet available (Reece-Hoyes and Walhout, 2012). Recently, Ji et al. (2014) developed a protein–DNA method, termed TF-centered Y1H, to identify the motifs recognized by a defined transcription factor. Unlike classic Y1H, this system utilizes a random short DNA sequence insertion library as prey and a transcription factor as the bait. With this method, these authors revealed novel DNA motifs specifically bound by bZip proteins from *Arabidopsis*.

2.4.5 Transient Assays to Analyze Protein–DNA Interactions *In Vivo*

As described earlier, identifying DNA-binding proteins that interact with a control region of interest is relatively simple. However, establishing definitively that a specific transcription factor directly regulates a target gene by binding to a defined element can be a difficult task. Hence, the hypothesis that an interaction is relevant can be tested by several different experiments of *in vivo* protein–DNA interaction, but it must be remembered that none of these experiments by themselves are conclusive. The information gained from each approach will contribute to determining whether and how the transcription factor in study regulates a subset of genes. To clarify whether a transcription factor activates a target gene, transient expression systems with reporter genes have been widely used (Figure 2.3D). There are many different transient expression methods that can be used in plants, like *Agrobacterium* infiltration, particle bombardment, and polyethylene glycol (PEG) mediated protoplast transformation (Yang et al., 2000; Ueki et al., 2009; Yoo et al., 2007). The common transient expression approach begins with cotransformation with a vector that drives overexpression of the DNA-binding protein and a reporter plasmid harboring a reporter gene regulated by the control region of interest (Figure 2.3D). A reporter activity assay is used to monitor the effect of the expressed protein. The reporter genes luciferase (*Luc*) and β-glucuronidase (*GUS*) are commonly used (Gallagher, 1992; Ow et al., 1986). In order that transformation efficiency (which would be expected to vary among independent transformations) can be normalized, a reference plasmid harboring another reporter gene driven by a constitutive promoter is cotransformed as well (Iwata et al., 2011). If overexpression results in activation of the reporter gene, a new experiment with a reporter plasmid containing a mutant-binding site can be used to demonstrate the requirement of the binding motif for the protein. A positive result with this type of experiment suggests that the DNA-binding protein can activate a gene under the control region of interest when both the DNA-binding protein and the control region are overproduced within the cell. However, it provides little evidence that the protein, when expressed at physiological concentrations, can regulate the endogenous target gene. Nevertheless, the hypothesis can be strengthened greatly by subjecting it to as many rigorous tests as possible, and after considering the data obtained from different assays it can be possible to establish whether a specific protein directly regulates a target gene or not.

Another strategy for testing the relevance of a protein–DNA interaction is an altered specificity experiment (Carey et al., 2012). Although difficult to design and perform, this method has the potential to provide more compelling evidence that a protein–DNA interaction is relevant. In the altered specificity experiment, the DNA-binding domain of the protein of interest is mutated so that it recognizes a different DNA sequence. On the other hand, this new sequence recognized by the altered protein is then inserted into the control region of interest in place of the sequence element recognized by the wild-type protein. The altered specificity DNA-binding protein is then expressed in cells containing an endogenous gene or reporter gene regulated by the altered control region, and its capacity to regulate transcription is monitored (Shah et al., 1997). The altered specificity strategy can provide compelling evidence that a DNA-binding protein acts at a particular target site.

2.5 ANALYSIS OF PROTEIN–PROTEIN INTERACTIONS

The vast majority of proteins do not operate alone but in complexes. Protein–protein interactions are important for coordinating cellular signaling events as well as metabolic functions in any cell. In the case of transcription factors, these proteins modulate transcription by interacting with other proteins, and their activities, DNA-binding capacities and subcellular localization are usually modulated by the action of other proteins or enzymes. Knowing which other proteins physically and functionally interact with a given transcription factor is essential to gaining insight into the molecular networks by which the transcription factor fulfills its function. Numerous techniques have been developed over the years to detect and study protein–protein interactions either *in vitro* or *in vivo*. In this section, the most common methods of protein–protein complex identification and validation are summarized.

2.5.1 Methods for Protein–Protein Complex Identification

A typical protein–protein interaction study usually starts with an initial screen for novel binding partners. One of the best known methods of purifying protein complexes is the tandem affinity purification (TAP) method, which allows purifying the partners of a protein of interest from a protein extract. Among the *in vivo* methods, the yeast two-hybrid assay has become immensely popular because it enables interacting proteins from a cDNA library to be identified. A number of *in vivo* assays are variations on the theme of fragment complementation (Hu et al., 2002; Subramaniam et al., 2001; Spotts et al., 2002).

2.5.1.1 *Yeast Two-Hybrid Assay (Y2H)*

For years, Y2H (Fields and Song, 1989) has been the predominant method to identify protein interactions. The Y2H assay relies on the modular nature of transcription activators, which consist of a DNA-binding domain (BD) and a transcriptional activation domain (AD) (Brent and Ptashne, 1985; Hope and Struhl, 1986; Keegan et al., 1986). The DNA-binding domain serves to target the activator behind the promoter of target genes and the activation domain contacts other proteins of the transcriptional machinery to enable transcription to occur. The two-hybrid system is based on the observation that the two domains of a yeast activator (e.g., GAL4) need not be covalently linked and can be brought together by the interaction of any two proteins. To analyze the interaction, two proteins X and Y, which are hybrid proteins where the GAL4 BD is fused to protein X and the GAL4 AD is fused to protein Y, must be constructed. In addition, these hybrids contain a yeast nuclear localization signal to carry the proteins into the nucleus. These two hybrids are expressed in yeast cells containing one or more reporter genes whose promoters contain elements recognized specifically by the BD of GAL4. If the X and Y proteins interact, they create a functional activator, by bringing the AD into close proximity with the BD, which is able to recognize and bring about expression of the reporter gene (Fields, 2009; Figure 2.4A). The two-hybrid system can be used to screen libraries of AD hybrids to identify proteins that bind to a protein of interest (de Folter and Immink, 2011). These screens result in the immediate availability of the cloned gene for any new protein identified. In addition, since multiple clones that encode overlapping regions of a protein are often identified, the minimal domain for interaction may be readily apparent from the initial screen (Iwabuchi et al., 1993; Vojtek, 1993).

Several Y2H versions exist that differ in the respective protein fragments used as activation and DNA-binding domains or reporter genes. As reporter genes, *E. coli lacZ* (Fields and Song, 1989) and selectable yeast genes such as *HIS3* (Durfee et al., 1993) and *LEU2* (Zervos et al., 1993) are employed.

However, even this elegant and powerful technique has its drawbacks. For example, the interaction is tested only in the nucleus of a yeast cell. Moreover, interactions that depend on cellular compartmentalization or on more than two interaction partners are not always detectable by this method. Proteins must be able to fold and exist stably in yeast cells and to retain activity as fusion proteins. The use of protein fusions also means that the site of interaction may be occluded by one of the transcription factor domains. Interactions dependent on a posttranslational modification that does not occur in yeast cells will not be

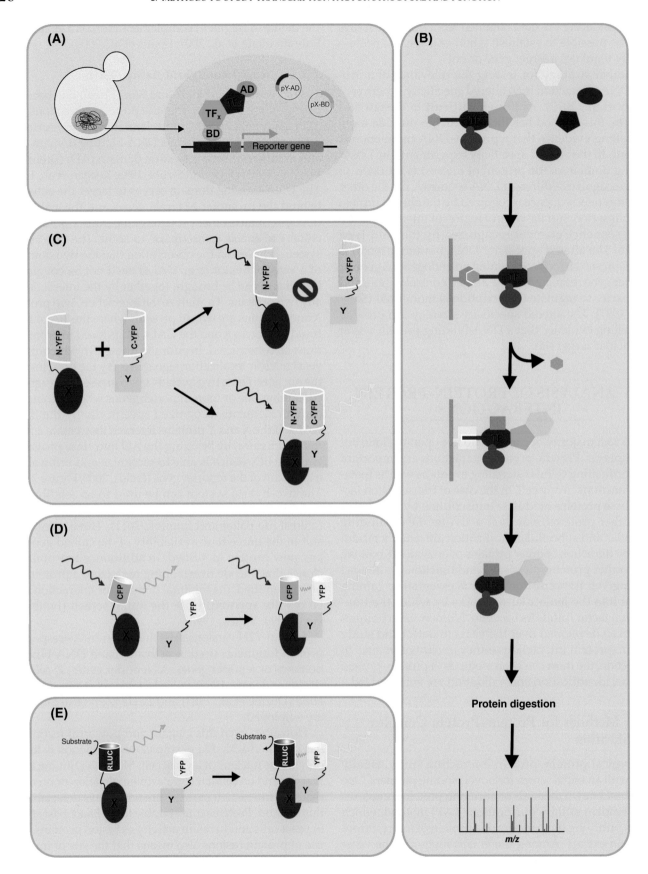

◀ FIGURE 2.4 **Methods to analyze protein–protein interactions.** (A) Yeast two-hybrid assay (Y2H). A yeast strain containing a reporter gene with a promoter recognized specifically by the yeast transcription factor GAL4 is transformed with two plasmids each encoding the proteins under study, TF_X and TF_Y, fused to the binding domain (BD; pink) or the activation domain (AD; green) of GAL4. If a physical interaction between TF_X and TF_Y occurs, the functionality of GAL4 is restored and the reporter gene is expressed. (B) Tandem affinity purification (TAP) for the purification of protein complexes from plants. The bait protein is expressed fused to two tags (e.g., protein A, green hexagon, and calmodulin-binding peptide, blue bar), separated by a protease cleavage site for the tobacco etch virus (TEV) protease. In the first round of purification, complexes are isolated through binding of the protein–A tag to an immobilized immunoglobulin G (IgG) and eluted by addition of the TEV protease. In the second round of purification, the complexes are isolated through binding of the remaining tag to a second affinity column. After elution, the proteins are digested and the peptides obtained are analyzed by mass spectrometry for protein identification. (C) Bimolecular fluorescence complementation (BiFC). The protein of interest (X, pink oval) and the potential partner (Y, green rectangle) are expressed fused to inactive N-terminal and C-terminal fragments of the yellow fluorescent protein (YFP). The interaction between X and Y brings the fragments together and reconstitutes the native fluorescent protein. If the proteins do not interact, the fluorescent signal is not detected. (D, E) In the fluorescence resonance energy transfer (FRET) (D) and the bioluminiscence resonance energy transfer (BRET) (E) methods a donor/acceptor pair of proteins with overlapping emission/absorption spectra is employed. If the proteins under study (X and Y) interact, the donor molecule transmits energy to the acceptor molecule and the emission energy of the acceptor is detected. In FRET (D) one of the proteins is fused to a fluorescent donor molecule (CFP, cyan cylinder) while in BRET (E) it is fused to the luciferase RLUC (blue cylinder) which emits luminescence in the presence of the substrate coelenterazine. The other candidate protein is fused to the acceptor fluorophore (YFP, yellow cylinder).

detected. Many proteins will activate transcription when fused to a DNA-binding domain, and this activation prevents a library screen from being performed. However, it is often possible to delete a small region of a protein that activates transcription and hence remove the activation function while retaining other properties of the protein. Finally, the yeast two-hybrid assay is based on reporter gene expression as an indirect readout. Despite this, several systematic and stringently controlled experimental and bioinformatics studies have demonstrated that Y2H can produce data of excellent quality, often surpassing that of other assays (Braun et al., 2009; Huang and Bader, 2009; Venkatesan et al., 2009; Yu et al., 2008, 2011). Especially important is the elimination of both constitutive and spontaneous autoactivators – usually constructs containing the DNA-binding domain that activate the reporter in the absence of an interaction – by fusing the cDNA library to the AD. The Y2H approach in plants has been summarized in several excellent reviews (Uhrig, 2006; Lalonde et al., 2008; Zhang et al., 2010).

2.5.1.2 Tandem and One-step Tag-Based Affinity Purification

While Y2H screening is the most utilized strategy to identify the partners of a protein of interest (Parrish et al., 2006; Tardif et al., 2007; Suter et al., 2008; Yu et al., 2008; Bonetta, 2010; *Arabidopsis* Interactome Mapping Consortium, 2011; Seo et al., 2011; Vernoux et al., 2011), this method is not applicable to the study of large multimeric protein complexes. One of the best-known methods for the isolation of protein complexes is TAP (Rohila et al., 2006; Van Leene et al., 2008; Rubio et al., 2005; Schoonheim et al., 2007). With this method, multicomponent protein complexes can be isolated from a cell lysate through affinity purification steps and then analyzed by mass spectrometry (MS) to identify purified proteins. The basic concept of TAP is the use of a so-called TAP tag, which is fused to a bait protein. The initial version of the TAP tag consisted of two sequential affinity tags separated by a tobacco etch virus (TEV) protease cleavage site that enabled two consecutive purification steps and reduced the amount of nonspecific binding (Figure 2.4B). Once expressed *in vivo* (stable or transient) the TAP-tagged protein associates to its endogenous targets and, after lysis of the cells, the TAP-tagged protein is allowed to bind via the first part of the TAP tag (e.g., protein A) to a specific column. Then the TEV protease is added and the TAP-tagged protein is cleaved, leaving the first affinity tag on the column. The bait protein, still fused to the second part of the TAP tag (e.g., a calmodulin-binding peptide) is then bound to a second column (e.g., calmodulin-coated beads), which is rinsed and eluted (e.g., by a buffer-containing EDTA) (Figure 2.4B). A platform with the GS tag, which combines two IgG-binding domains of protein G with a streptavidin-binding peptide separated by two tobacco etch virus cleavage sites, was developed to analyze protein complexes in *Arabidopsis thaliana* cell suspension cultures. This GS tag outperforms the traditional TAP tag in plant cells, regarding both specificity and complex yield (Van Leene et al., 2008). In addition to the possibility of isolating multimeric complexes, TAP has the advantage that complex formation occurs *in vivo* (*in planta*) and that posttranslationally modified proteins (e.g., phosphorylated, glycosylated, or oxidized/reduced proteins) are present. Even if TAP is highly sensitive and selective (Rohila et al., 2006; Rubio et al., 2005; Chang et al., 2009; Braun et al., 2013), weak protein–protein interactions may be lost during the purification steps. Another problem is that a relatively large amount of starting material is required, which can make purification and identification of low-abundance binding partners of transcription factors a difficult task.

Nowadays, many affinity tags are available for affinity purification followed by mass spectrometry methods (such as affinity purification and mass spectrometry, AP-MS) based on a single purification step. These systems employ the biotin peptide (Bio), the Flag epitope, c-Myc, His, HA, or the green fluorescent protein (GFP) as a tag (Fukao, 2012). In addition to GFP and its variants such as YFP and CFP (Cristea et al. 2005), magnetic beads conjugated to an anti-RFP antibody also work well for AP-MS analysis (Fukao, 2012). The interacting proteins detected by the fluorescent protein tag purification method can be confirmed by means of colocalization experiments in transgenic plants expressing fluorescent fusion proteins (see in later sections).

2.5.2 Methods for Verification of Protein–Protein Interactions

One of the major challenges when working with protein–protein interactions is to distinguish specific from unspecific binding. For example, MS-based protein identification has become so sensitive that any protein–protein interaction screen will result in a large number of identified contaminant proteins. Therefore, the general consensus is to confirm protein–protein interaction data using one or more independent approaches to arrive at an accurate evaluation. In some cases, the original method of screening can be used. Ideally, *in vivo* confirmation of the interaction is necessary. To do this, coimmunoprecipitation, fluorescence resonance energy transfer (FRET) and bimolecular fluorescence complementation (BiFC) methodologies can be used.

2.5.2.1 *Coimmunoprecipitation*

Coimmunoprecipitation is a technique in which a protein complex is purified from a cell lysate using an immobilized antibody against a known component of the complex and the interacting partners are analyzed by western blotting or MS (Ransone, 1995; Masters, 2004; Backstrom et al., 2007). Several strategies can be followed, using an antibody against an endogenous protein or a tagged fusion protein either transiently or stably expressed. Moreover, protein extracts can be incubated with recombinant bait proteins to trap interacting proteins (Swatek et al., 2011). As coimmunoprecipitation usually generates a significant background signal, it is important to perform appropriate negative controls (Ransone, 1995).

2.5.2.2 *In Vivo Split Methods*

In split methods, also called protein fragment complementation assays (PCA), the two proteins to be tested for interaction are fused to two fragments of a reporter protein, neither of which by itself has reporter activity. If the two proteins interact, they may bring the two split components together such that they can fold and the functionality of the reporter protein is restored. Among these reporters, the yellow fluorescent protein (YFP) and its variants have been commonly used (Figure 2.4C). This PCA method is referred to as bimolecular fluorescence complementation (BiFC; Kerppola, 2008). In BiFC, when two nonfluorescent fragments of an otherwise fluorescent protein are brought into close proximity by the interaction of two proteins fused to the fragments, complex formation can be visualized directly in a living cell by epifluorescence or confocal microscopy (Bhat et al., 2006; Ohad et al., 2007; Weinthal and Tzfira, 2009). Simultaneously, the subcellular location of the protein complex in the cell can be determined (Ohad et al., 2007; Weinthal and Tzfira, 2009). The BiFC assay can be performed by means of transient or stable expression of the fusion proteins (Waadt et al., 2014). Several binary plasmids for plant transformation have been developed for regular or gateway-based cloning (Gehl et al., 2009; Martin et al., 2009; Citovsky et al., 2008; Walter et al., 2004). Additionally, a series of multicolor BiFC (mcBiFC) vectors have been developed using spectral variants of the fluorescent proteins, which allows the interaction between a given "bait" protein and multiple "prey" proteins in living plant cells to be studied simultaneously (Lee et al., 2008; Waadt et al., 2008).

Another split reporter that can be used is luciferase. The luciferase split assay was first described in mammalian cells (Luker et al., 2004) and later adapted to plants (Fujikawa and Kato, 2007; Chen et al., 2008). It is particularly useful in plants because the autofluorescence of photosynthetic pigments is usually a problem in fluorescence-based assays. Split luciferase was shown to be the most sensitive protein–protein detection method; it is also able to detect protein dissociation (Li et al., 2011).

2.5.2.3 *Resonance Energy Transfer Methods*

Energy transfer techniques are additional strategies for assaying protein–protein interactions in living cells. The basic principle of these methods is that when the physical distance between two fluorescent proteins is small (<100 Å), energy can be passed from one to the other by a phenomenon called resonance energy transfer. In the case of the FRET method (Bhat et al., 2006), two fluorophores with overlapping emission/absorption spectra linked to proteins that might interact with each other are employed (Figure 2.4D). If the proteins interact and transition dipoles are appropriately oriented, the donor fluorophore is able to transfer its excited-state energy to the acceptor fluorophore. This energy transfer leads to a decrease in the donor's fluorescence intensity and a decreased lifetime of the excited state. As the acceptor molecule is a fluorophore, an increase in the acceptor's emission intensity is manifested. The efficiency of the resonance transfer depends upon the spectral overlap of the fluorophores, their relative orientation, as well as the distance between the donor and acceptor fluorophores (Pollok and Heim, 1999). Ideally, the acceptor should exhibit minimal excitation at the wavelength used to excite the donor fluorophore. Chromophore-mutated green fluorescent proteins (GFPs) with an excellent spectral overlap have been widely used in FRET studies as CFP/YFP and GFP/RFP variants (Pollok and Heim, 1999; Goedhart et al., 2007). The choice of fluorophores is a very important aspect in FRET. A comprehensive review article about this topic has been published (Shaner et al., 2005). On the other hand, the bioluminescence resonance energy transfer (BRET) method (Xu et al. 1999) uses a bioluminescent luciferase (RLUC), as a donor protein, which is genetically fused to one candidate protein, and a green fluorescent protein mutant (YFP) fused to the other protein of interest (Subramanian et al., 2004, Xu et al., 2007; Figure 2.4E). Interactions between the two fusion proteins can bring the luciferase and the fluorescent protein close enough for resonance energy transfer to occur, thus changing the color of the bioluminescent emission (Pfleger and Eidne, 2006).

Both FRET and BRET are nondestructive *in vivo* assays that allow real time measurements of protein–protein interactions. Even if FRET can be used to detect the molecular interactions between two proteins, it has been suggested as a method to analyze conformational alterations within a single polypeptide (Van Roessel and Brand, 2002; Janetopoulos et al., 2001). However, FRET has the disadvantage that it requires an excitation light source that may cause photobleaching or phototoxicity, as well as autofluorescence of plant pigments and unintended biological effects owing to cellular photoreceptors. Second, not only the photon donor (often CFP), but even the photon acceptor (often YFP) may be partially activated by the excitation light source, contributing to background signal (Bhat et al., 2006). In this sense, BRET does not need an excitation light source and is measured against a background of complete darkness. Thus, all photons emitted by the YFP acceptor originate from the luciferase and are indicative of BRET. BRET, on the other hand, requires a substrate. In the case of RLUC, this substrate is coelenterazine, which is nontoxic and membrane permeable (Subramanian et al., 2004).

A number of considerations must be taken into account before starting any fluorophore-based *in planta* protein–protein interaction assay (as FRET, BRET, or BiFC). First, these methods utilize tagged variants of the proteins of interest that can present alterations in their physiological parameters (such as subcellular localization, stability, and biological activity; Bhat et al., 2006). Thus, wherever possible, complementation of mutant phenotypes or analysis of protein activities *in vitro* must be carried out. Since proteins can be tagged either N- or C-terminally, all possible pairwise combinations should be tested when performing FRET or BiFC assays (Frank et al., 2005). Second, the type of promoter utilized to express the fusion proteins must be considered. Although expression is frequently driven by strong constitutive promoters, native gene promoters should be used to avoid possible artifacts that may either promote or inhibit the protein–protein interactions. Wherever possible, expression should take place in respective double-null mutants, since endogenous untagged copies of the proteins may interfere with the interaction assay. In summary, transgenic lines expressing both tagged proteins under control of their own promoters against a respective double-mutant genetic background must be used, whenever possible.

It is important to note that these methods determine the existence of a close physical proximity between the two tagged fusion proteins, but do not constitute proof of a true protein–protein interaction. To state this, evidence of direct interaction by *in vitro* assays using recombinant proteins or by Y2H is required. An interesting way to validate *in vivo* assays is by using mutant versions of the proteins in residues critical for the interaction as a control. If the loss of the interaction is coincident with an altered plant phenotype and the native proteins colocalize, then the biological relevance of the protein complex is demonstrated.

Acknowledgments

The authors acknowledge support from Consejo Nacional de Investigaciones Científicas y Técnicas (CONICET, Argentina), Agencia Nacional de Promoción Científica y Tecnológica (ANPCyT, Argentina), and Universidad Nacional del Litoral.

References

Abel, S., Theologis, A., 1994. Transient transformation of *Arabidopsis* leaf protoplasts: a versatile experimental system to study gene expression. Plant J. 5, 421–427.

Amendt, B.A., Sutherland, L.B., Russo, A.F., 1999. Transcriptional antagonism between Hmx1 and Nkx2.5 for a shared DNA-binding site. J. Biol. Chem. 274, 11635–11642.

Aoyama, T., Chua, N., 1997. A glucocorticoid-mediated transcriptional induction system in transgenic plants. Plant J. 11, 605–612.

Arabidopsis Interactome Mapping Consortium, 2011. Evidence for network evolution in an *Arabidopsis* interactome map. Science 333, 601–607.

Backstrom, S., Elfving, N., Nilsson, R., Wingsle, G., Bjorklund, S., 2007. Purification of a plant mediator from *Arabidopsis thaliana* identifies PFT1 as the Med25 subunit. Mol. Cell 26, 717–729.

Badis, G., Berger, M.F., Philippakis, A.A., Talukder, S., Gehrke, A.R., Jaeger, S.A., Chan, E.T., Metzler, G., Vedenko, A., Chen, X., Kuznetsov, H., Wang, C.F., Coburn, D., Newburger, D.E., Morris, Q., Hughes, T.R., Bulyk, M.L., 2009. Diversity and complexity in DNA recognition by transcription factors. Science 324, 1720–1723.

Balasubramanian, B., Pogozelski, W.K., Tullius, T.D., 1998. DNA strand breaking by the hydroxyl radical is governed by the accessible surface areas of the hydrogen atoms of the DNA backbone. Proc. Natl. Acad. Sci. USA 95, 9738–9743.

Berger, M.F., Bulyk, M.L., 2009. Universal protein-binding microarrays for the comprehensive characterization of the DNA-binding specificities of transcription factors. Nat. Protoc. 4, 393–411.

Berger, M.F., Philippakis, A.A., Qureshi, A.M., He, F.S., Estep 3rd, P.W., Bulyk, M.L., 2006. Compact, universal DNA microarrays to comprehensively determine transcription factor binding site specificities. Nat. Biotechnol. 24, 1429–1435.

Bhat, R.A., Lahaye, T., Panstruga, R., 2006. The visible touch: *in planta* visualization of protein–protein interactions by fluorophore-based methods. Plant Methods 2, 12.

Blackwell, T.K., Weintraub, H., 1990. Differences and similarities in DNA-binding preferences of MyoD and E2A protein complexes revealed by binding site selection. Science 250, 1104–1110.

Bonetta, L., 2010. Protein–protein interactions: interactome under construction. Nature 468, 851–854.

Braun, P., Aubourg, S., Van Leene, J., De Jaeger, G., Lurin, C., 2013. Plant protein interactomes. Annu. Rev. Plant Biol. 64, 61–87.

Braun, P., Tasan, M., Dreze, M., Barrios-Rodiles, M., Lemmens, I., Yu, H., Sahalie, J.M., Murray, R.R., Roncari, L., de Smet, A.S., Venkatesan, K., Rual, J.F., Vandenhaute, J., Cusick, M.E., Pawson, T., Hill, D.E., Tavernier, J., Wrana, J.L., Roth, F.P., Vidal, M., 2009. An experimentally derived confidence score for binary protein–protein interactions. Nat. Methods 6, 91–97.

Brent, R., Ptashne, M., 1985. A eukaryotic transcriptional activator bearing the DNA specificity of a prokaryotic repressor. Cell 43, 729–736.

Brooks, J.E., Blumenthal, R.M., Gingeras, T.R., 1983. The isolation and characterization of the *Escherichia coli* DNA adenine methylase (dam) gene. Nucleic Acids Res. 11, 837–851.

Bulyk, M.L., 2007. Protein binding microarrays for the characterization of DNA–protein interactions. Adv. Biochem. Eng. Biotechnol. 104, 65–85.

Bulyk, M.L., Gentalen, E., Lockhart, D.J., Church, G.M., 1999. Quantifying DNA–protein interactions by double-stranded DNA arrays. Nat. Biotechnol. 17, 573.

Bulyk, M.L., Huang, X., Choo, Y., Church, G.M., 2001. Exploring the DNA binding specificities of zinc fingers with DNA microarrays. Proc. Natl. Acad. Sci. USA 98, 7158–7163.

Caddick, M.X., Greenland, A.J., Jepson, I., Krause, K.-P., Qu, N., Riddell, K.V., Salter, M.G., Schuch, W., Sonnewald, U., Tomsett, A.B., 1998. An ethanol inducible gene switch for plants used to manipulate carbon metabolism. Nat. Biotechnol. 16, 177–180.

Carey, M.F., Peterson, C.L., Smale, S.T., 2012. Confirming the functional importance of a protein–DNA interaction. Cold Spring Harb. Protoc. 2012, 733–757.

Carey, M.F., Peterson, C.L., Smale, S.T., 2013. DNase I footprinting. Cold Spring Harb. Protoc. 2013, 469–478.

Chai, C., Xie, Z., Grotewold, E., 2011. SELEX (systematic evolution of ligands by exponential enrichment), as a powerful tool for deciphering the protein–DNA interaction space. Plant transcription factors. Methods Mol. Biol. 754 (Part 5), 249–258.

Chang, I.F., Curran, A., Woolsey, R., Quilici, D., Cushman, J.C., et al., 2009. Proteomic profiling of tandem affinity purified 14-3-3 protein complexes in *Arabidopsis thaliana*. Proteomics 9, 2967–2985.

Chen, H., Zou, Y., Shang, Y., Lin, H., Wang, Y., et al., 2008. Firefly luciferase complementation imaging assay for protein–protein interactions in plants. Plant Physiol. 146, 368–376.

Citovsky, V., Gafni, Y., Tzfira, T., 2008. Localizing protein–protein interactions by bimolecular fluorescence complementation in planta. Methods 45, 196–206.

Craft, J., Samalova, M., Baroux, C., Townley, H., Martinez, A., Jepson, I., Tsiantis, M., Moore, I., 2005. New pOp/LhG4 vectors for stringent glucocorticoid-dependent transgene expression in *Arabidopsis*. Plant J. 41, 899–918.

Cristea, I.M., Williams, R., Chait, B.T., Rout, M.P., 2005. Fluorescent proteins as proteomic probes. Mol. Cell Proteomics 4, 1933–1941.

Das, P.M., Ramachandran, K., vanWert, J., Singal, R., 2004. Chromatin immunoprecipitation assay. Biotechniques 37, 961–969.

de Folter, S., Immink, R.G., 2011. Yeast protein–protein interaction assays and screens. Methods Mol. Biol. 754, 145–165.

Dervan, P.B., 1986. Design of sequence specific DNA-binding molecules. Science 232, 464–471.

Doi, N., Takashima, H., Kinjo, M., Sakata, K., Kawahashi, Y., Oishi, Y., Oyama, R., Miyamoto-Sato, E., Sawasaki, T., Endo, Y., Yanagawa, H., 2002. Novel fluorescence labeling and high-throughput assay technologies for *in vitro* analysis of protein interactions. Genome Res. 12, 487–492.

Durfee, T., Becherer, K., Chen, P.L., Yeh, S.H., Yang, Y., Kilburn, A.E., Lee, W.H., Elledge, S.J., 1993. The retinoblastoma protein associates with the protein phosphatase type 1 catalytic subunit. Genes Dev. 7, 555–569.

Elliott, D.A., Brand, A.H., 2008. The GAL4 system: a versatile system for the expression of genes. Methods Mol. Biol. 420, 79–95.

Fields, S., 2009. Interactive learning: lessons from two hybrids over two decades. Proteomics 9, 5209–5213.

Fields, S., Song, O., 1989. A novel genetic system to detect protein–protein interactions. Nature 340, 245–246.

Frank, M., Thumer, L., Lohse, M.J., Bunemann, M., 2005. G protein activation without subunit dissociation depends on a G alpha(i)-specific region. J. Biol. Chem. 280, 24584–24590.

Fried, M., Crothers, D.M., 1981. Equilibria and kinetics of lac repressor-operator interactions by polyacrylamide gel electrophoresis. Nucleic Acids Res. 9, 6505–6525.

Fried, M., Daugherty, M.A., 1998. Electrophoretic analysis of multiple protein–DNA interactions. Electrophoresis 19, 1247–1253.

Fujikawa, Y., Kato, N., 2007. Split luciferase complementation assay to study protein–protein interactions in *Arabidopsis* protoplasts. Plant J. 52, 185–195.

Fukao, Y., 2012. Protein–protein interactions in plants. Plant Cell Physiol. 53, 617–625.

Galas, D.J., Schmitz, A., 1978. DNase footprinting: a simple method for the detection of protein–DNA binding specificity. Nucleic Acids Res. 5, 3157–3170.

Gallagher, S.R., 1992. GUS Protocols: Using the GUS Gene as a Reporter of Gene Expression. Academic Press, Boston, MA.

Garner, M.M., Revzin, A., 1981. A gel electrophoresis method for quantifying the binding of proteins to specific DNA regions: application to components of the *Escherichia coli* lactose operon regulatory system. Nucleic Acids Res. 9, 3047–3060.

Gatz, C., Frohberg, C., Wendenburg, R., 1992. Stringent repression and homogeneous de-repression by tetracycline of a modified CaMV 35S promoter in intact transgenic tobacco plants. Plant J. 2, 397–404.

Gehl, C., Waadt, R., Kudla, J., Mendel, R.R., Hansch, R., 2009. New GATEWAY vectors for high throughput analyses of protein–protein interactions by bimolecular fluorescence complementation. Mol. Plant 2, 1051–1058.

Gendrel, A.V., Lippman, Z., Yordan, C., Colot, V., Martienssen, R.A., 2002. Dependence of heterochromatic histone H3 methylation patterns on the *Arabidopsis* gene DDM1. Science 297, 1871–1873.

Germann, S., Gaudin, V., 2011. Mapping *in vivo* protein–DNA interactions in plants by DamID, a DNA adenine methylation-based method. Methods Mol. Biol. 754, 307–321.

Gerstle, J.T., Fried, M.G., 1993. Measurement of binding kinetics using the gel electrophoresis mobility shift assay. Electrophoresis 14, 725–731.

Gilmour, D.S., Lis, J.T., 1984. Detecting protein–DNA interactions *in vivo*: distribution of RNA polymerase on specific bacterial genes. Proc. Natl. Acad. Sci. USA 81, 4275–4279.

Gilmour, D.S., Lis, J.T., 1985. *In vivo* interactions of RNA polymerase II with genes of *Drosophila melanogaster*. Mol. Cell. Biol. 5, 2009–2018.

Godoy, M., Franco-Zorrilla, J.M., Pérez-Pérez, J., Oliveros, J.C., Lorenzo, O., Solano, R., 2011. Improved protein-binding microarrays for the identification of DNA-binding specificities of transcription factors. Plant J. 66, 700–711.

Goedhart, J., Vermeer, J.E., Adjobo-Hermans, M.J., van Weeren, L., Gadella, T.W., 2007. Sensitive detection of p65 homodimers using red-shifted and fluorescent protein-based FRET couples. PLoS ONE 2, e1011.

Gopinath, S.C., 2007. Methods developed for SELEX. Anal. Bioanal. Chem. 387, 171–182.

Grotewold, E., Drummond, B.J., Bowen, B., Peterson, T., 1994. The myb-homologous P gene controls phlobaphene pigmentation in maize floral organs by directly activating a flavonoid biosynthetic gene subset. Cell 76, 543–553.

Guyer, D., Tuttle, A., Rouse, S., Volrath, S., Johnson, M., Potter, S., Görlach, J., Goff, S., Crossland, L., Ward, E., 1998. Activation of latent transgenes in *Arabidopsis* using a hybrid transcription factor. Genetics 149, 633–639.

Harbison, C.T., Gordon, D.B., Lee, T.I., Rinaldi, N.J., Macisaac, K.D., Danford, T.W., Hannett, N.M., Tagne, J.B., Reynolds, D.B., Yoo, J., Jennings, E.G., Zeitlinger, J., Pokholok, D.K., Kellis, M., Rolfe, P.A., Takusagawa, K.T., Lander, E.S., Gifford, D.K., Fraenkel, E., Young, R.A., 2004. Transcriptional regulatory code of a eukaryotic genome. Nature 431, 99–104.

Hellman, L.M., Fried, M.G., 2007. Electrophoretic mobility shift assay (EMSA) for detecting protein–nucleic acid interactions. Nat. Protoc. 2, 1849–1861.

Helwa, R., Hoheisel, J.D., 2010. Analysis of DNA–protein interactions: from nitrocellulose filter binding assays to microarray studies. Anal. Bioanal. Chem. 398, 2551–2561.

Hiratsu, K., Matsui, K., Koyama, T., Ohme-Takagi, M., 2003. Dominant repression of target genes by chimeric repressors that include the EAR motif, a repression domain, in *Arabidopsis*. Plant J. 34, 733–739.

Hope, I.A., Struhl, K., 1986. Functional dissection of a eukaryotic transcriptional activator protein, GCN4 of yeast. Cell 46, 885–894.

Hu, C.D., Chinenov, Y., Kerppola, T.K., 2002. Visualization of interactions among bZIP and Rel family proteins in living cells using bimolecular fluorescence complementation. Mol. Cell 9, 789–798.

Huang, H., Bader, J.S., 2009. Precision and recall estimates for two-hybrid screens. Bioinformatics 25, 372–378.

Huang, H., Mizukami, Y., Hu, Y., Ma, H., 1993. Isolation and characterization of the binding sequences for the product of the *Arabidopsis* floral homeotic gene AGAMOUS. Nucleic Acids Res. 21, 4769–4776.

Iwabuchi, K., Li, B., Bartel, P., Fields, S., 1993. Use of the two-hybrid system to identify the domain of p53 involved in oligomerization. Oncogene 8, 1693–1696.

Iwata, Y., Lee, M.H., Koizumi, N., 2011. Analysis of a transcription factor using transient assay in *Arabidopsis* protoplasts. Methods Mol. Biol. 754, 107–117.

Jain, S., Tullius, T., 2008. Footprinting protein–DNA complexes using the hydroxyl radical. Nat. Protoc. 3, 1092–1100.

Janetopoulos, C., Jin, T., Devreotes, P., 2001. Receptor-mediated activation of heterotrimeric G-proteins in living cells. Science 291, 2408–2411.

Ji, X., Wang, L., Nie, X., He, L., Zang, D., Liu, Y., Zhang, B., Wang, Y., 2014. A novel method to identify the DNA motifs recognized by a defined transcription factor. Plant Mol. Biol. 86, 367–380.

Kagale, S., Rozwadowski, K., 2011. EAR motif-mediated transcriptional repression in plants: an underlying mechanism for epigenetic regulation of gene expression. Epigenetics 6, 141–146.

Kaufmann, K., Muino, J.M., Osterns, M., Farinelli, L., Krajewsli, P., Angenent, G.C., 2010. Chromatin immunoprecipitation (ChIP) of plant transcription factors followed by sequencing (ChIP-SEQ) or hybridization to whole genome arrays (ChIP-CHIP). Nat. Protoc. 5, 457–472.

Keegan, L., Gill, G., Ptashne, M., 1986. Separation of DNA binding from the transcription-activating function of a eukaryotic regulatory protein. Science 231, 699–704.

Kerppola, T.K., 2008. Bimolecular fluorescence complementation (BiFC) analysis as a probe of protein interactions in living cells. Annu. Rev. Biophys. 37, 465–487.

Kim, M.J., Lee, T.H., Pahk, Y.M., Kim, Y.H., Park, H.M., Choi, Y.D., Nahm, B.H., Yeon-Ki Kim, Y.K., 2009. Quadruple 9-mer-based protein binding microarray with DsRed fusion protein. BMC Mol. Biol. 10, 91–102.

Kuo, M.H., Allis, C.D., 1999. *In vivo* cross-linking and immunoprecipitation for studying dynamic protein: DNA associations in a chromatin environment. Methods 19, 425–433.

Lalonde, S., Ehrhardt, D.W., Loque, D., Chen, J., Rhee, S.Y., Frommer, W.B., 2008. Molecular and cellular approaches for the detection of protein–protein interactions: latest techniques and current limitations. Plant J. 53, 610–635.

Lee, T.I., Johnstone, S.E., Young, R.A., 2006. Chromatin immunoprecipitation and microarray-based analysis of protein location. Nat. Protoc. 1, 729–748.

Lee, T.I., Rinaldi, N.J., Robert, F., Odom, D.T., Bar-Joseph, Z., Gerber, G.K., Hannett, N.M., Harbison, C.T., Thompson, C.M., Simon, I., Zeitlinger, J., Jennings, E.G., Murray, H.L., Gordon, D.B., Ren, B., Wyrick, J.J., Tagne, J.B., Volkert, T.L., Fraenkel, E., Gifford, D.K., Young, R.A., 2002. Transcriptional regulatory networks in *Saccharomyces cerevisiae*. Science 298, 799–804.

Lee, L.Y., Fang, M.J., Kuang, L.Y., Gelvin, S.B., 2008. Vectors for multicolor bimolecular fluorescence complementation to investigate protein–protein interactions in living plant cells. Plant Methods 4, 24.

Li, J., Herskowitz, I., 1993. Isolation of the ORC6, a component of the yeast origin recognition complex by a one-hybrid system. Science 262, 1870–1874.

Li, J.F., Bush, J., Xiong, Y., Li, L., McCormack, M., 2011. Large-scale protein–protein interaction analysis in *Arabidopsis* mesophyll protoplasts by split firefly luciferase complementation. PLoS ONE 6, e27364.

Lloyd, A.M., Schena, M., Walbot, V., Davis, R.W., 1994. Epidermal cell fate determination in *Arabidopsis*: patterns defined by a steroid-inducible regulator. Science 266, 436–439.

Luker, K.E., Smith, M.C., Luker, G.D., Gammon, S.T., Piwnica-Worms, H., Piwnica-Worms, D., 2004. Kinetics of regulated protein–protein interactions revealed with firefly luciferase complementation imaging in cells and living animals. Proc. Natl. Acad. Sci. USA 101, 12288–12293.

Manfield, L.W., Stockley, P.G., 2009. Ethylation interference footprinting of DNA–protein complexes. Methods Mol. Biol. 543, 105–120.

Mardis, E.R., 2007. ChIP-seq: welcome to the new frontier. Nat. Methods 4, 613–614.

Martin, K., Kopperud, K., Chakrabarty, R., Banerjee, R., Brooks, R., Goodin, M.M., 2009. Transient expression in *Nicotiana benthamiana* fluorescent marker lines provides enhanced definition of protein localization, movement and interactions *in planta*. Plant J. 59, 150–162.

Martinez, A., Sparks, C., Hart, C.A., Thompson, J., Jepson, I., 1999. Ecdysone agonist inducible transcription in transgenic tobacco plants. Plant J. 19, 97–106.

Masters, S.C., 2004. Co-immunoprecipitation from transfected cells. Methods Mol. Biol. 261, 337–348.

Menges, M., Murray, J.A., 2002. Synchronous *Arabidopsis* suspension cultures for analysis of cell-cycle gene activity. Plant J. 30, 203–212.

Mitsuda, N., Matsui, K., Ikeda, M., Nakata, M., Oshima, Y., Nagatoshi, Y., Ohme-Takagi, M., 2011. CRES-T, an effective gene silencing system utilizing chimeric repressors. Methods Mol. Biol. 754, 87–105.

Moore, I., Gälweiler, L., Grosskopf, D., Schell, J., Palme, K., 1998. A transcription activation system for regulated gene expression in transgenic plants. Proc. Natl. Acad. Sci. USA 95, 376–381.

Moore, I., Samalova, M., Kurup, S., 2006. Transactivated and chemically inducible gene expression in plants. Plant J. 45, 651–683.

Moorman, C., Sun, L.V., Wang, J., de Wit, E., Talhout, W., Ward, L.D., Greil, F., Lu, X.J., White, K.P., Bussemaker, H.J., van Steensel, B., 2006. Hotspots of transcription factor colocalization in the genome of *Drosophila melanogaster*. Proc. Natl. Acad. Sci. USA 103, 12027–12032.

Mukherjee, S., Berger, M.F., Jona, G., Wang, X.S., Muzzey, D., Snyder, M., Young, R.A., Bulyk, M.L., 2004. Rapid analysis of the DNA-binding specificities of transcription factors with DNA microarrays. Nat. Genet. 36, 1331–1339.

Negre, N., Hennetin, J., Sun, L.V., Lavrov, S., Bellis, M., White, K.P., Cavalli, G., 2006. Chromosomal distribution of PcG proteins during *Drosophila* development. PLoS Biol. 4, e170.

Nole-Wilson, S., Krizek, B.A., 2000. DNA binding properties of the *Arabidopsis* floral development protein AINTEGUMENTA. Nucleic Acids Res. 28, 4076–4082.

Ohad, N., Shichrur, K., Yalovsky, S., 2007. The analysis of protein–protein interactions in plants by bimolecular fluorescence complementation. Plant Physiol. 145, 1090–1099.

Oliphant, A.R., Brandl, C.J., Struhl, K., 1989. Defining the sequence specificity of DNA-binding proteins by selecting binding sites from random-sequence oligonucleotides: analysis of yeast GCN4 protein. Mol. Cell. Biol. 9, 2944–2949.

Ouwerkerk, P.B., Meijer, A.H., 2001. Yeast one-hybrid screening for DNA–protein interactions. Curr. Protoc. Mol. Biol. 12, 12.1–12.22.

Ow, D.W., De Wet, J.R., Helinski, D.R., Howell, S.H., Wood, K.V., Deluca, M., 1986. Transient and stable expression of the firefly luciferase gene in plant cells and transgenic plants. Science 234, 856–859.

Padidam, M., Gore, M., Lu, D.L., Smirnova, O., 2003. Chemical inducible, ecdysone receptor-based gene expression system for plants. Transgenic Res. 12, 101–109.

Parrish, J.R., Gulyas, K.D., Finley, R.L., 2006. Yeast two-hybrid contributions to interactome mapping. Curr. Opin. Biotechnol. 17, 387–393.

Pfleger, K.D.G., Eidne, K.A., 2006. Illuminating insights into protein–protein interactions using bioluminescence resonance energy transfer (BRET). Nat. Methods 3, 165–174.

Pollok, B.A., Heim, R., 1999. Using GFP in FRET-based applications. Trends Cell Biol. 9, 57–60.

Ransone, L.J., 1995. Detection of protein–protein interactions by coimmunoprecipitation and dimerization. Methods Enzymol. 254, 491–497.

Reece-Hoyes, J.S., Walhout, A.J.M., 2012. Yeast one-hybrid assays: a historical and technical perspective. Methods 57, 441–447.

Ren, B., Robert, F., Wyrick, J.J., Aparicio, O., Jennings, E.G., Simon, I., Zeitlinger, J., Schreiber, J., Hannett, N., Kanin, E., Volkert, T.L., Wilson, C.J., Bell, S.P., Young, R.A., 2000. Genome-wide location and function of DNA binding proteins. Science 290, 2306–2309.

Rohila, J.S., Chen, M., Chen, S., Chen, J., Cerny, R., Dardick, C., Canlas, P., Xu, X., Gribskov, M., Kanrar, S., Zhu, J.K., Ronald, P., Fromm, M.E., 2006. Protein–protein interactions of tandem affinity purification-tagged protein kinases in rice. Plant J. 46, 1–13.

Rubio, V., Shen, Y.P., Saijo, Y., Liu, Y.L., Gusmaroli, G., Dinesh-Kumar, S.P., Deng, X.W., 2005. An alternative tandem affinity purification strategy applied to Arabidopsis protein complex isolation. Plant J. 41, 767–778.

Sadowski, I., Ma, J., Triezenberg, S., Ptashne, M., 1988. GAL4-VP16 is an unusually potent transcriptional activator. Nature 335, 563–564.

Saleh, A., Alvarez-Venegas, R., Avramova, Z., 2008. An efficient chromatin immunoprecipitation (ChIP) protocol for studying histone modifications in Arabidopsis plants. Nat. Protoc. 3, 1018–1025.

Schena, M., Lloyd, A.M., Walbot, V., Davis, R.W., 1991. A steroid inducible gene expression system for plant cells. Proc. Natl. Acad. Sci. USA 88, 10421–10425.

Schoonheim, P.J., Veiga, H., Pereira Dda, C., Friso, G., van Wijk, K.J., de Boer, A.H., 2007. A comprehensive analysis of the 14-3-3 interactome in barley leaves using a complementary proteomics and two-hybrid approach. Plant Physiol. 143, 670–683.

Seo, Y.S., Chern, M., Bartley, L.E., Han, M., Jung, K.H., Lee, I., et al., 2011. Towards establishment of a rice stress response interactome. PLoS Genet. 7, e1002020.

Shah, P.C., Bertolino, E., Singh, H., 1997. Using altered specificity Oct-1 and Oct-2 mutants to analyze the regulation of immunoglobulin gene transcription. EMBO J. 16, 7105–7117.

Shaner, N.C., Steinbach, P.A., Tsien, R.Y., 2005. A guide to choosing fluorescent proteins. Nat. Methods 2, 905–909.

Shaw, P.E., Stewart, A.F., 1994. Identification of protein–DNA contacts with dimethyl sulfate: methylation protection and methylation interference. Methods Mol. Biol. 30, 79–87.

Shaw, P.E., Stewart, A.F., 2009. Identification of protein/DNA contacts with dimethyl sulfate. Methylation protection and methylation interference. Methods Mol. Biol. 543, 97–104.

Sieweke, M., 2000. Detection of transcription factor partners with a yeast one hybrid screen. Methods Mol. Biol. 130, 59–77.

Spotts, J.M., Dolmetsch, R.E., Greenberg, M.E., 2002. Time-lapse imaging of a dynamic phosphorylation-dependent protein–protein interaction in mammalian cells. Proc. Natl. Acad. Sci. USA 99, 15142–15147.

Steiner, S., Pfannschmidt, T., 2009. Fluorescence-based electrophoretic mobility shift assay in the analysis of DNA-binding proteins. Methods Mol. Biol. 479, 273–289.

Subramaniam, R., Desveaux, D., Spickler, C., Michnick, S.W., Brisson, N., 2001. Direct visualization of protein interactions in plant cells. Nat. Biotechnol. 19, 769–772.

Subramanian, C., Xu, Y., Johnson, C.H., von Arnim, A.G., 2004. In vivo detection of protein–protein interaction in plant cells using BRET. Methods Mol. Biol. 284, 271–286.

Suter, B., Kittanakom, S., Stagljar, I., 2008. Two-hybrid technologies in proteomics research. Curr. Opin. Biotechnol. 19, 316–323.

Swatek, K.N., Graham, K., Agrawal, G.K., Thelen, J.J., 2011. The 14-3-3 isoforms chi and epsilon differentially bind client proteins from developing Arabidopsis seed. J. Proteome Res. 10, 4076–4087.

Tardif, G., Kane, N.A., Adam, H., Labrie, L., Major, G., Gulick, P., et al., 2007. Interaction network of proteins associated with abiotic stress response and development in wheat. Plant Mol. Biol. 63, 703–718.

Tolhuis, B., de Wit, E., Muijrers, I., Teunissen, H., Talhout, W., van Steensel, B., van Lohuizen, M., 2006. Genome-wide profiling of PRC1 and PRC2 Polycomb chromatin binding in Drosophila melanogaster. Nat. Genet. 38, 694–699.

Tompa, R., McCallum, C., Delrow, J., Henikoff, J., van Steensel, B., Henikoff, S., 2002. Genome-wide profiling of DNA methylation reveals transposon targets of CHROMOMETHYLASE3. Curr. Biol. 12, 65–68.

Tron, A.E., Comelli, R.N., Gonzalez, D.H., 2005. Structure of homeodomain–leucine zipper/DNA complexes studied using hydroxyl radical cleavage of DNA and methylation interference. Biochemistry 44, 16796–16803.

Tuerk, C., Gold, L., 1990. Systematic evolution of ligands by exponential enrichment: RNA ligands to bacteriophage T4 DNA polymerase. Science 249, 505–510.

Tullius, T.D., 1988. DNA footprinting with hydroxyl radical. Nature 332, 663–664.

Tullius, T.D., 1989. Physical studies of protein–DNA complexes by footprinting. Annu. Rev. Biophys. Biophys. Chem. 18, 213–237.

Tullius, T.D., Dombroski, B.A., 1986. Hydroxyl radical "footprinting": high-resolution information about DNA–protein contacts and application to lambda repressor and Cro protein. Proc. Natl. Acad. Sci. USA 83, 5469–5473.

Ueki, S., Lacroix, B., Krichevsky, A., Lazarowitz, S.G., Citovsky, V., 2009. Functional transient genetic transformation of Arabidopsis leaves by biolistic bombardment. Nat. Protoc. 4, 71–77.

Uhrig, J.F., 2006. Protein interaction networks in plants. Planta 224, 771–781.

Van Leene, J., Witters, E., Inze, D., De Jaeger, G., 2008. Boosting tandem affinity purification of plant protein complexes. Trends Plant Sci. 13, 517–520.

Van Roessel, P., Brand, A.H., 2002. Imaging into the future: visualizing gene expression and protein interactions with fluorescent proteins. Nat. Cell Biol. 4, E15–E20.

van Steensel, B., Henikoff, S., 2000. Identification of in vivo DNA targets of chromatin proteins using tethered dam methyltransferase. Nat. Biotechnol. 18, 424–428.

van Steensel, B., Delrow, J., Henikoff, S., 2001. Chromatin profiling using targeted DNA adenine methyltransferase. Nat. Genet. 27, 304–308.

Venkatesan, K., Rual, J.F., Vazquez, A., Stelzl, U., Lemmens, I., et al., 2009. An empirical framework for binary interactome mapping. Nat. Methods 6, 83–90.

Vernoux, T., Brunoud, G., Farcot, E., Morin, V., Van den Daele, H., Legrand, J., et al., 2011. The auxin signalling network translates dynamic input into robust patterning at the shoot apex. Mol. Syst. Biol. 7, 508.

Viola, I.L., Gonzalez, D.H., 2011. Footprinting and missing nucleoside analysis of transcription factor DNA complexes. Methods Mol. Biol. 754, 259–275.

Viola, I.L., Reinheimer, R., Ripoll, R., Uberti Manassero, N.G., Gonzalez, D.H., 2012. Determinants of the DNA binding specificity of Class I and Class II TCP transcription factors. J. Biol. Chem. 287, 347–356.

Vogel, M.J., Peric-Hupkes, D., van Steensel, B., 2007. Detection of in vivo protein–DNA interactions using DamID in mammalian cells. Nat. Protoc. 2, 1467–1478.

Vojtek, A.B., Hollenberg, S.M., Cooper, J.A., 1993. Mammalian Ras interacts directly with the serine/threonine kinase Raf. Cell 74, 205–214.

Waadt, R., Schmidt, L.K., Lohse, M., Hashimoto, K., Bock, R., Kudla, J., 2008. Multicolor bimolecular fluorescence complementation (mcBiFC) reveals simultaneous formation of alternative CBL/CIPK complexes in planta. Plant J. 56, 505–516.

Waadt, R., Schlücking, K., Schroeder, J.I., Kudla, J., 2014. Protein fragment bimolecular fluorescence complementation analyses for the in vivo study of protein–protein interactions and cellular protein complex localizations. Methods Mol. Biol. 1062, 629–658.

Walter, J., Dever, C.A., Biggin, M.D., 1994. Two homeodomain proteins bind with similar specificity to a wide range of DNA sites in *Drosophila* embryos. Genes Dev. 8, 1678–1692.

Walter, M., Chaban, C., Schutze, K., Batistic, O., Weckermann, K., et al., 2004. Visualization of protein interactions in living plant cells using bimolecular fluorescence complementation. Plant J. 40, 428–438.

Wang, Z., Gerstein, M., Snyder, M., 2009. RNA-Seq: a revolutionary tool for transcriptomics. Nat. Rev. Genet. 10, 57–63.

Warren, C.L., Kratochvil, N.C., Hauschild, K.E., Foister, S., Brezinski, M.L., Dervan, P.B., Phillips, Jr., G.N., Ansari, A.Z., 2006. Defining the sequence recognition profile of DNA-binding molecules. Proc. Natl. Acad. Sci. USA 103, 867–872.

Weinthal, D., Tzfira, T., 2009. Imaging protein–protein interactions in plant cells by bimolecular fluorescence complementation assay. Trends Plant Sci. 14, 59–63.

Wilde, R.J., Cooke, S.E., Brammar, W.J., Schuch, W., 1994. Control of gene expression in plant cells using a 434:VP16 chimeric protein. Plant Mol. Biol. 24, 381–388.

Winter, C.M., Austin, R.S., Blanvillain-Baufume, S., Reback, M.A., Monniaux, M., Wu, M.F., Sang, Y., Yamaguchi, A., Yamaguchi, N., Parker, J.E., Parcy, F., Jensen, S.T., Li, H., Wagner, D., 2011. LEAFY target genes reveal floral regulatory logic, *cis*-motifs, a link to biotic stimulus response. Dev. Cell 20, 430–443.

Wissmann, A., Hillen, W., 1991. DNA contacts probed by modification protection and interference studies. Methods Enzymol. 208, 365–379.

Wu, J., Smith, L.T., Plass, C., Huang, T.H., 2006. ChIP-chip comes of age for genome-wide functional analysis. Canc. Res. 66, 6899–6902.

Wu, M.F., Sang, Y., Bezhani, S., Yamaguchi, N., Han, S.K., Li, Z., Su, Y., Slewinski, T.L., Wagner, D., 2012. SWI2/SNF2 chromatin remodeling ATPases overcome polycomb repression and control floral organ identity with the LEAFY and SEPALLATA3 transcription factors. Proc. Natl. Acad. Sci. USA 109, 3576–3581.

Xu, Y., Piston, D.W., Johnson, C.H., 1999. A bioluminescence resonance energy transfer (BRET) system: application to interacting circadian clock proteins. Proc. Natl. Acad. Sci. USA 96, 151–156.

Xu, X., Soutto, M., Xie, Q., Servick, S., Subramanian, C., von Arnim, A.G., Johnson, C.H., 2007. Imaging protein interactions with bioluminescence resonance energy transfer (BRET) in plant and mammalian cells and tissues. Proc. Natl. Acad. Sci. USA 104, 10264–10269.

Yamaguchi, N., Wu, M.-F., Winter, C., Berns, M., Nole-Wilson, S., Yamaguchi, A., Coupland, G., Krizek, B., Wagner, D., 2013. A molecular framework for auxin-mediated initiation of floral promordia. Dev. Cell 24, 271–282.

Yamaguchi, N., Winter, C.M., Wu, M.F., Kwon, C.S., William, D.A., Wagner, D., 2014. PROTOCOLS: Chromatin Immunoprecipitation from *Arabidopsis* Tissues. Arabidopsis Book 12, e0170.

Yang, Y., Li, R., Qi, M., 2000. In vivo analysis of plant promoters and transcription factors by agroinfiltration of tobacco leaves. Plant J. 22, 543–551.

Yoo, S.D., Cho, Y.H., Sheen, J., 2007. *Arabidopsis* mesophyll protoplasts: a versatile cell system for transient gene expression analysis. Nat. Protoc. 2, 1565–1572.

Yu, H., Braun, P., Yildirim, M.A., Lemmens, I., Venkatesan, K., Sahalie, J., et al., 2008. High-quality binary protein interaction map of the yeast interactome network. Science 322, 104–110.

Yu, X., Ivanic, J., Memisevic, V., Wallqvist, A., Reifman, J., 2011. Categorizing biases in high-confidence high-throughput protein–protein interaction data sets. Mol. Cell. Proteomics 10, M111.012500.

Zervos, A.S., Gyuris, J., Brent, R., 1993. Mxi1, a protein that specifically interacts with Max to bind Myc–Max recognition sites. Cell 72, 223–232.

Zhang, Y., Gao, P., Yuan, J.S., 2010. Plant protein–protein interaction network and interactome. Curr. Genomics 11, 40–46.

Zuo, J., Niu, Q.W., Chua, N.H., 2000. An estrogen receptor-based transactivator XVE mediates highly inducible gene expression in transgenic plants. Plant J. 24, 265–273.

3

General Aspects of Plant Transcription Factor Families

Jong Chan Hong

Division of Life Science, Applied Life Science (BK21 Plus Program), Plant Molecular Biology and Biotechnology Research Center, Gyeongsang National University, Jinju, Korea; Division of Plant Sciences, University of Missouri, Columbia, MO, USA

OUTLINE

- 3.1 Introduction — 35
- 3.2 Overview of the Transcription Cycle in Eukaryotes — 36
  - 3.2.1 Core Promoter Elements and General Transcription Factors — 36
  - 3.2.2 Role of Nucleosome-Modifying Enzymes and Chromatin Remodelers — 36
  - 3.2.3 Productive Transcription Initiation and Elongation — 37
  - 3.2.4 Transcription Termination, and Reinitiation — 39
- 3.3 Components Involved in the Formation of the RNAPII Preinitiation Complex in Plants — 39
  - 3.3.1 GTFs and RNAPII — 39
  - 3.3.2 Mediator — 42
  - 3.3.3 Corepressors of the Plant Gro/Tup1 Family — 43
- 3.4 Plant Transcription Factor Families — 44
- 3.5 Major TF Families that are Conserved Across Eukaryotes — 44
  - 3.5.1 bZIP Transcription Factor Family — 44
  - 3.5.2 bHLH TF Family — 48
  - 3.5.3 MYB TF Family — 48
  - 3.5.4 HSF Family — 49
- 3.6 Plant-Specific TF Families — 49
  - 3.6.1 AP2/ERF Family — 49
  - 3.6.2 WRKY TF Family — 49
  - 3.6.3 NAC TF Family — 50
  - 3.6.4 TCP Family — 50
- 3.7 TFs without DBD but Interacting with DBD-Containing TFs — 50
  - 3.7.1 B-Box Zinc Finger Transcription Factors — 50
  - 3.7.2 Aux/IAA Family — 51
  - 3.7.3 JAZ Protein Family — 51
- 3.8 Conclusion — 52
- References — 52

3.1 INTRODUCTION

As sessile organisms, plants have evolved elaborate mechanisms to respond to various biotic and abiotic stresses and maintain optimal growth in a changing environment. Messenger RNA (mRNA) transcription is the most important control point for cellular activity to be regulated against a changing environment. Nuclear-encoded genes are transcribed by RNA polymerase II (RNAPII), which requires complex interactions with many proteins involved in successful formation of the preinitiation complex (PIC) at core promoter elements (CPEs). The proteins that regulate gene expression through specific binding to *cis*-acting elements in the promoter regions of target genes are called transcription factors (TFs). TFs that bind CPEs and are involved in basal transcription are called general transcription factors (GTFs). GTFs are multisubunit protein complexes involved in core promoter recognition, fundamental nucleation of the RNAPII transcriptional PIC, and initiation of transcription. GTFs have been named TFIIA, TFIIB, TFIID, TFIIE, TFIIF, and TFIIH according to their chromatographic elution profiles and order of discovery (Thomas and Chiang, 2006).

Eukaryotes have large genomes as compared to prokaryotes, and the eukaryotic genome is organized into compact chromatin with histone and nonhistone proteins. The basic unit of chromatin is the nucleosome, where DNA is wrapped around a histone octamer, which imposes the first level of complexity that the cellular transcriptional machinery encounters. As the transcription process requires a dynamic chromatin structure, diverse cellular factors, such as histone-modifying enzymes, adenosine triphosphate (ATP)-dependent chromatin remodelers, and histone chaperones, are involved. In addition, DNA-bound TFs that act as activators or repressors interact with cofactors such as the mediator (Med) complex and corepressors to either activate or repress target promoters, respectively. Thus, the formation of a functional PIC at the core promoter combines the action of TFs, chromatin-associated proteins, and cofactors. This chapter begins with an overview of the eukaryotic transcription cycle to introduce the protein components that are involved in transcriptional regulation according to their functional classes. Plant genomes have more than 2000 genes that encode the TFs of many different families. As an introductory chapter, a general overview of each TF family is presented according to the structural features and functional roles in plant development, hormone signaling, and biotic and abiotic stresses.

3.2 OVERVIEW OF THE TRANSCRIPTION CYCLE IN EUKARYOTES

3.2.1 Core Promoter Elements and General Transcription Factors

Eukaryotic transcription is a highly organized and tightly controlled process that exhibits regulation at multiple steps. It starts with the sequence-specific binding of transcription activators to upstream distant regulatory elements (DRE) and then promotes a sequential recruitment of GTFs and RNAPII to the target gene promoter (Thomas and Chiang, 2006) (Figure 3.1). The specific DNA elements that are located in regions most proximal to the transcription start site (TSS) and direct the accurate initiation of transcription are called CPEs. TSSs can be located at defined sites or dispersed over broad regions, giving rise to focused or dispersed initiation. Thus, CPEs are located upstream or downstream of the target gene start site; in rare cases, CPEs are also found in the coding region (Juven-Gershon and Kadonaga, 2010). CPEs are classified into TATA-containing and TATA-less promoters. The TATA sequence is bound by TATA-binding protein (TBP) which, along with several TBP-associated factors (TAFs), forms the TFIID complex. In addition to the TATA box, other major CPEs recognized by TFIID or TAFs include the initiator (INR), downstream promoter element (DPE), and motif ten element (MTE; Juven-Gershon et al., 2008). The sequence immediately flanking the TATA box can contain TFIIB recognition elements (BREs). In the majority of eukaryotic promoters that lack a canonical TATA box, so-called TATA-less promoters INR and DPE play major roles in promoter recognition by the transcription machinery. After binding of TFIID, TFIIB forms a stable ternary complex with TBP-bound DNA (Deng and Roberts, 2007). Next, TFIIF facilitates RNAPII binding to the core promoter, which is followed by TFIIE and TFIIH binding that further stabilize the formation of the RNAPII preinitiaton complex.

The role of TFs in the activation of transcription can be explained by several different mechanisms such as recruitment of coactivators and GTFs to promoters, modification of chromatin structure through chromatin-remodeling proteins, and enhancing steps that occur after formation of a functional PIC, a transition from a paused RNAPII to productive initiation and elongation (Hahn and Young, 2011). The recruitment mode of transcription initiation, which is the activation of transcription by recruitment of coactivator Meds and GTFs to CPEs, has been well documented (Ptashne and Gann, 1997). However, recent studies also recognized the significance of regulation after the initiation step, by pausing of RNAPII during transcription elongation at the promoter-proximal region (Adelman and Lis, 2012; Figure 3.2). The promoter-proximal pausing of RNAPII was initially described for the *c-fos* and *Hsp70* genes, both of which are rapidly induced in response to environmental signals. Further studies showed that the phenomenon of RNAPII pausing is prevalent in higher eukaryotes. Almost 30% of gene promoters in *Drosophila* and mammals harbor a paused RNAPII (Core et al., 2008). These paused RNAPII promoters predominantly belong to inducible and developmentally regulated genes (Nechaev and Adelman, 2011).

3.2.2 Role of Nucleosome-Modifying Enzymes and Chromatin Remodelers

The organization of eukaryotic DNA into chromatin poses a significant obstacle for binding TFs to their cognate sequences and hence in most scenarios negatively influences all steps of transcription (Li et al., 2007; Clapier and Cairns, 2009). The dynamics of chromatin structure are tightly regulated through multiple mechanisms

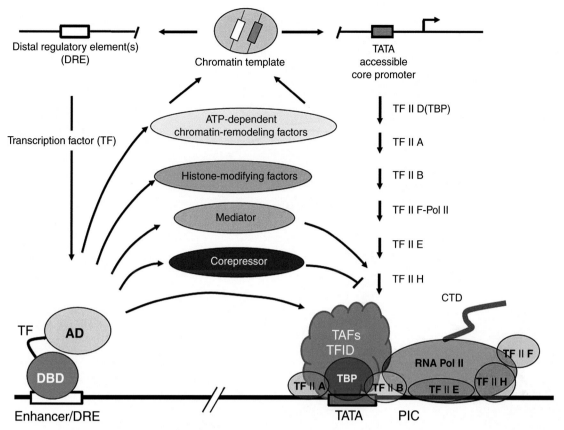

FIGURE 3.1 PIC assembly pathway for Class II genes with a TATA-containing core promoter and regulation by sequence-specific TFs and interacting cofactors. Assembly of a PIC containing RNAPII and GTFs is nucleated by binding TFIID to the TATA element of the core promoter. A model for the regulation of PIC assembly and function involves sequentially (1) binding of TFs to distal regulatory elements (DREs); (2) TF interactions with Meds or corepressors that facilitate or repress transcription; and (3) TF interactions with chromatin factors such as histone-modifying enzymes and chromatin-remodeling factors to facilitate recruitment of GTF and RNAPII transcription. TAFs, TBP-associated factors; DBD, DNA binding domain; AD, activation domain. *Adapted from Roeder (2005).*

including histone modification, chromatin remodeling, histone-variant incorporation, and histone eviction. TFs mediate the recruitment of histone-modifying enzymes such as acetyltransferases and methyltransferases to chromatin, which work in concert with nucleosome remodelers to reorganize the chromatin architecture. The enzymatic action of acetyltransferases adds acetyl groups to several residues of histones H3 (K9, K14) and H4 (K16) of the promoter-proximal nucleosomes. In addition, methyltransferases such as SET1 di- and trimethylate histone H3 at lysine 4 (H3K4) (Li et al., 2007). The occurrence of these histone marks signifies an open and permissive chromatin environment competent for transcription. In contrast, di- and trimethylation of H3K9 and H3K27 by SUV39h or G9a methyltransferases leads to the formation of a repressive chromatin state that is nonpermissive for transcription. Hyperacetylated or H3K4 trimethylated promoter-proximal nucleosomes then recruit ATP-dependent chromatin remodelers such as SWI/SNF (Clapier and Cairns, 2009). Chromatin remodelers recognize active epigenetic marks on nucleosomes with the aid of bromodomains (acetylated lysine recognition module) or chromodomains (methylated lysine binding module). This is followed by eviction or exchange of covalently modified as well as remodeled nucleosomes with the help of histone chaperones (ASF1, FACT) (Hsieh et al., 2013, Shandilya et al., 2007).

3.2.3 Productive Transcription Initiation and Elongation

Although the formation of a functional and stable PIC at gene promoters is foremostly important for transcription initiation, it does not guarantee productive transcription as it is evident from the observation of abortive transcription and stalled RNAPII. Transcripts of less than 5 nt (nucleotides) are unstable and result in a high frequency of abortive initiation. Productive initiation is achieved at a transcript length of about 25 nt, which succesfully form a stable transcription elongation complex (Saunders et al., 2006; Liu et al., 2011) (Figure 3.2).

The C-terminal domain (CTD) of RNAPII and proper modification of the N-terminal histone tails of nucleosomes are essential to creating a dynamic environment for

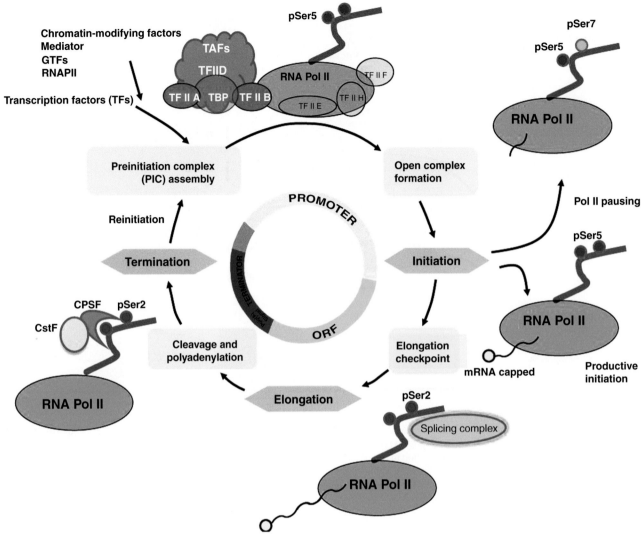

FIGURE 3.2 **RNAPII transcription cycle.** The main phases of the transcription cycle are colored in orange; important events of regulation are highlighted in yellow. The circle in the middle depicts the occurrence of the events in relation to the position on the gene. GTFs, general TFs; ORF, open reading frame. RNAPII C-terminal domain (CTD) phosphorylation is colored for Ser-5P (red), Ser-2P (blue), and Ser-7P (light green). *The transcription cycle is drawn according to Hahn (2004) and Shandilya and Roberts (2012).*

productive transcription initiation. The CTD of RNAPII is composed of tandem heptapeptide repeats of Tyr^1-Ser^2-Pro^3-Thr^4-Ser^5-Pro^6-Ser^7 ($Y^1S^2P^3T^4S^5P^6S^7$) (52 such repeats in humans, 25 repeats in yeast, 42 repeats in *Arabidopsis*; Egloff and Murphy, 2008). The differential phosphorylation status of serines at positions 2, 5, and 7 of the heptapeptide repeat play important roles in specific stages of transcription. To initiate transcription, RNAPII is loaded onto an active promoter in a hypophosphorylated state. Activity of TFIIH (CDK7/cyclin H) is required to begin promoter clearance and synthesis of a short nascent RNA by RNAPII. Initiation also requires phosphorylation of Ser-5 (Ser-5P) at the CTD by the CDK7 subunit of TFIIH upon PIC formation (Ohkuma and Roeder, 1994). Ser-5P at the CTD provokes conformational changes that allow binding of the capping enzyme that adds a methylguanosine cap to the 5′ end of the transcript. When ~25 nt of nascent RNA are synthesized, the cap structure is attached to its 5′ end.

After productive initiation there is a gradual loss of Ser-5P with an increase in Ser-2P in CTD repeats. The phosphorylated Ser-2 (Ser-2P), which is produced by CDK9 of the positive transcription elongation factor b (P-TEFb), peaks along the open reading frame (ORF) towards the 3′ terminal region of the genes (Peterlin and Price, 2006). Ser-2P recruits mRNA-splicing complexes during elongation, and termination complexes during 3′ processing and transcription termination (Nag et al., 2007). Thus, control of phosphorylation at Ser-2 of CTD repeats is critical for elongation checkpoint and mature mRNA production by RNAPII as it signals elongation, RNA processing, and termination. Phosphorylation of Ser-7 of CTD repeats has also been reported to be associated with RNAPII pausing (Chiba et al., 2010).

For rapidly induced genes such as *HSP70* and *c-fo*s, when RNAPII synthesizes around 50 nt, RNAPII is recognized by the negative elongation factor (NELF) and the DRB sensitivity-inducing factor (DSIF) causing its promoter-proximal pausing. To overcome the inhibitory effect of NELF/DSIF factors and to initiate productive elongation, P-TEFb kinase activity is required for the release of paused RNAPII through the phosphorylation of the RNAPII CTD, the NELF E subunit, and the SPT5 subunit of DSIF. The phosphorylation of SPT5 results in the release of DSIF from NELF, and DSIF proceeds to travel with RNAPII as a positive elongation factor for processivity. Thus, control of P-TEFb kinase activity is another important regulatory step after successful PIC formation as it determines the fate of nascent RNA transcripts, transcription abortion, or productive elongation from stalled RNAPII (Smith and Shilatifard, 2013, Kohoutek, 2009).

One of the key epigenetic marks associated with elongation is the methylation of histone H3 at K36 by the methyltransferase SET2. H2B monoubiquitination is also associated with elongation and is present both at promoters and within the ORF (Li et al., 2007). As the transcription bubble proceeds, the histones that were evicted during elongation are rapidly deposited on the DNA behind the elongating RNAPII with the help of histone chaperones such as Spt6, Spt16 (FACT), and ASF. Additionally, specific histone deacetylases (HDACs) such as Rpd3S recognize Set2 methylated histones and deacetylates them to maintain a hypoacetylated state of redeposited histones until the next round of transcription.

3.2.4 Transcription, Termination, and Reinitiation

Upon completion of transcription, the RNAPII and the newly synthesized RNA are released from the DNA template. The Ser-2P and Ser-5P of the CTD are selectively dephosphorylated by phosphatases (Ser-2P by Fcp1 and Ser-5P by Ssu72 in yeast/SCP1 in mammals). The released hypophosphorylated RNAPII is competent to initiate a fresh round of transcription. In the meantime, the GTFs still remain associated with the promoter and act as bookmarks long after RNAPII has escaped form the PIC (Sarge and Park-Sarge, 2005). Promoter-bound GTFs form a scaffold that allows reinitiation by RNAPII during successive rounds of transcription, which is termed as the "reinitiation scaffold". Activator-dependent GTF-stabilizing interactions are therefore critical for RNAPII recycling events.

It has emerged that the terminal and promoter regions of active genes can interact to facilitate reinitiation via a phenomenon known as gene looping. TFIIB is known to interact with the cleavage and polyadenylation specificity factor (CPSF) and cleavage stimulatory factor (CstF) components (Wang et al., 2010a). The phosphorylated TFIIB at Ser-65 by CDK7 of TFIIH is important not only in productive transcription initiation, but also in facilitating gene looping via its interaction with CstF.

3.3 COMPONENTS INVOLVED IN THE FORMATION OF THE RNAPII PREINITIATION COMPLEX IN PLANTS

In eukaryotes, the transcription of protein-coding genes is controlled by complex networks of many proteins of different functional classes. Sequence-specific TFs activate or repress transcription of their target genes by binding to *cis*-acting elements. TF binding to target promoters determines the formation of a macromolecular protein assembly, called the preinitiation complex (PIC), which includes GTFs, RNAPII, and transcription cofactors such as Meds and corepressors. Activation or repression of gene expression also requires chromatin remodeling in the vicinity of the gene to allow access for TFs and the general transcription machinery (GTFs, RNAPII, and Meds). The role of chromatin in gene expression is well described in Li et al. (2007), thus chromatin-associated proteins are not discussed in this chapter. A list of *Arabidopsis* core components involved in forming the RNAPII PIC is presented in Table 3.1.

3.3.1 GTFs and RNAPII

GTFs are multisubunit protein complexes involved in core promoter recognition, fundamental nucleation of the RNAPII transcriptional PIC, and the initiation of transcription. GTFs (TFIIA, TFIIB, TFIID, TFIIE, TFIIF, and TFIIH) are conserved in all eukaryotes including plants (Thomas and Chiang, 2006).

TFIID is a multisubunit protein complex that plays an important role in core promoter recognition and in nucleation of the PIC for RNAPII. Additionally, the subunits of TFIID also serve as recruitment targets for TFs bound to upstream promoter elements. TFIID is comprised of the TBP and 12–15 TAFs. In *Arabidopsis* there are two TBPs and 14 TAFs in 23 loci (Lawit et al., 2007, Lago et al., 2004). No TAF3 homolog has been observed in plants. As is the case in other eukaryotic organisms, plant TAFs, and other GTF subunits share a high level of sequence homology in conserved regions, as predicted from their conserved function (Table 3.1).

The role of TFIID in transcription activation is reflected by binding TAF1 to acetylated histone (H3K4), TAF3 interaction with trimethylated histone H3 (H3K4–me3), which is mediated by hSET-1 histone methylase, and TAF1 and TAF6 binding to the INR and DPE of CPEs, respectively (Thomas and Chiang, 2006; Lauberth et al., 2013; Figure 3.3). Several *Arabidopsis* TAFs have been reported to be involved in plant development and stress responses.

TABLE 3.1 List of *Arabidopsis* Core Components Involved in Forming the RNAPII Transcriptional PIC

Class	Module	Subunit	Locus	Class	Module	Subunit	Locus
RNA pol II	Core	Rpb1	At4g35800	Med	Head	Med6	At3g21350
		Rpb2	At4g21710			Med8	At2g03070
		Rpb3	At2g15430			Med11	At3g01435
		Rpb11	At3g52090			Med17	At5g20170
		Rpb6	At5g51940			Med18	At2g22370
		Rpb5	At2g41340			Med19_1	At5g12230
		Rpb8A	At1g54250			Med19_2	At5g19480
		Rpb8B	At3g59600			Med20_1	At2g28230
		Rpb10	At1g11475			Med20_2	At4g09070
		Rpb12	At5g41010			Med22_1	At1g07950
		Rpb9a	At3g16980			Med22_2	At1g16430
		Rpb9b	At4g16265			Med28	At3g52860
	Stalk	Rpb4	At3g28956			Med30	At5g63480
		Rpb7	At1g06790		Middle	Med4	At5g02850
						Med7_1	At5g03220
GTF	TFIIA	TFIIA-S	At4g24440			Med7_2	At5g03500
		TFIIA-L1	At1g07480			Med9	At1g55080
		TFIIA-L2	At1g07470			Med10_1	At1g26665
		TFIIA-L3	At5g59230			Med10_2	At5g41910
						Med21	At4g04780
	TFIIB	TFIIB1	At2g41630			Med31	At5g19910
		TFIIB2	At3g10330		Tail	Med2/29/32	At1g11760
		TFIIB3	At3g29380			Med3/27	At3g09180
		TFIIB4	At3g57370			Med33a	At3g23590
		TFIIB5	At4g36650			Med33b	At2g48110
		TFIIB6	At4g10680			Med14	At3g04740
						Med15_1	At1g15770
	TFIIE	TFIIEα1	At1g03280			Med15_2	At1g15780
		TFIIEα2	At4g20340			Med15_3	At2g10440
		TFIIEα3	At4g20810			Med16	At4g04920
		TFIIEβ1	At4g21010			Med23	At1g23230
		TFIIEβ2	At4g20330		Kinase	Med12	At4g00450
						Med13	At1g55325
	TFIIF	TFIIFα	At4g12610			Cdk8	At5g63610
		TFIIFβ1	At3g52270			CycC_1	At5g48630
		TFIIFβ2	At1g75510			CycC_2	At5g48640
	TFIID	TBP1	At1g13445		Unassigned	Med25	At1g25540
		TBP2	At1g55520			Med26_1	At3g10820
		TAF1	At1g32750			Med26_2	At5g05140

3.3 COMPONENTS INVOLVED IN THE FORMATION OF THE RNAPII PREINITIATION COMPLEX IN PLANTS

TABLE 3.1 List of *Arabidopsis* Core Components Involved in Forming the RNAPII Transcriptional PIC (cont.)

Class	Module	Subunit	Locus	Class	Module	Subunit	Locus
		TAF1b	At2g19040			Med26_3	At5g09850
		TAF2	At1g73960			Med34	At1g31360
		TAF4	At5g43130			Med35_1	At1g44910
		TAF4b	At1g27720			Med35_2	At3g19670
		TAF5	At5g25150			Med35_3	At3g19840
		TAF6	At1g04950			Med36_1	At4g25630
		TAF6b	At1g54360			Med36_2	At5g52470
		TAF7	At1g55300			Med37_1	At1g09080
		TAF8	At4g34340			Med37_2	At3g12580
		TAF9	At1g54140			Med37_3	At5g02490
		TAF10	At4g31720			Med37_4	At5g02500
		TAF11	At4g20280			Med37_5	At5g28540
		TAF11b	At1g20000			Med37_6	At5g42020
		TAF12	At3g10070	Corepressor	LUG/LUH	LUG	At3g01435
		TAF12b	At1g17440			LUH	At2g32700
		TAF13	At1g02680		TPL/TPR	TPL	At1g15750
		TAF14	At2g18000			TPR1	At1g80490
		TAF14b	At5g45600			TPR2	At3g16830
		TAF15	At1g50300			TPR3	At5g27030
		TAF15b	At5g58470			TPR4	At3g15880

Homozygous T-DNA insertion mutants of AtTAF6 are lethal, and heterozygotes show a markedly reduced rate of pollen tube growth (Lago et al., 2005). Overexpression of TAF10 causes morphological abnormalities and affects seed germination and salt stress responses (Gao et al., 2006b). In addition to TAF involvement in plant development and osmotic stress, a TAF subunit, TAF1b (gene product of HAF2), is involved in the integration

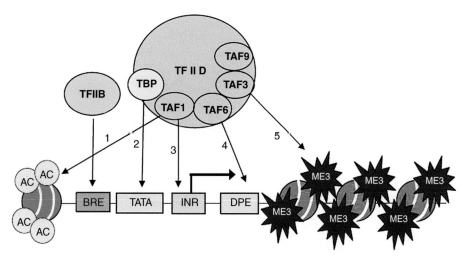

FIGURE 3.3 **A model of how TFIID subunits interact with CPEs to facilitate PIC formation.** TBP, TAF1, and TAF6 interact with TATA, INR, and DPE of CPEs, respectively. TAF1 and TAF3 also interact with acetylated histone H3 (yellow marks) and trimethylated histone H3 (H3K4me3, red asterisks) as well as enhancing recruitment of TFIID and RNAPII. *Adapted from Lauberth et al. (2013).*

of light signals in transcriptional regulation (Bertrand et al., 2005).

Arabidopsis RNAPII, the main enzyme for mRNA transcription, is a protein complex with 12 subunits. The largest and second largest subunits (Rpb1 and Rpb2) are the catalytic subunits that interact to form DNA entry and exit channels, the active site, and the RNA exit channel. In plants, there are two additional RNAPs, PolIV and PolV, which contain atypical largest subunits and second largest subunits (Pikaard et al., 2008; Ream et al., 2009). The atypical largest subunit genes of PolIV and PolV are NRPD1 and NRPE1, respectively. The second largest subunits of PolIV and PolV are designated by the synonymous names NRPD2 or NRPE2. PolIV and PolV are functionally distinct, with PolIV required for small interfering RNA (siRNA) production and PolV generating noncoding transcripts at target loci (Wierzbicki et al., 2008). Mutations in atypical catalytic subunit genes affect the short-range or long-distance spread of RNA-silencing signals, responses to biotic and abiotic stresses, and flowering time (Dunoyer et al., 2007; Smith et al., 2007).

3.3.2 Mediator

Med is a multiprotein complex that acts as a link between TFs and the RNAPII core complex and brings about either transcriptional activation or repression (Poss et al., 2013). Med facilitates transcription by increasing the efficiency/rate of RNAPII PIC formation at the promoter and activating transcription from promoters with stalled RNAPII (Lee et al., 2010). In yeast, Med was shown to participate in the transcription of a majority of genes (Bjorklund and Gustafsson, 2005). Yeast Med is a complex of 25 subunits, where 8 subunits form the head module, 7 constitute the middle module, and 6 constitute the tail. In addition, a fourth regulatory module comprising two Med subunits (Med12 and Med13) along with a cyclin-dependent kinase, CDK8, and its associated cyclin, CycC, collectively called the "kinase or CDK8/CycC module", is also part of the Med complex (Figure 3.4). Med serves as a central scaffold within the PIC and helps to regulate RNAPII activity (Poss et al., 2013). Med is also generally targeted by sequence-specific DNA-binding TFs that work to control gene

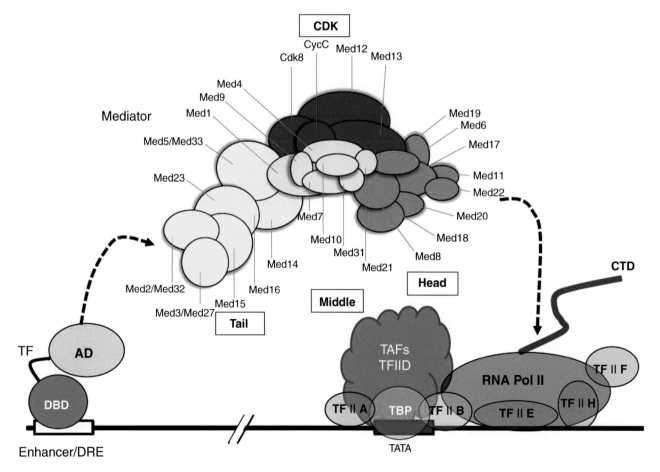

FIGURE 3.4 **Med components and the role of head and tail modules that link TFs to transcription activation.** Med transmits regulatory signals from gene-specific transcription activator proteins to the general transcription machinery, including RNAP II, and GTFs. The Tail interacts with a variety of activators/repressors and the regulatory signals are transferred via the Middle module to the Head that physically contacts RNAPII. The RNAPII CTD is indicated.

expression programs in response to developmental or environmental cues. At a basic level, Med functions by relaying signals from TFs directly to the RNAPII enzyme, thereby facilitating TF-dependent regulation of gene expression (Poss et al., 2013). Thus, Med is essential for converting biological inputs (communicated by TFs) to physiological responses (via changes in gene expression). The recruitment of RNAPII to CPEs is achieved by direct contacts between the head and the middle Med modules and the CTD of Rpb1, the largest RNAPII subunit (Kornberg, 2005; Fig. 3.4). Med stimulates TFIIH-dependent phosphorylation of the RNAPII C-terminal domain, an event that correlates with promoter escape (Boeing et al., 2010). The CDK8/cycC module phosphorylates Med, TFIIH, and the RNAPII C-terminal domain and negatively regulates transcription by disrupting Med–RNAPII interaction, a necessary event in transcription initiation and reinitiation. In addition, the CDK8/CycC module acts as a molecular switch to control the coactivator functions of Med (Knuesel et al., 2009).

Plant Meds have been isolated by biochemical purification from *Arabidopsis*, which allowed the identification of 21 conserved and 6 *Arabidopsis*-specific Med subunits (Backstrom et al., 2007). From sequence analysis of highly conserved regions across metazoa, fungi, and plants, Med was finally found to have 32 subunits (Med1–Med23, Med25, Med26, Med28, Med30, Med31, and Med34–Med37) along with the CDK8/CycC kinase module (Mathur et al., 2011; Table 3.1). This study identified STRUWWELPETER as Med14 and PHYTOCHROME AND FLOWERING TIME1 as Med25 subunits. Later studies identified SENSITIVE TO FREEZING6 (SFR6) as Med16, and two of the *Arabidopsis* Med33 subunits (AT2G48110 and AT3G23590) as REDUCED EPIDERMAL FLUORESCENCE4 (REF4) and REF4-RELATED1 (RFR1), respectively (Backstrom et al., 2007). The function of plant Med is diverse as it has demonstrated roles in plant development, flowering, hormone signaling, and biotic and abiotic stress tolerance (Kidd et al., 2011). The Med12–Med13 pair regulates pattern formation timing during embryogenesis in *Arabidopsis* (Gillmor et al., 2010). Dominant mutations in REF4 lead to reduced accumulation of phenylpropanoid end products and affect plant growth. Med25, also known as PHYTOCHROME AND FLOWERING TIME1 has diverse roles in flowering, hormone signaling, and abiotic and biotic stress responses (Kidd et al., 2011). *Arabidopsis* Med25 directly interacts with activator DREB2A through the activation interacting domain (ACID), which has sequence homology to the human Med25-ACID (hMed25) that interacts with the herpes simplex virus VP16 activator (Aguilar et al., 2014).

3.3.3 Corepressors of the Plant Gro/Tup1 Family

Corepressors are transcriptional regulators that are incapable of independent DNA binding, being recruited directly or indirectly by DNA-binding TFs to repress target gene expression. The Groucho (Gro)/Tup1 family, first identified in *Drosophila* and *Saccharomyces*, respectively, represents an archetypal class of corepressors that are recruited by a range of DNA-binding TFs to elicit a repressed chromosomal state, thereby shutting off gene expression (Chen and Courey, 2000; Liu and Karmarkar, 2008). In *Arabidopsis*, the Gro/Tup1-like proteins constitute a small family of 13 members (*http://smart.embl-heidelberg.de/*) that contain two representative protein family memberss, called LUG/LUH and TPL/TPR (Liu and Karmarkar, 2008).

In *Arabidopsis*, the first Gro/Tup1 family member to be isolated was LEUNIG (LUG) (Conner and Liu, 2000). LUG was identified due to its role in regulating the expression of the floral homeotic gene *AGAMOUS*. LUG and its close homolog LEUNIG_HOMOLOG (LUH) share characteristic C-terminal WD-40 repeats and the N-terminal LUFS domain. The LUFS domain contains the LisH (for Lis1-homologous) motif previously defined as the domain for protein–protein interaction in yeast (Cerna and Wilson, 2005). The LUFS domain interacts directly with an adaptor protein, called SEUSS, that serves to link LUG/LUH to a range of other TFs such as APETALA1 (AP1) and SEPALLATA3 (SEP3) (Sridhar et al., 2006).

TOPLESS (TPL) is another member of the Gro/Tup1 corepressor family that interacts directly with many TFs (Causier et al., 2012). TPL defines a family of five genes in *Arabidopsis* – TPL and TPL-related (TPR) 1 to 4. TPL and TPR4, previously known as WSIP1 and WSIP2, directly interact with the *Arabidopsis* homeodomain transcription factor WUSCHEL (WUS). The interaction is mediated by the N-terminal region of TPL/TPR, which also contains a LisH motif for protein interaction with WUS. Other targets characterized for TPL/TPR corepressors are (1) the auxin/indole-3-acetic acid (AUX/IAA) proteins involved in auxin signaling (Szemenyei et al., 2008b), (2) NINJA, which interacts with jasmonate ZIM-domain (JAZ) proteins in jasmonic acid (JA) signaling (Pauwels et al., 2010), and (3) SUPPRESSOR OF NPR1-1, CONSTITUTIVE1 (SNC1) involved in defense responses against pathogen infection (Zhu et al., 2010). The interaction of WUS, AUX/IAA, and NINJA with TPL/TPR corepressors is mediated by a small conserved protein motif known as the ERF-associated amphiphilic repression (EAR) domain, which has been identified in many transcription repressors (Kagale et al., 2010). High-throughput yeast two-hybrid screens using TPL/TPR corepressors as bait identified many TFs of at least 17 distinct families forming the TPL interactome (Causier et al., 2012). The interactome data established the role of TPL family members as general corepressors in diverse biological pathways such as auxin and JA hormone signaling, biotic and abiotic stress responses, and flower development. The Gro/Tup1 group of corepressors represses transcription by recruitment of HDAC, which induce chromatin condensation and gene silencing at target sites (Chen and Courey, 2000). *Arabidopsis* LUG and

TPL families have also been shown to function together with HDAC 19 (HDA19) (Gonzalez et al., 2007).

3.4 PLANT TRANSCRIPTION FACTOR FAMILIES

TFs are generally defined as proteins binding directly to the promoters of target genes in a sequence-specific manner to regulate expression of their target genes. TFs contain DNA-binding domains (DBDs) that recognize specific DNA sequences in promoters. However, there are regulatory proteins without a DBD that directly interact with DNA-binding TFs to form transcriptional complexes and they are also categorized as TFs.

The classification of TFs into different families is generally based on their characteristic DBDs (Riechmann et al., 2000). After completion of the *Arabidopsis* genome sequence, about 1500 TFs belonging to approximately 30 different TF families were initially identified (Riechmann et al., 2000). Subsequent analyses have recognized about 2000 TF genes in the *Arabidopsis* genome. *Arabidopsis* TF databases are used as the basis of classification in other plant species. There are four representative databases of *Arabidopsis* TFs: RARTF (*http://rarge.gsc.riken.jp/rartf/*; Iida et al., 2005), AGRIS (*http://arabidopsis.med.ohio-state.edu/AtTFDB/*; Davuluri et al., 2003), DATF (*http://datf.cbi.pku.edu.cn/*; Guo et al., 2005), and PlnTFDB (*http://plntfdb.bio.uni-potsdam.de/*; Riaño-Pachón et al., 2007). Four independent reports have shown that approximately 2000 genes encode the TFs in *Arabidopsis* (Table 3.2). Each database classifies TFs into families based on their own classification criteria, and the number of loci in each family is different among the four databases. A total of 51, 51, 64, and 67 families (72 families in total) and 1965, 1837, 1914, and 1949 loci, respectively, were identified. Altogether, 2620 loci are classified as TFs with a total of 1318 loci being recognized by all four databases (Mitsuda and Ohme-Takagi, 2009). These differences are mainly due to the different definition of a TF in each database. In addition to *Arabidopsis* TF databases, the availability of complete plant genome sequences and the development of various software tools have enabled scientists to identify TF sets from various plant species and compile their basic information in a number of species-specific TF databases, such as RARTF (Iida et al., 2005), DRTF (Gao et al., 2006a), Soybean TFDB (Mochida et al., 2009), SoyDB (Wang et al., 2010b), TOBFAC (Rushton et al., 2008), and wDBTF (Romeuf et al., 2010), or integrative databases, such as DATFAP (Fredslund, 2008), Grassius (Yilmaz et al., 2009), RiceSRTFDB (Priya and Jain, 2013), PlnTFDB (Perez-Rodriguez et al., 2010), LegumeTFDB (Mochida et al., 2010), GramineaeTFDB (Mochida et al., 2011), PlantTFDB (Zhang et al., 2011), and TreeTFDB (Mochida et al., 2013; Table 3.3). Plant TF databases also contain some proteins that can be categorized as other transcription regulator (TR) classes; for example, *Arabidopsis* polycomb group (PcG) and JumonjiC proteins are histone methylases and demethylases, respectively.

Plant TFs are characterized by the larger number of genes and by the variety of TF families when compared with those of *Drosophila melanogaster* or *Caenorhabditis elegans*. The number of TF genes in *Arabidopsis*, which is about 2000, is significantly larger than that found in these species (around 600), which have genomes of similar size (Riechmann et al., 2000). The ratio of TF genes to the total number of genes in *Arabidopsis* is 5–10% depending on databases, which is higher than that of *D. melanogaster* (4.7%) and *C. elegans* (3.6%), although it is comparable with that of humans (6.0%). In addition to the larger number of TF genes in *Arabidopsis*, there is a greater variety of TFs, with a greater diversity of DNA-binding specificities, compared with *D. melanogaster* or *Caenorhabditis elegans*. These characteristic features of *Arabidopsis* TFs suggest that transcriptional regulation may be more complex and diversified in plants than in animals. For example, zinc-finger TFs represent more than half of all TFs in *D. melanogaster* or *C. elegans*, whereas those in *Arabidopsis* represent around 20% (Riechmann et al., 2000). Around half of *Arabidopsis* TFs are plant specific and possess DBDs found only in plants (Riechmann et al., 2000). AP2-ERF, NAC, Dof, YABBY, WRKY, GARP, TCP, SBP, ABI3-VP1 (B3), EIL, and LFY are plant-specific TF families (Figure 3.5). The three-dimensional structures of several plant-specific DBDs (i.e., NAC, WRKY, SBP, EIL, B3 and AP2-ERF) have been determined (reviewed by Yamasaki et al., 2013; Chapter 4). Most *Arabidopsis* TFs form large families which share similar DBD structures. For example, AP2-ERF and NAC domain families contain >100 members each. MYB, MADS box, basic helix–loop–helix(bHLH), bZIP, and HB, which are not plant-specific families, also form large families. These families play important roles in the control of plant growth and development against changing environments.

3.5 MAJOR TF FAMILIES THAT ARE CONSERVED ACROSS EUKARYOTES

3.5.1 bZIP Transcription Factor Family

Basic region/leucine zipper (bZIP) TFs have a basic region (BR) that binds DNA and a leucine zipper region (ZR), a motif for protein dimerization. Yeast GCN4 and mammalian Jun/Fos or CREB have been extensively studied for bZIP domain–DNA interactions, homodimer formation, heterodimer formation, and TF posttranslational modifications (Ellenberger et al., 1992 Vinson et al., 1989). *Arabidopsis* has about four times as many bZIP genes as yeast, worms, and humans (Riechmann

TABLE 3.2 Comparison of *Arabidopsis* TF Databases

TF family	InterPro or GenBank	Riechmann	RARTF	AGRIS	DATF	PlnTFDB
ABI3/VP1*	CAA48241	14	51	11	60	56
ALFIN-like	AAA20093	7	47	7	7	7
AP2/EREBP	IPR001471	144	93	136	146	146
ARF	AAC49751	23	71	22	23	23
ARID	IPR001606	4	6	7	10	10
AS2		0	0	0	42	0
AUX/IAA	AAC39440	26	21	0	29	27
bHLH	IPR001092	139	157	162	127	134
bZIP	IPR001871	81	56	73	72	71
C2C2 (Zn)-BBX	A56133	33	51	30	37	17
C2C2 (Zn)-Dof	CAA66600	37	33	36	36	36
C2C2 (Zn)-GATA	IPR000679	28	37	30	26	29
C2C2 (Zn)-YABBY	AAD30526	6	5	6	5	6
C2H2 (Zn)	IPR000822	105	177	211	134	96
C3H-type (Zn)	IPR000232	33	47	165	59	67
CCAAT	A26771/P13434/Q02526	36	37	35	36	43
CPP (Zn)	CAA09028	8	8	8	8	8
E2F/DP	O00716/Q64163	8	8	8	8	7
EIL	AAC49750	6	6	6	6	6
GARP	AAD55941/BAA74528	56	51	55	53	52
GRAS	AAB06318	32	32	31	33	33
HB	IPR001356	89	97	91	87	91
HMG-box	IPR000910	10	11	0	11	11
HSF	IPR000232	26	27	21	23	23
JUMONJI	T30254	9	13	5	17	17
LFY	AAA32826	1	3	1	1	1
MADS	IPR002100	82	106	109	104	102
MYB	IPR001005/IPR000818	190	189	197	199	209
NAC	BAB10725	109	106	94	107	101
Nin-like	CAB61243	15	14	0	14	0
PCG		4	35	0	34	0
SBP	CAB56581	16	17	16	16	16
TCP	AAC26786	25	24	26	23	24
Trihelix	S39484	28	31	29	26	23
TUB	IPR000007	11	11	10	0	10
WRKY(Zn)	S72443	72	72	72	72	72
Others		20	215	127	231	375
Totals		1533	1965	1837	1922	1949

*The number of loci in each database is shown. Plant-specific TF families are in boldface.

TABLE 3.3 Databases and Websites for Plant TFs

Websites	Acronym	References	Website address	Plant species
Arabidopsis gene regulatory information server	AGRIS	Davuluri et al. (2003)	http://arabidopsis.med.ohio-state.edu/AtTFDB/	*Arabidopsis*
Plant TF databases	PlantTFDB	Jin et al. (2014)	http://planttfdb.cbi.pku.edu.cn/	Multiple species
Database of *Arabidopsis* TFs	DATF	Guo et al. (2005)	http://datf.cbi.pku.edu.cn/	*Arabidopsis*
Database of rice TFs	DRTF	Gao et al. (2006a)	http://drtf.cbi.pku.edu.cn/	Rice
Database of poplar TF	DPTF	Zhu et al. (2007)	http://dptf.cbi.pku.edu.cn/	Poplar
RIKEN *Arabidopsis* TF database	RARTF	Iida et al. (2005)	http://rarge.gsc.riken.jp/rartf/	*Arabidopsis*
Plant TF database	PlnTFDB	Riaño-Pachón et al., 2007	http://plntfdb.bio.uni-potsdam.de/v3.0/	Multiple species
Plant transcription associated protein database	PlanTAPDB	Richardt et al. (2007)	http://www.cosmoss.org/bm/plantapdb	Multiple species
Soybean database for TF	SoyDB	Wang et al. (2010b)	http://casp.rnet.missouri.edu/soydb/	Soybean
Soybean TF database	Soybean TFDB	Mochida et al. (2009)	http://soybeantfdb.psc.riken.jp/	Soybean
Tobacco TF database	TOBFAC	Rushton et al. (2008)	http://compsysbio.achs.virginia.edu/tobfac/	Tobacco
Database of wheat TF	wDBTF	Romeuf et al. (2010)	http://wwwappli.nantes.inra.fr:8180/wDBFT/	Wheat
Grass regulatory information services	Grassius	Yilmaz et al. (2009)	http://grassius.org/	Maize, sugarcane, sorghum, rice
Rice stress-responsive TF	RiceSRTFDB	Priya and Jain (2003)	http://www.nipgr.res.in/RiceSRTFDB.html	Rice
Legume TF database	LegumeTFDB	Mochida et al. (2010)	http://legumetfdb.psc.riken.jp/	Soybean, *Lotus*, *Medicago*
Gramineae TF database	GramineaeTFDB	Mochida et al. (2011)	http://gramineaetfdb.psc.riken.jp	*Brachypodium*, rice, sorghum, maize
Stress responsive TF database	STIFDB	Naika et al. (2013)	http://caps.ncbs.res.in/stifdb2/	*Arabidopsis*, rice
Tree TF database	TreeTFDB	Mochida et al. (2013)	http://treetfdb.bmep.riken.jp/index.pl	Jatropha, papaya, cassava, poplar castor bean grapevine
Interspecies TF function finder	IT3F	Bailey et al. (2008)	http://jicbio.bbsrc.ac.uk/IT3F/	*Arabidopsis*, rice

et al., 2000). Genetic and molecular analysis in *Arabidopsis* showed that bZIP TFs regulate diverse biological processes such as light signaling, abiotic stress responses, pathogen defense, seed maturation, and flower development (Jakoby et al., 2002). In *Arabidopsis*, 75 putative genes (generic name assigned as *AtbZIP1–AtbZIP75*) are present, encoding proteins with the bZIP signature motif. AtbZIP proteins are clustered into 10 groups (A–I and S), according to the sequence similarities of their basic region, additional conserved motifs, and sizes. Group-A bZIP proteins include ABA response element (ABRE) binding factors (ABFs) and ABA-responsive element binding proteins (AREBs). They function as important players in ABA signal transduction both in seeds and vegetative tissues, which is a key process of abiotic stress responses (drought, salt, and cold; Kim et al., 1997; Uno et al., 2000). Group-B bZIP proteins include AtbZIP17, 28, and 49 which play important roles in endoplasmic reticulum (ER) stress responses (Tajima et al., 2008). These proteins are membrane-bound ER localized bZIP proteins that localize to the nucleus after ER stress (Chapter 25). Group D contains 10 bZIP proteins that are key players in pathogen defense responses (Jakoby et al., 2002). TGA proteins of group-D

3.5 MAJOR TF FAMILIES THAT ARE CONSERVED ACROSS EUKARYOTES 47

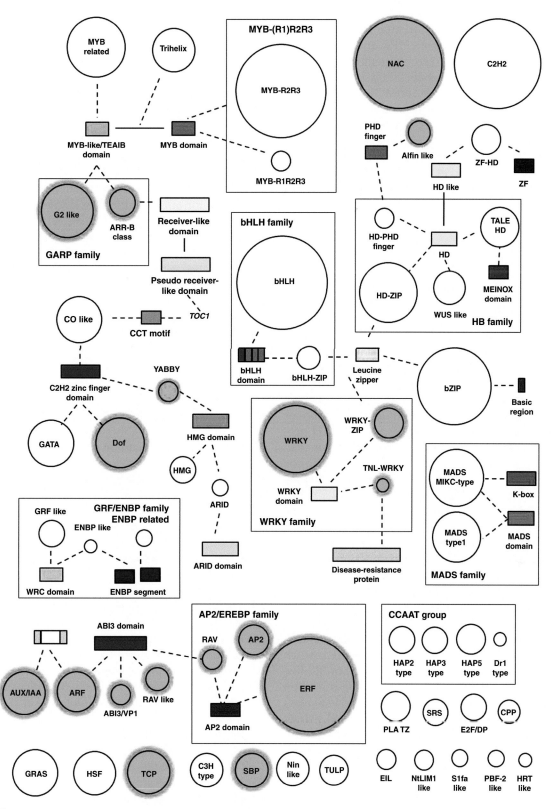

FIGURE 3.5 *Arabidopsis* **TF families.** TF families are represented by circles, whose size is proportional to the number of members in the family. Domains that have similar structural features that connect different groups of TFs are indicated with rectangles. Plant-specific TF families are indicated by blue circles. Dashed lines indicate that a given domain is a characteristic of the family or subfamily to which it is connected. *Adapted from Riechmann et al. (2000).*

A. GENERAL ASPECTS OF PLANT TRANSCRIPTION FACTORS

bZIPs interact with the nonexpressor of PR1 (NPR1). The salicylic acid (SA) synthesized after pathogen infection mediates TGA release from NPR and localization to the nucleus for binding to *cis*-elements that contain the TGACG motif (Johnson et al., 2003). ELONGATED HYPOCOTYL5 (HY5) and HY5 HOMOLOG (HYH) are group-H bZIP proteins that interact with G-boxes of many light-regulated genes (Chattopadhyay et al., 1998; Chapter 21). The stability of HY5 proteins is controlled by COP1 E3 ubiquitin ligase via interaction with the WD40 domain (Osterlund et al., 2000). These bZIP proteins do not activate target genes by themselves but through other interacting proteins such as STH2 and STH3 (Datta et al., 2007).

3.5.2 bHLH TF Family

Basic helix-loop-helix (bHLH) TFs are widely distributed in eukaryotes and constitute one of the largest TF families (Riechmann et al., 2000). bHLH TFs contain the bHLH domain of approximately 60 amino acids, with two functionally distinctive regions, the basic region and the HLH region. The 15 amino acid basic region at the N-terminus of the bHLH domain functions as a DNA-binding motif. The HLH region contains two amphipathic α-helices with a linking loop of variable length to/from homo- or heterodimmers (Atchley et al., 1999). Some bHLH proteins bind to sequences containing a consensus core element called the E-box (5′-CANNTG-3′), with the G-box (5′-CACGTG-3′) being the most common form (Atchley et al., 1999). Animal bHLHs are classified as six major functional and evolutionary lineages (Group A–F), while most bHLH proteins are classified as Group A or B. In *Arabidopsis* and rice, 162 and 167 bHLH-coding genes have been identified (Bailey et al., 2008; Li et al., 2006). Sequence analysis suggests that most plant bHLH proteins belong to Group B. The members of the bHLH TF family in both *Arabidopsis* and rice are divided into two major groups that contain a canonical bHLH or lack the basic region required for DNA binding. They are further classified into 25 subfamilies (Li et al., 2006). A group of bHLH TFs, which includes phytochrome interacting factor (PIF) or PIF-like (PIL) TFs, are key regulators in light signaling (Castillon et al., 2007; Chapter 21). These PIF/PIL bHLH subclass proteins have closely related bHLH domains and a small conserved N-terminal domain, called the "active phytochrome binding" (AFB) domain. The AFB domain is crucial for mediating the interaction with the Pfr form of phyB. A known function of PIFs is their role as negative regulators in phyB and phyA-mediated photomorphogenesis. PIF3, PIF4, and PIL6 act as negative regulators of phyB signaling. The bHLH TF MYC2 has been described as a master regulator of the crosstalk between the signaling pathways of JA and those of other phytohormones such as abscisic acid (ABA), salicylic acid (SA), gibberellins (GAs), auxins (IAA; Kazan and Manners, 2013), anthocyanin biosynthesis (GLABRA3 (GL3), ENHANCER OF GLABRA3 (EGL3), TRANSPARENT TESTA8 (TT8)), seed coat differentiation, and trichome/root hair formation (MYC1, GL3, EGL3, TT8).

Combinatorial interactions among bHLH TFs and MYB TFs are reported to play a key role in flavonoid biosynthesis in plants (Ramsay and Glover, 2005). The maize C1 bHLH protein interacts with the MYB R protein to activate maize flavonoid pathways. *Arabidopsis* GL3 and EGL3 interact with MYB factors GLABRA1 (GL1) or WEREWOLF (WER) to form trichomes and root hairs. TT8, GL3, and EGL3 interact with TRANSPARENT TESTA2 (TT2) and PRODUCTION OF ANTHOCYANIN PIGMENT1 (PAP1) and PAP2 for anthocyanin biosynthesis. The inclusion of TRANSPARENT TESTA GLABRA1 (TTG1), a WD40-repeat protein, in a ternary complex (called MBW for the MYB–bHLH–WD40 complex) activates genes involved in these cellular processes (Ramsay and Glover, 2005). Altogether, plant bHLH TFs are involved in a diverse array of biological processes such as hormone signaling, abiotic and biotic stress responses, cellular differentiation, and flavonoid biosynthesis.

3.5.3 MYB TF Family

MYB proteins integrate a superfamily of TFs that has the largest number of members of any *Arabidopsis* TF family: 197 members with a highly conserved DBD known as the MYB domain (Riechmann et al., 2000; Dubos et al., 2010; Katiyar et al., 2012). This domain generally consists of up to four imperfect amino acid sequence repeats (R) of about 52 amino acids, each forming a helix–turn–helix structure that intercalates in the major groove of the DNA. MYB proteins can be divided into different classes depending on the number of MYB repeats (one to four). The three repeats of the prototypic MYB protein c-Myb are referred to as R1, R2, and R3, and repeats from other MYB proteins are named according to their similarity to R1, R2, or R3 of c-Myb. Plant MYB proteins were classified as three major groups: R2R3-MYB, with two adjacent repeats; R1R2R3-MYB, with three adjacent repeats; and a heterogeneous group collectively referred to as MYB-related proteins, which usually contain a single MYB repeat. In plants, the MYB family has selectively expanded, particularly through the large family of R2R3-MYB (Dubos et al., 2010). In the *Arabidopsis* genome, 138 are R2R3-MYB, 5 are R1R2R3-MYB, 52 are MYB-related, and 2 are atypical MYB genes (Yanhui et al., 2006; Katiyar et al., 2012).

R2R3-MYB proteins have been further subdivided into 25 subgroups and found to be involved in the control of a variety of plant-specific processes including plant secondary metabolism (PAP1/2 and PFG1-3 for anthocyanin and flavonol biosynthesis, respectively), cell fate, and identity regulation (GL1 for trichome, WER for root hair development), and responses to biotic and abiotic stresses (Dubos et al., 2010, Stracke et al., 2001). MYB-related proteins include TRIPTYCHON (TRY) and CAPRICE (CPC), which are involved in the control of cellular morphogenesis, CIRCADIAN CLOCK ASSOCIATED1 (CCA1) and LATE ELONGATED HYPOCOTYL (LHY) are core components of the circadian oscillator, and the GARP family including the KANADI and GOLDEN2-LIKE proteins involved in organ morphogenesis and chloroplast development, respectively (Du et al., 2013, Katiyar et al., 2012).

3.5.4 HSF Family

Heat stress transcription factors (HSFs) mediate the rapid accumulation of heat shock proteins (HSPs) in response to both heat stress and many chemical stressors (Von Koskull-Doring et al., 2007; Nover et al., 1996). HSPs play a central role not only in the protection against stress damage, but also in the folding, intracellular distribution, and degradation of proteins (Wang et al., 2004). HSFs recognize conserved binding motifs, so-called heat stress elements (HSE: 5'-AGAAnnTTCT3') that exist in the promoters of *Hsp* genes (Nover et al., 1996). HSFs have a modular structure with a DBD and an oligomerization domain (OD). In addition, depending on the subfamily, they contain a nuclear localization signal (NLS), a nuclear export signal (NES), and an activator motif (AHA motif). The OD (or HR-A/B region) is connected to the N-terminal DBD by a flexible linker of variable length (15–80 amino acid residues). A heptad pattern of hydrophobic residues in the HR-A/B region leads to the formation of a coiled structure that allows leucine zipper-type protein interactions. Plant HSFs are classified into three classes, A, B, and C, based on the peculiarities of their ODs. The HR-A/B region of plant Class B HSFs is compact, whereas Class A and C HSFs have extended HR-A/B regions with insertions of 21 (Class A) or 7 (Class C) amino acid residues between the A and B parts. Plants have many more HSF-encoding genes than other eukaryotes (21 in *Arabidopsis*, 4 in vertebrates, and 1 in *Drosophila*). Of the 21 HSFs in *Arabidopsis*, 15 belong to Class A, 5 to Class B, and 1 to Class C (Scharf et al., 2012, Guo et al., 2008).

Functional studies show that HsfA1a has a unique function as master regulator of acquired thermotolerance and trigger of the HS response and that later on, by interaction with HsfA2 and B1 in a functional triad, affects different aspects of HS response and recovery (Hahn et al., 2011). HsfA2 is the most highly induced HSF in stressed plants and plays a role in thermotolerance and a broader role for expression of general stress-related nonchaperone-encoding genes like *APX2* (*ASCORBATE PEROXIDASE2*; Nishizawa et al., 2006). Class B HSFs function as active repressors of the transcription of HS-inducible genes. The LFGV-tetrapeptide motif found in the repressor domain of Class B HSFs (LKLFGVWL) is known to have repressor functions in other plant TFs (Ikeda and Ohme-Takagi, 2009). Class B HSFs are also necessary for suppression of the general HS response under non-HS conditions and in the attenuating period.

3.6 PLANT-SPECIFIC TF FAMILIES

3.6.1 AP2/ERF Family

The AP2/ERF family is a large group of plant-specific TFs containing AP2/ERF-type DBDs, which consist of about 60–70 amino acids and are involved in DNA binding (Mizoi et al., 2012; Nakano et al., 2006). AP2/ERF family members are encoded by 145 loci in *Arabidopsis* and 167 loci in rice. Members of the AP2/ERF family can be divided into three groups based on overall structure: the AP2, RAV, and ERF subfamilies. Members of the AP2 subfamily (14 members in *Arabidopsis*) contain double AP2/ERF domains, ERF family (125 members) proteins contain a single AP2/ERF domain, and RAV family (6 members) proteins contain an AP2/ERF domain and an additional B3 domain, which is a DNA-binding domain conserved in other plant-specific TFs, including VP1/ABI3. The ERF family is sometimes further classified two major subfamilies, the ERF subfamily and the CBF/DREB subfamily (Sakuma et al., 2002). The AP2 domain was first identified as a repeated motif within the *Arabidopsis* homeotic gene *APETALA 2* (*AP2*) involved in flower development. The ERF domain was first found in tobacco ethylene-responsive element binding proteins (EREBPs) as a conserved DNA-binding motif. ERF subfamily members are mainly involved in responses to biotic stresses by recognition of the GCC box (5'AGCCGCC3'), which is a DNA sequence involved in ethylene-responsive gene transcription (Ohme-Takagi and Shinshi, 1995). CBF/DREB subfamily members play crucial roles in abiotic stresses by recognizing a dehydration-responsive element (DRE) with a core motif 5'A/GCCGAC3'. Members of the DREB1/CBF subgroup (DREB1A/CBF3, DREB1B/CBF1, and DREB1C/CBF2) are cold inducible and are major regulators of cold stress responses, while those of the DREB2 subgroup (DREB2A and DREB2B) play important roles in dehydration and heat stress responses (Mizoi et al., 2012).

3.6.2 WRKY TF Family

The WRKY TF family is one of the best-studied plant-specific TF families and comprises 74 members in *Arabidopsis* (Rushton et al., 2010; Ulker and Somssich, 2004; Chapter 11). The WRKY protein family owes its name to the highly conserved 60 amino acid long WRKY domain, which contains a conserved amino acid sequence motif WRKYGQK at the N-terminus and a novel zinc-finger-like motif at the C-terminus. These two motifs are vital for binding to the consensus *cis*-acting element termed the W-box (TTGACT/C). Based on both the number of WRKY domains and features of the the zinc-finger motif, WRKY proteins are categorized into three subfamilies: the first Group I has two WRKY domains, group II has one WRKY domain with the same Cys2–His2 zinc-finger motif, and

Group III has one WRKY domain containing a different Cys2–His2 zinc-finger motif. Group II WRKYs are further divided into a–e based on additional conserved motifs outside the WRKY domain.

WRKY TFs are well-known as participants in various biotic stress responses (Pandey and Somssich, 2009) and abiotic stress responses (Chen et al., 2012). WRKY TFs are also involved in several developmental and physiological processes such as embryogenesis, seed coat and trichome development, anthocyanin biosynthesis, and hormone signaling (Rushton et al., 2010). AtWRKY52/RRS1, a member of Group III that contains TIR–NBS–LRR (TNL) and WRKY domains, confers immunity on the bacterial pathogen *Ralstonia solanacearum* by nuclear interaction with the type III bacterial effector PopP2 (Deslandes et al., 2003). The AtWRKY52 also interacts with the R protein RPS4 to provide dual resistance towards fungal and bacterial pathogens (Narusaka et al., 2009). This suggests that TNL-WRKYs may allow a shortcut to the effector-triggered immunity (ETI) pathway, leading to defense gene activation. Group IId WRKY proteins contain a calmodulin (CaM)-binding domain, designated the C-motif (DxxVxKFKxVISLxxxR), suggesting possible regulation by CaM and Ca^{2+} fluxes (Park et al., 2005). The leucine zipper containing Group IIa WRKY proteins (AtWRKY18, 40 and 60) form homo- or heterodimeric complexes.

3.6.3 NAC TF Family

NAC TFs are plant-specific TFs that have been shown to function in relation to plant development as well as in abiotic and biotic stress responses (Ooka et al., 2003b; Nakashima et al., 2012; Chapter 13). The NAC domain was identified based on consensus sequences from Petunia NAM and *Arabidopsis* ATAF1/2 and CUC proteins (Aida et al., 1997). NAC proteins contain a highly homologous region containing the DNA-binding NAC domain in the N-terminus. The NAC domain is approximately 150 amino acid long and contains 5 conserved regions (A–E). X-ray structure analysis of *Arabidopsis* ANAC019 revealed a new type of TF fold consisting of a twisted β-sheet that is surrounded by a few helical elements (Ernst et al., 2004). The N-terminal NAC domain contains a NLS, and the highly variable C-terminal region contains a transactivation domain.

Many NAC proteins, including *Arabidopsis* CUC2, have important functions in plant development (Chapter 15). Some NAC genes are upregulated during wounding and bacterial infections, whereas others mediate viral resistance (Nakashima et al., 2012). Many NAC proteins are thought to be transcriptional activators as *Arabidopsis* ATAF1/2 and AtNAM (NARS2) function as activators in a yeast system. There are 75 predicted NAC proteins in *Arabidopsis* and 105 NACs in rice (Ooka et al., 2003a).

Phylogeny analysis of conserved amino acid residues and the NAC domain revealed that they are broadly classified into six major groups. As many NAC TFs are induced by various stress treatments, stress-responsive NAC genes are grouped into the SNAC (stress-responsive NAC) group which has three subgroups (*SNAC-A, B,* and *C*; Nuruzzaman et al., 2010). *Arabidopsis* has 17 NAC proteins containing a transmembrane domain (TM) motif at the C-terminus, which are called the "NTL subfamily" (Kim et al., 2007; Chapter 25).

3.6.4 TCP Family

TCP TFs constitute a small family of plant-specific TFs represented by the first three identified members TEOSINTE BRANCHED1 in maize (*Zea mays*), CYCLOIDEA in snapdragon (*Antirrhinum majus*), and PROLIFERATING CELL FACTOR1 (PCF) in rice (*Oryza sativa*; Kosugi and Ohashi, 1997; Chapter 9). The *Arabidopsis thaliana* genome encodes 24 TCP TFs, which are divided into Class I and Class II TCPs based on sequence similarities (Martín-Trillo and Cubas, 2010). All TCP TFs share the TCP domain, a 59-amino-acid-long, noncanonical bHLH domain responsible for nuclear targeting, DNA binding, and mediating protein–protein interactions. The TCP domain mediates the binding of TCP proteins to GC-rich DNA sequence motifs *in vitro* (Kosugi and Ohashi, 2002). These motifs have been identified as *cis*-elements in many plant genes coding for proteins such as CYCLIN, PCNA, and CCA1. TCP proteins are important regulators of plant growth and development and control multiple traits in diverse plant species, including flower and petal development, plant architecture, leaf morphogenesis and senescence, embryo growth, and circadian rhythm (Chapter 16). At the cellular level, most TCP genes modulate cell proliferation in the axillary organs of plants either by directly controlling the transcription of cyclin or by bringing about cell differentiation (Martín-Trillo and Cubas, 2010).

3.7 TFs WITHOUT DBD BUT INTERACTING WITH DBD-CONTAINING TFs

3.7.1 B-Box Zinc Finger Transcription Factors

B-box (BBX) proteins are a class of Zn finger TFs that contain an N-terminal B-box domain and sometimes a CONSTANS, CO-like, and TOC1 (CCT) domain in the C-terminus (Gangappa and Botto, 2014, Khanna et al., 2009). BBX proteins share a 40-amino-acid-long B-box motif that can be divided into two types, B-box1 and B-box2, based on their consensus sequence and the spacing of zinc-binding residues. Conserved residues in the B-box motif have been shown to be crucial in mediating protein–protein interactions (Datta et al., 2008). The CCT domain is a basic motif of 42–43 amino acids with functional roles in transcriptional regulation and nuclear protein transport in some BBX proteins. In *Arabidopsis*, there are 32 BBX proteins encoded in the genome. BBX proteins are divided into five structural groups (Groups I–V) depending on the type or number of BBX motifs and the presence of the CCT domain. Groups I (BBX1–6) and II (BBX7–13) have two B-boxes and a CCT domain, whereas Group III (BBX14–17) has one B-box and a CCT domain. Group IV contains two tandem repeat B-box motifs in the N-terminus but lacks the CCT domain and is also called the "double B-box" (DBB) subfamily. Group V (BBX27-32) has only a short N-terminal B-box domain with one or two B-box motifs.

BBX proteins play a key role in a variety of plant regulatory networks controlling growth and developmental processes that include seedling photomorphogenesis, photoperiodic regulation of flowering, shade avoidance, and responses to biotic and abiotic stresses (Gangappa and Botto, 2014). One of the best-studied BBX proteins is BBX1/CONSTANS (CO), which contains two BBXs and a CCT domain. It promotes flowering of *Arabidopsis* in response to day length (Putterill et al., 1995). CO regulates flowering time by positively regulating expression of FLOWERING LOCUS T (FT). Members of Group IV act as positive (BBX21/STH2, BBX22/STH3) or negative (BBX24/STO, BBX25/STH) regulators of photomorphogenesis by interacting with HY5 bZIP proteins through their BBX motifs (Gangappa et al., 2013). These BBX proteins are involved in seedling de-etiolation, hypocotyl growth, anthocyanin production, chlorophyll accumulation, lateral root growth, and cotyledon unfolding. The diverse roles of BBX proteins can be attributed to the BBX domain, as it mediates homo- and heterodimer formation both within and outside the BBX protein family and interactions with members of other TF families.

3.7.2 Aux/IAA Family

The AUX/IAA family represents a class of proteins interacting with auxin response factors (ARFs). They act as repressors of auxin-regulated genes (Dharmasiri and Estelle, 2004; Overvoorde et al., 2005). Aux/IAA genes are rapidly induced by auxin and represent a major class of primary auxin-responsive genes after hormone treatment. Canonical Aux/IAA proteins share four conserved amino acid sequence motifs known as Domains I, II, III, and IV. Domain I is a repressor domain that contains a conserved leucine repeat motif (LxLxLx) similar to the so-called EAR (ethylene-responsive element-binding factor-associated amphiphilic repression) domain (Tiwari et al., 2004). Domain I is also required for the recruitment of the transcriptional corepressor TOPLESS (Szemenyei et al., 2008a).

Domain II confers protein instability, leading to rapid degradation of Aux/IAA through interaction with the F-box protein TIR1 (a component of the SCFTIR1 ubiquitin ligase complex; Dharmasiri et al., 2005). C-terminal Domains III and IV are shared with ARF proteins, and are known to promote homo- and heterodimerization of Aux/IAA polypeptides, as well as interaction between Aux/IAAs and ARFs (Overvoorde et al., 2005).

There are 23 ARFs, most of which contain a conserved N-terminal DBD, a variable middle transcriptional regulatory region (MR), and a carboxy-terminal dimerization domain (CTD). The DBD of ARFs specifically binds to the conserved auxin response element (AuxRE, TGTCTC) in promoter regions of primary or early auxin-responsive genes. The structure of the transcriptional regulatory region of each ARF determines whether the ARF acts as an activator or repressor. The ARF C-terminal domain is modular with amino acid sequences related to Domains III and IV of Aux/IAA proteins, and functions as a dimerization domain among ARF CTDs or with several Aux/IAA proteins. The binding of auxin to its receptors leads to the degradation of Aux/IAA proteins (Dharmasiri et al., 2005). This auxin-dependent proteolysis releases ARFs from Aux/IAA repressor binding and facilitates the activation of auxin-responsive genes.

3.7.3 JAZ Protein Family

JASMONATE ZIM-DOMAIN (JAZ) proteins act as jasmonate (JA) coreceptors and transcriptional repressors in JA signaling. In *Arabidopsis* there are 12 members (JAZ1–JAZ12) of this family (Pauwels and Goossens, 2011). JAZ proteins are composed of two major domains, the TIFY/ZIM and the Jas domain, with the latter mediating the interaction between transcriptional repressor JAZ proteins and the receptor F-box protein CORONATINE INSENSITIVE1 (COI1; Melotto et al., 2008). JAZ proteins repress the JA response by interacting with bHLH TFs MYC2 and MYC3, which bind to DNA sequences directly and regulate downstream gene expression (Pauwels and Goossens, 2011). On perception of JA-Ile, a bioactive JA, JAZ proteins are degraded and JA-Ile dependent gene expression is activated by MYC2. *Arabidopsis* JAZ proteins recruit the Groucho/Tup1-type corepressor TOPLESS (TPL) and TPL related proteins (TPRs) through an adaptor protein, designated NOVEL INTERACTOR of JAZ (NINJA; Pauwels et al., 2010). NINJA acts as a transcriptional corepressor whose activity is mediated by a functional TPL-binding EAR repression motif.

JAZ proteins interact with many TFs of different families and hence provide possible mechanisms to understand the role of JA in plant development, defense, and hormone signaling (Kazan and Manners, 2012). The search for proteins that interact with JAZ identified two other MYC2-related bHLH TFs, MYC3 and MYC4, with functionally overlapping yet distinct roles with MYC2 in JA signaling. In addition, JAZ proteins, through their conserved Jas domain, interact with three other bHLH TFs, TT8, GL3, and EGL3, and two R2R3 MYB TFs, PAP1/MYB75 and GLABRA1 (GL1). These five TFs belong to the MYB -bHLH -WD40 complex which has well-established roles in JA-mediated anthocyanin accumulation and trichome initiation. JAZ repressors also interact with and suppress the transcriptional activities of the two other MYBs, MYB21 and MYB24, required for stamen development. The identification of DELLA (REPRESSOR of GA1-3 [RGA]1 and RGA2) proteins as JAZ targets further extend the role of JAZ in JA and GA hormone signaling.

3.8 CONCLUSION

The transcription of eukaryotic genes requires complex interaction between many regulatory proteins of different functional classes. TFs are a group of regulatory proteins that control gene expression by binding to DNA and in so doing either activate or repress mRNA transcription. TF binding to promoter DNA facilitates recruitment of the Med complex. It further facilitates recruitment of GTFs to CPEs for successful formation of PIC and transcription initiation. The promoter-proximal pausing of RNAPII found in inducible and developmentally regulated genes suggests the importance of regulation after transcription initiation. Proper modification of the CTD of RNAPII and N-terminal histone tails is essential for successful transcription initiation and elongation.

Plant TFs are characterized by a large number of genes and by the variety of TF gene families. Plants also have additional RNA polymerases (RNAPIV and V), plant-specific Med subunits, and a large number of plant-specific TF families. This reflects the variety of plant-specific cellular responses to various biotic and abiotic stresses that plants confront as sessile organisms. Each plant hormone shows specific modes of signal transduction from the receptor to the TFs being used. Genetic and biochemical studies have identified the protein–protein interactions of TFs of different families, which signifies the importance of combinatorial regulation of gene expression. Genetic and molecular studies of *Arabidopsis* TFs have provided valuable information on a variety of plant-specific responses, such as plant defense to pathogen infection, responses to light, and environmental stresses such as cold, drought, high salinity, and plant development.

Acknowledgments

The author acknowledges support provided by grants from the Next Generation Biogreen 21 Program (SSAC grant PJ01107102) and the Basic Science Research Program funded by the Ministry of Education (NRF-2013R1A1A2010131).

References

Adelman, K., Lis, J.T., 2012. Promoter-proximal pausing of RNA polymerase II: emerging roles in metazoans. Nat. Rev. Genet. 13, 720–731.

Aguilar, X., Blomberg, J., Brannstrom, K., Olofsson, A., Schleucher, J., Bjorklund, S., 2014. Interaction studies of the human and *Arabidopsis thaliana* Med25–ACID proteins with the herpes simplex virus VP16- and plant-specific Dreb2a transcription factors. PLoS ONE 9, e98575.

Aida, M., Ishida, T., Fukaki, H., Fujisawa, H., Tasaka, M., 1997. Genes involved in organ separation in *Arabidopsis*: an analysis of the cup-shaped cotyledon mutant. Plant Cell Online 9, 841–857.

Atchley, W.R., Terhalle, W., Dress, A., 1999. Positional dependence, cliques, and predictive motifs in the bHLH protein domain. J. Mol. Evol. 48, 501–516.

Backstrom, S., Elfving, N., Nilsson, R., Wingsle, G., Bjorklund, S., 2007. Purification of a plant mediator from *Arabidopsis thaliana* identifies PFT1 as the Med25 subunit. Mol. Cell 26, 717–729.

Bailey, P.C., Dicks, J., Wang, T.L., Martin, C., 2008. IT3F: a web-based tool for functional analysis of transcription factors in plants. Phytochemistry 69, 2417–2425.

Bertrand, C., Benhamed, M., Li, Y.F., Ayadi, M., Lemonnier, G., Renou, J.P., Delarue, M., Zhou, D.X., 2005. *Arabidopsis* HAF2 gene encoding TATA-binding protein (TBP)-associated factor TAF1 is required to integrate light signals to regulate gene expression and growth. J. Biol. Chem. 280, 1465–1473.

Bjorklund, S., Gustafsson, C.M., 2005. The yeast Mediator complex and its regulation. Trends Biochem. Sci. 30, 240–244.

Boeing, S., Rigault, C., Heidemann, M., Eick, D., Meisterernst, M., 2010. RNA polymerase II C-terminal heptarepeat domain Ser-7 phosphorylation is established in a mediator-dependent fashion. J. Biol. Chem. 285, 188–196.

Castillon, A., Shen, H., Huq, E., 2007. Phytochrome interacting factors: central players in phytochrome-mediated light signaling networks. Trends Plant Sci. 12, 514–521.

Causier, B., Ashworth, M., Guo, W., Davies, B., 2012. The TOPLESS interactome: a framework for gene repression in *Arabidopsis*. Plant Physiol. 158, 423–438.

Cerna, D., Wilson, D.K., 2005. The structure of Sif2p, a WD repeat protein functioning in the SET3 corepressor complex. J. Mol. Biol. 351, 923–935.

Chattopadhyay, S., Ang, L.-H., Puente, P., Deng, X.-W., Wei, N., 1998. *Arabidopsis* bZIP protein HY5 directly interacts with light-responsive promoters in mediating light control of gene expression. Plant Cell Online 10, 673–683.

Chen, G., Courey, A.J., 2000. Groucho/TLE family proteins and transcriptional repression. Gene 249, 1–16.

Chen, L., Song, Y., Li, S., Zhang, L., Zou, C., Yu, D., 2012. The role of WRKY transcription factors in plant abiotic stresses. Biochim. Biophys. Acta 1819, 120–128.

Chiba, K., Yamamoto, J., Yamaguchi, Y., Handa, H., 2010. Promoter-proximal pausing and its release: molecular mechanisms and physiological functions. Exp. Cell Res. 316, 2723–2730.

Clapier, C.R., Cairns, B.R., 2009. The biology of chromatin remodeling complexes. Annu. Rev. Biochem. 78, 273–304.

Conner, J., Liu, Z., 2000. LEUNIG, a putative transcriptional corepressor that regulates AGAMOUS expression during flower development. Proc. Natl. Acad. Sci. USA 97, 12902–12907.

Core, L.J., Waterfall, J.J., Lis, J.T., 2008. Nascent RNA sequencing reveals widespread pausing and divergent initiation at human promoters. Science 322, 1845–1848.

Datta, S., Hettiarachchi, C., Johansson, H., Holm, M., 2007. SALT TOLERANCE HOMOLOG2, a B-box protein in *Arabidopsis* that activates transcription and positively regulates light-mediated development. Plant Cell Online 19, 3242–3255.

Datta, S., Johansson, H., Hettiarachchi, C., Irigoyen, M.L., Desai, M., Rubio, V., Holm, M., 2008. LZF1/SALT TOLERANCE HOMOLOG3, an *Arabidopsis* B-box protein involved in light-dependent development and gene expression, undergoes COP1-mediated ubiquitination. Plant Cell Online 20, 2324–2338.

Davuluri, R.V., Sun, H., Palaniswamy, S.K., Matthews, N., Molina, C., Kurtz, M., Grotewold, E., 2003. AGRIS: *Arabidopsis* gene regulatory information server, an information resource of *Arabidopsis* cis-regulatory elements and transcription factors. BMC Bioinform. 4, 25.

Deng, W., Roberts, S.G., 2007. TFIIB and the regulation of transcription by RNA polymerase II. Chromosoma 116, 417–429.

Deslandes, L., Olivier, J., Peeters, N., Feng, D.X., Khounlotham, M., Boucher, C., Somssich, I., Genin, S., Marco, Y., 2003. Physical interaction between RRS1-R, a protein conferring resistance to bacterial wilt, and PopP2, a type III effector targeted to the plant nucleus. Proc. Natl. Acad. Sci. 100, 8024–8029.

Dharmasiri, N., Estelle, M., 2004. Auxin signaling and regulated protein degradation. Trends Plant Sci. 9, 302–308.

Dharmasiri, N., Dharmasiri, S., Estelle, M., 2005. The F-box protein TIR1 is an auxin receptor. Nature 435, 441–445.

Du, H., Wang, Y.B., Xie, Y., Liang, Z., Jiang, S.J., Zhang, S.S., Huang, Y.B., Tang, Y.X., 2013. Genome-wide identification and evolutionary and expression analyses of MYB-related genes in land plants. DNA Res. 20, 437–448.

Dubos, C., Stracke, R., Grotewold, E., Weisshaar, B., Martin, C., Lepiniec, L., 2010. MYB transcription factors in *Arabidopsis*. Trends Plant Sci. 15, 573–581.

Dunoyer, P., Himber, C., Ruiz-Ferrer, V., Alioua, A., Voinnet, O., 2007. Intra- and intercellular RNA interference in *Arabidopsis thaliana* requires components of the microRNA and heterochromatic silencing pathways. Nat. Genet. 39, 848–856.

Egloff, S., Murphy, S., 2008. Cracking the RNA polymerase II CTD code. Trends Genet. 24, 280–288.

Ellenberger, T.E., Brandl, C.J., Struhl, K., Harrison, S.C., 1992. The GCN4 basic region leucine zipper binds DNA as a dimer of uninterrupted alpha helices: crystal structure of the protein–DNA complex. Cell 71, 1223–1237.

Ernst, H.A., Nina Olsen, A., Skriver, K., Larsen, S., Lo Leggio, L., 2004. Structure of the conserved domain of ANAC, a member of the NAC family of transcription factors. EMBO Rep. 5, 297–303.

Fredslund, J., 2008. DATFAP: a database of primers and homology alignments for transcription factors from 13 plant species. BMC Genomics 9, 140.

Gangappa, S.N., Botto, J.F., 2014. The BBX family of plant transcription factors. Trends Plant Sci. 19, 460–470.

Gangappa, S.N., Holm, M., Botto, J.F., 2013. Molecular interactions of BBX24 and BBX25 with HYH, HY5 HOMOLOG, to modulate *Arabidopsis* seedling development. Plant Signal. Behav. 8.

Gao, G., Zhong, Y., Guo, A., Zhu, Q., Tang, W., Zheng, W., Gu, X., Wei, L., Luo, J., 2006a. DRTF: a database of rice transcription factors. Bioinformatics 22, 1286–1287.

Gao, X., Ren, F., Lu, Y.T., 2006b. The *Arabidopsis* mutant stg1 identifies a function for TBP-associated factor 10 in plant osmotic stress adaptation. Plant Cell Physiol. 47, 1285–1294.

Gillmor, C.S., Park, M.Y., Smith, M.R., Pepitone, R., Kerstetter, R.A., Poethig, R.S., 2010. The MED12–MED13 module of Mediator regulates the timing of embryo patterning in *Arabidopsis*. Development 137, 113–122.

Gonzalez, D., Bowen, A.J., Carroll, T.S., Conlan, R.S., 2007. The transcription corepressor LEUNIG interacts with the histone deacetylase HDA19 and mediator components MED14 (SWP) and CDK8 (HEN3) to repress transcription. Mol. Cell Biol. 27, 5306–5315.

Guo, A., He, K., Liu, D., Bai, S., Gu, X., Wei, L., Luo, J., 2005. DATF: a database of *Arabidopsis* transcription factors. Bioinformatics 21, 2568–2569.

Guo, J., Wu, J., Ji, Q., Wang, C., Luo, L., Yuan, Y., Wang, Y., Wang, J., 2008. Genome-wide analysis of heat shock transcription factor families in rice and *Arabidopsis*. J. Genet. Genomics 35, 105–118.

Hahn, S., 2004. Structure and mechanism of the RNA polymerase II transcription machinery. Nat. Struct. Mol. Biol. 11, 394–403.

Hahn, A., Bublak, D., Schleiff, E., Scharf, K.-D., 2011. Crosstalk between Hsp90 and Hsp70 chaperones and heat stress transcription factors in tomato. Plant Cell Online 23, 741–755.

Hahn, S., Young, E.T., 2011. Transcriptional regulation in *Saccharomyces cerevisiae*: transcription factor regulation and function, mechanisms of initiation, and roles of activators and coactivators. Genetics 189, 705–736.

Hsieh, F.K., Kulaeva, O.I., Patel, S.S., Dyer, P.N., Luger, K., Reinberg, D., Studitsky, V.M., 2013. Histone chaperone FACT action during transcription through chromatin by RNA polymerase II. Proc. Natl. Acad. Sci. USA 110, 7654–7659.

Iida, K., Seki, M., Sakurai, T., Satou, M., Akiyama, K., Toyoda, T., Konagaya, A., Shinozaki, K., 2005. RARTF: database and tools for complete sets of *Arabidopsis* transcription factors. DNA Res. 12, 247–256.

Ikeda, M., Ohme-Takagi, M., 2009. A novel group of transcriptional repressors in *Arabidopsis*. Plant Cell Physiol. 50, 970–975.

Jakoby, M., Weisshaar, B., Droge-Laser, W., Vicente-Carbajosa, J., Tiedemann, J., Kroj, T., Parcy, F., bZIP Research Group, 2002. bZIP transcription factors in *Arabidopsis*. Trends Plant. Sci. 7, 106–111.

Jin, J.P., Zhang, H., Kong, L., Gao, G., Luo, J.C., 2014. PlantTFDB 3.0: a portal for the functional and evolutionary study of plant transcription factors. Nucleic Acids Res. 42 (D1), D1182–D1187.

Johnson, C., Boden, E., Arias, J., 2003. Salicylic acid and NPR1 induce the recruitment of trans-activating TGA factors to a defense gene promoter in *Arabidopsis*. Plant Cell Online 15, 1846–1858.

Juven-Gershon, T., Hsu, J.Y., Theisen, J.W., Kadonaga, J.T., 2008. The RNA polymerase II core promoter – the gateway to transcription. Curr. Opin. Cell Biol. 20, 253–259.

Juven-Gershon, T., Kadonaga, J.T., 2010. Regulation of gene expression via the core promoter and the basal transcriptional machinery. Dev. Biol. 339, 225–229.

Kagale, S., Links, M.G., Rozwadowski, K., 2010. Genome-wide analysis of ethylene-responsive element-binding factor-associated amphiphilic repression motif-containing transcriptional regulators in *Arabidopsis*. Plant Physiol. 152, 1109–1134.

Katiyar, A., Smita, S., Lenka, S.K., Rajwanshi, R., Chinnusamy, V., Bansal, K.C., 2012. Genome-wide classification and expression analysis of MYB transcription factor families in rice and *Arabidopsis*. BMC Genomics 13, 544.

Kazan, K., Manners, J.M., 2012. JAZ repressors and the orchestration of phytohormone crosstalk. Trends Plant Sci. 17, 22–31.

Kazan, K., Manners, J.M., 2013. MYC2: the master in action. Mol. Plant 6, 686–703.

Khanna, R., Kronmiller, B., Maszle, D.R., Coupland, G., Holm, M., Mizuno, T., Wu, S.H., 2009. The *Arabidopsis* B-box zinc finger family. Plant Cell Online 21, 3416–3420.

Kidd, B.N., Cahill, D.M., Manners, J.M., Schenk, P.M., Kazan, K., 2011. Diverse roles of the Mediator complex in plants. Semin. Cell Dev. Biol. 22, 741–748.

Kim, S., Chung, H., Thomas, T., 1997. Isolation of a novel class of bZIP transcription factors that interact with ABA-responsive and embryo-specification elements in the Dc3 promoter using a modified yeast one-hybrid system. Plant J. Cell Mol. Biol. 11, 1237.

Kim, S.Y., Kim, S.G., Kim, Y.S., Seo, P.J., Bae, M., Yoon, H.K., Park, C.M., 2007. Exploring membrane-associated NAC transcription factors in *Arabidopsis*: implications for membrane biology in genome regulation. Nucleic Acids Res. 35, 203–213.

Knuesel, M.T., Meyer, K.D., Donner, A.J., Espinosa, J.M., Taatjes, D.J., 2009. The human CDK8 subcomplex is a histone kinase that requires Med12 for activity and can function independently of mediator. Mol. Cell Biol. 29, 650–661.

Kohoutek, J., 2009. P-TEFb – the final frontier. Cell Div. 4, 19.

Kornberg, R.D., 2005. Mediator and the mechanism of transcriptional activation. Trends Biochem. Sci. 30, 235–239.

Kosugi, S., Ohashi, Y., 1997. PCF1 and PCF2 specifically bind to cis elements in the rice proliferating cell nuclear antigen gene. Plant Cell Online 9, 1607–1619.

Kosugi, S., Ohashi, Y., 2002. DNA binding and dimerization specificity and potential targets for the TCP protein family. Plant J. 30, 337–348.

Lago, C., Clerici, E., Dreni, L., Horlow, C., Caporali, E., Colombo, L., Kater, M.M., 2005. The *Arabidopsis* TFIID factor AtTAF6 controls pollen tube growth. Dev. Biol. 285, 91–100.

Lago, C., Clerici, E., Mizzi, L., Colombo, L., Kater, M.M., 2004. TBP-associated factors in *Arabidopsis*. Gene 342, 231–241.

Lauberth, S.M., Nakayama, T., Wu, X., Ferris, A.L., Tang, Z., Hughes, S.H., Roeder, R.G., 2013. H3K4me3 interactions with TAF3 regulate preinitiation complex assembly and selective gene activation. Cell 152, 1021–1036.

Lawit, S.J., O'Grady, K., Gurley, W.B., Czarnecka-Verner, E., 2007. Yeast two-hybrid map of *Arabidopsis* TFIID. Plant Mol. Biol. 64, 73–87.

Lee, S.K., Fletcher, A.G., Zhang, L., Chen, X., Fischbeck, J.A., Stargell, L.A., 2010. Activation of a poised RNAPII-dependent promoter requires both SAGA and mediator. Genetics 184, 659–672.

Li, B., Carey, M., Workman, J.L., 2007. The role of chromatin during transcription. Cell 128, 707–719.

Li, X., Duan, X., Jiang, H., Sun, Y., Tang, Y., Yuan, Z., Guo, J., Liang, W., Chen, L., Yin, J., Ma, H., Wang, J., Zhang, D., 2006. Genome-wide analysis of basic/helix–loop–helix transcription factor family in rice and *Arabidopsis*. Plant Physiol. 141, 1167–1184.

Liu, X., Bushnell, D.A., Silva, D.A., Huang, X., Kornberg, R.D., 2011. Initiation complex structure and promoter proofreading. Science 333, 633–637.

Liu, Z., Karmarkar, V., 2008. Groucho/Tup1 family co-repressors in plant development. Trends Plant. Sci. 13, 137–144.

Martín-Trillo, M., Cubas, P., 2010. TCP genes: a family snapshot ten years later. Trends Plant Sci. 15, 31–39.

Mathur, S., Vyas, S., Kapoor, S., Tyagi, A.K., 2011. The Mediator complex in plants: structure, phylogeny, and expression profiling of representative genes in a dicot (*Arabidopsis*) and a monocot (rice) during reproduction and abiotic stress. Plant Physiol. 157, 1609–1627.

Melotto, M., Mecey, C., Niu, Y., Chung, H.S., Katsir, L., Yao, J., Zeng, W., Thines, B., Staswick, P., Browse, J., Howe, G.A., He, S.Y., 2008. A critical role of two positively charged amino acids in the Jas motif of *Arabidopsis* JAZ proteins in mediating coronatine- and jasmonoyl isoleucine-dependent interactions with the COI1 F-box protein. Plant J. 55, 979–988.

Mitsuda, N., Ohme-Takagi, M., 2009. Functional analysis of transcription factors in *Arabidopsis*. Plant Cell. Physiol. 50, 1232–1248.

Mizoi, J., Shinozaki, K., Yamaguchi-Shinozaki, K., 2012. AP2/ERF family transcription factors in plant abiotic stress responses. Biochim. Biophys. Acta 1819, 86–96.

Mochida, K., Yoshida, T., Sakurai, T., Yamaguchi-Shinozaki, K., Shinozaki, K., Tran, L.S., 2009. In silico analysis of transcription factor repertoire and prediction of stress responsive transcription factors in soybean. DNA Res. 16, 353–369.

Mochida, K., Yoshida, T., Sakurai, T., Yamaguchi-Shinozaki, K., Shinozaki, K., Tran, L.S., 2010. LegumeTFDB: an integrative database of *Glycine max*, *Lotus japonicus* and *Medicago truncatula* transcription factors. Bioinformatics 26, 290–291.

Mochida, K., Yoshida, T., Sakurai, T., Yamaguchi-Shinozaki, K., Shinozaki, K., Tran, L.S., 2011. In silico analysis of transcription factor repertoires and prediction of stress-responsive transcription factors from six major Gramineae plants. DNA Res. 18, 321–332.

Mochida, K., Yoshida, T., Sakurai, T., Yamaguchi-Shinozaki, K., Shinozaki, K., Tran, L.S., 2013. TreeTFDB: an integrative database of the transcription factors from six economically important tree crops for functional predictions and comparative and functional genomics. DNA Res. 20, 151–162.

Nag, A., Narsinh, K., Martinson, H.G., 2007. The poly(A)-dependent transcriptional pause is mediated by CPSF acting on the body of the polymerase. Nat. Struct. Mol. Biol. 14, 662–669.

Naika, M., Shameer, K., Mathew, O.K., Gowda, R., Sowdhamini, R., 2013. STIFDB2: an updated version of plant stress-responsive transcrIption factor database with additional stress signals, stress-responsive transcription factor binding sites and stress-responsive genes in arabidopsis and rice. Plant Cell Physiol. 54, e8.

Nakano, T., Suzuki, K., Fujimura, T., Shinshi, H., 2006. Genome-wide analysis of the ERF gene family in *Arabidopsis* and rice. Plant Physiol. 140, 411–432.

Nakashima, K., Takasaki, H., Mizoi, J., Shinozaki, K., Yamaguchi-Shinozaki, K., 2012. NAC transcription factors in plant abiotic stress responses. Biochim. Biophys. Acta 1819, 97–103.

Narusaka, M., Shirasu, K., Noutoshi, Y., Kubo, Y., Shiraishi, T., Iwabuchi, M., Narusaka, Y., 2009. RRS1 and RPS4 provide a dual resistance-gene system against fungal and bacterial pathogens. Plant J. 60, 218–226.

Nechaev, S., Adelman, K., 2011. Pol II waiting in the starting gates: regulating the transition from transcription initiation into productive elongation. Biochim. Biophys. Acta 1809, 34–45.

Nishizawa, A., Yabuta, Y., Yoshida, E., Maruta, T., Yoshimura, K., Shigeoka, S., 2006. *Arabidopsis* heat shock transcription factor A2 as a key regulator in response to several types of environmental stress. Plant J. 48, 535–547.

Nover, L., Scharf, K.D., Gagliardi, D., Vergne, P., Czarnecka-Verner, E., Gurley, W.B., 1996. The Hsf world: classification and properties of plant heat stress transcription factors. Cell Stress Chaperones 1, 215–223.

Nuruzzaman, M., Manimekalai, R., Sharoni, A.M., Satoh, K., Kondoh, H., Ooka, H., Kikuchi, S., 2010. Genome-wide analysis of NAC transcription factor family in rice. Gene 465, 30–44.

Ohkuma, Y., Roeder, R.G., 1994. Regulation of TFIIH ATPase and kinase activities by TFIIE during active initiation complex formation. Nature 368, 160–163.

Ohme-Takagi, M., Shinshi, H., 1995. Ethylene-inducible DNA binding proteins that interact with an ethylene-responsive element. Plant Cell Online 7, 173–182.

Ooka, H., Satoh, K., Doi, K., Nagata, T., Otomo, Y., Murakami, K., Matsubara, K., Osato, N., Kawai, J., Carninci, P., 2003a. Comprehensive analysis of NAC family genes in *Oryza sativa* and *Arabidopsis thaliana*. DNA Res. 10, 239–247.

Ooka, H., Satoh, K., Doi, K., Nagata, T., Otomo, Y., Murakami, K., Matsubara, K., Osato, N., Kawai, J., Carninci, P., Hayashizaki, Y., Suzuki, K., Kojima, K., Takahara, Y., Yamamoto, K., Kikuchi, S., 2003b. Comprehensive analysis of NAC family genes in *Oryza sativa* and *Arabidopsis thaliana*. DNA Res. 10, 239–247.

Osterlund, M.T., Hardtke, C.S., Wei, N., Deng, X.W., 2000. Targeted destabilization of HY5 during light-regulated development of *Arabidopsis*. Nature 405, 462–466.

Overvoorde, P.J., Okushima, Y., Alonso, J.M., Chan, A., Chang, C., Ecker, J.R., Hughes, B., Liu, A., Onodera, C., Quach, H., 2005. Functional genomic analysis of the AUXIN/INDOLE-3-ACETIC ACID gene family members in *Arabidopsis thaliana*. Plant Cell Online 17, 3282–3300.

Pandey, S.P., Somssich, I.E., 2009. The role of WRKY transcription factors in plant immunity. Plant Physiol. 150, 1648–1655.

Park, C.Y., Lee, J.H., Yoo, J.H., Moon, B.C., Choi, M.S., Kang, Y.H., Lee, S.M., Kim, H.S., Kang, K.Y., Chung, W.S., 2005. WRKY group IId transcription factors interact with calmodulin. FEBS Lett. 579, 1545–1550.

Pauwels, L., Barbero, G.F., Geerinck, J., Tilleman, S., Grunewald, W., Prez, A.C., Chico, J.M., Bossche, R.V., Sewell, J., Gil, E., 2010. NINJA connects the co-repressor TOPLESS to jasmonate signalling. Nature 464, 788–791.

Pauwels, L., Goossens, A., 2011. The JAZ proteins: a crucial interface in the jasmonate signaling cascade. Plant Cell. 23, 3089–3100.

Perez-Rodriguez, P., Riaño-Pachón, D.M., Correa, L.G., Rensing, S.A., Kersten, B., Mueller-Roeber, B., 2010. PlnTFDB: updated content and new features of the plant transcription factor database. Nucleic Acids Res. 38, D822–D827.

Peterlin, B.M., Price, D.H., 2006. Controlling the elongation phase of transcription with P-TEFb. Mol. Cell 23, 297–305.

Pikaard, C.S., Haag, J.R., Ream, T., Wierzbicki, A.T., 2008. Roles of RNA polymerase IV in gene silencing. Trends Plant. Sci. 13, 390–397.

Poss, Z.C., Ebmeier, C.C., Taatjes, D.J., 2013. The Mediator complex and transcription regulation. Crit. Rev. Biochem. Mol. Biol. 48, 575–608.

Priya, P., Jain, M., 2013. RiceSRTFDB: a database of rice transcription factors containing comprehensive expression, cis-regulatory element and mutant information to facilitate gene function analysis. Database (Oxford) 2013, bat027.

Ptashne, M., Gann, A., 1997. Transcriptional activation by recruitment. Nature 386, 569–577.

Putterill, J., Robson, F., Lee, K., Simon, R., Coupland, G., 1995. The CONSTANS gene of Arabidopsis promotes flowering and encodes a protein showing similarities to zinc finger transcription factors. Cell 80, 847–857.

Ramsay, N.A., Glover, B.J., 2005. MYB–bHLH–WD40 protein complex and the evolution of cellular diversity. Trends Plant Sci. 10, 63–70.

Ream, T.S., Haag, J.R., Wierzbicki, A.T., Nicora, C.D., Norbeck, A.D., Zhu, J.K., Hagen, G., Guilfoyle, T.J., Pasa-Tolic, L., Pikaard, C.S., 2009. Subunit compositions of the RNA-silencing enzymes Pol IV and Pol V reveal their origins as specialized forms of RNA polymerase II. Mol. Cell 33, 192–203.

Riaño-Pachón, D.M., Ruzicic, S., Dreyer, I., Mueller-Roeber, B., 2007. PlnTFDB: an integrative plant transcription factor database. BMC Bioinform. 8, 42.

Richardt, S., Lang, D., Reski, R., Frank, W., Rensing, S.A., 2007. PlanTAPDB - A phylogeny-based comprehensive resource of plant transcription associated proteins. Plant Physiol. 143, 1452–1466.

Riechmann, J.L., Heard, J., Martin, G., Reuber, L., Jiang, C., Keddie, J., Adam, L., Pineda, O., Ratcliffe, O.J., Samaha, R.R., Creelman, R., Pilgrim, M., Broun, P., Zhang, J.Z., Ghandehari, D., Sherman, B.K., Yu, G., 2000. Arabidopsis transcription factors: genome-wide comparative analysis among eukaryotes. Science 290, 2105–2110.

Roeder, R.G., 2005. Transcriptional regulation and the role of diverse coactivators in animal cells. FEBS Lett. 579, 909–915.

Romeuf, I., Tessier, D., Dardevet, M., Branlard, G., Charmet, G., Ravel, C., 2010. wDBTF: an integrated database resource for studying wheat transcription factor families. BMC Genomics 11, 185.

Rushton, P.J., Bokowiec, M.T., Laudeman, T.W., Brannock, J.F., Chen, X., Timko, M.P., 2008. TOBFAC: the database of tobacco transcription factors. BMC Bioinform. 9, 53.

Rushton, P.J., Somssich, I.E., Ringler, P., Shen, Q.J., 2010. WRKY transcription factors. Trends Plant. Sci. 15, 247–258.

Sakuma, Y., Liu, Q., Dubouzet, J.G., Abe, H., Shinozaki, K., Yamaguchi-Shinozaki, K., 2002. DNA-binding specificity of the ERF/AP2 domain of Arabidopsis DREBs, transcription factors involved in dehydration- and cold-inducible gene expression. Biochem. Biophys. Res. Commun. 290, 998–1009.

Sarge, K.D., Park-Sarge, O.K., 2005. Gene bookmarking: keeping the pages open. Trends. Biochem. Sci. 30, 605–610.

Saunders, A., Core, L.J., Lis, J.T., 2006. Breaking barriers to transcription elongation. Nat. Rev. Mol. Cell Biol. 7, 557–567.

Scharf, K.D., Berberich, T., Ebersberger, I., Nover, L., 2012. The plant heat stress transcription factor (Hsf) family: structure, function and evolution. Biochim. Biophys. Acta 1819, 104–119.

Shandilya, J., Gadad, S., Swaminathan, V., Kundu, T.K., 2007. Histone chaperones in chromatin dynamics: implications in disease manifestation. Subcell Biochem. 41, 111–124.

Shandilya, J., Roberts, S.G., 2012. The transcription cycle in eukaryotes: from productive initiation to RNA polymerase II recycling. Biochim. Biophys. Acta 1819, 391–400.

Smith, E., Shilatifard, A., 2013. Transcriptional elongation checkpoint control in development and disease. Genes Dev. 27, 1079–1088.

Smith, L.M., Pontes, O., Searle, I., Yelina, N., Yousafzai, F.K., Herr, A.J., Pikaard, C.S., Baulcombe, D.C., 2007. An SNF2 protein associated with nuclear RNA silencing and the spread of a silencing signal between cells in Arabidopsis. Plant Cell 19, 1507–1521.

Sridhar, V.V., Surendrarao, A., Liu, Z., 2006. APETALA1 and SEPALLATA3 interact with SEUSS to mediate transcription repression during flower development. Development 133, 3159–3166.

Stracke, R., Werber, M., Weisshaar, B., 2001. The R2R3–MYB gene family in Arabidopsis thaliana. Curr. Opin. Plant. Biol. 4, 447–456.

Szemenyei, H., Hannon, M., Long, J.A., 2008a. TOPLESS mediates auxin-dependent transcriptional repression during Arabidopsis embryogenesis. Science 319, 1384–1386.

Szemenyei, H., Hannon, M., Long, J.A., 2008b. TOPLESS mediates auxin-dependent transcriptional repression during Arabidopsis embryogenesis. Science 319, 1384–1386.

Tajima, H., Iwata, Y., Iwano, M., Takayama, S., Koizumi, N., 2008. Identification of an Arabidopsis transmembrane bZIP transcription factor involved in the endoplasmic reticulum stress response. Biochem. Biophys. Res. Commun. 374, 242–247.

Thomas, M.C., Chiang, C.M., 2006. The general transcription machinery and general cofactors. Crit. Rev. Biochem. Mol. Biol. 41, 105–178.

Tiwari, S.B., Hagen, G., Guilfoyle, T.J., 2004. Aux/IAA proteins contain a potent transcriptional repression domain. Plant Cell 16, 533–543.

Ulker, B., Somssich, I.E., 2004. WRKY transcription factors: from DNA binding towards biological function. Curr. Opin. Plant Biol. 7, 491–498.

Uno, Y., Furihata, T., Abe, H., Yoshida, R., Shinozaki, K., Yamaguchi-Shinozaki, K., 2000. Arabidopsis basic leucine zipper transcription factors involved in an abscisic acid-dependent signal transduction pathway under drought and high-salinity conditions. Proc. Natl. Acad. Sci. 97, 11632–11637.

Vinson, C.R., Sigler, P.B., McKnight, S.L., 1989. Scissors-grip model for DNA recognition by a family of leucine zipper proteins. Science 246, 911–916.

Von Koskull-Doring, P., Scharf, K.D., Nover, L., 2007. The diversity of plant heat stress transcription factors. Trends Plant Sci. 12, 452–457.

Wang, W., Vinocur, B., Shoseyov, O., Altman, A., 2004. Role of plant heat-shock proteins and molecular chaperones in the abiotic stress response. Trends Plant Sci. 9, 244–252.

Wang, Y., Fairley, J.A., Roberts, S.G., 2010a. Phosphorylation of TFIIB links transcription initiation and termination. Curr. Biol. 20, 548–553.

Wang, Z., Libault, M., Joshi, T., Valliyodan, B., Nguyen, H.T., Xu, D., Stacey, G., Cheng, J., 2010b. SoyDB: a knowledge database of soybean transcription factors. BMC Plant Biol. 10, 14.

Wierzbicki, A.T., Haag, J.R., Pikaard, C.S., 2008. Noncoding transcription by RNA polymerase Pol IVb/Pol V mediates transcriptional silencing of overlapping and adjacent genes. Cell 135, 635–648.

Yamasaki, K., Kigawa, T., Seki, M., Shinozaki, K., Yokoyama, S., 2013. DNA-binding domains of plant-specific transcription factors: structure, function, and evolution. Trends Plant Sci. 18, 267–276.

Yanhui, C., Xiaoyuan, Y., Kun, H., Meihua, L., Jigang, L., Zhaofeng, G., Zhiqiang, L., Yunfei, Z., Xiaoxiao, W., Xiaoming, Q., Yunping, S., Li, Z., Xiaohui, D., Jingchu, L., Xing-Wang, D., Zhangliang, C., Hongya, G., Li-Jia, Q., 2006. The MYB transcription factor superfamily of Arabidopsis: expression analysis and phylogenetic comparison with the rice MYB family. Plant Mol. Biol. 60, 107–124.

Yilmaz, A., Nishiyama, Jr., M.Y., Fuentes, B.G., Souza, G.M., Janies, D., Gray, J., Grotewold, E., 2009. GRASSIUS: a platform for comparative regulatory genomics across the grasses. Plant Physiol. 149, 171–180.

Zhang, H., Jin, J., Tang, L., Zhao, Y., Gu, X., Gao, G., Luo, J., 2011. PlantTFDB 2.0: update and improvement of the comprehensive plant transcription factor database. Nucleic Acids Res. 39, D1114–D1117.

Zhu, Q.H., Guo, A.Y., Gao, G., Zhong, Y.F., Xu, M., Huang, M.R., Luo, J.C., 2007. DPTF: a database of poplar transcription factors. Bioinformatics 23, 1307–1308.

Zhu, Z., Xu, F., Zhang, Y., Cheng, Y.T., Wiermer, M., Li, X., Zhang, Y., 2010. *Arabidopsis* resistance protein SNC1 activates immune responses through association with a transcriptional corepressor. Proc. Natl. Acad. Sci. USA 107, 13960–13965.

CHAPTER

4

Structures, Functions, and Evolutionary Histories of DNA-Binding Domains of Plant-Specific Transcription Factors

Kazuhiko Yamasaki

Biomedical Research Institute, National Institute of Advanced Industrial Science and Technology (AIST), Tsukuba, Japan; RIKEN Quantitative Biology Center, Tsurumi-ku, Yokohama, Japan

OUTLINE

4.1 Introduction — 57	4.2.6 Other DBDs of Plant-Specific Transcription Factors — 65
4.2 Description of Respective DBDs — 59	4.3 Evolutionary History of Plant-Specific TFs — 65
4.2.1 AP2/ERF Domain — 59	4.3.1 Endonucleases as Origins of DBDs — 65
4.2.2 B3 Domain — 60	4.3.2 Lineage-Specific Expansion — 68
4.2.3 WRKY Domain — 62	References — 69
4.2.4 NAC Domain — 63	
4.2.5 SBP Domain — 64	

4.1 INTRODUCTION

After the genome sequence of *Arabidopsis thaliana* was completed, ~1500 probable transcription factors (TFs) were identified and classified into ~30 families, according to conserved sequence motifs that are mostly basic and correspond to the DNA-binding domains (DBDs) (Riechmann et al., 2000). Several plant TF databases developed afterwards increased the numbers of *Arabidopsis* TFs and families to ~2000 and 50–70, respectively (Mitsuda and Ohme-Takagi, 2009; see Chapter 3). About half of these TF families were considered plant-specific because the relevant motifs are distinct from those in TFs of prokaryotes or other eukaryotic lineages. Among the plant-specific TF families, major ones are APETALA2 (AP2)/ethylene-responsive element binding factor (ERF); *NO APICAL MERISTEM, ATAF1/2, CUP-SHAPED COTYLEDON 2* (NAC); WRKY; *ABSCISIC ACID INSENSITIVE 3* (ABI3)/*VIVIPAROUS1* (VP1); auxin response factor (ARF), and *SQUAMOSA*-promoter binding protein (SBP),

containing 16–150 family members (Figure 4.1). TFs in AP2/ERF, NAC, SBP, and WRKY families possess DBDs called the same names as the respective families, while ABI3/VP1 and ARF family members share a DBD called B3, resulting in classification into the B3 superfamily (Swaminathan et al., 2008).

The conserved sequence motifs are typically related to structures and/or functions. Namely, the relevant amino acid residues are involved in formation of the structure core of the domain or binding to the target DNA. For example, in the WRKYGQK sequence motif of the WRKY family, the first Trp is involved in the structure core, while the other residues are involved in contact to DNA (see Section "WRKY Domain"). These were clearly revealed by the three-dimensional structure of the protein domain and that of the protein–DNA complex.

Moreover, three-dimensional structures are much more sensitive to the relationship between the different proteins than primary sequences. This is because comparison of the structures is virtually the same as that focusing on

Plant Transcription Factors. http://dx.doi.org/10.1016/B978-0-12-800854-6.00004-X
Copyright © 2016 Elsevier Inc. All rights reserved.

(A) AP2/ERF (AP2/ERF family)

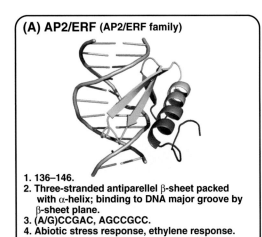

1. 136–146.
2. Three-stranded antiparellel β-sheet packed with α-helix; binding to DNA major groove by β-sheet plane.
3. (A/G)CCGAC, AGCCGCC.
4. Abiotic stress response, ethylene response.
5. <u>1gcc</u>, 2gcc, 3gcc.

(B) B3 (B3 superfamily: ABI3/VP1 family, ARF family, etc.)

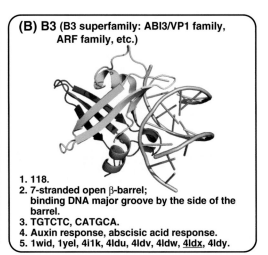

1. 118.
2. 7-stranded open β-barrel; binding DNA major groove by the side of the barrel.
3. TGTCTC, CATGCA.
4. Auxin response, abscisic acid response.
5. 1wid, 1yel, 4i1k, 4ldu, 4ldv, 4ldw, <u>4ldx</u>, 4ldy.

(C) WRKY (WRKY family)

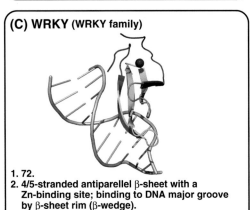

1. 72.
2. 4/5-stranded antiparellel β-sheet with a Zn-binding site; binding to DNA major groove by β-sheet rim (β-wedge).
3. TTGAC(C/T).
4. Biotic and abiotic stress response.
5. 1wj2, 2ayd, <u>2lex</u>.

(D) NAC (NAC family)

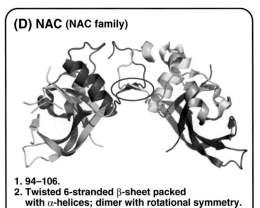

1. 94–106.
2. Twisted 6-stranded β-sheet packed with α-helices; dimer with rotational symmetry.
3. AN$_5$TCN$_7$ACACGCATGT.
4. Development, biotic and abiotic stress response.
5. 1ul4, <u>1ul7</u>, 3swm, 3swp, 3ulx, 4dul.

(E) SBP (SBP family)

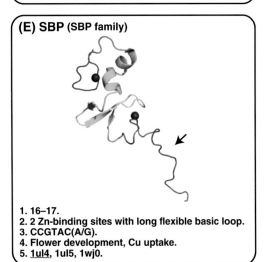

1. 16–17.
2. 2 Zn-binding sites with long flexible basic loop.
3. CCGTAC(A/G).
4. Flower development, Cu uptake.
5. <u>1ul4</u>, 1ul5, 1wj0.

FIGURE 4.1 **Structures of the DNA-binding domains (DBDs) that characterize the major families of plant-specific transcription factors (TFs).** Shown are (A) the AP2/ERF domain in complex with DNA, (B) the B3 domain in complex with DNA, (C) the WRKY domain in complex with DNA, (D) the NAC domain, and (E) the SBP domain. Proteins are shown in the colors of the rainbow from blue (N-terminus) to red (C-terminus), while DNAs are in gray. In (C) and (E) the red spheres show Zn ions. In (D) the red oval indicates the antiparallel β-sheet forming the dimerization interface. In (E) the arrow indicates the C-terminal basic tail. Information on (1) number of proteins in *Arabidopsis thaliana*, including those predicted in genome, (2) brief structural description, (3) representative recognition sequences, (4) representative functions of family members, and (5) Protein Data Bank (PDB) entry codes (those used in the figure are underlined) are also provided. The numbers of proteins are given in ranges because they differ from one database to another except for the WRKY family (Mitsuda and Ohme-Takagi, 2009). The B3 superfamily includes the number by Swaminathan et al. (2008). The molecular figures were produced using PyMOL (Shrödinger, LLC). *This figure is a modification of that used in a previous review paper (Yamasaki et al., 2013), as reproduced with permission of Elsevier.*

residues important in structure formation located at the equivalent positions; the contributions of other residues such as those in variable loops are discounted. For example, structure comparison revealed unexpected similarity between the B3 domain and DBDs of prokaryotic restriction endonucleases, after which similarity in primary sequence of equivalent positions could also be detected (see Section "B3 Domain"). This result provides an important clue to evolutionary history.

Thus, three-dimensional structures make profound impacts on the understanding of functional mechanisms and evolution. In this chapter, the author summarizes these implications in terms of the DBDs of plant-specific TF families. Based on facts observed for respective DBDs, general aspects regarding the evolution of plant-specific TFs are also described. The topics in this chapter largely cover the contents of two previous review papers (Yamasaki et al., 2008, 2013), but update the results and add more detail to the discussions.

4.2 DESCRIPTION OF RESPECTIVE DBDs

4.2.1 AP2/ERF Domain

The AP2/ERF domain was originally identified as a highly conserved DBD of transcription factors related to floral development and ethylene response (Jofuku et al., 1994; Ohme-Takagi and Shinshi, 1995). Approximately 150 genes on the *Arabidopsis* genome share this highly basic motif encompassing ~60 amino acids, which comprise the largest of the plant-specific TF families (Figure 4.1). The family is further classified into four subfamilies; AP2, ERF, dehydration-responsive element-binding protein (DREB), and *RELATED TO ABI3/VP1* (RAV). Functions of the AP2/ERF TFs cover a variety of plant-specific reactions (i.e., responses to abiotic stresses, such as cold, dehydration, heat shock, and mechanical stress, ethylene response, and the development of flowers, roots, embryos, and seeds; Jofuku et al., 1994; Ohme-Takagi and Shinshi, 1995; Fowler and Thomashow, 2002; Aida et al., 2004; Chandler et al., 2007; Kagaya and Hattori, 2009; Dietz et al., 2010; Mizoi et al., 2012).

DREB and ERF subfamily members possess a single AP2/ERF domain, whereas AP2 subfamily members possess two repeated domains. RAV subfamily members possess an N-terminal AP2/ERF domain and a C-terminal B3 domain (see Section "B3 Domain"; Kagaya et al., 1999). The CBF1 protein in the DREB subfamily also requires amino acids outside the AP2/ERF domain for DNA binding (Canella et al., 2010). Variations in the combination of DBDs, as well as in sequences inside and outside the AP2/ERF domain, bring about differences in the DNA sequences recognized by the proteins. Namely, DREB and ERF TFs recognize similar but slightly different sequences, such that DREB TFs recognize the dehydration-responsive element (DRE) sequence, 5'-(A/G)CCGAC-3', while ERF TFs recognize the GCC-box sequence, 5'-AGCCGCC-3' (Ohme-Takagi and Shinshi, 1995; Sakuma et al., 2002; Hao et al., 2003). TFs of the AP2 subfamily recognize longer sequences with a consensus of 5'-GCAC(A/G)N(A/T)TCCC(A/G)ANG(C/T)-3' (Nole-Wilson and Krizek, 2000). The AP2/ERF domain of the RAV subfamily TF recognizes the 5'-CAACA-3' sequence independently of the B3 domain, which recognizes the 5'-CACCTG-3' sequence (Kagaya et al., 1999).

The structure of the complex of the AP2/ERF domain of AtERF1 and the GCC-box DNA has been determined by NMR spectroscopy (Figures 4.1A and 4.2A), where the AP2/ERF domain was described as a "GCC-box binding domain" in the original report (Allen et al., 1998). The

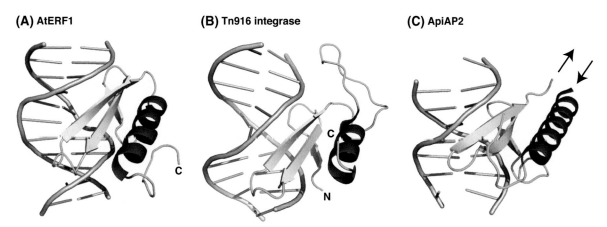

FIGURE 4.2 **Structures of the AP2/ERF and related domains.** Shown are (A) AP2/ERF domain of AtERF1 in complex with DNA (Allen et al., 1998; PDB code 1gcc), (B) noncatalytic DBD of Tn*916* integrase in complex with DNA (Wojciak et al., 1999; PDB code 1b69), and (C) DBD of *Plasmodium falciparum* PF14_0633 in complex with DNA (Lindner et al., 2010; PDB code 3igm). The regions of the α-helix, β-sheet, and loop are colored in red, yellow, and green, respectively. The N- and C-termini of the regions presented are indicated when appropriate. In (C) the monomeric half of the secondary structure unit-swapped dimer is shown, where the α-helix and β-sheet belong to different chains; the connection points and the N to C direction are indicated by arrows. *The figures were drawn using PyMOL (Shrödinger, LLC).*

structure of the AP2/ERF domain consists of a three-stranded antiparallel β-sheet and an α-helix. The β-sheet moiety fits into the major groove of the DNA, in which the plane of the sheet is nearly parallel to the helical axis of the DNA. This DNA-binding mode is considered atypical because the majority of DBDs, such as the Zn finger domain in combination with an antiparallel β-sheet and an α-helix, place the α-helix in the major groove of the DNA. For sequence recognition, three arginine and two tryptophan residues, which are highly conserved in the sequence motif, directly contact bases.

Determinant residues to illustrate the difference in sequence specificity between subfamilies have been analyzed mainly by site-directed mutagenesis experiments, at least partly in terms of the AtERF1 structure (Hao et al., 2002; Sakuma et al., 2002; Krizek, 2003; Liu et al., 2006; Yang et al., 2009). The data revealed that determinant residues between DREB and ERF TFs, which recognize slightly different sequences, are not necessarily those in direct contact with bases; those likely to structurally influence the orientation of base-contacting residues can be the determinant for specificity. For the AP2 domain, which recognizes greatly different DNA sequences from other subfamilies, the DNA-binding interface was suggested to be very different from that of AtERF1 (Krizek, 2003). The DNA-binding mode shown in Figure 4.2A is applicable to ERF domains, but not strictly speaking necessarily to AP2 domains (as the name of the domain "AP2/ERF" is typically used in this chapter).

Although the DNA-binding mode of AP2/ERF is atypical, three-dimensional structures and DNA-binding modes strikingly similar to those of AtERF1 have been reported for the noncatalytic DBDs of Tn916-integrase and λ-integrase (Wojciak et al., 1999; Fadeev et al., 2009); they also possess a three-stranded antiparallel β-sheet and an α-helix, where the β-sheet fits into the DNA major groove (Figure 4.2B). A similarity to the DNA-binding mode regarding the homing endonuclease I-PpoI was also pointed out (Wojciak et al., 1999). What stands out is that all these proteins are endonucleases of bacterial or primitive eukaryotic origins and are associated with transposable elements (see also Section for "Endonucleases as Origins of DBDs").

Inspired more or less by these unexpected structural similarities, sequence searches on genomic DNAs have indeed identified genes coding for homing endonucleases that possess regions homologous to the AP2/ERF domain in ciliate, cyanobacteria, and bacteriophage (Magnani et al., 2004; Wuitschick et al., 2004). Moreover, a large number of probable TFs containing the motif of the AP2/ERF domain have been identified in apicomplexa (Balaji et al., 2005; Painter et al., 2011). The crystal structure of the complex of an apicomplexan AP2/ERF-like domain and DNA was also reported. This structure possesses the same structural motif and DNA-binding mode as that of AtERF1 and that of the bacterial integrases. This apicomplexan domain, however, differs in that it was found in a secondary structure unit-swapped homodimer where the α-helix of one monomer folds with the β-sheet of the other. The recognition sequences of apicomplexan AP2/ERF TFs were found to be diverse and dissimilar to those of plant AP2/ERF TFs (De Silva et al., 2008; Painter et al., 2011).

4.2.2 B3 Domain

The B3 domain was named for the third basic region of approximately 110 amino acids in ABI3/VP1 proteins (Giraudat et al., 1992). The most likely TFs to share the B3 domain are classified as ARF, *LEAFY COTYLEDON2-ABI3-VAL* (LAV; including the ABI3/VP1 family), RAV (note that this RAV family is larger than the RAV subfamily of the AP2/ERF family; i.e., the former also contains proteins from outside the AP2/ERF family, lacking the AP2/ERF domain), and REPRODUCTIVE MERISTEM (REM) families. These TFs further comprise the B3 superfamily including ~120 members in *Arabidopsis* (Swaminathan et al., 2008; Figure 4.1B). The functions of representative members of the ARF and LAV families are responses to auxin and abscisic acid (McCarty et al., 1989; Giraudat et al., 1992; Ulmasov et al., 1997), while those of the RAV and REM families members are related to stress responses and vernalization, respectively (Fowler and Thomashow, 2002; Kagaya and Hattori, 2009; Levy et al., 2002).

Members of the ARF and LAV families possess a single B3 domain, whereas those of the REM family possess up to six repeats (Swaminathan et al., 2008). As described earlier, representative members of the RAV family possess an AP2/ERF domain in addition to the B3 domain (Kagaya et al., 1999). Recognition sequences have been identified for ARF family members as 5'-TGTCTC-3' (Ulmasov et al., 1997), for ABI3/VP1 members of the LAV family as 5'-CATGCA-3' (Suzuki et al., 1997), and for the B3 domain of RAV family members as 5'-CACCTG-3' (Kagaya et al., 1999). In contrast, VERNALIZATION1 (VRN1) of the REM family shows nonspecific DNA-binding ability (Levy et al., 2002), and At1g16640 of the REM family lacks DNA-binding activity (Waltner et al., 2005).

The structures of the B3 domains of RAV1, At1g16640 (REM family), VRN1 (C-terminal of one of the two B3 domains), ARF1, and ARF5 have been determined by NMR and crystallography, in which the structure of ARF1 was further determined in complex with DNA (Yamasaki et al., 2004a; Waltner et al., 2005; King et al., 2013; Boer et al., 2014). B3 domain structures consist of a seven-stranded β-sheet arranged in an open barrel (barrel-like structure without closing the sheet; or pseudobarrel), which is accompanied by two short α-helices located at the two ends of the barrel in question (shown at the top and bottom in Figure 4.3A) and a very short helical turn

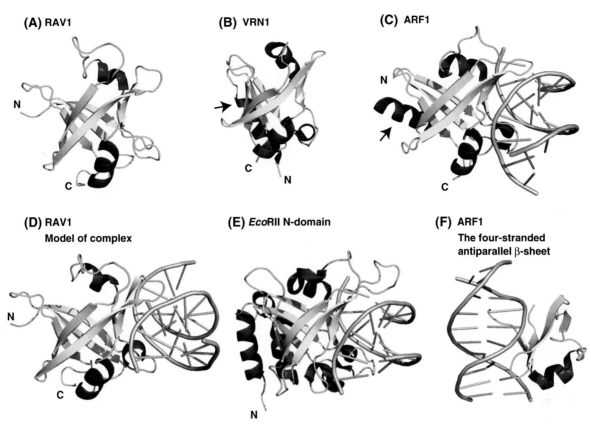

FIGURE 4.3 **Structures of the B3 and related domains.** Shown are (A) the B3 domain of *Arabidopsis* RAV1 (Yamasaki et al., 2004a; PDB code 1wid), (B) the C-terminal B3 domain of *Arabidopsis* VRN1 (King et al., 2013; PDB code 4i1k), (C) the B3 domain of *Arabidopsis* ARF1 in complex with DNA (Boer et al., 2014; PDB code 4ldx), (D) model of the RAV1-B3–DNA complex (Yamasaki et al., 2013), (E) the noncatalytic DBD of *Escherichia coli* restriction enzyme *Eco*RII in complex with DNA (Golovenko et al., 2009; PDB code 3hqf), and (F) a four-stranded antiparallel β-sheet region of ARF1 and DNA, isolated from the structure shown in (C), for comparison with Figure 4.2 or Figure 4.1C (see text). The N- and C-termini are indicated when appropriate. Arrows in B and C indicate additional α-helices (see text).

in the DNA-contacting region (Figure 4.3C). In addition, the B3 domain of VRN1 possesses a long α-helix at the N-terminus, which is tightly packed at the side of the pseudobarrel (Figure 4.3B). Likewise for ARF1, there is a long helix belonging to the N-terminal region, one of the two regions relevant for forming symmetric dimer (Boer et al., 2014), contacting a similar position of the pseudobarrel (Figure 4.3C).

For the RAV1 B3 domain, the first B3 structure determined, several residues in a basic patch of the surface were picked up as DNA-contacting residues, based on an NMR titration experiment evaluating the effects of DNA binding on the NMR crosspeaks assigned to individual amino acid residues (Yamasaki et al., 2004a). The equivalent area of the VRN1 B3 domain was also shown to be important in DNA binding by site-directed mutagenesis, even though the VRN1 lacks sequence specificity (King et al., 2013). These are consistent with the structure of the ARF1/DNA complex (e.g., for Leu200 and Trp245 of RAV1, assigned as the base-contacting residues as a result of NMR and modeling; Yamasaki et al., 2004a, 2013). Equivalent residues, His136 and Arg181, respectively, of ARF1 are indeed base-contacting residues (Boer et al., 2014). These base-contacting residues, however, are not highly conserved among the B3 domains of different families, which has to do with variation in the recognition sequences and the property regarding DNA-binding activity (Yamasaki et al., 2004a; King et al., 2013).

A striking similarity was found between structures of the B3 domains and those of noncatalytic DBDs of the restriction endonucleases *Eco*RII of *Escherichia coli* and *Bfi*I of *Bacillus firmus* (Zhou et al., 2004; Grazulis et al., 2005; Golovenko et al., 2009, 2014; Figure 4.3E). The B3-like structures of these restriction endonucleases are slightly larger than those of the B3 domains and consist of an eight-stranded open β-barrel and six helices. Seven of the eight strands match those of the B3 domains. The helices are also in similar positions to those of the B3 domains, including the extra helices of ARF1 and VRN1. Once structural similarity has been identified, sequence similarities were also detected between the B3-like domain of *Eco*RII and the B3 domains, with pairwise amino acid identities of up to ~20% (~60% when permitting conservative changes; e.g., leucine and isoleucine), which shows that they are likely to be evolutionarily related (Yamasaki et al., 2004a). This is indeed analogous to the case of AP2/ERF and Tn*916* and the λ-integrases described in the previous section, in that the DBDs of plant-specific TFs are related to the noncatalytic DBDs of prokaryotic endonucleases.

Based on the structure of the *Eco*RII DBD/DNA complex, a model of the RAV1–B3/DNA complex has been constructed using a computational approach (Yamasaki et al., 2013; Figure 4.3D). It is clear that the binding mode and molecular interface are very similar to those of ARF1 (Boer et al., 2014; Figure 4.3C), likewise forming a helical turn in the DNA-contacting area. Note that another model of the RAV1–B3/DNA complex previously constructed only from NMR titration data possesses similar molecular interfaces, although the relative orientation of the protein and the helical axis of DNA differs by approximately 90° (Yamasaki et al., 2004a, 2008).

The seven-stranded pseudobarrel of the B3 domain can be divided into three- and four-stranded antiparallel β-sheets, which connect to each other by a relatively short area in a parallel mode. The four-stranded β-sheet forms the interface with DNA (Figure 4.3F). Associated with an α-helix, this manner of binding is similar to that of the AP2/ERF domain (Figure 4.2) when three of the four strands are compared. This may reflect a distant relation between these domains, presumably relating to their origins (see Section "Endonucleases as Origins of DBDs").

4.2.3 WRKY Domain

The WRKY domain was originally identified in factors relating to the transcription control of amylase (Ishiguro and Nakamura, 1994; Rushton et al., 1995). The DBD of ~60 amino acids is characterized by the invariant WRKYGQK motif, for which the domain and family have been named, and the C—X_{4-5}—C—X_{22-23}—H—X_1—H (C_2—H_2) or C—X_7—C—X_{23}—H—X_1—C (C_2—HC) motif that composes a Zn-binding site (Eulgem et al., 2000; see Chapter 11). More than 70 TFs of *Arabidopsis* possess the WRKY domain (Figure 4.1C). WRKY family members are classified as three groups: typically, those possessing two copies of the WRKY domain (Group I); those possessing a single WRKY domain with the C_2—H_2 motif (Group II); and those possessing a single WRKY domain with the C_2—HC motif (Group III). The representative functions of WRKY family TFs are related to responses to biotic and abiotic stresses, including bacterial or fungal pathogens, and pathogen-related hormone salicylic acid (reviewed in Rushton et al., 2010; Agarwal et al., 2011). WRKY TFs recognize the 5′-TTGAC(C/T)-3′ sequence termed the W-box. For Group I WRKY proteins, the C-terminal WRKY domain (WRKY-C), not the N-terminal one, is responsible for recognizing the W-box (Ishiguro and Nakamura, 1994; Eulgem et al., 1999).

The structures of the WRKY-C domain of AtWRKY1 and AtWRKY4, both of which belong to Group I, and that of the AtWRKY4–WRKY-C/DNA complex have been determined by NMR and crystallography (Yamasaki et al., 2005a, 2012; Duan et al., 2007; Figure 4.1C). These consist of a four- or five-stranded antiparallel β-sheet with a Zn-binding site formed by Cys and His residues in the C_2—H_2 motif. The β-sheet binds to the major groove of DNA, with the plane of the sheet placed nearly perpendicular to the helical axis of DNA, although the contacting strand tilts from the sheet and fits into the groove. The conserved WRKYGQK motif is located in this contacting strand; the first Trp (W) residue orients the bulky side chain to the other strands and forms the core of the structure, whereas all the other residues (RKYGQK), mostly basic, are directly involved in DNA binding. The Gly residue causes this strand to bend extensively, which allows penetration into the DNA groove. The W-box sequence is recognized mainly through hydrophobic contacts with the methyl groups of the thymine bases of the DNA strand, for which the Gly, Tyr, Arg, and Lys residues of the WRKYGQK motif are involved. Note that Arg and Lys possess long hydrophobic side chains with hydrophilic ends. The Arg and Lys residues are involved in extensive hydrogen bonding and electrostatic interactions with DNA phosphates at the same time.

The three-dimensional structure and DNA-binding mode of the WRKY domain are similar to those of the GCM domain (Figure 4.4B; Cohen et al., 2003). The GCM domain is a Zn-containing DNA-binding domain found in the TFs of animals, ranging from insects to mammals. It contains two different Zn-binding units: the WRKY-like Zn-binding unit and a Zn ribbon unit, where the N- and C-termini of the latter are inserted in a loop between the β-strands of the former (only the former is shown in Figure 4.4B; here, "Zn-ribbon" refers to the Zn-binding unit formed by antiparallel β-strands and extended loops). The β-strands of the GCM domain, especially the one that enters the DNA groove, are significantly shorter than those in the WRKY domain. To gain deep entry into the groove, the GCM domain employs the short strand and the subsequent loop regions, instead of bending a longer strand as in the WRKY domain. Moreover, the Zn-binding site is formed by residues at equivalent positions to those in the WRKY domain in a conserved manner. For the two His residues forming the Zn-binding site, it is common that the Nδ atom of the N-terminal and the Nε atom of the C-terminal, separated by two terminals, are involved in the coordinate bonds. The common DNA-binding mode of the WRKY domain and WRKY-like unit of the GCM domain are termed the β-wedge (Yamasaki et al., 2012). An extensive database search reveals that WRKY, GCM, and other Zn-binding proteins have been grouped and classified in the WRKY–GCM1 superfamily (Babu et al., 2006).

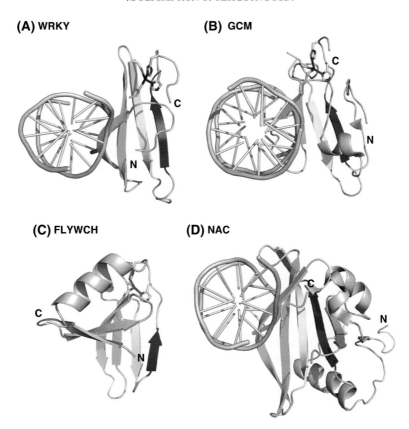

FIGURE 4.4 **Structures of the WRKY and related domains.** Shown are (A) the C-terminal WRKY domain of AtWRKY4 in complex with DNA (Yamasaki et al., 2012; PDB code 2lex), (B) the WRKY-like unit in the GCM domain of mouse GCM1 (Cohen et al., 2003; PDB code 1odh), (C) the fifth FLYWCH domain of human FLYWCH-type Zn finger-containing protein 1 (Enomoto et al., manuscript not available; PDB code 2rpr), and (D) the NAC domain of *Arabidopsis* ANAC in complex with DNA (Welner et al., 2012; PDB code 3swp). The β-strands located at topologically equivalent regions are shown in the same colors. In (A)–(C) the Zn-binding sites are shown in stick representation.

The group clearly contains the DBDs of transposases, such as animal *phantom* transposase (Marquez and Pritham, 2010). The DBD motif of *phantom* is a FLYWCH Zn-binding domain. The structure of another FLYWCH Zn-binding domain is registered in the Protein Data Bank (PDB code: 2rpr; manuscript not available). Indeed, there are close similarities to the WRKY and GCM domains regarding the arrangement of the four β-strands in a six-stranded β-sheet and the binding geometry of Zn (Figure 4.4C). Simple modeling indicated that the two β-strands that are not aligned to those of the WRKY domain also fit into the major groove of DNA when the β-wedge binding mode was adopted (data not shown). The similarity between the WRKY and FLYWCH domains is again an analog to the cases of AP2/ERF and B3, in that the noncatalytic DBDs of endonucleases are related to the DBDs of plant-specific TFs.

Some copies of WRKY proteins are also identified in nonphotosynthetic primitive eukaryotes (i.e., the protist *Giardia lamblia* and the slime mold *Dictyostelium discoideum*, which belong to Group I and possess two WRKY domains; Ulker and Somssich, 2004). It follows that the two WRKY domains came about as a result of duplication from a single WRKY domain of an ancestral WRKY protein. Despite the C-terminal WRKY domain solely being considered responsible for specific DNA binding, most residues directly bound to DNA are highly conserved in the N-terminal WRKY domain as well (Yamasaki et al., 2005a, 2012). Therefore, it is unlikely that the N-terminal domain has no DNA-binding activity at all, either in a specific or nonspecific manner; the DNA-contacting residues are still under positive selection pressure during evolution. Indeed, an *in vivo* binding experiment showed that a full-length Group I WRKY protein showed approximately four times higher activity than its truncated fragment containing only the C-terminal WRKY domain (Eulgem et al., 1999).

4.2.4 NAC Domain

The NAC domain is a relatively large DBD encompassing approximately 150 amino acids, which further forms a homo- or heterodimer, originally identified in transcription factors related to embryonic and floral development (Souer et al., 1996; Aida et al., 1997; Olsen et al., 2005). Approximately 100 proteins that share the NAC domain are identified in *Arabidopsis*, which comprises the second largest plant-specific TF family (Riechmann et al., 2000; Figure 4.1D). The proteins are involved in a wide range of plant-specific phenomena, such as development of plant-specific organs and responses to abiotic and biotic stresses, including responses to stress-related abscisic acid (Souer et al., 1996; Aida et al., 1997; Tran et al., 2004; Nakashima et al., 2012; Puranik et al., 2012; see Chapter 13). The recognition DNA sequences differ among the family members, and base requirements tend to be weak in the respective positions (Puranik et al., 2012). The sequences possess palindromic properties with different distances such as $AN_5TCN_7\underline{ACACGCATGT}$ with a consecutive pseudopalindromic core (underlined) for abscisic acid-responsive NAC (ANAC) proteins (Tran et al., 2004) and $(T/A)NN(C/T)(T/C/G)TN_7\underline{A(A/C)GN(A/C/T)(A/T)}$ with a pseudopalindromic sequence (underlined) and a seven-base gap for SND1 and VND7 (Zhong et al., 2010), where the palindromic propensity appears to be related to dimerization of the NAC domain.

The dimeric structures of the NAC domains of *Arabidopsis* ANAC and rice (*Oryza sativa*) stress-responsive NAC1 (SNAC1) have been determined by crystallography (Ernst et al., 2004; Chen et al., 2011; Figure 4.1D). Furthermore, the structure of the NAC/DNA complex based on low-resolution X-ray crystallography and small-angle X-ray scattering has been reported (Welner et al., 2012; Figure 4.4D). The structural core of NAC consists of a largely twisted six-stranded antiparallel β-sheet packed with two α-helices. The dimer interface is formed by a short intermolecular antiparallel β-sheet (shown as a red oval in Figure 4.1D), which is assisted by hydrogen bonds/salt bridges between the Arg and Glu side chains. This dimer interface region is connected to the cores of the two monomers through hinge-like regions, thus allowing the monomers to have flexible relative positions. The relative orientation of monomers differs between ANAC and SNAC1 by ~30° (Yamasaki et al., 2013), which may allow diversity in the gap between the pseudopalindromic recognition sequences.

There is a similarity between NAC and WRKY structures; all (four or five) β-strands of the WRKY domain have equivalent strands in NAC domains (Figure 4.4). The β-strand located on the edge of the NAC domain, which corresponds to the WRKYGQK sequence of the WRKY domain (colored cyan in Figure 4.4), bends in the ANAC protein (much like WRKY) or breaks in the middle in the SNAC1 protein, resulting in convex curvature of the β-sheet edge. Moreover, this β-strand possesses a WKATGXD[K/R] sequence resembling the WRKYGQK sequence. In complex with DNA, this strand enters the DNA major groove (Welner et al., 2012) and the β-wedge mode of DNA binding is adopted. Considering this structural similarity, accompanied by local sequence similarity, an evolutionary link between NAC and WRKY DBDs is highly likely. Also, the structural similarity between the NAC and GCM domains has already been demonstrated (Olsen et al., 2005). This is the reason NAC proteins have been included in the WRKY–GCM1 superfamily (Babu et al., 2006).

4.2.5 SBP Domain

SBPs (*SQUAMOSA* promoter-binding proteins) were originally identified as a group of TFs that bind to the promoter DNA of the floral meristem identity gene *SQUAMOSA* (Klein et al., 1996). The DBD termed the SBP domain a highly basic region of approximately 80 amino acids characterized by 10 Cys or His residues. In *Arabidopsis*, 16–17 TFs possess SBP domains, the representative function of which is regulation of flower development and developmental transitions (Cardon et al., 1999; see Chapter 18). The SBP domain recognizes a consensus of 5′-(C)(C)GTAC(A/G)-3′ as shown by random selection experiments (Birkenbihl et al., 2005; Liang et al., 2008).

The structures of *Arabidopsis* SPL4 and SPL7 (SPL: *SQUAMOSA* promoter-binding protein-like) have been determined by NMR. They are atypical in that they consist of two very short α-helices, a short three-stranded β-sheet, and a long flexible C-terminal loop in common (Yamasaki et al. 2004b; Figure 4.1E). The structure contains two Zn-binding sites formed by eight Cys or His residues in a $Cys_3HisCys_2HisCys$ or $Cys_6HisCys$ sequence motif in which the first four residues coordinate to one Zn ion and the last four coordinate to the other. Despite the atypical structure, the N- and C-terminal Zn-coordinating units show similarities to the classical Cys_2His_2 Zn finger (C_2H_2 ZnF) and the Zn ribbon, respectively, suggesting their evolutionary origins (Yamasaki et al., 2004b). Analyses of a truncated fragment of *Arabidopsis* SPL12 showed that the N-terminal Zn-binding site, but not the C-terminal site, is necessary to maintain the tertiary structure of protein (Yamasaki et al. 2006). Mutational analyses showed that the C-terminal site is important in DNA binding (Birkenbihl et al., 2005), probably through guiding the basic C-terminal loop (shown by an arrow in Figure 4.1E) to correctly fit into the DNA groove. Although a DNA-binding mode of SBP has been proposed by NMR titration analysis (Yamasaki et al., 2004b), a determined structure of the SBP/DNA complex by NMR or crystallography is still awaited.

An SBP of the green alga *Chlamydomonas reinhardtii* Cu response regulator 1 (CRR1) and its *Arabidopsis* ortholog SPL7 are related to the regulation of Cu uptake and probably behave as Cu sensors (Kropat et al., 2005; Yamasaki et al., 2009). For *Chlamydomonas* CRR1, it has been hypothesized that Cu binds to the SBP domain and thereby impairs DNA-binding activity (Kropat et al., 2005). Considering the elastic property of the C-terminal

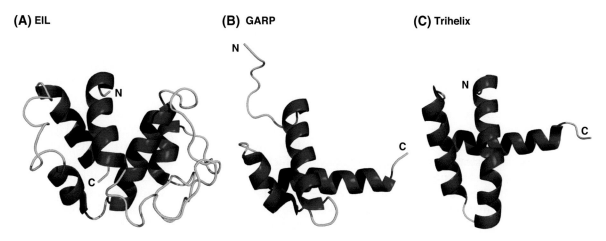

FIGURE 4.5 **Structures of the DBDs of other plant-specific TFs.** Shown are (A) the major DBD of *Arabidopsis* EIL3 (EIL family; Yamasaki et al., 2005b; PDB code 1wij), (B) B-motif DBD of *Arabidopsis* ARR10 (GARP family; Hosoda et al., 2002; PDB code 1riz), and (C) trihelix DBD of *Arabidopsis* GT-1 (Nagata et al., 2010; PDB code 2jmw).

Zn-binding site regarding structural folding, the site may accept Cu ions as well. Because the coordination geometry of Cu is different from that of Zn, binding of Cu may interfere with DNA binding, without appropriately guiding the C-terminal basic tail to the DNA groove.

4.2.6 Other DBDs of Plant-Specific Transcription Factors

Other than the above, the structures of the major DBD of ethylene-insentive3 (EIN3)-like protein3 (EIL3) belonging to the EIL family (Yamasaki et al., 2005b), the B-motif DBD of *Arabidopsis* type-B response regulator10 (ARR10) belonging to the GARP family (Hosoda et al., 2002), and the trihelix DBD of GT-1 belonging to the trihelix family (Nagata et al., 2010), which are considered plant-specific TFs, have been determined by NMR spectroscopy (Figure 4.5).

The EIL family includes the EIN3 and EIL proteins that function as key TFs in the ethylene signaling pathway (Chao et al., 1997). Six EILs are identified in *Arabidopsis* (Mitsuda and Ohme-Takagi, 2009). The longest of the basic regions of the protein encompasses approximately 130 amino acids, containing the key mutational site (*ein3-3* site) for DNA binding and signaling. This region, the major DBD, shows DNA-binding activity when isolated and possesses a structure containing five α-helices with a novel fold (Figure 4.5A; Yamasaki et al. 2005b). To understand the DNA-binding mechanism, however, the structure of the complex with DNA needs to be determined.

GARP family and trihelix family TFs are considered plant-specific, although they show weak sequence similarity to Myb transcription factors, which are also seen in animals (Riechmann et al., 2000). In *Arabidopsis*, 40–50 GARP family TFs and approximately 30 trihelix family TFs have been identified (Mitsuda and Ohme-Takagi, 2009).

The structures are indeed similar to Myb, containing three α-helices (Figure 4.5B, C; Hosoda et al., 2002; Nagata et al., 2010). The structure of trihelix DBDs is atypical in that the two α-helices from the N-terminus are significantly longer, with a kink in the middle of the first. NMR titration experiments and computational modeling reveal that the C-terminal helix acts as the recognition helix, in much the same way as Myb, both for the trihelix and GARP. In addition, the long N-terminal loop is also involved in DNA binding, which is typical of homeodomains of homeobox TFs (Gehring et al., 1994), but not necessarily of Myb. Comparison of amino acid sequences also indicates that GARP is more related to a homeodomain than Myb (Hosoda et al., 2002). However, Myb is known to be related to a homeodomain and shares the same structural fold and DNA-binding mode. Thus, these two families are strongly related to TFs that are not specific to higher plants and may be more appropriately classified as Myb-related or homeobox families.

4.3 EVOLUTIONARY HISTORY OF PLANT-SPECIFIC TFs

4.3.1 Endonucleases as Origins of DBDs

This section looks at the evolutionary histories of the DBDs of major plant-specific TFs (i.e., AP2/ERF, B3, WRKY, NAC, and SBP; Figure 4.6). These DBDs, with the exception of NAC, have been identified in the unicellular green alga *C. reinhardtii*, which is often modeled as an ancestor of land plants. Moreover, the proteins that show similarities to the AP2/ERF, WRKY, or B3 domains in amino acid sequences, three-dimensional structures, and DNA-binding mode have now been identified in a wide range of organisms (i.e., animals, protists, slime mold, bacteria, and viruses, as described in the previous sections). Very interestingly, the closest similarities between AP2/ERF and Tn*916* integrase or λ integrase, between WRKY and FLYWCH, and between B3 and restriction endonuclease *Eco*RII or *Bfi*I are those between the DBDs of plant-specific TFs and the noncatalytic DBDs of endonucleases. The endonucleases in the former two, in particular, are those included in transposable elements. As mentioned, a link has been shown between Tn*916* integrase and homing endonucleases, which are also tightly coupled with transposable elements (i.e., intron and intein; Wojciak et al., 1999). These are likely to be highly exposed to recombination with other genes of multiple hosts as a result of integration, self-splicing, and propagation of transposable elements. When the gene of the host genome is also an endonuclease, such as a restriction endonuclease or a colicin, the survival rate of the recombinant gene should be higher, since it may still contain endonuclease activity. Indeed, regarding the structures and sequence motifs of catalytic domains, many restriction endonucleases and homing endonucleases are classified in the same superfamilies, such as HNH, GIY-YIG, and PD-(D/E)xK, showing firm evolutionary links and/or extensive exchange of domains (Orlowski and Bujnicki, 2008; Taylor and Stoddard, 2012). The exchangeability of domains is also seen between restriction enzymes *Eco*RII and *Bfi*I, where they share the noncatalytic B3-like DBD, but possess different types of catalytic domains (Zhou et al., 2004; Grazulis et al., 2005). This might be brought about through the medium of transposable elements.

As shown in the above sections, most DBDs of the major families of plant-specific TFs utilize β-sheets in order to contact DNA (Figure 4.1), which used to be considered atypical. Coincidently, however, most of the homing endonucleases contain β-sheets that bind to DNA (Taylor and Stoddard, 2012; see structures of protein/DNA complexes in the upper panel of Figure 4.6). PI-*Sce*I of the LAGLIDADG family of homing endonuclease contains a three-stranded β-sheet packed with a C-terminal adjacent α-helix, where the β-sheet deeply penetrates the DNA major groove (Moure et al., 2002; "ERF-like" in Figure 4.6) in a manner similar to AtERF1 and related DBDs (Figure 4.2). I-*Hmu*I of the HNH family also possesses a β-sheet region that contacts DNA similarly ("ERF-like," in Figure 4.6), although the packing between the β-sheet and the C-terminal adjacent α-helix is disrupted (another four-stranded β-sheet intervenes in the middle; Shen et al., 2004). I-*Ppo*I of the His–Cys family also possesses such a corresponding region, where the DNA-contacting three-stranded β-sheet and the α-helix are tightly packed against each other ("ERF-like," in Figure 4.6), although this α-helix is located N-terminal to the β-sheet in the primary structure (blue in the figure; instead a helical turn without packing to the β-sheet exists; Flick et al., 1998). Thus, among these homing endonuclease structures, the ERF-like region of PI-*Sce*I most closely resembles the structure of AP2/ERF and related domains regarding the topology of the secondary structure unit. However, the β-sheet is long and packed tightly with other α-helices and a β-sheet; it appears very difficult to isolate the domain as it is. Alternatively, the ciliate, cyanobacteria, and bacteriophage-homing endonucleases containing AP2/ERF-like regions possess the HNH-type catalytic domain (Magnani et al., 2004; Wuitschick et al., 2004), which suggests a link to I-*Hmu*I. However, domain rearrangement will still be necessary as the orders of the two regions in the primary sequences are different between the two groups of endonucleases. Moreover, the ERF-like region is tightly packed with other regions including the HNH catalytic domain. In contrast, the ERF-like region of I-*Ppo*I is somewhat isolated from other secondary structure units. Considering the helix-swapped structure of ApiAP2 (Lindner et al., 2010), replacing the N-terminal helix by a C-terminal one in the primary sequence may well not be unlikely during evolution. Therefore, whereas it is not easy to determine which homing endonuclease is most related to the ancestor of AP2/ERF and related domains, the processes involved in significant rearrangement can be used to isolate them as noncatalytic DBDs.

The catalytic domain of HNH and His–Cys family endonucleases binds a catalytic metal through a β-strand/β-strand/α-helix structure motif (ββα-metal fold), which is shared with some bacterial colicin nucleases and restriction enzymes (Figure 4.6; Taylor and Stoddard, 2012). This fold is the same as that in C_2H_2 ZnF and the metal is exchangeable; colicin is known to hold various divalent cations including Zn when His residues are involved in Zn binding (Ku et al., 2002). Therefore, there may be an evolutionary link between the nuclease catalytic domain and C_2H_2 ZnF. A presumable link between C_2H_2 ZnF and WRKY has also been suggested but it requires significant rearrangement of secondary structure units including substitution of a helix by a strand (Babu et al., 2006). Because the Zn-binding site of WRKY is formed by residues on antiparallel β-strands and extended loops, a link to the Zn ribbon should also be considered; a Zn ribbon-like structure is seen in I-*Ppo*I (Figure 4.6; Flick et al., 1998). What is more, the SBP domain appears to be a structural merge of a C_2H_2 ZnF and a Zn ribbon-like structure (Yamasaki et al. 2004b), which may be taken from homing endonucleases, either independently or simultaneously. Moreover, when the structures of B3 domains (Figure 4.3) and a half-monomer of the I-*Ppo*I dimer (Figure 4.6, upper right) are compared, it is tempting to remark that if the three β-sheets of the latter, containing seven strands in total, are bundled to a pseudobarrel, the locations of two α-helices and the interface of DNA binding are indeed similar to those of the former when it comes to partial similarity between AP2/ERF and B3 (Figure 4.3F).

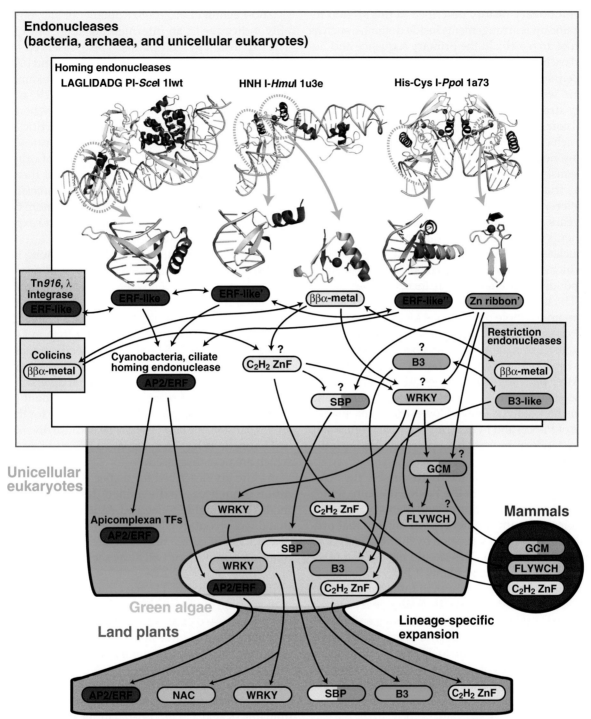

FIGURE 4.6 Probable evolutionary histories for the DBDs of plant-specific TFs and related DBDs. The evolutionary direction is shown from top to bottom with most DBDs originating from parts of homing endonucleases. Squares in the upper region encircle groups of endonucleases, whereas the colored large figures with other shapes in the lower region indicate groups of organisms. The domains are shown by small oval icons, in which those that share evolutionary histories are colored in the same scheme; the icons containing two colors indicate that the domain is evolutionarily related to two different domains (e.g., the N-terminal part of SBP is related to C_2H_2 ZnF and the C-terminal part is related to the Zn ribbon). The arrows indicate probable/proposed evolutionary connections and the directions between proteins possessing the DBDs. The question marks accompanying the domain icons show that they are in proteins presumed to exist temporarily during evolution. For example, the SBP-domain that is included in a homing endonuclease is not identified in existing organisms. The entire structures of homing endonucleases (PDB codes are shown above the structures) are indicated at the top where helices, strands, and loops are colored red, yellow, and green, respectively. The red spheres indicate metal ions (i.e., Mn for I-*Hmu*I and Mg or Zn for I-*Ppo*I). The parts encircled by cyan broken ovals are shown larger scale, colored in the colors of the rainbow from blue (N-terminus) to red (C-terminus). *This figure is a modification of that used in a previous review paper (Yamasaki et al., 2013), as reproduced with permission of Elsevier.*

The evolutionary history described in this section requires significant rearrangements inside domains, such as relocation of an α-helix in the primary sequence and helix/strand exchange, which generally speaking are unlikely to happen; such rearrangements by protein-engineering approaches would hardly be successful. This is because secondary structure elements are necessary to form the structural core and rearrangements unfold the structures easily. To achieve stable folding after rearrangements, all relevant amino acids should be changed consistently and simultaneously. This might have happened, however, in numerous trials though recombination and mutations when bacteria and early eukaryotes predominated billions of years ago. As mentioned previously, transposable elements are likely to facilitate the recombination of endonucleases. Intein homing endonucleases, such as PI-*SceI*, which can be embedded in other proteins forming covalent bonds, would have greatly increased the number of trials. The large structures of homing endonucleases, which contain multiple domains packing closely against one another, such as PI-*SceI* (Figure 4.6, upper left), may also facilitate rearrangements as "incubators"; the enzyme may keep functioning even in the face of a partial unstable moiety. After rearrangement, it would also have taken time to isolate domains that can fold independently and function as noncatalytic DBDs. However, comparing the structures of PI-*SceI* and I-*HmuI*, for example (Figure 4.6), such degradation appears to have proceeded in the latter more than in the former.

In contrast to the DBDs already discussed, the NAC domain may originate from the WRKY domain, which was born during the evolution of land plants presumably by recombination with another type of domain that possesses the property of dimerization by forming antiparallel β-sheets (Figure 4.1D). This should have taken place during the evolution of land plants 0.4–0.5 billion years ago, which is much more recent than for other DBDs.

It should be pointed out that the evolutionary history of eukaryotic TFs that originated from homing endonuclease is not necessarily plant-specific. Because the MH1 domain of the mammalian Smad TF shows close structural similarity to a part of I-*PpoI*, it is considered to have originated from homing endonucleases (Grishin, 2001).

4.3.2 Lineage-Specific Expansion

The DBDs of plant-specific TFs, with the exception of SBP, experienced extensive higher plant-specific expansion. For example, only one B3-containing protein and one WRKY-containing protein have been identified in *Chlamydomonas* (Ulker and Somssich, 2004; Riaño-Pachón et al., 2008; Romanel et al., 2009), although 118 and 72, respectively, were identified in *Arabidopsis* (Figure 4.1). The numbers of B3 and WRKY TFs in moss (*Physcomitrella patens*) are approximately half those of higher plants (Riaño-Pachón et al., 2008; Romanel et al., 2009), probably indicative of an intermediate stage of expansion. The expansion in land plants is also observed for other TF families, such as Myb, bHLH, MADS, and bZIP, even though they are not plant-specific. For example, approximately 200 Myb superfamily (Myb-R2R3 and Myb-related) TFs are identified in *Arabidopsis*, whereas only 3–10 copies are identified in other eukaryotic lineages (Riechmann et al., 2000). The expansion rates of plant TFs are significantly higher than those of other plant genes (Shiu et al., 2005). Considering that the genome size of *Chlamydomonas* is similar to that of *Arabidopsis* and that the number of genes of the former is more than half that of the latter (Merchant et al., 2007), the expansions seen in TFs are indeed special.

Genome-wide comparative analyses among Brassicaceae species and *Arabidopsis thaliana* strains revealed the dynamic equilibrium between duplication, transposon-mediated modulation, *de novo* formation, and deletion of genes, where the resultant generation of new genes in *Arabidopsis* is nearly three times faster than in *Drosophila* (Rutter et al., 2012). The lineage-specific expansion of TFs is a result of repeated duplication and survival. For duplicated genes to survive, functional diversification is required, since loss of genes with a redundant function causes no change in phenotype. The reason for the extensive expansion of plant TFs may be because diversification in TF functions has been achieved in a number of ways, which are now discussed.

As already described for respective DBDs, the recognition sequences are diversified despite DNA-binding modes being highly conserved. Mutations in the residues that directly contact DNA bases and those that affect the orientation of contacting residues also varied the recognition sequence of DBDs, as seen in AP2/ERFs (Hao et al., 2002; Liu et al., 2006).

In addition, many TFs have multiple DBDs resulting in elongation of recognition sequences; a single DBD recognizes only 4–6 base pairs, as already described, which may not be sufficient for the selection of proper target sites, especially when the numbers of TFs and target genes are on the increase. For example, AP2 TFs possess a tandem repeat of AP2/ERF domains (Jofuku et al., 1994) and RAV proteins are a combination of AP2/ERF and B3 domains (Kagaya et al., 1999). Duplication and recombination may have yielded such functional diversification. Moreover, ARF, NAC, and some WRKY TFs form homo- and heterodimers (Ernst et al., 2004; Xu et al., 2006; Boer et al., 2014) when regions outside the DBDs are essential for dimerization in the cases of ARF and WRKY. Therefore, mutations in dimerization regions, affecting interaction partner selectivity, may also affect DNA-binding activity.

Like other proteins, TFs are involved in extensive protein–protein interaction networks. If a protein-binding domain were to switch to another the interaction partner would change, resulting in alteration to the network. For example, TFs with the same DBD can behave as activators or repressors of transcription, depending on short motifs, as a result of which TFs interact with activator proteins or repressor proteins. The ERF-associated amphiphilic repression (EAR) motif identified in most TF families is representative of such motifs, which makes the TFs behave as transcription repressors, probably by specifying interaction partners to the repression machinery (Ohta et al., 2001; Hiratsu et al., 2003; Kagale et al., 2010). Moreover, the functions of ARFs are controlled by interacting with negative regulator proteins, such as Aux/IAA (auxin–indole acetic acid) proteins, which possess a dimerization domain equivalent to that of ARFs but lack a DNA-binding domain (Ulmasov et al., 1997). It has been shown that changing interaction partners (rewiring the interaction network) can indeed be a survival force for duplicated genes (Arabidopsis Interactome Mapping Consortium, 2011).

Finally, the transcription of TFs can also be controlled by other TFs as a result of their binding to the promoter region. Thus, sequence changes in the promoter region of TFs may also alter their positions in the transcription network and effectively diversify the functions of TFs.

Once newly acquired TFs are integrated into key regulatory systems, such as stress responses and the development of organs, they become essential and survive. Thus, the number of TFs dramatically increased during the evolution of higher plants, accompanied by generation of a complex system. Roughly half the TFs of higher plants were gained from the TFs in moss for adaptation to the terrestrial environment, and the remainder were for improved systems, such as vascular bundle and flower.

Acknowledgments

The author acknowledges fellow collaborators in the structural genomics project on plant transcription factors, headed by Professors S. Yokoyama and K. Shinozaki, from which six original and two review papers are cited. These works were brought about by the highly efficient cell-free system for protein synthesis developed by Dr. T. Kigawa and coworkers and the library of Arabidopsis thaliana full-length cDNAs constructed by Dr. M. Seki and coworkers. The project was supported by the RIKEN Structural Genomics/Proteomics Initiative (RSGI) and the National Project on Protein Structural and Functional Analyses, Ministry of Education, Culture, Sports, Science and Technology.

References

Agarwal, P., Reddy, M.P., Chikara, J., 2011. WRKY: its structure, evolutionary relationship, DNA-binding selectivity, role in stress tolerance and development of plants. Mol. Biol. Rep. 38, 3883–3896.

Aida, M., Beis, D., Heidstra, R., Willemsen, V., Blilou, I., Galinha, C., Nussaume, L., Noh, Y.S., Amasino, R., Scheres, B., 2004. The PLETHORA genes mediate patterning of the Arabidopsis root stem cell niche. Cell 119, 109–120.

Aida, M., Ishida, T., Fukaki, H., Fujisawa, H., Tasaka, M., 1997. Genes involved in organ separation in Arabidopsis: an analysis of the cup-shaped cotyledon mutant. Plant Cell 9, 841–857.

Allen, M.D., Yamasaki, K., Ohme-Takagi, M., Tateno, M., Suzuki, M., 1998. A novel mode of DNA recognition by a β-sheet revealed by the solution structure of the GCC-box binding domain in complex with DNA. EMBO J 17, 5484–5496.

Arabidopsis Interactome Mapping Consortium, 2011. Evidence for network evolution in an Arabidopsis interactome map. Science 333, 601–607.

Babu, M.M., Iyer, L.M., Balaji, S., Aravind, L., 2006. The natural history of the WRKY-GCM1 zinc fingers and the relationship between transcription factors and transposons. Nucleic Acids Res. 34, 6505–6520.

Balaji, S., Babu, M.M., Iyer, L.M., Aravind, L., 2005. Discovery of the principal specific transcription factors of apicomplexa and their implication for the evolution of the AP2–integrase DNA binding domains. Nucleic Acids Res. 33, 3994–4006.

Birkenbihl, R.P., Jach, G., Saedler, H., Huijser, P., 2005. Functional dissection of the plant-specific SBP-domain: overlap of the DNA-binding and nuclear localization domains. J. Mol. Biol. 352, 585–596.

Boer, D.R., Freire-Rios, A., van den Berg, W.A.M., Saaki, T., Manfield, I.W., Kepinski, S., Lopez-Vidrieo, I., Manuel Franco-Zorrilla, J., De Vries, S.C., Solano, R., Weijers, D., Coll, M., 2014. Structural basis for DNA binding specificity by the auxin-dependent ARF transcription factors. Cell 156, 577–589.

Canella, D., Gilmour, S.J., Kuhn, L.A., Thomashow, M.F., 2010. DNA binding by the Arabidopsis CBF1 transcription factor requires the PKKP/RAGRxKFxETRHP signature sequence. Biochim. Biophys. Acta Gene Regul. Mechanisms 1799, 454–462.

Cardon, G., Hohmann, S., Klein, J., Nettesheim, K., Saedler, H., Huijser, P., 1999. Molecular characterisation of the Arabidopsis SBP-box genes. Gene 237, 91–104.

Chandler, J.W., Cole, M., Flier, A., Grewe, B., Werr, W., 2007. The AP2 transcription factors DORNRÖSCHEN and DORNRÖSCHEN-LIKE redundantly control Arabidopsis embryo patterning via interaction with PHAVOLUTA. Development 134, 1653–1662.

Chao, Q.M., Rothenberg, M., Solano, R., Roman, G., Terzaghi, W., Ecker, J.R., 1997. Activation of the ethylene gas response pathway in Arabidopsis by the nuclear protein ETHYLENE-INSENSITIVE3 and related proteins. Cell 89, 1133–1144.

Chen, Q., Wang, Q., Xiong, L., Lou, Z., 2011. A structural view of the conserved domain of rice stress-responsive NAC1. Protein Cell 2, 55–63.

Cohen, S.X., Moulin, M., Hashemolhosseini, S., Kilian, K., Wegner, M., Muller, C.W., 2003. Structure of the GCM domain–DNA complex: a DNA-binding domain with a novel fold and mode of target site recognition. EMBO J. 22, 1835–1845.

De Silva, E.K., Gehrke, A.R., Olszewski, K., Leon, I., Chahal, J.S., Bulyk, M.L., Llinas, M., 2008. Specific DNA-binding by apicomplexan AP2 transcription factors. Proc. Natl. Acad. Sci. USA 105, 8393–8398.

Dietz, K.-J., Vogel, M.O., Viehhauser, A., 2010. AP2/EREBP transcription factors are part of gene regulatory networks and integrate metabolic, hormonal and environmental signals in stress acclimation and retrograde signalling. Protoplasma 245, 3–14.

Duan, M.-R., Nan, J., Liang, Y.-H., Mao, P., Lu, L., Li, L., Wei, C., Lai, L., Li, Y., Su, X.-D., 2007. DNA binding mechanism revealed by high resolution crystal structure of Arabidopsis thaliana WRKY1 protein. Nucleic Acids Res. 35, 1145–1154.

Ernst, H.A., Olsen, A.N., Skriver, K., Larsen, S., Lo Leggio, L., 2004. Structure of the conserved domain of ANAC, a member of the NAC family of transcription factors. EMBO Rep. 5, 297–303.

Eulgem, T., Rushton, P.J., Schmelzer, E., Hahlbrock, K., Somssich, I.E., 1999. Early nuclear events in plant defence signalling: rapid gene activation by WRKY transcription factors. EMBO J. 18, 4689–4699.

Eulgem, T., Rushton, P.J., Robatzek, S., Somssich, I.E., 2000. The WRKY superfamily of plant transcription factors. Trends Plant Sci. 5, 199–206.

Fadeev, E.A., Sam, M.D., Clubb, R.T., 2009. Nmr structure of the amino-terminal domain of the lambda integrase protein in complex with DNA: immobilization of a flexible tail facilitates beta-sheet recognition of the major groove. J. Mol. Biol. 388, 682–690.

Flick, K.E., Jurica, M.S., Monnat, R.J., Stoddard, B.L., 1998. DNA binding and cleavage by the nuclear intron-encoded homing endonuclease I-*Ppo*I. Nature 394, 96–101.

Fowler, S., Thomashow, M.F., 2002. *Arabidopsis* transcriptome profiling indicates that multiple regulatory pathways are activated during cold acclimation in addition to the CBF cold response pathway. Plant Cell 14, 1675–1690.

Gehring, W.J., Qian, Y.Q., Billeter, M., Furukubotokunaga, K., Schier, A.F., Resendezperez, D., Affolter, M., Otting, G., Wuthrich, K., 1994. Homeodomain–DNA recognition. Cell 78, 211–223.

Giraudat, J., Hauge, B.M., Valon, C., Smalle, J., Parcy, F., Goodman, H.M., 1992. Isolation of the *Arabidopsis ABI3* gene by positional cloning. Plant Cell 4, 1251–1261.

Golovenko, D., Manakova, E., Tamulaitiene, G., Grazulis, S., Siksnys, V., 2009. Structural mechanisms for the 5′-CCWGG sequence recognition by the N- and C-terminal domains of EcoRII. Nucleic Acids Res. 37, 6613–6624.

Golovenko, D., Manakova, E., Zakrys, L., Zaremba, M., Sasnauskas, G., Grazulis, S., Siksnys, V., 2014. Structural insight into the specificity of the B3 DNA-binding domains provided by the co-crystal structure of the C-terminal fragment of BfiI restriction enzyme. Nucleic Acids Res. 42, 4113–4122.

Grazulis, S., Manakova, E., Roessle, M., Bochtler, M., Tamulaitiene, G., Huber, R., Siksnys, V., 2005. Structure of the metal-independent restriction enzyme BfiI reveals fusion of a specific DNA-binding domain with a nonspecific nuclease. Proc. Natl. Acad. Sci. USA 102, 15797–15802.

Grishin, N.V., 2001. MH1 domain of Smad is a degraded homing endonuclease. J. Mol. Biol. 307, 31–37.

Hao, D.Y., Ohme-Takagi, M., Yamasaki, K., 2003. A modified sensor chip for surface plasmon resonance enables a rapid determination of sequence specificity of DNA-binding proteins. FEBS Lett 536, 151–156.

Hao, D.Y., Yamasaki, K., Sarai, A., Ohme-Takagi, M., 2002. Determinants in the sequence specific binding of two plant transcription factors, CBF1 and NtERF2, to the DRE and GCC motifs. Biochemistry 41, 4202–4208.

Hiratsu, K., Matsui, K., Koyama, T., Ohme-Takagi, M., 2003. Dominant repression of target genes by chimeric repressors that include the EAR motif, a repression domain, in *Arabidopsis*. Plant J. 34, 733–739.

Hosoda, K., Imamura, A., Katoh, E., Hatta, T., Tachiki, M., Yamada, H., Mizuno, T., Yamazaki, T., 2002. Molecular structure of the GARP family of plant Myb-related DNA binding motifs of the *Arabidopsis* response regulators. Plant Cell 14, 2015–2029.

Ishiguro, S., Nakamura, K., 1994. Characterization of a cDNA encoding a novel DNA-binding protein, SPF1, that recognizes SP8 sequences in the 5′ upstream regions of genes coding for sporamin and beta-amylase from sweet potato. Mol. Gen. Genet 244, 563–571.

Jofuku, K.D., Denboer, B.G.W., Vanmontagu, M., Okamuro, J.K., 1994. Control of *arabidopsis* flower and seed development by the homeotic gene *APETALA2*. Plant Cell 6, 1211–1225.

Kagale, S., Links, M.G., Rozwadowski, K., 2010. Genome-wide analysis of ethylene-responsive element binding factor-associated amphiphilic repression motif-containing transcriptional regulators in *Arabidopsis*. Plant Physiol. 152, 1109–1134.

Kagaya, Y., Hattori, T., 2009. *Arabidopsis* transcription factors, RAV1 and RAV2, are regulated by touch-related stimuli in a dose-dependent and biphasic manner. Genes Genet. Syst. 84, 95–99.

Kagaya, Y., Ohmiya, K., Hattori, T., 1999. RAV1, a novel DNA-binding protein, binds to bipartite recognition sequence through two distinct DNA-binding domains uniquely found in higher plants. Nucleic Acids Res. 27, 470–478.

King, G.J., Chanson, A.H., McCallum, E.J., Ohme-Takagi, M., Byriel, K., Hill, J.M., Martin, J.L., Mylne, J.S., 2013. The *Arabidopsis* B3 domain protein VERNALIZATION1 (VRN1) is involved in processes essential for development, with structural and mutational studies revealing its DNA-binding surface. J. Biol. Chem. 288, 3198–3207.

Klein, J., Saedler, H., Huijser, P., 1996. A new family of DNA binding proteins includes putative transcriptional regulators of the *Antirrhinum majus* floral meristem identity gene *SQUAMOSA*. Mol. Gen. Genet 250, 7–16.

Krizek, B.A., 2003. AINTEGUMENTA utilizes a mode of DNA recognition distinct from that used by proteins containing a single AP2 domain. Nucleic Acids Res. 31, 1859–1868.

Kropat, J., Tottey, S., Birkenbihl, R.P., Depege, N., Huijser, P., Merchant, S., 2005. A regulator of nutritional copper signaling in *Chlamydomonas* is an SBP domain protein that recognizes the GTAC core of copper response element. Proc. Natl. Acad. Sci. USA 102, 18730–18735.

Ku, W.Y., Liu, Y.W., Hsu, Y.C., Liao, C.C., Liang, P.H., Yuan, H.S., Chak, K.F., 2002. The zinc ion in the HNH motif of the endonuclease domain of colicin E7 is not required for DNA binding but is essential for DNA hydrolysis. Nucleic Acids Res. 30, 1670–1678.

Levy, Y.Y., Mesnage, S., Mylne, J.S., Gendall, A.R., Dean, C., 2002. Multiple roles of *Arabidopsis VRN1* in vernalization and flowering time control. Science 297, 243–246.

Liang, X., Nazarenus, T.J., Stone, J.M., 2008. Identification of a consensus DNA-binding site for the *Arabidopsis thaliana* SBP domain transcription factor, AtSPL14, and binding kinetics by surface plasmon resonance. Biochemistry 47, 3645–3653.

Lindner, S.E., De Silva, E.K., Keck, J.L., Llinas, M., 2010. Structural determinants of DNA binding by a *P. falciparum* ApiAP2 transcriptional regulator. J. Mol. Biol. 395, 558–567.

Liu, Y., Zhao, T.J., Liu, J.M., Liu, W.Q., Liu, Q., Yan, Y.B., Zhou, H.M., 2006. The conserved Ala37 in the ERF/AP2 domain is essential for binding with the DRE element and the GCC box. FEBS Lett. 580, 1303–1308.

Magnani, E., Sjolander, K., Hake, S., 2004. From endonucleases to transcription factors: evolution of the AP2 DNA binding domain in plants. Plant Cell 16, 2265–2277.

Marquez, C.P., Pritham, E.J., 2010. *Phantom*, a new subclass of *Mutator* DNA transposons found in insect viruses and widely distributed in animals. Genetics 185, 1507–1582.

McCarty, D.R., Carson, C.B., Stinard, P.S., Robertson, D.S., 1989. Molecular analysis of *viviparous-1*: an abscisic acid-insensitive mutant of maize. Plant Cell 1, 523–532.

Merchant, S.S., Prochnik, S.E., Vallon, O., Harris, E.H., Karpowicz, S.J., Witman, G.B., Terry, A., Salamov, A., Fritz-Laylin, L.K., Marechal-Drouard, L., Marshall, W.F., Qu, L.H., Nelson, D.R., Sanderfoot, A.A., Spalding, M.H., Kapitonov, V.V., Ren, Q., Ferris, P., Lindquist, E., Shapiro, H., et al., 2007. The *chlamydomonas* genome reveals the evolution of key animal and plant functions. Science 318, 245–250.

Mitsuda, N., Ohme-Takagi, M., 2009. Functional analysis of transcription factors in *arabidopsis*. Plant Cell Physiol. 50, 1232–1248.

Mizoi, J., Shinozaki, K., Yamaguchi-Shinozaki, K., 2012. AP2/ERF family transcription factors in plant abiotic stress responses. Biochim. Biophys. Acta Gene Regul. Mechanisms 1819, 86–96.

Moure, C.M., Gimble, F.S., Quiocho, F.A., 2002. Crystal structure of the intein homing endonuclease PI-*Sce*I bound to its recognition sequence. Nat. Struct. Biol. 9, 764–770.

Nagata, T., Niyada, E., Fujimoto, N., Nagasaki, Y., Noto, K., Miyanoiri, Y., Murata, J., Hiratsuka, K., Katahira, M., 2010. Solution structures of the trihelix DNA-binding domains of the wild-type and a phosphomimetic mutant of *Arabidopsis* GT-1: mechanism for an increase in DNA-binding affinity through phosphorylation. Prot. Struct. Funct. Bioinfo. 78, 3033–3047.

REFERENCES

Nakashima, K., Takasaki, H., Mizoi, J., Shinozaki, K., Yamaguchi-Shinozaki, K., 2012. NAC transcription factors in plant abiotic stress responses. Biochim. Biophys. Acta Gene Regul. Mechanisms 1819, 97–103.

Nole-Wilson, S., Krizek, B.A., 2000. DNA binding properties of the *Arabidopsis* floral development protein AINTEGUMENTA. Nucleic Acids Res. 28, 4076–4082.

Ohme-Takagi, M., Shinshi, H., 1995. Ethylene-inducible DNA-binding proteins that interact with an ethylene-responsive element. Plant Cell 7, 173–182.

Ohta, M., Matsui, K., Hiratsu, K., Shinshi, H., Ohme-Takagi, M., 2001. Repression domains of class II ERF transcriptional repressors share an essential motif for active repression. Plant Cell 13, 1959–1968.

Olsen, A.N., Ernst, H.A., Lo Leggio, L., Skriver, K., 2005. NAC transcription factors: structurally distinct, functionally diverse. Trends Plant Sci. 10, 79–87.

Orlowski, J., Bujnicki, J.M., 2008. Structural and evolutionary classification of Type II restriction enzymes based on theoretical and experimental analyses. Nucleic Acids Res. 36, 3552–3569.

Painter, H.J., Campbell, T.L., Llinas, M., 2011. The apicomplexan AP2 family: integral factors regulating *plasmodium* development. Mol. Biochem. Parasitol. 176, 1–7.

Puranik, S., Sahu, P.P., Srivastava, P.S., Prasad, M., 2012. NAC proteins: regulation and role in stress tolerance. Trends. Plant Sci. 17, 369–381.

Riaño-Pachón, D.M., Correa, L.G., Trejos-Espinosa, R., Mueller-Roeber, B., 2008. Green transcription factors: a *Chlamydomonas* overview. Genetics 179, 31–39.

Riechmann, J.L., Heard, J., Martin, G., Reuber, L., Jiang, C.Z., Keddie, J., Adam, L., Pineda, O., Ratcliffe, O.J., Samaha, R.R., Creelman, R., Pilgrim, M., Broun, P., Zhang, J.Z., Ghandehari, D., Sherman, B.K., Yu, C.L., 2000. *Arabidopsis* transcription factors: genome-wide comparative analysis among eukaryotes. Science 290, 2105–2110.

Romanel, E.A.C., Schrago, C.G., Counago, R.M., Russo, C.A.M., Alves-Ferreira, M., 2009. Evolution of the B3 DNA binding superfamily: new insights into REM family gene diversification. PLoS One, 4.

Rushton, P.J., Macdonald, H., Huttly, A.K., Lazarus, C.M., Hooley, R., 1995. Members of a new family of DNA-binding proteins bind to a conserved *cis*-element in the promoters of α-*Amy2* genes. Plant Mol. Biol. 29, 691–702.

Rushton, P.J., Somssich, I.E., Ringler, P., Shen, Q.J., 2010. WRKY transcription factors. Trends Plant Sci. 15, 247–258.

Rutter, M.T., Cross, K.V., Van Woert, P.A., 2012. Birth, death and subfunctionalization in the *Arabidopsis* genome. Trends Plant Sci. 17, 204–212.

Sakuma, Y., Liu, Q., Dubouzet, J.G., Abe, H., Shinozaki, K., Yamaguchi-Shinozaki, K., 2002. DNA-binding specificity of the ERF/AP2 domain of *Arabidopsis* DREBs, transcription factors involved in dehydration- and cold-inducible gene expression. Biochim. Biophys. Res. Commun. 290, 998–1009.

Shen, B.W., Landthaler, M., Shub, D.A., Stoddard, B.L., 2004. DNA binding and cleavage by the HNH homing endonuclease I-HmuI. J. Mol. Biol. 342, 43–56.

Shiu, S.H., Shih, M.C., Li, W.H., 2005. Transcription factor families have much higher expansion rates in plants than in animals. Plant Physiol. 139, 18–26.

Souer, E., Vanhouwelingen, A., Kloos, D., Mol, J., Koes, R., 1996. The *No Apical Meristem* gene of petunia is required for pattern formation in embryos and flowers and is expressed at meristem and primordia boundaries. Cell 85, 159–170.

Suzuki, M., Kao, C.Y., McCarty, D.R., 1997. The conserved B3 domain of VIVIPAROUS1 has a cooperative DNA binding activity. Plant Cell 9, 799–807.

Swaminathan, K., Peterson, K., Jack, T., 2008. The plant B3 superfamily. Trends Plant Sci. 13, 647–655.

Taylor, G.K., Stoddard, B.L., 2012. Structural, functional and evolutionary relationships between homing endonucleases and proteins from their host organisms. Nucleic Acids Res. 40, 5189–5200.

Tran, L.S.P., Nakashima, K., Sakuma, Y., Simpson, S.D., Fujita, Y., Maruyama, K., Fujita, M., Seki, M., Shinozaki, K., Yamaguchi-Shinozaki, K., 2004. Isolation and functional analysis of *Arabidopsis* stress-inducible NAC transcription factors that bind to a drought-responsive *cis*-element in the early responsive to dehydration stress 1 promoter. Plant Cell 16, 2481–2498.

Ulker, B., Somssich, I.E., 2004. WRKY transcription factors: from DNA binding towards biological function. Curr. Opin. Plant Biol. 7, 491–498.

Ulmasov, T., Hagen, G., Guilfoyle, T.J., 1997. ARF1, a transcription factor that binds to auxin response elements. Science 276, 1865–1868.

Waltner, J.K., Peterson, F.C., Lytle, B.L., Volkman, B.F., 2005. Structure of the B3 domain from *Arabidopsis thaliana* protein At1g16640. Protein Sci. 14, 2478–2483.

Welner, D.H., Lindemose, S., Grossmann, J.G., Mollegaard, N.E., Olsen, A.N., Helgstrand, C., Skriver, K., Lo Leggio, L., 2012. DNA binding by the plant-specific NAC transcription factors in crystal and solution: a firm link to WRKY and GCM transcription factors. Biochem. J. 444, 395–404.

Wojciak, J.M., Connolly, K.M., Clubb, R.T., 1999. NMR structure of the Tn*916* integrase–DNA complex. Nat. Struct. Biol. 6, 366–373.

Wuitschick, J.D., Lindstrom, P.R., Meyer, A.E., Karrer, K.M., 2004. Homing endonucleases encoded by germ line-limited genes in *Tetrahymena thermophila* have APETELA2 DNA binding domains. Eukaryot. Cell 3, 685–694.

Xu, X.P., Chen, C.H., Fan, B.F., Chen, Z.X., 2006. Physical and functional interactions between pathogen-induced *Arabidopsis* WRKY18, WRKY40, and WRKY60 transcription factors. Plant Cell 18, 1310–1326.

Yamasaki, K., Kigawa, T., Inoue, M., Tateno, M., Yamasaki, T., Yabuki, T., Aoki, M., Seki, E., Matsuda, T., Tomo, Y., Hayami, N., Terada, T., Shirouzu, M., Osanai, T., Tanaka, A., Motoaki, S., Shinozaki, K., Yokoyama, S., 2004a. Solution structure of the B3 DNA binding domain of the *Arabidopsis* cold-responsive transcription factor RAV1. Plant Cell 16, 3448–3459.

Yamasaki, K., Kigawa, T., Inoue, M., Tateno, M., Yamasaki, T., Yabuki, T., Aoki, M., Seki, E., Matsuda, T., Nunokawa, E., Ishizuka, Y., Terada, T., Shirouzu, M., Osanai, T., Tanaka, A., Seki, M., Shinozaki, K., Yokoyama, S., 2004b. A novel zinc-binding motif revealed by solution structures of DNA-binding domains of *Arabidopsis* SBP-family transcription factors. J. Mol. Biol. 337, 49–63.

Yamasaki, K., Kigawa, T., Inoue, M., Tateno, M., Yamasaki, T., Yabuki, T., Aoki, M., Seki, I., Matsuda, T., Tomo, Y., Hayami, N., Terada, T., Shirouzu, M., Tanaka, A., Seki, M., Shinozaki, K., Yokoyama, S., 2005a. Solution structure of an *Arabidopsis* WRKY DNA binding domain. Plant Cell 17, 944–956.

Yamasaki, K., Kigawa, T., Inoue, M., Tateno, M., Yamasaki, T., Yabuki, T., Aoki, M., Seki, E., Matsuda, T., Tomo, Y., Terada, T., Shirouzu, M., Tanaka, A., Seki, M., Shinozaki, K., Yokoyama, S., 2005b. Solution structure of the major DNA-binding domain of *Arabidopsis thaliana* ethylene-insensitive3-like3. J. Mol. Biol. 348, 253–264.

Yamasaki, K., Kigawa, T., Inoue, M., Tateno, M., Yamasaki, T., Yabuki, T., Aoki, M., Seki, E., Matsuda, T., Tomo, Y., Terada, T., Shirouzu, M., Tanaka, A., Seki, M., Shinozaki, K., Yokoyama, S., 2006. An *Arabidopsis* SBP-domain fragment with a disrupted c-terminal zinc-binding site retains its tertiary structure. FEBS Lett. 580, 2109–2116.

Yamasaki, K., Kigawa, T., Inoue, M., Watanabe, S., Tateno, M., Seki, M., Shinozaki, K., Yokoyama, S., 2008. Structures and evolutionary origins of plant-specific transcription factor DNA-binding domains. Plant Physiol. Biochem. 46, 394–401.

Yamasaki, H., Hayashi, M., Fukazawa, M., Kobayashi, Y., Shikanai, T., 2009. SQUAMOSA promoter binding protein-like7 is a central regulator for copper homeostasis in *Arabidopsis*. Plant Cell 21, 347–361.

Yamasaki, K., Kigawa, T., Watanabe, S., Inoue, M., Yamasaki, T., Seki, M., Shinozaki, K., Yokoyama, S., 2012. Structural basis for sequence-specific DNA recognition by an *Arabidopsis* WRKY transcription factor. J. Biol. Chem. 287, 7683–7691.

Yamasaki, K., Kigawa, T., Seki, M., Shinozaki, K., Yokoyama, S., 2013. DNA-binding domains of plant-specific transcription factors: structure, function, and evolution. Trends Plant Sci. 18, 267–276.

Yang, S., Wang, S., Liu, X., Yu, Y., Yue, L., Wang, X., Hao, D., 2009. Four divergent *Arabidopsis* ethylene-responsive element-binding factor domains bind to a target DNA motif with a universal CG step core recognition and different flanking bases preference. FEBS J 276, 7177–7186.

Zhong, R., Lee, C., Ye, Z.H., 2010. Global analysis of direct targets of secondary wall NAC master switches in *Arabidopsis*. Mol. Plant 3, 1087–1103.

Zhou, X.Y.E., Wang, Y.J., Reuter, M., Mucke, M., Kruger, D.H., Meehan, E.J., Chen, L.Q., 2004. Crystal structure of type IIE restriction endonuclease *Eco*RII reveals an autoinhibition mechanism by a novel effector-binding fold. J. Mol. Biol. 335, 307–319.

CHAPTER 5

The Evolutionary Diversification of Genes that Encode Transcription Factor Proteins in Plants

Toshifumi Nagata, Aeni Hosaka-Sasaki, Shoshi Kikuchi

Plant Genome Research Unit, Agrogenomics Research Center, National Institute of Agrobiological Sciences (NIAS), Tsukuba, Ibaraki, Japan

OUTLINE

5.1 Introduction – Distinctive Features of TF Genes in Plants (*Arabidopsis* and Rice)	73	
5.2 A Comparative Analysis of TF Genes Between Plants and Animals	76	
5.3 A Comparative Analysis of Transcription Factor Genes in 32 Diverse Organisms	76	
5.3.1 The Evolution of Prokaryotes to Unicellular Eukaryotes (Bacteria, Photosynthetic Bacteria, Unicellular Animals, and Unicellular and Colony-Type Plants)	76	
5.3.2 The Evolutionary Characteristics of Slime Mold, Yeast, and Fungi TF Genes	89	
5.3.3 The Emergence of Multicellular Organisms and the Evolution of Invertebrates (Sea Anemones, Nematodes, Segmental Worms, Sea Urchins, and Flies)	89	
5.3.4 The Evolution of Vertebrates (Fish, Amphibians, Birds, Mice, and Humans)	90	
5.3.5 The Evolution of Terrestrial Plants (Ferns and Mosses)	91
5.3.6 The Evolution of Seed Plants (Woody Plants, Dicots, and Monocots)	91
5.3.6.1 Woody Plants	91
5.3.6.2 Dicotyledonous Plants	92
5.3.6.3 Monocotyledonous Plants	92
5.4 The Appearance of New TF Gene Members During Evolution	92
5.5 The Different Evolutionary Methods of TF Genes in Animals and Plants	93
5.6 TF Gene Evolution and Its Biological Function	94
5.7 Conclusion: The Regulatory Role of Individual Transcription Factors	96
References	96

5.1 INTRODUCTION – DISTINCTIVE FEATURES OF TF GENES IN PLANTS (ARABIDOPSIS AND RICE)

Genome-wide comparative structural analyses of genes that encode for transcription factor (TF) proteins have occurred for many organisms because of the significant progress of genome analysis. Structural and functional genome analyses have investigated how many genes contribute to the formation of the gene network system and how they participate in developmental processes and responses to environmental stresses. TF genes have key roles in the gene network in facilitating many complex regulatory systems.

Complex regulatory systems have increasingly evolved to facilitate adaptation in many organisms. Evolutionary development has also generally coincided with an increase in gene number and regulatory networks for adaptation to various environmental stimuli. Consequently, the increasing complexity of gene regulation depends on an increase in the number and function of genes that encode TF genes. Many studies (Putnam et al., 2007; Sea Urchin Genome Sequencing Consortium et al., 2006; International Chicken Genome Sequencing Consortium, 2004; Merchant et al., 2007; Eichinger et al., 2005; Rensing et al., 2008; Ming et al., 2008; Tuskan et al., 2006) have identified and characterized species-specific individual regulatory species of TF genes, and recent genomic sequencing (Riechmann et al., 2000; Xiong et al., 2005; Davuluri et al., 2003) has provided complete analyses of TF genes in many organisms. These studies have provided an understanding of the various features of gene regulation networks. In the animal kingdom, these analyses have shown that (1) TFs involved in the early patterning and development of bilaterians are present in cnidarian genomes and are active in development; (2) nearly all subclasses of important vertebrate hematopoietic and immune TFs are present in the sea urchin genome; and (3) the Krüppel-associated box (KRAB) domain containing the C2H2 Zn finger gene has not expanded in chickens as it has in mammals (humans and mice) (Putnam et al., 2007; Sea Urchin Genome Sequencing Consortium et al., 2006; International Chicken Genome Sequencing Consortium, 2004) (Figure 5.1A).

In the plant kingdom (1) genes that constitute the light-regulated transcription network of angiosperms and their

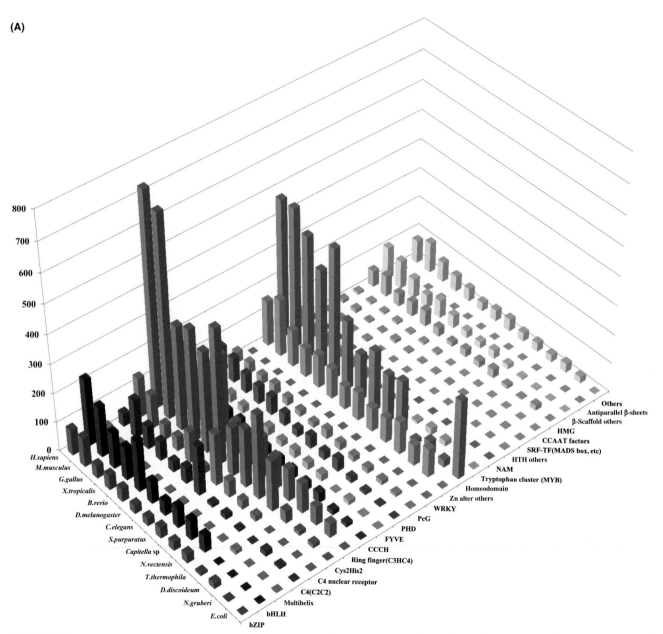

FIGURE 5.1 **The evolution of transcription factor genes in animals and plants.** (A) The evolution of transcription factor genes in animals.

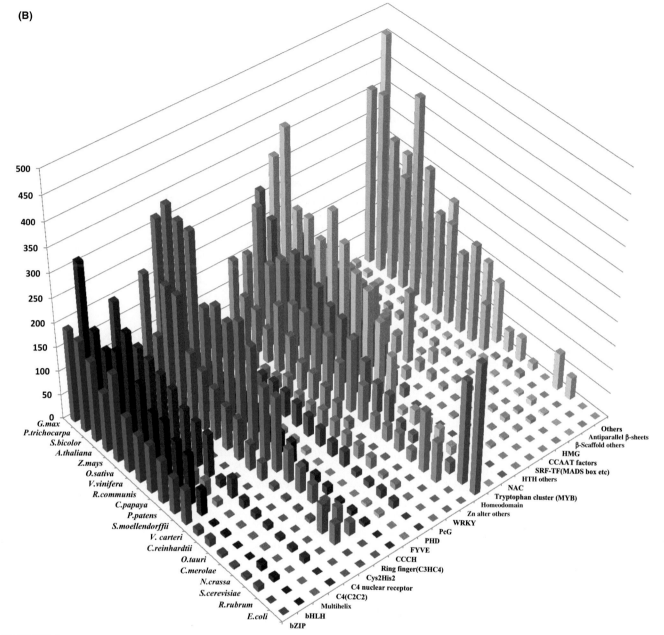

FIGURE 5.1 *(Cont.)* (B) The evolution of transcription factor genes in plants. bZIP, leucine zipper factor; GRAS, GRAS family; GeBP, GeBP family; bHLH, basic helix–loop–helix family; TCP, TCP family; Myc, Myc family; AP-2, AP-2 family; Nin-like, Nin-like (=RWP-RK) family; CXC, CPP (=CXC) family; Trihelix, trihelix family; SAND, SAND (=ULT) family; EIL, EIL family; GATA, GATA (=C2C2-GATA) family; Dof, Dof (=C2C2-Dof) family; Co-like, Co-like (=C2C2-Co-like) family; ZIM, ZIM family; YABBY, YABBY (=C2C2-YABBY) family; C4 nuclear receptor, C4 (= C2C2) nuclear receptor family; GRF, GRF family; C2H2, C2H2 family; SWIM, SWIM family; Ring finger, Ring finger (=C3HC4, C3H) family; C3HC, C3HC family; CCCH, CCCH family; TAZ, TAZ family; FYVE, FYVE family; PHD, PHD family; PcG, PcG family; CW, CW family; WRKY, WRKY family; UBR1, UBR1 family; TRAF, TRAF family; HRT, HRT family; VOZ, VOZ-9 (=VOZ) family; AN1, AN1-like; DHHC, DHHC family; Jmj, jumonji (=JUMONJI) family; A20, A20 family; MIZ, MIZ family; BED, BED family; LSD, LSD1 family; NF-X1, NF-X1 family; PLATZ, PLATZ family; HD, homeodomain (HB) family; ZF-HD, ZF-HD family; DDT, DDT family; folk head, folkhead-like (E2F-DP) (=E2F-DP) family; winged helix, winged helix (E2F-DP) (=E2F-DP) family; HSF, HSF (E2F-DP) (=HSF) family; LIM, LIM family; POU, POU family; ETS, ETS family; Myb, Myb + Myb-related family; GARP, GARP (ARR-B + G2-like) family; NAC, NAC family; SRF, SRF (=MADS) family; CCAAT factors, CCAAT (= CCAAT-Dr1 + CCAAT-Hap2 + CCAAT-Hap3 + CCAAT-Hap5) family; HMG, HMG family; ARID, ARID family; AP2/EREBP, AP2/EREBP (AP2-EREBP) family; AUX_IAA, AUX_IAA (=AUX-IAA = AUX/IAA) family; ABI3/VP1, ABI3/VP1 (=ABI3VP1 = ABI3-VP1); ARF, ARF family; TUBBY (TLP), TUBBY (TLP) (=TUB = TLP) family; BBR/BPC, BBR/BPC (=BBR-BPC) family; BZR, BZR (BES) (=BES1) family; CAMTA, CAMTA family; GIF, GIF family; AS2, AS2 family; FLO, FLO/LEAFY (=LFY) family; MBF, MBF1 family; SAP, SAP family; SBP, SBP family; PAZ, PAZ family.

A. GENERAL ASPECTS OF PLANT TRANSCRIPTION FACTORS

algal orthologs were established in *Chlamydomonas*; (2) most of the ubiquitous families of TFs are represented in *Dictyostelium* with the exception of ubiquitous basic helix–loop–helix (HLH) proteins; (3) duplicated TF genes are retained in the *Physcomitrella patens* lineage at a lower rate than those in the flowering plant lineage but at a higher rate than in algae; (4) most TFs are represented by fewer genes in papaya than in *Arabidopsis*, although the RWP-RK, MADS-box, Scarecrow, TCP, and Jumonji family gene numbers are greater in papaya; and (5) most TF domains in *Populus* were 40–80% higher compared to *Arabidopsis*, and ARF and ERF proteins (a subfamily of the AP2/EREBP family) are more highly diverged in poplars than in *Arabidopsis* (Merchant et al., 2007; Eichinger et al., 2005; Rensing et al., 2008; Ming et al., 2008; Tuskan et al., 2006) (Figure 5.1B).

Arabidopsis and rice databases have provided genomic data for a comprehensive analysis of plant TF genes (Riechmann et al., 2000; Xiong et al., 2005; Davuluri et al., 2003). These studies have shown that rice and *Arabidopsis* contain 1500–1600 TF genes, comprising 4–6% of the total genes. Plant-specific TF genes include AP2/ERF (/DREB/EREBP), NAC, WRKY (Zn finger), GARP, Dof, CO homolog, YABBY, GRAS, Trihelix, TCP, ARF, C3H-TYPE2 (Zn finger), SBP, Nin homolog, ABI3/VPI, Alfin homolog, EIL, and LFY, which comprise 45–50% of the total TF genes. By contrast, animal-specific TF genes include NHR (C8) (Zn), Adf-1, T-BOX, ETS, DM (Zn), PAIRED (W/0 HB), Runt/CBFa, NF-kB/REL/dorsal, and Smad, which comprise 15–45% of the total TF genes (Chapter 3). *Arabidopsis* has more MADS and REM TF genes than rice, whereas rice has more NAC, WRKY, and GRAS family TF genes. Large clusters of REM, GRAS, MADS, NAC, and WRKY genes in rice have been previously reported (Riechmann et al., 2000; Xiong et al., 2005). However, analyses of individual TF gene families (GATA, Dof, Co-like, homeodomain, MYB, NAC, MADS, HMG, bZIP, bHLH, helix-turn-helix (HTH), AP2/EREBP, etc.) have also provided much information on TF evolution (Reyes et al., 2004; Umemura et al., 2004; Griffiths et al., 2003; Banerjee-Basu and Baxevanis, 2001; Yanhui et al., 2006; Ooka et al., 2003; Becker and Theissen, 2003; Soullier et al., 1999; Amoutzias et al., 2007; Simionato et al., 2007; Aravind et al., 2005; Magnani et al., 2004).

5.2 A COMPARATIVE ANALYSIS OF TF GENES BETWEEN PLANTS AND ANIMALS

A general analysis of the TF gene diversity encoded in the genomes of 32 organisms indicates the following: (1) there were more than four large diversifications (animal, plants, bacteria, fungi, etc.) of TFs; (2) plant genomes encode more TF genes than animal genomes; (3) TF genes in animals appear to have expanded when they became multicellular organisms, and in plants, the expansion appears to have occurred when they became multicellular terrestrial organisms; (4) in animal genomes the TF genes encoding C2H2 ZF, C4 receptor ZF, Ring finger, homeodomain, and HMG have increased during evolution, whereas in plant genomes, the TF genes encoding bZIP, C2H2 ZF, C4 nonreceptor-type ZF, Ring finger, homeodomain, MYB, SRF-TF (MADS), and antiparallel β-sheet factors (AP2/EREBP, AUX_IAA, ABI3/VP1, ARF) have increased; (5) the diversification of animals and plants is different and evolved following the unicellular organism stage; (6) increased folds in the total TF gene round numbers among plants (3400/160 = 21-fold) is more than twice that among animals (2700/300 = ninefold); (7) the number of vertebrate TF genes is similar to that in angiosperms; and (8) the number of TF genes in multicellular invertebrates is similar to that in ferns and mosses (Figure 5.2).

5.3 A COMPARATIVE ANALYSIS OF TRANSCRIPTION FACTOR GENES IN 32 DIVERSE ORGANISMS

TF gene family numbers were compared (Table 5.1) for *Escherichia coli* K12-MG1655, *Arabidopsis thaliana*, *Oryza sativa*, *Caenorhabditis elegans*, *Drosophila melanogaster*, *Homo sapiens* NCBI, *Neurospora crassa* 74-OR23-IVA, *Saccharomyces cerevisiae* S228C, *Mus musculus*, *Gallus gallus*, *Xenopus tropicalis*, *Danio rerio*, *Stronglyocentrotus purpuratus*, *Capitella capitata* sp. I, *Nematostella vectensis*, *Tetrahymena thermophila*, *Dictyostelium discoideum*, *Naegleia gruberi*, *Rhodospirillum rubrum*, *Cyanidioschyzon merolae*, *Ostreococcus tauri*, *Chlamydomonas reinhardtii*, *Volvox carteri*, *Selaginella moellendorffii*, *Physcomitrella patens*, *Carica papaya*, *Ricinus communis*, *Vitis vinifera*, *Sorghum bicolor*, *Zea mays* spp. *mays*, *Glycine max*, and *Populus trichocarpa*. The detailed descriptions of the TF gene patterns in these organisms are presented in Figures 5.3A–F.

5.3.1 The Evolution of Prokaryotes to Unicellular Eukaryotes (Bacteria, Photosynthetic Bacteria, Unicellular Animals, and Unicellular and Colony-Type Plants)

Large evolutionary changes in TF genes can be observed in the transition from prokaryote to eukaryote (summarized in Figure 5.4). In bacteria, most of the TF genes (>95%) belong to the group containing HTH structures. In addition, bacteria have a few prototypes of eukaryotic TF genes (σ-factor is similar to TFA, Zur has the Zn(II)-binding domain, PhoB has an additional β-sheet α-helix complex, Lac repressor has a bZIP-like domain, etc.). Bacterial HTH TFs are the origin of homeodomain proteins and similar

5.3 A COMPARATIVE ANALYSIS OF TRANSCRIPTION FACTOR GENES IN 32 DIVERSE ORGANISMS

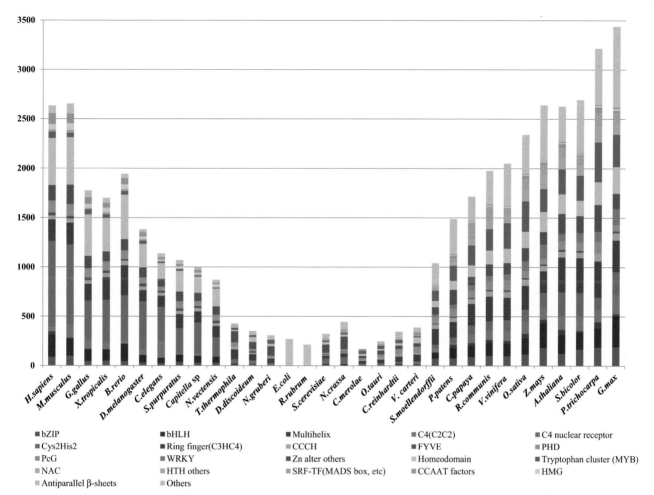

FIGURE 5.2 **A comparison of transcription factor genes in 32 diverse organisms.** Leucine zipper factors, leucine zipper (=bZIP) factor; GRAS, GRAS family; GeBP, GeBP family; bZIP others, bZIP other families; helix–loop–helix factors, basic helix–loop–helix family; TCP (HLH), TCP family; bHLH others, bHLH other families; cell cycle (Myc), Myc family; AP-2, AP-2 family; Nin-like, Nin-like (=RWP-RK) family; CPP (CXC), CPP (=CXC) family; trihelix, trihelix family; SAND (ULT), SAND (=ULT) family; EIL, EIL family; GATA, GATA (=C2C2-GATA) family; Dof, Dof (=C2C2-Dof) family; Co-like, Co-like(=C2C2-Co-like) family; ZIM, ZIM family; YABBY, YABBY (=C2C2-YABBY) family; nuclear receptor, C4 (= C2C2) nuclear receptor family; GRF, GRF family; C4 others, C4 (=C2C2) other families; C2H2, C2H2 family; SWIM, SWIM family; Ring finger (C3HC4), Ring finger (=C3HC4, C3H) family; C3HC, C3HC family; CCCH, CCCH family; TAZ, TAZ family; FYVE, FYVE family; PHD, PHD family; Alfin, Alfin (=Alfin-like) family; PcG, PcG family; CW, CW family; WRKY, WRKY family; RAS, RAS family; UBR1, UBR1 family; UBP, UBP family; TRAF, TRAF family; CHY, CHY family; ZPR1, ZPR1 family; HRT, HRT family; VOZ-9, VOZ-9 (=VOZ) family; Zz, Zz family; AN1-like, AN1-like family; DHHC, DHHC family; jumonji: jumonji (=JUMONJI) family; A20, A20 family; MIZ, MIZ family; BED, BED family; LSD1, LSD1 family; NF-X1, NF-X1 family; PLATZ, PLATZ family; Zn alter others, Zn-altered other families; homeodomain, homeodomain (HB) family; ZF-HD, ZF-HD family; DDT, DDT family; folkhead-like (E2F-DP), folkhead-like (E2F-DP) (=E2F-DP) family; FHA (E2F-DP), FHA (E2F-DP) family; winged helix (E2F-DP), winged helix (E2F-DP) (=E2F-DP) family; HSF (E2F-DP), HSF (E2F-DP) (=HSF) family; LIM, LIM family; POU, POU family; prokaryotic regulator, prokaryotic regulator (=bacterial helix–turn–helix factors) family; HD-ZIP, HD-ZIP family; ETS, ETS family; Myb, Myb + Myb-related family; GARP (ARR-B), GARP (ARR-B) (=ARR-B) family; GARP(G2-like), GARP (G2-like) (=G2-like) family; GARP others, GARP other families; NAC, NAC family; HTH others, helix–turn–helix other families; SRF (MADS), SRF (=MADS) family; CCAAT factors, CCAAT(= CCAAT-Dr1 + CCAAT-Hap2 + CCAAT-Hap3 + CCAAT-Hap5) family; HMG, HMG family; TBP (TFIID), TBP (=TFIID) family; ARID, ARID family; β-scaffold others, β-scaffold other families; AP2/EREBP, AP2/EREBP (AP2 EREBP) family; AUX_IAA, AUX_IAA(=AUX-IAA = AUX/IAA) family; ABI3/VP1, ABI3/VP1 (=ABI3VP1 = ABI3-VP1) family; ARF, ARF family; others (B3), others (B3) family; TUBBY (TLP), TUBBY (TLP) (=TUB = TLP) family; S1Fa-like, S1Fa-like family; BBR/BPC, BBR/BPC (=BBR-BPC) family; BZR(BES), BZR (BES) (=BES1) family; CAMTA, CAMTA family; WHIRLY, WHIRLY (=Whirly) family; GIF, GIF family; AS2, AS2 family; FLO/LEAFY, FLO/LEAFY (=LFY) family; LUG, LUG family; MBF1, MBF1 family; NZZ, NZZ family; SAP, SAP family; SBP, SBP family; PAZ, PAZ family; Sigma70-like, Sigma70-like family; others, other families.

species. Photosynthetic bacteria have similar components to those of nonphotosynthetic bacteria. Therefore, the development of photosynthetic capabilities did not require substantial transcription regulation differences in bacteria. The most significant diversification in TFs occurred during the evolution from prokaryotes to eukaryotes. The evolution of transcription systems from prokaryotes to eukaryotes has resulted in functional changes in general TFs, which has manifested in the loss of DNA-binding ability and the absence of helicase function. In addition,

TABLE 5.1 Number of Sequence Specific Transcription Factor Genes in 32 Organisms

DNA-binding structure	Super family	Gene family	H. sapiens	M. musculus	G. gallus	X. tropicalis	B. rerio	D. melanogaster	C. elegans	S. purpuratus	Capitella sp	N. vectensis	T. thermophila
Basic domain	bZIP	Leucine zipper factors	78	81	40	44	28	24	23	33	25	27	9
		GRAS	0	0	0	0	0	0	0	0	0	0	0
		GeBP	0	0	0	0	0	0	0	0	0	0	0
		bZIP others	10	19	4	7	17	5	4	4	4	4	0
	bHLH	Helix–loop–helix factors	229	174	123	113	177	74	44	68	69	58	0
		TCP (HLH)	0	0	0	0	0	0	0	0	0	0	0
		bHLH others	0	0	0	0	0	2	5	4	0	0	0
	bHLH-ZIP	Cell cycle (Myc)	2	6	4	1	1	0	0	1	1	1	0
		AP-2	11	10	6	5	14	1	4	2	1	1	0
		Nin-like	0	0	0	0	0	0	0	0	0	0	0
	Multihelix	CPP (CXC)	7	1	1	2	1	2	2	2	1	1	13
		Trihelix	0	0	0	0	0	0	0	0	0	0	0
		SAND (ULT)	22	8	3	2	1	2	4	0	1	1	0
		EIL	0	0	0	0	0	0	0	0	0	0	0
Zn-coordinating	C4	GATA	30	29	16	15	22	10	20	9	12	3	0
		Dof	0	0	0	0	0	0	0	0	0	0	0
		Co-like	0	0	0	0	0	0	0	0	0	0	0
		ZIM	0	0	0	0	0	0	0	0	0	0	0
		YABBY	0	0	0	0	0	0	0	0	0	0	0
		Nuclear receptor	137	116	51	48	105	23	178	56	29	18	0
		GRF	9	8	7	4	3	2	1	2	3	3	0
		C4 others	6	95	55	46	10	44	149	0	0	0	1
	C2H2	Cys2His2	728	688	352	379	344	463	164	183	270	166	40
		SWIM	6	7	7	7	4	1	2	19	22	8	2
	Zn alter	Ring finger (C3HC4)	218	222	168	230	305	113	112	142	113	112	101
		C3HC	1	0	0	0	1	0	1	0	0	0	0
		CCCH	39	41	37	40	47	37	31	31	33	29	35
		TaZ	4	2	3	2	4	2	5	2	1	1	0
		FYVE	30	27	28	24	15	27	19	27	23	24	2
		PHD	86	84	65	60	63	43	27	42	27	43	18
		Alfin	0	0	0	0	0	0	0	0	0	0	0
		PcG	36	36	29	36	26	20	16	31	24	19	17
		zf-CW	3	7	3	4	3	0	0	3	3	4	0

A. GENERAL ASPECTS OF PLANT TRANSCRIPTION FACTORS

5.3 A COMPARATIVE ANALYSIS OF TRANSCRIPTION FACTOR GENES IN 32 DIVERSE ORGANISMS

D. discoideum	N. gruberi	E. coli	R. rubrum	S. cerevisiae	N. crassa	C. merolae	O. tauri	C. reinhardtii	V. carteri	S. moellendorffi	P. patens	C. papaya	R. communis	V. vinifera	O. sativa	Z. mays	A. thaliana	S. bicolor	P. trichocarpa	G. max
19	3	0	0	7	5	4	7	14	12	23	38	45	50	48	84	119	75	89	84	110
1	0	0	0	0	0	0	0	0	0	50	38	39	46	47	24	60	32	73	98	73
0	0	0	0	0	0	0	0	0	0	0	0	4	4	3	8	4	16	6	7	9
0	0	0	0	3	1	0	0	1	0	0	0	1	0	0	0	0	0	0	0	0
0	0	0	0	8	12	1	1	5	3	48	100	91	112	110	143	206	162	146	159	260
0	0	0	0	0	0	0	0	0	0	6	6	21	21	19	22	44	24	27	33	53
0	0	0	0	1	1	0	0	0	0	0	0	0	0	0	2	0	6	0	1	0
0	0	0	0	0	0	0	0	0	0	0	0	0	0	0	2	0	4	0	0	0
0	0	0	0	0	0	0	0	0	0	0	0	0	0	0	1	0	1	0	0	0
2	0	0	0	0	0	1	4	14	10	2	8	5	10	8	13	9	14	12	15	9
1	1	0	0	0	0	2	2	2	2	4	6	4	6	6	12	4	8	8	12	5
0	0	0	0	0	0	0	0	0	0	1	25	23	22	18	23	24	36	?	43	?
0	0	0	0	0	0	0	0	0	0	0	0	0	0	0	1	0	2	0	2	0
0	0	0	0	0	0	0	0	0	0	6	2	4	4	4	7	8	6	6	6	15
18	4	0	0	10	6	6	4	7	10	8	10	23	20	19	30	35	32	29	36	50
0	0	0	0	0	0	0	2	1	1	5	0	19	22	26	34	30	36	29	41	65
0	0	0	0	0	0	0	3	1	4	13	5	26	28	30	19	25	31	32	14	39
0	0	0	0	0	0	0	0	0	0	12	0	13	14	15	17	31	15	19	16	29
0	0	0	0	0	0	0	0	0	0	0	0	8	6	7	10	9	6	7	13	13
0	0	0	0	0	0	0	0	0	0	0	0	0	0	0	0	0	0	0	0	0
0	4	0	0	1	2	1	1	3	3	3	2	1	5	3	5	1	7	29	13	5
0	0	1	0	0	0	0	0	0	0	0	0	0	0	1	6	0	14	1	0	0
9	7	0	0	43	60	7	4	5	10	33	55	69	82	70	108	119	130	89	113	179
5	6	0	0	0	2	0	0	0	0	4	0	18	10	38	16	22	110	151	3	47
63	48	0	0	27	36	28	39	54	70	158	157	219	250	228	241	218	352	353	367	322
0	1	0	0	1	0	0	1	1	1	2	1	2	2	4	2	3	2	6	3	5
15	12	0	0	7	10	7	16	19	18	45	44	49	54	59	65	57	59	56	81	74
0	0	0	0	0	0	0	1	2	2	3	5	6	5	4	10	9	9	5	8	1
15	3	0	0	7	6	0	0	1	2	8	11	11	13	15	13	15	19	18	0	18
9	4	0	0	16	17	12	11	32	30	37	43	49	53	54	49	37	56	63	70	64
0	0	0	0	0	0	0	1	1	2	7	3	6	4	13	5	7	6	9	9
16	9	0	0	2	11	8	23	27	30	36	39	30	30	47	28	23	34	25	49	3
0	0	0	0	0	0	1	1	1	2	0	2	5	4	6	6	14	7	7	13

(Continued)

TABLE 5.1 Number of Sequence Specific Transcription Factor Genes in 32 Organisms *(cont.)*

DNA-binding structure	Super family	Gene family	Organisms										
			H. sapiens	*M. musculus*	*G. gallus*	*X. tropicalis*	*B. rerio*	*D. melanogaster*	*C. elegans*	*S. purpuratus*	*Capitella* sp	*N. vectensis*	*T. thermophila*
		WRKY	0	0	0	0	0	0	0	0	0	0	0
		UBR1	5	5	10	6	7	6	8	6	6	3	7
		UBP	12	11	12	10	10	9	5	9	5	7	6
		TRAF	5	7	12	7	5	3	1	7	7	17	7
		CHY	2	0	0	1	2	1	0	1	2	4	3
		ZPR1	1	1	2	0	2	1	1	4	1	2	1
		HRT	0	0	0	0	0	0	0	0	0	0	0
		VOZ-9	0	0	0	0	0	0	0	0	0	0	0
		ZZ	20	18	18	14	14	20	10	21	11	15	4
		AN1-like	8	6	6	5	5	6	4	2	4	5	2
		DHHC	27	23	23	20	22	26	18	17	23	14	35
		Jumonji	12	9	4	4	6	1	1	0	1	1	0
		A20	9	5	5	3	8	2	2	6	3	3	0
		MIZ	12	14	7	5	6	4	4	6	2	2	3
		BED	6	2	1	0	3	5	4	4	4	0	1
		LSD1	0	0	0	0	0	0	0	0	0	0	0
		NF-X1	6	2	2	3	1	3	1	4	3	2	1
		PLATZ	0	0	0	0	0	0	0	0	0	0	0
		Zn alter others	23	85	17	15	14	14	14	5	5	5	41
Helix–turn–helix	HD	Homeodomain	219	244	232	189	234	144	79	123	95	110	1
		zf-HD	0	0	0	0	0	0	0	0	0	0	0
		DDT	9	3	4	3	3	3	2	5	1	3	0
		Folkhead-like (E2FTDP)	76	69	37	52	66	28	27	38	35	30	0
		Winged helix (E2FTDP)	15	15	12	11	11	4	5	7	5	10	7
		HSF (E2FTDP)	9	5	11	3	8	5	2	2	1	3	3
		LIM	67	80	80	46	69	39	29	10	8	6	2
		POU	32	16	15	11	22	8	4	7	6	6	0
		Prokaryotic regulator	0	0	0	0	0	0	0	0	0	0	0
		HD-ZIP	0	0	0	0	0	0	0	0	0	0	0
		ETS	50	49	31	26	37	7	11	17	13	13	0
		Myb	58	43	49	35	37	34	12	18	13	8	22
		GARP (ARR-B)	0	0	0	0	0	0	0	0	0	0	0
		GARP (G2-like)	0	0	0	0	0	0	0	0	0	0	0

D. discoideum	N. gruberi	E. coli	R. rubrum	S. cerevisiae	N. crassa	C. merolae	O. tauri	C. reinhardtii	V. carteri	S. moellendorffii	P. patens	C. papaya	R. communis	V. vinifera	O. sativa	Z. mays	A. thaliana	S. bicolor	P. trichocarpa	G. max
1	0	0	0	0	0	0	2	1	1	19	37	50	58	60	96	141	72	93	101	158
8	0	0	0	2	3	1	0	0	0	4	1	2	2	4	1	1	3	3	2	4
3	3	0	0	4	1	2	3	2	2	5	4	5	6	5	3	2	9	4	9	9
22	1	0	0	0	2	1	4	1	2	3	2	0	1	1	4	7	1	3	1	1
2	3	0	0	2	3	1	1	1	1	3	6	6	8	7	9	4	10	7	9	12
1	1	0	0	1	1	0	0	0	0	1	1	1	1	2	1	1	6	1	1	2
0	0	0	0	0	0	0	0	0	0	0	7	1	1	1	1	1	2	?	1	?
0	0	0	0	0	1	0	0	0	0	0	2	1	1	1	2	1	2	?	4	?
7	11	0	0	2	3	5	5	3	3	5	16	5	9	8	5	4	18	9	5	8
3	2	0	0	2	1	1	0	1	1	6	10	4	8	6	11	4	15	13	0	15
15	4	0	0	7	5	6	10	6	7	14	21	16	23	24	20	23	37	27	41	26
0	0	0	0	1	1	4	5	7	7	5	11	3	6	5	1	10	7	10	20	11
2	3	0	0	0	0	1	0	1	1	3	6	4	4	5	7	5	7	9	10	8
3	1	0	0	2	2	1	1	0	0	2	4	1	2	4	3	3	3	4	3	3
0	0	0	0	1	0	0	0	0	0	0	0	0	0	5	11	4	19	29	59	2
0	0	0	0	0	0	0	0	1	1	3	3	4	6	5	6	7	5	7	14	8
1	1	0	0	0	0	0	0	1	1	2	2	2	2	3	2	3	2	3	3	3
0	0	0	0	0	0	1	1	3	1	10	13	11	11	10	6	11	10	14	20	14
2	66	0	0	59	105	0	6	1	1	2	30	5	11	4	7	13	14	4	43	2
6	11	52	41	8	6	5	6	1	0	27	43	55	64	73	89	117	110	79	127	165
0	0	0	0	0	0	0	0	0	1	7	8	10	11	12	18	12	15	12	23	26
1	0	0	0	2	1	1	0	1	10	4	2	7	5	8	5	7	7	8	7	6
18	10	123	100	5	4	3	7	12	6	9	15	13	22	13	16	9	17	13	18	11
2	3	0	0	0	0	5	3	6	2	4	10	6	6	5	9	19	12	10	10	12
1	0	0	0	5	3	3	1	2	1	7	8	18	20	20	25	30	24	24	31	25
22	1	0	0	4	5	2	1	1	0	5	3	8	10	14	5	9	13	9	13	20
0	0	0	0	0	0	0	0	0	0	0	0	0	0	0	0	0	0	0	0	0
0	0	95	74	0	0	0	0	0	0	0	0	0	0	0	0	0	0	0	0	0
0	0	1	0	0	0	0	0	0	0	0	1	2	2	1	0	1	0	1	2	7
0	0	0	0	0	0	0	0	0	0	0	0	0	0	0	0	0	0	0	0	0
18	41	0	0	15	18	33	27	26	30	100	107	201	217	259	253	229	203	254	306	327
0	0	0	0	0	0	0	1	1	1	0	5	0	0	0	8	0	10	0	15	0
0	0	0	0	0	0	0	2	3	2	0	41	0	0	0	46	0	43	0	66	0

(Continued)

A. GENERAL ASPECTS OF PLANT TRANSCRIPTION FACTORS

TABLE 5.1 Number of Sequence Specific Transcription Factor Genes in 32 Organisms (cont.)

DNA-binding structure	Super family	Gene family	H. sapiens	M. musculus	G. gallus	X. tropicalis	B. rerio	D. melanogaster	C. elegans	S. purpuratus	Capitella sp	N. vectensis	T. thermophila
		Others	6	0	0	0		0	0	0	0	0	0
	NAM	NAM	0	0	0	0	0	0	0	0	0	0	0
	Others	HTH others	0	7	0	6	0	0	0	0	0	0	0
β-scaffold		SRF-TF (MADS box etc)	11	12	5	4	11	5	3	4	4	3	1
		CCAAT factors	9	8	7	9	5	8	7	1	2	2	1
		HMG	56	71	44	49	54	30	15	25	21	30	11
		TBP (TFIID)	3	4	6	3	3	6	2	1	2	1	2
		ARID	13	16	16	12	19	6	4	7	7	6	2
		β-scaffold others	94	76	45	30	39	31	18	14	19	8	2
Others	Antiparallel β-sheets	AP2/EREBP	0	0	0	0	0	0	0	0	0	0	0
		AUX_IAA (B3)	0	0	0	0	0	0	0	0	0	0	0
		ABI3/VP1 (B3)	0	0	0	0	0	0	0	0	0	0	0
		ARF (B3)	0	0	0	0	0	0	0	0	0	0	0
		Others (B3)	0	0	0	0	0	0	0	0	0	0	0
		TUBBY (TLP)	10	9	5	4	6	2	2	4	2	2	2
		S1Fa-like	0	0	0	0	0	0	0	0	0	0	0
		BBR/BPC	0	0	0	0	0	0	0	0	0	0	0
		BZR (BES)	0	0	0	0	0	0	0	0	0	0	0
		CAMTA	6	6	0	4	1	1	1	1	0	0	0
		WHIRLY	0	0	0	0	0	0	0	0	0	0	0
		GIF	0	0	0	0	0	0	0	0	0	0	0
		AS2	0	0	0	0	0	0	0	0	0	0	0
		FLO/LEAFY	0	0	0	0	0	0	0	0	0	0	0
		LUG	0	0	0	0	0	0	0	0	0	0	0
		MBF1	8	1	1	3	1	4	1	2	6	1	1
		NZZ	0	0	0	0	0	0	0	0	0	0	0
		SAP	38	24	20	22	17	11	7	11	10	14	1
		SBP	0	0	0	0	0	0	0	0	0	0	0
		PAZ	6	8	9	9	4	10	27	10	6	6	10
		Others	10	56	34	12	15	1	0	14	5	7	10

5.3 A COMPARATIVE ANALYSIS OF TRANSCRIPTION FACTOR GENES IN 32 DIVERSE ORGANISMS

D. discoideum	N. gruberi	E. coli	R. rubrum	S. cerevisiae	N. crassa	C. merolae	O. tauri	C. reinhardtii	V. carteri	S. moellendorffii	P. patens	C. papaya	R. communis	V. vinifera	O. sativa	Z. mays	A. thaliana	S. bicolor	P. trichocarpa	G. max
0	0	0	0	0	0	1	1	2	2	4	2	0	0	0	0	0	0	0	15	0
0	0	0	0	0	2	0	0	0	0	20	32	68	92	81	123	149	113	112	167	109
0	0	0	0	5	4	4	4	2	2	0	21	4	92	7	0	6	0	11	17	20
4	2	0	0	4	2	1	1	2	2	17	22	149	37	59	105	92	106	77	111	90
1	1	0	0	4	9	6	9	8	9	1	27	5	6	5	31	24	34	9	59	21
4	17	0	0	7	6	0	5	7	6	8	9	11	12	12	11	11	10	9	16	20
1	1	0	0	1	1	1	1	1	1	2	1	1	3	1	4	4	4	4	2	5
2	1	0	0	1	2	4	1	2	3	5	7	5	9	4	3	1	7	3	8	8
0	0	0	0	2	8	0	0	0	0	0	0	0	0	0	1	0	1	0	0	0
0	0	0	0	0	0	0	10	12	16	60	147	95	116	128	181	253	145	157	211	191
0	0	0	0	0	0	0	0	0	0	4	2	18	19	23	38	37	29	24	33	36
0	0	0	0	0	0	0	0	1	1	26	30	42	56	43	13	88	20	87	66	101
0	0	0	0	0	0	0	0	0	0	7	13	11	18	18	33	34	23	24	37	34
0	0	0	0	0	0	0	0	0	0	0	10	0	0	0	20	0	16	0	20	0
0	1	0	0	0	0	0	1	3	3	6	6	6	7	15	26	22	11	12	11	22
0	0	0	0	0	0	0	0	0	0	0	1	1	1	2	3	2	3	0	2	2
0	0	0	0	0	0	0	0	0	0	1	0	2	4	2	3	4	7	5	16	7
0	0	0	0	0	0	0	0	0	0	5	6	6	7	6	2	7	6	8	12	13
0	0	0	0	0	0	0	0	0	0	6	1	4	4	4	3	9	6	7	7	9
0	0	0	0	0	0	0	1	1	1	3	2	5	4	3	10	5	3	6	3	9
1	0	0	0	0	0	0	0	1	1	3	3	2	2	4	10	2	3	3	5	3
0	0	0	0	0	0	0	0	0	0	15	14	25	28	44	16	34	42	36	49	57
0	0	0	0	0	0	0	0	0	0	2	2	1	1	1	1	1	1	1	1	3
0	0	0	0	0	0	0	0	0	0	0	1	2	1	2	4	1	2	2	5	2
2	0	0	0	5	4	1	0	1	1	1	3	3	23	3	2	1	3	2	6	3
0	0	0	0	0	0	0	0	0	0	0	0	0	0	0	0	0	1	0	0	0
15	8	0	0	5	6	0	9	10	10	9	9	5	8	8	10	6	1	13	1	6
0	0	0	0	0	0	0	0	21	15	11	14	11	15	18	9	21	15	18	29	27
6	5	0	0	2	1	0	0	1	1	12	12	10	13	20	12	12	19	22	18	16
7	8	0	0	33	65	0	4	11	10	54	77	2	13	90	0	23	1	115	36	278

A. GENERAL ASPECTS OF PLANT TRANSCRIPTION FACTORS

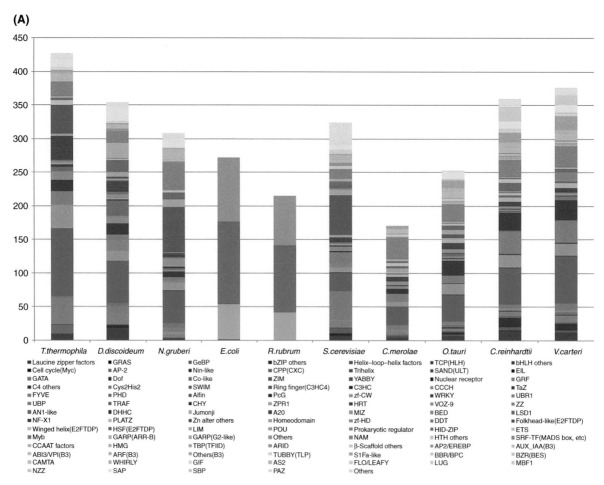

FIGURE 5.3 A comparison of transcription factor genes in 32 diverse organisms (*E. coli* K12-MG1655, *A. thaliana*, *O. sativa*, *C. elegans*, *D. melanogaster*, *H. sapiens* NCBI, *N. crassa* 74-OR23-IVA, *S. cerevisiae* S228C, *M. musculus*, *G. gallus*, *X. tropicalis*, *D. rerio*, *S. purpuratus*, *C. capitata* sp. I, *N. vectensis*, *T. thermophila*, *D. discoideum*, *N. gruberi*, *R. rubrum*, *C. merolae*, *O. tauri*, *C. reinhardtii*, *V. carteri*, *S. moellendorffii*, *P. patens*, *C. papaya*, *R. communis*, *V. vinifera*, *S. bicolor*, *Z. mays* ssp. *mays*, *G. max*, and *P. trichocarpa*). (A) Pattern of prokaryotes, unicellular organisms, and colonized algae.

cell volume has increased more than 10-fold, which is required for packaging naked DNA strands into condensed molecules using nucleosomes. Gene expression control in eukaryotes requires a more complex system because of the increase in the total gene number, multiplication of transcription templates (chromosomes), and the evolution of the monocistronic system from the prokaryotic polycistronic system. Therefore, in eukaryotes, sequence-specific TFs have stronger DNA-binding affinity and more precise specificity than observed in prokaryotes. The original domain – the basic region that binds to the major groove of DNA and supports the HTH structure – has improved DNA-binding specificity and affinity and acquired additional functions, such as enzymatic activity and protein–protein interactions, in eukaryotes. Additionally, the nuclear-binding domains that originated in enzymatic proteins have been customized for transcription regulation. Examples include GreA (transcript cleavage factor) for the basic domain factor and HNH (homing endonuclease) for the antiparallel β-sheet factor (Kulish et al., 2000). Zn finger nuclear-binding domains have also originated in certain species of bacteria (alpha bacterium). These new categories of DNA-binding structures have been identified in general eukaryotic TF functional units (basic domain = TFIIB; bHLH = TFIIE-β; Zn finger = TFIIIA; β-scaffold = TBP). Thus, eukaryotes have evolved by adapting DNA-binding structures for transcription regulation.

Gene regulatory networks are less complicated, and genome sizes are smaller in prokaryotes (bacteria) and unicellular eukaryotes (amoebas, protozoa, and algae) than in higher eukaryotes. Consequently, the number of TF genes (170–440) is 10–25% of that observed in higher eukaryotes, and prokaryotes and unicellular eukaryotes have fewer TF components and fewer species of TF genes (Bouhouche et al., 2000).

The unicellular red alga, *C. merolae*, has few eukaryotic features: no cell wall; a single nucleus; one mitochondrion; one plastid; one microbody (peroxisome); one Golgi apparatus with two cisternae and coated vesicles; one ER; a few lysosome-like structures; and a small volume of cytosol.

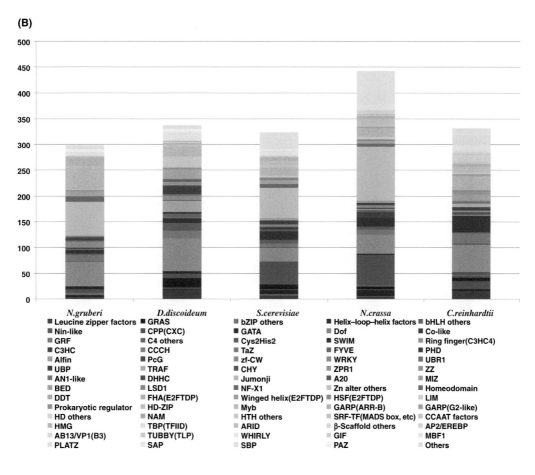

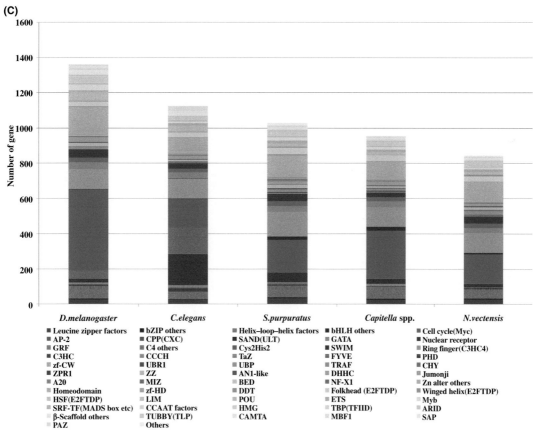

FIGURE 5.3 *(Cont.)* (B) Pattern of yeast and fungi. (C) Pattern of invertebrate animals.

A. GENERAL ASPECTS OF PLANT TRANSCRIPTION FACTORS

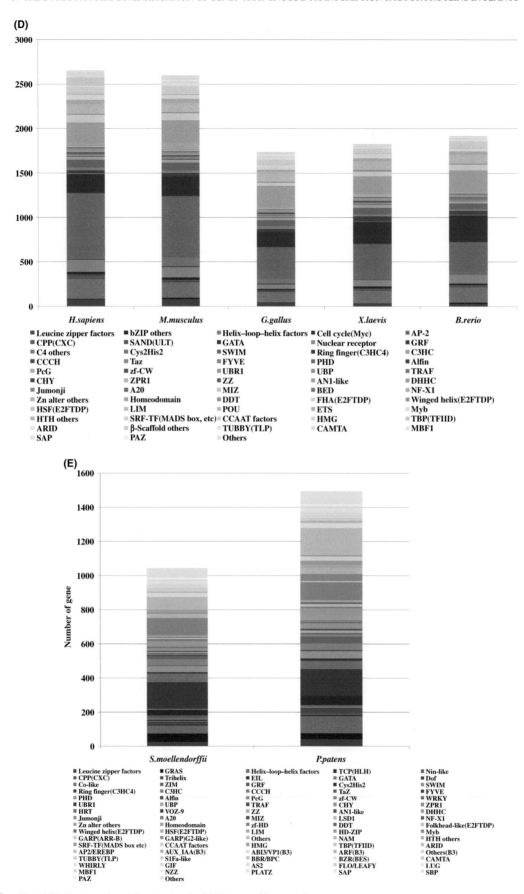

FIGURE 5.3 *(Cont.)* (D) Pattern of vertebrate animals. (E) Pattern of ferns and mosses.

A. GENERAL ASPECTS OF PLANT TRANSCRIPTION FACTORS

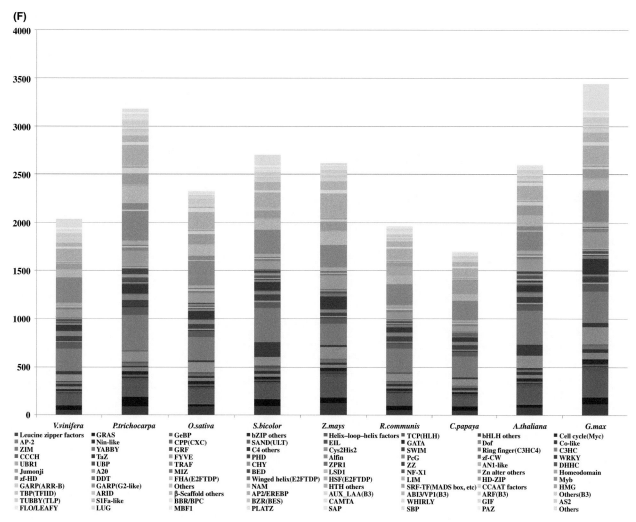

FIGURE 5.3 *(Cont.)* (F) Pattern of seed plants.

C. merolae has evolved the TF components of a typical lower plant. For example, Myb is the most adapted HTH TF gene, and the Ring finger and other altered Zn finger genes are highly diverged in *C. merolae*. New TF genes in *C. merolae* include bZIP, bHLH, Nin-like, CXC, GATA, GRF, C2H2 Zn finger, Ring finger, PHD, PcG, PLATZ and other altered Zn fingers, HSF, LIM, SRF-TF(MADS), CCAAT factor, ARID, and MBF. Notably, the total number of TF genes in *C. merolae* are fewer than those observed in bacteria, which might have resulted because of the efficient division of the eukaryotic transcription regulation system that recycles the common general TF complex unit involved in the regulation of most genes. With the evolution from red algae to green algae and the acquisition of chloroplasts and establishment of complex cellular systems, unicellular plants gradually increased the number of TF gene family components by adding newly adapted families (Dof, Co-like, TAZ, FYVE, Alfin, CW, WRKY, GARP, HMG, AP2/EREBP, TUBBY, WHIRLY, GIF, SAP, SBP, and PAZ).

O. tauri is a phytoplankton that lives in the ocean. *O. tauri* belongs to an early diverging class (Prasinophyceae) of the green plant lineage. The most striking feature of *O. tauri* and related species is their minimal cellular organization: a naked, approximately 1 μm cell; the absence of flagella; a single chloroplast; and a single mitochondrion. *O. tauri* has 1.5 times more TF genes than *C. merolae* because of a gradual increase in TF components and the addition of new families (Dof, Co-like, GARP, HMG, TUBBY, WHIRLY, etc.). Only the Myb family is smaller (10%) than *C. merolae*, and antiparallel β-sheet group factors (AP2/EREBP) emerged at this stage. There are fewer total TF genes in Prasinophyceae than in bacteria. The chlorophytes (*Chlamydomonas* and *Ostreococcus*) diverged from the streptophytes (land plants and their close relatives) over one billion years ago. These lineages are a component of the green plant lineage (Viridiplantae), which previously diverged from opisthokonts (animals, fungi, and Choanozoa).

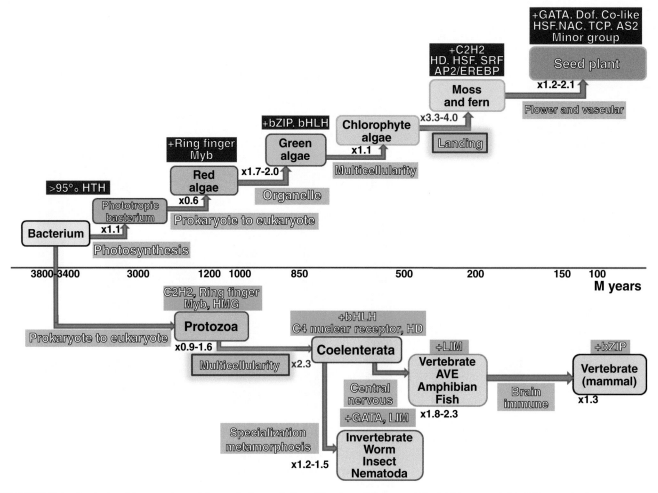

FIGURE 5.4 A relationship overview of the evolution and diversification of TF genes. The fold increase (×) in TF genes is shown.

C. reinhardtii is a unicellular (10 μm) soil-dwelling green alga with multiple mitochondria, two anterior flagella for motility and mating, and a chloroplast that houses the photosynthetic apparatus and critical metabolic pathways. The total number of TF genes is 1.4 times larger than in *O. tauri*. New groups of altered Zn finger genes (FYVE, A20, LSD1, and NF-X1) have emerged, and PHD, Nin-like, and SBP members have increased. The Myb gene families remain fewer than in *C. merolae* and ABI3/VP1, and GIF and PAZ genes are new additions at this stage. The general patterns of TF genes in *C. merolae*, *O. tauri* and *C. reinhardtii* are similar to those of higher plants.

Volvox is composed of 1000–3000 flagellate cells similar to those observed in *Chlamydomonas*: interconnected and arranged in a sphere (coenobium). Although *Volvox* is equivalent to the multicellular complex of *Chlamydomonas*, the total number of TF genes is only 10% higher and there are no new TF families. Therefore, this form of multicellular organization has not caused a substantial change in plants. However, there was a substantial increase in TF genes that accompanied multicellularization in animals (Figure 5.3D). This large difference between animals and plants may be related to their different systems of multicellular structures. Multicellular animal cells are highly specialized and precisely regulated to conserve a limited external energy supply. However, plant cells have retained their own energy production system, and the component cells of plant organs are not as highly regulated as in animals. In plants, the drastic adaptation of new TF genes has occurred with the development and differentiation of the vascular system of terrestrial plants.

TF genes in protozoan animals show different patterns from those of higher eukaryotic animals. *N. gruberi* is an amoeba-flagellate and typically undergoes a cellular differentiation process where it changes from a crawling amoeba to a streamlined swimming flagellate. *N. gruberi* previously possessed organelles identical to those of higher animals but did not have multicellular organism systems (nervous system, digestive organs, body structure). After evolving from bacteria, unicellular animals acquired many TF genes, including bZIP, CXC, GATA, GIF, C2H2 Zn finger, Ring finger, most of the altered Zn finger families, LIM, Myb, SRF-TF, CCAAT factor, ARID, HMG, TUBBY, SAP, and PAZ. The Myb and altered Zn finger TF genes are the main components of the amoeba-flagellate TF genes. Additionally, *N. gruberi* also has a small ratio of C2H2 Zn finger and homeodomain proteins.

T. thermophila is a free-living ciliate protozoa with nuclear dimorphism. Each cell has two nuclei – the micronucleus (MIC) and macronucleus (MAC) – which contain distinct but closely related genomes. An analysis of the complete MAC genome indicated that it is amplified threefold (104 Mb) and is larger than the nematode genome (94 Mb). The MAC genome is estimated to code for 27,000 proteins, which is equivalent to seed plants and vertebrates. The total number of TF genes is 543 or approximately 1.5 times the number observed in amoeba-flagellates. The general pattern of the TFs in *Tetrahymena* is similar to amoeba-flagellates, but Ring finger and other altered Zn finger genes have increased and Myb has decreased to levels lower than those observed in amoeba. C2H2 Zn finger genes have also increased but not to the levels observed in higher animals. The numbers of C2H2 Zn finger and homeodomain protein genes have not increased in either unicellular animal. The altered Zn-coordinating factor genes comprise >60% of the total TF genes in these unicellular animals.

5.3.2 The Evolutionary Characteristics of Slime Mold, Yeast, and Fungi TF Genes

Some primitive organisms developed unique life systems that have been maintained in other (higher) organisms. These organisms are highly specialized not only in their biological features but also in the composition of their TF genes (Figure 5.3B).

The soil-dwelling social amoeba (*D. discoideum*) prompts solitary cells to aggregate and develop as a true multicellular organism, thus producing a fruiting body comprised of a cellular cellulosic stalk that supports a bolus of spores. Therefore, it has evolved mechanisms that direct how a homogeneous population of cells differentiate into distinct cell types. The total number of TF genes in *D. discoideum* is identical to *C. reinhardtii* and 1.1 times that of *N. gruberi*. Consistent with its biological features, the patterns of TF genes indicate a mixed pattern of animal and plant TF genes. In *D. discoideum*, the majority of altered Zn finger, LIM, and homeodomain protein TF genes correspond with animal genes and fewer correspond with plant genes. The opposite composition was observed for Myb, bZIP, and GATA genes. Orthologs of some plant-specific genes (GRAS, Nin-like, and WRKY) have been detected, but orthologs of animal-specific TF genes have not been detected. Therefore, slime mold TF genes are slightly more similar to plants than animals.

Yeast and fungi have similar numbers of TF genes, which are 1.4 times the number of genes observed in slime mold, and have unique patterns differing from animals and plants. Although their life cycles (generation of 2n and n spans, lengths of multicellular periods) are different, both *N. crassa* (red bread mold) and *S. cerevisiae* (budding yeast) belong to Ascomycota. *N. crassa* has an approximately three times larger genome, 1.5 times the total genes, and 1.5 times the TF genes of *S. cerevisiae*, but the patterns of the TF genes are similar. Both organisms have a large proportion of C2H2 Zn finger genes and have evolved other unique Zn finger genes, including those observed in higher organisms. These organisms also have many Myb genes, and the number of bHLH genes is similar to that observed in higher eukaryotes. There are no plant-unique TFs (Dof, Co-like, TAZ, Alfin, CW, WRKY, GARP, AP2/EREBP, TUBBY, WHIRLY, GIF, and SBP), as observed in unicellular plants, and no animal-specific TFs, but there are yeast- or fungi-specific TFs. Statistical comparison of DNA-binding structures shows that most species (>60%) have adapted Zn-coordinating factors. Based on these patterns, yeast and fungi appear to have adapted the TF gene contents of animals rather than plants. Yeast, fungi, and *D. discoideum* not only contain primitive eukaryote TFs but also have more organism-specific genes. These organisms have continuously evolved a transcription regulation system through competition with higher organisms and/or the horizontal import of genes from other species.

5.3.3 The Emergence of Multicellular Organisms and the Evolution of Invertebrates (Sea Anemones, Nematodes, Segmental Worms, Sea Urchins, and Flies)

The evolution of unicellular organisms to multicellular organisms requires cellular regulatory systems. The development of a complex multicellular body and system integration that respond to environmental variables are new adaptations (Figure 5.3C). Therefore, in addition to an increase in the total gene numbers, the genes that control development (homeodomain protein, etc.) and hormonal receptors have increased.

The organisms in this group have approximately 800–1400 TF genes, which is one-third to one-half the number observed in higher animals. These organisms have nearly all of the TF gene species but fewer members within the gene families, particularly for certain TF genes (C2H2 Zn finger, C4 nuclear receptor, homeodomain protein, Ring finger, etc.) which have increased considerably in higher animals. Compared to unicellular organisms, the total number of TF genes has more than doubled, and the relative proportion of the gene families has also changed substantially. The bHLH genes have increased considerably, C2H2 Zn finger genes have increased, and the homeodomain protein numbers have remained at the levels observed in bacteria. The general patterns of gene groups are similar, even in the absence of increased homeodomain protein and C2H2 Zn finger genes. However, the C4-type Zn finger genes (ligand receptor, etc.) show greater divergence in nematodes. Coincident with the increasing complexity of these organisms, the bZIP genes have increased gradually, the altered Zn finger genes (TAZ, FYVE, CW, etc.) have expanded or evolved new members, the altered homeodomain proteins (folk head, LIM, POU), other HTH (ETS) and HMG genes have expanded, and newly adapted CAMTA genes have evolved. By this stage of evolution, more than 90% of the TF gene families observed in higher vertebrates had emerged.

N. vectensis (starlet sea anemone) constitutes the oldest eumetazoan phylum, the Cnidaria, and has simplified features of higher animals such as flagellated sperm, development through gastrulation, multiple germ layers, true epithelia covering a basement membrane, a lined gut (enteron), a neuromuscular system, multiple sensory systems, and a fixed body axis. Compared to unicellular amoebas, *N. vectensis* has a 5- to 10-fold larger genome and 2–3 times the number of TF genes. The expansion of C2H2 Zn finger and homeodomain protein genes and the emergence of the bHLH family changed the distribution pattern within the total TF genes. The Myc, AP2, SAND(ULT), C4 nuclear receptor, GIF, CW, Jumonji, A20, DDT, POU, and ETS genes evolved from the initial stages of multicellular organisms. However, the CXC, DHHC, and Myb genes have apparently decreased.

Capitella sp. *I* (segmented worm) is in the phylum Annelida and superphylum Lophotrochozoa. This species is a small benthic marine worm that has a segmented body, centralized nervous system, displays continuous adult growth by adding body segments from a posterior growth zone, regenerative abilities, a holoblastic spiral cleavage program, and an indirect life cycle. The genome size of *Capitella* sp. *I* is one-half and the TF gene number is 1.1 times that of a sea anemone. Most of the increase in TF genes occurred from the expansion of C2H2 Zn finger genes. bHLH, GATA, C4 nuclear receptor, SWIM, and DHHC genes have also increased, while homeodomain protein, PHD, TRAF, Zz, HMG, and SAP genes have decreased slightly.

S. purpuratus (purple sea urchin) has a radial adult body, an endoskeleton, aqueous vascular system, and nonadaptive immune system. Compared to sea anemones, the genome is 1.8 times larger and the number of TF genes is 1.2 times larger. The general pattern of TF gene numbers is similar to that of sea anemones and segmented worms. Increased amounts of bHLH, homeodomain protein, C2H2 Zn finger, C4 nuclear receptor, Ring finger, LIM, and ETS genes are the main differences between sea anemones and urchins. The sea urchin has more divergently altered Zn-coordinating factor, C4-type Zn finger, and homeodomain protein genes, and segmented worms have more C2H2 Zn finger genes.

C. elegans (a free-living nematode) is an unsegmented, vermiform, and bilaterally symmetrical roundworm with a cuticle integument, four main epidermal cords and a fluid-filled pseudocoelomate cavity. *C. elegans* has a mouth, pharynx, intestine, gonads, and collagenous cuticle. It has a small genome (94 Mb) and approximately the identical number of TF genes as sea anemones. The nematode genes for the C4 nuclear receptor, GATA, LIM, and PAZ are particularly expanded, whereas bHLH, homeodomain protein, and HMG genes are lower.

D. melanogaster (fruit fly) has a small genome (120 Mb) but approximately 1.5 times the number of TFs compared to *N. vectensis*. The C2H2 Zn finger gene family has expanded and bHLH, homeodomain protein, LIM, Myb, and CCAAT factor genes have gradually increased compared to *N. vectensis*. The general pattern of the TFs of *Drosophila*, particularly the large ratio of C2H2 Zn finger genes, is similar to that of mammals.

5.3.4 The Evolution of Vertebrates (Fish, Amphibians, Birds, Mice, and Humans)

In the animal kingdom, invertebrates and vertebrates evolved completely different tactics for survival. Invertebrates have simplified and shortened life styles and life spans, whereas vertebrates have complicated and longer life styles and life spans which are accompanied by additional developmental and signal transduction systems and an increase in regulatory genes (Figure 5.3D).

The acquisition of a central nervous system made it possible to have more variety in the complexity of the organs and in response to environmental stimuli. Vertebrates have 1.5–3 times the number of TF genes (approximately 1800–2600) compared with multicellular invertebrates (Figures 5.3C,D). The general patterns of vertebrate TF gene family compositions are similar to higher invertebrate multicellular organisms (flies, sea urchins, etc.). Many of the TF genes have increased 1.1- to 1.3-fold, and C2H2 Zn finger, Ring finger, C4 nuclear receptor, bHLH, and homeodomain protein genes have expanded in vertebrates.

The general patterns of TF genes among vertebrates are highly similar to most of the differences caused by the degree of expansion of the C2H2 Zn finger, C4 nuclear receptor-type Zn finger, bHLH, bZIP, Ring finger, and homeodomain protein gene families. The expansion of the C2H2 Zn finger gene family is the main reason for the differences in total gene numbers among vertebrates and the primary difference between animal and plant TF genes. Among vertebrates, fish have 1.2 times the genes compared to amphibians (frogs), frogs have 1.2 times the genes of avian species (chickens), and mammals (approximately 2500–2600 TF genes) have approximately 1.3–1.4 times more genes than the other orders (1700–1900). The many genes in fish may have occurred by total gene duplication after divergence from other vertebrates (Kasahara et al., 2007). Chickens (Aves) have a compressed genome structure, and its gene regulatory systems might have similar characteristics. The large proportion of Zn-coordinating TFs is a feature of mammals. Apparently, mammals have also increased the number of basic DNA-binding structure families (bHLH, bZIP). This large ratio of non-Zn finger-type TF genes is unique to vertebrates.

If animals represent the most advanced evolutionary state, then the most recently evolved transcription regulation system contains approximately 2500–2700 TF genes.

5.3.5 The Evolution of Terrestrial Plants (Ferns and Mosses)

The red alga (*C. merolae*) and the moss (*P. patens*) are the most primitive land plants. These organisms have a simple vascular system, leaf-like tissues, and sporangia. Therefore, these organisms previously acquired the basic systems of homeostasis and reproduction. Although both organisms are not adapted to survive in drought conditions, they inhabit large areas of land. Their evolution required the development of a vascular transport system, durable body structures, DNA repair functions, and signal transduction capabilities. Compared to mosses, ferns are more similar to seed plants in their body structure, tissue complexity, and reproduction system. The hormonal signal transduction system is particularly more evolved in ferns. Auxin, ABA, and cytokinin systems are present in mosses and ferns, but gibberellin (GA), jasmonic acid (JA), and brassinosteroid (BR) systems are only present in ferns. The development of multicellular terrestrial plants was accompanied by an expansion of TF genes (Figure 5.3E). Ferns and mosses have approximately 1000–1350 TF genes, respectively, or one-third to one-half the number observed in seed plants. Ferns and mosses have nearly all the species of TF genes observed in seed plants but fewer components, and some specific TF genes may not have increased.

Multicellular land plants have approximately 3–4 times the number of TF genes observed in aquatic unicellular organisms. Ferns and mosses have similar ratios of TF genes that show pattern continuity with seed plants. Compared to aquatic unicellular plants, mosses and ferns have larger proportions of antiparallel β-sheet family (AP2/EREBP, AUX_IAA, ARF, etc.), Myb family, and basic domain family (bZIP, GRAS, bHLH) genes. The ratio of Zn-coordinating factor genes has decreased. New families (GRAS, EIL, ZIM, ZF-HD, NAC, AUX_IAA, ARF, BBR/BPC, BZR, CAMTA, AS2, and FLO/LEAFY) have been added and present TF genes (bHLH, WRKY, and Myb) have expanded. Detailed comparisons have shown differences between ferns and mosses. Mosses have 1.3 times more TF genes and large proportions of AP2/EREBP, WRKY, and bHLH genes. The Myb, Ring finger, and GRAS genes are more diverged in ferns. HD-ZIP, S1Fa-like, and LUG genes are absent in ferns, and mosses have no Dof, ZIM, SWIM, CW, or BBR/BPC genes. Although the total number of TF genes is smaller in ferns, Zn-coordinating factor genes are more divergent than in mosses.

5.3.6 The Evolution of Seed Plants (Woody Plants, Dicots, and Monocots)

Land plant development required adaptation to conditions of limited water. In seed plants, the flowering tissues developed to protect gametes from drought stress. Therefore, the angiosperms have more specifically differentiated flowering tissues and systems to control flowering (Figure 5.3F). Seed plants have 1.2–3.3 times more TF genes (1600–3500) than ferns and mosses. All the remaining TF gene species (GeBP, YABBY) have emerged in seed plants, and many TF genes (bZIP, GRAS, bHLH, SWIM, Ring finger, WRKY, BED, HD, HSF, Myb, NAC, SRF-TF, HMG, CCAAT factor, AP2/EREBP, AUX_IAA, ARF, TUBBY, and AS2) have gradually increased or expanded. Although the total number of TF genes varies by more than twofold, the general patterns of TF genes are similar among seed plants. Detailed analyses have been completed for woody plant, dicot, and monocot groups.

5.3.6.1 Woody Plants

Grapes (*V. vinifera*) belong to the order Vitales and family Vitaceae. The poplar (*P. trichocarpa*) belongs to the order Malpighiales and family Salicaceae. Although their genome size (poplar is nine times larger than grape), total gene number (poplar is 1.5 times larger than grape) and TF gene number (poplar is 1.3 times larger than grape) are different, and the phylogenic distance is not close, the patterns of TF genes are similar, which may be because of their common body structure (woody-type plant). Therefore, we analyzed these two plants as representatives of the woody plant group. The GRAS, BED, Myb, AS2, and SBP TF genes were slightly more divergent in woody plants than in other seed plants. The C2H2 Zn finger, SWIM, Ring finger, and WRKY genes had fewer members than other seed plants. Therefore, non-Zn-coordinating TF genes have diverged considerably in woody plants. There was a large difference in the total numbers (2000 and 3200) of TF genes in grapes and poplars. Grapes had higher divergence rates for Myb and altered Zn finger genes than poplars. However, poplars had higher divergence rates for GRAS, NAC, AP2/EREBP, and AUX_IAA genes.

5.3.6.2 Dicotyledonous Plants

The papaya (*C. papaya*) belongs to the order Brassicales and family Caricaceae. The castor bean (*R. communis*) belongs to the order Malpighiales, family Euphorbiaceae, and tribe Acalypheae. *Arabidopsis* (mouse-ear cress) belongs to the order Brassicales and family Brassicaceae. The soybean (*G. max*) belongs to the order Fabales and family Fabaceae. The bHLH, C2H2 Zn finger, SWIM, Ring finger, and homeodomain protein genes are more divergent in dicots than in other seed plants. The NAC and AP2/EREBP genes are reduced compared to other seed plants. Therefore, dicots diverged more Zn-coordinating factor genes than other seed plants. There is a wide range (1600–3500) in the total number of TF genes in these dicots. Among these dicots, papaya has a large proportion of SRF-TF (MADS) genes, castor bean has more NAC genes, *Arabidopsis* has more Ring finger genes and soybean has more GRAS, bHLH, C2H2 Zn finger, WRKY, homeodomain protein, Myb, AP2/EREBP, and AUX_IAA genes.

5.3.6.3 Monocotyledonous Plants

Rice (*Oryza sativa*), maize (*Zea mays*), and sorghum (*S. bicolor*) all belong to the order Poales and family Poaceae. Therefore, the common pattern for the Poaceae family has been more highly analyzed than other groups of plants. The bZIP, WRKY, NAC, AP2/EREBP, and ABI3/VP1 TF genes are slightly more divergent in monocots than other seed plants. The Ring finger, homeodomain protein, SRF-TF, and AS2 genes are less abundant in monocots than other seed plants. Among these three monocots, rice has more Myb and SRF-TF (MADS) genes, maize has more bZIP, bHLH, WRKY, AP2/EREBP, and AUX_IAA genes, and sorghum has more GRAS and Ring finger genes.

There are many small differences among seed plants, but their TF gene patterns are similar (Figure 5.3F). Notably, monocots and dicots are highly similar in their TF gene patterns. Therefore, physiological differences between the two groups may be associated with a few TF genes.

5.4 THE APPEARANCE OF NEW TF GENE MEMBERS DURING EVOLUTION

The total number of TF genes and the diversity of TF species change with the evolution of the organism. Therefore, the relationship between the evolution and emergence of new TFs can be overviewed by means of comparison analysis of the statistical data (Figure 5.4).

Large evolutionary changes in TF genes can be observed in the transition from prokaryote to eukaryote. In bacteria, most of the TF genes (>95%) belong to the group containing HTH structures. In addition, bacteria have a few prototypes of eukaryotic TF genes (σ-factor bears similarity to TFA, Zur has the Zn(II)-binding domain, PhoB has an additional β-sheet α-helix complex, Lac repressor also has a bZIP-like domain, etc.). Bacterial HTH TFs are the origin of homeodomain proteins and similar species (Madan Babu and Teichmann, 2003). Unicellular animals have TF genes for bZIP, GATA, C2H2 Zn finger, altered Zn-coordinating factors (CCCH, PHD, UBP, etc.), LIM, SRT TF, CCAAT factor, and TUBBY genes, whereas many Ring finger, Myb, and HMG genes are present. Multicellular aquatic animals have newly adapted bHLH, Myc, SAND, C4 nuclear receptor, Taz, Jmj, DDT, POU, and ETS genes and expanded the number of bHLH, bZIP, C4 nuclear receptor, C2H2 Zn finger, and homeodomain factor genes. A few genes (CW, C3HC) initially appeared in vertebrate animals, and many Zn-coordinating factor (C2H2 Zn finger, C4 nuclear receptor, etc.), bHLH, homeodomain protein, and HMG genes have increased in number.

In plants, however, unicellular aquatic organisms have all the TF gene species present in unicellular animals in addition to bHLH, Nin-like, Dof, Co-like, WRKY, AP2/EREBP, and WHILRY genes. Myb, Ring finger, Nin-like, and PHD gene families have expanded in unicellular aquatic plants. Multicellular aquatic organisms do not have new TF genes but have similar total amounts. Multicellular terrestrial organisms have many new TF genes (TCP, EIL, SWIM, NAM, ZF-HD, AUX_IAA, ARF, BBR/BPC, BZR, CAMTA, AS2, and FLO). The GRAS, bHLH, WRKY, homeodomain protein, Myb, NAM, SRF-TF, and AP2/EREBP gene families have expanded in both ferns and mosses. A few TF genes (GeBP, YABBY, Myc, AP-2, and SAND) are new in seed plants, and various TF genes (bZIP, bHLH, C2H2 Zn finger, Myb) have significantly increased in number.

In animals, the largest emergence of new TF genes occurred during the development of multicellular organisms. Following that evolutionary stage, no new genes have diverged, instead, the present genes have expanded, as observed for HTH, Zn finger, HMG, basic domain factor, and other genes at the vertebrate and mammal stages. Despite several evolutionary steps, no new TF gene types occurred, whereas the number of previously present gene types increased. This may well be the reason for optimization of the C2H2 and Zn finger types in animal gene regulation. In plants, the largest diversification occurred during the development of terrestrial plants. During the formation of seed plants, new genes continued to evolve and present genes also increased. There was also a difference in the timing of divergence periods between animals and plants. The most recent and most drastic emergence of new TF genes occurred in animals approximately 800 million years ago, while TF expansion occurred 500–300 million years ago.

In animals, the emergence of new structural genes occurred early, and small modifications of the structural domains were adapted more recently (~680 million years ago). In plants, however, the largest change occurred during the development of multicellular land plants approximately 400 million years ago, but newly categorized gene families have also continuously emerged in plants. All categories of DNA-binding genes have continued to expand in seed plants. Therefore, there are differences in the evolutionary time period and duration of TF gene diversification between animals and plants. The period containing the largest diversification of plant TF genes is 300 million years later than that for animals, and plant TF genes have continuously adapted more divergent changes throughout their evolution.

5.5 THE DIFFERENT EVOLUTIONARY METHODS OF TF GENES IN ANIMALS AND PLANTS

The relative proportions of genes that encode different DNA-binding structures have changed during evolution (Figure 5.5). These differences might have depended on the evolutionary relationship of the organism and gene regulation systems. However, the structure of the helix-turn-helix (HTH) gene is shorter and simpler than the eukaryotic HTH gene in both photosynthetic and non-photosynthetic bacteria, and HTH (bacterial-type) is the predominant TF gene, comprising >95% of all TF genes. Lower eukaryotes have evolved a large proportion of Zn-coordinating factor genes in lower animals, which have

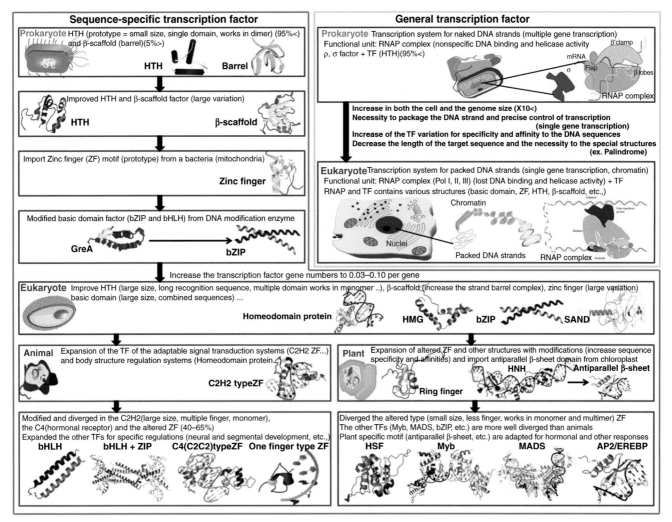

FIGURE 5.5 **Hypothetical scheme for depicting the evolution of transcription factors categorized according to their functional domains for DNA binding.** HTH, helix–turn–helix; barrel, barrel structure domain; β-scaffold, β-scaffold factor; GreA, GreA transcription elongation factor; homeo box, homeodomain protein; HMG, high-mobility gene; bZIP, leucine zipper; multihelix, multihelix structure; C2H2-type ZF, Cys2His2-type Zn finger; C3H4-type ZF, C2H4-type Zn finger (Ring finger); HNH, HNH family of endonuclease (homing endonuclease); bHLH, basic helix–loop–helix; bHLH+ ZIP, basic helix–loop–helix + leucine zipper; C4-type ZF, C4(C2C2)-type Zn finger; one finger-type ZF, one finger-type Zn finger; HSF, heat shock factor; Myb, Myb; MADS, SRF (MADS) transcription factor; antiparallel β-sheet, antiparallel β-sheet factor.

60–70% ZF and 20–30% HTH genes, and plants, which have 50–60% ZF and 30–40% HTH genes.

Most ZF genes have been maintained throughout animal evolution although the HTH family gene numbers increased and the adaptation of basic domain TF genes occurred during the development of multicellular organisms. Higher animals have TF genes for the basic domain, the β-scaffold factor, and other new structures; however, their total proportion is less than 15% and most are ZF and HTH genes. In mammals, the expansion of the C2H2 family of Zn finger genes has considerably increased the total number of Zn-coordinating factor genes.

In plants, the proportion of ZF genes declined from the 60% of lower plants to 30% because of an increase in non-ZF genes during terrestrial plant evolution. During this time the ratio of ZF genes decreased to less than 30% from 50–60% of total TF genes, HTH genes remained at 30%, basic domain genes comprised 15%, β-scaffold factor genes were <10%, and antiparallel β-sheet structure genes comprised approximately 15%. Basic domain factors are relatively few in ferns and mosses. After the development of land plants, the relative proportions of different TF gene families have not changed significantly although the number of family members has gradually increased. Therefore, each type of TF has increased accordingly. Various TF gene families have expanded in seed plants: papaya (SRF-TF = β-scaffold); castor bean (MBF1 = others); grape (Myb = HTH); rice (Myb = HTH, SRF-TF = β-scaffold); maize (bZIP and bHLH = basic domain, WRKY = Zn-coordinating, AP2/EREBP, and AUX_IAA = antiparallel β-sheet); *Arabidopsis* (Ring finger = Zn-coordinating); sorghum (GRAS = basic domain, Ring finger); poplar (GRAS, NAM, AP2/EREBP, and AUX_IAA), and soybean (GRAS, WRKY, homeodomain, Myb, AP2/EREBP, and AUX_IAA). Therefore, various forms of DNA-binding structures have accommodated the increasingly complex regulatory networks observed in higher plants.

5.6 TF GENE EVOLUTION AND ITS BIOLOGICAL FUNCTION

In Figure 5.6 we illustrate our proposed evolutionary relationships for TF gene families. The similarities and differences of these families are discussed relative to the evolution of animals and plants. In animals, Zn-coordinating factor genes contribute most of the diversity among the TF genes because C2H2-type Zn finger genes continually changed throughout evolution. During the evolution from prokaryote to single-celled eukaryote, completely new species of DNA-binding structures have emerged and are represented by at least a few members. During this stage, several gene families (e.g., C2H2 Zn finger, Myb, HMG, and altered Zn fingers) occur. During the change from single to multiple-celled animals, the TF gene ratios changed most drastically, the C2H2 Zn finger and HMG families expanded further, and many new bHLH and C4 nuclear receptor Zn finger genes were added. Homeodomain factors also comprised many of the total genes at this stage. Therefore, we propose that the newly adapted groups might have regulated cell differentiation, signal transduction, and body formation. Originating from coelenterates, the animal kingdom significantly bifurcated into vertebrates and invertebrates. Vertebrates have a spinal cord in the central nervous system and actively respond to external stimuli. The complexity of body structure, life span, and habitat has increased for vertebrates. Altered Zn finger and LIM families were added in this evolutionary phase. Mammals, as the most developed group of vertebrates, have many C2H2 Zn finger and divergent bZIP genes. The diversification of the leucine zipper protein is unique in the animal kingdom and might regulate epidermal cell (= nerve tissue) development. Invertebrate groups have evolved passive tactics to respond to their environment. The invertebrate life cycle may be highly preprogrammed and less flexible and less capable of independent responses. Reproduction is the top priority; therefore, the life span is shorter, the habitat is smaller, and the body structure is less complex than for vertebrates. Invertebrate evolution was accompanied by a divergence of Zn cocoordinating factors (GATA, C4 nuclear receptor, and C2H2) and HTH (LIM and Myb). Compared to vertebrate evolution, bHLH, HMG, and homeodomain protein genes are less divergent, while Myb genes are more divergent. The small proportion of basic domain factor (bZIP and bHLH) and homeodomain protein TF genes is the primary difference between invertebrates and vertebrates. The genes might be involved in the regulatory networks of sensory and nervous systems and complex body development. During the evolution of higher seed plants, the TF gene families in plants, in contrast, have continuously diverged in DNA-binding species and the number of genes within groups. Similar to lower animals, the evolution of lower plants was accompanied by the emergence of all species of DNA-binding structures represented by at least a few member genes. Many Myb and altered Zn finger genes emerged at this stage. Differing from animals, the C2H2 Zn finger and HMG gene numbers are low, and the proportion of basic domain factor genes is large in early unicellular plants. Organelle development was accompanied by an increase in basic domain factor (bZIP and bHLH), Myb, and altered Zn finger genes. As land plants developed, many new groups of TF genes appeared (C2H2 Zn finger, altered Zn finger, homeodomain factor, HSF, SRF-TF, AP2/EREBP). For hormonal signal transduction, plants diverged AP2/EREBP instead of C4 nuclear receptors similar to animals and have SRF-TF genes for body form instead of animal homeodomain factor genes. HMG genes

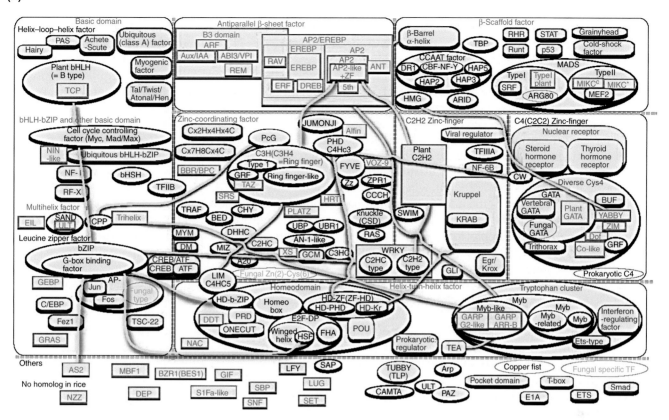

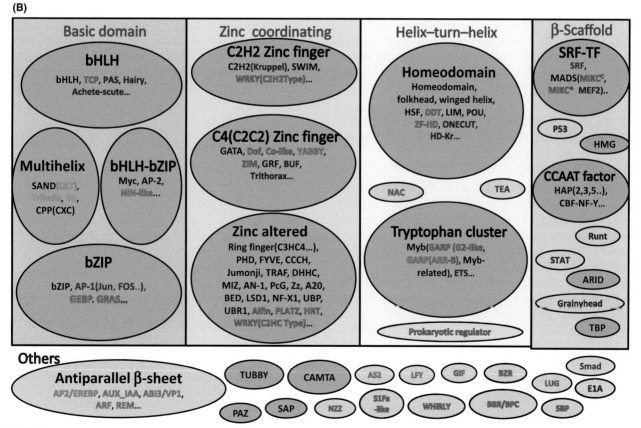

FIGURE 5.6 **A relationship map of the main transcription factors (TFs) categorized according to DNA-binding domain structure.** The gene family groups are presented in color: black, common among animals and plants; red, animal specific; green, plant specific; yellow, specific to other organisms. Families that have similar DNA-binding structures are connected by a line. (A) Relationship map of the main TFs. (B) Simplified relationship map of main TFs.

A. GENERAL ASPECTS OF PLANT TRANSCRIPTION FACTORS

have been relatively stable throughout all stages of plant evolution. Seed plants have a well-developed vascular system and a variety of flowering tissues. The evolution from ferns and mosses to seed plants added various TF genes (GATA, Dof, Co-like, NAC, HSF, Myb, SRF, TCP, and AS2), many of which were plant specific. Although seed plants diverged in many TF genes, there are similar relative proportions of their members among plants. Compared to mammals, seed plants have similar total numbers of TF genes. Mammals are particularly divergent in Zn-coordinating factors (58% of total), whereas, most plant TF genes are observed in the non-Zn-coordinating DNA-binding domains (HTH, basic domain, β-scaffold, antiparallel β-sheet, and multihelix factor). Zn finger TFs have the following characteristics: (1) control of DNA-binding affinity and target sequence specificity according to the number of fingers; (2) typically function as a monomer; and (3) do not need special structures (palindrome, high GC content, etc.) for target sequences. Therefore, higher animals (humans) have particularly expanded the C2H2-type (multifinger and monomer types) Zn finger genes to more than 700 (>30% of total TFs; Schmidt and Durrett, 2004). Given the characteristics of Zn finger TFs, why do plants not have as many Zn finger genes (or as large a proportion) as humans? There was no C2H2-type Zn finger gene expansion or emergence of C4 nuclear receptor-type Zn finger genes in plants. However, plants have adapted many different types of structures for DNA-binding in their TFs. Plants evolutionarily show greater divergence in major TF structures and additional minor specific domains for TFs.

5.7 CONCLUSION: THE REGULATORY ROLE OF INDIVIDUAL TRANSCRIPTION FACTORS

In animals, one TF group with a DNA-binding domain determines biological function. For example, C2H2 Zn factors and β-scaffold TFs are associated with the regulation of housekeeping (metabolic enzymes, homeostasis) genes; bZIP factors activate the regulation of signal transduction (ligand, hormone, synaptic transduction) genes by phosphorylation; bZIP and bHLH are associated with stress (heat, UV, active oxygen species) response genes; C4 Zn factors (dimer) + cofactor are associated with nuclear receptor (hormone, vitamin) genes; and HTH (homeobox), HLH, bZIP, and ZF genes are associated with development and pattern formation. Tissue-specific gene expression shows the following associations: epidermis (bZIP and bHLH), sense organ (bZIP), nerve (bHLH and HTH homeobox), muscle (bHLH and HTH homeobox), and viscous tissue (bHLH and HTH homeobox). Other associations relate to virus infection (HTH), the immunity system (β-scaffold), and cell growth factors (bZIP, ZF, HTH, and β-scaffold). However, the gene specificity shown by TFs is less obvious in plants, and only a few cases have been reported: SRF (MADS) controls flowering tissue differentiation, AP2/EREBP regulates responses to plant hormones (auxin, gibberellin, ethylene, brassinosteroid, jasmonic acid, etc.) and stress, B3 regulates auxin responses, and NAC regulates both stress responses and development (William et al., 2004; Kofuji et al., 2003; Shigyo et al., 2006; Waltner et al., 2005; Nuruzzaman et al., 2010). A variety of phenomena are regulated by more highly divergent TFs: Myb, Ring finger, bZIP, bHLH, and homeodomain. The adaptation of entire TF structures appears to have occurred equally in plants. Plants also have many lineage-specific gene family expansions, and TF families have had much higher expansion rates in plants than in animals. This larger ratio of multiplied genes may be caused by the higher viability of plants, relative to changed regulations, compared to animals. There is an obvious developmental role and definition in animal organs and tissues. This obvious identification was caused by necessary signal transduction cascades and/or the hierarchy of the regulatory systems of gene expression, tissue differentiation, and development. Therefore, regulatory systems have a tendency to consist of identical gene families to synchronize regulatory evolution. However, plant tissues have high variability and/or differentiation potentials compared to many other organs and tissues. Therefore, selection for TF species might have been less severe. This multiplication resulted in various structures, particularly plant TFs. Plants also evolved new structures and modified DNA-binding domains that responded to hormonal, developmental, and disease resistance signal transductions.

Acknowledgments

The authors acknowledge financial support of the research grant from the Ministry of Agriculture, Forestry, and Fisheries's Rice Genome Project. English proofreading was performed by American Journal Experts.

References

Amoutzias, G.D., Veron, A.S., Weiner, 3rd., J., Robinson-Rechavi, M., Bornberg-Bauer, E., et al., 2007. One billion years of bZIP transcription factor evolution: conservation and change in dimerization and DNA-binding site specificity. Mol. Biol. Evol. 24, 827–835.

Aravind, L., Anantharaman, V., Balaji, S., Babu, M.M., Iyer, L.M., 2005. The many faces of the helix–turn–helix domain: transcription regulation and beyond. FEMS Microbiol. Rev. 29, 231–262.

Banerjee-Basu, S., Baxevanis, A.D., 2001. Molecular evolution of the homeodomain family of transcription factors. Nucleic Acids Res. 29, 3258–3269.

Becker, A., Theissen, G., 2003. The major clades of MADS–box genes and their role in the development and evolution of flowering plants. Mol. Phylogenet. Evol. 29, 464–489.

Bouhouche, N., Syvanen, M., Kado, C.I., 2000. The origin of prokaryotic C2H2 zinc finger regulators. Trends Microbiol. 8, 77–81.

Davuluri, R.V., Sun, H., Palaniswamy, S.K., Matthews, N., Molina, C., et al., 2003. AGRIS: *Arabidopsis* gene regulatory information server, an information resource of *Arabidopsis cis*-regulatory elements and transcription factors. BMC Bioinform. 4, 25.

Eichinger, L., Pachebat, J.A., Glöckner, G., Rajandream, M.A., Sucgang, R., et al., 2005. The genome of the social amoeba *Dictyostelium discoideum*. Nature 35, 43–57.

Griffiths, S., Dunford, R.P., Coupland, G., Laurie, D.A., 2003. The evolution of CONSTANS-like gene families in barley, rice, and *Arabidopsis*. Plant Physiol. 131, 1855–1867.

International Chicken Genome Sequencing Consortium, 2004. Sequence and comparative analysis of the chicken genome provide unique perspectives on vertebrate evolution. Nature 432, 695–716.

Kasahara, M., et al., 2007. The medaka draft genome and insights into vertebrate genome evolution. Nature 447, 714–719.

Kofuji, R., Sumikawa, N., Yamasaki, M., Kondo, K., Ueda, K., et al., 2003. Evolution and divergence of the MADS–box gene family based on genome-wide expression analyses. Mol. Biol. Evol. 20, 1963–1977.

Kulish, D., et al., 2000. The functional role of basic patch, a structural element of *Escherichia coli* transcript cleavage factors GreA and GreB. J. Biol. Chem. 275, 12789–12798.

Madan Babu, M., Teichmann, S.A., 2003. Evolution of transcription factors and the gene regulatory network in *Escherichia coli*. Nucleic Acids Res. 31, 1234–1244.

Magnani, E., Sjolander, K., Hake, S., 2004. From endonucleases to transcription factors: evolution of the AP2 DNA binding domain in plants. Plant Cell 16, 2265–2277.

Merchant, S.S., Prochnik, S.E., Vallon, O., Harris, E.H., Karpowicz, S.J., et al., 2007. The *Chlamydomonas* genome reveals the evolution of key animal and plant functions. Science 318, 245–250.

Ming, R., Hou, S., Feng, Y., Yu, Q., Dionne-Laporte, A., et al., 2008. The draft genome of the transgenic tropical fruit tree papaya (*Carica papaya* Linnaeus). Nature 452, 991–996.

Nuruzzaman, M1., Manimekalai, R., Sharoni, A.M., Satoh, K., Kondoh, H., Ooka, H., Kikuchi, S., 2010. Genome-wide analysis of NAC transcription factor family in rice. Gene 465(1-2): 30–44.

Ooka, H., Satoh, K., Doi, K., Nagata, T., Otomo, Y., et al., 2003. Comprehensive analysis of NAC family genes in *Oryza sativa* and *Arabidopsis thaliana*. DNA Res. 10, 239–247.

Putnam, N.H., Srivastava, M., Hellsten, U., Dirks, B., Chapman, J., et al., 2007. Sea anemone genome reveals ancestral eumetazoan gene repertoire and genomic organization. Science 317, 86–94.

Rensing, S.A., Lang, D., Zimmer, A.D., Terry, A., Salamov, A., et al., 2008. The *Physcomitrella* genome reveals evolutionary insights into the conquest of land by plants. Science 319, 64–69.

Reyes, J.C., Muro-Pastor, M.I., Florencio, F.J., 2004. The GATA family of transcription factors in *Arabidopsis* and rice. Plant Physiol. 134, 1718–1732.

Riechmann, J.L., Heard, J., Martin, G., Reuber, L., Jiang, C.Z., et al., 2000. *Arabidopsis* transcription factors: genome-wide comparative analysis among eukaryotes. Science 290, 2105–2110.

Schmidt, D., Durrett, R., 2004. Adaptive evolution drives the diversification of zinc-finger binding domains. Mol. Biol. Evol. 12, 2326–2339.

Sea Urchin Genome Sequencing Consortium, et al., 2006. The genome of the sea urchin *Strongylocentrotus purpuratus*. Science 314, 941–952.

Shigyo, M., Hasebe, M., Ito, M., 2006. Molecular evolution of the AP2 subfamily. Gene 366, 256–265.

Simionato, E., Ledent, V., Richards, G., Thomas-Chollier, M., Kerner, P., et al., 2007. Origin and diversification of the basic helix–loop–helix gene family in metazoans: insights from comparative genomics. BMC Evol. Biol. 7, 33.

Soullier, S., Jay, P., Poulat, F., Vanacker, J.M., Berta, P., Laudet, V., 1999. Diversification pattern of the HMG and SOX family members during evolution. J. Mol. Evol. 48, 517–527.

Tuskan, G.A., Difazio, S., Jansson, S., Bohlmann, J., Grigoriev, I., et al., 2006. The genome of black cottonwood *Populus trichocarpa* (Torr. & Gray). Science 313, 1596–1604.

Umemura, Y., Ishiduka, T., Yamamoto, R., Esaka, M., 2004. The Dof domain, a zinc finger DNA-binding domain conserved only in higher plants, truly functions as a Cys2/Cys2 Zn finger domain. Plant J. 37, 741–749.

Waltner, J.K., Peterson, F.C., Lytle, B.L., Volkman, B.F., 2005. Structure of the B3 domain from *Arabidopsis thaliana* protein At1g16640. Protein Sci. 14, 2478–2483.

William, D.A., Su, Y., Smith, M.R., Lu, M., Baldwin, D.A., Wagner, D., 2004. Genomic identification of direct target genes of LEAFY. Proc. Natl. Acad. Sci. USA. 101(6): 1775–80.

Xiong, Y., Liu, T., Tian, C., Sun, S., Li, J., Chen, M., 2005. Transcription factors in rice: a genome-wide comparative analysis between monocots and eudicots. Plant Mol. Biol. 59, 191–203.

Yanhui, C., Xiaoyuan, Y., Kun, H., Meihua, L., Jigang, L., et al., 2006. The MYB transcription factor superfamily of *Arabidopsis*: expression analysis and phylogenetic comparison with the rice MYB family. Plant Mol. Biol. 60, 107–124.

SECTION B

EVOLUTION AND STRUCTURE OF DEFINED PLANT TRANSCRIPTION FACTOR FAMILIES

6	Structure and evolution of plant homeobox genes	101
7	Homeodomain–Leucine Zipper transcription factors: structural features of these proteins, unique to plants	113
8	Structure and evolution of plant MADS domain transcription factors	127
9	TCP transcription factors: evolution, structure, and biochemical function	139
10	Structure and evolution of plant GRAS family proteins	153
11	Structure and evolution of WRKY transcription factors	163
12	Structure, function, and evolution of the Dof transcription factor family	183
13	NAC transcription factors: from structure to function in stress-associated networks	199

CHAPTER 6

Structure and Evolution of Plant Homeobox Genes

Ivana L. Viola, Daniel H. Gonzalez

Instituto de Agrobiotecnología del Litoral (CONICET-UNL), Cátedra de Biología Celular y Molecular, Facultad de Bioquímica y Ciencias Biológicas, Universidad Nacional del Litoral, Santa Fe, Argentina

OUTLINE

6.1 Introduction	101
6.2 Structure of the Homeodomain	102
6.3 Specific Contacts with DNA	102
6.4 Plant Homeodomain Families	104
6.4.1 HD-ZIP Superclass	105
6.4.2 The TALE Superclass	105
6.4.3 The PINTOX Class	106
6.4.4 WOX Class	106
6.4.5 DDT Class	107
6.4.6 PLINC Zn Finger Class	107
6.4.7 PHD Finger Homeodomain Family	107
6.4.8 NDX Class	108
6.4.9 SAWADEE Class	108
6.4.10 LD Class	108
6.5 The Evolution of Plant Homeobox Genes	108
References	110

6.1 INTRODUCTION

Homeobox proteins are a large family of transcription factors (TFs) that contain a highly conserved DNA-binding domain of 60 amino acids known as the homeodomain. The first homeobox genes were identified in the fruit fly *Drosophila melanogaster* from a study of homeotic mutants that substitute one corporal segment by another different one. The structural analysis of these genes and sequence comparisons with other homeotic genes subsequently isolated allowed the discovery of a highly conserved DNA sequence of 180 bp (base pairs), the homeobox, which encodes the homeodomain (McGinnis et al., 1984a, b). Soon after, it was proposed that the homeodomain is a DNA-binding domain involved in recognizing specific DNA sequences and that homeodomain proteins act as TFs. Genes with homeoboxes were discovered in invertebrates, vertebrates, fungi, and plants and in all cases were related to developmental processes.

Homeodomain proteins are classified into different families and subfamilies according to the conservation of amino acid sequences inside and outside the homeodomain (Derelle et al., 2007). In addition, members of each family share other conserved domains, named codomains, which mediate protein–protein or protein–DNA interactions. These additional domains confer a higher degree of regulatory specificity through the combined action of homeodomain proteins with proteins bound to nearby DNA sequences or by the coupling of homeodomains to additional DNA-binding domains (Jiang et al., 1991; Keleher et al., 1989; Stern et al., 1989; Hayashi and Scott, 1990). As described further, variants of the homeodomain that have insertions or unusual substitutions of amino acids were found. This divides homeodomain proteins into "typical" and "atypical". Nevertheless, the structural features that define the homeodomain are conserved in atypical homeodomains.

6.2 STRUCTURE OF THE HOMEODOMAIN

The first structural study of a homeodomain was carried out by NMR spectroscopy of the Antennapedia (Ant) homeodomain and X-ray crystallography of the engrailed (en) homeodomain, both from *D. melanogaster* (Otting et al., 1990; Kissinger et al., 1990). The results indicated that the homeodomain adopts a structure of three α-helices connected by a loop and a turn, where helices I, II, and III comprise residues 10–22, 28–37, and 42–58, respectively (Figure 6.1A). In particular, in the Ant homeodomain helix III comprises residues 42–52 and is continued in a more flexible fourth helix formed by amino acids 53–60. Helices I and II are antiparallel and helix III is positioned perpendicular to the other two. Helices II and III are connected by a tight turn and form a helix–turn–helix (HTH) motif similar to the HTH DNA-binding motif found in prokaryotic repressors. Helix III, named the "recognition helix", fits into the major groove of DNA, and a disordered N-terminal arm (amino acids 1–9 of the homeodomain) establishes specific contacts in the minor groove (Figure 6.1A). Apart from helix III and the N-terminal arm, only three residues, at positions 25 (from the loop between helices I and II), 28, and 31 (from helix II), make contact with the sugar phosphate backbone allowing proper alignment of the homeodomain on DNA (Wolberger et al., 1991; Gehring et al., 1994). The homeodomain is a compact globular structure with a hydrophobic core that allows helix III and the N-terminal arm to establish specific contacts with adjacent bases in the bound DNA sequence. Therefore, the overall arrangement of the three helices is critical for DNA-binding specificity. In this sense the hydrophobic core is constituted by several nonpolar, invariant, or highly conserved amino acids: leucines 16, 38, and 40, isoleucine or valine 45, tryptophan 48, and phenylalanine 49. By comparing the homeodomain sequences of different species, a consensus sequence defining a homeodomain was determined, and it was found that the most highly conserved amino acids are those involved in stabilization of the folded structure and those localized in the DNA-binding region of helix III (Laughon, 1991). Thereby, two positions are nearly perfectly conserved (Trp48 and Asn51), two positions predominantly contain one amino acid (positions 16 and 49), and position 53 is almost always an arginine.

6.3 SPECIFIC CONTACTS WITH DNA

While the hydrophobic surface of helix III and helices I and II, form the inside of the homeodomain, the hydrophilic surface of helix III interacts with bases and the sugar phosphate backbone in the major groove of DNA. In spite of the functional diversity found for homeodomain proteins, many of them recognize DNA sequences that contain a TAAT core but differ in nucleotides 3' to it. Binding to a conserved core agrees with the high degree of conservation of the residues involved in establishing specific contacts with DNA. A key residue in the interaction of helix III with DNA is asparagine at position 51, which forms two hydrogen bonds with an adenine of the target DNA sequence (TA<u>A</u>T; Figure 6.1B). Since Asn51 is almost invariable, the core sequence recognized by the homeodomain can be defined around this adenine (Wilson et al., 1996; Tron et al., 2005; Viola and Gonzalez, 2006). The residue located at position 47, usually isoleucine or valine, is also important for the interaction of helix III with DNA; this residue establishes van der Waals contacts with a timline located next (3') to the adenine contacted by Asn51 (TAA<u>T</u>; Figure 6.1B).

In general, positions 5 and 6 of the target sequence are contacted by the amino acid located at position 50 of the homeodomain (Figure 6.1B). It has been shown that Lys50 imposes strong preferences for cytosines at positions 5 and 6, and that higher flexibility is observed when position 50 is occupied by a small amino acid, which can establish indirect contacts mediated by water molecules (Mathias et al., 2001). In several studies, it has been demonstrated that changes in residue 50 generate changes in DNA-binding preferences at positions 5 and 6 of the target site (Treisman et al., 1989; Wilson et al., 1996; Percival-Smith et al., 1990; Schier and Gehring, 1992). Another residue involved in determining DNA-binding specificity is the one present at position 54, which generally contacts bases 4, 5, and/or 6 of the target DNA sequence. Furthermore, it has been reported that the combinatorial action of amino acids 50 and 54 of helix III determines sequence preferences at positions 5 and 6 (Damante et al., 1996; Pellizzari et al., 1997; Kissinger et al., 1990; Dave et al., 2000; Koizumi et al., 2003; Viola and Gonzalez, 2006).

The amino acid at position 55 is usually basic and makes contact with the sugar phosphate backbone. However, arginine at position 55 determines the presence of a guanine 5' complementary to the adenine contacted by Asp51 (T<u>G</u>AT) in some homeodomains (Figure 6.1B). In the MATa1/MATα2 heterodimer, Arg55 of the a1 homeodomain forms two hydrogen bonds with this guanine as well as a hydrogen bond with the thymine complementary to the adenine contacted by Asn51 (Li et al., 1995).

Positions 1 and 2 of the core sequence are contacted by the N-terminal arm of the homeodomain through arginine residues present at variable positions, depending on the protein (Figure 6.1B). This flexibility is related to the disordered structure of this arm. The removal of the N-terminal arm considerably decreases the binding affinity of homeodomains for their target sites (Percival-Smith et al., 1990; Shang et al., 1994).

With the isolation of new homeodomains, it became evident that many of them are longer than 60 amino acids

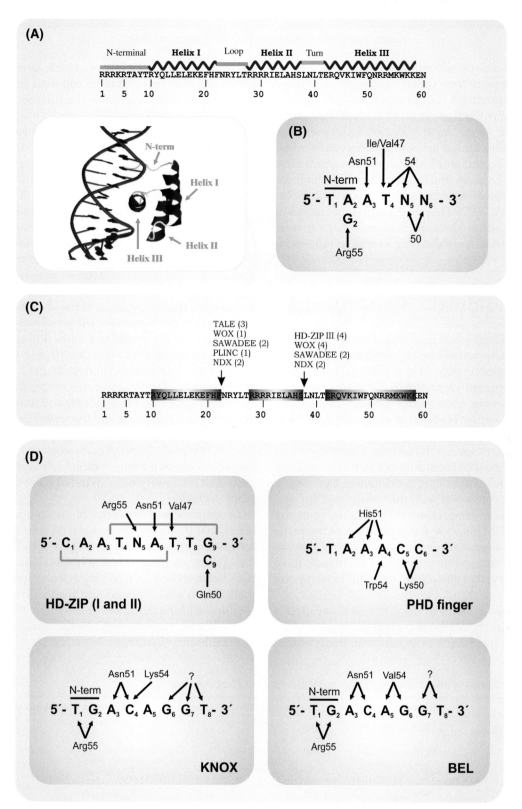

FIGURE 6.1 **Structure of the homeodomain.** (A) Primary, secondary, and tridimensional structure of the homeodomain. The 60-amino-acid homeodomain consensus sequence is presented in the upper part, along with a scheme of the homeodomain secondary structure. In the lower part, the tridimensional structure of the engrailed (en) homeodomain–DNA complex is shown. (B) Diagram of the interactions established by Antennapedia-like homeodomains with the sequence 5′-TAATNN-3′, as deduced from structural and mutagenesis studies. (C) Several classes of plant homeodomain proteins contain atypical homeodomains with insertions in either the loop or the turn. Arrows indicate the site of amino acid insertion; the number of extra amino acids is shown in brackets. (D) Diagram of DNA contacts established by plant homeodomains. Panels show the proposed models of interaction of HD-ZIP (I and II), TALE (KNOX and BEL), and PHD finger homeodomains with their consensus binding sequences deduced from SELEX experiments. Assignment of the contacts is based on mutagenesis and footprinting studies and on comparisons with known animal HD–DNA complexes (Sessa et al., 1993, 1997; Tron et al., 2005; Viola and Gonzalez, 2007; Tioni et al., 2005; Viola and Gonzalez, 2009). The blue square brackets show the two partially overlapping 5′-TNATTG-3′ sequences that compose the binding site of HD-ZIP proteins. See Section 6.4.1 for details.

B. EVOLUTION AND STRUCTURE OF DEFINED PLANT TRANSCRIPTION FACTOR FAMILIES

due to the presence of insertions; as a consequence, these homeodomains were then called "atypical". The three amino acid loop extension (TALE) superclass of homeodomain proteins was the first group of atypical homeodomains identified. As the name indicates, TALE members contain three extra amino acids in the loop connecting helices I and II. Typical and TALE homeobox genes were found in plants, animals, and fungi, suggesting that these two types of homeoboxes arose early in eukaryotic evolution (Derelle et al., 2007). Despite this difference, atypical homeodomains adopt a similar tridimensional structure and contact DNA in a similar way to typical homeodomains.

6.4 PLANT HOMEODOMAIN FAMILIES

The first plant homeobox gene was identified in mutant maize plants characterized by the presence of knot-like structures in leaves (Vollbrecht et al., 1991). These alterations in development were due to the ectopic expression of the homeobox gene *knotted-1* (Hake, 1992; Smith et al., 1992; see Chapter 14). Soon after, increasingly more genes encoding homeodomain proteins were isolated from different plant species. It has been determined that the function of most of them is to act as transcriptional regulators primarily controlling developmental processes. This function is consistent with that postulated for animal homeodomains, suggesting that, although they are not necessarily involved in the regulation of the same developmental events, transcriptional control mechanisms in which homeodomain proteins participate have remained conserved in eukaryotes over millions of years of evolution.

Homeodomain proteins of the different plant species have been classified over the years into different families and subfamilies. In addition to sequence similarity within the homeodomain, members of each family share other conserved domains and have common features such as protein size and location of the homeodomain in the molecule. Based on exhaustive phylogenetics analysis of over 1000 sequences of homeodomain proteins encoded in the genomes of a variety of phyla, including flowering plants, *Selaginella moellendorffii* (Tracheophyta), mosses (Bryophyta), unicellular green algae (Chlorophyta), and red algae (Rhodophyta), we now know that plant homeodomain proteins can be grouped into 14 distinct classes (Mukherjee et al., 2009). This classification is supported by the patterns of sequence conservation among full-length homeodomain proteins, conservation of codomains and other motifs, and conservation of class-specific intron positions within all 14 homeodomain gene classes. In addition, this analysis led to the identification of more homeobox genes than reported previously. The number of homeodomain proteins encoded in different plant genomes varies from only a few (five in the green alga *Chlamydomonas reinhardtii* and seven in *Ostreococcus lucimarinus* and *Ostreococcus tauri*) to more than 100 (110 in *Arabidopsis* and rice, 148 in poplar and maize) (Table 6.1). The clubmoss *Selaginella* and the moss *Physcomitrella patens* encode 45 and 66 homeodomain proteins, respectively (Mukherjee et al., 2009). A comparison of the number of homeobox genes present in different plant genomes suggests that the family has expanded considerably during plant evolution.

In comparison with animal homeodomains, homeodomains that do not fit the classical 60 amino acids pattern are quite common in plants (Figure 6.1C). However, although several groups of plant homeodomains have insertions between helices, when the extra regions are removed their amino acid sequences fit well with the consensus established for animal homeodomains (Bürglin, 1994). The following sections describe the features of each class of plant homeodomain proteins.

TABLE 6.1 Distribution of Plant Homeobox Genes and Pseudogenes in Plant Genomes

Division	Organism	Number of homeobox genes	Number of homeobox pseudogenes
Eudicots	*A. thaliana*	110	–
	Poplar	148	1
Monocots	Rice	110	7
	Maize	148	1
Lycopodiophyta	*Selaginella*	45	1
Moss	*P. patens*	66	1
Unicellular green algae	*C. reinhardtii*	5	–
	O. lucimarinus	7	–
	O. tauri	7	–
Unicellular red algae	*C. merolae*	7	–

6.4.1 HD-ZIP Superclass

The HD-ZIP superclass is composed of classes HD-ZIP I, HD-ZIP II, HD-ZIP III, and HD-ZIP IV (Sessa et al., 1994), all of which are characterized by the presence of a leucine zipper dimerization motif located in the carboxyl terminal to the homeodomain (see Chapter 7). HD-ZIP genes comprise approximately 50% of plant homeobox genes in mosses and flowering plants. The name of this family comes from the analogy to b-Zip transcription factors, which possess a leucine zipper associated with a basic DNA-binding motif. In HD-Zip proteins the homeodomain is located in the amino terminal half of the protein. The HD-ZIP II class presents a conserved "CPSCE" motif downstream of the leucine zipper (Chan et al., 1998). The HD-ZIP III and HD-ZIP IV classes are characterized by the START (steroidogenic acute regulatory protein–related lipid transfer; Ponting and Aravind, 1999) and HD-SAD (START-associated conserved domain; Schrick et al., 2004) domains in the C-terminal of the protein. Nevertheless, two characteristics divide the HD-ZIP III and IV classes: HD-ZIP III proteins contain a C-terminal domain, the MEKHLA domain, which is absent in the HD-ZIP IV class and have four extra residues inserted between helices II and III of the homeodomain (Figure 6.1D; Mukherjee and Burglin, 2006; Mukherjee et al., 2009). A more detailed description of the HD-ZIP family is presented in Chapter 7.

HD-ZIP I and HD-ZIP II proteins bind a pseudopalindromic DNA sequence of the type 5′-CAATNATTG-3′, where N can be A/T or G/C for HD-Zip I or II homeodomains, respectively. These different preferences have been attributed to a distinct orientation of the side chain of Arg55 that is involved in the recognition of the central position of the target sequence (Sessa et al., 1993). Otherwise, this consensus sequence can be viewed as two partially overlapping 5′-TNATTG-3′ sequences, similar to those bound by animal homeodomains. It has been suggested that, in the HD-ZIP-DNA complex, Arg55 of one of the two HD-ZIP monomers recognizes the central position of the sequence, while Arg55 of the other monomer may interact with the sugar-phosphate backbone (Sessa et al., 1997). As HD-ZIP I and HD-ZIP II proteins contain the almost invariant Asn51, the highly conserved Ile/Val47, and Gln50, as in Ant and en, a mode of DNA interaction similar to that of animal homeodomains can be assumed (Figure 6.1D). However, HD-ZIP proteins seem to have lost the ability to make base contacts in the minor groove of DNA through the N-terminal arm, probably to prevent steric hindrance between the two N-terminal arms bound to the central portion of the target sequence (Sessa et al., 1998). This lack of interaction through the N-terminal arm may be responsible for the inability of these proteins to bind DNA as monomers. In addition, it has been suggested that changes in the DNA-binding residues present in the loop located between helices I and II are also responsible for this inability (Tron et al., 2004).

On the other hand, HD-ZIP III proteins have four additional amino acids between the homeodomain and the leucine zipper. They also bind the extended sequence 5′-GTAAT(G/C)ATTAC-3′ (Sessa et al., 1998). Members of the HD-ZIP IV group bind target sequences characterized by two TAAA cores (Tron et al., 2001; Ohashi et al., 2003). Tron et al. (2001) showed that the amino acids present at positions 47 and 54 of the homeodomain act in combination to determine the preferential recognition of A or T at the fourth position of the half-sequence recognized by HD-ZIP proteins (TAA<u>A</u> or TAA<u>T</u>).

6.4.2 The TALE Superclass

The TALE superclass constitutes a large family of homeodomain proteins found in plants, fungi, and animals. They are characterized by a homeodomain with a three-amino-acid insertion in the loop that connects helices I and II (Bürglin, 1997). In animals and fungi, TALE proteins are divided into two families, PBC and MEIS. In plants, TALE homeodomain proteins are divided into two classes, KNOX and BEL (Bharathan et al., 1997; see Chapter 14). KNOX and BEL genes make up one of the oldest classes of plant homeobox genes and have representatives in unicellular red and green algae.

The KNOX class contains the KNOX domain, a conserved region located N-terminal to the homeodomain, which is composed of two blocks (KNOX A and KNOX B), and a shorter motif adjacent to the homeodomain named ELK (Vollbrecht et al., 1991). This family of proteins has been divided into two classes, KNOX I and II, according to sequence conservation within and outside the homeodomain and intron positions in the genes (Kerstetter et al., 1994). Furthermore, these groups differ in their expression patterns. The BEL class is characterized by two conserved domains located upstream of the homeodomain, the SKY and BEL domains (Bellaoui et al., 2001). According to predictions based on amino acid sequences, the SKY box and the BEL domain form α-helices, which would confer the ability to establish protein–protein contacts (Bellaoui et al., 2001; Chen et al., 2003). In addition, the VSLTLGL box, a conserved element whose function is unknown, is present in a position C-terminal to the BEL homeodomain, and another motif of 10 amino acids was detected at the N- and C-terminal ends of BEL proteins (Mukherjee et al., 2009).

In animals, MEIS and PBC TALE proteins form complexes that increase the affinity of the respective proteins for their target sites on DNA. Likewise, plant KNOX and BEL proteins heterodimerize through their conserved N-terminal codomains (Müller et al., 2001; Bellaoui et al., 2001; Smith et al., 2002; Byrne et al., 2003; Chen et al., 2003; Smith and Hake, 2003; Bhatt et al., 2004; Hackbusch et al., 2005; Kanrar et al., 2006; Viola and Gonzalez, 2006; Lin et al., 2013). MEIS and PBC TALE proteins also interact with non-TALE homeodomain proteins in animals, but this has not been observed for plant TALE proteins.

Binding site selection and mutagenesis studies indicated that KNOX proteins bind DNA sequences containing a TGAC core (Figure 6.1D; Krusell et al., 1997; Smith et al., 2002; Tioni et al., 2005). Similar studies with the BEL protein ATH1 from *Arabidopsis* showed that its homeodomain recognizes the sequence TGACAGGT (Viola and Gonzalez, 2006), identical to the consensus sequence of KN1, a KNOX I protein from maize (Figure 6.1D). Although the DNA-binding properties of ATH1, and probably all BEL proteins, show similarities to those presented by KNOX proteins, some differences exist in binding affinity and selectivity. These different properties, mainly attributable to the amino acid present at position 54 of the homeodomain (valine in BEL and lysine in KNOX proteins), are probably essential for the respective functions of these TFs, considering the conservation observed at this position within each protein family (Viola and Gonzalez, 2006).

It has been reported that the interaction of several KNOX and BEL proteins from maize, potato, and *Arabidopsis* produces an increase in affinity for DNA sequences containing one or two TGAC cores (Smith et al., 2002, Viola and Gonzalez, 2009; Lin et al., 2013). The sequences involved in protein–protein interactions are necessary and sufficient to increase binding to DNA, and complex formation is not correlated with the establishment of new detectable contacts as deduced from missing nucleoside experiments (Viola and Gonzalez, 2009). Even though the BEL proteins ATH1 and BEL1 from *Arabidopsis* are able to interact *in vivo* in yeast with a single TGAC core (TAGACAGGT) and to produce complexes with the KNOX protein STM, which have similar binding properties to those observed with individual proteins. However, single proteins were unable to bind, or did so only very weakly, to sequences containing tandem TGAC sites (TGACTTGACAGGT) *in vivo*, and a pronounced increase in binding efficiency was obtained when the two proteins were coexpressed. This behavior may reflect the existence of different regulatory modes by KNOX and BEL proteins, either alone or in combination, depending on the specific sequences present in their target genes (Viola and Gonzalez, 2009).

6.4.3 The PINTOX Class

The proteins from this class are named after one of its components, the plant interactor homeobox protein GF14c-int. from rice, which has been isolated for its interaction with GF14-c in a two-hybrid screen (Cooper et al., 2003). This class of proteins form a clade characterized by substitution of the almost invariable Asn51 by aspartic acid. Upstream of the homeodomain, these TFs contain a conserved basic domain of about 70 amino acids, the PINTOX domain, together with a conserved acidic domain named Acid Pint located further upstream. The N-terminal of these proteins has conserved hydrophobic and basic residues (Mukherjee et al., 2009). Phylogenetic studies indicated that PINTOX genes originated before the divergence of *Cholorophyta* and *Streptophyta*. Although it has been reported that the PINTOX gene *OCP3* from *Arabidopsis* participates in adaptive responses to drought stress and necrotrophic fungal infections (Ramírez et al., 2009), until now the function of most of the members of this class remains unknown.

6.4.4 WOX Class

The WUSCHEL-related homeobox (WOX) is a large group of TFs specifically found in plants. WOX members contain an atypical homeodomain with one or two extra residues between helices I and II, and four to five extra residues between helices II and III. In addition, they contain a distinctive conserved motif, the WUS Box, which gives the name to the family (Haecker et al., 2004; see Chapter 14). Phylogenetic analyses indicated that the WOX family can be divided into three clades based on the time of their first appearance in the plant kingdom: ancient, intermediate, and modern (also called the WUS clade). The ancient clade contains members from green algae, moss, and vascular plants; the intermediate clade has members from vascular plant only; the WUS clade only contains members from seed plants, confirming the evolutionary relationship of WOX proteins. No members of the WOX class were identified in red algae.

In addition to the homeodomain and the WOX domain, most of the members of the same clade share one or more common motifs. To date, a total of seven motifs have been identified in these members, the function of most of which is still unknown (Lian et al., 2014). In the ancient class certain members from green algae, mosses, and angiosperms contain motif 2 at the N-terminus of the WOX domain, but others only have the homeodomain (Lian et al., 2014), suggesting that they may represent the ancestral form of WOXs. The members of the intermediate clade are more diversified and seven additional motifs were observed, while all members of the WUS clade contain only the homeodomain and the WUS box. This suggests that the acquisition of motifs is associated with subfunctionalization or neofunctionalization of WOX homeodomain genes. In the intermediate clade, two extra motifs are located at the N-terminus of the WOX domain, while the other four are at the C-terminus. In addition, two subgroups can be identified, named WOX8/9 and WOX11/12, which were brought about as a result of a duplication of the ancestor of this clade. Recently, a systematic analysis of the origins and evolutionary history of WOX family members was performed employing 350 WOX sequences from 50 plant species (Lian et al., 2014).

6.4.5 DDT Class

A group of homeodomain proteins that are present in unicellular green algae and land plants is characterized by the presence of the DDT domain downstream of the homeodomain (Doerks et al., 2001). This domain is also found in members of the HOX family of animal homeodomain proteins and in nuclear proteins from plants and animals not belonging to the homeodomain family, such as the BAZ family, BPTF transcription-factor-like proteins and PHD-domain-containing proteins. The DDT domain generally has 60 amino acids and contains regions of conserved phenylalanine and leucine residues. It has been suggested that the DDT domain consists of three α-helices that are capable of DNA binding. However, this hypothesis must be experimentally determined. In plants the DDT class is divided into three subclasses, D-TOX1, D-TOX2, and D-TOX3 (for DDT homeobox class domain; Mukherjee et al., 2009). D-TOX3 genes are specific of eudicots and present a conserved domain named D-TOX A. D-TOX1 and D-TOX2 members contain six additional conserved domains distributed throughout the entire length of the protein named D-TOX B to G. The D-TOX F motif is a Zn finger motif that has the pattern $C-X_2-C-X_{10}-C-X_2-C$, common among glucocorticoid hormone receptors and GATA transcription factors. A typical characteristic of the Cys2/Cys2-type Zn-finger domain is its ability to interact with DNA as well as with proteins, allowing it to play an essential role in chromatin rearrangement (Matsushita et al., 2007). The D-TOX E motif shows high sequence similarity with a gene (WSTF) associated with the human Williams–Beuren syndrome (Cus et al., 2006). The WSTF gene was found to play an important role in chromatin remodeling (Bozhenok et al., 2002).

6.4.6 PLINC Zn Finger Class

This class of TF, first named ZF-HD for the presence of two highly conserved Zn-finger-like motifs upstream of the homeodomain, is present in flowering plants, *Selaginella*, and mosses but not in unicellular red or green algae (Mukherjee et al., 2009). This suggests that this class originated in Streptophyta and was already present in the common ancestor of mosses and vascular plants. The first and second putative Zn fingers have the consensus sequences $C-X_3-H-X_9-D-X_1-C$ and $C-X_2-C-X_1-C-H-X_3-H$, respectively. In addition, most members of this family are characterized by the presence of a methionine at position 49 of the homeodomain, instead of the conserved phenylalanine, which is essential for proper binding to the core sequence TAAT. This binding site preference, similar to that of canonical homeodomains, can be explained by the fact that Phe49 is not thought to directly contact DNA; however, it is important for proper folding of the HD, particularly in stabilizing the hydrophobic core of the HD structure bound to DNA (Kissinger et al., 1990; Wolberger et al., 1991; Hirsch and Aggarwal, 1995). PLINC Zn finger proteins are also characterized by a four-amino-acid insertion between helices I and II of the homeodomain (Windhövel et al., 2001). Yeast two-hybrid analysis of PLINC class proteins from *Arabidopsis* suggested that these proteins form homo and heterodimers through sequences located in the Zn finger region, with a general tendency to form heterodimeric interactions (Tan and Irish, 2006). In *Arabidopsis*, these proteins are expressed in flowers, and it has been suggested that they play an important role in abiotic stress responses (Tran et al., 2007).

6.4.7 PHD Finger Homeodomain Family

The family of PHD finger proteins is characterized by the presence of a $Cys_4HisCys_3$ Zn finger motif of 50–80 amino acids in the amino terminus of the protein. This motif, the PHD finger (for plant homeodomain finger), presents a region of cysteine residues regularly arranged in a similar way to that observed in metal binding-domains and is one of the most abundant modules present in nuclear proteins (Bienz, 2006). After its identification in the homeodomain protein HAT3.1 from *Arabidopsis* (Schindler et al., 1993), numerous PHD fingers were identified in proteins from animals, yeasts, and plants. However, the PHD finger only appears associated with a homeodomain in plants (Aasland et al., 1995; Bienz, 2006).

The PHD finger domain has histone-binding activities (Musselman and Kutateladze, 2011) and was first characterized in 2006 as a reader of histone H_3 trimethylation on lysine 4 (H3K4me3; Wysocka et al., 2006; Shi et al., 2006; Pena et al., 2006; Li et al., 2006). Most proteins with PHD finger domains are involved in the control of gene transcription and chromatin dynamics.

The proteins of the PHD finger homeodomain class constitute a small family of TFs (e.g., there are two members in *Arabidopsis* and rice, four in poplar, and five in maize), which is absent from green or red algae. These proteins present a conserved motif of 90 amino acids rich in charged residues, called PEX-PHD, N-terminal to the PHD finger. Between the PHD finger and the homeodomain there are other conserved motifs (Mukherjee et al., 2009). In *Arabidopsis* the two genes that encode PHD finger homeodomain proteins, HAT3.1 (Schindler et al., 1993) and PRHA (Plesch et al., 1997), are representative members of the two subclasses in which this family is divided. Both groups of proteins are present in flowering plants and the PRHA subclass is also present in mosses and *Selaginella* (Mukherjee et al., 2009).

The homeodomains of many PHD finger proteins have an unusual feature: they differ markedly in the nature of residues that contact DNA. However, the positions that constitute the hydrophobic core are conserved, and it can be assumed that their three-dimensional structure is similar to that of other homeodomains (Viola and Gonzalez, 2007). From this, the question of whether the homeodomain of PHD finger proteins is able to act as a DNA-binding motif and, if so, whether it recognizes specific sequences emerges. At least for ZmHox2a from maize and HAT3.1 from *Arabidopsis*, this seems to be the case. ZmHox2a contains two functional homeodomains, HD1 and HD2, which recognize the sequences TCCT and GATC, respectively (Kirch et al., 1998). HAT3.1 binds the consensus sequence T(A/G)(A/C)ACCA, with a preference for A at positions 2 and 3 (Viola and Gonzalez, 2007). All these sequences differ from the classical TAAT core. For HAT3.1, this can be attributed to the presence of a tryptophan residue at position 54 of the homeodomain. This tryptophan would be involved in interacting with position 4 of the core and would be one of the determinants of the TAAA-binding sequence, rather than TAAT as for most homeodomains (Figure 6.1D). In addition, the presence of Lys50 would be responsible for the preference of a CC dinucleotide at positions 5 and 6, as observed for animal homeodomain proteins like bicoid (Baird-Titus et al., 2006). HAT3.1 also possesses a histidine residue, rather than the conserved asparagine, at position 51, which recognizes nucleotides 2 to 4 of the binding sequence (Viola and Gonzalez, 2007).

6.4.8 NDX Class

NDX homeodomain proteins represent a small family of TFs with atypical and highly divergent homeodomains with an insertion of six amino acids between helices II and III. The homeodomain is located close to the C-terminal of these proteins. The genomes of *Arabidopsis*, maize, rice, and *Selaginella* encode only one NDX gene, while two are encoded in poplar and mosses. Studies in soybean and *Lotus japonicus* showed that NDX genes are expressed during root nodule formation and in the apical root and shoot meristems. It has been suggested that these genes participate in the development of structural nodule features in *L. japonicus* (Jorgensen et al., 1999; Gronlund et al., 2003). NDX proteins have two additional motifs located upstream and downstream of the homeodomain, named NDX-A and NDX-B, respectively (Mukherjee et al., 2009). Phylogenetic studies suggested that the NDX class appeared early in the evolution of the land plant clade, before the separation of mosses and vascular plants because NDX genes are present in flowering plants, mosses, and *Selaginella*, but not in unicellular red and green algae.

6.4.9 SAWADEE Class

SAWADEE genes encode a class of homeodomain proteins of unknown function present in Streptophyta but not in Chlorophyta or Rhodophyta. These genes encode atypical homeodomains characterized by the presence of a longer loop between helices I and II and 10 extra amino acids between helices II and III. At the C-terminal to the homeodomain there is a conserved region of 130–140 amino acids known as the SAWADEE domain (Mukherjee et al., 2009). The SAWADEE domain is also detected in nonhomeodomain proteins from eudicots and monocots and has several conserved cysteine and histidine residues that may be involved in metal binding.

6.4.10 LD Class

LD proteins possess a homeodomain that contains unusual substitutions in residues highly conserved in other homeodomains, such as Phe48, instead of tryptophan and a nonconserved residue at position 51. These proteins were identified in maize and *Arabidopsis* (van Nocke et al., 2000; Gronlund et al., 2003), and later in other flowering plants, *Selaginella*, and mosses through phylogenetic analysis (Mukherjee et al., 2009). Apart from the homeodomain, LD class proteins contain five conserved codomains named LD1–LD5. The LD3 domain is also conserved in other groups of plant TFs and was named LUMI by Mukherjee et al. (2009). In *Arabidopsis* the *LD* (*LUMINIDEPENDENS*) gene appears to be involved in the regulation of floral transition (Domagalska et al., 2007).

6.5 THE EVOLUTION OF PLANT HOMEOBOX GENES

Homeodomain proteins have been classified over the years into different families and subfamilies. In addition to similarities within the homeodomain, the members of each family share other conserved domains as well as common features, such as protein size and location of the homeodomain in the molecule. In this manner, animal homeobox genes were previously classified into about 30 groups (Bürglin, 1994) and then, with the identification of novel homeobox genes (e.g., PBC, MEIS, PKNOX/PREP, TGIF, and IRO; Bürglin, 1997), into at least 49 different families (Bürglin, 2005). Plant homeobox genes were first classified into seven families (Bharathan et al., 1997; Bürglin, 1997). Afterwards, with the availability of sequenced genomes from different species representative of distinct phyla, a classification into 14 classes was established (Mukherjee et al., 2009). Studies based on parsimony and neighbor-joining analyses of primary amino acid sequences from angiosperm, metazoan, and fungal homeodomains suggested that there are two major groups of homeodomain proteins (Bharathan et al., 1997). Phylogenetic analysis showed that the last common ancestor of angiosperms, fungi, and metazoa contained these two groups of homeodomain proteins, one with typical (or

Ant-like) homeodomains of 60 amino acids and the other one with atypical homeodomains with an insertion between helices I and II (Figure 6.2; Mukherjee et al., 2009). The results support the hypothesis that duplication of the atypical homeodomain brought about the two subgroups of TALE proteins, which exhibit codomains that are similar between plants and animals. One of the copies acquired an intron and brought about the BEL and PBC

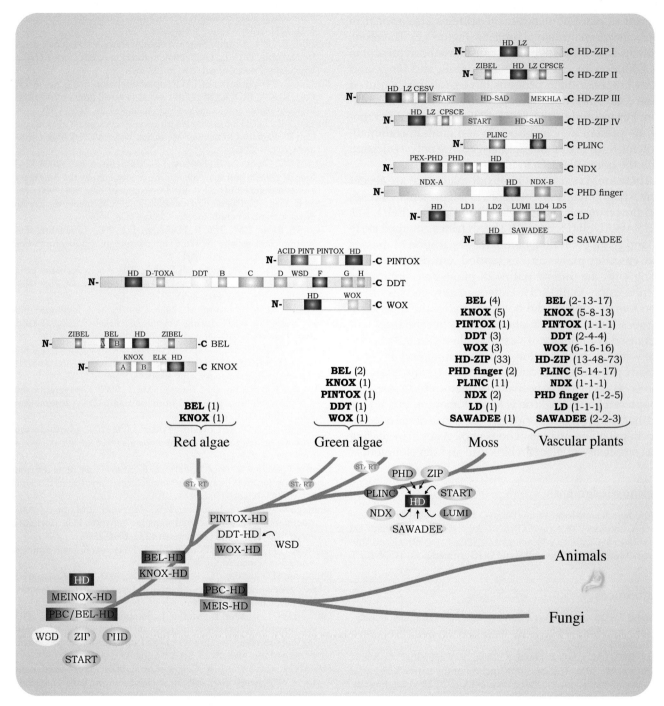

FIGURE 6.2 **Evolution of plant homeodomain proteins and their codomains.** The tree shows that incorporation of codomains and diversification of homeodomain proteins took place at different stages of plant evolution. Squares represent homeodomain proteins and ovals the codomains or motifs that are also present in nonhomeodomain proteins. The classes of homeodomain proteins present in each lineage are listed in bold and the numbers in parentheses indicate the number of homeodomain proteins of each class present in representative species (red algae, *Cyanodioschyzon merolae*; green algae, *O.s lucimarinus*; mosses, *P. patens*; vascular plants, *S moellendorffii*; *Arabidopsis thaliana*, and maize, respectively). At the top of the diagram are the schematic structures of the 14 classes of plant homeodomain proteins, showing the presence of codomains and other conserved motifs. Typical and atypical homeodomains are in blue and pink, respectively.

B. EVOLUTION AND STRUCTURE OF DEFINED PLANT TRANSCRIPTION FACTOR FAMILIES

families from plants and animals, respectively. The other copy brought about the MEINOX protein, the ancestor of KNOX (plant) and MEIS (animal) homeodomain proteins. However, none of the non-TALE (typical) proteins from plants can be associated with an animal homeodomain protein. In addition, it is interesting that some codomains present in plant homeodomain proteins also appear in nonhomeodomain animal proteins (Figure 6.2). Therefore, proteins with these codomains must have been present in the common ancestor of plants, animals, and fungi, and plant homeodomain proteins have probably acquired these codomains from other proteins during plant evolution. For example, the START, PHD, ZIP, and WSD motifs are present in homeodomain and nonhomeodomain proteins from plants and animals, but are only associated with the homeodomain in plants.

There are six classes of homeodomains that are absent from Rhodophyta and Chlorophyta but present in mosses and flowering plants: HD-ZIP, PLINC, NDX, PHD, LD, and SAWADEE (Figure 6.2). These classes emerged early in land plant evolution, before the separation of mosses and vascular plants and expanded and differentiated along with the differentiation of plants into organisms of increasing complexity. However, new classes of homeodomain proteins did not appear after the separation of mosses and vascular plants. The differentiation and increased complexity of vascular plants seem to correlate with the proliferation and diversification of genes already present in their ancestors. While certain classes of homeobox genes have been under intense study (see, e.g., Chapters 14 and 22), little is known about many others. Further studies may provide new evidence of the effect of functional differentiation of homeobox genes in increasing the complexity of plant architecture and development.

Acknowledgments

The authors acknowledge support from Consejo Nacional de Investigaciones Científicas y Técnicas (CONICET, Argentina), Agencia Nacional de Promoción Científica y Tecnológica (ANPCyT, Argentina), and Universidad Nacional del Litoral. ILV and DHG are members of CONICET.

References

Aasland, R., Gibson, T.J., Stewart, A.F., 1995. The PHD finger: implications for chromatin-mediated transcriptional regulation. Trends Biochem. Sci. 20, 56–59.

Baird-Titus, J.M., Clark-Baldwin, K., Dave, V., Caperelli, C.A., Ma, J., Rance, M., 2006. The solution structure of the native K50 bicoid homeodomain bound to the consensus TAATCC DNA-binding site. J. Mol. Biol. 356, 1137–1151.

Bellaoui, M., Pidkowich, M.S., Samach, A., Kushalappa, K., Kohalmi, S.E., et al., 2001. The Arabidopsis BELL1 and KNOX TALE homeodomain proteins interact through a domain conserved between plants and animals. Plant Cell 13, 2455–2470.

Bharathan, G., Janssen, B.J., Kellogg, E.A., Sinha, N., 1997. Did homeodomain proteins duplicate before the origin of angiosperms, fungi, and metazoa? Proc. Natl. Acad. Sci. USA 94, 13749–13753.

Bhatt, A.M., Etchells, J.P., Canales, C., Lagodienko, A., Dickson, H., 2004. VAAMANA a BEL1-like homeodomain protein interacts with KNOX proteins BP and STM and regulates inflorescence stem growth in Arabidopsis. Gene 328, 103–111.

Bienz, M., 2006. The PHD finger, a nuclear protein–interaction domain. Trends Biochem. Sci. 31, 35–40.

Bozhenok, L., Wade, P.A., Varga-Weisz, P., 2002. WSTF-ISWI chromatin remodeling complex targets heterochromatic replication foci. EMBO J. 21, 2231–2241.

Bürglin, T., 1994. A comprehensive classification of homeobox genes. In: Duboule, D. (Ed.), Guidebook to the Homeobox Genes. Oxford University Press, Oxford, UK, pp. 27–269.

Bürglin, T., 1997. Analysis of TALE superclass homeobox genes (MEIS, PBC, KNOX, IROQUOIS, TGIF) reveals a novel domain conserved between plants and animals. Nucl. Acids Res. 25, 4173–4180.

Bürglin, T.R., 2005. Homeodomain proteins. Encyclopedia of Molecular Cell Biology and Molecular Medicine. Wiley-VCH Verlag GmbH and Co, 179–222.

Byrne, M.E., Groover, A.T., Fontana, J.R., Martienssen, R.A., 2003. Phyllotactic pattern and stem cell fate are determined by the Arabidopsis homeobox gene BELLRINGER. Development 130, 3941–3950.

Chan, R., Gago, G., Palena, C., González, D., 1998. Homeoboxes in plant development. Biochim. Biophys. Acta 93134, 1–19.

Chen, H., Rosin, F.M., Prat, S., Hannapel, D.J., 2003. Interacting transcription factors from the TALE superclass regulate tuber formation. Plant Physiol. 132, 1391–1404.

Cooper, B., Clarke, J.D., Budworth, P., et al., 2003. A network of rice genes associated with stress response and seed development. Proc. Natl. Acad. Sci. USA 100, 4945–4950.

Cus, R., Maurus, D., Kuhl, M., 2006. Cloning and developmental expression of WSTF during Xenopus laevis embryogenesis. Gene Expr. Patterns 6, 340–346.

Damante, G., Pellizzari, L., Esposito, G., Fogolari, F., Viglino, P., Fabbro, D., et al., 1996. A molecular code dictates sequence-specific DNA recognition by homeodomains. EMBO J. 15, 4992–5000.

Dave, V., Zhao, C., Yang, F., Tung, C., Ma, J., 2000. Reprogrammable recognition codes in bicoid homeodomain–DNA interaction. Mol. Cell Biol. 20, 7673–7684.

Derelle, R., Lopez, P., Le Guyader, H., Manuel, M., 2007. Homeodomain proteins belong to the ancestral molecular tool kit of eukaryotes. Evol. Dev. 9, 212–219.

Doerks, T., Copley, R., Bork, P., 2001. DDT – a novel domain in different transcription and chromosome remodeling factors. Trends Biochem. Sci. 26, 145–146.

Domagalska, M.A., Schomburg, F.M., Amasino, R.M., et al., 2007. Attenuation of brassinosteroid signaling enhances FLC expression and delays flowering. Development 134, 2841–2850.

Gehring, W.J., Affolter, M., Bürglin, T.R., 1994. Homeodomain proteins. Annu. Rev. Biochem. 63, 487–526.

Gronlund, M., Gustafsen, C., Roussis, A., et al., 2003. The Lotus japonicus ndx gene family is involved in nodule function and maintenance. Plant Mol. Biol. 52, 303–316.

Hackbusch, J., Richter, K., Müller, J., Salamini, F., Uhrig, J.F., 2005. A central role of Arabidopsis thaliana ovate family proteins in networking and subcellular localization of 3 aa loop extension homeodomain proteins. Proc. Natl. Acad. Sci. USA 102, 4908–4912.

Haecker, A., Gross-Hardt, R., Geiges, B., et al., 2004. Expression dynamics of WOX genes mark cell fate decisions during early embryonic patterning in Arabidopsis thaliana. Development 131, 657–668.

Hake, S., 1992. Unraveling the knots in plant development. Trends Genet. 8, 109–114.

Hayashi, S., Scott, M.P., 1990. What determines the specificity of action of Drosophila homeodomain proteins? Cell 63, 883–894.

Hirsch, J.A., Aggarwal, A.K., 1995. Structure of the even-skipped homeodomain complexed to AT-rich DNA: new perspectives on homeodomain specificity. EMBO J. 14, 6280–6291.

Jiang, J., Hoey, T., Levine, M., 1991. Autoregulation of a segmentation gene in *Drosophila*: combinatorial interaction of the even-skipped homeobox protein with a distal enhancer element. Genes Dev. 5, 265–277.

Jorgensen, J.E., Gronlund, M., Pallisgaard, N., et al.,1999. A new class of plant homeobox genes is expressed in specific regions of determinate symbiotic root nodules. Plant Mol. Biol. 40, 65–77.

Kanrar, S., Onguka, O., Smith, H.M., 2006. *Arabidopsis* inflorescence architecture requires the activities of KNOX–BELL homeodomain heterodimers. Planta 224, 1163–1173.

Keleher, C.A., Passmore, S., Johnson, A.D., 1989. Yeast repressor alpha 2 binds to its operator cooperatively with yeast protein Mcm1. Mol. Cell. Biol. 9, 5228–5230.

Kerstetter, R., Vollbrecht, E., Lowe, B., Veit, B., Yamaguchi, J., Hake, S., 1994. Sequence analysis and expression patterns divide the maize knotted1-like homeobox genes into two classes. Plant Cell 6, 1877–1887.

Kirch, T., Bitter, S., Kisters-Woike, B., Werr, W., 1998. The two homeodomains of the ZmHox2a gene from maize originated as an internal gene duplication and have evolved different target site specificities. Nucl. Acids Res. 26, 4714–4720.

Kissinger, C., Liu, B., Martin-Blanco, E., Kornberg, T., Pabo, C., 1990. Crystal structure of an engrailed homeodomain–DNA complex at 2,8 Å resolution: a framework for understanding homeodomain–DNA interactions. Cell 63, 579–590.

Koizumi, K., Lintas, C., Nirenberg, M., Maeng, J., Ju, J., Mack, J., et al., 2003. Mutations that affect the ability of the vnd/Nk-2 homeoprotein to regulate gene expression: transgenic alterations and tertiary structure. Proc. Natl. Acad. Sci. USA 100, 3119–3124.

Krusell, L., Rasmussen, I., Gausing, K., 1997. DNA binding sites recognised *in vitro* by a knotted class I homeodomain protein encoded by the *hooded* gene, *k*, in barley (*Hordeum vulgare*). FEBS Lett. 408, 25–29.

Laughon, A., 1991. DNA binding specificity of homeodomains. Biochemistry 30, 11357–11367.

Li, H., Ilin, S., Wang, W., Duncan, E.M., Wysocka, J., Allis, C.D., Patel, D.J., 2006. Molecular basis for site-specific read-out of histone H3K4me3 by the BPTF PHD finger of NURF. Nature 442, 91–95.

Li, T., Stark, M.R., Johnson, A.D., Wolberger, C., 1995. Crystal structure of the MATa1/MATa2 homeodomain heterodimer bound to DNA. Science 270, 262–269.

Lian, G., Ding, Z., Wang, Q., Zhang, D., Xu, J., 2014. Origins and evolution of WUSCHEL-related homeobox protein family in plant kingdom. Sci. World J., 534140.

Lin, T., Sharma, P., Gonzalez, D.H., Viola, I.L., Hannapel, D.J., 2013. The impact of the long-distance transport of a BEL1-like messenger RNA on development. Plant Physiol. 161, 760–772.

Mathias, J., Zhong, H., Jin, Y., Vershon, A., 2001. Altering the DNA-binding specificity of the yeast Matα2 homeodomain protein. J. Biol. Chem. 276, 32696–32703.

Matsushita, A., Sasaki, S., Kashiwabara, Y., et al., 2007. Essential role of GATA2 in the negative regulation of thyrotropin beta gene by thyroid hormone and its receptors. Mol. Endocrinol. 21, 865–884.

McGinnis, W., Garber, R.L., Wirz, J., Kuroiwa, A., Gehring, W.J., 1984a. A homologous protein-coding sequence in *Drosophila* homeotic genes and its conservation in other metazoans. Cell 37, 403–408.

McGinnis, W., Levine, M.S., Hafen, E., et al., 1984b. A conserved DNA sequence in homoeotic genes of the *Drosophila* Antennapedia and bithorax complexes. Nature 308, 428–433.

Mukherjee, K., Burglin, T.R., 2006. MEKHLA, a novel domain with similarity to PAS domains, is fused to plant homeodomain–leucine zipper III proteins. Plant Physiol. 140, 1142–1150.

Mukherjee, K., Brocchieri, L., Burglin, T.R., 2009. A comprehensive classification and evolutionary analysis of plant homeobox genes. Mol. Biol. Evol. 26, 2775–2794.

Müller, J., Wang, Y., Franzen, R., Santi, L., Salamini, F., Rohde, W., 2001. *In vitro* interactions between barley TALE homeodomain proteins suggest a role for protein–protein associations in the regulation of Knox gene function. Plant J. 27, 13–23.

Musselman, C.A., Kutateladze, T.G., 2011. Handpicking epigenetic marks with PHD fingers. Nucl. Acids Res. 39, 9061–9071.

Ohashi, Y., Oka, A., Pousada, R.R., Possenti, M., Ruberti, I., Morelli, G., Aoyama, T., 2003. Modulation of phospholipid signaling by GLABRA2 in root-hair pattern formation. Science 300, 1427–1430.

Otting, G., Qian, Y., Billeter, M., Muller, M., Affolter, M., Gehring, W., Wuthrich, K., 1990. Protein–DNA contacts in the structure of a homeodomain–DNA complex determined by nuclear magnetic resonance spectroscopy in solution. EMBO J. 9, 3085–3092.

Pellizzari, L., Tell, G., Fabbro, D., Pucillo, C., Damante, G., 1997. Functional interference between contacting amino acids of homeodomains. FEBS Lett. 407, 320–324.

Pena, P.V., Davrazou, F., Shi, X., Walter, K.L., Verkhusha, V.V., Gozani, O., Zhao, R., et al., 2006. Molecular mechanism of histone H3K4me3 recognition by plant homeodomain of ING2. Nature 442, 100–103.

Percival-Smith, A., Müller, M., Affolter, M., Gehring, W., 1990. The interaction with DNA of wild-type and mutant fushi tarazu homeodomains. EMBO J. 9, 3967–3974.

Plesch, G., Stormann, K., Torres, J.T., Walden, R., Somssich, I.E., 1997. Developmental and auxin-induced expression of the *Arabidopsis* PRHA homeobox gene. Plant J. 12, 635–647.

Ponting, C.P., Aravind, L., 1999. START: a lipid-binding domain in StAR, HD-ZIP and signalling proteins. Trends Biochem. Sci. 24, 130–132.

Ramírez, V., Coego, A., López, A., Agorio, A., Flors, V., Vera, P., 2009. Drought tolerance in *Arabidopsis* is controlled by the OCP3 disease resistance regulator. Plant J. 58, 578–591.

Schier, A., Gehring, W., 1992. Direct homeodomain–DNA interaction in the autoregulation of the fushi tarazu gene. Nature 356, 804–807.

Schindler, U., Beckmann, H., Cashmore, A.R., 1993. HAT3.1, a novel *Arabidopsis* homeodomain protein containing a conserved cysteine-rich region. Plant J. 4, 137–150.

Schrick, K., Nguyen, D., Karlowski, W.M., et al., 2004. START lipid/sterol-binding domains are amplified in plants and are predominantly associated with homeodomain transcription factors. Genome Biol. 5, R41.

Sessa, G., Morelli, G., Ruberti, I., 1993. The Athb-1 and -2 HD-Zip domains homodimerize forming complexes of different DNA binding specificities. EMBO J. 12, 3507–3517.

Sessa, G., Carabelli, M., Ruberti, I., 1994. Identification of distinct families of HD-Zip proteins in *Arabidopsis thaliana*. Plant Mol. Biol., 412–426.

Sessa, G., Morelli, G., Ruberti, I., 1997. DNA-binding specificity of the homeodomain–leucine zipper domain. J. Mol. Biol. 274, 303–309.

Sessa, G., Steindler, C., Morelli, G., Ruberti, I., 1998. The *Arabidopsis* ATHB-8,-9 and -14 genes are members of a small gene family coding for highly related HD-Zip proteins. Plant Mol. Biol. 38, 609–622.

Shang, Z., Ebright, Y., Iler, N., Pendergrast, P., Echelard, Y., McMahon, A., Ebright, R., Abate, C., 1994. DNA affinity cleaving analysis of homeodomain–DNA interaction: identification of homeodomain consensus sites in genomic DNA. Proc. Natl. Acad. Sci. USA 91, 118–122.

Shi, X.B., Hong, T., Walter, K.L., Ewalt, M., Michishita, E., Hung, T., Carney, D., et al., 2006. ING2 PHD domain links histone H3 lysine 4 methylation to active gene repression. Nature 442, 96–99.

Smith, H., Boschke, I., Hake, S., 2002. Selective interaction of plant homeodomain proteins mediates high DNA-binding affinity. Proc. Natl. Acad. Sci. USA 99, 9579–9584.

Smith, H., Hake, S., 2003. The interaction of two homeobox genes, BREVIPEDICELLUS and PENNYWISE, regulates internode patterning in the *Arabidopsis* inflorescence. Plant Cell 15, 1717–1727.

Smith, L., Greene, B., Veit, B., Hake, S., 1992. A dominant mutation in the maize homeobox gene, Knotted-1, causes its ectopic expression in leaf cells with altered fates. Development 116, 21–30.

Stern, S., Tanaka, M., Herr, W., 1989. The Oct-1 homoeodomain directs formation of a multiprotein–DNA complex with the HSV transactivator VP16. Nature 341, 624–630.

Tan, Q.K., Irish, V.F., 2006. The *Arabidopsis* zinc finger-homeodomain genes encode proteins with unique biochemical properties that are coordinately expressed during floral development. Plant Physiol. 140, 1095–1108.

Tioni, M.F., Viola, I.L., Chan, R.L., Gonzalez, D.H., 2005. Site-directed mutagenesis and footprinting analysis of the interaction of the sunflower KNOX protein HAKN1 with DNA. FEBS J. 272, 190–202.

Tran, L.S., Nakashima, K., Sakuma, Y., et al., 2007. Co-expression of the stress-inducible zinc finger homeodomain ZFHD1 and NAC transcription factors enhances expression of the ERD1 gene in *Arabidopsis*. Plant J. 49 (1), 46–63.

Treisman, J., Gonczy, P., Vashishtha, M., Harris, E., Desplan, C., 1989. A single amino acid can determine the DNA binding specificity of homeodomain proteins. Cell 59, 553–562.

Tron, A.E., Bertoncini, C.W., Palena, C.M., Chan, R.L., Gonzalez, D.H., 2001. Combinatorial interactions of two amino acids with a single base pair define target site specificity in plant dimeric homeodomain proteins. Nucl. Acids Res. 29, 4866–4872.

Tron, A.E., Welchen, E., Gonzalez, D.H., 2004. Engineering the loop region of a homeodomain–leucine zipper protein promotes efficient binding to a monomeric DNA binding site. Biochemistry 43, 15845–15851.

Tron, A.E., Comelli, R.N., Gonzalez, D.H., 2005. Structure of homeodomain–leucine zipper/DNA complexes studied using hydroxyl radical cleavage of DNA and methylation interference. Biochemistry 44, 16796–16803.

van Nocke, S., Muszynski, M., Briggs, K., Amasino, R.M., 2000. Characterization of a gene from *Zea mays* related to the *Arabidopsis* flowering-time gene LUMINIDEPENDENS. Plant Mol. Biol. 44, 107–122.

Viola, I.L., Gonzalez, D.H., 2006. Interaction of the BELL-like protein ATH1 with DNA. A role of homeodomain residue 54 in specifying the different binding properties of BELL and KNOX proteins. J. Biol. Chem. 387, 31–40.

Viola, I.L., Gonzalez, D.H., 2007. Interaction of the PHD finger homeodomain protein HAT3.1 from *Arabidopsis thaliana* with DNA. Specific DNA binding by a homeodomain with histidine at position 51. Biochemistry 46, 7416–7425.

Viola, I.L., Gonzalez, D.H., 2009. Binding properties of the complex formed by the *Arabidopsis* TALE homeodomain proteins STM and BLH3 to DNA containing single and double target sites. Biochimie 91, 974–981.

Vollbrecht, E., Veit, B., Sinha, N., Hake, S., 1991. The developmental gene Knotted-1 is a member of a maize homeobox gene family. Nature 350, 241–243.

Wilson, D., Sheng, G., Jun, S., Desplan, C., 1996. Conservation and diversification in homeodomain–DNA interactions: a comparative genetic analysis. Proc. Natl. Acad. Sci. USA 93, 6886–6891.

Windhövel, A., Hein, I., Dabrowa, R., Stockhaus, J., 2001. Characterization of a novel class of plant homeodomain proteins that bind to the C4 phosphoenolpyruvate carboxylase gene of *Flaveria trinervia*. Plant Mol. Biol. 45, 201–214.

Wolberger, C., Vershon, A., Liu, B., Johnson, A., Pabo, C., 1991. Crystal structure of a MATα2 homeodomain–operator complex suggests a general model for homeodomain–DNA interaction. Cell 67, 517–528.

Wysocka, J., Swigut, T., Xiao, H., Milne, T.A., Kwon, S.Y., Landry, J., Kauer, M., et al., 2006. A PHD finger of NURF couples histone H3 lysine 4 trimethylation with chromatin remodelling. Nature 442, 86–90.

CHAPTER 7

Homeodomain–Leucine Zipper Transcription Factors: Structural Features of These Proteins, Unique to Plants

Matías Capella, Pamela A. Ribone, Agustín L. Arce, Raquel L. Chan

Instituto de Agrobiotecnología del Litoral, Universidad Nacional del Litoral, CONICET, Santa Fe, Argentina

OUTLINE

7.1 Homeoboxes and Homeodomains in Eukaryotic Kingdoms	114
7.2 Plant Homeoboxes	114
7.3 The Plant Homeodomain Superfamily	114
7.4 Different Domains Present in Homeodomain Transcription Factors	115
7.4.1 PHD Finger (Plant Homeodomain Associated to a Finger Domain)	115
7.4.2 ZF-HD (Conserved Zinc Finger-Like Motifs Associated with a Homeodomain)	115
7.4.3 WOX (Wuschel-Related Homeobox)	115
7.4.4 KNOX (Knotted-Related Homeobox)	115
7.4.5 BELL (Named for the Distinctive Bell Domain)	117
7.4.6 DDT (Named for the Presence of a DDT Domain Downstream of the Homeodomain)	117
7.4.7 NDX (Nodulin Homeobox Genes)	117
7.4.8 LD (Luminidependens Homeobox Genes)	117
7.4.9 PINTOX	117
7.4.10 SAWADEE (Homeodomain Associated to a Sawadee Domain)	117
7.5 The HD-Zip Family	117
7.5.1 HD-Zip Subfamily I	118
7.5.2 HD-Zip Subfamily II	120
7.5.3 HD-Zip Subfamily III	121
7.5.4 HD-Zip Subfamily IV	121
7.6 Target Sequences Recognized by the HD-Containing Transcription Factors	122
7.7 What do we Know About the Target Sequences of the HD-Zip Proteins?	123
7.8 Concluding Remarks	124
References	124

Plant Transcription Factors. http://dx.doi.org/10.1016/B978-0-12-800854-6.00007-5
Copyright © 2016 Elsevier Inc. All rights reserved.

7.1 HOMEOBOXES AND HOMEODOMAINS IN EUKARYOTIC KINGDOMS

In the cell, the regulation of gene transcription is based on specific interactions between regulatory proteins, known as transcription factors (TFs), and their target genes. TFs typically consist at least of two domains. The first one is a DNA-binding domain, which recognizes specific sequences in the regulatory regions of their target genes, and the second domain is able to establish protein–protein interactions. TFs can activate or repress their targets expression, thereby changing the cell transcriptome.

A homeobox (HB) encodes one of the most widespread and best characterized DNA-binding domains in all eukaryotic organisms: the homeodomain (HD). HBs were named after the homeotic effect that their mutation or ectopic expression produces: a developmental anomaly in which a segment of the body develops in the likeness of another. HDs were originally identified as protein regions encoded by several genes in *Drosophila melanogaster* development (Gehring, 1987). Dominant mutations at the *Antennapedia* locus, for example, lead to the substitution of the antennae by legs on the head of *Drosophila* (Garber et al., 1983). Another example is found in *bithorax* mutants, where the small halters which are used as balancers during flight are transformed into a second pair of wings (Bender et al., 1983).

The knowledge about the mechanisms of sequence-specific DNA binding by HDs was provided by three-dimensional structures of individual protein–DNA complexes coupled with directed mutagenesis and biochemical analysis (Ades and Sauer, 1995; Gehring et al., 1994; Wolberger, 1996). Now we know the HD consists of a conserved 60-amino-acid motif present in TFs mainly involved in developmental processes. The HD folds into a bundle of three α-helices, named I, II and III, joined by two loops and a turn, and is able to bind DNA as a monomer with high affinity through the interactions established by helix III (called the recognition helix) with the major groove of the target DNA while the disordered N-terminal arm, located upstream the first helix, interacts with the DNA minor groove (Gehring et al., 1994; Chapter 6). Since their discovery in *Drosophila*, HDs were found in TFs from evolutionary distant organisms, such as humans, nematodes, fungi, and plants. The high degree of conservation of this type of domain among proteins from different kingdoms indicates that this structure is crucial to maintain its functionality and that the role played by the HD is essential (Moens and Selleri, 2006).

7.2 PLANT HOMEOBOXES

In plants, the HB-containing genes were first identified by Erik Vollbrecht and coworkers in 1991 (Vollbrecht et al., 1991), analyzing a *Zea mays* mutant obtained by transposon tagging exhibiting striking "knotted" leaves. The analysis of such a mutant resulted in the isolation of *KNOTTED1*, the first HB-containing gene of plants. Since then, many HB-containing genes have been identified and isolated from a wide variety of plant species, including mono- and dicots. Nonetheless, unlike animals, neither of the known plant HB-containing genes has a canonical homeotic effect; so the legacy of the family name (homeobox) seems rather inadequate (Ariel et al., 2007).

7.3 THE PLANT HOMEODOMAIN SUPERFAMILY

Since the discovery of the maize *KNOTTED1* gene, a large number of genes encoding HD-containing TFs have been identified, but, among them, the majority have not been characterized so far. Members of this HD superfamily participate in a wide variety of developmental processes, such as identification and maintenance of meristematic cells, embryogenesis, pollen maturation, regulation of floral development, and response to environmental conditions. The full description of several plant genomes, such as those of rice (*Oryza sativa*), maize (*Z. mays*), *Arabidopsis thaliana*, and poplar (*Populus trichocarpa*), among others, increased the knowledge about plant HB genes and enabled the classification of the plant HD-containing proteins in several families, according to the similarity in HD sequence, protein size, HD location within the TF, association with other domains or conserved motifs, and also the genes structures including intron positions within the *Arabidopsis* genes (Ariel et al., 2007; Mukherjee et al., 2009). According to Mukherjee et al. (2009), these families are: PHD finger, ZF-HD, WOX, the TALE superclass (which includes BELL and KNOX families), DDT, NDX, LD, PINTOX, SAWADEE, and HD-Zip (composed of four subfamilies named I, II, III, and IV). The analysis of the HD sequences from different, and unrelated, organisms of different kingdoms suggested that some plant HD-TFs are more related to animal and fungi HD-TFs than to other plant HDs belonging to different families. Therefore, HDs seem to have diverged before the separation of the branches leading to plants, animals, and fungi (Ariel et al., 2007).

Aiming at illustrating the scenario constituted by plant HD containing TFs, Table 7.1 summarizes the available information about the number of members from families constituting the plant HD superfamily in model species and their main functions. Additionally, Figure 7.1 shows an evolutionary tree of plant HD-TFs and a schematic representation of their main known structural features.

TABLE 7.1 Classification and Main Physiological Events of HB-Containing Genes from Model Plants

Family	At[a]	Pt[b]	Os[c]	Zm[d]	Main functions
HD-Zip I	17	22	14	24	Response to abiotic stress
HD-Zip II	10	16	13	17	Cell expansion and proliferation
HD-Zip III	5	8	9	8	Organ and vascular development
HD-Zip IV	16	15	12	21	Epidermal cell differentiation
KNOX	8	15	12	13	Initiation and maintenance of the shoot apical meristem
BELL	13	19	14	17	Ovule development
ZF-HD	14	17	11	17	Regulation of floral development
WOX	16	19	15	16	Determination of cell fate
PHD-finger	2	4	2	5	Unknown
DDT	4	7	3	4	Maintenance of plant vegetative state
LD	1	2	1	1	Regulation of flowering time
PINTOX	1	1	1	1	Response to necrotrophic pathogens
SAWADEE	2	1	3	3	Transcriptional silencing through DNA methylation
NDX	1	2	1	1	Nodule development

[a] A. thaliana; [b] P. trichocarpa *(poplar)*; [c] O. sativa *(rice)*; [d] Z. mays *(maize)*

7.4 DIFFERENT DOMAINS PRESENT IN HOMEODOMAIN TRANSCRIPTION FACTORS

7.4.1 PHD Finger (Plant Homeodomain Associated to a Finger Domain)

The pathogenesis-related HB gene A (PRHA) and the HAT3.1 gene (Schindler et al., 1993; Plesch et al., 1997) from *A. thaliana* have a PHD finger domain, located several hundred amino acids upstream the HD. These PHD are able to activate the transcription in yeast, plant, and animal cells (Halbach et al., 2000). Based on the position of the unique histidine residue, the cysteine scaffold of the PHD finger (Cys_4–His–Cys_3) is clearly distinct from RING fingers (Cys_3–His–Cys_4), LIM domains (Cys_2–His–Cys_5), or DRIL domains, in which two RING finger motifs are closely linked. Adjacent to the PHD domain proteins of this class there are an additional 90 amino acids conserved motif rich in charged residues, called PEX-PHD (Mukherjee et al., 2009).

7.4.2 ZF-HD (Conserved Zinc Finger-Like Motifs Associated with a Homeodomain)

The members of this family present two highly conserved zinc (Zn)-finger-like motifs upstream of the HD. The first putative Zn finger has a consensus sequence C—X_3—H—X_9—D—X_1—C, where H is not always conserved, and the second one exhibits the conserved pattern C—X_2—C—X_1—C—H—X_3—H (Mukherjee et al., 2009).

7.4.3 WOX (Wuschel-Related Homeobox)

The *Arabidopsis* WUS (AtWUS) protein contains three functional domains: the acidic region, the WUS box, and the EAR-like motif, which contributes significantly to the WUS function as a TF (Ikeda et al., 2009). The EAR-like motif was also found in other members of the WOX family. However, the AtWUS functional acidic region (10 amino acids) upstream of the WUS box was not detectable in other plant WUS proteins, so it was speculated that this acidic region maybe a key domain only in AtWUS (Zhang et al., 2010b).

7.4.4 KNOX (Knotted-Related Homeobox)

The members of this family are characterized by the presence of a KNOX domain upstream of the HD, composed of two blocks (KNOX A and KNOX B) separated by a variable region (Bürglin, 1997), as well as a shorter motif situated between the KNOX and the HD, named ELK (Vollbrecht et al., 1991). The KNOX domain shares similarities with the MEIS and PBC domains of animal TALE HD proteins, suggesting that these domains derived from a common ancestral domain that has been named MEINOX (MEINOX being derived from MEIS and KNOX; Bürglin, 1997; Bürglin, 1998).

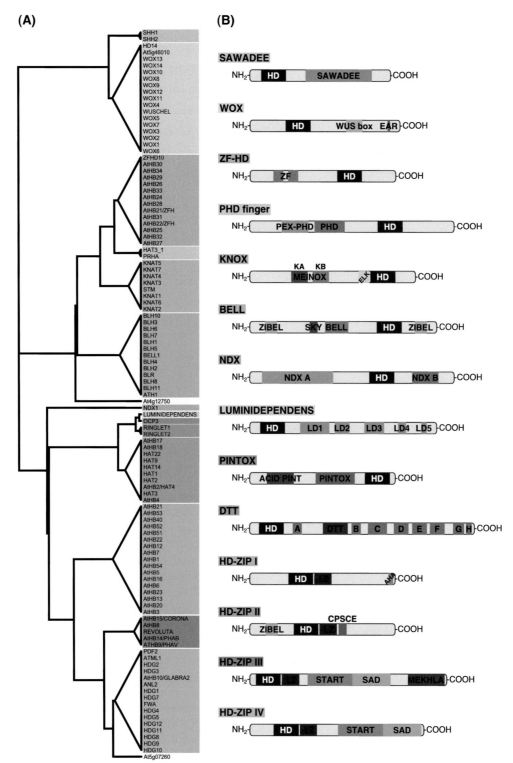

FIGURE 7.1 (A) Phylogenetic tree of all *Arabidopsis* HD-containing TFs based on the alignment of full-length protein sequences by using MAFFT v7, PhyML 3.1, and FigTree 1.4.2. (B) Schematic representation of the distinctive domains exhibited by each family of HB-containing genes. AHA, aromatic and large hydrophobic residues embedded in an acidic context; EAR, amphiphilic repression motif; ELK motif, named for the three conserved amino acids Glu, Leu, and Lys; HD, homeodomain; MEINOX, derived from association of MEIS and KNOX domains; MEKHLA domain, named for the highly conserved amino acids Met, Glu, Lys, His, Leu, and Ala; LZ, leucine zipper; PEX-PHD, motif rich in charged residues; PHD, plant homeodomain; SAD, START adjacent domain; START, steroidogenic acute regulatory protein-related lipid transfer domain; ZF, Zn finger.

7.4.5 BELL (Named for the Distinctive Bell Domain)

The BELL class of proteins was identified and characterized by the presence of two conserved domains upstream of the HD, called SKY and BEL (Bellaoui et al., 2001; Becker et al., 2002). Later, Mukherjee et al. (2009) were able to identify a third highly conserved 10-amino-acid motif repeated at both the C-terminal and N-terminal ends of the BELL proteins. These short motifs were named ZIBEL.

7.4.6 DDT (Named for the Presence of a DDT Domain Downstream of the Homeodomain)

This group is characterized by the presence of the DDT (named for DNA-binding HD and different TFs) domain downstream of the HD. This DDT domain is exclusively associated with DNA-binding domains and is generally about 60 amino acids in length. Multiple sequence alignment revealed several particular conserved regions, including conserved charged residues, N-terminal phenylalanines, and C-terminal leucines (Doerks et al., 2001). Members of this family also have seven other conserved motifs, distributed throughout the entire protein, named D-TOX A–G (DDT HB class domain). One of them, the D-TOX F motif, is a Zn finger motif, highly conserved between monocots and eudicots, characterized by the pattern C—X_2—C—X_{10}—C—X_2—C (Mukherjee et al., 2009).

7.4.7 NDX (Nodulin Homeobox Genes)

The first report about the presence of HB genes expressed in the nodule (nodulin HB genes, NDX) of soybean and *Lotus japonicus* was informed by Jørgensen and coworkers (1999). From the alignment of moss and flowering plant sequences, newly additional motifs were identified, termed NDX-A and NDX-B. NDX-A is a 540-amino-acid domain located upstream the HD while NDX-B has 80 amino acids and is located at the C-terminal of the protein. Both NDX-A and -B are highly conserved among flowering plants and mosses (Mukherjee et al., 2009).

7.4.8 LD (Luminidependens Homeobox Genes)

A single copy of a HB gene called LD (*LUMINIDEPENDENS*) has been identified in *Arabidopsis*, maize, rice, *Selaginella*, and moss while two copies of the gene have so far been identified in poplar (van Nocke et al., 2000; Mukherjee et al., 2009). These TFs are characterized by the presence of five conserved codomains downstream the HD, which were named LD1–LD5 (Mukherjee et al., 2009).

7.4.9 PINTOX

Members of this family were named after one of the genes of this class, the *Plant Interactor Homeobox* rice gene GF14c-int. The main structural feature of these TFs is a highly conserved basic domain of about 70 amino acids (named the PINTOX domain) located upstream of the HD. On the other hand, upstream of the PINTOX domain, there is a conserved acidic domain, Acid Pint, whereas the N-terminal region exhibits conserved hydrophobic and basic residues (Mukherjee et al., 2009).

7.4.10 SAWADEE (Homeodomain Associated to a Sawadee Domain)

These genes are characterized by a segment of 130 amino acids located at the C-terminal of the HD, termed the SAWADEE domain. The SAWADEE has several conserved cysteine and histidine residues, suggesting that it could coordinate a metal (Mukherjee et al., 2009).

7.5 THE HD-ZIP FAMILY

Members of the HD-Zip family are characterized by the presence of a plant-specific leucine zipper (LZ) domain, termed the HALZ (HB associated leucine zipper), immediately downstream of the HD. The LZ folds into an α-helix having a lysine in each seventh position on the same side of the helix. This three-dimensional structure allows the formation of hetero- or homodimers through hydrophobic interactions between both peptides (Landschulz et al., 1988; Sessa et al., 1993). HDs and LZs can be found in TFs from other eukaryotic kingdoms, but their combination in a single protein is unique to plants (Schena and Davis, 1992). The HD-Zip domain is highly conserved in proteins from *Physcomitrella patens*, mono- and dicots. The HD is responsible for recognition and specific DNA binding, whereas the LZ acts as a dimerization motif. It is noteworthy that the removal of the LZ absolutely abolishes the binding ability of these TFs, indicating that the relative orientation of the monomers, driven by dimerization, is crucial for efficient DNA recognition (Tron et al., 2004).

The existence of an atypical HD-Zip-containing TF in vertebrates, designated HOMEZ (HD leucine zipper encoding gene), has been reported (Bayarsaihan et al., 2003). However, a sequence comparison between this protein and plant HD-Zip proteins, as well as with other LZ-containing proteins, revealed that the structure of the LZ in HOMEZ significantly differs from a canonical LZ (Ariel et al., 2007).

The classification of HD-Zip proteins in four subfamilies is supported by four distinguishing characteristics: (a) sequence conservation of the HD-Zip domain determining DNA-binding specificities; (b) the intron/exon

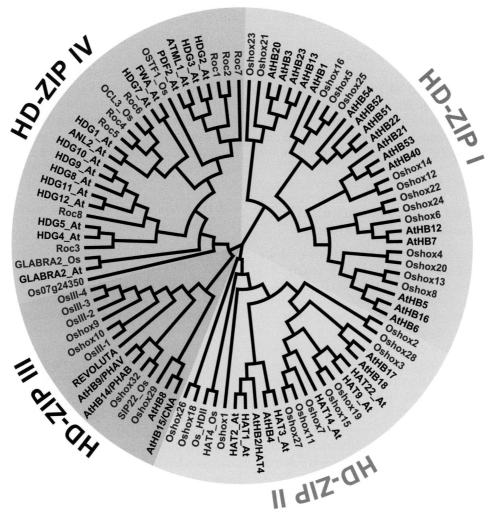

FIGURE 7.2 Circular representation of the evolutionary tree of *A. thaliana* (black) and *O. sativa* (blue) HD-Zip-containing TFs. The tree, which should be considered unrooted, was obtained based on the alignment of full-length protein sequences using MAFFT v7, PhyML 3.1, and FigTree 1.4.2.

patterns of the encoding genes; (c) the additional conserved motifs outside the HD-Zip domain; and (d) the pathways in which they participate. A phylogenetic tree that resolves HD-Zip proteins from *A. thaliana* and *O. sativa* is shown in Figure 7.2. Additionally, a schematic representation of the putative transcriptional mechanisms of HD-Zip proteins characterizing each subfamily is summarized in Figure 7.3.

7.5.1 HD-Zip Subfamily I

The HD-Zip subfamily I is composed of 17 members in *A. thaliana* and of 14 members in *O. sativa* (Ariel et al., 2007; Agalou et al., 2008). The encoded proteins have a size of about 35 kDa and exhibit a highly conserved HD and a less conserved LZ (Chan et al., 1998). Genes encoding members of the HD-Zip I subfamily have introns in helix I and in the LZ. Four different intron/exon patterns can be found: (a) AtHB52 and AtHB54 lack introns; (b) AtHB21, AtHB40, AtHB53, AtHB22, and AtHB51 have only one intron downstream the codon for the fourth leucine in the LZ domain; (c) AtHB3, AtHB20, AtHB13, AtHB23, AtHB5, AtHB6, AtHB16, AtHB1, AtHB7, and AtHB12 have only one intron located after the codon for the fifth leucine; and (d) AtHB1 has an additional intron within the region corresponding to helix I (Henriksson et al., 2005).

The 17 members of *Arabidopsis* have been classified in six groups according to phylogenetic relationships and intron/exon patterns (Henriksson et al., 2005). More recently, a new phylogenetic reconstruction, taking HD-Zip subfamily I proteins from different species, resolved these TFs into six groups (named I–VI). This recent tree considered the presence and distribution of uncharacterized conserved motifs in the carboxy terminal region (CTR; Arce et al., 2011).

AHA motifs were found among the conserved motifs in the CTRs. These motifs exhibit a characteristic pattern of aromatic and large hydrophobic residues embedded in

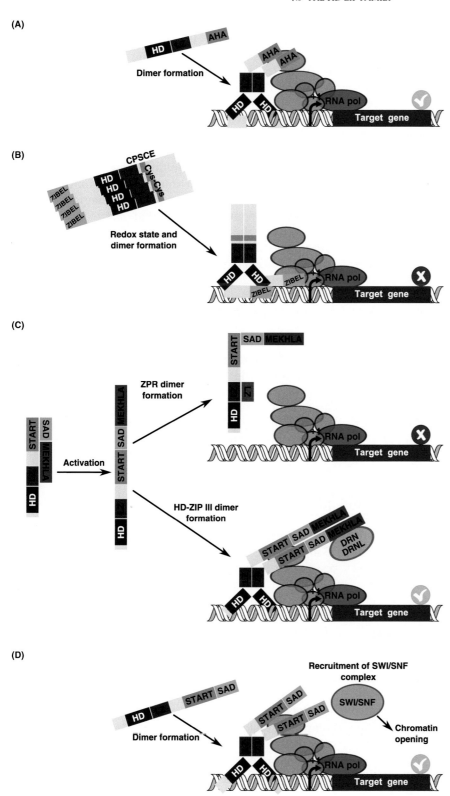

FIGURE 7.3 Schematic representation of putative action mechanisms of HD-Zip proteins. (A) HD-Zip I recognize and bind DNA by forming hetero- or homodimers and then activate transcription through interaction of their AHA motif with components of the basal transcription machinery. (B) In an oxidant environment, HD-Zip II form multimers through intermolecular disulfide bonds and cannot be imported to the nucleus. When the redox state of the cell changes, HD-Zip II could hetero- or homodimerize and repress the transcriptional expression of their target genes by means of the ZIBEL domain (with contains a LxLxL motif). (C) Normally, the MEKHLA domain inhibits HD-Zip III dimerization in a steric fashion. After activation by an unknown condition, HD-Zip III are allowed to form hetero- or homodimers and activate transcription, or they can interact with DORNRÖSCHEN (DRN) or DRN-LIKE (DRNL) and regulate the transcription of target genes. However, HD-Zip III can also interact with little zipper proteins (ZPR), and these dimers are unable to bind DNA. (D) As dimers, HD-Zip IV recruit the SWI/SNF complex to the target gene, leading to chromatin opening and activation of transcription. HD-Zip IV could also activate target expression through its START domain.

an acidic context, and were first described as activation motifs present in tomato HSF TFs (heat stress factors) by Treuter et al. (1993). Several years later, Döring et al. (2000) analyzed the activation motifs present in tomato HSFA1 and HSFA2. They concluded that the AHAs form an amphipathic and negatively charged helix to contact components of the basal transcription complex.

Although transactivation activity was experimentally demonstrated for several HD-Zip I proteins (Meijer et al., 2000; Henriksson et al., 2005; Wang et al., 2005), the role of the CTRs in such a transactivation function was experimentally confirmed only for AtHB1, AtHB12 (Lee et al., 2001; Arce et al., 2011), and more recently for AtHB7, AtHB13, Oshox22, HvHox2, and Vrs1 (Zhang et al., 2012; Sakuma et al., 2013; Capella et al., 2014). Consistently, conserved motifs fitting putative AHAs were informatically detected in the C-terminal of HD-Zip TFs (Arce et al., 2011). These motifs were demonstrated to be functional in the *Arabidopsis* AtHB1, AtHB7, AtHB12, and AtHB13 in both plants and yeasts. Additionally, tryptophans within each AHA-like motif play different roles but these amino acids seem to be very important in all cases (Capella et al., 2014).

The importance of HD-Zip I AHA motifs *in planta* has also been indirectly demonstrated. Sakuma et al. (2010) identified *HvHox2*, a putative paralog of the gene *VRS1* known for being responsible for the six-rowed phenotype in barley plants bearing the recessive allele *vrs1*. The HD-Zip I HvHox2 exhibits 14 additional amino acids in its CTR, conserved in proteins from other species and matching the characteristics of an AHA motif (Capella et al., 2014), while VRS1 does not. Another example is the HD-Zip I Tendril less (TL) from the garden pea. A TL mutant plant has leaflets instead of tendrils. The same phenotype has been observed when TL encodes a mutant protein lacking 12 amino acids in its CTR (Hofer et al., 2009); these 12 amino acids form part of a putative AHA-like motif (Arce et al., 2011).

It was demonstrated that several *Arabidopsis* HD-Zip I TFs interact with components of the basal transcriptional machinery. For example, AtHB1 was able to interact with AtTBP2, AtHB12 with AtTFIIB, AtHB7 with both, AtTBP and AtTFIIB, while AtHB13 showed weak interactions with all of them (Capella et al., 2014). The interaction of AtHB1 with AtTBP2 was further analyzed and it was shown that the deletion of the segment comprised between positions 259 and 272, containing the AHA motif, or the mutation of tryptophan residues 269 and 271 to alanine, completely abolished their interaction (Capella et al., 2014). This observation indicated that transcriptional activation exerted by these HD-Zip TFs occurs when recruiting the basal transcriptional machinery to the promoter of a target gene and/or stimulating transcription. This ability has also been attributed to several transcriptional activators (Pugh, 1996).

One member of the *Arabidopsis* subfamily, AtHB6, interacts with the protein phosphatase 2C ABI1 (<u>a</u>bscisic <u>a</u>cid <u>i</u>nsensitive 1), which is a key signal transduction factor that regulates various ABA responses (Himmelbach et al., 2002). The analysis of the interaction of ABI1 and AtHB6 in a yeast two-hybrid system revealed that such interaction needs a functional domain in ABI1. In addition, other features are crucial for the interaction; among them, a serine residue within the HD of AtHB6. Nevertheless, the dephosphorylation of AtHB6 by ABI1 was not experimentally demonstrated. On the other hand, *in vitro* assays in which AtHB6 was phosphorylated by PKA revealed that such phosphorylation impaired the ability of AtHB6 to bind DNA (Himmelbach et al., 2002). In a more recent work, it was shown that AtHB6 is also recognized by BPM3, an *Arabidopsis* member of the MATH-BTB (Meprin and TRAF homology – Bric-a-brac, Tramtrack, and Broad Complex) family, but in this case the recognized domain was the LZ. When this interaction occurs, CRL3 (Cullin-RING E3 ubiquitin ligases 3) recognizes the complex and facilitates the transfer of ubiquitin moieties to AtHB6, as a preparative step for its degradation by the 26S proteasome. It was not demonstrated but rather remains a possibility that other HD-Zip I members were also targeted by these E3 ubiquitin ligases. In this sense, in yeast cells AtHB5 and AtHB16 also interact with BPMs (Lechner et al., 2011).

Recently, it was shown that AtHB23 interacts with phytochrome B (phyB) in the nucleus and is involved in the phyB-mediated light-signaling pathway in *Arabidopsis* by an unclear mechanism (Choi et al., 2013). Nevertheless, up to now the domain involved in AtHB23 and phyB interaction has not been mapped.

HD-Zip I members are able to homo- or heterodimerize through the LZ; however, the interaction is selective. For example, *in vitro* assays demonstrated that AtHB5 is able to form heterodimers with AtHB6, AtHB7, and AtHB12 but interacts weakly with AtHB16 and does not form heterodimers at all with AtHB1 (Johannesson et al., 2001).

7.5.2 HD-Zip Subfamily II

HD-Zip subfamily II has 10 members in *Arabidopsis* and 13 in rice (Ariel et al., 2007; Jain et al., 2008). These proteins are similar in size compared with those of subfamily I. In addition, like members of subfamily I, HD-Zip II TFs can form heterodimers through interactions with the LZ domain. For example, using a yeast two-hybrid system, Meijer et al. (2000) showed that the rice HD-Zip II Oshox7 can interact with Oshox1, Oshox2, and Oshox3, but the interaction between Oshox2 and Oshox3 is weak.

Remarkably, all the *Arabidopsis* HD-Zip II members have two introns within the HD-Zip encoding region, and the location of such introns is conserved even when comparing distantly related genes. The first intron is located within the HD second helix-encoding region, while the second one splits the HD from the LZ. This particular exon/intron organization observed in the HD-Zip II proteins is not found in any of the HD-Zip I members (Ciarbelli et al., 2008).

HD-Zip II proteins show high sequence conservation in the HD-Zip domain, as well as in two additional regions: the CPSCE (named after the conserved amino acids cysteine, proline, serine, cysteine, and glutamate) motif adjacent to and downstream of the LZ, and an

N-terminal consensus sequence named ZIBEL (Mukherjee et al., 2009). The CPSCE domain is able to sense the cell redox state. In oxidant conditions, these TFs form high molecular weight multimers through intermolecular Cys–Cys bridges. In this state of high-molecular-weight macromolecules, HD-Zip II proteins could not be imported to the nucleus to exert their role (Tron et al., 2002). It was speculated that through the ZIBEL motif, also present at both N-terminal and C-terminal termini of the BELL class, the HD-Zip II and BELL HD proteins may interact with each other or with the same target proteins (Mukherjee et al., 2009). In addition, within this motif, there is an LxLxL motif in several HD-Zip II proteins, important for conferring a transcriptional repression function to the AUX/IAA factors, involved in the regulation of auxin response (Tiwari et al., 2004).

Furthermore, it was reported that the HD of HAT1, an *Arabidopsis* member of this subfamily, mediates the interaction of this TF with BES1/BZR1 (BRI1-EMS SUPPRESOR 1/BRASSINAZOLE RESISTANT 1), a TF involved in brassinosteroid responses, and thus cooperatively represses BR-repressed gene expression (Zhang et al., 2014). HAT1 interacts also with GSK3 (GLYCOGEN SYNTHASE KINASE 3)-like kinase BIN2 (BRASSINOSTEROID-INSENSITIVE 2), a negative regulator of the brassinosteroid pathway, through its LZ domain and can be phosphorylated by it, stabilizing HAT1 (Zhang et al., 2014).

7.5.3 HD-Zip Subfamily III

The HD-Zip subfamily III is formed by nine members in rice and five in *Arabidopsis* (Ariel et al., 2007; Jain et al., 2008). There are 17 internal introns, all inside the coding region, in three of the five *Arabidopsis* genes. *AtHB14* (also known as PHABULOSA) lacks the internal intron 6 and *AtHB8* lacks intron 15. Most genomic sequences from other land plants have 17 introns and the splicing sites map within, or between, the same codons in most cases (Floyd et al., 2006).

HD-Zip III proteins are characterized by the presence of four additional amino acids between the HD and the LZ. Among these domains, more than a half of the amino acids are conserved, and downstream from the HD-Zip, these proteins exhibit an additional domain termed START (steroidogenic acute regulatory protein-related lipid transfer) followed by an adjacent conserved region called SAD (START-adjacent domain). Although many START-containing proteins found in the animal kingdom have been well characterized, up to now no lipid ligands have been identified in plants (Schrick et al., 2004). However, the high conservation of this motif achieved throughout evolution indicates that it is likely to play a significant role. Additionally, all members have a conserved domain in the C-terminus, called MEKHLA, which shares a significant similarity with the PAS domain, found in a superfamily of proteins directly or indirectly involved in signal transduction (Mukherjee and Bürglin, 2006). The PAS-like domain in HD-Zip III can interact with DORNRÖSCHEN (DRN) (also known as ENHANCER OF SHOOT REGENERATION1; ESR1) and DRN-LIKE (DRNL; also known as ESR2), two linked paralogs encoding AP2 domain-containing proteins. This dimerization could control the transcription of target genes (Chandler et al., 2007). It is also known that in REVOLUTA (an *Arabidopsis* member of this subfamily), the MEKHLA domain can sterically prevent its dimerization, probably by blocking access to the LZ domain, acting as a negative regulator of its own activity. This inhibition could be relieved by an up to now unknown cellular signal (Magnani and Barton, 2011).

HD-Zip III TFs can interact with members of a protein family called ZPR (little zipper proteins), characterized by the presence of a small LZ domain similar to that found in HD-Zip III proteins. These heterodimers are unable to bind DNA, so this interaction can provide an important point of regulation, controlling the ratio of inactive heterodimers to active homodimers (Wenkel et al., 2007). This kind of inhibition was also shown for ZF-HD TFs, which interact with mini Zn finger proteins (MIF); such heterodimers inhibit the nuclear localization of ZF-HD or in the nucleus have a lower DNA-binding activity compared to ZF-HD homodimers (Hong et al., 2011).

7.5.4 HD-Zip Subfamily IV

The HD-Zip subfamily IV constitutes a large subfamily of genes, composed of 16 members in *Arabidopsis* and 12 in rice (Ariel et al., 2007; Jain et al., 2008). Like HD-Zip III TFs, HD-Zip IV exhibit START and SAD domains; however, HD-Zip IV binding and dimerization are closer to those of subfamilies I and II. The most distinguishable features of HD-Zip IV TFs are the presence of a loop in the middle of the LZ domain and the absence of the MEKHLA motif. The proteins belonging to this subfamily have also been named HD-Zip GL2 or simply the GL2 family after its founding member, the *Arabidopsis* GLABRA2 protein (Palena et al., 1997; Nakamura et al., 2006).

As was pointed out earlier, one distinguishable feature of HD-Zip subfamilies III and IV is the presence of a conserved region downstream of the START domain, the SAD domain, and little is known about its function. However, both the START and SAD domains of the cotton HD-Zip IV TF GbML1 are necessary for binding to the C-terminal domain of GbMYB25, a TF shown to be a key regulator of cotton fiber initiation (Zhang et al., 2010a).

Members of this subfamily act as transcriptional activators, and it looks like this ability is located at the N-terminal of the START domain (Depege-Fargeix et al., 2011; Wei et al., 2012). Moreover, it was suggested that binding of lipids by the START domain might be necessary for nuclear transport of these TFs in a way similar to the model proposed for the glucocorticoid receptors in mammals (Chew et al., 2013).

Transcriptional activation is accompanied by changes in chromatin structure giving TFs access to their target regulatory regions, and this step often requires multiprotein complexes capable of manipulating the nucleosome architecture. Interestingly, the maize SWITCH complex protein 3C1 (ZmSWI3C1), one of the core subunits of the SWITCH/SUCROSE NONFERMENTING (SWI/SNF) ATP-dependent chromatin-remodeling complex in yeast and animals (Sarnowski et al., 2005), was identified and confirmed as an interacting partner of maize HD-Zip IV outer cell layer 1 (ZmOCL1). However, the motif responsible for this interaction has not yet been mapped (Depege-Fargeix et al., 2011). ATP-dependent chromatin-remodeling complexes, such as SWI/SNF, temporarily inactivate histone interactions with DNA, allowing DNA to be more accessible to TFs. However, the SWI/SNF complex binds DNA in a nonspecific manner. Thus, it is likely that the interaction between ZmOCL1 and ZmSWI3C1 could recruit the SWI/SNF complex to a particular promoter region(s) of OCL1 target genes, leading to chromatin opening and active transcription (Depege-Fargeix et al., 2011).

7.6 TARGET SEQUENCES RECOGNIZED BY THE HD-CONTAINING TRANSCRIPTION FACTORS

In general, DNA sequences recognized and efficiently bound by HD-containing proteins include the ATTA core, or TAAT in the complementary strand (Ariel et al., 2007; Chapter 6), with the sequence NNATTA being the canonical binding site for most HDs with the specificity given by the first two bases (Fraenkel and Pabo, 1998; Conolly et al., 1999). In fact, several experiments in mammals proved that the binding of TAATCC by the Bicoid class of HDs is promoted by the residue Lys50, instead of the TAAT(T/G)(A/G) recognized by the Gln50-containing Antennapedia and Engrailed classes (Hanes and Brent, 1989; Treisman et al., 1989; Percival-Smith et al., 1990).

Both structural data and *in vitro/in vivo* analysis suggest that the primary sequence determinants within the N-terminal arm may help to define sequence preferences, but intramolecular or intermolecular interactions can also influence recognition (Noyes et al., 2008). Nonetheless, notwithstanding a common DNA-binding architecture, it has been shown that there is a meaningful variation in the sequence composition recognized by the HD superfamily (Noyes et al., 2008). For instance, two superclasses of HDs, denoted as typical and atypical (Banerjee-Basu and Baxevanis, 2001; Mukherjee and Bürglin, 2007), with low residue conservation and significantly different DNA-binding motifs, share a nearly identical docking with the DNA (Kissinger et al., 1990; Wolberger et al., 1991). A recent study on 327 human TFs and 84 mouse TFs using high-throughput SELEX found that the HD superfamily, for which almost 150 proteins were analyzed, was among few superfamilies that included large subsets of proteins that bound a single motif (Jolma et al., 2013).

Animal HD-containing proteins can bind DNA as monomers, homo- or heterodimers, or higher order complexes; in some cases, the preferred recognition sequence of monomers in these complexes may even be modified (Noyes et al., 2008). For instance, the binding specificity of *Drosophila* HD proteins Hox and Exd as monomers is remarkably different from that of the complex Hox–Exd (Joshi et al., 2007), raising the prospect that the monomeric binding preferences may not always be relevant to targeting *in vivo*. Nevertheless, other evidence suggests that cofactor alterations to binding specificities of the monomer are likely to be e exception rather than the rule. Carr and Biggin (1999) and Berger et al. (2008) separately showed that there is good correlation between monomer binding *in vitro* and the genomic fragments bound *in vivo* for several mammal HD-containing proteins, supporting the biological relevance of the binding preferences of HD monomers. These evidences could indicate that monomer-binding preferences are likely to be a component of targeting mechanisms in general (Berger et al., 2008).

In the plant kingdom, the DNA-HD interaction has been less studied than in animals. However, the binding specificity of several plant HD TFs has been analyzed. One example is the maize HB-containing gene *KNOTTED1*. The DNA-binding sequence of this protein, TGACAG(C/G)T, was biochemically identified, and it was shown that KNOTTED1 can interact with knotted interacting protein (KIP), a BEL1-like TALE HD protein. *In vitro* assays proved that both KNOTTED1 and KIP could recognize and bind specifically this motif, but this interaction has low affinity. Nevertheless, the KNOTTED1–KIP complex binds the DNA sequence with high affinity, indicating that the association of these two proteins could have a function in transcriptional regulation (Smith et al., 2002).

On the other hand, to orchestrate the expression of direct target genes, WUS recognizes at least two divergent DNA sequence motifs, TCAYRTGA (which is similar to a G-Box) and TTAATGG, with different binding affinities (Lohmann et al., 2001; Busch et al., 2010). This DNA-binding affinity difference suggests that WUS might regulate distinct sets of targets in a concentration-dependent manner (Busch et al., 2010).

In the case of the ZF-HD members, the DNA sequence CACTAAATTGTCAC was determined, using *in vitro* and one-hybrid assays, to be specifically recognized by the *A. thaliana* ZFHD1 (Tran et al., 2006). Within these 14 nucleotides recognized by ZFHD1, there is a sequence (TAATT) that resembles the binding sequence TGTAATT of the α2-HD (Wolberger et al., 1991). Additionally, it was shown that AtHB33, belonging to the *Arabidopsis* Zn finger HD family, bound to the DNA region (AGTGTCTTGTA-ATTAAA), in which the NNATTA consensus sequence is present (Tan and Irish, 2006). The recognition site of ZFHD1 seems to contain this consensus (underlined), except for the last nucleotide, which is G instead of A. This last nucleotide is absent in the α2-HD DNA-binding sequence. Altogether, this could indicate that this last residue is not essential (Tran et al., 2006).

7.7 WHAT DO WE KNOW ABOUT THE TARGET SEQUENCES OF THE HD-ZIP PROTEINS?

Several *in vitro* assays, consisting of PCR-assisted binding site selection and footprinting assays, demonstrated that proteins belonging to the HD-Zip I subfamily recognize the pseudopalindromic sequence CAAT(A/T)ATTG forming hetero- or homodimers (Sessa et al., 1993; Chan et al., 1998; Palena et al., 1999; Johannesson et al., 2001; Palena et al., 2001). The *in vivo* recognition of this sequence was demonstrated for *MtHB1*, a *Medicago truncatula* HD-Zip I gene involved in lateral root emergence (Ariel et al., 2007), and for *Oshox4*, a TF with a negative function in rice gibberellin responses (Dai et al., 2008). The only member that is apparently unable to bind to DNA *in vitro* is AtHB7 (Johannesson et al., 2001). Several *Arabidopsis* members of this subfamily (*AtHB5*, *AtHB7*, *AtHB12*, *AtHB21*, and *AtHB40*) and one member of *Rosa hybrida* (*RhHB1*) have been shown to bind *cis*-elements in the promoter regions of their target genes that differ from the preferred pseudopalindromic sequence found *in vitro* (Comelli et al., 2012; Valdés et al., 2012; De Smet et al., 2013; Lü et al., 2014). Nevertheless, all regulatory regions have at least the central core of this site, so it looks like the interaction with target gene promoters *in vivo* may be mediated by *cis*-elements related to the consensus HD-bound sequence.

Biochemical *in vitro* analysis performed with HaHB4, a sunflower member of this subfamily, indicated that positively charged residues at the N-terminal flexible arm plays an essential role in determining the affinity of DNA–protein interaction, without changing the specificity (Palena et al., 2001). On the other hand, the amino acidic composition of the loop located between helixes I and II of HD-Zip proteins must be regarded as one of the segments responsible for the inability of these TFs to bind DNA in their monomeric form as most of HD-containing proteins do (Tron et al., 2004).

Proteins encoded by genes of the HD-Zip subfamily II also form dimers that bind to a specific pseudopalindromic sequence similar to that bound by members of subfamily I, but they prefer different nucleotides at the central position of the recognized sequence. HD-Zip II TFs can recognize the nucleotide C/G instead of A/T, meaning that the target sequence is CAAT(C/G)ATTG. The binding specificity of the central nucleotide in the pseudopalindromic sequence seems to be conferred in part by differences in the relative orientations of their monomers, in which the amino acids 46 and 56 of helix III are involved (Ala and Trp in HD-Zip I; Glu and Thr in HD-Zip II), together with a different spatial orientation of the conserved Arg55 in both proteins. Arg55 would be directly responsible for the interaction (Tron et al., 2005).

Little is known about the interaction between DNA and HD-Zip III proteins, since this binding is less studied than in the case of the other subfamilies (Ariel et al., 2007). Nonetheless, GTAAT(G/C)ATTAC was determined as the extended sequence for which *AtHB9* has the highest affinity *in vitro* (Sessa et al., 1998). In *A. thaliana* it was also shown that REVOLUTA recognizes the intron region in the 5′ untranslated region of *ZPR3* and positively regulates its expression (Magnani and Barton, 2011). Within this region, the sequence TAATCATTAC could be found, which resembles the one recognized by AtHB9.

Finally, proteins belonging to subfamily IV do not have a consensus target site, since they show a binding preference for alternative sequences (Ariel et al., 2007). For instance, the *Helianthus annuus* member HAHR1 could recognize, as a preferential target, the sequence CATT(A/T)AATG (Tron et al., 2001). On the other hand, the results of binding site selection experiments using HDG7, HDG9, ATML1, and PDF2 recombinant proteins revealed a GCATT(A/T)AATGC consensus sequence (Nakamura et al., 2006). This sequence overlaps the L1 box sequence TAAATG(C/T)A recognized *in vitro* by ATML1 and *in vivo* by GL2 (Abe et al., 2001; Khosla et al., 2014). Together, it looks like the sequences recognized by HD-Zip IV proteins are all characterized by a TAAA core. This motif is in fact present in a target site of GL2 within the phospholipase Dξ1 gene promoter (Ohashi et al., 2003).

A recent work employed protein-binding microarrays to determine the binding specificities of 63 plant TFs (Franco-Zorrilla et al., 2014). The results showed that although proteins from the same subfamily display highly similar binding preferences, some minor differences could be found in the DNA recognition patterns. Unexpectedly, many TFs also bound secondary sequences with similar or lower affinity. Nonetheless, the three HD-Zip proteins studied, AtHB12 and AtHB51 from subfamily I and ICU4/CORONA from subfamily III, did not present this ability. The binding sequences determined for these TFs were in agreement with previous studies on proteins from each group, except for the preference of AtHB12 for a G/C in the central position of the pseudopalindrome, usually an A/T for proteins of subfamily I (Franco-Zorrilla et al., 2014).

7.8 CONCLUDING REMARKS

Throughout evolution, the HB gene family has proliferated and diversified in accordance with the growth in structural and developmental complexity of the organisms in which they were expressed. Since its first discovery in *Drosophila*, many functional and DNA-binding studies have been carried out, and we are now starting to understand the functionality of these genes and classes, but many questions remain unanswered.

In particular, HD-Zip proteins have been more deeply analyzed in the past few years thanks to novel techniques and the availability of sequences, especially from nonmodel plant species. However, there are more open questions to be answered in the next future. In particular, which are the targets of the HD-Zip TFs? Which are their partners? Which roles are the conserved motifs playing outside the HD-Zip? Do they suffer posttranslational regulations? How is the crosstalk between these TFs, phytohormones, and other molecules to regulate plant responses?

It is expected that answers to these, other, and new questions will be found in the future, allowing the scientific community to understand the complex regulatory mechanisms in which plant TFs are involved.

References

Abe, M., Takahashi, T., Komeda, Y., 2001. Identification of a *cis*-regulatory element for L1 layer-specific gene expression, which is targeted by an L1-specific homeodomain protein. Plant J. 26, 487–494.

Ades, S.E., Sauer, R.T., 1995. Specificity of minor-groove and major-groove interactions in a homeodomain–DNA complex. Biochemistry 34, 14601–14608.

Agalou, A., Purwantomo, S., Overnas, E., Johannesson, H., Zhu, X., Estiadi, A., de Kam, R.J., Engstrom, P., Slamet-Loedin, I.H., Zhu, Z., Wang, M., Xiong, L., Meijer, A.H., Ouwerkerk, P.B.F., 2008. A genome-wide survey of HD-Zip genes in rice and analysis of drought-responsive family members. Plant Mol. Biol. 66, 87–103.

Arce, A.L., Raineri, J., Capella, M., Cabello, J.V., Chan, R.L., 2011. Uncharacterized conserved motifs outside the HD-Zip domain in HD-Zip subfamily I transcription factors; a potential source of functional diversity. BMC Plant Biol. 11, 42.

Ariel, F.D., Manavella, P.A., Dezar, C.A., Chan, R.L., 2007. The true story of the HD-Zip family. Trends Plant Sci. 12, 419–426.

Ariel, F., Diet, A., Verdenaud, M., Gruber, V., Frugier, F., Chan, R., Crespi, M., 2010. Environmental regulation of lateral root emergence in Medicago truncatula requires the HD-Zip I transcription factor HB1. Plant Cell 22, 2171–2183.

Banerjee-Basu, S., Baxevanis, A.D., 2001. Molecular evolution of the homeodomain family of transcription factors. Nucleic Acids Res. 29, 3258–3269.

Bayarsaihan, D., Enkhmandakh, B., Majeyev, A., Greally, J.M., Leckman, J.F., Ruddle, F.H., 2003. Homez, a homeobox leucine zipper gene specific to the vertebrate lineage. Proc. Natl. Acad. Sci. USA 100, 10358–10363.

Becker, A., Bey, M., Bürglin, T.R., Saedler, H., Theissen, G., 2002. Ancestry and diversity of BEL1-like homeobox genes revealed by gymnosperm (*Gnetum gnemon*) homologs. Dev. Genes Evol. 212, 452–457.

Bellaoui, M., Pidkowich, M.S., Samach, A., Kushalappa, K., Kohalmi, S.E., Modrusan, Z., Crosby, W.L., Haughn, G.W., 2001. The *Arabidopsis* BELL1 and KNOX TALE homeodomain proteins interact through a domain conserved between plants and animals. Plant Cell 13, 2455–2470.

Bender, W., Akam, M., Karch, F., Beachy, P.A., Peifer, M., Spierer, P., Lewis, E.B., Hogness, D.S., 1983. Molecular genetics of the bithorax complex in *Drosophila melanogaster*. Science 221, 23–29.

Berger, M.F., Badis, G., Gehrke, A.R., Talukder, S., Philippakis, A.A., Pena-Castillo, L., Alleyne, T.M., Mnaimneh, S., Botvinnik, O.B., Chan, E.T., Khalid, F., Zhang, W., Newburger, D., Jaeger, S.A., Morris, Q.D., Bulyk, M.L., Hughes, T.R., 2008. Variation in homeodomain DNA binding revealed by high-resolution analysis of sequence preferences. Cell 133, 1266–1276.

Bürglin, T.R., 1997. Analysis of TALE superclass homeobox genes (MEIS, PBC, KNOX, Iroquois. TGIF) reveals a novel domain conserved between plants and animals. Nucleic Acids Res. 25, 4173–4180.

Bürglin, T.R., 1998. The PBC domain contains a MEINOX domain: coevolution of Hox and TALE homeobox genes? Dev. Genes Evol. 208, 113–116.

Busch, W., Miotk, A., Ariel, F.D., Zhao, Z., Forner, J., Daum, D., Suzaki, T., Schuster, C., Schulthesis, S.J., Leibfried, A., Heubeib, S., Ha, N., Chan, R.L., Lohmann, J.U., 2010. Transcriptional control of a plant stem cell niche. Dev. Cell 18, 841–853.

Capella, M., Ré, D.A., Arce, A.L., Chan, R.L., 2014. Plant homeodomain–leucine zipper I transcription factors exhibit different functional AHA motifs that selectively interact with TBP or/and TFIIB. Plant Cell Rep. 33, 955–967.

Carr, A., Biggin, M.D., 1999. A comparison of *in vivo* and *in vitro* DNA-binding specificities suggests a new model for homeoprotein DNA binding in *Drosophila* embryos. EMBO J. 18, 1598–1608.

Chan, R.L., Gago, G.M., Palena, C.M., Gonzalez, D.H., 1998. Homeoboxes in plant development. Biochim. Biophys. Acta 1442, 1–19.

Chandler, J.W., Cole, M., Flier, A., Grewe, B., Werr, W., 2007. The AP2 transcription factors DORNRÖSCHEN and DORNRÖSCHEN-LIKE redundantly control *Arabidopsis* embryo patterning via interaction with PHAVOLUTA. Development 134, 1653–1662.

Chew, W., Hrmova, M., Lopato, S., 2013. Role of homeodomain leucine zipper (HD-Zip) IV transcription factors in plant development and plant protection from deleterious environmental factors. Int. J. Mol. Sci. 14, 8122–8147.

Choi, H., Jeong, S., Kim, D.S., Na, H.J., Ryu, J.S., Lee, S.S., Nam, H.G., Lim, P.O., Woo, H.R., 2013. The homeodomain–leucine zipper ATHB23, a phytochrome B-interacting protein, is important for pythochrome B-mediated red light signaling. Plant Physiol. 150, 308–320.

Ciarbelli, A.R., Ciolfi, A., Salvucci, S., Ruzza, V., Possenti, M., Carabelli, M., Fruscalzo, A., Sessa, G., Morelli, G., Ruberti, I., 2008. The *Arabidopsis* homeodomain–leucine zipper II gene family: diversity and redundancy. Plant Mol. Biol. 68, 465–478.

Comelli, R.N., Welchen, E., Kim, H.J., Hong, J.C., Gonzalez, D.H., 2012. Delta subclass HD-Zip proteins and a B-3 AP2/ERF transcription factor interact with promoter elements required for expression of the *Arabidopsis cytochrome c oxidase 5b-1* gene. Plant Mol. Biol. 80, 157–167.

Conolly, J.P., Augustine, J.G., Francklyn, C., 1999. Mutational analysis of the engrailed homeodomain recognition helix by phage display. Nucleic Acids Res. 27, 1182–1189.

REFERENCES

Dai, M., Hu, Y., Ma, Q., Zhao, Y., Zhou, D.X., 2008. Functional analysis of rice *HOMEOBOX4* (*Oshox4*) gene reveals a negative function in gibberellin responses. Plant Mol. Biol. 66, 289–301.

De Smet, I., Lau, S., Ehrismann, J.S., Axiotis, I., Kolb, M., Kientz, M., Weijers, D., Jürgens, G., 2013. Transcriptional repression of *BODENLOS* by HD-ZIP transcription factor HB5 in *Arabidopsis thaliana*. J. Exp. Bot. 64, 3009–3019.

Depege-Fargeix, N., Javelle, M., Chambrier, P., Frangne, N., Gerentes, D., Perez, P., Rogowsky, P.M., Vernoud, V., 2011. Functional characterization of the HD-ZIP IV transcription factor OCL1 from maize. J. Exp. Bot. 62, 293–305.

Doerks, T., Copley, R., Bork, P., 2001. DDT – a novel domain in different transcription and chromosome remodeling factors. Trends Biochem. Sci. 26, 145–146.

Döring, P., Treuter, E., Kistner, C., Lyck, R., Chen, A., Nover, L., 2000. The role of AHA motifs in the activator function of tomato heat stress transcription factors HsfA1 and HsfA2. Plant Cell 12, 265–278.

Floyd, S.K., Zalewski, C.S., Bowman, J.L., 2006. Evolution of class III homeodomain–leucine zipper genes in streptophytes. Genetics 173, 373–388.

Fraenkel, E., Pabo, C.O., 1998. Comparison of X-ray and NMR structures for the Antennapedia homeodomain–DNA complex. Nat. Struct. Biol. 5, 692–697.

Franco-Zorrilla, J.M., López-Vidriero, I., Carrasco, J.L., Godoy, M., Vera, P., Solano, R., 2014. DNA-binding specificities of plant transcription factors and their potential to define target genes. Proc. Natl. Acad. Sci. USA 111, 2367–2372.

Garber, R.L., Kuroiwa, A., Gehring, W.J., 1983. Genomic and cDNA clones of the homeotic locus *Antennapedia* in *Drosophila*. EMBO J. 2, 2027–2036.

Gehring, W.J., 1987. Homeoboxes in the study of development. Science 236, 1245–1252.

Gehring, W.J., Qian, Y.Q., Billeter, M., Furukubo-Tokunaga, K., Schier, A.F., Resendez-Perez, D., Affolter, M., Otting, D., Wüthrich, K., 1994. Homeodomain–DNA recognition. Cell 78, 211–223.

Halbach, T., Scheer, N., Werr, W., 2000. Transcriptional activation by the PHD finger is inhibited through an adjacent lecune zipper that binds 14-3-3 proteins. Nucleic Acids Res. 28, 3542–3550.

Hanes, S.D., Brent, R., 1989. DNA specificity of the bicoid activator protein is determined by homeodomain recognition helix residue 9. Cell 57, 1275–1283.

Henriksson, E., Olsson, A.S.B., Johannesson, H., Johansson, H., Hanson, J., Engstrom, P., Söderman, E., 2005. Homeodomain leucine zipper class I genes in *Arabidopsis*. Expression patterns and phylogenetic relationships. Plant Physiol. 139, 509–518.

Himmelbach, A., Hoffmann, T., Leube, M., Hohener, B., Grill, E., 2002. Homeodomain protein ATHB6 is a target of the protein phosphatase ABI1 and regulates hormone responses in *Arabidopsis*. EMBO J. 21, 3029–3038.

Hofer, J., Turner, L., Moreau, C., Ambrose, M., Isaac, P., Butcher, S., Weller, J., Dupin, A., Dalmais, M., Le Signor, C., Bendahmane, A., Ellis, N., 2009. Tendril-less regulates tendril formation in pea leaves. Plant Cell 21, 420–428.

Hong, S.Y., Kim, O.K., Kim, S.G., Yang, M.S., Park, C.M., 2011. Nuclear import and DNA binding of the ZHD5 transcription factor is modulated by a competitive peptide inhibitor in *Arabidopsis*. J. Biol. Chem. 286, 1659–1668.

Ikeda, M., Mitsuda, N., Ohme-Takagi, M., 2009. Arabidopsis WUSCHEL is a bifunctional transcription factor that acts as a repressor in stem cell regulation and as an activator in floral patterning. Plant Cell 21, 3493–3505.

Jain, M., Tyagi, A.K., Khurana, J.P., 2008. Genome-wide identification, classification, evolutionary expansion and expression analyses of homeobox genes in rice. FEBS J. 275, 2845–2861.

Johannesson, H., Wang, Y., Engström, P., 2001. DNA-binding and dimerization preferences of *Arabidopsis* homeodomain–leucine zipper transcription factors *in vitro*. Plant Mol. Biol. 45, 63–73.

Jolma, A., Yan, J., Whitington, T., Toivonen, J., Nitta, K.R., Rastas, P., Morgunova, E., Enge, M., Taipale, M., Wei, G., Palin, K., Vaquerizas, J.M., Vincentelli, R., Luscombe, N.M., Hughes, T.R., Lemaire, P., Ukkonen, E., Kivioja, T., Taipale, J., 2013. DNA-binding specificities of human transcription factors. Cell 152, 327–339.

Jørgensen, J.E., Grønlund, M., Pallisgaard, N., Larsen, K., Marcker, K.A., Jensen, E.O., 1999. A new class of plant homeobox genes is expressed in specific regions of determinate symbiotic root nodules. Plant Mol. Biol. 40, 65–77.

Joshi, R., Passner, J.M., Rohs, R., Jain, R., Sosinsky, A., Crickmore, M.A., Jacob, V., Aggarwal, A.K., Honig, B., Mann, R.S., 2007. Functional specificity of a **Hox** protein mediated by the recognition of minor groove structure. Cell 131, 530–543.

Khosla, A., Paper, J.M., Boehler, A.P., Bradley, A.M., Neumann, T.R., Schrick, K., 2014. HD-Zip proteins GL2 and HDG11 have redundant functions in *Arabidopsis* trichomes, and GL2 activates a positive feedback loop via MYB23. Plant Cell 26, 2184–2200.

Kissinger, C.R., Liu, B.S., Martin-Blanco, E., Kornberg, T.B., Pabo, C.O., 1990. Crystal structure of an engrailed homeodomain–DNA complex at 2.8 A resolution: a framework for understanding homeodomain–DNA interactions. Cell 63, 579–590.

Landschulz, W.H., Johnson, P.F., McKnight, S.L., 1988. The leucine zipper: a hypothetical structure common to a new class of DNA binding proteins. Science 240, 1759–1764.

Lechner, E., Leonhardt, N., Eisler, H., Parmentier, Y., Alioua, M., Jacquet, H., Leung, J., Genschik, P., 2011. MATH/BTB CRL3 receptors target the homeodomain–leucine zipper ATHB6 to modulate abscisic acid signaling. Dev. Cell 21, 1116–1128.

Lee, Y.H., Oh, H.S., Cheon, C.I., Hwang, I.T., Kim, Y.J., Chun, J.Y., 2001. Structure and expression of the *Arabidopsis thaliana* homeobox gene Athb-12. Biochem. Biophys. Res. Commun. 284, 133–141.

Lohmann, J.U., Hong, R.L., Hobe, M., Busch, M.A., Parcy, F., Simon, R., Weigel, D., 2001. A molecular link between stem cell regulation and floral patterning in *Arabidopsis*. Cell 105, 793–803.

Lü, P., Zhang, C., Liu, J., Liu, X., Jiang, G., Jiang, X., Khan, M.A., Wang, L., Hong, B., Gao, J., 2014. RhHB1 mediates the antagonism of gibberellins to ABA and ethylene during rose (*Rosa hybrida*) petal senescence. Plant J. 78, 578–590.

Magnani, E., Barton, M.K., 2011. A Per-ARNT-Sim-like sensor domain uniquely regulates the activity of the homeodomain leucine zipper transcription factor REVOLUTA in *Arabidopsis*. Plant Cell 23, 567–582.

Meijer, A.H., de Kam, R.J., d'Erfurth, I., Shen, W., Hoge, J.H.C., 2000. HD-Zip proteins of families I and II from rice: interactions and functional properties. Mol. Gen. Genet. 263, 12–21.

Moens, C.B., Selleri, L., 2006. Hox cofactors in vertebrate development. Dev. Biol. 291, 193–206.

Mukherjee, K., Bürglin, T.R., 2006. MEKHLA, a novel domain with similarity to PAS domains, is fused to plant homeodomain–leucine zipper III proteins. Plant Physiol. 140, 1142–1150.

Mukherjee, K., Bürglin, T.R., 2007. Comprehensive analysis of animal TALE homeobox genes: new conserved motifs and cases of accelerated evolution. J. Mol. Evol. 65, 137–153.

Mukherjee, K., Brocchieri, L., Bürglin, T.R., 2009. A comprehensive classification and evolutionary analysis of plant homeobox genes. Mol. Biol. Evol. 26, 2775–2794.

Nakamura, M., Katsumata, H., Abe, M., Yabe, N., Komeda, Y., Yamamoto, K.T., Takahashi, T., 2006. Characterization of the class IV homeodomain–leucine zipper (HD-Zip IV) gene family in *Arabidopsis*. Plant Physiol. 141, 1363–1375.

Noyes, M.B., Christensen, R.G., Wakabayashi, A., Stormo, G.D., Brodsky, M.H., Wolfe, S.A., 2008. Analysis of homeodomain specificities allows the family-wide prediction of preferred recognition sites. Cell 133, 1277–1289.

Ohashi, Y., Oka, A., Rodrigues-Pousada, R., Possenti, M., Ruberti, I., Morelli, G., Aoyama, T., 2003. Modulation of phospholipid signaling by GLABRA2 in root-hair pattern formation. Science 300, 1427–1430.

Palena, C.M., Chan, R.L., Gonzalez, D.H., 1997. A novel type of dimerization motif, related to leucine zippers, is present in plant homeodomain proteins. Biochim. Biophys. Acta 1352, 203–212.

Palena, C.M., Gonzalez, D.H., Chan, R.L., 1999. A monomer–dimer equilibrium modulates the interaction of the sunflower homeodomain–leucine zipper protein HAHB-4 with DNA. Biochem. J. 341, 81–87.

Palena, C.M., Tron, A.E., Bertoncini, C.W., Gonzalez, D.H., Chan, R.L., 2001. Positively charged residues at the N-terminal arm of the homeodomain are required for efficient DNA binding by homeodomain–leucine zipper proteins. J. Mol. Biol. 308, 39–47.

Percival-Smith, A., Müller, M., Affolter, M., Gehring, W.J., 1990. The interaction with DNA of wild-type and mutant fushi tarazu homeodomains. EMBO J. 9, 3967–3974.

Pugh, B.F., 1996. Mechanisms of transcription complex assembly. Curr. Opin. Cell Biol. 8, 303–311.

Plesch, G., Störmann, K., Torres, J.T., Walden, R., Somssich, I.E., 1997. Developmental and auxin-induced expression of the Arabidopsis prha homeobox gene. Plant J. 12:635–647.

Sakuma, S., Pourkheirandish, M., Matsumoto, T., Koba, T., Komatsuda, T., 2010. Duplication of a well-conserved homeodomain–leucine zipper transcription factor gene in barley generates a copy with more specific functions. Funct. Integr. Genomics 10, 123–133.

Sakuma, S., Pourkheirandish, M., Hensel, G., Kumlehn, J., Stein, N., Tagiri, A., Yamaji, N., Ma, J.F., Sassa, H., Koba, T., Komatsuda, T., 2013. Divergence of expression pattern contributed to neofunctionalization of duplicated HD-Zip I transcription factor in barley. New Phytol. 197, 939–948.

Sarnowski, T.J., Ríos, G., Jásik, J., Świeżewskia, S., Kaczanowskia, S., Lib, Y., Kwiatkowskac, A., Pawlikowskaa, K., Koźbiałc, M., Koźbiałc, P., Koncs, C., Jerzmanowski, A., 2005. SWI3 subunits of putative SWI/SNF chromatin-remodeling complexes play distinct roles during Arabidopsis development. Plant Cell 17, 2454–2472.

Schena, M., Davis, R.W., 1992. HD-Zip proteins members of Arabidopsis homeodomain protein superfamily. Proc. Natl. Acad. Sci. USA 89, 3894–3898.

Schindler, U., Beckmann, H., Cashmore, A.R., 1993. HAT3.1, a novel Arabidopsis homeodomain protein containing a conserved cysteine-rich region. Plant J. 4:137–150.

Schrick, K., Nguyen, D., Karlowski, W.M., Mayer, K.F.X., 2004. START lipid/sterol-binding domains are amplified in plants and are predominantly associated with homeodomain transcription factors. Genome Biol. 5, R41.

Sessa, G., Morelli, G., Ruberti, I., 1993. The Athb-1 and -2 HD-Zip domains homodimerize forming complexes of different DNA binding specificities. EMBO J. 12, 3507–3517.

Sessa, G., Steindler, C., Morelli, G., Ruberti, I., 1998. The Arabidopsis ATHB-8, -9 and -14 genes are members of a small gene family coding for highly related HD-Zip proteins. Plant Mol. Biol. 38, 609–622.

Smith, H.M.S., Boschke, I., Hake, S., 2002. Selective interaction of plant homeodomain proteins mediates high DNA-binding affinity. Proc. Natl. Acad. Sci. USA 99, 9579–9584.

Tan, K.-G.Q., Irish, F.V., 2006. The Arabidopsis zinc finger-homeodomain genes encode proteins with unique biochemical properties that are coordinately expressed during floral development. Plant Physiol. 140, 1095–1108.

Tiwari, S.B., Hagen, G., Guilfoyle, T.J., 2004. Aux/IAA proteins contain a potent transcriptional repression domain. Plant Cell 16, 533–543.

Tran, L.S.P., Nakashima, K., Sakuma, Y., Osakabe, Y., Qin, F., Simpson, S.D., Maruyama, K., Fujita, Y., Shinozaki, K., Yamaguchi-Shinozaki, K., 2006. Co-expression of the stress-inducible zinc finger homeodomain ZFHD1 and NAC transcription factors enhances expression of the ERD1 gene in Arabidopsis. Plant J. 49, 46–63.

Treisman, J., Gonczy, P., Vashishtha, M., Harris, E., Desplan, C., 1989. A single amino acid can determine the DNA binding specificity of homeodomain proteins. Cell 59, 553–562.

Treuter, E., Nover, L., Ohme, K., Scharf, K.-D., 1993. Promoter specificity and deletion analysis of three heat stress transcription factors of tomato. Mol. Gen. Genet. 240, 113–125.

Tron, A.E., Bertoncini, C.W., Palena, C.M., Chan, R.L., Gonzalez, D.H., 2001. Combinatorial interactions of two amino acids with a single base pair define target site specificity in plant dimeric homeodomain proteins. Nucleic Acids Res. 29, 4866–4872.

Tron, A.E., Bertoncini, C.W., Chan, R.L., Gonzalez, D.H., 2002. Redox regulation of plant homeodomain transcription factors. J. Biol. Chem. 277, 34800–34807.

Tron, A.E., Welchen, E., Gonzalez, D.H., 2004. Engineering the loop region of a homeodomain–leucine zipper protein promotes efficient binding to a monomeric DNA binding site. Biochemistry 43, 15845–15851.

Tron, A.E., Comelli, R.N., Gonzalez, D.H., 2005. Structure of homeodomain–leucine zipper/DNA complexes studied using hydroxyl radical cleavage of DNA and methylation interference. Biochemistry 44, 16796–16803.

Valdés, A.E., Övernäs, E., Johansson, H., Rada-Iglesias, A., Engström, P., 2012. The homeodomain–leucine zipper (HD-Zip) class I transcription factors ATHB7 and ATHB12 modulate abscisic acid signalling by regulating protein phosphatase 2C and abscisic acid receptor gene activities. Plant Mol. Biol. 80, 405–418.

van Nocke, S., Muszynski, M., Briggs, K., Amasino, R.M., 2000. Characterization of a gene from Zea mays related to the Arabidopsis flowering-time gene LUMINIDEPENDENS. Plant Mol. Biol. 44, 107–122.

Vollbrecht, E., Veit, B., Sinha, N., Hake, S., 1991. The developmental gene Knotted-1 is a member of a maize homeobox gene family. Nature 350, 241–243.

Wang, Y.J., Li, Y.D., Luo, G.Z., Tian, A.G., Wang, H.W., Zhang, J.S., Chen, S.Y., 2005. Cloning and characterization of an HDZip I gene GmHZ1 from soybean. Planta 221, 831–843.

Wei, Q., Kuai, B.K., Hu, P., Ding, Y., 2012. Ectopic-overexpression of an HD-Zip IV transcription factor from Ammopiptanthus mongolicus (Leguminosae) promoted upward leaf curvature and non-dehiscent anthers in Arabidopsis thaliana. Plant Cell Tissue Organ Cult. 110, 299–306.

Wenkel, S., Emery, J., Hou, B.-H., Evans, M.M.S., Barton, M.K., 2007. A feedback regulatory module formed by LITTLE ZIPPER and HD-ZIPIII genes. Plant Cell 19, 3379–3390.

Wolberger, C., 1996. Homeodomain interactions. Curr. Opin. Struct. Biol. 6, 62–68.

Wolberger, C., Vershon, A.K., Liu, B., Johnson, A.D., Pabo, C.O., 1991. Crystal structure of a MATα2 homeodomain–operator complex suggests a general model for homeodomain–DNA interactions. Cell 67, 517–536.

Zhang, F., Zuo, K., Zhang, J., Liu, X., Zhang, L., Sun, X., Tang, K., 2010a. An L1 box binding protein, GbML1, interacts with GbMYB25 to control cotton fibre development. J. Exp. Bot. 61, 3599–3613.

Zhang, X., Zong, J., Liu, J., Yin, J., Zhang, D., 2010b. Genome-wide analysis of WOX gene family in rice, sorghum, maize, Arabidopsis and poplar. J. Integr. Plant Biol. 52, 1016–1026.

Zhang, S., Haider, I., Kohlen, W., Jiang, L., Bouwmeester, H., Meijer, A.H., Schluepmann, H., Liu, C.-M., Ouwerkerk, P.B.F., 2012. Function of the HD-Zip I gene Oshox22 in ABA-mediated drought and salt tolerances in rice. Plant Mol. Biol. 80, 571–585.

Zhang, D., Ye, H., Guo, H., Johnson, A., Zhang, M., Lin, H., Yin, Y., 2014. Transcription factor HAT1 is phosphorylated by BIN2 kinase and mediates brassinosteroid repressed gene expression in Arabidopsis. Plant J. 77, 59–70.

CHAPTER 8

Structure and Evolution of Plant MADS Domain Transcription Factors

Günter Theißen, Lydia Gramzow

Department of Genetics, Friedrich Schiller University Jena, Jena, Germany

OUTLINE

8.1 Introduction: Who Cares About MADS Domain Transcription Factors? 127
8.2 The Structure of MADS Domain Proteins 128
   8.2.1 Defining the Subject of Interest: What is a MADS Domain? 128
   8.2.2 Structural Features and Diversity of MADS Domain Proteins 130
8.3 Evolution of MADS Domain Transcription Factors 132
   8.3.1 Origin and Early Evolution of MADS Domain Proteins 132
8.3.2 The Great Expansion of the MADS World in Land Plants 132
8.3.3 Functional Aspects of MADS Box Gene Evolution 134
   8.3.3.1 Type I Proteins in Plants 134
   8.3.3.2 Type II Proteins in Plants 134
8.4 Concluding Remarks 136
References 137

8.1 INTRODUCTION: WHO CARES ABOUT MADS DOMAIN TRANSCRIPTION FACTORS?

As revealed by mutant analysis, often including spectacular phenotypes such as filled flowers or wood formation in a tiny herb (Yanofsky et al., 1990; Galimba et al., 2012; Melzer et al., 2008), MCM1/AGAMOUS/DEFICIENS/SRF (MADS; serum response factor, (SRF)) domain transcription factors are involved in controlling many developmental processes of angiosperms (flowering plants). First identified as the products of homeotic genes specifying organ identity during flower development (Sommer et al., 1990; Yanofsky et al., 1990), it soon became clear that plant MADS domain proteins have functions far beyond that. In fact, it eventually turned out that in the model plant thale cress (*Arabidopsis thaliana*) MADS domain transcription factors act throughout the whole lifecycle of the plant, including gametophyte (pollen and embryo sac) formation, seed development, vegetative growth, the transition to flowering, and fruit development (for a recent review, see Gramzow and Theißen, 2010; Smaczniak et al., 2012). *A. thaliana* is especially amenable to comprehensive mutant (forward and reverse genetic) analysis, but also in several other angiosperms such as snapdragon (*Antirrhinum majus*), spectacular and often similar mutant phenotypes have been found (see, e.g., Sommer et al., 1990; Schwarz-Sommer et al., 1990), corroborating the considerable importance of MADS domain transcription factors for angiosperm development. MADS domain transcription factors thus have intrigued many students of flowering plant development and evolution, and they continue to do so.

However, MADS domain transcription factors originated much earlier than flowering plants, and actually much earlier than the green plant lineage. Due to experimental limitations there is, unfortunately, only very limited knowledge based on mutant analysis about the MADS box gene function in green plants other than angiosperms. However, several lines of circumstantial evidence suggest that MADS domain proteins are also important transcriptional regulators of development in all other green plant lineages including algae, bryophytes (liverworts, hornworts, and mosses), lycophytes, monilophytes (ferns and their allies), and gymnosperms. MADS domain proteins underwent an impressive increase in number and diversity during the evolution of land plants (embryophytes), and the origin and diversification of some clades of genes encoding MADS domain transcription factors is closely correlated to the origin of evolutionary novelties such as seeds, flowers, and fruits (Theißen et al., 2000; Becker and Theißen, 2003). Even though causality is notoriously difficult to demonstrate in such cases, did the diversification of MADS domain proteins facilitate the evolution of plant diversity, or was it the other way around? MADS domain transcription factors have been intensively used as model systems to investigate the interrelationship between the evolution of gene repertoire and morphological innovations. So, for anyone interested in the molecular mechanisms behind the development and evolution of plants, MADS domain transcription factors are difficult to ignore.

8.2 THE STRUCTURE OF MADS DOMAIN PROTEINS

8.2.1 Defining the Subject of Interest: What is a MADS Domain?

The defining feature of MADS domain proteins is the presence of a MADS domain (Figure 8.1). It is about 58-amino-acid residues long (with minor variation depending on the publication you look at) and is usually encoded by just one exon (Henschel et al., 2002). The approximate 174 nucleotide residues of DNA encoding the MADS domain have been termed the MADS box; therefore, the genes they are part of are known as MADS box genes (Schwarz-Sommer et al., 1990).

The first MADS box gene ever isolated was *ARG80* from *Saccharomyces cerevisiae* (brewer's, baker's, or budding yeast; Dubois et al., 1987), but at the time it was not recognized as the representative of a large, important, and fascinating class of genes encoding transcription factors. Thus, the acronym "MADS" and the term "MADS box protein" was coined later after four subsequently characterized genes, <u>M</u>INICHROMOSOME MAINTENANCE FACTOR 1 (MCM1) from *S. cerevisiae*, <u>A</u>GAMOUS (AG) from *A. thaliana*, <u>D</u>EFICIENS (DEF) from *A. majus*, and <u>S</u>ERUM RESPONSE FACTOR (SRF) from *Homo sapiens* (human) and some animal species, had been isolated and defined as encoding the "four founding proteins" (Schwarz-Sommer et al., 1990). *ARG80* may have been ignored as part of the acronym because it was assumed to correspond to the *MCM1* gene of *S. cerevisiae* (Sommer et al., 1990). In reality, however, it is an ancient paralog (Theißen et al., 1996; Baker et al., 2013). In this chapter we use the terms "box" and "domain" for DNA and protein sequences, respectively.

The presence of MADS boxes in organisms as distantly related as human beings and snapdragon suggested that genes encoding a MADS domain were widely distributed throughout eukaryotes. Since MCM1 and SRF had been known as transcription factors it appeared immediately likely that plant genes also encode transcription factors (Sommer et al., 1990).

Transcription factors bind to specific DNA sequences that act as *cis*-regulatory elements in the promoters or enhancers of the genes whose transcription they control. The binding of transcription factors to these "target genes" occurs via a DNA-binding domain, which is almost always the part of the transcription factors most highly conserved during evolution. This conservation explains why the DNA-binding domain is so suitable for defining families of transcription factors and recognizing its members.

The analysis of conceptional sequences of thousands of MADS domain proteins has revealed that the MADS domain is by far the most highly conserved part of all MADS domain proteins. It does not come as a surprise that the MADS domain indeed represents their DNA-binding domain; however, it also has functions in protein dimerization and nuclear localization (Treisman, 1995).

MADS domain proteins bind to DNA in a sequence-specific way, but they do so not as single polypeptides, but as homo- or heterodimers, or even multimers (Kaufmann et al., 2005). As observed for most transcription factors, binding of MADS domain proteins to DNA also occurs in a sequence-specific way. Biochemical studies of diverse MADS domain proteins revealed that the amino-terminal (N-terminal) basic half of the MADS domain is important for sequence-specific binding, whereas the carboxy-terminal (C-terminal) half is part of a dimerization surface (Figure 8.1; Treisman, 1995). Several basic stretches – the most prominent being the motif KR[K/R]X$_4$KK at positions 22–30 of the MADS domain (Figure 8.1A) – constitute the nuclear localization signal (NLS) facilitating transport of transcription factors from the location of translation (cytoplasm) to the location of main function (nucleus; Gramzow and Theißen, 2010).

Different MADS domain protein dimers recognize slightly different sequences of DNA, but typically they resemble the so-called "serum response element (SRE)-type CArG box (C-A-rich-G-box)" (5'-CC(A/T)$_6$GG-3'), or similar sequences such as the "N10-type CArG box" or "MEF2 consensus binding site" (loosely defined as 5'-C(A/T)$_8$G-3; Shore and Sharrocks, 1995; Kaufmann et al., 2005; Verelst et al., 2007a; Zobell et al., 2010; Liu et al., 2013).

8.2 THE STRUCTURE OF MADS DOMAIN PROTEINS

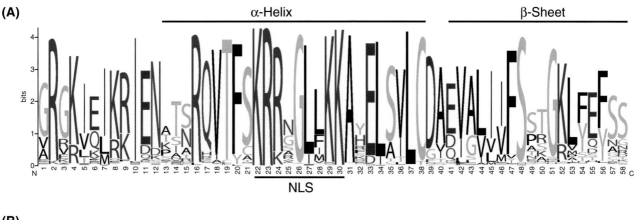

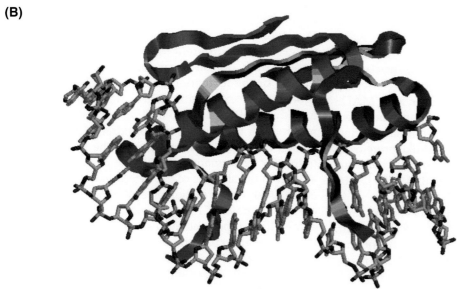

FIGURE 8.1 **Sequence and structure of the MADS domain.** (A) Sequence logo of the MADS domain based on an alignment of 867 plant MADS domain sequences including sequences of both Type I and Type II (MIKC-type) proteins. The logo was generated as generally described (Gramzow and Theißen, 2010). It shows the frequencies of amino acids at each position of the MADS domain by the relative heights of letters, along with the degree of conservation as the total height of a stack of letters, measured in bits of information. The conserved motif, which serves as part of the nuclear localization signal (NLS), is underlined. Parts of the sequence forming an α-helix and a β-sheet are indicated. (B) X-ray crystal structure of a dimer of the MADS domain of human serum response factor (SRF) bound to DNA (PDB 1SRS; note that no crystal structure exists for any plant MADS domain protein). The DNA is shown in wireframe representation. The two MADS domains of the dimer are colored blue and red, respectively. The α-helix is represented by a spring-like structure whereas the β-strands forming a β-sheet are shown as arrows.

The three-dimensional structure of any plant MADS domain has not been determined yet. However, X-ray crystal structures of the MADS domains of SRF, MCM1, and human MYOCYTE ENHANCER FACTOR 2 (MEF2) bound to DNA are available (Pellegrini et al., 1995; Tan and Richmond, 1998; Santelli and Richmond, 2000); given the extreme sequence conservation within the MADS domain, it appears quite safe to assume that the structure of MADS domains from plant proteins will be very similar.

Studies on the dimeric MADS domains of proteins from animals and fungi bound to CArG boxes revealed that the stretch of amino acids of the MADS domain folds into an N-terminal extension of 14 amino acids, followed by a long amphipathic α-helix and two β-strands (Figure 8.1B). The protein dimers constitute a stratified structure in which two structural units, stacked one above the other in the monomer, interact with the same unit in its partner (Treisman, 1995; Pellegrini et al., 1995). The base of the structure is given by a long antiparallel coil, constituted by the interaction of the α-helices encoded by the central part of the MADS domain. On top of this sits a four-stranded antiparallel β-sheet, constituted by the C-terminal part of each MADS domain. A remarkable feature of MADS domains bound to DNA is that the N-terminal extensions make extensive interactions with the minor groove of the DNA. Interactions include the insertion of the side chain of a highly conserved arginine (R) at position 2 (Figure 8.1A) into a region of the minor groove where it is narrowed (Pellegrini et al., 1995; Treisman, 1995; Santelli and Richmond, 2000). Narrowed minor groove recognition by a conserved arginine side chain appears to be a widely used mechanism of sequence-specific DNA binding by shape recognition (Rohs et al., 2009) that probably originated several times independently during evolution.

The coil formed by the α-helices of the MADS domain sits almost parallel to, and on top of, the narrowed DNA minor groove and makes extensive contacts with the phosphate backbone (Figure 8.1B; Pellegrini et al., 1995; Treisman, 1995). Some MADS domain proteins, such as SRF, bend the DNA strongly upon binding, while others, such as MEF2, do so only moderately (Pellegrini et al., 1995; Santelli and Richmond, 2000).

8.2.2 Structural Features and Diversity of MADS Domain Proteins

MADS domain transcription factors are structurally diverse; the defining MADS domain is the only conserved sequence element shared by all of them. In most MADS domain proteins it is located at the N-terminal, but there are several exceptions such as SRF in animal species, MCM1 in fungi, many AGAMOUS (AG)-like proteins in seed plants, and some MADS domain proteins of ferns (Purugganan et al., 1995; Muenster et al., 1997; Hasebe et al., 1998; Theißen et al., 1996, 2000).

Even though there are no other conserved domains present in all members of the family of MADS domain proteins, some subfamilies share regions of high sequence and structural similarity. The MEF2-like proteins of animals, for example, have a 28-residue MEF2 domain following immediately C-terminal to the MADS domain, which is not found in any other known MADS domain protein. This is the reason MEF2 proteins form DNA-binding dimers with themselves but not with other MADS domain proteins (Molkentin et al., 1996). Likewise, SRF, ARG80, MCM1, and their close relatives share a SAM domain immediately downstream of the MADS domain (Theißen et al., 1996). Neither the MEF2 nor the SAM domain is found in any plant MADS domain transcription factor, however.

All the MADS box genes that had been identified by "classical" forward genetics based on mutant phenotypes, including the widely known floral organ identity genes, encode proteins with a characteristic, highly conserved domain structure. In addition to the DNA-binding MADS domain (M) this domain structure includes an intervening domain (I), a keratin-like domain (K), and a C-terminal domain (C), in that order (Kaufmann et al., 2005). Transcription factors with this domain structure have therefore been termed MADS-domain/intervening-domain/keratin-like-domain/C-terminal domain (MIKC)-type MADS domain proteins (Muenster et al., 1997). The K domain is, besides the MADS domain, the most highly conserved domain of MIKC-type proteins. It is usually encoded by three exons (Figure 8.2A). The K domain shows only superficial similarity to keratin but is very likely not homologous to it. It has a length of approximately 70 amino acids (Henschel et al., 2002) and has been subdivided into three subdomains, K1, K2, and K3 (Kaufmann et al., 2005). Each of these subdomains is characterized by a heptad repeat [abcdefg]$_n$ where positions a and d usually contain hydrophobic amino acids, preferably leucine (Figure 8.2B; Yang et al., 2003; Yang and Jack, 2004). The subdomains were predicted to form amphipathic α-helices that constitute coils and thereby mediate protein–protein interactions of MADS domain proteins (Yang et al., 2003; Yang and Jack, 2004; Kaufmann et al., 2005). In line with this hypothesis it has been speculated that at least part of the K domain structure resembles a parallel leucine zipper (Yang et al., 2003). Several lines of experimental evidence suggest that in different MIKC-type proteins different parts of the K domain are involved in protein dimer formation or the formation of higher order (multimeric) protein complexes (Kaufmann et al., 2005). For example, a study on SEPALLATA3 (SEP3), a transcription factor required for the specification of floral organ identity, revealed that the C-terminal part of the K domain is required for the formation of multimeric protein complexes, at least in the absence of the C-terminal domain (Melzer et al., 2009). The formation of tetrameric complexes (such as "floral quartets") is a remarkable feature of at least some MIKC-type proteins, especially those involved in the specification of floral organ identity (Egea-Cortines et al., 1999; Theißen, 2001; Theißen and Saedler, 2001; Melzer et al., 2009; Melzer and Theißen, 2009; Gramzow and Theißen, 2010; Smaczniak et al., 2012; Mendes et al., 2013; Jetha et al., 2014). However, whether the K-domain was preferentially involved in establishing dimeric or multimeric protein contacts remained speculation for a while (Kaufmann et al., 2005).

The first X-ray crystal structure of the K domain of a MIKC-type protein, SEP3 again, was reported recently (Puranik et al., 2014). It revealed a remarkable structure that is clearly involved in both dimerization and tetramer formation (Figure 8.2C). It turned out that the K domain is involved in the formation of two (rather than three) amphipathic α-helices that are separated by a rigid kink. The second helix roughly corresponds to helices 2 and 3 of the previous models. The kink between the two helices prevents intramolecular association and presents separate dimerization and tetramerization interfaces made up of predominantly hydrophobic patches (Puranik et al., 2014). By presenting two amphipathic α-helices that are capable of forming intermolecular coils, MIKC-type transcription factors have obviously obtained a versatile oligomerization interface. It allows for homodimer, heterodimer, and tetramer formation with the K domains of other MIKC-type proteins. In theory, tetramers of MIKC-type MADS domain proteins should be capable of binding

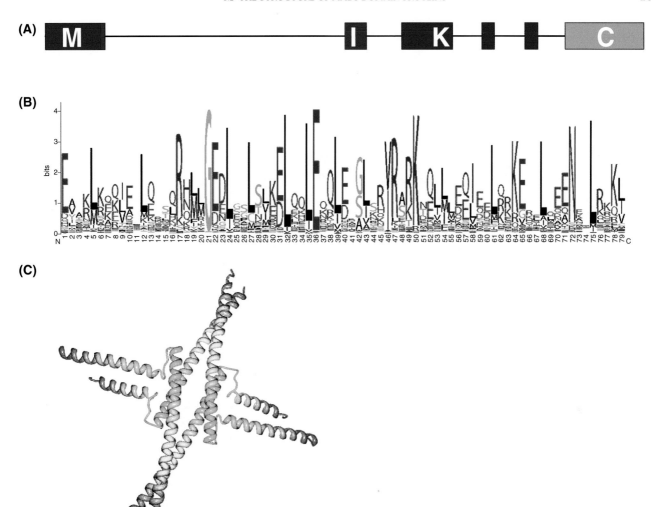

FIGURE 8.2 **Exon–intron structure of MIKC-type MADS box genes and sequence and structure of the K domain.** (A) Exon–intron structure of a typical MIKC-type MADS box gene where *ABS* from *A. thaliana* is shown as a representative here. Exons are shown as boxes and introns are indicated by lines. The part of the exon(s) encoding for the MADS (M), intervening (I), keratin-like (K), and C-terminal (C) domain are colored blue, red, purple, and green, respectively. Note that only protein-coding exons are shown. (B) Sequence logo of the K domain based on an alignment of 457 K domain sequences of MIKC-type proteins. The logo was generated in general as described for the MADS domain. (C) X-ray crystal structure of a tetramer of the K domain of SEP3 from *A. thaliana* (PDB 4OX0) (Puranik et al., 2014). The α-helices are shown as spring-like structures.

to two different CArG boxes by looping the DNA between the two binding sites (Egea-Cortines et al., 1999; Theißen, 2001; Theißen and Saedler, 2001). Several lines of experimental evidence have meanwhile corroborated this hypothesis (Melzer et al., 2009; Melzer and Theißen, 2009; Mendes et al., 2013; Jetha et al., 2014; Puranik et al., 2014).

The I domains and the C-terminal domains of MIKC-type proteins are much less well conserved than the MADS and K domains (Parenicova et al., 2003). The I domain is involved in determining the specificity of dimeric protein–protein interactions (Kaufmann et al., 2005).

The C-terminal domain is encoded by a variable number of exons, and is all in all the least conserved region of MIKC-type proteins (Kaufmann et al., 2005). Within subclades of MIKC-type proteins, however, conservation of specific motifs is recognizable, some of which may have originated by translational frameshift mutations (Vandenbussche et al., 2003). There is evidence that some C domains are involved in protein multimer formation or even just in dimerization; some C domains contain glutamine-rich or acidic stretches and function as transcriptional activation domains (reviewed by Kaufmann et al., 2005).

In plants, the first non-MIKC-type MADS box genes were only identified based on sequence analyses, especially during the course of the *A. thaliana* genome project. Absence or presence of a K box or domain has often been used as a decisive criterion to distinguish between Type I and Type II (or MIKC-type) genes and proteins of plants (Alvarez-Buylla et al., 2000). Apart from the

MADS domain, no other conserved domain unites all Type I MADS domain proteins. Several motifs specific to different groups of Type I proteins of plants have been identified, however (Parenicova et al., 2003; De Bodt et al., 2003). They do not belong to any otherwise described motifs, and no three-dimensional structures and specific functions have been assigned to these motifs (reviewed by Gramzow and Theißen, 2010). Type I genes in plants have typically just 1–2 exons and thus have a much simpler exon–intron structure than plant Type II (MIKC-type) genes, which usually have 6–8 exons (Figure 8.2A; Gramzow et al., 2010).

8.3 EVOLUTION OF MADS DOMAIN TRANSCRIPTION FACTORS

8.3.1 Origin and Early Evolution of MADS Domain Proteins

MADS box genes encoding MADS domain transcription factors have been found in almost all eukaryotes that have been investigated so far, including animals, fungi, plants, and protists; some parasitic excavates (*Trichomonas vaginalis* and *Giardia lamblia*) represent remarkable exceptions (Gramzow et al., 2010). In contrast, a convincing MADS box has never been identified in prokaryotes (eubacteria and archaea), despite the many genomes of diverse prokaryotes that have been sequenced (Theißen et al., 2000; Gramzow et al., 2010). Nevertheless, remote homology detection methods have provided evidence suggesting that the MADS box has prokaryotic roots. Specifically, it is believed that the MADS box probably originated from a DNA sequence encoding a region of subunit A of topoisomerase IIA (TOPOIIA-A; Gramzow et al., 2010). Topoisomerases bind to DNA while exerting their essential functions in DNA replication, transcription, recombination, and chromosome compaction and segregation, but they do so with only very limited sequence specificity. Thus, during the evolution of MADS domain proteins, the MADS domain must have acquired mutations that gained specificity in the recognition of CArG boxes (Gramzow et al., 2010).

Phylogeny reconstructions suggested that the *SRF*-like genes of animals, the *ARG80/MCM1*-like genes of fungi, and the plant MADS box genes without a K box are relatively closely related, hence they were all summarized as Type I genes; likewise, animal and fungal *MEF2*-like genes and *A. thaliana* MIKC-type genes appeared to be closely related, so they were all considered to be Type II genes (Alvarez-Buylla et al., 2000). Further analyses employing whole genome sequence information from the main eukaryotic lineages provided evidence that the most recent common ancestor of extant eukaryotes already possessed at least one type I (*SRF*-like) and one Type II (*MEF2*-like) MADS box gene. Thus, the ancestral MADS box gene that had originated from a TopoIIA-A gene, by gene duplication and sequence divergence, must have duplicated again in the lineage that led to extant eukaryotes, giving rise to Type I and Type II genes (Gramzow et al., 2010).

8.3.2 The Great Expansion of the MADS World in Land Plants

According to the scenario outlined above all lineages of extant eukaryotes may have inherited both types of MADS box genes from the most recent common ancestor (MRCA) of extant eukaryotes. During the phylogeny of the eukaryotes, however, the family of MADS box genes evolved in remarkably different ways in different eukaryotic lineages. In most of the major lineages of eukaryotes, the number of MADS box genes remained quite low. For example, only 1–4 MADS box genes are typically present in extant fungi (e.g., 4 in *S. cerevisiae*) and 2–8 in animals (metazoans; e.g., 2 in *Drosophila melanogaster* and 5 in humans), but despite their low numbers these genes encode transcription factors with important functions (e.g., in pheromone response or in cell proliferation and differentiation; reviewed by Gramzow et al., 2010).

The genomes of some chlorophyte green algae revealed only one MADS box gene (Figure 8.3). The conceptual proteins lack a K domain and may represent Type I proteins, thus suggesting that Type II genes were lost in the lineage that led to these green algae (Gramzow et al., 2012). However, more analyses are required to test this hypothesis.

In contrast to the minimal number of MADS box genes in chlorophytes, the number of MADS box genes increased considerably during the evolution of land plants (embryophytes). This can be inferred from the number of MADS box genes in extant species, as revealed by the analysis of whole genome sequences. Specifically, 23 MADS box genes have been identified in the moss *Physcomitrella patens* (Barker and Ashton, 2013) and 20 in the spikemoss *Selaginella moellendorffii* (Figure 8.3; Banks et al., 2011; Gramzow et al., 2012). In seed plants, the number of MADS box genes is 5–10 times higher. There are roughly about 100 MADS box genes in angiosperm genomes (Figure 8.3), exhibiting some variance, ranging in recent studies from 60 in *Ricinus communis* to 262 in *Carica papaya* (Gramzow and Theißen, 2013). In the genomes of conifers such as *Picea abies*, *Picea glauca*, and *Pinus taeda*, representing gymnosperms, an even higher number of MADS box genes has been found (Figure 8.3), but more investigations are required to determine how many of them (if any) are pseudogenes (Nystedt et al., 2013; Gramzow et al., 2014). Roughly half of the MADS box genes in angiosperms are Type I and

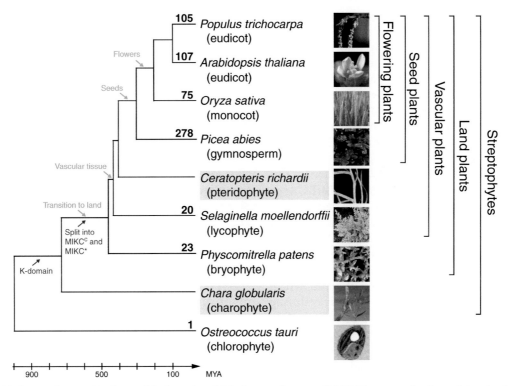

FIGURE 8.3 **Phylogenetic tree of major model plants for which the complete set of MADS box genes has been determined based on whole genome sequence data.** Numbers on terminal branches depict numbers of MADS box genes in the respective genomes. For pteridophytes (monilophytes; ferns and their allies) and charophytes whole genome data are not available yet, but their positions are shown for orientation. Arrows pointing to internal branches indicate major genetic (purple) or morphological innovations (green). The arrow at the bottom represents a time scale. MYA, million years ago.

Type II, again with considerable variance (Parenicova et al., 2003; Gramzow and Theißen, 2013), whereas the number of Type I genes is underrepresented in gymnosperms (Gramzow et al., 2014).

Why did the number of genes encoding MADS domain transcription factors increase so much in land plants, especially in flowering plants? There is no easy answer to this question, for several reasons. First, one should note that quite a number of gene families encoding transcription factors have undergone a much more dramatic expansion in plants than in other eukaryotes. The reason for this is not simply a higher duplication rate of plant genomes because there is also a higher degree of expansion for genes encoding transcription factors compared with other plant genes (Shiu et al., 2005). It has been argued that a special propensity of transcription factors for sub- and neofunctionalization may account for the higher expansion rate (Melzer and Theißen, 2011). For example, simple changes in expression domains of duplicate genes specifying cell or organ identity may suffice to bring about sub- or neofunctionalization. In line with this notion, duplication of Type II (MIKC-type) MADS box genes may have contributed to the origin of evolutionary innovations, and hence neofunctionalization, in an especially pronounced way in land plants.

Well-known examples are the morphological novelties of seed and flowering plants such as ovules and seeds, and flowers and fruits whose development evolved under the control of novel clades of MADS box gene subfamilies not present in nonseed plants (Theißen et al., 2000; Gramzow et al., 2014).

In case of genes encoding MADS domain transcription factors any explanation is necessarily especially complex because Type I and Type II (MIKC-type) genes had significantly different evolutionary dynamics (or "opposing life styles"; Nam et al., 2004; Gramzow and Theißen, 2013). Type II genes increased in numbers because they were preferentially retained after whole genome duplications, thus constituting ancient, conserved, and taxonomically widely distributed clades of paralogous genes. In comparison, Type I genes of plants have much higher birth and death rates, thus not constituting ancient clades, but young clusters of closely related paralogs with limited taxonomic distribution (Nam et al., 2004; Gramzow and Theißen, 2013).

The mechanisms which led to the differences in the evolutionary trajectories of Type I and Type II genes in plants are only partially understood. Differences in purifying selection, positive selection, and mutation have already been considered as potential, but not mutually exclusive, reasons for the differences between Type I and Type II gene evolution in plants.

According to the "gene balance hypothesis" genes that are more connected (e.g., due to protein–protein interactions of their gene products) are more likely retained after whole genome or large-scale duplications than after tandem or other small-scale duplications (Edger and Pires, 2009). Type II MADS domain proteins are probably involved in a higher number of complexes and in more complex multimers than Type I proteins (de Folter et al., 2005; Melzer and Theißen, 2009). One reason for this being the presence of the K domain as part of MIKC-type proteins. Therefore, Type II genes may have had a higher likelihood than Type I genes to be retained after whole genome duplications and to be lost after small-scale duplications due to purifying selection trying to keep gene products in balance.

Second, the duplication and retention of genes, which are involved in coevolutionary scenarios, such as during reproductive processes, may be under positive selection after small-scale duplications. Type I genes have not been comprehensively studied, but the few Type I genes for which functions are known are involved in female gametophyte, embryo, and seed development and hence quite directly in reproduction (Masiero et al., 2011). Thus, the differences in the evolutionary dynamics between Type I and Type II genes may also be caused by selection for diversity and neofunctionalization of Type I genes (Gramzow and Theißen, 2013).

Type I and Type II genes may also differ in mutation patterns. It has been hypothesized that Type I MADS box genes may have been highjacked by transposons, or may even have an intrinsic transposon activity (De Bodt et al., 2003), and, thus, might duplicate more often than "regular" genes such as Type II MADS box genes. Type I genes are also much shorter and have less exons than Type II genes; the chance of being duplicated as functional units thus might be higher for Type I than for Type II genes (Gramzow and Theißen, 2013).

Even though the story of Type II MADS box gene evolution appears to be largely one of conservation of ancient paralogs and expansion of copy number, it is clear that genes also quite frequently get lost during evolution. One may wonder, however, whether whole clades of ancient genes have ever been lost. A recent analysis employing whole genome sequence data has shown that one clade of MIKC-type genes, which already existed in the MRCA of extant seed plants (TM8-like genes) and another that was already established in the MRCA of extant angiosperms (FLC-like genes), was lost several times independently in different lineages of flowering plants. One species, *Spirodela polyrhiza*, revealed the loss of several clades of MADS box genes, a phenomenon that might be linked to the highly simplified body plan of this duckweed species (Gramzow and Theißen, 2015).

8.3.3 Functional Aspects of MADS Box Gene Evolution

8.3.3.1 Type I Proteins in Plants

Based on phylogeny reconstruction and the analysis of conserved domains the Type I MADS domain proteins of *A. thaliana* have been subdivided into three groups, Mα, Mβ, and Mγ (Parenicova et al., 2003; Nam et al., 2004; Arora et al., 2007). More recent phylogeny reconstructions based on a broader sampling, in terms of genes and species, recognized just two ancient clades of Type I genes, one containing all Mα genes and the other containing all previously identified Mβ and Mγ genes (Gramzow et al., 2012). There is evidence that both an Mα and an Mβ/γ gene already existed in the MRCA of extant mosses and vascular plants about 480 million years ago (MYA; Gramzow et al., 2012).

Type I MADS box genes in plants did not show up in forward genetics analyses, and they are only weakly expressed in all plant genomes that have been analyzed so far, suggesting that they are functionally redundant or otherwise of only limited functional importance (Gramzow et al., 2010). However, phenotypes of a few angiosperm mutants affected in Mα or Mβ/γ genes revealed that at least some Type I MADS box genes acquired functions in female gametophyte, embryo sac, and seed development (reviewed by Gramzow et al., 2010; Masiero et al., 2011). The function of most plant Type I proteins remains elusive, however.

8.3.3.2 Type II Proteins in Plants

Type II genes encoding MIKC-type proteins have been detected in all streptophytes (charophycean green algae + land plants), but never in chlorophytes (green algae; the sister group of the streptophytes; Tanabe et al., 2005). Thus, the K domain probably joined the MADS domain in the lineage that led to extant streptophytes more than 700 MYA (Figure 8.3; Kaufmann et al., 2005). The presence of MIKC-type genes thus seems to be a valuable "synapomorphy" (a shared, derived trait) of streptophytes.

Two different kinds of MIKC-type genes have been defined, the so-called "classic" MIKC-group (MIKC^C-group) and the MIKC*-group genes (Henschel et al., 2002). MIKC*-group genes probably originated more than 450 MYA in the lineage that led to extant land plants by duplication of an ancestral MIKC^C-group gene (Figure 8.3), followed by either an elongation of the I region or a duplication of parts of the K box (Henschel et al., 2002; Kwantes et al., 2012).

8.3.3.2.1 MIKC*-GROUP PROTEINS

Much like Type I genes, MIKC*-group genes were also detected later than MIKC^C-group genes (Henschel et al., 2002), probably due to their functional redundancy and limited importance for plant development. The K domain of MIKC*-group proteins deviates considerably from that of MIKC^C-group proteins and is more difficult to detect; in fact, due to this seeming absence of the K domain, MIKC*-group proteins have sometimes been confused with Type I proteins and described, for example, as Mδ proteins (Parenicova et al., 2003).

MIKC*-group genes have been identified in representatives of all major groups of land plants, including bryophytes, lycophytes, ferns, gymnosperms, basal angiosperms, eudicots, and monocots (Henschel et al., 2002; Zobell et al., 2010; Gramzow et al., 2012; Kwantes et al., 2012). In ferns and seed plants two different monophyletic lineages of MIKC*-group genes have been identified, termed P and S clade genes, which neither exist in bryophytes nor lycophytes (Nam et al., 2004; Gramzow et al., 2012; Kwantes et al., 2012). MIKC*-type proteins of ferns and seed plants (euphyllophytes) form – sometimes exclusively – interclade heterodimers with one interaction partner from the P clade and one from the S clade (Verelst et al., 2007a; Kwantes et al., 2012; Liu et al., 2013). Such dimers of MIKC*-group proteins preferentially bind to the N10-type rather than the SRE-type CArG box, which is in contrast to the behavior of typical MIKC^C-group protein dimers (Verelst et al., 2007a; Zobell et al., 2010).

In all land plants that have been investigated so far, MIKC*-type genes are expressed predominantly in the gametophytic generation, suggesting an ancestral function during the haploid phase of the plant life cycle (Kwantes et al., 2012). This expression domain was progressively restricted to the male gametophyte (pollen) in the lineage leading to flowering plants (Kwantes et al., 2012; Liu et al., 2013).

Mutant phenotypes obtained by reverse genetics revealed the function of MIKC*-type genes in two distantly related flowering plants, *A. thaliana* (a eudicot) and *Oryza sativa* (rice; a monocot). They showed that in both species MIKC*-type genes are required for proper pollen maturation and germination, which nicely reflects the expression of these genes during the late developmental stages of pollen (Verelst et al., 2007a, b; Adamczyk and Fernandez, 2009; Liu et al., 2013). The available data indicate a highly similar role played by MIKC*-type transcription factors in monocots and eudicots, one that must have been conserved for about at least 150 million years of flowering plant evolution. Redundancy of the MIKC*-group gene function is higher in *A. thaliana* than in *O. sativa*, however, since single loss-of-function mutants show an aberrant phenotype in rice, but not in thale cress (Verelst et al., 2007a; Adamczyk and Fernandez, 2009; Liu et al., 2013).

8.3.3.2.2 MIKC^C-GROUP PROTEINS

MIKC^C-group MADS box genes are involved in the control of many developmental processes in flowering plants, and, by inference, also in all other land plants. The phylogeny of MIKC^C-group genes is characterized by the formation of ancient paralogs, many of which originated by whole genome duplications, preferential gene retention, and sequence divergence linked to sub- and neofunctionalization. Radiations of genes occurred independently in mosses, lycophytes, ferns, and seed plants (Figure 8.4). The MIKC^C-group genes of seed plants are unique in that they are all members of 11 seed plant-specific superclades that were present in the MRCA of extant seed plants about 300 MYA, something that did not exist in the MRCA of monilophytes (ferns and their allies such as horsetails) and seed plants (gymnosperms + angiosperms) about 400 MYA. These genes evolved into 17 clades that had already been established in the MRCA of extant angiosperms (Gramzow et al., 2014).

In the fern *Ceratopteris richardii* and the moss *P. patens*, MIKC^C-group genes are expressed (and by inference, exert their function) in both gametophytes and sporophytes (Muenster et al., 1997; Hasebe et al., 1998; Theißen et al., 2000). In flowering plants such as *A. thaliana*, however, their expression and function are more focused on the sporophyte, where they control various aspects of development (Gramzow and Theißen, 2010; Smaczniak et al., 2012). Most well known are the floral organ identity genes that provide the class A, B, C, D, and E homeotic functions. Combinatorial interactions among these functions determine floral organ identities, with A + E specifying sepals, A + B + E specifying petals, B + C + E specifying stamens, C + E specifying carpels, and D + E specifying ovules (for a review see Theißen, 2001; Krizek and Fletcher, 2005; Causier et al., 2010; Gramzow and Theißen, 2010; Smaczniak et al., 2012). The different floral homeotic functions are provided by transcription factors encoded by members of different clades of MADS box genes (Figure 8.4). For example, stamen development requires the combinatorial interaction of DEF- and GLO-like proteins (class B) with an AG-like protein (C) and an AGL2- or AGL9-like protein (E) (Figure 8.4). The combinatorial interactions are probably physically realized by tetramerization and DNA binding of the respective MIKC^C-group proteins. In the case of floral homeotic proteins, these DNA-bound tetramers are called "floral quartets" (Theißen, 2001; Theißen and Saedler, 2001).

There is evidence that only some MIKC^C-group proteins possess transcription activation domains; it might be one function of tetramerization, therefore, to provide transcription activation potential to proteins that lack this capacity (Honma and Goto, 2001). Moreover, quartet formation with DNA binding is highly cooperative, as one dimer bound to a CArG box assists the other one to bind to another CArG box (Melzer et al., 2009). Due to this cooperativity small changes in protein concentration may suffice to change the occupancy of target DNA sites substantially. Thus, cooperativity in DNA binding might

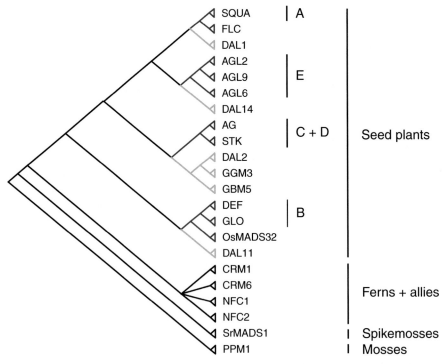

FIGURE 8.4 Phylogeny of some major clades of MIKC^c-group MADS box genes in land plants. Terminal clades ("gene subfamilies") are shown as triangles; gene clades from angiosperms are shown in red, clades from gymnosperms in blue. Gene names given represent clades (e.g., DEF = DEF-like genes), not individual genes. A, B, C, D, and E indicate floral homeotic functions represented by members of the respective gene clades. Note that only a fraction of the known clades is shown. For a more comprehensive overview, see Gramzow et al. (2014).

enable some MADS domain proteins to function as "molecular switches" (Theißen and Melzer, 2007).

However, the importance of the principle of quartet formation involving MIKC^C-group proteins may go well beyond the specification of floral organ identity, since MIKC^C-group genes are also involved in developmental processes that occur before and after floral organ determination. In *A. thaliana*, for example, the flowering time genes *AGAMOUS-LIKE24* (*AGL24*) and *SUPRESSOR OF OVEREXPRESSION OF CONSTANS1* (*SOC1*) are involved in the switch from vegetative to reproductive development. They activate meristem identity genes such as *APETALA1* (*AP1*) and *CAULIFLOWER* (*CAL*) that then activate the floral organ identity genes. Targets of floral homeotic genes again include MIKC^C-group genes such as *ARABIDOPSIS BSISTER* (*ABS*), *SEEDSTICK* (*STK*), *SHATTERPROOF1* (*SHP1*), and *SHATTERPROOF2* (*SHP2*) that are involved in ovule and fruit development. In fact, every stage of reproductive development is controlled, at least in part, by intricate gene regulatory networks (GRN) involving interactions between MIKC^C-group genes and the transcription factors they encode (Theißen and Melzer, 2013).

Many MIKC^C-group transcription factors participate in positive and negative autoregulatory feedback loops which in addition to their capacity of combinatorial multimerization and cooperative DNA binding might have been of critical importance for these proteins to become key regulators in the molecular networks controlling plant development (Gramzow et al., 2010; Theißen and Melzer, 2013).

8.4 CONCLUDING REMARKS

MADS domain transcription factors are important players in both plant development and evolution, and one may wonder about the molecular basis of their suitability for these respective processes. The common denominator of all MADS domain transcription factors and thus the usual suspect for any special feature of these proteins would be the DNA-binding MADS domain. We feel, however, that this is only part of the answer. The developmentally most important genes, as revealed by their mutant phenotypes, do not only share the MADS domain, but a whole domain structure termed MIKC, including a highly intriguing K domain. The dimerization and tetramerization interfaces presented by this domain enable efficient combinatorial and cooperative multimerization of MIKC-type MADS domain proteins and thus may explain their suitability as developmental switches. Plant-specific combinations of DNA-binding and protein–protein interaction domains may have originated more than once. For example, HD-Zip transcription factors exhibit a combination of a homeodomain (HD) with a leucine zipper (Zip) immediately downstream of the HD. HD-Zip proteins bind to DNA as dimers, with the Zip sequence acting as a dimerization motif (Chapter 7). They function in a wide variety of processes, including stress response and development (reviewed by Melzer and Theißen, 2011; Chapter 22). Nevertheless, with their ability to constitute tetrameric protein complexes composed of paralogous proteins in a cooperative way on looped DNA MIKC-type proteins appear really special, certainly when compared with Type I MADS domain proteins, but even when compared with HD-Zip proteins.

Acknowledgments

Günter Theißen acknowledges the considerable patience of Daniel H. Gonzalez and thanks him for his invitation to write this chapter.

References

Adamczyk, B.J., Fernandez, D.E., 2009. MIKC* MADS domain heterodimers are required for pollen maturation and tube growth in Arabidopsis. Plant Physiol. 149, 1713–1723.

Alvarez-Buylla, E.R., Pelaz, S., Liljegren, S.J., Gold, S.E., Burgeff, C., Ditta, G.S., de Pouplana, L.R., Martinez-Castilla, L., Yanofsky, M.F., 2000. An ancestral MADS-box gene duplication occurred before the divergence of plants and animals. Proc. Natl. Acad. Sci. USA 97, 5328–5333.

Arora, P., Ray, S., Singh, A.K., Singh, V.P., Tyagi, A.K., Kapoor, S., 2007. MADS-box gene family in rice: genome-wide identification, organization and expression profiling during reproductive development and stress. BMC Genomics 8, 242.

Baker, C.R., Hanson-Smith, V., Johnson, A.D., 2013. Following gene duplication, paralog interference constrains transcriptionl circuit evolution. Science 342, 104–108.

Banks, J.A., Nishiyama, T., Hasebe, M., Bowman, J.L., Gribskov, M., dePamphilis, C., Albert, V.A., Aono, N., Aoyama, T., Ambrose, B.A., Ashton, N.W., Axtell, M.J., Barker, E., Barker, M.S., Bennetzen, J.L., Bonawitz, N.D., Chapple, C., Cheng, C., Correa, L.G.G., Dacre, M., DeBarry, J., Dreyer, I., Elias, M., Engstrom, E.M., Estelle, M., Feng, L., Finet, C., Floyd, S.K., Frommer, W.B., Fujita, T., Gramzow, L., Gutensohn, M., Harholt, J., Hattori, J., Heyl, A., Hirai, T., Hiwatashi, Y., Ishikawa, M., Iwata, M., Karol, K.G., Koehler, B., Kolukisaoglu, U., Kubo, M., Kurata, T., Lalonde, S., Li, K., Li, Y., Litt, A., Lyons, E., Manning, G., Maruyama, T., Michael, T.P., Mikami, K., Miyazaki, S., Morinaga, S.-I., Murata, T., Mueller-Roeber, B., Nelson, D.R., Obara, M., Oguri, Y., Olmstead, R.G., Onodera, N., Petersen, B.L., Pils, B., Prigge, M., Rensing, S.A., Riaño-Pachón, D.M., Roberts, A.W., Sato, Y., Scheller, H.V., Schulz, B., Schulz, Ch., Shakirov, E.V., Shibagaki, N., Shinohara, N., Shippen, D.E., Sørensen, I., Sotooka, R., Sugimoto, N., Sugita, M., Sumikawa, N., Tanurdzic, M., Theißen, G., Ulvskov, P., Wakazuki, S., Weng, J.K., Willats, W.W.G.T., Wipf, D., Wolf, P.G., Yang, L., Zimmer, A.D., Zhu, Q., Mitros, T., Hellsten, U., Loqué, D., Otillar, R., Salamov, A., Schmutz, J., Shapiro, H., Lindquist, E., Lucas, S., Rokhsar, D., Grigoriev, I.V., 2011. The Selaginella genome identifies genetic changes associated with the evolution of vascular plants. Science 332, 960–963.

Barker, E.I., Ashton, N.W., 2013. A parsimonious model of lineage-specific expansion of MADS-box genes in Physcomitrella patens. Plant Cell Rep. 32, 1161–1177.

Becker, A., Theißen, G., 2003. The major clades of MADS-box genes and their role in the development and evolution of flowering plants. Mol. Phylogenet. Evol. 29, 464–489.

Causier, B., Schwarz-Sommer, Z., Davies, B., 2010. Floral organ identity: 20 years of ABCs. Semin. Cell. Dev. Biol. 21, 73–79.

De Bodt, S., Raes, J., Florquin, K., Rombauts, S., Rouzé, P., Theißen, G., Van de Peer, Y., 2003. Genomewide structural annotation and evolutionary analysis of the type I MADS-box genes in plants. J. Mol. Evol. 56, 573–586.

de Folter, S., Immink, R.G.H., Kieffer, M., Pařenicová, L., Henz, S.R., Weigel, D., Busscher, M., Kooiker, M., Colombo, L., Kater, M.M., Davies, B., Angenent, G.C., 2005. Comprehensive interaction map of the Arabidopsis MADS box transcription factors. Plant Cell 17, 1424–1433.

Dubois, E., Bercy, J., Descamps, F., Messenguy, F., 1987. Characterization of two new genes essential for vegetative growth in Saccharomyces cerevisiae: nucleotide sequence determination and chromosome mapping. Gene 55, 265–275.

Edger, P.P., Pires, J.C., 2009. Gene and genome duplications: the impact of dosage-sensitivity on the fate of nuclear genes. Chromosome Res. 17, 699–717.

Egea-Cortines, M., Saedler, H., Sommer, H., 1999. Ternary complex formation between the MADS-box proteins SQUAMOSA, DEFICIENS and GLOBOSA is involved in the control of floral architecture in Antirrhinum majus. EMBO J. 18, 5370–5379.

Galimba, K.D., Tolkin, T.R., Sullivan, A.M., Melzer, R., Theißen, G., Di Stilio, V.S., 2012. Loss of deeply conserved C-class floral homeotic gene function and C- and E-class protein interaction in a double-flowered ranunculid mutant. Proc. Natl. Acad. Sci. USA, E2267–E2275.

Gramzow, L., Theißen, G., 2010. A hitchhiker's guide to the MADS world of plants. Genome Biol. 11, 214.

Gramzow, L., Theißen, G., 2013. Phylogenomics of MADS-box genes in plants – two opposing life styles in one gene family. Biology 2, 1150–1164.

Gramzow, L., Theißen, G., 2015. Phylogenomics reveals surprising sets of essential and dispensable clades of MIKC^C-type MADS-box genes in flowering plants. J. Exp. Zool. B. doi: 10.1002/jez.b.22598

Gramzow, L., Ritz, M.S., Theißen, G., 2010. On the origin of MADS-domain transcription factors. Trends Genet. 26, 149–153.

Gramzow, L., Barker, E., Schulz, C., Ambrose, B., Ashton, N., Theißen, G., Litt, A., 2012. Selaginella genome analysis – entering the "homoplasy heaven" of the MADS world. Front. Plant Sci. 3, 214.

Gramzow, L., Weilandt, L., Theißen, G., 2014. MADS goes genomic in conifers: towards determining the ancestral set of MADS-box genes in seed plants. Ann. Bot. 114, 1407–1429.

Hasebe, M., Wen, C.K., Kato, M., Banks, J.A., 1998. Characterization of MADS homeotic genes in the fern Ceratopteris richardii. Proc. Natl. Acad. Sci. USA 95, 6222–6227.

Henschel, K., Kofuji, R., Hasebe, M., Saedler, H., Münster, T., Theißen, G., 2002. Two ancient classes of MIKC-type MADS-box genes are present in the moss Physcomitrella patens. Mol. Biol. Evol. 19, 801–814.

Honma, T., Goto, K., 2001. Complexes of MADS-box proteins are sufficient to convert leaves into floral organs. Nature 409, 525–529.

Jetha, K., Theißen, G., Melzer, R., 2014. Arabidopsis SEPALLATA proteins differ in cooperative DNA-binding during the formation of floral quartet-like complexes. Nucleic Acids Res. 42, 10927–10942.

Kaufmann, K., Melzer, R., Theißen, G., 2005. MIKC-type MADS-domain proteins: structural modularity, protein interactions and network evolution in land plants. Gene 347, 183–198.

Krizek, B.A., Fletcher, J.C., 2005. Molecular mechanisms of flower development: an armchair guide. Nature Rev. Genet. 6, 688–698.

Kwantes, M., Liebsch, D., Verelst, W., 2012. How MIKC* MADS-box genes originated and evidence for their conserved function throughout the evolution of vascular plant gametophytes. Mol. Biol. Evol. 29, 293–302.

Liu, Y., Cui, S., Wu, F., Yan, S., Lin, X., Du, X., Chong, K., Schilling, S., Theißen, G., Meng, Z., 2013. Functional conservation of MIKC*-type MADS box genes in Arabidopsis and rice pollen maturation. Plant Cell 25, 1288–1303.

Masiero, S., Colombo, L., Grini, P.E., Schnittger, A., Kater, M.M., 2011. The emerging importance of Type I MADS box transcription factors for plant reproduction. Plant Cell 23, 865–872.

Melzer, R., Theißen, G., 2009. Reconstitution of "floral quartets" in vitro involving class B and class E floral homeotic proteins. Nucleic Acids Res. 37, 2723–2736.

Melzer, R., Theißen, G., 2011. MADS and more: transcription factors that shape the plant. In: Yuan, L., Perry, S.E. (Eds.), Plant Transcription Factors – Methods and Protocols; Methods in Molecular Biology Series, Vol. 754, Springer Protocols, Humana Press, New York, pp. 3–18.

Melzer, R., Verelst, W., Theißen, G., 2009. The class E floral homeotic protein SEPALLATA3 is sufficient to loop DNA in "floral quartet"-like complexes in vitro. Nucleic Acids Res. 37, 144–157.

Melzer, S., Lens, F., Gennen, J., Vanneste, S., Rohde, A., Beeckman, T., 2008. Flowering-time genes modulate meristem determinacy and growth form in Arabidopsis thaliana. Nat. Genet. 40, 1489–1492.

Mendes, M.A., Guerra, R.F., Berns, M.C., Manzo, C., Masiero, S., Finziu, L., Kater, M.M., Colombo, L., 2013. MADS domain transcription factors mediate short-range DNA looping that is essential for target gene expression in Arabidopsis. Plant Cell 25, 2560–2572.

Molkentin, J.D., Black, B.L., Martin, J.F., Olson, E.N., 1996. Mutational analysis of the DNA binding, dimerization, and transcriptional activation domains of MEF2C. Mol. Cell Biol. 16, 2627–2636.

Muenster, T., Pahnke, J., Di Rosa, A., Kim, J.T., Martin, W., Saedler, H., Theissen, G., 1997. Floral homeotic genes were recruited from homologous MADS-box genes preexisting in the common ancestor of ferns and seed plants. Proc. Natl. Acad. Sci. USA 94, 2415–2420.

Nam, J., Kim, J., Lee, S., An, G., Ma, H., Nei, M., 2004. Type I MADS-box genes have experienced faster birth-and-death evolution than type II MADS-box genes in angiosperms. Proc. Natl. Acad. Sci. USA 101, 1910–1915.

Nystedt, B., Street, N.R., Wetterbom, A., Zuccolo, A., Lin, Y.C., Scofield, D.G., Vezzi, F., Delhomme, N., Giacomello, S., Alexeyenko, A., Vicedomini, R., Sahlin, K., Sherwood, E., Elfstrand, M., Gramzow, L., Holmberg, K., Hällmann, J., Keech, O., Klasson, L., Koriabine, M., Kucukoglu, M., Käller, M., Luthman, J., Lysholm, F., Nittylä, T., Olson, A., Rilakovic, N., Ritland, C., Rosselló, J.A., Sena, J., Svensson, T., Talavera-López, C., Theißen, G., Tuominen, H., Vanneste, K., Wu, Z.Q., Zhang, B., Zerbe, P., Arvestad, L., Bhalerao, R., Bohlmann, J., Bousquet, J., Garcia Gil, R., Hvidsten, T.R., de Jong, P., Mackay, J., Morgante, M., Ritland, K., Sundberg, B., Lee Thompson, S., Van de Peer, Y., Andersson, B., Nilsson, O., Ingvarsson, P.K., Lundeberg, J., Jansson, S., 2013. The Norway spruce genome sequence and conifer genome evolution. Nature 497, 579–584.

Parenicova, L., de Folter, S., Kieffer, M., Horner, D.S., Favalli, C., Busscher, J., et al., 2003. Molecular and phylogenetic analyses of the complete MADS-box transcription factor family in Arabidopsis: new openings to the MADS world. Plant Cell 15, 1538–1551.

Pellegrini, L., Song, T., Richmond, T.J., 1995. Structure of serum response factor core bound to DNA. Nature 376, 490–498.

Puranik, S., Acajjaoui, S., Conn, S., Costa, L., Conn, V., Vial, A., Marcellin, R., Melzer, R., Brown, E., Hart, D., Theißen, G., Silva, S.S., Parcy, F., Dumas, R., Nanao, M., Zubieta, C., 2014. Structural basis for the oligomerization of the MADS domain transcription factor SEPALLATA3 in Arabidopsis. Plant Cell 26, 3603–3615.

Purugganan, M.D., Rounsley, S.D., Schmidt, R.J., Yanofsky, M.F., 1995. Molecular evolution of flower development: diversification of the plant MADS-box regulatory gene family. Genetics 140, 345–356.

Rohs, R., West, S.M., Sosinsky, A., Liu, P., Mann, R.S., Honig, B., 2009. The role of DNA shape in protein–DNA recognition. Nature 461, 1248–1254.

Santelli, E., Richmond, T.J., 2000. Crystal structure of MEF2A core bound to DNA at 1.5 angstrom resolution. J. Mol. Biol. 297, 437–449.

Schwarz-Sommer, Z., Huijser, P., Nacken, W., Saedler, H., Sommer, H., 1990. Genetic-control of flower development by homeotic genes in Antirrhinum majus. Science 250, 931–936.

Shiu, S.H., Shih, M.C., Li, W.H., 2005. Transcription factor families have much higher expansion rates in plants than in animals. Plant Physiol. 139, 18–26.

Shore, P., Sharrocks, A.D., 1995. The MADS-box family of transcription factors. Eur. J. Biochem. 229, 1–13.

Smaczniak, C., Immink, R.G.H., Angenent, G.C., Kaufmann, K., 2012. Developmental and evolutionary diversity of plant MADS-domain factors: insights from recent studies. Development 139, 3081–3098.

Sommer, H., Beltrán, J.P., Huijser, P., Pape, H., Lönnig, W.E., Saedler, H., Schwarz-Sommer, Z., 1990. Deficiens, a homeotic gene involved in the control of flower morphogenesis in Antirrhinum majus: the protein shows homology to transcription factors. EMBO J. 9, 605–613.

Tan, S., Richmond, T.J., 1998. Crystal structure of the yeast MAT alpha 2/MCM1/DNA ternary complex. Nature 391, 660–666.

Tanabe, Y., Hasebe, M., Sekimoto, H., Nishiyama, T., Kitani, M., Henschel, K., Münster, T., Theissen, G., Nozaki, H., Ito, M., 2005. Characterization of MADS-box genes in charophycean green algae and its implication for the evolution of MADS-box genes. Proc. Natl. Acad. Sci. USA 102, 2436–2441.

Theißen, G., 2001. Development of floral organ identity: stories from the MADS house. Curr. Opin. Plant Biol. 4, 75–85.

Theißen, G., Melzer, R., 2007. Molecular mechanisms underlying origin and diversification of the angiosperm flower. Ann. Bot. 100, 603–619.

Theißen, G., Melzer, R., 2013. Flower development, genetics of. In: Maloy, S., Hughes, K. (Eds.), Brenner's Online Encyclopedia of Genetics, vol. 3, second ed. Academic Press, San Diego, pp. 67–71.

Theißen, G., Saedler, H., 2001. Floral quartets. Nature 409, 469–471.

Theißen, G., Kim, J., Saedler, H., 1996. Classification and phylogeny of the MADS-box multigene family suggest defined roles of MADS-box gene subfamilies in the morphological evolution of eukaryotes. J. Mol. Evol. 43, 484–516.

Theißen, G., Becker, A., Di Rosa, A., Kanno, A., Kim, J.T., Münster, T., Winter, K.U., Saedler, H., 2000. A short history of MADS-box genes in plants. Plant Mol. Biol. 42, 115–149.

Treisman, R., 1995. Inside the MADS box. Nature 376, 468–469.

Vandenbussche, M., Theißen, G., van de Peer, Y., Gerats, T., 2003. Structural diversification and neo-functionalization during floral MADS-box gene evolution by C-terminal frameshift mutations. Nucleic Acids Res. 31, 4401–4409.

Verelst, W., Saedler, H., Münster, T., 2007a. MIKC* MADS–protein complexes bind motifs enriched in the proximal region of late pollen-specific Arabidopsis promoters. Plant Physiol. 143, 447–460.

Verelst, W., Twell, D., de Folter, S., Immink, R., Saedler, H., Münster, T., 2007b. MADS-complexes regulate transcriptome dynamics during pollen maturation. Genome Biol. 8, R249.

Yang, Y., Fanning, L., Jack, T., 2003. The K domain mediates heterodimerization of the Arabidopsis floral organ identity proteins APETALA3 and PISTILLATA. Plant J. 33, 47–59.

Yang, Y., Jack, T., 2004. Defining subdomains of the K domain important for protein–protein interactions of plant MADS proteins. Plant Mol. Biol. 55, 45–59.

Yanofsky, M.F., Ma, H., Bowman, J.L., Drews, G.N., Kenneth, A., Feldmann, K.A., Meyerowitz, E.M., 1990. The protein encoded by the Arabidopsis homeotic gene agamous resembles transcription factors. Nature 346, 35–39.

Zobell, O., Faigl, W., Saedler, H., Münster, T., 2010. MIKC* MADS-box proteins: conserved regulators of the gametophytic generation of land plants. Mol. Biol. Evol. 27, 1201–1211.

CHAPTER

9

TCP Transcription Factors: Evolution, Structure, and Biochemical Function

Eduardo González-Grandío, Pilar Cubas

Department of Plant Molecular Genetics,
Centro Nacional de Biotecnología (CNB-CSIC), Madrid, Spain

OUTLINE

9.1 Introduction	139
9.2 Evolution of TCP Proteins	139
9.3 The TCP Domain: Structure and Function	142
9.3.1 TCP Proteins Form Homo- and Heterodimers	143
9.3.2 The TCP Domain Basic Region Binds DNA	144
9.3.3 TCP Proteins Recognize Specific DNA Sequences	144
9.3.4 TCP Proteins Bind DNA as Dimers	145
9.4 Activation and Repression Domains	147
9.5 TCP Factors as Intrinsically Disordered Proteins	148
9.6 Posttranslational Modifications of TCP	148
9.7 Concluding Remarks	149
References	149

9.1 INTRODUCTION

The Teosinte branched1/Cincinnata/proliferating cell factor (TCP) family is a group of plant-specific genes encoding transcription factors that share the so-called TCP domain, a 59-amino-acid stretch predicted to form a noncanonical basic helix–loop–helix (bHLH) structure (Cubas et al., 1999). The TCP domain was named after the first family members identified, four apparently unrelated proteins with different roles in the control of plant development and growth, CYCLOIDEA (CYC) from snapdragon (*Antirrhinum majus*), involved in the control of floral bilateral symmetry (Luo et al., 1996), Teosinte branched1 (TB1) from maize (*Zea mays*), a key determinant of maize apical dominance (Doebley et al., 1997), and the proliferating cell factors 1 and 2 (PCF1, PCF2) from rice (*Oryza sativa*), transcription factors that bind the promoter of the PROLIFERATING CELL NUCLEAR ANTIGEN (PCNA) gene (Kosugi and Ohashi, 1997). Since their discovery, a vast amount of information has been obtained on their evolution across the plant kingdom, from green algae to eudicots, as well as on the biochemical function of TCP proteins as transcriptional regulators. In this chapter, we summarize the most recent work studying the evolution of the TCP family in angiosperms as well as the biochemical role of TCP proteins as transcriptional regulators. We will focus specifically on the role of different motifs and specific residues of the TCP domain in protein–DNA and protein–protein interactions, and will discuss the involvement of other TCP protein regions in the posttranslational regulation and function of these proteins.

9.2 EVOLUTION OF TCP PROTEINS

Since its discovery in 1999 (Cubas et al., 1999), the TCP domain has been found in an increasingly large number of proteins of distinct plant species, from charophyte green algae, mosses, lycophytes, and ferns to higher plants. It has not been found in the chlorophyte *Chlamydomonas*

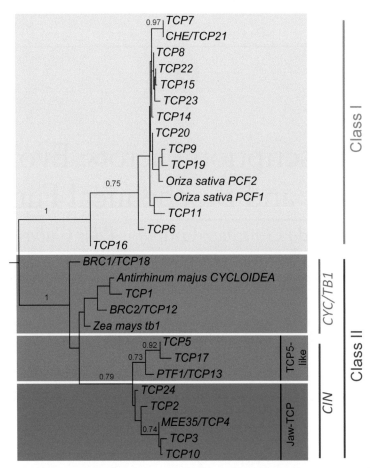

FIGURE 9.1 Maximum likelihood (ML) phylogenetic tree of Arabidopsis TCP factors. PCF1 and PCF2 from rice, CYC from *Antirrhinum*, and TB1 from maize are also included. The tree was generated using DNA sequences encoding the TCP domain and the PhyML program (www.phylogeny.fr; Dereeper et al., 2008). The reliability of internal branching was assessed using the bootstrap method with 100 bootstrap pseudoreplicates. Branches with support values over 70 are indicated.

reinhardtii nor outside the plant kingdom suggesting that TCPs originated about 700 million years ago, at the base of streptophytes (Martín-Trillo and Cubas, 2010). The TCP domain is highly conserved throughout plant species; more than 80% of the residues in this domain are shared by >50% of TCP proteins (Aggarwal et al., 2010).

Even the most basal plant species have two TCP protein types, class I (including PCF1 and PCF2) and class II (including CYC and TB1), distinguishable by conserved sets of residues within the TCP domain (Figures 9.1 and 9.3). As both classes are detected in even the simplest plant species, it has not been possible to determine which corresponds to the ancestral type. Class I is comprised of closely related proteins, whereas class II proteins fall into two well-defined groups, *CINCINNATA* (*CIN*) and CYC/TB1 clades, identifiable by further subsets of conserved amino acids (Figures 9.1 and 9.3). Outside the TCP domain, there are a few additional conserved motifs in some class II lineages, often used to refine phylogenetic analyses. Proteins of the CYC/TB1 clade share a poorly conserved glutamic acid–cysteine–glutamic acid stretch (the ECE motif; Howarth and Donoghue, 2006). A group of *CIN* genes (jaw–TCP genes, Figure 9.1, Palatnik et al., 2003; Danisman et al., 2013) contains a target for the microRNA *miR319* (Palatnik et al., 2003). Finally, CYC/TB1 factors (and a few CIN-like proteins) have a conserved 18–20-amino-acid arginine-rich motif predicted to form an α-helix, the R domain (arginine-rich domain), which appears to be ancient, as it is also found in class II proteins from basal species such as *Selaginella* (Cubas et al., 1999; Martín-Trillo and Cubas, 2010).

The remaining protein regions are very divergent, even among orthologous proteins of closely related species, which indicates rapid evolution within each lineage. Nuclear localization signals, transactivation domains (TAD), and coiled–coiled regions involved in protein–protein interactions have been detected in these regions (Kosugi and Ohashi, 1997; Martín-Trillo et al., 2011; Aggarwal et al., 2010; Tähtiharju et al., 2012; Valsecchi et al., 2013). Many of these peptides are predicted to be intrinsically disordered regions (IDR, Valsecchi et al., 2013), which are frequent targets of posttranslational modifications

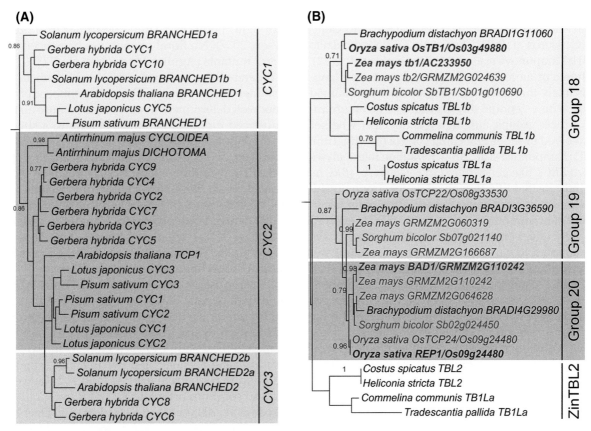

FIGURE 9.2 (A) ML phylogenetic tree of CYC/TB1 genes from dicotyledonous species. The three CYC subclades are highlighted. (B) ML phylogenetic tree of CYC/TB1-like genes of monocotyledonous species. In both cases the tree was generated as described in Figure 9.1. In (B) maize genes are highlighted in red, rice genes in purple, and sorghum genes in blue and in bold, the best characterized genes. Group nomenclature is according to Mondragon-Palomino and Trontin (2011) and Bartlett and Specht (2011).

(e.g., phosphorylation) and are often involved in protein–protein interactions mediating transactivation and multimerization (Valsecchi et al., 2013). Thus, the rapid evolution of regions outside the TCP and R domains might not be due to relaxation of selective constraints, but due to adaptive evolution, specialization, and functional diversification for each transcription factor.

In angiosperms, class I proteins are closely related and seem to have been amplified recently; nonetheless, there are no exhaustive phylogenetic analyses to determine their degree of conservation across plant lineages. In contrast, class II proteins can be clearly subdivided into two clades, their TCP domains are conserved in distantly related groups, and are usually associated to well-defined gene functions (Chapter 16).

The CIN clade contains genes mainly involved in lateral organ development. This clade could be more ancient than the CYC/TB1 clade, because all class II members are of the CIN type in basal groups such as mosses and spikemosses. The CIN clade has two further subclades, jaw–TCP genes (Palatnik et al., 2003) and TCP5-like genes (Danisman et al., 2013, Figure 9.1). Jaw–TCP genes have a target for microRNA *miR319* in all angiosperms analyzed to date, whereas TCP5-like genes do not have such a motif. This *miR319* target might have been a late acquisition by jaw–TCP genes after their separation from TCP5-like genes. Although *miR319* is detected in mosses, CIN genes do not have this sequence in the moss *Physcomitrella* (Arazi et al., 2005; Floyd and Bowman, 2007). As in the case of class I genes, in-depth phylogenetic studies that focus on the CIN clade evolution are still needed.

The CYC/TB1 clade is a major focus of many evolutionary/developmental and phylogenetic studies. This might be because it includes several genes with important, well-conserved developmental functions in angiosperms, specifically, the growth control of lateral meristems that give rise to flowers or lateral shoots. This small clade underwent independent amplification in monocots and dicots. Phylogenetic analyses of dicots indicate that three major CYC/TB1 gene subclades (CYC1, CYC2, CYC3, Figure 9.2A) arose in a series of duplication events at the base of core eudicots (Howarth and Donoghue, 2006). The first duplication led to CYC1 and the ancestor of CYC2 and CYC3. Functional studies suggest that at least two of these subclades became specialized in distinct roles in plant development.

CYC1 genes are involved mainly in the control of lateral branch outgrowth. A representative is the *Arabidopsis* single copy gene *BRANCHED1* (*BRC1*) (Aguilar-Martínez et al., 2007; Finlayson, 2007). *BRC1* duplications are reported in Solanaceae, Populus, Dipsacaceae, and Compositae (Carlson et al., 2011; Chen et al., 2013; Martín-Trillo et al., 2011; Parapunova et al., 2014; Tähtiharju et al., 2012), in which two or more *BRC1*-like genes have been identified. It is likely that these gene copies were maintained following whole genome duplication events before radiation of these groups, although it is not clear whether all are involved in shoot-branching control (Martín-Trillo et al., 2011; Parapunova et al., 2014). Evolution rate analyses of the two *BRC1*-like genes found in *Solanum* sp. (*BRC1a*, *BRC1b*) indicate that *BRC1b* evolved under strong purifying selection, consistent with a conservation of its role in the control of branch outgrowth, whereas *BRC1a* evolved more rapidly under positive selection. *BRC1a* might have become adapted to a new or more specialized, but still undetermined, role (Martín-Trillo et al., 2011).

CYC2 genes, expressed mainly in developing flowers, control the growth patterns of flower meristems and floral organs in many species. The most representative of these is *CYC* from *A. majus* (Luo et al., 1996), which had a key role in the evolution of floral zygomorphy (bilateral symmetry). Zygomorphy is thought to have evolved many times independently in dicots. In many of these groups CYC2 genes seem to have been recruited to control bilateral flower symmetry (reviewed in Busch and Zachgo, 2009; Hileman, 2014a, Hileman, 2014b). Indeed, *CYC2* genes expanded by duplication in groups in which floral zygomorphy evolved and became more complex, such as Leguminosae, Lamiales, and Asteraceae (Citerne et al., 2003; Juntheikki-Palovaara et al., 2014; Song et al., 2009; Tähtiharju et al., 2012; Hileman 2014a, 2014b). Amplification of the CYC2 subclade is particularly prominent in Asteraceae and Dipsaceae, two families in which capitula (complex, highly compressed inflorescence structures formed by small florets, usually with distinct morphologies) evolved independently. In these families there are 5–6 CYC2 genes (Broholm et al., 2008; Carlson et al., 2011; Chapman et al., 2008, Tähtiharju et al., 2012, Figure 9.2A), associated with the evolution of floral morphologies in the capitula, especially with the differences between ray and disk florets. For instance, in Dipsacaceae, the number of CYC2-like genes in discoid species (with capitula formed only by disk florets) is smaller than in radiate species (whose capitula have both disk and ray florets; Carlson et al., 2011). Amplification and evolution of *CYC2* genes in dicots and their impact on the emergence of flower zygomorphy and complex inflorescences has been reviewed extensively (e.g., Hileman, 2014b; Broholm et al., 2014).

CYC3 (Figure 9.2A), the sister clade of CYC2, comprises genes expressed in developing flowers and in vegetative buds of several species (Aguilar-Martínez et al., 2007; Howarth and Donoghue, 2006); their roles in development are nonetheless poorly characterized. *BRC2* loss of function causes a mild phenotype of increased shoot branching, particularly in short-day photoperiods or in poor light conditions (Aguilar-Martínez et al., 2007; Finlayson, 2007). No floral phenotypes have been detected in *brc2* mutants (Aguilar-Martínez et al., 2007).

Whereas three well-defined CYC/TB1 subclades have been identified in dicots, CYC/TB1 monocot phylogeny is not well determined. In part this is because all the fully sequenced monocot genomes belong to a single family, the Poaceae (grasses, such as *Z. mays*, *Brachypodium distachyon*, and *Sorghum bicolor*, Goff et al., 2002; International Brachypodium Initiative, 2010; Paterson et al., 2009; Schnable et al., 2009). Nonetheless, phylogenetic analyses of CYC/TB1 genes in grasses indicate several subclades, which evolved independently from those of the dicots (Bai et al., 2012; Bartlett and Specht, 2011; Mondragon-Palomino and Trontin, 2011). According to the nomenclature of Mondragon-Palomino and Trontin (2011), these genes comprise groups 18, 19, and 20 (Figure 9.2B). From their phylogenetic data, it can be inferred that the first duplication at the base of the monocots (or the grasses) led to group 18 and to the ancestor of groups 19 and 20. A single representative of each group is found in rice, *B. distachyon* and *S. bicolor*, and two in maize (Figure 9.2B).

Group 18 contains maize *Tb1* (and *Tb2*), as well as the rice and sorghum orthologs *OsTB1* (*FINE CULM1*, *FC1*; Takeda et al., 2003) and *SbTB1* (Lukens and Doebley, 2001, Figure 9.2B). *Tb1* and its orthologs are involved in lateral shoot development (Chapter 16). Maize *Tb1* also controls the morphology of female inflorescences (Doebley et al., 1997). Group 19 contains genes of uncharacterized function. Group 20 contains rice *RETARDED PALEA1* (*REP1*), which controls palea development and floral zygomorphy (Yuan et al., 2009), and maize *BRANCH ANGLE DEFECTIVE 1* (*BAD1*), necessary for normal tassel branch angle (Bai et al., 2012, Figure 9.2B).

Several monocot groups are formed mainly by species with zygomorphic flowers, such as Zingiberales and Orchidaceae. It would be of great interest to determine whether zygomorphy is also associated to amplification of some CYC/TB1 subclades, as observed for the CYC2 subclade in dicots; there is preliminary evidence that this could be the case for group 18 in the Zingiberales (Bartlett and Specht, 2011).

In summary, independent amplification of CYC/TB1 genes has generated subclades specialized in the control of shoot branching and/or flower development in monocots and in dicots.

9.3 THE TCP DOMAIN: STRUCTURE AND FUNCTION

The TCP domain is predicted to form a noncanonical bHLH secondary structure (Cubas et al., 1999; Kosugi and Ohashi, 1997). With the availability of a large number

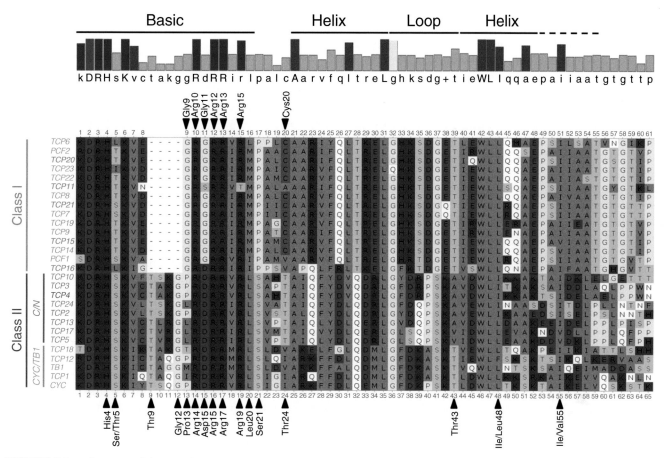

FIGURE 9.3 Alignment of the TCP domain of the 24 Arabidopsis TCP proteins, *A. majus* CYCLOIDEA, *Z. mays* TB1, and *O. sativa* PCF1 and PCF2. TCP proteins used for biochemical studies, cited in the text, are in bold. Amino acids with potential functional relevance are marked with arrowheads. Positively charged amino acids are highlighted in red, negatively charged in blue, phosphorylatable in green, hydrophobic in gray, α-helix breaking residues (proline, glycine) in yellow, cysteine in pink. The consensus amino acid sequence is shown on top. Capital letters denote 100% conservation, "+" symbol denotes two amino acids with the highest frequency. Vertical bars represent the degree of conservation. Colored bars indicate the most highly conserved amino acids.

of sequences, the TCP domain has been delimited to 62 and 58 amino acids for class I and II, respectively (Aggarwal et al., 2010). The basic domain of class I proteins is 4 amino acids shorter, and helix II is 6–7 amino acids longer than that of class II proteins (Figure 9.3). Diagnostic residues strongly conserved within each class have been identified, some of which might act as specificity determining positions (SDP, de Juan et al., 2013) that define class-type DNA-binding affinity and dimerization preferences (Section 9.3.1).

9.3.1 TCP Proteins Form Homo- and Heterodimers

Class I and II proteins form dimers in solution (Aggarwal et al., 2010; Kosugi and Ohashi, 2002) and there is considerable evidence that the TCP domain has a central function in this interaction. In *Arabidopsis* TCP4, the TCP domain is necessary and sufficient for dimer formation in yeast two-hybrid (Y2H) assays (Aggarwal et al., 2010); in rice PCF, this domain is necessary but not sufficient for dimerization, suggesting that other regions help stabilize the interactions (Kosugi and Ohashi, 1997).

Several studies in *Arabidopsis*, tomato, and rice have tested the ability of TCP proteins to interact in Y2H assays, and found that not all homodimerize, and they only form specific heterodimers, usually between proteins of the same class (Aggarwal et al., 2010; Arabidopsis Interactome Mapping Consortium, 2011; Danisman et al., 2013; Kosugi and Ohashi, 1997; Kosugi and Ohashi, 2002; Parapunova et al., 2014; Tähtiharju et al., 2012; Valsecchi et al., 2013; Viola et al., 2011). These assays detected a few examples of interclass dimerization in *Arabidopsis* and tomato, but their biological significance is not clear (Arabidopsis Interactome Mapping Consortium, 2011; Danisman et al., 2013; Parapunova et al., 2014). High-throughput Y2H assays with *Arabidopsis* TCPs indicated that, of 61 interactions detected, 7 were homo- and 54 heterodimerizations, of which

92.6% were among members of the same class (Danisman et al., 2013). CIN-like proteins of the jaw–TCP subclade preferred partners of the TCP5-like gene subclade, and vice versa. Interactions between CYC/TB1 proteins were not tested, as they were self-activating in Y2H assays (Danisman et al., 2013). After removal of their TAD, these *Arabidopsis* proteins were tested in an independent study, as well as the corresponding *Gerbera* and *Helianthus* CYC/TB1 proteins (Tähtiharju et al., 2012). Much as the case of the CIN proteins, homo- and heterodimerizations took place within (and were limited to) the group. Comparison of the heterodimers formed between *Gerbera* proteins and those formed among *Helianthus* proteins indicated conservation of some heteromeric complexes, which suggests functional relevance (Tähtiharju et al., 2012).

TCP proteins are thus able to homo- and heterodimerize, mainly between members of the same class and only rarely among members of different classes; nonetheless, these interactions have so far been characterized only in Y2H assays. Dimerization can only take place if the proteins accumulate in the same cells. Gene expression studies, essential to assess these interactions, are being carried out in some species (Aguilar-Martínez et al., 2007; Danisman et al., 2013; Koyama et al., 2007; Martín-Trillo et al., 2011; Parapunova et al., 2014; Tähtiharju et al., 2012; Ma et al., 2014). Bimolecular fluorescent complementation (BiFC) assays have been done for some of these proteins (TCP8:TCP8 and TCP8:TCP15, Valsecchi et al., 2013). Further BiFC assays, protein immunolocalizations, *in vivo* coimmunoprecipitations, and fluorescence resonance energy transfer (FRET) analyses will help to confirm the biological relevance of these interactions.

Few studies have identified amino acids essential for dimerization. The TCP amphipathic helices I and II are proposed to facilitate protein–protein interactions (Cubas et al., 1999). These types of helices tend to hide their hydrophobic surface from the hydrophilic environment, which promotes interaction with other hydrophobic surfaces, as described for HLH- and leucine-zipper-containing proteins. Indeed, helix II seems necessary for dimer formation, as a truncated TCP4 protein that lacks this helix does not dimerize (Aggarwal et al., 2010). Two hydrophobic residues in AtTCP4 helix II (Ile^{48} and Val^{55}, Figure 9.3) are essential for dimer formation; when either is substituted, the resulting mutant proteins do not dimerize (Aggarwal et al., 2010).

Preferential protein–protein interactions between members of each class seem to indicate that class-specific residues must participate in selective dimerization. Helix II, which is divergent in class I and class II proteins, might contribute to this selectivity. Studies are needed to assess the role of this and other TCP domain regions in dimerization preferences.

9.3.2 The TCP Domain Basic Region Binds DNA

Assays using *Arabidopsis* TCP4 proteins with partial deletions of this domain showed that the TCP domain, and in particular the basic region, was essential for DNA binding (Aggarwal et al., 2010). In contrast, in rice PCF1 and PCF2, this region was necessary but not sufficient for DNA binding to the PCNA promoter (Kosugi and Ohashi, 1997). Electrophoretic mobility shift assay (EMSA) of AtTCP4 in the presence of chemicals that discriminate between the major and minor grooves of the DNA double-helix further suggested that, like canonical bHLH proteins, the basic region binds the DNA major groove (Aggarwal et al., 2010). This 16–20-amino-acid region is composed of two short peptides of highly conserved basic residues (<u>K</u>D<u>RH</u>(S/T)<u>K</u> and <u>RXRRXR</u>) separated by a linker. The linker is 3 and 7 amino acids long in class I and II, respectively, and the sequence is poorly conserved (Figure 9.3). Secondary structure predictions suggest an extended conformation for the basic region, due to the glycine 9/12 (Gly$^{9/12}$, for class I and class II, respectively) and proline 13 (Pro^{13} for class II) residues in the linker, which prevent helix formation (Aggarwal et al., 2010; Cubas et al., 1999; Figure 9.3). Indeed, the TCP domain basic region is described as one of the most disordered DNA-binding domains (DBD) together with the AT hook domains (Liu et al., 2006; Valsecchi et al., 2013), and contrasts with the basic regions of other transcription factors such as bHLH and basic leucine zipper (bZIP), which form α-helical structures. This extended structure with an internal loop appears to be necessary for DNA binding in *Arabidopsis* TCP4, as mutations in the Gly^{12} or Pro^{13} of the loop completely abolish DNA binding (Aggarwal et al., 2010; Figure 9.3). Recent models suggest that parts of the basic region might acquire a helical structure after DNA binding, as observed for bHLH proteins (Aggarwal et al., 2010; Carroll et al., 1997; Ferre-D'Amare et al., 1994).

9.3.3 TCP Proteins Recognize Specific DNA Sequences

Available data indicates that TCP factors recognize 6–10 base pair (bp) motifs containing a GGNCC or GGNNCC core sequence. Kosugi and Ohashi (2002) reported that rice PCF2 (class I) and PCF5 (class II) have high affinity for distinct but overlapping motifs, PCF2 for GGNCCCAC and PCF5 for G(T/C)GGNCC. This raised the possibility that DNA sequences such as G(T/C)GGNCCCAC are recognized by both types of factors, which would lead to competition or cooperation. Information obtained mainly *in vitro* or in yeast assays such as random binding site selection, EMSA, systematic evolution of ligands by exponential enrichment (SELEX), protein-binding microarrays, yeast one-hybrid assays

(Trémousaygue et al., 2003; Li et al., 2005; Costa et al., 2005; Franco-Zorrilla et al., 2014; Kosugi and Ohashi, 2002; Schommer et al., 2008; Viola et al., 2012; Viola et al., 2011; Weirauch et al., 2014) confirmed this consensus sequence, although variations were also detected in length and composition of the flanking nucleotides. Motifs reported to be class I or class II protein targets are summarized in Table 9.1. However, some exceptions to class-type specificity have been reported. *Antirrhinum* class II CYC binds preferentially to class I motif GGNCCCNC (Costa et al., 2005), and *Arabidopsis* class I TCP16 recognizes class II motifs GTGGNCCC and GTGGaCCCa (Franco-Zorrilla et al., 2014; Viola et al., 2012). In the case of TCP16, this divergence is attributed to a specific amino acid change in the basic region, in a position proposed to interact with DNA (Viola et al., 2012). This indicates that class sequence specificity is not absolute and increases the possibilities for competition among TCP factors.

In vivo support for TCP–DNA binding is scanty, but in a few cases chromatin immunoprecipitation confirmed *in vitro* observations: *Arabidopsis* TCP20 binds a GCCCR motif in the promoters of *CYCB1;1*, *PCNA2*, *RPL24B*, *RPS15aD*, and *RPS27aB* (Li et al., 2005) and a TGGGCC motif in the *LOX2* promoter (Danisman et al., 2012); TCP3 binds a GGnCCC motif in several promoters (*miR164*, *AS1*, *IAA3/SHY2*, *SAUR65*; Koyama et al., 2010); CHE (TCP21) recognizes a GGTCCCAC motif in the *CCA1* promoter (Pruneda-Paz et al., 2009); TCP15 and TCP14 bind GGNCC motifs in the *CYCA2:3* and *RBR1* promoters (Li et al., 2012; Davière et al., 2014), and TCP14 binds other class I consensus motifs in the *CYCB1;1*, *PCNA2* promoters (Davière et al., 2014).

Several studies have tested which amino acids in the basic region participate in recognition of these DNA-binding sites and which provide specificity for class I or II motifs. Using a combined approach of mutagenesis, yeast functional assays, and structural models, Aggarwal et al. (2010) proposed that in *Arabidopsis* TCP4 (class II), His^4, Arg^{17}, and Leu^{20} establish direct contact with DNA (Figure 9.3). More recently, point mutant studies to test TCP20 and TCP11 (class I) binding abilities, indicated that Arg^{15} (almost completely conserved in this group; Figure 9.3) determines the preference of most TCPs for a G:C pair at positions 3 and 8 of the consensus-binding site (GTGGGCCCAC), and probably for a G:C pair at central positions 5 and 6 (GTGGGCCCAC, Viola et al., 2011).

Two amino acid positions in the basic domain are almost completely conserved in each class, but differ between classes; these are Gly^{11}/Asp^{15} (for class I and II, respectively) and the 4-amino-acid indel at positions 9–12 (Figure 9.3). These might be specificity-determining positions (SDPs; de Juan et al., 2013), which could be essential for defining the binding properties of each class. To study this, Viola et al. (2012) tested the DNA-binding preferences of *Arabidopsis* TCP20 and TCP16 (class I) and TCP4 (class II), as well as mutants and chimeras in these residues. They confirmed that Gly^{11}/Asp^{15} determined DNA-binding preferences for class I or II target sites. In contrast, the 4-amino-acid indel did not greatly affect DNA-binding properties, as elegantly demonstrated using TCP16, a class I protein with the typical class I deletion as well as an Asp^{15} typical of class II (Figure 9.3). TCP16 had binding properties typical of class II proteins but Asp^{11}>Gly^{11} mutations improved its affinity for class I sites. Conversely, Gly^{11}>Asp^{11} mutations in TCP20 increased affinity for class II sites. The Gly^{11}/Asp^{15} position might not establish direct contact with DNA, but could influence orientation and contact with the DNA bases of the four adjacent, fully conserved arginines (Arg$^{10/14}$, Arg$^{12/16}$, Arg$^{13/17}$, Arg$^{15/19}$-numbering for class I/II, Figure 9.3; Aggarwal et al., 2010; Viola et al., 2012).

Posttranslational modifications of the TCP domain might be used to modulate DNA-binding intensity. For instance, it would not be surprising that alteration of the potentially phosphorylatable Ser/Thr^5 and (class II) Ser^{21} of the basic domain could affect DNA-binding intensity or specificity (Martín-Trillo et al., 2011). Position Cys^{20}/Thr^{24} (class I/class II) near the basic region (Figure 9.3), proposed to be at the dimer interface (Aggarwal et al., 2010), appears to play an important role for efficient DNA binding in both class I and II proteins, perhaps by participation in protein dimerization. Indeed, mutations in TCP4 Thr^{24} abolish DNA binding (Aggarwal et al., 2010), and posttranslational modifications of Cys^{20} of TCP15, TCP20, and TCP21 lead to reduced DNA affinity (Viola et al., 2013).

9.3.4 TCP Proteins Bind DNA as Dimers

TCP proteins dimerize alone and in the presence of their target DNA. Dimerization seems to be necessary for DNA binding, as protein deletions that prevent dimer formation also abolish DNA binding (Aggarwal et al., 2010; Kosugi and Ohashi, 2002). Even point mutations in residues at the dimer interface that suppress dimerization in Y2H assays (e.g., helix II Ile/Leu^{48} and Ile/Val^{55} of class II TCPs; Figure 9.3) prevent TCP4 DNA-binding in EMSA assays (Aggarwal et al., 2010).

Rice PCF1:PCF2 and *Arabidopsis* TCP11:TCP15 heterodimers have greater affinity for DNA-binding sites than the corresponding homodimers of these proteins (Kosugi and Ohashi, 1997; Kosugi and Ohashi, 2002; Viola et al., 2011). This could be compatible with the observed preference of many TCP proteins for heterodimer formation in Y2H assays.

Canonical bHLH factors, which also act as dimers, bind the palindromic E-box motif CANNTG. Each monomer binds with similar affinity to DNA, and recognizes a half-site of the symmetrical E-box. If TCPs bind DNA in a similar fashion, it would be predicted that the binding sites of TCP homodimers would be perfect palindromes,

TABLE 9.1 List of Consensus Sequences Recognized by TCP Proteins

Species	TCP	Name	Class	Recognized motif	Method(s)	References
Rice	PCF2		I	GGNCCCAC	SELEX	Kosugi and Ohashi (2002)
Arabidopsis	TCP6		I	TGGGC(C/T)	Y1H	Giraud et al. (2010)
Arabidopsis	TCP7		I	TGGGC(C/T)	Y1H	Giraud et al. (2010)
Arabidopsis	TCP9		I	TGGGC(C/T)	Y1H	Giraud et al. (2010)
Arabidopsis	TCP11		I	(T/G)GTGGGCC	SELEX	Viola et al. (2011)
Arabidopsis	TCP11		I	TGGGC(C/T)	Y1H	Giraud et al. (2010)
Arabidopsis	TCP14		I	TGGGC(C/T)	Y1H	Giraud et al. (2010)
Arabidopsis	TCP14		I	GTGGGCCCAC	EMSA	Davière et al. (2014)
Arabidopsis	TCP14		I	GAGGGACCCT	EMSA, ChIP	Davière et al. (2014)
Arabidopsis	TCP14		I	TTGGGCCATT	EMSA, ChIP	Davière et al. (2014)
Arabidopsis	TCP14		I	TTGGGACCTC	EMSA, ChIP	Davière et al. (2014)
Arabidopsis	TCP14		I	GTGGGAACCA	EMSA, ChIP	Davière et al. (2014)
Arabidopsis	TCP14		I	GTGGGCCGTT	EMSA, ChIP	Davière et al. (2014)
Arabidopsis	TCP14		I	TTGGGCCAAA	EMSA, ChIP	Davière et al. (2014)
Arabidopsis	TCP14		I	GTGGGCCCAAA	EMSA, ChIP	Davière et al. (2014)
Arabidopsis	TCP15		I	GTGGGNCCgN	SELEX	Viola et al. (2011)
Arabidopsis	TCP15		I	gGGgCCCAC	PBM	Franco-Zorrilla et al. (2014)
Arabidopsis	TCP15		I	GGNCC	ChIP	Li et al. (2012)
Arabidopsis	TCP15		I	TGGGCC	Y1H	Giraud et al. (2010)
Arabidopsis	TCP15		I	GTGGGACC	EMSA	Uberti-Manassero et al. (2013)
Arabidopsis	TCP19		I	TGGGC(C/T)	Y1H	Giraud et al. (2010)
Arabidopsis	TCP19		I	GTGGGcCCc	PBM	Weirauch et al. (2014)
Arabidopsis	TCP20		I	GCCCR	EMSA, ChIP	Li et al. (2005)
Arabidopsis	TCP20		I	TGGGCC	EMSA	Trémousaygue et al. (2003)
Arabidopsis	TCP20		I	GTGGGNCCcN	SELEX	Viola et al. (2011)
Arabidopsis	TCP20		I	GTGGGACCGG	EMSA	Viola et al. (2012)
Arabidopsis	TCP20		I	TGGGC(C/T)	Y1H	Giraud et al. (2010)
Arabidopsis	TCP20		I	TGGGCC	ChIP	Danisman et al. (2012)
Arabidopsis	TCP20		I	GTGGGSCC	PBM	Weirauch et al. (2014)
Arabidopsis	TCP21	CHE	I	GGTCCCAC	EMSA, Y1H, ChIP	Pruneda-Paz et al. (2009)
Arabidopsis	TCP21	CHE	I	TGGGC(C/T)	Y1H	Giraud et al. (2010)
Arabidopsis	TCP23		I	gGGgCCCACa	PBM	Franco-Zorrilla et al. (2014)
Arabidopsis	TCP23		I	TGGGC(C/T)	Y1H	Giraud et al. (2010)
Antirrhinum	CYC		II	GGTCCCAC	EMSA	Costa et al. (2005)
Antirrhinum	CYC		II	GGNCCCNC	SELEX	Costa et al. (2005)
Rice	OsTB1		II	TGGGC(C/T)	EMSA	Lu et al. 2013
Rice	PCF5		II	G(T/c)GGNCCC	SELEX	Kosugi and Ohashi (2002)
Arabidopsis	TCP1		II	TGGGCC	Y1H	Giraud et al. (2010)
Arabidopsis	TCP2		II	TGGGC(C/T)	Y1H	Giraud et al. (2010)

TABLE 9.1 List of Consensus Sequences Recognized by TCP Proteins (cont.)

Species	TCP	Name	Class	Recognized motif	Method(s)	References
Arabidopsis	TCP2		II	tGGKMCCa	PBM	Weirauch et al. (2014)
Arabidopsis	TCP3		II	TGGGC(C/T)	Y1H	Giraud et al. (2010)
Arabidopsis	TCP3		II	GGnCCC	ChIP	Koyama et al. 2010
Arabidopsis	TCP3		II	GTGGTCCC	PBM	Weirauch et al. (2014)
Arabidopsis	TCP4	MEE35	II	GGACCA	SELEX	Schommer et al. (2008)
Arabidopsis	TCP4	MEE35	II	TGGGC(C/T)	Y1H	Giraud et al. (2010)
Arabidopsis	TCP4	MEE35	II	GTGGTCCC	EMSA	Aggarwal et al. (2010)
Arabidopsis	TCP4	MEE35	II	GTGGKCCC	PBM	Weirauch et al. (2014)
Arabidopsis	TCP5		II	TGGGCT	Y1H	Giraud et al. (2010)
Arabidopsis	TCP5		II	RTGGTCCC	PBM	Weirauch et al. (2014)
Arabidopsis	TCP10		II	TGGGC(C/T)	Y1H	Giraud et al. (2010)
Arabidopsis	TCP12	BRC2	II	TGGGC(C/T)	Y1H	Giraud et al. (2010)
Arabidopsis	TCP13	PTF1	II	TGGGCT	Y1H	Giraud et al. (2010)
Arabidopsis	TCP16		II	GTGGNCCCNN	SELEX, EMSA	Viola et al. (2012)
Arabidopsis	TCP16		II	tGGGtCCAC	PBM	Franco-Zorrilla et al. (2014)
Arabidopsis	TCP16		II	GTGGacCCgg	PBM	Weirauch et al. (2014)
Arabidopsis	TCP17		II	TGGGC(C/T)	Y1H	Giraud et al. (2010)
Arabidopsis	TCP18	BRC1	II	TGGGC(C/T)	Y1H	Giraud et al. (2010)
Arabidopsis	TCP24		II	TGGGC(C/T)	Y1H	Giraud et al. (2010)
Arabidopsis	TCP24		II	TGGKGCC	PBM	Weirauch et al. (2014)

The technique used to determine it is indicated.
EMSA, electrophoretic mobility shift assay; SELEX, systematic evolution of ligands by exponential enrichment; Y1H, yeast one-hybrid; PBM, protein binding matrix; ChIP, chromatin immunoprecipitation.

GTGGGCCCAC. In this way, each TCP monomer would interact with half a binding site, GTGGG. However, this is not the case; TCP-binding sites are not always perfect palindromes. SELEX experiments using *Arabidopsis* and rice TCPs, as well as hydroxyl radical foot-printing assays carried out with *Arabidopsis* proteins, showed that base selection is more efficient in one half of the palindromic sequence than in the other (Kosugi and Ohashi, 2002; Viola et al., 2011). This indicates that one of the monomers establishes more specific contacts with DNA and directs DNA recognition specificity.

9.4 ACTIVATION AND REPRESSION DOMAINS

Typical transcription factors not only have DBDs, but also TADs that promote interaction with the transcriptional machinery, directly or through coactivator proteins. TADs have been identified in some but not all TCP proteins when tested as baits in Y2H assays. As a rule, the most strongly self-activating TCPs are class II factors. In maize, *Arabidopsis*, *Gerbera*, and *Helianthus*, most proteins of the CYC/TB1 subclade (e.g., TB1, AtTCP1, BRC1, BRC2, GhCYC, HaCYC) and many of the *CIN* subclade (e.g., *Arabidopsis* TCP2, TCP4, TCP10, TCP24) have TADs, as identified in Y2H assays (Aggarwal et al., 2010; Danisman et al., 2013; Giraud et al., 2010; Kosugi and Ohashi, 2002; Tähtiharju et al., 2012). These TADs are located either in the amino- or the carboxy-terminal (N-t, C-t) part of the proteins and are not conserved in distantly related species. It is likely that these domains evolved independently in different clades.

In contrast, none of the rice class I proteins tested so far is self-activating, and only a few *Arabidopsis* ones are weakly self-activating (TCP11 and TCP20; Giraud et al., 2010). For others, the self-activating behavior reported is not consistent in the literature (e.g., TCP8, TCP14, and TCP15 are self-activating for Valsecchi et al. (2013) and Steiner et al. (2012), but not for Giraud et al. (2010) or Danisman et al. (2013)). Recently, Davière et al. (2014) showed that TCP14 is able to transcriptionally activate a *pCYCB1;1:GUS* reporter in *Nicotiana* leaves but it cannot be ruled out that this action requires interaction with other *Nicotiana* factors.

This could indicate that class II proteins are more often self-sufficient transcription factors able to bind DNA and trigger transcriptional activation. In contrast, most class I TCP factors could be nonautonomous transcriptional regulators that require additional interacting proteins for their activity. It is also possible that such TCP proteins could function as activators or repressors, depending on their interactions with other proteins (Herve et al., 2009). Indeed, TCP-binding sites are often associated with other cis-acting elements, indicating that some TCP could act as part of multimeric regulatory modules (Martín-Trillo and Cubas, 2010; Uberti-Manassero et al., 2013). This would be consistent with the proposal that the function of some class I TCPs (e.g., TCP14 and TCP15) is context dependent (Kieffer et al., 2011; Kim et al., 2014; Li et al., 2012; Rueda-Romero et al., 2012; Davière et al., 2014).

Remarkably, *Arabidopsis* class II TCP3 and TCP4 also have transcriptional repressor EAR (ERF amphiphilic repressor; Hiratsu et al., 2003) domains (LRLSL) and these factors (as well as TCP2) have been found to interact with TOPLESS (TPL) and TOPLESS-RELATED (TPR) transcriptional corepressors in Y2H assays (Causier et al., 2012). This supports a potential dual function of some TCPs as transcriptional activators and repressors. In addition, class I TCP8, TCP14, TCP16, and TCP23 have been reported to interact with TPR proteins in Y2H assays, although only partial EAR domains have been found in their protein sequences (Causier et al., 2012).

9.5 TCP FACTORS AS INTRINSICALLY DISORDERED PROTEINS

Bioinformatic analyses showed that the amino acid composition of TCP proteins is rich in disorder-promoting residues typical of intrinsically disordered proteins (IDPs; Dunker et al., 2001; Valsecchi et al., 2013). The IDPs are characterized by low compactness, low secondary structure content, and high flexibility (Uversky, 2002). These proteins have long IDRs prone to form coil-coiled (CC) structures, often involved in protein–protein interactions, and frequently modified by phosphorylation (Iakoucheva et al., 2004). IDRs appear to be common in transcription factors, and they participate in the recognition of various protein partners. Valsecchi et al. (2013) determined that class I TCPs are predicted to be more disordered than class II TCPs. The most potentially disordered *Arabidopsis* TCP proteins are TCP8, TCP9, TCP22, and TCP23. TCP8, predicted to be the most disordered factor, has a very disordered C-t IDR with a high probability to form a CC domain (Valsecchi et al., 2013). Biochemical and molecular biology experiments showed that this C-t region contributes to TCP8 self-assembly in trimers and higher-order multimers. TCP20, TCP15, and perhaps TCP4 are also described to form multimers through regions outside the TCP domain (Viola et al., 2013). IDPs are predicted to be promiscuous proteins, consistent with some observations made for TCPs. High-throughput Y2H assays showed that TCP15, TCP19, and TCP21 can interact, respectively, with 40, 29, and 27 proteins represented in the assay. TCP14, the most promiscuous TCP protein described to date, showed interaction with >150 proteins in these Y2H assays (Arabidopsis Interactome Mapping Consortium, 2011) and was found to interact with additional proteins in other yeast screenings (e.g., Rueda-Romero et al., 2012; Davière et al., 2014; Resentini et al., 2015). This promiscuity is not exclusive to class I TCPs; for instance, class II TCP13 interacts with 74 partners (Arabidopsis Interactome Mapping Consortium, 2011).

Y2H screenings using individual *Arabidopsis* proteins revealed further TCP potential partners. Several TCPs interact with core components of the circadian clock, indicating that these factors participate in the circadian regulation of gene expression (Pruneda-Paz et al., 2009; Giraud et al., 2010). More recently TCPs have also been reported to interact, in Y2H and *Nicotiana* BiFC assays, with the flowering regulators FLOWERING LOCUS T and SISTER OF FT (Mimida et al., 2011; Ho and Weigel, 2014; Niwa et al., 2013) and *in vivo* with DELLA proteins (Davière et al., 2014; Resentini et al., 2015). Of particular relevance for their function as transcriptional regulators is the previously mentioned interaction of some TCPs with TPL/TPR proteins in YTH assays (Causier et al., 2012), and *in vivo* with the EAR domain protein TCP INTERACTING PROTEIN1 (TIE1; Tao et al., 2013). These interactions are likely to transform TCP factors into transcriptional repressors thus reversing the response of their regulatory pathways and increasing the complexity of their functional activities. Interaction of TCPs with yet additional proteins has been reviewed elsewhere (Martín-Trillo and Cubas, 2010; Uberti-Manassero et al., 2013).

9.6 POSTTRANSLATIONAL MODIFICATIONS OF TCP

Posttranslational modifications of the TCP proteins, especially at the TCP domain, might control their DNA-binding intensity and their activity as transcriptional regulators. For instance, modification of the potentially phosphorylatable Ser and Thr in the TCP domain could affect DNA-binding intensity or specificity, protein dimerization, or all of these. Indeed, several strong mutations affect conserved Ser and Thr residues: the strong mutant alleles of maize class II gene *BAD1*, *bad1-1*, and *bad1-2*, are a $Thr^9 > Met^9$ and a $Ser^{21} > Phe^{21}$ mutation in the basic region, respectively (Bai et al., 2012; Figure 9.3). The strong mutant of the pea class II *BRC1* gene, *PsBRC1*, $psbrc1^{Cam}$ has a $Thr^{43} > Ile^{43}$ mutation in helix II (Figure 9.3; Braun et al., 2012).

Viola et al. (2013) found that the conserved Cys^{20} in class I TCP15, TCP20, and TCP21 (Figure 9.3), located near the basic region at the dimer interface is sensitive to redox conditions, such that proteins with oxidized Cys^{20} showed reduced DNA binding. Class II TCP4 has a Thr^{24} in this position (Figure 9.3) and mutations in this residue also abolish DNA binding (Aggarwal et al., 2010). O-linked N-acetylglucosamine (O-GlcNAc) modifications also regulate the posttranslational fate of target proteins. Indirect genetic evidence indicates that Arabidopsis TCP14 and TCP15 could be activated by posttranslational modification, through the O-GlcNAc transferase (OGT) SPINDLY (SPY), which would lead to induction of the cytokinin pathway (Steiner et al., 2012; Chapter 16).

9.7 CONCLUDING REMARKS

Over the last years, knowledge has been gained about the evolution of the TCP gene family and the molecular function of proteins containing the TCP domain. TCP transcription factors seem to act frequently as heterodimers formed by proteins of the same class that recognize specific GC-rich motifs in DNA. Well-defined lineages of class II TCP proteins have been associated with important developmental gene functions. Moreover, the evolution of complex morphological traits, such as the capitulum inflorescence, correlates with a remarkable amplification of certain class II clades. This is consistent with likely activity of class II proteins as autonomous transcription factors that not only have DBDs but often TADs to activate the transcriptional machinery. In contrast, most class I factors lack a TAD and could therefore have different functionalities depending on their tissue-specific interaction with other factors. The high promiscuity of many of these TCPs would facilitate this functional flexibility. Class I and class II factors could also act as transcriptional repressors through interactions with TPL/TPR and TIE1-related proteins. So far we still have an incomplete knowledge of how different amino acids and motifs contribute to elicit or modulate the function of TCP proteins as transcriptional regulators. Additional biochemical studies using mutant proteins combined with in vivo studies will help elucidate these questions.

Acknowledgments

The authors acknowledge support from the Spanish Ministerio de Educación y Ciencia (BIO2008-00581 and CSD2007-00057), Ministerio de Ciencia y Tecnología (BIO2011-25687), and the Fundación Ramón Areces.

References

Aggarwal, P., Das Gupta, M., Joseph, A.P., Chatterjee, N., Srinivasan, N., Nath, U., 2010. Identification of specific DNA binding residues in the TCP family of transcription factors in Arabidopsis. Plant Cell 22, 1174–1189.

Aguilar-Martínez, J.A., Poza-Carrion, C., Cubas, P., 2007. Arabidopsis BRANCHED1 acts as an integrator of branching signals within axillary buds. Plant Cell 19, 458–472.

Arabidopsis Interactome Mapping Consortium., 2011. Evidence for network evolution in an Arabidopsis interactome map. Science 333, 601–607.

Arazi, T., Talmor-Neiman, M., Stav, R., Riese, M., Huijser, P., Baulcombe, D.C., 2005. Cloning and characterization of micro-RNAs from moss. Plant J. 43, 837–848.

Bai, F., Reinheimer, R., Durantini, D., Kellogg, E.A., Schmidt, R.J., 2012. TCP transcription factor, BRANCH ANGLE DEFECTIVE 1 (BAD1), is required for normal tassel branch angle formation in maize. Proc. Natl. Acad. Sci. USA 109, 12225–12230.

Bartlett, M.E., Specht, C.D., 2011. Changes in expression pattern of the teosinte branched1-like genes in the Zingiberales provide a mechanism for evolutionary shifts in symmetry across the order. Am. J. Bot. 98, 227–243.

Braun, N., de Saint Germain, A., Pillot, J.P., Boutet-Mercey, S., Dalmais, M., Antoniadi, I., Li, X., Maia-Grondard, A., Le Signor, C., Bouteiller, N., et al., 2012. The pea TCP transcription factor PsBRC1 acts downstream of strigolactones to control shoot branching. Plant Physiol. 158, 225–238.

Broholm, S.K., Tähtiharju, S., Laitinen, R.A., Albert, V.A., Teeri, T.H., Elomaa, P., 2008. A TCP domain transcription factor controls flower type specification along the radial axis of the Gerbera (Asteraceae) inflorescence. Proc. Natl. Acad. Sci. USA 105, 9117–9122.

Broholm, S.K., Teeri, T.H., Elomaa, P., 2014. The molecular genetics of floral transition and flower development. In: Advances in Botanical Research. Academic Press, Elsevier, Oxford, pp. 297–333.

Busch, A., Zachgo, S., 2009. Flower symmetry evolution: towards understanding the abominable mystery of angiosperm radiation. BioEssays 31, 1181–1190.

Carlson, S.E., Howarth, D.G., Donoghue, M.J., 2011. Diversification of CYCLOIDEA-like genes in Dipsacaceae (Dipsacales): implications for the evolution of capitulum inflorescences. BMC Evol. Biol. 11, 325.

Carroll, A.S., Gilbert, D.E., Liu, X., Cheung, J.W., Michnowicz, J.E., Wagner, G., Ellenberger, T.E., Blackwell, T.K., 1997. SKN-1 domain folding and basic region monomer stabilization upon DNA binding. Genes Dev. 11, 2227–2238.

Causier, B., Ashworth, M., Guo, W., Davies, B., 2012. The TOPLESS interactome: a framework for gene repression in Arabidopsis. Plant Physiol. 158, 423–438.

Chapman, M.A., Leebens-Mack, J.H., Burke, J.M., 2008. Positive selection and expression divergence following gene duplication in the sunflower CYCLOIDEA gene family. Mol. Biol. Evol. 25, 1260–1273.

Chen, X., Zhou, X., Xi, L., Li, J., Zhao, R., Ma, N., Zhao, L., 2013. Roles of DgBRC1 in regulation of lateral branching in chrysanthemum (Dendranthema xgrandiflora cv Jinba). PLoS ONE 8, e61717.

Citerne, H., Luo, D., Pennington, R., Coen, E., Cronk, Q., 2003. A phylogenomic investigation of CYCLOIDEA-like TCP genes in the Leguminosae. Plant Physiol. 131, 1042–1053.

Costa, M.M., Fox, S., Hanna, A.I., Baxter, C., Coen, E., 2005. Evolution of regulatory interactions controlling floral asymmetry. Development 132, 5093–5101.

Cubas, P., Lauter, N., Doebley, J., Coen, E., 1999. The TCP domain: a motif found in proteins regulating plant growth and development. Plant J. 18, 215–222.

Danisman, S., van der Wal, F., Dhondt, S., Waites, R., de Folter, S., Bimbo, A., van Dijk, A.D., Muino, J.M., Cutri, L., Dornelas, M.C., et al., 2012. Arabidopsis class I., class II TCP transcription factors regulate jasmonic acid metabolism and leaf development antagonistically. Plant Physiol. 159, 1511–1523.

Danisman, S., van Dijk, A.D., Bimbo, A., van der Wal, F., Hennig, L., de Folter, S., Angenent, G.C., Immink, R.G., 2013. Analysis of functional redundancies within the Arabidopsis TCP transcription factor family. J. Exp. Bot. 64, 5673–5685.

Davière, J.M., Wild, M., Regnault, T., Baumberger, N., Eisler, H., Genschik, P., Achard, P., 2014. Class I TCP-DELLA interactions in inflorescence shoot apex determine plant height. Curr. Biol. 24, 1923–1928.

de Juan, D., Pazos, F., Valencia, A., 2013. Emerging methods in protein co-evolution. Nat. Rev. Genet. 14, 249–261.

Dereeper, A., Guignon, V., Blanc, G., Audic, S., Buffet, S., Chevenet, F., Dufayard, J.F., Guindon, S., Lefort, V., Lescot, M., et al., 2008. Phylogeny.fr: robust phylogenetic analysis for the non-specialist. Nucleic Acids Res. 36, 465–469.

Doebley, J., Stec, A., Hubbard, L., 1997. The evolution of apical dominance in maize. Nature 386, 485–488.

Dunker, A.K., Lawson, J.D., Brown, C.J., Williams, R.M., Romero, P., Oh, J.S., Oldfield, C.J., Campen, A.M., Ratliff, C.M., Hipps, K.W., et al., 2001. Intrinsically disordered protein. J. Mol. Graph. Model. 19, 26–59.

Ferre-D'Amare, A.R., Pognonec, P., Roeder, R.G., Burley, S.K., 1994. Structure and function of the b/HLH/Z domain of USF. EMBO J. 13, 180–189.

Finlayson, S.A., 2007. *Arabidopsis TEOSINTE BRANCHED1-LIKE 1* regulates axillary bud outgrowth and is homologous to monocot *teosinte branched1*. Plant Cell Physiol. 48, 667–677.

Floyd, K.S., Bowman, J.L., 2007. The ancestral developmental tool kit of land plants. Int. J. Plant Sci. 168, 1–35.

Franco-Zorrilla, J.M., Lopez-Vidriero, I., Carrasco, J.L., Godoy, M., Vera, P., Solano, R., 2014. DNA-binding specificities of plant transcription factors and their potential to define target genes. Proc. Natl. Acad. Sci. USA 111, 2367–2372.

Giraud, E., Ng, S., Carrie, C., Duncan, O., Low, J., Lee, C.P., Van Aken, O., Millar, A.H., Murcha, M., Whelan, J., 2010. TCP transcription factors link the regulation of genes encoding mitochondrial proteins with the circadian clock in *Arabidopsis thaliana*. Plant Cell 22, 3921–3934.

Goff, S.A., Ricke, D., Lan, T.H., Presting, G., Wang, R., Dunn, M., Glazebrook, J., Sessions, A., Oeller, P., Varma, H., et al., 2002. A draft sequence of the rice genome (*Oryza sativa* L. ssp. *japonica*). Science 296, 92–100.

Herve, C., Dabos, P., Bardet, C., Jauneau, A., Auriac, M.C., Ramboer, A., Lacout, F., Tremousaygue, D., 2009. *In vivo* interference with *AtTCP20* function induces severe plant growth alterations and deregulates the expression of many genes important for development. Plant Physiol. 149, 1462–1477.

Hileman, L.C., 2014a. Bilateral flower symmetry – how, when and why? Curr. Opin. Plant Biol. 17, 146–152.

Hileman, L.C., 2014b. Trends in flower symmetry evolution revealed through phylogenetic and developmental genetic advances. Philos. Trans. R. Soc. Lond. B. Biol. Sci. 369, doi: 10.1098/rstb.2013.0348

Hiratsu, K., Matsui, K., Koyama, T., Ohme-Takagi, M., 2003. Dominant repression of target genes by chimeric repressors that include the EAR motif, a repression domain, in *Arabidopsis*. Plant J. 34, 733–739.

Ho, W.W., Weigel, D., 2014. Structural features determining flower-promoting activity of *Arabidopsis* FLOWERING LOCUS T. Plant Cell 26, 552–564.

Howarth, D.G., Donoghue, M.J., 2006. Phylogenetic analysis of the "ECE" (CYC/TB1) clade reveals duplications predating the core eudicots. Proc. Natl. Acad. Sci. USA 103, 9101–9106.

Iakoucheva, L.M., Radivojac, P., Brown, C.J., O'Connor, T.R., Sikes, J.G., Obradovic, Z., Dunker, A.K., 2004. The importance of intrinsic disorder for protein phosphorylation. Nucleic Acids Res. 32, 1037–1049.

International Brachypodium Initiative, 2010. Genome sequencing and analysis of the model grass *Brachypodium distachyon*. Nature 463, 763–768.

Juntheikki-Palovaara, I., Tähtiharju, S., Lan, T., Broholm, S.K., Rijpkema, A.S., Ruonala, R., Kale, L., Albert, V.A., Teeri, T.H., Elomaa, P., 2014. Functional diversification of duplicated CYC2 clade genes in regulation of inflorescence development in *Gerbera hybrida* (Asteraceae). Plant J. 79, 783–796.

Kieffer, M., Master, V., Waites, R., Davies, B., 2011. TCP14 and TCP15 affect internode length and leaf shape in *Arabidopsis*. Plant J. 68, 147–158.

Kim, S.H., Son, G.H., Bhattacharjee, S., Kim, H.J., Nam, J.C., Nguyen, P.D., Hong, J.C., Gassmann, W., 2014. The *Arabidopsis* immune adaptor SRFR1 interacts with TCP transcription factors that redundantly contribute to effector-triggered immunity. Plant J. 78, 978–989.

Kosugi, S., Ohashi, Y., 1997. PCF1 and PCF2 specifically bind to *cis* elements in the rice *proliferating cell nuclear antigen* gene. Plant Cell 9, 1607–1619.

Kosugi, S., Ohashi, Y., 2002. DNA binding and dimerization specificity and potential targets for the TCP protein family. Plant J. 30, 337–348.

Koyama, T., Furutani, M., Tasaka, M., Ohme-Takagi, M., 2007. TCP transcription factors control the morphology of shoot lateral organs via negative regulation of the expression of boundary-specific genes in *Arabidopsis*. Plant Cell 19, 473–484.

Koyama, T., Mitsuda, N., Seki, M., Shinozaki, K., Ohme-Takagi, M., 2010. TCP transcription factors regulate the activities of ASYMMETRIC LEAVES1 and miR164, as well as the auxin response, during differentiation of leaves in *Arabidopsis*. Plant Cell 22, 3574–3588.

Li, C., Potuschak, T., Colón-Carmona, A., Gutiérrez, R.A., Doerner, P., 2005. *Arabidopsis* TCP20 links regulation of growth and cell division control pathways. Proc. Natl. Acad. Sci. USA 102, 12978–12983.

Li, Z.Y., Li, B., Dong, A.W., 2012. The *Arabidopsis* transcription factor AtTCP15 regulates endoreduplication by modulating expression of key cell-cycle genes. Mol. Plant 5, 270–280.

Liu, J., Perumal, N.B., Oldfield, C.J., Su, E.W., Uversky, V.N., Dunker, A.K., 2006. Intrinsic disorder in transcription factors†. Biochemistry 45, 6873–6888.

Lu, Z., Yu, H., Xiong, G., Wang, J., Jiao, Y., Liu, G., Jing, Y., Meng, X., Hu, X., Qian, Q., Fu, X., Wang, Y., Li, J., 2013. Genome-wide binding analysis of the transcription activator IDEAL PLANT ARCHITECTURE1 reveals a complex network regulating rice plant architecture. Plant Cell 25, 3743–3759.

Lukens, L., Doebley, J., 2001. Molecular evolution of the teosinte branched gene among maize and related grasses. Mol. Biol. Evol. 18, 627–638.

Luo, D., Carpenter, R., Vincent, C., Copsey, L., Coen, E., 1996. Origin of floral asymmetry in *Antirrhinum*. Nature 383, 794–799.

Ma, J., Wang, Q., Sun, R., Xie, F., Jones, D.C., Zhang, B., 2014. Genome-wide identification and expression analysis of TCP transcription factors in *Gossypium raimondii*. Sci. Rep. 4, doi: 10.1038/srep06645

Martín-Trillo, M., Cubas, P., 2010. TCP genes: a family snapshot ten years later. Trends Plant Sci. 15, 31–39.

Martín-Trillo, M., Grandio, E.G., Serra, F., Marcel, F., Rodriguez-Buey, M.L., Schmitz, G., Theres, K., Bendahmane, A., Dopazo, H., Cubas, P., 2011. Role of tomato BRANCHED1-like genes in the control of shoot branching. Plant J. 67, 701–714.

Mimida, N., Kidou, S.-I., Iwanami, H., Moriya, S., Abe, K., Voogd, C., Varkonyi-Gasic, E., Kotoda, N., Näsholm, T., 2011. Apple FLOWERING LOCUS T proteins interact with transcription factors implicated in cell growth and organ development. Tree Physiol. 31 (5), 555–566.

Mondragon-Palomino, M., Trontin, C., 2011. High time for a roll call: gene duplication and phylogenetic relationships of TCP-like genes in monocots. Ann. Bot. 107, 1533–1544.

Niwa, M., Daimon, Y., Kurotani, K., Higo, A., Pruneda-Paz, J.L., Breton, G., Mitsuda, N., Kay, S.A., Ohme-Takagi, M., Endo, M., et al., 2013. BRANCHED1 interacts with FLOWERING LOCUS T to repress the floral transition of the axillary meristems in *Arabidopsis*. Plant Cell 25, 1228–1242.

Palatnik, J.F., Allen, E., Wu, X.L., Schommer, C., Schwab, R., Carrington, J.C., Weigel, D., 2003. Control of leaf morphogenesis by microRNAs. Nature 425, 257–263.

Parapunova, V., Busscher, M., Busscher-Lange, J., Lammers, M., Karlova, R., Bovy, A.G., Angenent, G.C., de Maagd, R.A., 2014. Identification, cloning and characterization of the tomato TCP transcription factor family. BMC Plant Biol. 14, 157.

Paterson, A.H., Bowers, J.E., Bruggmann, R., Dubchak, I., Grimwood, J., Gundlach, H., Haberer, G., Hellsten, U., Mitros, T., Poliakov, A., et al., 2009. The *Sorghum bicolor* genome and the diversification of grasses. Nature 457, 551–556.

Pruneda-Paz, J.L., Breton, G., Para, A., Kay, S.A., 2009. A functional genomics approach reveals CHE as a component of the *Arabidopsis* circadian clock. Science 323, 1481–1485.

Resentini, F., Felipo-Benavent, A., Colombo, L., Blázquez, M.A., Alabadí, D., Masiero, S., 2015. TCP14 and TCP15 mediate the promotion of seed germination by gibberellins in Arabidopsis thaliana. Mol. Plant 8 (3), 482–485.

Rueda-Romero, P., Barrero-Sicilia, C., Gomez-Cadenas, A., Carbonero, P., Onate-Sanchez, L., 2012. *Arabidopsis thaliana* DOF6 negatively affects germination in non-after-ripened seeds and interacts with TCP14. J. Exp. Bot. 63, 1937–1949.

Schnable, P.S., Ware, D., Fulton, R.S., Stein, J.C., Wei, F., Pasternak, S., Liang, C., Zhang, J., Fulton, L., Graves, T.A., et al., 2009. The B73 maize genome: complexity, diversity, and dynamics. Science 326, 1112–1115.

Schommer, C., Palatnik, J.F., Aggarwal, P., Chetelat, A., Cubas, P., Farmer, E.E., Nath, U., Weigel, D., 2008. Control of jasmonate biosynthesis and senescence by miR319 targets. PLoS Biol. 6, 1991–2001.

Song, C.F., Lin, Q.B., Liang, R.H., Wang, Y.Z., 2009. Expressions of ECE-CYC2 clade genes relating to abortion of both dorsal and ventral stamens in *Opithandra* (Gesneriaceae). BMC Evol. Biol. 9, 244.

Steiner, E., Efroni, I., Gopalraj, M., Saathoff, K., Tseng, T.S., Kieffer, M., Eshed, Y., Olszewski, N., Weiss, D., 2012. The *Arabidopsis* O-linked N-acetylglucosamine transferase SPINDLY interacts with class I TCPs to facilitate cytokinin responses in leaves and flowers. Plant Cell 24, 96–108.

Tähtiharju, S., Rijpkema, A.S., Vetterli, A., Albert, V.A., Teeri, T.H., Elomaa, P., 2012. Evolution and diversification of the CYC/TB1 gene family in Asteraceae – a comparative study in *Gerbera* (Mutisieae) and sunflower (Heliantheae). Mol. Biol. Evol. 29, 1155–1166.

Takeda, T., Suwa, Y., Suzuki, M., Kitano, H., Ueguchi-Tanaka, M., Ashikari, M., Matsuoka, M., Ueguchi, C., 2003. The *Ostb1* gene negatively regulates lateral branching in rice. Plant J. 33, 513–520.

Tao, Q., Guo, D., Wei, B., Zhang, F., Pang, C., Jiang, H., Zhang, J., Wei, T., Gu, H., Qu, L.J., et al., 2013. The TIE1 transcriptional repressor links TCP transcription factors with TOPLESS/TOPLESS-RELATED corepressors and modulates leaf development in *Arabidopsis*. Plant Cell 25, 421–437.

Trémousaygue, D., Garnier, L., Bardet, C., Dabos, P., Hervé, C., Lescure, B., 2003. Internal telomeric repeats and "TCP domain" protein-binding sites co-operate to regulate gene expression in *Arabidopsis thaliana* cycling cells. Plant J. 33, 957–966.

Uberti-Manassero, N.G., Viola, I.L., Welchen, E., Gonzalez, D.H., 2013. TCP transcription factors: architectures of plant form. Biomol. Concepts 4, 111–127.

Uversky, V.N., 2002. Natively unfolded proteins: a point where biology waits for physics. Protein Sci. 11, 739–756.

Valsecchi, I., Guittard-Crilat, E., Maldiney, R., Habricot, Y., Lignon, S., Lebrun, R., Miginiac, E., Ruelland, E., Jeannette, E., Lebreton, S., 2013. The intrinsically disordered C-terminal region of *Arabidopsis thaliana* TCP8 transcription factor acts both as a transactivation and self-assembly domain. Mol. Biosyst. 9, 2282–2295.

Viola, I.L., Uberti Manassero, N.G., Ripoll, R., Gonzalez, D.H., 2011. The *Arabidopsis* class I TCP transcription factor AtTCP11 is a developmental regulator with distinct DNA-binding properties due to the presence of a threonine residue at position 15 of the TCP domain. Biochem. J. 435, 143–155.

Viola, I.L., Reinheimer, R., Ripoll, R., Manassero, N.G., Gonzalez, D.H., 2012. Determinants of the DNA binding specificity of class I and class II TCP transcription factors. J. Biol. Chem. 287, 347–356.

Viola, I.L., Guttlein, L.N., Gonzalez, D.H., 2013. Redox modulation of plant developmental regulators from the class I TCP transcription factor family. Plant Physiol. 162, 1434–1447.

Weirauch, M.T., Yang, A., Albu, M., Cote, A.G., Montenegro-Montero, A., Drewe, P., Najafabadi, H.S., Lambert, S.A., Mann, I., Cook, K., et al., 2014. Determination and inference of eukaryotic transcription factor sequence specificity. Cell 158, 1431–1443.

Yuan, Z., Gao, S., Xue, D.W., Luo, D., Li, L.T., Ding, S.Y., Yao, X., Wilson, Z.A., Qian, Q., Zhang, D.B., 2009. *RETARDED PALEA1* controls palea development and floral zygomorphy in rice. Plant Physiol. 149, 235–244.

CHAPTER

10

Structure and Evolution of Plant GRAS Family Proteins

Cordelia Bolle

Department Biologie I, Lehrstuhl für Molekularbiologie der Pflanzen (Botanik),
Biozentrum der LMU München, Planegg-Martinsried, Germany

OUTLINE

10.1 Presence of GRAS Proteins in Plants and other Organisms	153	10.3.1 GRAS Domain	157
		10.3.2 N-Terminal Domain	159
10.2 Genomic Organization (intron/exon)	156	10.3.3 Additional Domains	160
		10.4 Conclusion	160
10.3 Structure of GRAS Proteins	157	References	160

GRAS proteins are named after the first three proteins identified in *Arabidopsis thaliana*, GIBBERELLIC ACID INSENSITIVE (GAI), REPRESSOR OF GA1 (RGA), and SCARECROW (SCR; Di Laurenzio et al., 1996; Peng et al., 1997; Silverstone et al., 1998). In the last two decades several GRAS proteins have been biologically and functionally characterized and have been shown to play diverse roles in plant signaling and development (Chapter 19). The DELLA proteins, a subbranch of GRAS proteins, have been identified as regulators of gibberellin (GA) signaling but also integrate several environmental stimuli and endogenous programs (Schwechheimer, 2011; Sun, 2011; Chapter 20). SCR and SHORT-ROOT (SHR) have been shown to be predominantly involved in root patterning (Pauluzzi et al., 2012). The HAIRY MERISTEM (HAM) proteins are important for apical meristem maintenance, and LATERAL SUPRESSORs are important for axial meristem development (Schumacher et al., 1999; Engstrom, 2012). Nodulation signaling pathway (NSP)1, NSP2, and REQUIRED FOR ARBUSCULAR MYCORRHIZATION (RAM)1 are important for strigolactone signaling and nodulation and/or mycorrhiza colonization (Liu et al., 2011; Gutjahr and Parniske, 2013), whereas PAT1 and members of this subbranch are involved in phytochrome signaling (Bolle et al., 2000; Torres-Galea et al., 2013).

GRAS proteins are about 400–770 amino acids long, and the GRAS domain proper, which resides in their C-Terminus, is usually the major part of the protein (Bolle, 2004). The GRAS domain has been defined by its highly conserved amino acids, sometimes arranged in motifs such as VHIID, PFYRE, and SAW, which always occur in the same order (Pysh et al., 1999; Figure 10.1). Furthermore, the VHIID motif is flanked by two leucine-rich repeats (LRI and LRII). In contrast to the conserved GRAS domain, the N-terminal region of GRAS proteins is highly variable both in sequence and length.

10.1 PRESENCE OF GRAS PROTEINS IN PLANTS AND OTHER ORGANISMS

So far, GRAS proteins could be identified in all sequenced genomes of higher plants, even in lower plants such as the lycophyte *Selaginella moellendorffii* and the bryophyte *Physcomitrella patens* (Yasumura et al., 2007; Engstrom, 2011). This led to the fact that GRAS proteins are usually described as plant specific. No GRAS

Plant Transcription Factors. http://dx.doi.org/10.1016/B978-0-12-800854-6.00010-5
Copyright © 2016 Elsevier Inc. All rights reserved.

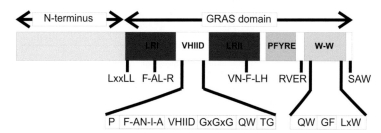

FIGURE 10.1 Schematic representation of the different domains of GRAS proteins. Highly conserved motifs are indicated (see Table 10.1). LR, leucine repeat.

proteins could be detected in the genomes of algae such as *Ostreococcus*, *Volvox*, and *Chlamydomonas*. Nevertheless, GRAS proteins are present in the aquatic alga *Spirogyra pratensis*, belonging to the Zygnematales (Engstrom, 2011). The origin of land plants can be traced back to the Charophytes, to which the Zygnematales belong. Sequencing data have yet to determine how far GRAS proteins are distributed between the Charophytes. No GRAS proteins have been identified in any fungi or metazoa.

A recent bioinformatic study showed that the structure prediction of GRAS domains resembles the Rossmann fold of methyltransferases, a superfamily known from bacteria, and that bacterial ancestors of GRAS proteins could be identified (Zhang et al., 2012). This comparison suggests that the GRAS domain resembles most closely various N- and O-methylases of small molecules, although variant from previously identified O-methylases in plants (Lam et al., 2007). The first conserved sequence block of Rossmann-fold SAM-dependent methyltransferases includes in its C-terminal part the consensus GxGxG, which is considered the main SAM-binding site. Furthermore, a conserved acidic residue (D or E) is localized close by, perhaps so that a water molecule could be polarized (Kozbial and Mushegian, 2005). This residue could be the aspartic acid within the VHIID motif. Data presented by Zhang et al. (2012) suggests that although this GxGxG motif is in most cases present in plant GRAS proteins other SAM-binding residues could be lost compared to the bacterial forms, suggesting that they can no longer act as methyltransferases. So far, no methyltransferase activity has been reported for GRAS proteins. This would mean that the original function could be lost in plants and a new one could evolve. Moreover, proteins with a Rossman fold from other organisms have been reported that lack methyltransferase activity (Kozbial and Mushegian, 2005). Nevertheless, plant GRAS proteins should still be able to bind to their substrates which are suggested to be small molecules such as GA by Zhang et al. (2012). On the other hand, besides GA for the DELLA proteins and perhaps strigolactones for NSP1/2, small molecules cannot be tied directly to the function of GRAS proteins analyzed to date, suggesting that binding small molecules might not be a unifying mode of action for GRAS proteins.

It has been suggested by Zhang et al. (2012) that a single transfer of a GRAS domain from a bacterial source to the common ancestor of land plants took place. Bacterial proteins with homology to GRAS domains can be identified in several Myxococcales (*Hyalangium minutum*, *Corallococcus coralloides*, *Myxococcus stipitatus*, *Stigmatella aurantiaca*, *Cystobacter violaceus*, and *Sorangium cellulosum*), in the cyanobacteria *Fischerella* sp. (PCC 9339 and *Microcystis aeruginosa*), in two Actinomycetales (*Microbispora* sp. ATCC PTA-5024, *Streptosporangium amethystogenes*), in the Bacteroidetes (*Alkaliflexus imshenetskii*, *Runella* sp., *Pontibacter* sp., and *Nafulsella turpanensis*), and in the Bacillales (*Paenibacillus mucilaginosus* and *Cohnella panacarvi*; Zhang et al., 2012; Bolle unpublished; Figure 10.2). It is only in *Runella slithyformis* that two homologous proteins can be detected; otherwise, only one *GRAS* gene seems to be present per genome. The bacterial GRAS proteins clearly define a new clade and do not fall into the clades defined in plants. The GRAS protein of *S. cellulosum* shares the highest homology with plant proteins compared to other bacterial GRAS proteins, but not with the other GRAS proteins found in Myxoccocales. Interestingly, proteins resembling GRAS proteins are not found in all bacteria of a single order but sometimes only in a single group of close relatives. However, they seem to be spread in a wide variety of bacteria. In the well-studied Synecoccales (Cyanophyceae) no protein with a clear GRAS domain could be identified, but only a distant homolog in which the conserved amino acids of the GRAS domain are not present. Many of the bacteria that contain GRAS domains also seem to be growing on decaying plant material, either demonstrating an advantage for the presence of a GRAS protein under these growing conditions or perhaps not ruling out that horizontal gene transfer from plants to bacteria could also be possible.

In higher plants the number of clades for the GRAS protein family varies depending on the analysis, and usually 8–15 monophyletic clades have been identified (Figure 10.2). The two *S. pratensis* GRAS proteins already identified fall into the monophyletic clades established in higher plants (Engstrom, 2011). On the other hand, the *Spirogyra* sequence derived from an expressed sequence tag (EST) clone identified by the authors seems very basal and clusters with bacterial GRAS proteins in a maximum likelihood phylogenetic tree (Figure 10.2). Interestingly,

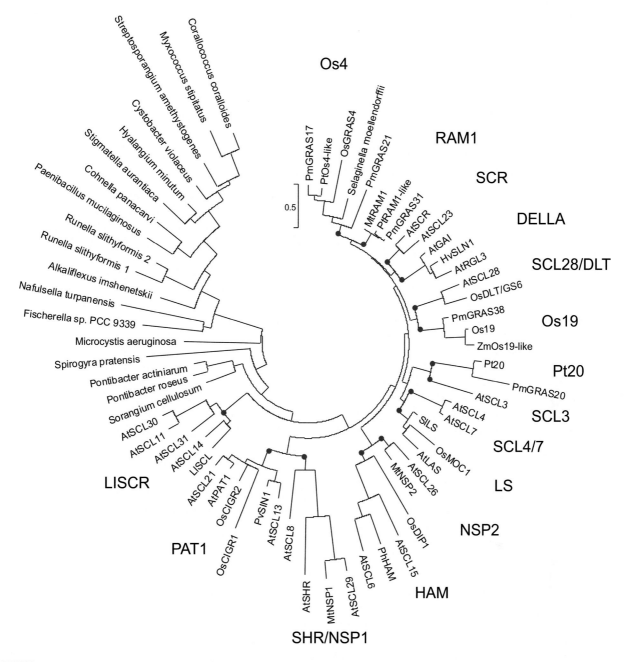

FIGURE 10.2 Unrooted phylogenetic tree with representatives of all GRAS clades identified to date including bacterial GRAS proteins. At, *A. thaliana*; Hv, *Hordeum vulgare*; Ll, *Lilium longiflorum*; Mt, *Medicago trunculata*; Os, *O. sativa*; Ph, *Petunia x hybrida*; Pv, *Phaseolus vulgaris*; Sl, *Solanum lycopersicum*; Pm, *P. mume*; Pt, *P. trichocarpa*; Zm, *Z. mays*. Red dots are branching points that define clades. Alignment was carried out using ClustalW of Lasergene 11 (DNASTAR) in which only conserved GRAS domains were employed, and the tree was generated using Mega6 (maximum likelihood).

GRAS proteins are not found in chlorophyta suggesting that perhaps GRAS proteins were important for adaptation to land life. It seems that every clade in higher plants also includes representatives of lower plants such as *Physcomitrella* and *Selaginella*, suggesting that diversification into the different clades took place very early in evolutionary terms. Therefore, after the Zygnematales, which contain only few GRAS proteins, rapid diversification must have taken place just around the time of the transition of aquatic life forms to land growth (Engstrom et al., 2011), as at least 12 clades can already be seen in lower plants and the *S. moellendorffii* genome contains more *GRAS* genes relative to genome size than rice or *Arabidopsis* (Song et al., 2014). This shows that diversification precedes the evolution of proper roots and leaves, in which the processes that GRAS proteins play in higher plants are important. On the other hand, phytochrome and GA signaling, in which PAT1 and DELLA proteins are involved, have already been demonstrated to be present in lower plants (Vandenbussche et al., 2007; Duanmu et al., 2014). Learning more about the biological role of GRAS proteins in lower plants will enhance our understanding of the importance of GRAS proteins throughout evolution and their adaptation to different functions.

In all sequenced higher plant genomes the family of GRAS proteins has been identified; nevertheless, the number of GRAS genes varies widely. In most species 30–60 genes have been identified, and even Physcomitrella (43 genes) and Selaginella (54 genes) are within this range (Engstrom, 2011). Brassicaceae usually seem to have low numbers of GRAS genes: *Brassica napus* with 37 genes, *Brassica rapa* ssp. *pekinensis* with 48 genes, and *A. thaliana* with 34 genes (Bolle, 2004; Tian et al., 2004; Lee et al., 2008; Liu and Widmer, 2014; Song et al., 2014). On the other hand, several species have over 100 family members such as *Glycine max* (139 genes) and *Gossypium raimondii* (113 genes). In trees an increase of GRAS genes can usually be observed such as in *Malus x domestica* (127 genes), *Populus trichocarpa* (106 genes), *Eukalyptus grandis* (94 genes), with some exceptions being in *Citrus* (48–49) and *Prunus mume* (46 genes; Liu and Widmer, 2014; Lu et al., 2015; Plant Transcription Factor Database, http://planttfdb.cbi.pku.edu.cn/family.php?fam=GRAS). The number of genes in monocots varies, with *Zea mays* and *Sorghum bicolor* at the upper end (184/112 genes), but *Oryza sativa* (60 genes), *Brachypodium distachyon* (48 genes), and *Triticum aestivum* (56 genes) with fewer copies (Tian et al., 2004; Liu and Widmer, 2014; Song et al., 2014; Plant Transcription Factor Database). In some cases, additional (or fewer) GRAS genes might be identified as newer genome annotations are developed.

When the number of GRAS genes is compared to the genome size, there is an underrepresentation in *M. x domestica* (factor 0.144) and *O. sativa* (0.161), but similar values in *S. moellendorffii* (0.254), *P. trichocarpa* (0.241), *M. truncatula* (0.291), and *A. thaliana* (0.244), suggesting that the number of GRAS proteins is mostly related to the genome size and duplication events of species (Song et al., 2014). It is known that gene families arise from whole genome duplications, segmental duplications, and gene duplications. For *Populus*, which has an expanded GRAS protein family, it can be shown that both segmental duplications and tandem duplications played a role in the expansion of the gene family (Liu and Widmer, 2014). Twelve tandem clusters of GRAS genes with 2–6 genes per cluster can be identified in *Populus*, which means that 39% of the genes arose from tandem duplication, whereas in *Arabidopsis* 4 genes (12%) and in rice 15 genes (25%) were located in tandem clusters (Liu and Widmer, 2014). Segmental duplications in *Populus* gave rise to 73% of the genes, whereas in *Arabidopsis* 17 genes (50%) and in rice 24 genes (40%) derive from segmental duplications (Liu and Widmer, 2014).

The distinct clades were first established for the *A. thaliana* gene versions and are named after the representatives that have been analyzed in detail (see Figure 10.2). These are LlSCR, SHR/NSP1, HAM, PAT1, SCL26/NSP2, SCL4/7, LS, DELLA, SCR, SCL28/DLT, and SCL3. Furthermore, some clades are not present in *Arabidopsis*, such as Os4 (OsGRAS4), Os19 (OsGRAS19), RAM1 (MtRAM1), and Pt20 (Tian et al., 2004; Liu and Widmer, 2014; Lu et al., 2015). Pt20 has been identified as a species-specific clade for *P. trichocarpa*, and no orthologs can be found in *Arabidopsis* or rice, but have been identified in *P. mume*. Representatives of the rice clade Os4 can also be found in *Populus*, *Prunus*, and *Selaginella*. In contrast, for the Os19 clade mainly monocot representatives can be identified, but a *P. mume* GRAS protein also clusters to this branch (PmGRAS38). A RAM1 homolog can not only be found as expected in other legumes but also in *P. trichocarpa* and *P. mume*. The three new GRAS clades mentioned in *P. mume* (Group IX, X, and PmGRAS38) therefore all find homologs in *P. trichocarpa* (Liu and Widmer, 2014; Lu et al., 2015). This suggests that the lack of these branches in *A. thaliana* and *B. rapa* might be due to a loss of the clade in the Brassicaceae family, not a specific diversification in monocots or *Populus*.

Clades that are specific to certain species, and the species-dependent expansion of clades, suggest that adaptive evolution took place here. Several plant species possess only a single DELLA gene, such as rice or tomato, while in the Brassicaceae five (*Arabidopsi thaliana*) to six (*Brassica napus*) genes can be found (Yasumura et al., 2007). In contrast, one *SCL3* gene can be found in *Arabidopsis*, but three in *P. trichocarpa*, three in *P. mume*, and seven in rice (Liu and Widmer, 2014). GRAS proteins that diversified in *Populus* or *Prunus* might have an important role for lignification and tree-like growth, whereas the expansion of the SCL3 or LlSCR clade in rice might be important for functions specific to monocots.

The function within the orthologs of one clade, as far as determined until now, can, but does not have to be similar. DELLA proteins are highly conserved in their sequence and function, and similarly the orthologs of SCR, SHR, and HAM seem to have related functions in all higher plant species. On the other hand, in the PAT1 branch, proteins important for phytochrome signaling can be found, but also CIGR1 and 2, which could play a role in defense, and SIN1, which is important for the number and size of nodules formed upon inoculation of *Phaseolus vulgaris* with *Rhizobium etli* (Day et al., 2003; Torres-Galea et al., 2006, 2013; Battaglia et al., 2014).

10.2 GENOMIC ORGANIZATION (INTRON/EXON)

One indication that GRAS proteins could be of bacterial origin is that they have in most cases no, or only one, intron, which is a characteristic feature of genes derived from prokaryotic genomes (Jain et al., 2008). The proportions of intronless genes in *Arabidopsis*, rice, and *Populus* genomes are 21.7%, 19.9%, and 18.9%, respectively (Jain et al., 2008). For *GRAS* genes, 67.6% (*Arabidopsis*), 55% (rice), and 54.7% (*Populus*) are intronless (Tian et al., 2004; Liu and Widmer, 2014).

In *Arabidopsis* only one GRAS gene has an intron within the coding region, *PAT1*, but several *GRAS* genes have introns in the 5′ untranslated region (e.g., *PAT1*, *SCL21*, and *SCL13*). Only on rare occasions are two or more introns detected in *GRAS* genes. However, this is more often the case in rice (11%) and *P. trichocarpa* (13%) (*Arabidopsis*: AtPAT1; rice: OsGRAS3, 16, 40, 42, 43, 45, 53; *Populus*: PtGRAS6, 11, 25, 26, 31, 49, 67, 68, 75, 79, 82, 92, PtSCL1.1, PtSCL1.2; *P. mume*: PmGRAS38; Tian et al., 2004; Liu and Widmer, 2014). Members of the same subbranch have a mostly conserved intron/exon structure and exons of similar length.

10.3 STRUCTURE OF GRAS PROTEINS

As no 3D structure is currently available for GRAS proteins, the folding of the molecule relies on bioinformatical predictions. Song et al. (2014) calculated the instability index of GRAS proteins in *B. rapa*, which indicates that GRAS proteins generally belong to a group of unstable proteins. As GRAS proteins act as regulators, this instability might be important for adapting a signaling pathway to the presence and absence of a stimulus. The presence and abundance of GRAS proteins seems in many cases tightly regulated and related to their biological function. This has been demonstrated for DELLA proteins and MO-CI, which are both degraded by the ubiquitin pathway as part of their biological role (Dill et al., 2001; Fu et al., 2002; Lin et al., 2012). Degradation of DELLA proteins requires dephosphorylation at a serine or threonine residue, and for RGA it was shown that phosphatase TOPP4 is necessary to increase degradation (Qin et al., 2014). A rice casein kinase I (EL1) was isolated and phosphorylated SLR1 both at the N- and the C-terminus, thereby stabilizing it and negatively regulating GA signaling (Dai and Xue 2010). Without the phosphorylation of EL1 the GA-dependent degradation of SLR1 is enhanced.

Furthermore, phosphorylation sites seem to play an important role in the regulation of GRAS protein activity. Kinases can phosphorylate the rice DELLA protein SLR1 within the polyS/T/V, DELLA, and TVHYNP domains (Itoh et al., 2005). However, the phosphorylation of SLR1 is not necessary for the interaction of SLR1 with the GID2/F-box protein. Reversible phosphorylation is required for plant stress-responsive induction of a tobacco GRAS protein, NtGRAS1. Two rice chitin-inducible gibberellin-responsive proteins (CIGR1 and CIGR2), from the AtPAT1 subfamily, were induced by GA signaling depending on phosphorylation and dephosphorylation events (Czikkel and Maxwell, 2007; Day et al., 2003). Furthermore, a calcium- and calmodulin-dependent protein kinase (CCaMK) acts upstream of MtNSP1 and MtNSP2, but direct phosphorylation has not yet been shown.

10.3.1 GRAS Domain

In most cases the GRAS domain is downstream of an N-terminal extension, contains about 350 AA, and is hydrophilic in nature. In some cases it has been reported that proteins contain multiple GRAS domains (e.g., OsGRAS39 with two domains and OsGRAS54 with three domains), but some of these annotations are derived from the misinterpretation of tandem duplication events (as former SCL33 in *A. thaliana* split into AT2G29060 and AT2G29065). In *Populus* it has been shown that some members that group with DELLA proteins start directly with a GRAS domain followed by either another functional domain (e.g., PtGRAS25, 26) or none in the C-terminal part (e.g., PtGRAS81, 83). Most of the bacterial homologs of GRAS proteins only comprise the GRAS domain. Secondary structure prediction revealed a regular succession of α-helices and β-strands within this domain (Zhang et al., 2012).

Conserved amino acids and motives have already been described within the GRAS domain, but for most of them neither a biological nor biochemical role could be attributed (Pysh et al., 1999; Figure 10.1; Table 10.1). The two leucine-rich repeats (also called leucine heptad repeats, LRI and LRII) flanking the VHIID motif are about 100 amino acid residues in length and enriched in leucines. In most cases the leucine residues do not occur as heptad repeats, therefore, they do not resemble classical leucine zippers. Nonetheless, the presence of conserved leucines suggests that these domains could be important for protein–protein interactions and several interactions have been mapped to this area. One example is the LRI domain of RGA which is essential for the interaction with AtIDD3 (Yoshida et al., 2014). An LXXLL sequence occurs in several GRAS proteins at the beginning of the first leucine-rich domain, but is not as highly conserved. This motif matches the consensus sequence, which has been demonstrated to mediate the binding of steroid receptor coactivator complexes to cognate nuclear receptors in mammalians (Heery et al., 1997). The significance of this motif in plants is not yet known. Within the LRI domain several leucine residues are at conserved positions, as is the motif Fx_2ALx_2R. The 20 amino acids surrounding the VHIID domain also contain some conserved residues, mainly an invariant proline and the motif $Fx_4ANx_2Ix_2A$. The VHIID sequence is present in all members of the family, but only histidine and aspartic acid residues are highly conserved. In bacterial GRAS proteins only aspartic acid is conserved. It can be shown that BnSCL1 interacts with a

TABLE 10.1 Motifs Identified in GRAS Domains

Identifier	LxxLL	P	F-AN-I-A	VHIID	Rossmann	QW	TG	VN-F-LH	PFYRE P	E	RF	R/Kx2-3D/Ex2-3Y	F	GREI	RVER	W-W	GF	LxW	SAW
Motif	LxxLL	F-AL-R	P	F-AN-I-A	VHIID	GxGxG	QW	TG	VN-F-LH	P	E	RF	R/Kx2-3D/Ex2-3Y	F	GREI	RVERHE	QW	GF	LxW	SAW
AA (ref: AtGAI)	169	213	248	252	271	277	283	304	361	393	400	413	421	425	444	459	468	477	515	526
AtGAI	LxxxLL	F-FT-Q			VHVID	SxSxG							RxxExxxY		GKQI					
AtPAT1	LxxDL	L-GL-Q				QxGxG							RxxExxxY		ARDV		KW			SCAW
PhHAM	—	F-AL-P		F-SN-I-A	IHIID	DxGxG		TA	IN-F-IW			HL	—Y	L	LPSI	RLRSPD				LTW
AtSCL3				V-TN-I-A	VHVID	DxSxP			VS-L-LH			RL	RxxExxxxY		GEEI	RRERHE	KW		IxW	
AtSCL14		F-SL-R		A-AN-M-F	IHIID	GxSxG			VN-F-FR		I		RxxExxxxY			RVERPE			QxW	SLW
AtSCR		F-AM-R				DxMxG			VH-Q-LY				RxxExxxY		SKEI	RSGEVK	SW			
AtSHR	DxxLL				IHIVD	SxTxC		TT	IN-G-MH			GF	RxxExxxxF		GRAI	STERRE	KW			
SlLS	IxxLL				IHIVD	DxNxG			IN-F-LH				RxxExxxY			RKERHE	SW			
S. pratensis	nd			F-SS-I-A	VHVLE	GxWxT	NW	TA	IC-L-SH	D			RxxxxxxxF	H		nd	nd	nd	nd	SSW
S. cellulosum					IHVID								RxxExxxxY		GQEV		AW		LxR	
P. actiniarum	LxxIL	L-QV-Q			VYLLD				VN-L-LH			RL	RxxExxxxY				AW	GY		TAW
M. stipitatus	KxxLL	F-AV-R		T-AN-L-L	ATLVD		QA	VG	VN-F-MH				RxxxxxxxF		GREV	RCERHE	TW			LAV
Fischerella sp.	SxxLL	F-AL-K		Y-GN-L-Y	VAYMD		QA	IG	IN-A-LH				RxxxCxxxY			RHERLE	VW		VxV	YDF
R. slithyformis	—	L-AM-H		L-TN-L-A	PVLMD		QV	VG		I			RxxCxxxY		SREI	RVEKHY			TxF	LFW
S. aurantiaca	KxxLL			M-AN-L-F	ITVLD		QE	FA	AT-F-LH				RxxAxxxY		SREV	RSERHE	AW		LxY	LCA
P. mucilaginosus	—			D-AN-L-A	VRLLE		QI	HA				RM	RxxCxxxF			RCERHE	NW		FxS	LCA

Variation of the motifs in selected GRAS proteins is shown. The position of the amino acids (AA) is given according to AtGAI. Blank cells indicate 100% identity to the motif; bold AAs indicate variant AAs within conserved motifs; nd, could not be determined; —, motif not present; x or - within the motif indicates not conserved AA within a motif.

histone deacetylase (AtHDA19) via its VHIID motif (Gao et al., 2004). The aspartic acid residue is also the acidic residue localized close to the GxGxG, which is considered the main SAM-binding site of the Rossmann fold of SAM-dependent methyltransferases. Not all glycine residues need to be conserved and replacements are typically by residues with small side chains with a disposition for bending (Kozbial and Mushegian, 2005). At least one glycine is conserved in most GRAS proteins (besides AtSCL3), whereas all three are usually conserved in bacterial GRAS proteins. Moreover, the following 30 amino acids are conserved, especially motifs QW and TG. This shows that the 100 amino acids surrounding the VHIID domain are highly conserved and probably important for GRAS function. In the following LRII domain, several leucine residues are again conserved and another motif can be found, VNx_2FxLH.

In contrast to the motifs mentioned above, the conserved amino acids of the PFYRE motif are spaced within 50 amino acids and localized after the LRII. Most notably the P, Y, and F are conserved in most GRAS proteins, as well as in bacterial GRAS proteins. A consensus sequence for a tyrosine phosphorylation site (R/K $x_{(2,3)}$ D/E $x_{(2,3)}$ Y; Patschinsky et al., 1982) is present in several members of the family, overlapping with the tyrosine in the PFYRE motif. Its function as a phosphorylation site, however, has yet to be demonstrated. For the DELLA protein SLN1, intragenic mutations in the PFYRE motif were shown to compensate for the *sln1* dwarf phenotype (Chandler and Harding, 2013). Also, mutations in the PFYRE domain of SLN1 led to a more slender phenotype and might be important to form secondary interactions with GID1 (Hirano et al., 2010). The RVER motif defines another stretch of highly conserved amino acids, both in plants and bacteria.

The C-terminal part of the GRAS domain (50 AA) contains several invariant tryptophans and a GF motif. It seems that modifications in this area lead to problems in functionality as plants with mutations in DELLA proteins that affect this domain are more slender (Hirano et al., 2010; Chandler and Harding, 2013). For the binding of the DELLA protein GAI to the transcription factor GAF1, belonging to the IDD family, this C-terminal domain also seems to be important and sufficient (Fukazawa et al., 2014). The deletion of the C-terminal domain of the PAT1 protein leads to a more pronounced phenotype in the *pat1-1* mutant and overexpression lines compared to a loss-of-function mutant, suggesting a complete block of the signaling pathway by this shorter protein perhaps due to inappropriate interaction (Bolle et al., 2000; Torres-Galea et al., 2013). Most GRAS proteins contain, at their very C-terminal part, the SAW motif, containing a third conserved tryptophan residue, but this domain is lacking in bacteria.

Comparing all the motifs within the GRAS domain of plants, the relative to front (RF) amino acids within the PFYRE motif and the RVER motif show most variance, suggesting that perhaps these domains are important for specificity toward different interacting proteins. The *S. pratensis* sequence shows significant changes from plant GRAS proteins; notably the conserved aspartate in the VHIID motif is changed to a glutamate, which can otherwise only be observed in *P. mucilaginosus*. On the other hand, *Spirogyra* retains the histidine residue similar to all plant GRAS proteins and *S. cellulosum*. Due to the fact that this sequence is derived from an EST, single amino acid substitutions have to be considered with caution.

10.3.2 N-Terminal Domain

The N-terminal domain is not conserved in sequence and length. On average there are 100–250 amino acids, but examples of very short N-terminal domains (AtSCL21 and AtSCL3, with 35–50 amino acids) or extremely long domains with over 300 amino acids (e.g., AtSCL14, LlSCR, PhHAM) exist. In no case has homology to other known proteins been found in this domain in plants. Therefore, the N-terminal domain has often been referred to as the domain that gives specificity to the GRAS domain. This is especially true for DELLA proteins that contain conserved motifs within their N-termini (namely the DELLA motif and (T)VHYNP(S)) which are crucial for their biological function (Willige et al., 2007). The LxCxE motif in the N-terminus of SCR proteins is highly conserved in SCR orthologs and is important for the interaction with RBR (Cruz-Ramirez et al., 2012). *In vitro* and *in vivo* assays established that many GRAS proteins contain a transactivation domain (e.g., OsGAI/SLR1, NSP1, NSP2, and LlSCR) and that this domain can often be found in the N-terminal region (Morohashi et al., 2003; Hirsch et al., 2009; Fukozawa et al., 2014). In addition, the nuclear localization signal is often found in the N-terminal domain. This shows that in many cases the N-terminal domain adds specificity to the GRAS domain probably via protein–protein interaction, localization, or modification.

Many N-termini of GRAS proteins contain homopolymeric stretches of amino acid residues such as prolines, glutamines, tyrosines, serines, glycines, aspartic acids, or alanines. Serine, threonine, and sometimes tyrosine residues could also indicate that phosphorylation/dephosphorylation events could take place in this region. Protein phosphorylation is predicted to occur within intrinsically disordered regions, probably due to easier steric access of kinases and phosphatases (Iakoucheva et al., 2004). Sun et al., 2011 have shown that the N-terminal domain of GRAS proteins contains molecular recognition features, which are short segments that are usually located within intrinsically disordered proteins (IDPs). These segments can be bound by different partners as they can change in structure due to their disorder-to-order transitions. This would give the N-terminal domain a high degree of binding plasticity which may be linked to functional versatility. Possessing an intrinsically disordered N-terminal domain, Sun et al., (2011) claim that GRAS proteins constitute the first functionally required unfoldome from the plant kingdom. IDPs have been shown to be important for many cellular functions, especially in cell signaling and transcriptional regulation (Dunker et al., 2014).

10.3.3 Additional Domains

Only in rare cases do GRAS proteins possess an additional domain that can be identified. In *P. trichocarpa* it has been shown that the GRAS domain is associated with an additional DNA-binding domain, namely a BED-type zinc finger (PtGRAS75, 79, and 92, all clustering to the DELLA clade). In some instances in *Populus* it can also be observed that a peptidase C48 domain is fused C-terminally to the GRAS domain (PtGRAS26, 26; DELLA clade; Liu and Widmer, 2014).

In bacteria, GRAS-like proteins from three different *Pontibacter* species contain an N-terminal histidinol phosphate aminotransferase domain. Similarly, for one of the *R. slithyformis* GRAS proteins a similar protein can be found in the reading frame after the GRAS protein (Zhang et al., 2012). The significance of this association with aminotransferases in bacteria for the function of the protein has still to be established.

10.4 CONCLUSION

Plant GRAS proteins diversified early in evolution, probably close to the time land was colonized. Nevertheless, the GRAS domain is highly conserved between all members, even when compared to lower plants, which is reflected in the conserved amino acids of the GRAS domain and their invariant order of occurence. Several of these conserved amino acids can also be found in bacterial GRAS proteins. This suggests that proteins have been optimized for their function early in evolution and since then duplications have brought about enlargement of the family. Adaptation has led to amplifications of specific subbranches of the GRAS protein family in individual species or plant families. In several cases it has been shown that different GRAS proteins interact with several members of another protein family (Chapter 19), and this coevolution could have brought about conservation of the secondary structure. On the other hand, several areas are not conserved and are highly variable between GRAS proteins, especially in the N-terminal region. The specificity of individual GRAS proteins to certain interaction partners, and pathways, could be brought about in these regions.

References

Battaglia, M., Ripodas, C., Clua, J., Baudin, M., Aguilar, O.M., Niebel, A., Zanetti, M.E., Blanco, F.A., 2014. A nuclear factor Y interacting protein of the GRAS family is required for nodule organogenesis, infection thread progression, and lateral root growth. Plant Physiol. 164, 1430–1442.

Bolle, C., 2004. The role of GRAS proteins in plant signal transduction and development. Planta 218, 683–692.

Bolle, C., Koncz, C., Chua, N.H., 2000. PAT1, a new member of the GRAS family, is involved in phytochrome A signal transduction. Genes Dev. 14, 1269–1278.

Chandler, P.M., Harding, C.A., 2013. 'Overgrowth' mutants in barley and wheat: new alleles and phenotypes of the 'Green Revolution' DELLA gene. J. Exp. Bot. 64, 1603–1613.

Cruz-Ramirez, A., Diaz-Trivino, S., Blilou, I., Grieneisen, V.A., Sozzani, R., Zamioudis, C., Miskolczi, P., Nieuwland, J., Benjamins, R., Dhonukshe, P., Caballero-Perez, J., Horvath, B., Long, Y., Mahonen, A.P., Zhang, H., Xu, J., Murray, J.A., Benfey, P.N., Bako, L., Maree, A.F., Scheres, B., 2012. A bistable circuit involving SCARECROW-RETINOBLASTOMA integrates cues to inform asymmetric stem cell division. Cell 150, 1002–1015.

Czikkel, B.E., Maxwell, D.P., 2007. NtGRAS1, a novel stress-induced member of the GRAS family in tobacco, localizes to the nucleus. J. Plant Physiol. 164, 1220–1230.

Dai, C., Xue, H.W., 2010. Rice early flowering1, a CKI, phosphorylates DELLA protein SLR1 to negatively regulate gibberellin signalling. EMBO J. 29, 1916–1927.

Day, R.B., Shibuya, N., Minami, E., 2003. Identification and characterization of two new members of the GRAS gene family in rice responsive to N-acetylchitooligosaccharide elicitor. Biochim. Biophys. Acta 1625, 261–268.

Di Laurenzio, L., Wysocka-Diller, J., Malamy, J.E., Pysh, L., Helariutta, Y., Freshour, G., Hahn, M.G., Feldmann, K.A., Benfey, P.N., 1996. The SCARECROW gene regulates an asymmetric cell division that is essential for generating the radial organization of the *Arabidopsis* root. Cell 86, 423–433.

Dill, A., Jung, H.S., Sun, T.P., 2001. The DELLA motif is essential for gibberellin-induced degradation of RGA. Proc. Natl. Acad. Sci. USA 98, 14162–14167.

Duanmu, D., Bachy, C., Sudek, S., Wong, C.H., Jimenez, V., Rockwell, N.C., Martin, S.S., Ngan, C.Y., Reistetter, E.N., van Baren, M.J., Price, D.C., Wei, C.L., Reyes-Prieto, A., Lagarias, J.C., Worden, A.Z., 2014. Marine algae and land plants share conserved phytochrome signaling systems. Proc. Natl. Acad. Sci. USA 111, 15827–15832.

Dunker, A.K., Bondos, S.E., Huang, F., Oldfield, C.J., 2015. Intrinsically disordered proteins and multicellular organisms. Semin. Cell Dev. Biol. 37, 44–55.

Engstrom, E.M., 2011. Phylogenetic analysis of GRAS proteins from moss, lycophyte and vascular plant lineages reveals that GRAS genes arose and underwent substantial diversification in the ancestral lineage common to bryophytes and vascular plants. Plant Signal. Behav. 6, 850–854.

Engstrom, E.M., 2012. HAM proteins promote organ indeterminacy: but how? Plant Signal. Behav. 7, 227–234.

Fu, X., Richards, D.E., Ait-Ali, T., Hynes, L.W., Ougham, H., Peng, J., Harberd, N.P., 2002. Gibberellin-mediated proteasome-dependent degradation of the barley DELLA protein SLN1 repressor. Plant Cell 14, 3191–3200.

Fukazawa, J., Teramura, H., Murakoshi, S., Nasuno, K., Nishida, N., Ito, T., Yoshida, M., Kamiya, Y., Yamaguchi, S., Takahashi, Y., 2014. DELLAs function as coactivators of GAI-ASSOCIATED FACTOR1 in regulation of gibberellin homeostasis and signaling in *Arabidopsis*. Plant Cell 26, 2920–2938.

Gao, M.J., Parkin, I., Lydiate, D., Hannoufa, A., 2004. An auxin-responsive SCARECROW-like transcriptional activator interacts with histone deacetylase. Plant Mol. Biol. 55, 417–431.

Gutjahr, C., Parniske, M., 2013. Cell and developmental biology of arbuscular mycorrhiza symbiosis. Ann. Rev. Cell Dev. Biol. 29, 593–617.

Heery, D.M., Kalkhoven, E., Hoare, S., Parker, M.G., 1997. A signature motif in transcriptional co-activators mediates binding to nuclear receptors. Nature 387, 733–736.

Hirano, K., Asano, K., Tsuji, H., Kawamura, M., Mori, H., Kitano, H., Ueguchi-Tanaka, M., Matsuoka, M., 2010. Characterization of the molecular mechanism underlying gibberellin perception complex formation in rice. Plant Cell 22, 2680–2696.

Hirsch, S., Kim, J., Munoz, A., Heckmann, A.B., Downie, J.A., Oldroyd, G.E., 2009. GRAS proteins form a DNA binding complex to induce gene expression during nodulation signaling in Medicago truncatula. Plant Cell 21, 545–557.

Iakoucheva, L.M., Radivojac, P., Brown, C.J., O'Connor, T.R., Sikes, J.G., Obradovic, Z., Dunker, A.K., 2004. The importance of intrinsic disorder for protein phosphorylation. Nucleic Acids Res. 32, 1037–1049.

Itoh, H., Sasaki, A., Ueguchi-Tanaka, M., Ishiyama, K., Kobayashi, M., Hasegawa, Y., Minami, E., Ashikari, M., Matsuoka, M., 2005. Dissection of the phosphorylation of rice DELLA protein, SLENDER RICE1. Plant Cell Physiol. 46, 1392–1399.

Jain, M., Khurana, P., Tyagi, A.K., Khurana, J.P., 2008. Genome-wide analysis of intronless genes in rice and Arabidopsis. Funct. Integr. Genomics 8, 69–78.

Kozbial, P.Z., Mushegian, A.R., 2005. Natural history of S-adenosylmethionine-binding proteins. BMC Struct. Biol. 5, 19.

Lam, K.C., Ibrahim, R.K., Behdad, B., Dayanandan, S., 2007. Structure, function, and evolution of plant O-methyltransferases. Genome 50, 1001–1013.

Lee, M.H., Kim, B., Song, S.K., Heo, J.O., Yu, N.I., Lee, S.A., Kim, M., Kim, D.G., Sohn, S.O., Lim, C.E., Chang, K.S., Lee, M.M., Lim, J., 2008. Large-scale analysis of the GRAS gene family in Arabidopsis thaliana. Plant Mol. Biol. 67, 659–670.

Lin, Q., Wang, D., Dong, H., Gu, S., Cheng, Z., Gong, J., Qin, R., Jiang, L., Li, G., Wang, J.L., Wu, F., Guo, X., Zhang, X., Lei, C., Wang, H., Wan, J., 2012. Rice APC/C(TE) controls tillering by mediating the degradation of MONOCULM 1. Nat. Commun. 3, 752.

Liu, X., Widmer, A., 2014. Genome-wide comparative analysis of the GRAS gene family in Populus, Arabidopsis and rice. Plant Mol. Biol. Rep. 32, 1129–1145.

Liu, W., Kohlen, W., Lillo, A., Op den Camp, R., Ivanov, S., Hartog, M., Limpens, E., Jamil, M., Smaczniak, C., Kaufmann, K., Yang, W.C., Hooiveld, G.J., Charnikhova, T., Bouwmeester, H.J., Bisseling, T., Geurts, R., 2011. Strigolactone biosynthesis in Medicago truncatula and rice requires the symbiotic GRAS-type transcription factors NSP1 and NSP2. Plant Cell 23, 3853–3865.

Lu, J., Wang, T., Xu, Z., Sun, L., Zhang, Q., 2015. Genome-wide analysis of the GRAS gene family in Prunus mume. Mol. Genet. Genomics 290, 303–317.

Morohashi, K., Minami, M., Takase, H., Hotta, Y., Hiratsuka, K., 2003. Isolation and characterization of a novel GRAS gene that regulates meiosis-associated gene expression. J. Biol. Chem. 278, 20865–20873.

Patschinsky, T., Hunter, T., Esch, F.S., Cooper, J.A., Sefton, B.M., 1982. Analysis of the sequence of amino acids surrounding sites of tyrosine phosphorylation. Proc. Natl. Acad. Sci. USA 79, 973–977.

Pauluzzi, G., Divol, F., Puig, J., Guiderdoni, E., Dievart, A., Perin, C., 2012. Surfing along the root ground tissue gene network. Dev. Biol. 365, 14–22.

Peng, J., Carol, P., Richards, D.E., King, K.E., Cowling, R.J., Murphy, G.P., Harberd, N.P., 1997. The Arabidopsis GAI gene defines a signaling pathway that negatively regulates gibberellin responses. Genes Dev. 11, 3194–3205.

Pysh, L.D., Wysocka-Diller, J.W., Camilleri, C., Bouchez, D., Benfey, P.N., 1999. The GRAS gene family in Arabidopsis: sequence characterization and basic expression analysis of the SCARECROW-LIKE genes. Plant J. 18, 111–119.

Qin, Q., Wang, W., Guo, X., Yue, J., Huang, Y., Xu, X., Li, J., Hou, S., 2014. Arabidopsis DELLA protein degradation is controlled by a type-one protein phosphatase, TOPP4. PLoS Genet. 10, e1004464.

Schumacher, K., Schmitt, T., Rossberg, M., Schmitz, G., Theres, K., 1999. The lateral suppressor (Ls) gene of tomato encodes a new member of the VHIID protein family. Proc. Natl. Acad. Sci. USA 96, 290–295.

Schwechheimer, C., 2011. Gibberellin signaling in plants – the extended version. Front. Plant Sci. 2, 107.

Silverstone, A.L., Ciampaglio, C.N., Sun, T., 1998. The Arabidopsis RGA gene encodes a transcriptional regulator repressing the gibberellin signal transduction pathway. Plant Cell 10, 155–169.

Song, X.M., Liu, T.K., Duan, W.K., Ma, Q.H., Ren, J., Wang, Z., Li, Y., Hou, X.L., 2014. Genome-wide analysis of the GRAS gene family in Chinese cabbage (Brassica rapa ssp. pekinensis). Genomics 103, 135–146.

Sun, T.P., 2011. The molecular mechanism and evolution of the GA-GID1-DELLA signaling module in plants. Curr. Biol. 21, R338–R345.

Sun, X., Xue, B., Jones, W.T., Rikkerink, E., Dunker, A.K., Uversky, V.N., 2011. A functionally required unfoldome from the plant kingdom: intrinsically disordered N-terminal domains of GRAS proteins are involved in molecular recognition during plant development. Plant Mol. Biol. 77, 205–223.

Tian, C., Wan, P., Sun, S., Li, J., Chen, M., 2004. Genome-wide analysis of the GRAS gene family in rice and Arabidopsis. Plant Mol. Biol. 54, 519–532.

Torres-Galea, P., Huang, L.F., Chua, N.H., Bolle, C., 2006. The GRAS protein SCL13 is a positive regulator of phytochrome-dependent red light signaling, but can also modulate phytochrome A responses. Mol. Genet. Genomics 276, 13–30.

Torres-Galea, P., Hirtreiter, B., Bolle, C., 2013. Two GRAS proteins, SCARECROW-LIKE21 and PHYTOCHROME A SIGNAL TRANSDUCTION1, function cooperatively in phytochrome A signal transduction. Plant Physiol. 161, 291–304.

Vandenbussche, F., Fierro, A.C., Wiedemann, G., Reski, R., Van Der Straeten, D., 2007. Evolutionary conservation of plant gibberellin signalling pathway components. BMC Plant Biol. 7, 65.

Willige, B.C., Ghosh, S., Nill, C., Zourelidou, M., Dohmann, E.M., Maier, A., Schwechheimer, C., 2007. The DELLA domain of GA INSENSITIVE mediates the interaction with the GA INSENSITIVE DWARF1A gibberellin receptor of Arabidopsis. Plant Cell 19, 1209–1220.

Yasumura, Y., Crumpton-Taylor, M., Fuentes, S., Harberd, N.P., 2007. Step-by-step acquisition of the gibberellin-DELLA growth-regulatory mechanism during land-plant evolution. Curr. Biol. 17, 1225–1230.

Yoshida, H., Hirano, K., Sato, T., Mitsuda, N., Nomoto, M., Maeo, K., Koketsu, E., Mitani, R., Kawamura, M., Ishiguro, S., Tada, Y., Ohme-Takagi, M., Matsuoka, M., Ueguchi-Tanaka, M., 2014. DELLA protein functions as a transcriptional activator through the DNA binding of the indeterminate domain family proteins. Proc. Natl. Acad. Sci. USA 111, 7861–7866.

Zhang, D., Iyer, L.M., Aravind, L., 2012. Bacterial GRAS domain proteins throw new light on gibberellic acid response mechanisms. Bioinformatics 28, 2407–2411.

CHAPTER 11

Structure and Evolution of WRKY Transcription Factors

Charles I. Rinerson*, Roel C. Rabara*, Prateek Tripathi**, Qingxi J. Shen†, Paul J. Rushton*

*Texas A and M AgriLife Research and Extension Center, Dallas, TX, USA
**Molecular and Computational Biology Section, Dana and David Dornsife College of Letters, Arts and Sciences, University of Southern California, Los Angeles, CA, USA
†School of Life Sciences, University of Nevada, at Las Vegas, NV, USA

OUTLINE

11.1 Introduction	163
11.2 The Structure of the WRKY Domain	164
11.3 The Evolution of WRKY Genes	164
11.3.1 Distribution of WRKY Genes	164
11.3.2 Nonplant WRKY Genes	165
11.3.3 WRKY Genes in Unicellular Green Algae	169
11.3.4 WRKY Genes in Multicellular Green Algae	169
11.3.5 WRKY Genes in Mosses and Spikemosses	169
11.3.6 WRKY Genes in Flowering Plants	171
11.3.7 The Relationship with FLYWCH, GCM1, and BED Proteins	171
11.4 R Protein–WRKY Genes	173
11.5 Conclusion: A Reevaluation of WRKY Evolution	177
References	179

11.1 INTRODUCTION

As of the time of writing (2015) there has been 20 years of research into WRKY transcription factors. Over those 20 years, we have learned a lot about these transcription factors but there is still much that remains unknown. Some of the early suggestions have proved to be correct, such as the conserved cysteines and histidines in the WRKY domain forming a novel Zn-finger-like motif and the WRKY amino acid sequence binding directly to the W box (TTGACC/T) DNA-binding site. However, an increase in the number of available sequenced plant genomes has shown that some other early suggestions concerning the evolution of WRKY transcription factors were incorrect. There has indeed been a lineage-specific expansion of WRKY transcription factors in plants but they are not found exclusively in plants. Group III WRKY genes were not the last to evolve and the six well-defined groups of WRKY genes found in flowering plants are not the only groups of WRKY genes in the plant kingdom. Other more recent suggestions such as the Group III WRKY genes being the youngest, Group IIb genes being descendents of Group IIa WRKY genes, and Group IId WRKY genes also evolving from Group IIa (Brand et al., 2013) are now apparently incorrect. The recent availability of the first genome sequence from a member of the Charophyta (the filamentous terrestrial alga *Klebsormidium flaccidum*; Hori et al., 2014) fills in a gap in the evolutionary history of WRKY genes associated with the colonization of land by plants and reveals some unexpected new insights into WRKY evolution.

11.2 THE STRUCTURE OF THE WRKY DOMAIN

As soon as the WRKY domain was characterized, it was suggested that it contained a novel Zn finger structure and the first evidence to support this came from studies with 2-phenanthroline that chelates zinc ions. Addition of 2-phenenthroline to gel retardation assays that contained *Escherichia-coli*-expressed WRKY proteins resulted in a loss of binding to the W box target sequence (Rushton et al., 1995). The other suggestion was that the WRKY signature amino acid sequence at the N-terminus of the WRKY domain directly binds to the W box sequence in the DNA of target promoters. These suggestions were shown to be correct by publication of the solution structure of the C-terminal WRKY domain of the *Arabidopsis* WRKY4 protein. The WRKY domain was found to form a four-stranded β-sheet (Yamasaki et al., 2005; see Chapter 4). Soon afterwards, a crystal structure of the C-terminal WRKY domain of the *Arabidopsis* WRKY1 protein was reported. This showed a similar result to the solution structure except that it may contain an additional β-strand at the N-terminus of the domain (Duan et al., 2007). From these two studies it appears that the conserved WRKYGQK signature amino acid sequence enters the major groove of the DNA to bind to the W box. The first structural determination of the WRKY domain complexed with a W Box was reported in 2012 (Yamasaki et al., 2012). The nuclear magnetic resonance (NMR) solution structure of the WRKY DNA-binding domain of *Arabidopsis* WRKY4 in complex with W box DNA revealed that part of a four-stranded β-sheet enters the major groove of DNA in an atypical mode that the authors named the β-wedge, where this sheet is almost perpendicular to the DNA helical axis. As initially predicted, amino acids in the conserved WRKYGQK signature motif contact the W box DNA bases mainly through extensive apolar contacts with thymine methyl groups (Yamasaki et al., 2012). These structural data explain the conservation of both the WRKY signature sequence at the N-terminus of the WRKY domain and the conserved cysteine and histidine residues. It also provides the molecular basis for the previously noted remarkable conservation of both the WRKY amino acid signature sequence and the W box DNA sequence (Eulgem et al., 2000).

11.3 THE EVOLUTION OF WRKY GENES

There have been several major contributions to our understanding of the evolution of WRKY transcription factor genes. The first was the definition of the seven major groups of WRKY genes found in flowering plants (Groups I, IIa, IIb, IIc, IId, IIe, and III) by Eulgem et al., (2000). This classification was only partly based on phylogenetic analyses but has proven over time to be an accurate representation of the major groups of WRKY genes in flowering plants (Eulgem et al., 2000; Rushton et al., 2010). In 2005, Zhang and Wang used the availability of an increasing number of plant genome sequences to produce a hypothesis of the evolution of WRKY genes in plants (Zhang and Wang, 2005). They hypothesized that a proto-WRKY gene with a single WRKY domain underwent domain duplication to produce Group I WRKY genes. Subsequent loss of the N-terminal WRKY domain led to Group IIc genes from which all other WRKY genes evolved, Group III genes being the last (Zhang and Wang, 2005). Since this paper, the first genome sequences of a moss (Rensing et al., 2008), a spike moss (Banks et al., 2011), and most significantly, a filamentous green alga from the Charophytes (Hori et al., 2014) have been published. These data rewrite the evolution of WRKY transcription factors and based on this larger data set a new picture emerges (Rinerson et al., 2015).

Another major contribution came from Babu et al. (2006). The authors looked for WRKY-like genes outside of the plant kingdom and showed that WRKY domains share a similar Zn finger domain and four-strand fold with GCM1 and FLYWCH domains and that they may be derived from a BED finger and ultimately a C_2H_2 Zn finger domain (Babu et al., 2006). This has proved to be a controversial proposal but appears to be at least partly true. The Zn finger structures of these proteins do appear to have some similarities at the primary amino acid level, suggesting that they may be derived from a BED finger-like structure. However, there are no similarities in the WRKY signature portion of the domains at the primary amino acid level, and phylogenetic software such as ClustalW in the MEGA4 package cannot align these different classes of proteins as they are too dissimilar (Rinerson et al., 2015). The most likely conclusion is that the Zn finger portions of WRKY, GCM1, and FLYWCH proteins do share a common ancestor but that any common structural features other than the Zn finger are unrelated in evolutionary terms as they share no similarities at the primary amino acid sequence level. We suggest that the "classical" WRKY transcription factors evolved early in plant evolution from a Zn-finger-like structure and instances of WRKY genes outside of the plant kingdom are through lateral gene transfer (see Section "Distribution of WRKY genes").

11.3.1 Distribution of WRKY Genes

Some WRKY genes are found outside of the plant kingdom. This includes genes in diplomonads, social amoebae and other amoebozoa, and fungi *incertae sedis* (Figure 11.1). This is an unusual distribution and not easy

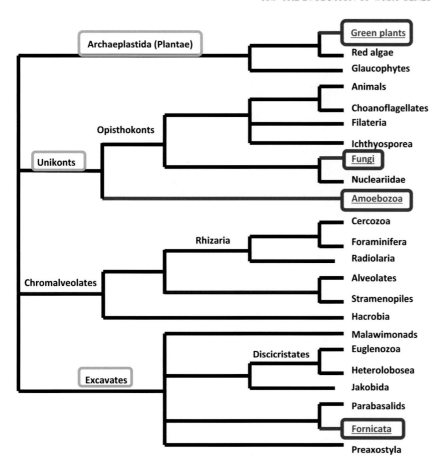

FIGURE 11.1 **The distribution of WRKY genes in the tree of life.** Red-boxed names indicate the presence of WRKY genes. *Modified from Rinerson et al. (2015).*

to explain. How, for instance, can there be WRKY genes in the distantly related Fornicata but not in red algae?

The most striking genes outside the plant kingdom are WRKY genes from the early diverging fungal lineages such as *Mortierella verticillata, Mucor circinelloides, Absidia idahoensis, Lichtheimia corymbifera, Rhizopus delemar*, and *Rhizophagus irregularis*. These genes clearly contain WRKY domains based on evidence from BLAST searches and protein sequence searches against profile–Hidden Markov Model (HMM) databases, but they do not belong to any of the groups of WRKY genes from flowering plants (Figure 11.2). They are some of the most divergent WRKY genes.

So how can we explain this patchy phyletic distribution of WRKY genes in the plant lineage, plus certain diplomonads, social amoebae, fungi *incertae sedis*, and amoebozoa? Such a patchy distribution is a feature of lateral gene transfer (Fitzpatrick, 2012). It cannot easily be explained by multiple losses of WRKY genes in multiple independent lineages. Other features of lateral gene transfer include finding similar genes shared amongst unrelated species that share a specific niche/geographical location (Kunin et al., 2005). This would indeed appear to be the case with the diplomonads, social amoebae, fungi *incertae sedis*, and amoebozoa that contain WRKY genes. It

is striking that almost all of the organisms outside of the plant kingdom that contain WRKY genes can be found in one of two ecological niches. Either they live in the soil in proximity to plant roots and/or they are parasites of humans/animals, often in the gut. Both niches would put them close to plant material (either alive or rotting in the soil or being digested in the digestive system). Further support for this hypothesis comes from studies of the nonplant organisms themselves. Diplomonads are known to be hotbeds of lateral gene transfer (Andersson et al., 2003) and studies of the genome of *Acanthamoeba castellanii* highlighted extensive lateral gene transfer (Clarke et al., 2013). Plant–fungi lateral gene transfers appear to be both rare and ancient (Richards et al., 2009), but at least four examples have been described, suggesting that the transfer of a WRKY gene from an alga to an early fungus is indeed possible.

11.3.2 Nonplant WRKY Genes

Fungal WRKY genes appear the most divergent compared with higher plant WRKY genes as the WRKY signature amino acid sequence is present as a conserved WKNNGNT rather than WRKYGQK (Figure 11.3). In addition, the Zn finger motif is a conserved

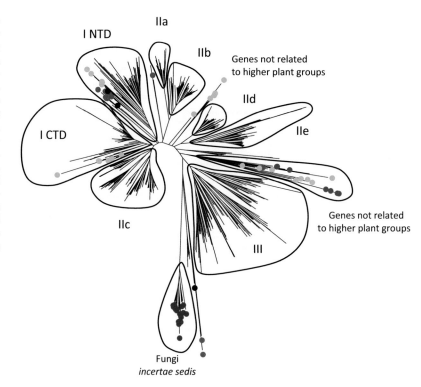

FIGURE 11.2 Neighbor-joining phylogenetic tree derived from a MUSCLE alignment of WRKY domains from the complete WRKY gene family from the following species: *A. thaliana, G. max, B. distachyon, S. moellendorffii, P. patens, C. reinhardtii, Chlorella variabilis, Coccomyxa subellipsoidea, M. pusilla, O. lucimarinus, Ostreococcus tauri, V. carteri, K. flaccidum, Bathycoccus prasinos, Dictyostelium discoideum, Polysphondylium pallidum, Dictyostelium fasciculatum, Fonticula alba, A. castellanii, G. lamblia, G. intestinalis, Dictyostelium purpureum, Auxenochlorella protothecoides, S. salmonicida, M. circinelloides, R. delemar, A. idahoensis, L. corymbifera, R. irregularis,* and *M. verticillata*. Fungal genes are marked with a red dot, unicellular green algae green, diplomonads blue, amoebozoa black, and social amoebae purple. The higher plant WRKY groups are marked I–III. I NTD and I CTD denote the N-terminal and C-terminal domains from Group I proteins. The tree was produced using MEGA 6.

C—X6—C—H—X3—C. This spacing of C- and H-residues is unique among WRKY proteins. Fungal WRKY TFs also contain only one WRKY domain. Until now (2015), the ancestral form of WRKY proteins has been suggested to be similar to the Group I WRKY TFs with two domains (N- and C-terminal), but it has been clear that a proto-WRKY with a single domain was likely present before domain duplication occurred. As plant–fungi lateral gene transfers appear to be both rare and ancient (Richards et al., 2009), this suggests that fungal-type WRKY genes may have descended from a gene closer to the original ancestral single domain-type WRKY gene than Group I genes.

Fungi such as the fungi *incertae sedis* are ancient, probably over 1000 million years old (Hedges et al., 2004), and have been suggested as playing an important role in the evolution of land plants (Simon et al., 1993). *Rhizopus microsporus* is a widely distributed soil fungus that can cause mucormycosis in immunocompromised humans and seedling blight in rice (Lackner and Hertweck, 2011) and is one of the very few fungi that harbor bacterial endosymbionts (Partida-Martinez et al., 2007). Fungi *incertae sedis* species such as *R. irregularis* (formerly *Glomus intraradices*; Tisserant et al., 2013) are mycorrhizal fungi and actually penetrate plant cells. It is therefore possible that an ancient WRKY gene was transferred from a plant cell early during the evolution of land plants and that this gene has given rise to the fungal type of WRKY gene. Fossil evidence shows arbuscular mycorrhizal symbiosis to be at least as old as the earliest land plants (470–480 million years ago) and to predate plant roots (Schüßler et al., 2001). It is likely that colonization of the land by plants was therefore dependent on fungal provision of inorganic nutrients and water (Schüßler et al., 2001). It is possible that these first terrestrial symbioses with fungal cells led to an early lateral gene transfer of a WRKY gene to a nonplant host. If so, then the single domain fungal type of WRKY gene may reflect the single WRKY domain present in the oldest form of WRKY transcription factors from unicellular organisms.

Other evidence for ancient lateral gene transfer of WRKY genes comes from the genes present in diplomonad species (*Giardia lamblia, Giardia intestinalis,* and *Spironucleus salmonicida*; Figure 11.2). Diplomonads are free-living flagellates that are often common in stagnant fresh water, but most are commensal in the intestines of animals. Some are parasitic and cause disease. They include *G. lamblia*, which causes giardiasis in humans. Diplomonad WRKY genes are not members of flowering plant groups and are found in a distinct clade with several unicellular green algae WRKY genes (Figure 11.2). This suggests that the diplomonads obtained a unicellular green alga WRKY gene and that this group of WRKY genes is not represented in modern day flowering plants. Again this argues in favor of an ancient lateral gene transfer event or events. The majority of the diplomonad WRKY genes contain two WRKY domains but these WRKY domains are not closely related to the two WRKY domains in Group I WRKY proteins from flowering plants. This suggests that either WRKY domain duplication was a very early event and that two domain proteins from diplomonads and higher plants are now

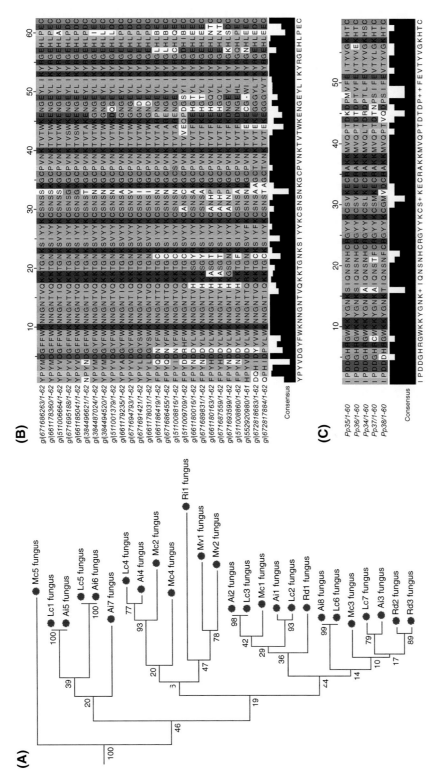

FIGURE 11.3 Fungal and *Physcomitrella patens* Group III WRKY proteins. (A) Neighbor-joining phylogenetic tree derived from a MUSCLE alignment of WRKY domains from *M. circinelloides*, *R. delemar*, *A. idahoensis*, *L. corymbifera*, *R. irregularis*, and *M. verticillata*. Numbers indicate bootstrap values from 1000 replicates. (B) ClustalW2 multiple sequence alignment and consensus sequence of WRKY domains from fungi *incertae sedis*. (C) ClustalW2 multiple sequence alignment and consensus sequence of WRKY domains from *P. patens* Group III proteins (PpWRKY35-38). *Modified from Rinerson et al. (2015).*

only distantly related to each other, or that domain duplication has occurred on more than one occasion in the history of WRKY transcription factors, a suggestion that has previously been made (Brand et al., 2013).

The third group of nonplant organisms that contains WRKY genes in their genomes includes some social amoebae and other amoebozoa (Figure 11.2). Most of these amoebozoa genes are Group I-like and contain two WRKY domains. We have performed phylogenetic analyses of these amoebozoa WRKY genes together with WRKY genes from unicellular green algae and higher plants using neighbor-joining (NJ), maximum likelihood (ML), minimum evolution, and maximum parsimony (MP) methods (Rinerson et al., 2015). In several cases (most notably NJ trees), the two WRKY domains from the amoebozoa genes both appear to be more related to Group I N-terminal domains rather than C-terminal domains (Figures 11.2 and 11.4). These observations are consistent with the following hypothesis of WRKY gene evolution. A duplication event of the N-terminal WRKY domain occurred early in WRKY gene evolution and the second N-terminal domain began to evolve to become the C-terminal in unicellular green algae. Before the C-terminal WRKY domain had evolved into the higher plant form, a lateral gene transfer event occurred that resulted in an amoebozoa species containing an intermediate Group I form of WRKY gene with the C-terminal WRKY domain sharing some N-terminal features. Consistent with this is the observation that the amoebozoa genes lack the PR intron that is present in the Group I WRKY genes of modern day multicellular plants from filamentous

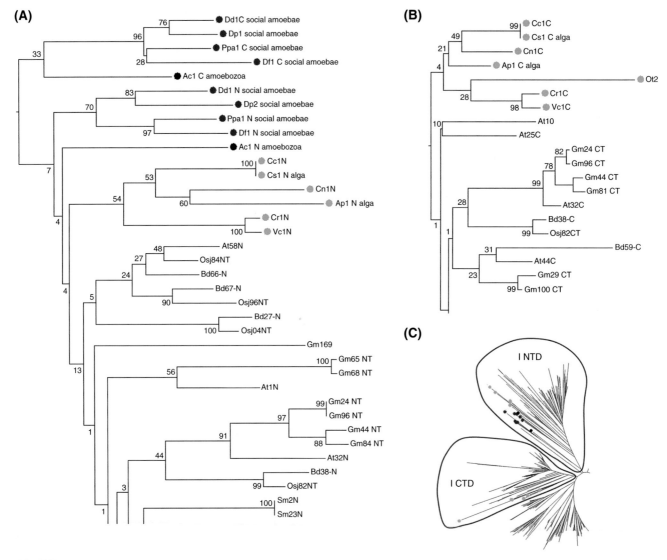

FIGURE 11.4 **Group I-like WRKY proteins from social amoebae and other amoebozoa.** (A) Part of the N-terminal WRKY domain clade from a neighbor-joining phylogenetic tree derived from a MUSCLE alignment of WRKY domains from the species described in Figure 11.2. Unicellular green algae WRKY domains are marked with a green dot, amoebozoa black, and social amoebae purple. (B) C-terminal WRKY domains. (C) Unrooted version of the same tree in (A) and (B). *Modified from Rinerson et al. (2015).*

green algae onward but absent in unicellular green alga genes. This suggests that lateral gene transfer occurred from a unicellular green alga-like gene and not from a multicellular plant gene.

There is one further observation concerning the distribution of WRKY genes outside the plant kingdom. Some searches of expressed sequence tag (EST) databases show WRKY genes in further nonplant species. It is possible that other gene transfer events have occurred in addition to the three that have been documented (Rinerson et al., 2015) but until such genes are shown to be present in the genomes of other types of organism, these EST sequences may also be a result of contamination with plant material.

11.3.3 WRKY Genes in Unicellular Green Algae

The WRKY genes in unicellular green algae fall into three groups based on phylogenetic analyses (Figure 11.2). One group corresponds to the Group I genes found in flowering plants. These genes have been postulated to be ancestral to all higher plant WRKY genes, largely because the only WRKY gene present in the unicellular green alga *Chlamydomonas reinhardtii* is of this type. However, analyses suggest that this situation is far from clear and another possibility is that Groups IIa and IIb genes evolved directly from an ancestral single domain WRKY gene and not from a Group I WRKY gene (Rinerson et al., 2015). Consistent with this suggestion, there is a second type of WRKY gene that is found in many unicellular green algae, such as *Volvox carteri*, *Ostreococcus lucimarinus*, and *Micromonas pusilla*, which appears to have little similarity to any WRKY groups from higher plants. These genes appear to fall into at least two different clades (Figure 11.2). Interestingly, the diplomonad WRKY genes cluster with this second group of unicellular green alga WRKY genes (Figures 11.2 and 11.5), suggesting that it was lateral gene transfer from this class of algal genes that led to diplomonad genes (Figure 11.2). This second type of algal WRKY gene does not appear to be represented in higher plant genomes and this suggests that these WRKY genes have no counterparts in higher plants. However, the presence of these single-domain WRKY genes in unicellular green algae suggests that Group I genes with two WRKY domains were not the only early WRKY genes that could have given rise to WRKY groups in higher plants.

11.3.4 WRKY Genes in Multicellular Green Algae

Until the early 2010s, there had been a large gap in the available genome sequences in the plant kingdom between unicellular green algae, such as *C. reinhardtii*, which typically have 1–3 WRKY genes, and mosses, such as *Physcomitrella patens*, which have 30–40 genes. This situation changed with the publication of the *K. flaccidum* genome sequence (Hori et al., 2014). The *K. flaccidum* genome contains just two WRKY genes (Rinerson et al., 2015; Figure 11.6). The first is a Group I gene (kfl00096) that contains two WRKY domains similar to the single *C. reinhardtii* gene. Unexpectedly, the second gene (kfl00189) is a Group IIb gene. Phylogenetic analyses show that kfl00189 clusters with other Group IIb WRKY genes (Figure 11.6). The amino acid sequence of the kfl00189 WRKY domain also has hallmarks of Group IIb proteins from flowering plants, such as the C—X5—C spacing in the Zn finger motif, the sequence QVQR in the middle of the finger, and the sequence DGCx immediately before the WRKY amino acid signature (Figure 11.6). All of these primary amino acid sequences are features of Group IIb WRKY proteins (Eulgem et al., 2000). Strikingly, kfl00189 also contains the conserved QVQR type intron that flowering plant Groups IIb and IIa genes possess rather than the PR-type intron shared by all other flowering plant WRKY genes (Eulgem et al., 2000).

These observations necessitate a re-evaluation of the current view of WRKY gene evolution. Previously, it had been assumed that Group IIb genes evolved from Group IIc-like genes later in the evolution of plants (Zhang and Wang, 2005). Now it is clear that only Group I genes clearly predate them. However, the new information showing an early evolution of Group IIb genes poses a new question. Did Group IIb genes evolve from a Group I gene or did they evolve independently from a single WRKY-domain-containing unicellular green alga gene? We have called these two different possibilities the "IIa + b Separate Hypothesis" and the "Group I Hypothesis" (Rinerson et al., 2015). The "IIa + b Separate Hypothesis" suggests that Group IIa and IIb WRKY genes did not evolve from Group I genes whereas the "Group I Hypothesis" suggests that all WRKY genes in higher plants evolved from a Group I gene. Group I WRKY genes from unicellular green algae do not appear to have the conserved PR intron, and both the PR intron and the QVQR intron first appear in filamentous green algae. Both hypotheses are therefore possible and only additional genomic information might possibly answer this question.

11.3.5 WRKY Genes in Mosses and Spikemosses

The further evolution of WRKY genes in multicellular plants is rather clearer. The first available genome sequence of a moss (*P. patens*) showed that it contains Group I, Group IIb, Group IIc, Group IId, and Group III-like genes (Rensing et al., 2008). The newly evolved Group

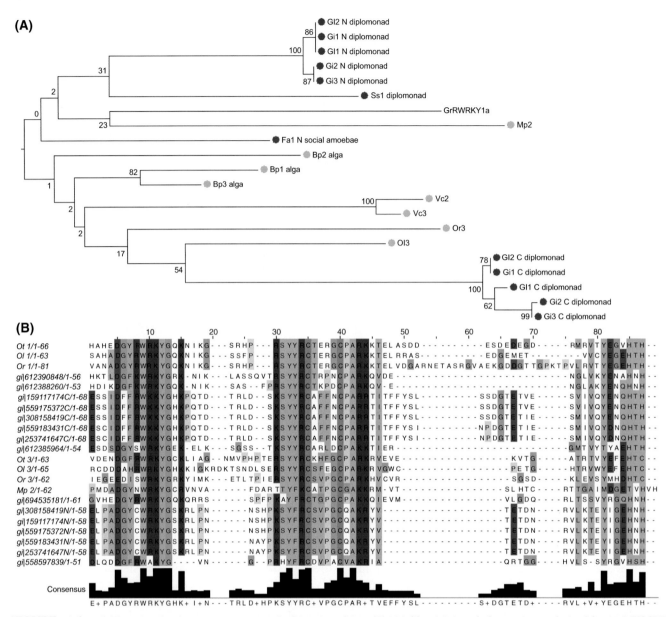

FIGURE 11.5 **WRKY genes that have no counterparts in flowering plants.** (A) Neighbor-joining phylogenetic tree derived from a MUSCLE alignment of WRKY domains from unicellular green algae (green), diplomonads (blue), and social amoebae (purple). (B) ClustalW2 multiple sequence alignment and consensus sequence of WRKY domains from WRKY genes that have no counterparts in flowering plants. Names are indicated by the abbreviations in (A) or by accession numbers.

IIc WRKY genes appear to have evolved from Group I genes by loss of the N-terminal domain. It is likely that both Group IId and Group III genes evolved from Group IIc-like/Group I C-terminal domain genes based on the presence of the conserved PR intron.

P. patens has Group III-like genes which show distinct features that are not present in Group III genes from more advanced plants. In some NJ trees, *P. patens* Group III WRKY proteins are found adjacent to, but not included in, the higher plant Group III clade. In ML trees, *P. patens* genes were found within the higher plant Group III clade. Additionally, the single domain genes from fungi appear to be closer phylogenetically to Group III genes than any other group (Figure 11.2). The consensus amino acid sequence of the WRKY signature from these variant moss Group III proteins is WKNNGNT, compared with WKKYGNK in fungal genes, and WRKYGQK in flowering plant Group III genes (Figure 11.3). In filamentous green algae, there appear (based on the single available genome) to be only Group I and Group IIb genes. It is therefore extremely likely that Group III genes evolved from Group I genes and not IIb genes because Group I and

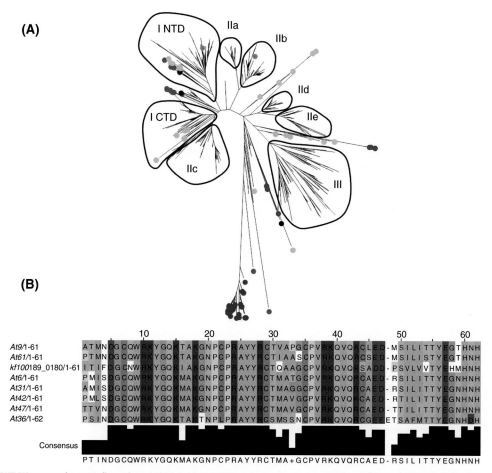

FIGURE 11.6 **WRKY genes from *K. flaccidum*.** (A) Neighbor-joining phylogenetic tree similar to Figure 11.2 but also containing the three WRKY domains from *K. flaccidum* (orange). (B) ClustalW2 multiple sequence alignment and consensus sequence of WRKY domains from *Arabidopsis* Group IIb genes and the IIb gene from *K. flaccidum*.

Group III share the PR intron. It is now clear that previous suggestions that Group III genes were the last group to evolve are certainly incorrect as Group III genes predate Group IIa and Group IIe genes (Figure 11.2).

The genome sequence of the spikemoss *Selaginella moellendorffii* (Banks et al., 2011) provides a view of the WRKY gene family in a primitive vascular plant. The approximately 40 million years of evolution that separates the mosses from *S. moellendorffii*, has seen two major changes in the WRKY gene family. First, the appearance of Group IIe genes and, second, vascular plants starting with *S. moellendorffii* have Group III WRKY genes that are similar to those in flowering plants with a similar Zn finger structure, WRKY amino acid sequence, and intron/exon boundaries.

11.3.6 WRKY Genes in Flowering Plants

All of the main groups of WRKY genes that are present in flowering plants are present in *S. moellendorffii* except for Group IIa genes, which were therefore the last group to evolve and appear to have arisen from Group IIb genes. Group IIa genes are the group with the smallest number of members but nevertheless appear to play many important roles in regulating stress responses (both biotic and abiotic).

11.3.7 The Relationship with FLYWCH, GCM1, and BED Proteins

As previously mentioned, Babu et al. (2006) have suggested that WRKY domains share a similar Zn finger domain and four-strand fold with GCM1 and FLYWCH domains and that they may be derived from a BED finger and ultimately a C2H2 Zn finger domain (Babu et al., 2006). FLYWCH proteins are "too divergent to be aligned" in MEGA4 using CLUSTALW with our data set of WRKY domains (Rinerson et al., 2015). The same is true with GCM1-like sequences. Inspection of the domains show that the Zn finger structures share some amino acid similarities with the WRKY domain, even though the spacing between cysteine and histidine

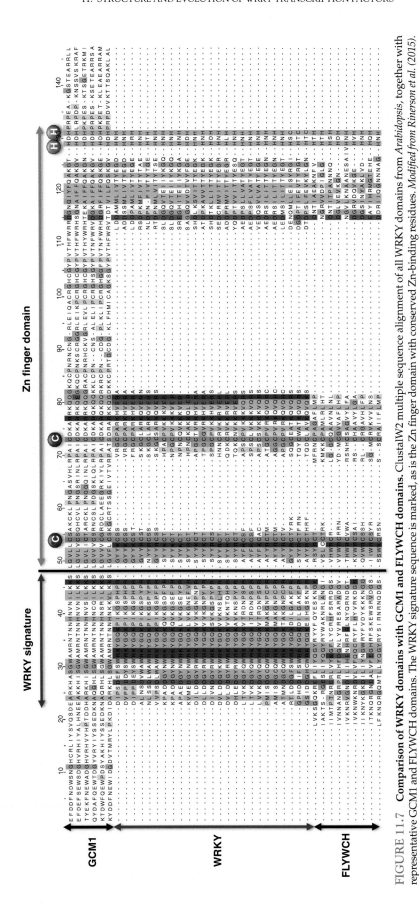

FIGURE 11.7 **Comparison of WRKY domains with GCM1 and FLYWCH domains.** ClustalW2 multiple sequence alignment of all WRKY domains from *Arabidopsis*, together with representative GCM1 and FLYWCH domains. The WRKY signature sequence is marked, as is the Zn finger domain with conserved Zn-binding residues. *Modified from Rinerson et al. (2015).*

B. EVOLUTION AND STRUCTURE OF DEFINED PLANT TRANSCRIPTION FACTOR FAMILIES

residues varies greatly, but that appears to be the limit of the similarity (Figure 11.7; Rinerson et al., 2015). The primary amino acid sequences of the N-terminal part of the domains that contain the WRKY signature show no similarities and cannot be aligned against each other. This lack of similarity is not found with the WRKY proteins in the diplomonads, social amoebae, fungi *incertae sedis*, and amoebozoa as they all share amino acid similarities in the WRKY signature part of the domain that binds directly to DNA (Rinerson et al., 2015). It is possible that the Zn finger portion of the WRKY domain does share a common ancestry with FLYWCH, GCM1, MULE, and BED proteins that ultimately derives from an ancestral C2H2 Zn finger motif. However, "classical" WRKY transcription factors are too divergent to be considered part of a larger family with these proteins and the examples of "classical" WRKY proteins in a small number of nonplant genomes is therefore consistent with lateral gene transfer.

11.4 R PROTEIN–WRKY GENES

One of the most unusual features of the WRKY gene family in flowering plants is the existence of chimeric proteins comprising domains typical for both intracellular-type resistance (R) proteins (nucleotide-binding site–leucine-rich repeat, NBS–LRR proteins) and WRKY transcription factors (Rushton et al., 2010). With the sequencing of the *Arabidopsis thaliana* genome, three such R protein–WRKY genes were found (*AtWRKY16*, *AtWRKY19*, and *AtWRKY52*) and it seemed likely that R protein–WRKY-like genes were a feature of most plant genomes. With the completed sequencing of many more plant genomes and extensive EST collections it has become clear that the reality is considerably more complex. We have searched available plant genome sequences and EST collections to build an atlas of R protein–WRKY-like genes (Figure 11.8; Rinerson et al., 2015). This has been a complicated undertaking because it was necessary

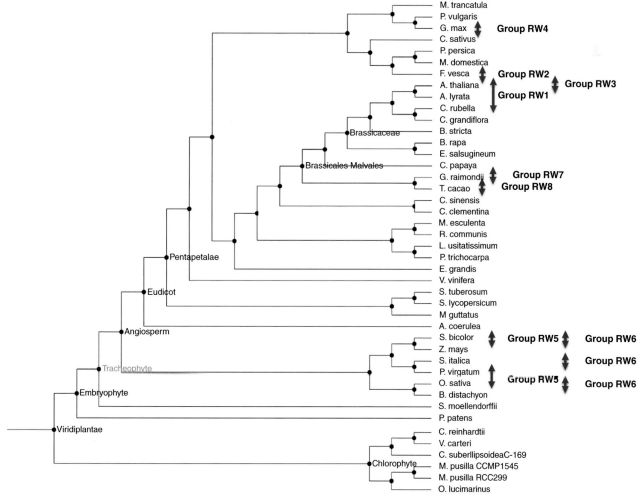

FIGURE 11.8 **Distribution of the eight R protein–WRKY families.** A phylogenetic tree of sequenced plant genomes is presented together with the distribution of the eight R protein–WRKY families. *Based partly on phylogenetic analysis at http://phytozome.jgi.doe.gov. Modified from Rinerson et al. (2015).*

to search for the presence of both NBS–LRR-like domains (or similar) and at least one WRKY domain in a single gene. Simple BLAST (Basic Local Alignment Search Tool) searches often yielded results with only NBS–LRR domains or WRKY domains. Nevertheless, it was possible to establish not only that many plant genomes contain no R protein–WRKY-like genes but also that a considerable number of plant genomes do indeed contain such genes. More interestingly, the combinations of domains and domain architectures found in R protein–WRKY proteins are novel so that many of the proteins are not NBS–LRR–WRKY proteins, but all contain domains typical of R proteins together with one or more WRKY domains.

The majority of plant R genes encode a class of innate immune receptors (NLRs) with nucleotide-binding and leucine-rich repeat domains. R-gene evolution is thought to be facilitated by the formation of R-gene clusters, which permit sequence exchanges via recombinatorial mispairing and generate high haplotypic diversity. This pattern of evolution may also generate diversity at other loci that contribute to the R-complex (Friedman and Baker, 2007).

Table 11.1 shows a list of R protein–WRKY-like genes. These genes are present in *A. thaliana*, *Arabidopsis lyrata*, *Sorguhum bicolor*, *Capsella rubella*, *Oryza sativa* ssp. *japonica*, *O. sativa* ssp. *indica*, *Fragaria vesca*, *Aegilops tauschii*, *Glycine*

TABLE 11.1 NBS–LRR–WRKY Genes

New name	Species	Gene model and other names	Genomic position
AtRWRKY52	*A. thaliana*	AtWRKY52	Chr5: 18326203–18332609
AtRWRKY16	*A. thaliana*	AtWRKY16	Chr5: 18176914–18181805
AtRWRKY19	*A. thaliana*	AtWRKY19	Chr4: 7201656–7208766
AlRWRKY1	*A. lyrata*	AL915586	Scaffold 8: 2126189–2132181
AlRWRKY2	*A. lyrata*	AL915663	Scaffold 8: 2768079–2773021
AlRWRKY3	*A. lyrata*	AL915648	Scaffold 8: 2623905– 2628796
SbRWRKY1	*S. bicolor*	SOBIC.002G104400	ChrO2: 12369911–12381876
SbRWRKY2	*S. bicolor*	Sobic.008G174100	ChrO8: 53517353–53522939
SbRWRKY3	*S. bicolor*	Sobic.002G168300	ChrO2: 52695615–52704484
CrWRKY1	*C. rubella*	CARUBV10025744M	Scaffold 8: l365382–1370221
CrWRKY2	*C. rubella*	CARUBV10025742M	Scaffold 8: 1163667–1169265
OsjRWRKY1	*O. sativa* ssp. *japonica*	LOC_Os07gl7230. FgenesH prediction different	Chr7: 10149830– 10,159,829
OsiRWRKY1	*O. sativa* ssp. *indica*	BGIOSGA035675	Chromosome 11: 21830082–21837218
OsjRWRKY2	*O. sativa* ssp. *japonica*	Retrotransposon at 3 prime end gi 1108864659	Chrll: 27783900– 27793499
FvRWRKY1	*F. vesca*	MRNA21370	LG7: 18263740–18277966
FvRWRKY2	*F. vesca*	MRNA13368	LG7:22236162– 22242036
FvRWRKY3	*F. vesca*	MRNA03900ALT	LG7: 9804380– 9813690
FvRWRKY4	*F. vesca*	MRNA16678alt	LG6: 802048–808355
AtaRWRKY1	*A. tauschii*	R7VZB5	Scaffold 2: 19315
GmRWRKY1	*G. Max*	GlymaO5g29921.1	GmO5: 35364051– 35374699
GrRWRKY1	*G. raimondii*	Gorai.008G201000	ChrO8: 48587929– 48599304
GrRWRKY2	*G. raimondii*	Gorai.008G200800	ChrO8: 48660141– 48668419
TcRWRKY1	*T. cacao*	ThecclEG006109	Scaffold 2: 820964– 828678
TcRWRKY2	*T. cacao*	ThecclEG006103	Scaffold 2: 805474– 814178
TcRWRKY3	*T. cacao*	ThecclEG006116tl	Scaffold 2: 845249– 851848
HvRWRKY1	*H. vulgare*	MLOC 74974.5	Chr5: 483720745–483727238
SiRWRKY1	*S. italica*	Si028710m.g	Scaffold 2: 26415481 –26421105
PvRWRKY1	*P. virgatum*	Pavir.J20878	sg0.contig22731/9–CL19939Contigl
FvWRKY5	*F. vesca*	MRNA21370. Tandem repeat with FvRWRKY1	LG7: 18263740–18277966

max, Gossypium raimondii, Theobroma cacao, Hordeum vulgare, Setaria italica, and *Panicum virgatum*. In other sequenced plant genomes such as *Zea mays, Brachypodium distachyon,* and *Medicago truncatula* they appear to be lacking. Interestingly, R protein–WRKY-like genes appear to have evolved on multiple independent occasions in the plant kingdom but current information suggests that they are confined to higher plants (Figure 11.8). It appears that at least eight independent genomic rearrangements resulting in NBS–LRR–WRKY (or similar) genes have occurred in the genomes of currently sequenced higher plant species (Figure 11.9). Representatives are present that contain WRKY domains from Group I, Group IIb, Group IId, Group IIe, and Group III. Phylogenetic analyses and inspection of the architecture of the R protein–WRKY-like proteins reveals that there are at least eight types (Figure 11.9) and that many of these proteins have not only novel arrangements of WRKY domains, but also contain novel combinations of other protein domains (Figure 11.10). The groups have been classified as R protein–WRKYs (RWs) by Rinerson et al. (2015) as follows:

Group RW1: TIR–NB ARC–LRR–WRKY (IIe). Found in *Capsella* and *Arabidopsis*.

Group RW2: TIR–NB ARC–LRR–WRKY (III) – [WRKY (III)]. May have one or two Group III WRKY domains and may therefore be two groups. Found in strawberry.

Group RW3: PAH–WRKY (I NT)–WRKY (I CT)–NB ARC. AtWRKY19 is the only member of this family. Found in *Arabidopsis*.

Group RW4: TIR–NB ARC–LRR–WRKY (III). Has the same domains as Group RW2 but different architecture and does not cluster with Group RW2 proteins. Found in soybean.

Group RW5: [B3]–LRR–NB ARC–LRR–WRKY (IIe). Some, but not all, have a B3 DNA-binding domain. Found in the monocots rice, sorghum, and switch grass.

Group RW6: NB ARC–LRR–WRKY (III)–WRKY (III). These are found in the monocots sorghum, barley, rice, foxtail millet, and Tausch's goatgrass. One of the proteins has only one WRKY domain and another has an additional NAC DNA-binding domain.

Group RW7: LRR–WRKY(III)–WRKY (IId)–calmodulin-binding domain–WRKY (IIc). The two members of this group are found in *G. raimondii* (a possible progenitor species of tetraploid cotton). The WRKY domains from GrRWRKY1 are truncated and difficult to classify.

Group RW8: WRKY (III)–NB ARC–LRR. Found in cacao.

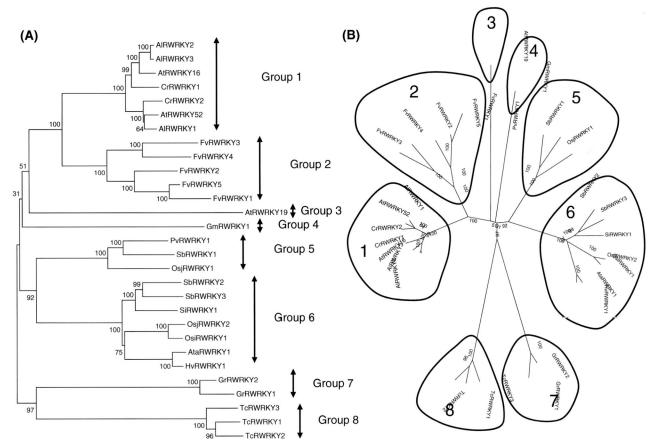

FIGURE 11.9 **Phylogenetic analyses of the R protein–WRKY (RW) families.** (A) Neighbor-joining phylogenetic tree derived from a MUSCLE alignment of full-length R protein–WRKY proteins. Numbers indicate bootstrap values from 1000 replicates. (B) Nonrooted version of the same tree as presented in (A). *Modified from Rinerson et al. (2015).*

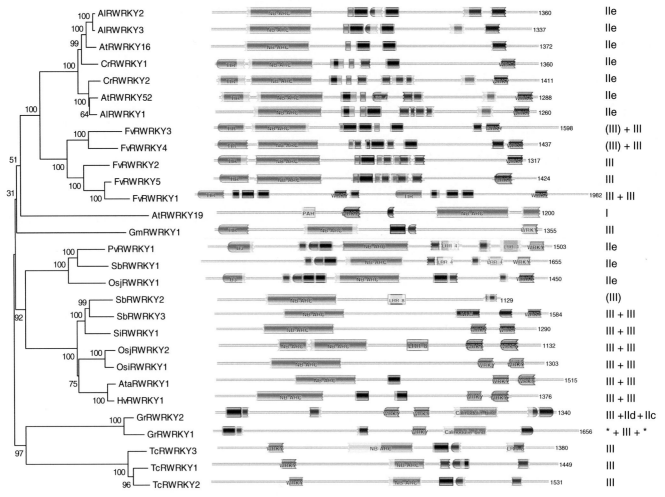

FIGURE 11.10 HMMER analyses of the R protein–WRKY families. Next to each predicted protein in the phylogenetic tree is the HMMER-derived overview of protein architecture with protein domains shown. WRKY domains are shown in reddish purple, TIR domains in green, leucine-rich repeat domains in blue or black, NB–ARC domains in lilac, calmodulin-binding domains in yellowish green, NAC domains in dark purple, and B3 domains in green. The number of WRKY domains and their groups are shown to the right of the proteins. *Modified from Rinerson et al. (2015).*

TIR stands for toll/interleukin-1 receptor; ARC for APAF-1, R proteins, and CED-4 domain; LRR stands for leucine-rich repeat; PAH for paired amphipathic helix; NB for nucleotide binding; R for resistance protein.

It is clear that these genomic rearrangements are associated with specific plant lineages and appear therefore to be relatively recent events. For example, Groups RW2 and RW4 are found in the Fabidae, and RW5 and RW6 in the grasses (Figure 11.8). Other groups such as RW2, RW7, and RW8 have only been found in a single species and even considering the limited availability of plant genome sequences, it is likely that they are present in only a small number of related species. This suggests that the formation of many of these R protein–WRKY-like genes are recent events and this is consistent with information showing that many R-genes are fast evolving and characterized by chimeric structures resulting from frequent sequence exchanges among group members (Kuang et al., 2005).

Sequence exchange between R-gene paralogs is considered to be the dominant mechanism for generating variations of type I resistance genes (Kuang et al., 2004). In addition, it has been known for some time that novel disease resistance specificities result from sequence exchange between tandemly repeated genes (Parniske et al., 1997). It may be significant that many of the R protein–WRKY genes contain one or more Group III WRKY domains because we have previously shown that tandem repeats of Group III WRKY genes exist in species such as *B. distachyon* (Tripathi et al., 2012). It is possible that the existence of R protein–WRKY genes reflects the frequent recombination associated with some R-genes but it may also reflect a high level of recombination at some WRKY gene loci, especially tandem repeats. The strawberry R protein–WRKY genes *FvRWRKY1* and *FvRWRKY5* illustrate the relative instability of these genes in the genome (Figures 11.8–11.10). *FvRWRKY1* and *FvRWRKY5* are found on linkage group

7 between 18245853 and 18295852. FGENESH predictions are of a single large polypeptide of 2854 amino acids. However, this predicted polypeptide contains what could be two very similar proteins with a TIR–NBS–LRR–WRKY (III) structure. The proteins are similar but not identical with blocks of similarity separated by dissimilar regions. Strikingly, an N-terminal segment of 186 amino acids from the first TIR–NBS–LRR–WRKY protein from amino acid 12 onward is present as an identical 186 amino acid in the second TIR–NBS–LRR–WRKY protein (Rinerson et al., 2015). Clearly, there has been a genomic rearrangement and duplication of some TIR–NBS–LRR–WRKY sequences. This illustrates that novel R protein–WRKY combinations appear to be formed through rearrangements including duplications.

It is possible that, once formed, some R protein–WRKY genes are under positive selection as they combine different components of signaling pathways that may either create new diversity in signaling or accelerate signaling by short-circuiting signaling pathways. In favor of this hypothesis are the identities of other domains that have been incorporated in R protein–WRKY proteins. These domains do not seem to be random segments of protein coding genes but rather other signaling components such as B3 and NAC DNA-binding domains and calmodulin-binding domains (Figure 11.10).

It has also been observed that many transposable elements are found at R-gene loci, including retrotransposons, transposons, and miniature inverted transposable elements (Kuang et al., 2005). This may provide one mechanism by which R-gene loci are rearranged. Transposable elements are found next to at least one of the R protein–WRKY genes that have been described by Rinerson et al. (2015; *OsjRWRKY2*) and transposable elements may therefore play a role in the creation of some R protein–WRKY genes.

It is possible that a small number of the predicted R protein–WRKY genes do not actually form chimeric proteins that contain all of the domains predicted by gene predictions such as FGENESH and HMMs. Further research will be required to determine the exact protein architecture produced from each individual gene. However, it is clear from studies of the *Arabidopsis* NBS–LRR–WRKY genes that these genes do indeed encode chimeric proteins and that the WRKY domain is indeed functional and binds to DNA (Noutoshi et al., 2005).

11.5 CONCLUSION: A REEVALUATION OF WRKY EVOLUTION

The availability of increasing numbers of sequenced plant genomes has facilitated a reevaluation of the evolution of the WRKY transcription factor family. In particular, the publication of the first charophyte genome sequence from *K. flaccidum* (Hori et al., 2014) filled a large gap in the available genome sequences in the plant kingdom between unicellular green algae, such as *C. reinhardtii*, which typically have 1–3 WRKY genes, and mosses ,such as *P. patens*, which have 30–40 genes. The *K. flaccidum* genome contains just two WRKY genes but the presence of a Group IIb WRKY gene was unexpected and necessitates a rewriting of the evolution of WRKY transcription factors. Similarly, the presence of WRKY transcription factor genes outside of the plant lineage in some diplomonads, social amoebae, fungi *incertae sedis*, and amoebozoa also sheds new light on the early evolution of WRKY genes. Although there are still some gaps in our knowledge, the following hypothesis for how WRKY genes have evolved, including two possible origins for Group IIa + b genes, best fits the available data (Figure 11.11).

Early in the green lineage, a BED-finger-like C2H2 Zn finger domain evolved into a WRKY domain by the addition of a WRKY-like motif N-terminal to the Zn finger. This single-domain WRKY transcription factor served as the progenitor for all other WRKY genes. Early during the colonization of land by plants the first terrestrial interactions with fungal cells, including symbioses, led to an early lateral gene transfer of a WRKY gene to a nonplant host. These single-domain fungal type WRKY genes from fungi *incertae sedis* are ancient and reflect the single WRKY domain present in the oldest form of WRKY transcription factors from unicellular organisms. Other lateral gene transfers also occurred to diplomonads, social amoebae, and other amoebozoa and these WRKY genes are either from algal types of WRKY genes that are not present in higher plants or early Group I-like genes. In the early multicellular terrestrial algae, Group IIb genes evolved. This is an important observation that causes a reevaluation of WRKY transcription factor evolution because previously Group IIb genes were thought to have evolved later and postdate Group IIc genes. The early appearance of Group IIb genes now poses a new question: Did they evolve from a Group I gene or did they evolve independently from a single WRKY domain-containing green alga gene? With limited available information, it is currently difficult to differentiate between these two different possibilities which we have called the "IIa + b Separate Hypothesis" and the "Group I Hypothesis" (Rinerson et al., 2015). The "IIa + b Separate Hypothesis" suggests that Group IIa and IIb WRKY genes did not evolve from Group I genes whereas the "Group I Hypothesis" suggests that all WRKY genes evolved from a Group I gene. Group I WRKY genes from unicellular green algae do not appear to have the conserved PR intron, and both the PR intron and the QVQR intron first appear in filamentous green algae. Both hypotheses therefore appear possible. The conserved PR intron was not present in the DNA coding for the first Group I C-terminal domains but had evolved

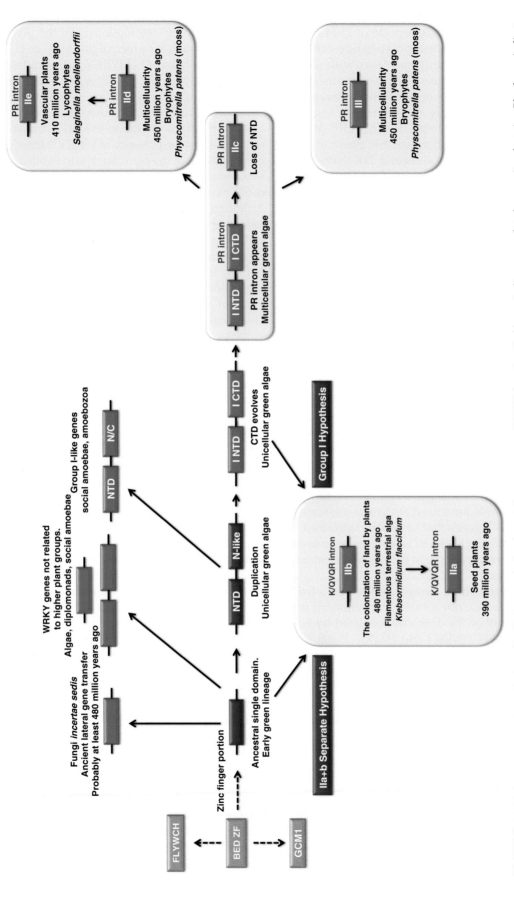

FIGURE 11.11 Overview of the evolution of WRKY transcription factors. Boxes represent WRKY domains. Red boxes indicate postulated progenitor domains. Blue boxes indicate WRKY domains from present day species. Green boxes indicate FLYWCH, GCM1, and BED Zn finger domains. Conserved introns are shown in red lettering. The four major flowering plant WRKY lineages are shown in light blue boxes. The direction of evolution is shown by arrows.

B. EVOLUTION AND STRUCTURE OF DEFINED PLANT TRANSCRIPTION FACTOR FAMILIES

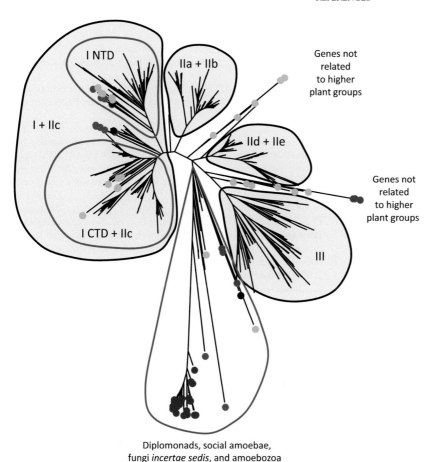

FIGURE 11.12 **Maximum likelihood phylogenetic tree of the WRKY gene family.** The four major flowering plant WRKY lineages are shown in light blue boxes and these boxes correspond to the boxes in Figure 11.11.

by the time plants had started to colonize the land. This conserved intron is found in all other WRKY groups and suggests that Groups IIc, IId, IIe, and III evolved from the C-terminal WRKY domains of Group I genes. It appears from our phylogenetic analyses and genomic searches that there are four major WRKY transcription factor lineages in flowering plants (Figures 11.11 and 11.12), Groups I + IIc, Groups IIa + IIb, Groups IId + IIe, and Group III and, while it is possible that Groups IIa + IIb evolved independently from an ancestral single domain WRKY gene, it seems clear that all other groups of flowering plant WRKY genes evolved from the C-terminal domain of Group I genes.

During the evolution of flowering plants, one other type of WRKY genes evolved that contain domains typical for both intracellular-type R proteins and WRKY transcription factors. These R protein–WRKY (RW) proteins are not found in all plant genomes but have evolved many times and with differing domain structures (Figures 11.8–11.10). The formation of these R protein–WRKY genes is recent with classes being restricted to specific flowering plant lineages. Once formed, R protein–WRKY genes may be selected for as they combine different components of signaling pathways which may either create new diversity in signaling or accelerate signaling by short-circuiting signaling pathways.

Acknowledgments

The authors would like to thank Mani Kant Choudhary, Marissa Miller, Naveen Kumar, Malini Rao, Deena Rushton, and Nikhil Kesarla in the Rushton Lab. We would also like to thank all of those that have contributed to the field of WRKY transcription factor evolution, especially Yuanji Zhang, Liangjiang Wang, Thomas Eulgem, Imre Somssich, Luise Brand, Dierk Wanke, Madan Babu, and Lakshminarayanan Aravind. This project was supported in part by National Research Initiative grants 2008-35100-04519 and 2008-35100-05969 from the USDA National Institute of Food and Agriculture.

References

Andersson, J.O., Sjogren, A.M., Davis, L.A., Embley, T.M., Roger, A.J., 2003. Phylogenetic analyses of diplomonad genes reveal frequent lateral gene transfers affecting eukaryotes. Curr. Biol. 13, 94–104.

Babu, M.M., Iyer, L.M., Balaji, S., Aravind, L., 2006. The natural history of the WRKY–GCM1 zinc fingers and the relationship between transcription factors and transposons. Nucleic Acid Res. 34, 6505–6520.

Banks, J.A., Nishiyama, T., Hasebe, M., Bowman, J.L., Gribskov, M., dePamphilis, C., Albert, V.A., Aono, N., Aoyama, T., Ambrose, B.A., Ashton, N.W., Axtell, M.J., Barker, E., Barker, M.S., Bennetzen, J.L., Bonawitz, N.D., Chapple, C., Cheng, C., Correa, L.G., Dacre, M.,

DeBarry, J., Dreyer, I., Elias, M., Engstrom, E.M., Estelle, M., Feng, L., Finet, C., Floyd, S.K., Frommer, W.B., Fujita, T., Gramzow, L., Gutensohn, M., Harholt, J., Hattori, M., Heyl, A., Hirai, T., Hiwatashi, Y., Ishikawa, M., Iwata, M., Karol, K.G., Koehler, B., Kolukisaoglu, U., Kubo, M., Kurata, T., Lalonde, S., Li, K., Li, Y., Litt, A., Lyons, E., Manning, G., Maruyama, T., Michael, T.P., Mikami, K., Miyazaki, S., Morinaga, S., Murata, T., Mueller-Roeber, B., Nelson, D.R., Obara, M., Oguri, Y., Olmstead, R.G., Onodera, N., Petersen, B.L., Pils, B., Prigge, M., Rensing, S.A., Riaño-Pachón, D.M., Roberts, A.W., Sato, Y., Scheller, H.V., Schulz, B., Schulz, C., Shakirov, E.V., Shibagaki, N., Shinohara, N., Shippen, D.E., Sorensen, I., Sotooka, R., Sugimoto, N., Sugita, M., Sumikawa, N., Tanurdzic, M., Theissen, G., Ulvskov, P., Wakazuki, S., Weng, J.K., Willats, W.W., Wipf, D., Wolf, P.G., Yang, L., Zimmer, A.D., Zhu, Q., Mitros, T., Hellsten, U., Loque, D., Otillar, R., Salamov, A., Schmutz, J., Shapiro, H., Lindquist, E., Lucas, S., Rokhsar, D., Grigoriev, I.V., 2011. The *Selaginella* genome identifies genetic changes associated with the evolution of vascular plants. Science 332, 960–963.

Brand, L.H., Fischer, N.M., Harter, K., Kohlbacher, O., Wanke, D., 2013. Elucidating the evolutionary conserved DNA-binding specificities of WRKY transcription factors by molecular dynamics and *in vitro* binding assays. Nucleic Acid Res. 21, 9764–9778.

Clarke, M., Lohan, A.J., Liu, B., Lagkouvardos, I., Roy, S., Zafar, N., Bertelli, C., Schilde, C., Kianianmomeni, A., Burglin, T.R., Frech, C., Turcotte, B., Kopec, K.O., Synnott, J.M., Choo, C., Paponov, I., Finkler, A., Heng Tan, C.S., Hutchins, A.P., Weinmeier, T., Rattei, T., Chu, J.S., Gimenez, G., Irimia, M., Rigden, D.J., Fitzpatrick, D.A., Lorenzo-Morales, J., Bateman, A., Chiu, C.H., Tang, P., Hegemann, P., Fromm, H., Raoult, D., Greub, G., Miranda-Saavedra, D., Chen, N., Nash, P., Ginger, M.L., Horn, M., Schaap, P., Caler, L., Loftus, B.J., 2013. Genome of *Acanthamoeba castellanii* highlights extensive lateral gene transfer and early evolution of tyrosine kinase signaling. Genome Biol. 14, R11.

Duan, M.R., Nan, J., Liang, Y.H., Mao, P., Lu, L., Li, L., Wei, C., Lai, L., Li, Y., Su, X.D., 2007. DNA binding mechanism revealed by high resolution crystal structure of *Arabidopsis thaliana* WRKY1 protein. Nucleic Acid Res. 35, 1145–1154.

Eulgem, T., Rushton, P.J., Robatzek, S., Somssich, I.E., 2000. The WRKY superfamily of plant transcription factors. Trends Plant Sci. 5, 199–206.

Fitzpatrick, D.A., 2012. Horizontal gene transfer in fungi. FEMS Microbiol. Lett. 329, 1–8.

Friedman, A.R., Baker, B.J., 2007. The evolution of resistance genes in multi-protein plant resistance systems. Curr. Opin. Genet. Dev. 17, 493–499.

Hedges, S.B., Blair, J.E., Venturi, M.L., Shoe, J.L., 2004. A molecular timescale of eukaryote evolution and the rise of complex multicellular life. BMC Evol. Biol. 4, 279–284.

Hori, K., Maruyama, F., Fujisawa, T., Togashi, T., Yamamoto, N., Seo, M., Sato, S., Yamada, T., Mori, H., Tajima, N., Moriyama, T., Ikeuchi, M., Watanabe, M., Wada, H., Kobayashi, K., Saito, M., Masuda, T., Sasaki-Sekimoto, Y., Mashiguchi, K., Awai, K., Shimojima, M., Masuda, S., Iwai, M., Nobusawa, T., Narise, T., Kondo, S., Saito, H., Sato, R., Murakawa, M., Ihara, Y., Oshima-Yamada, Y., Ohtaka, K., Satoh, M., Sonobe, K., Ishii, M., Ohtani, R., Kanamori-Sato, M., Honoki, R., Miyazaki, D., Mochizuki, H., Umetsu, J., Higashi, K., Shibata, D., Kamiya, Y., Sato, N., Nakamura, Y., Tabata, S., Ida, S., Kurokawa, K., Ohta, H., 2014. *Klebsormidium flaccidum* genome reveals primary factors for plant terrestrial adaptation. Nature Commun. 5, 3978.

Kuang, H., Woo, S.-S., Meyers, B.C., Nevo, E., Michelmore, R.W., 2004. Multiple genetic processes result in heterogeneous rates of evolution within the major cluster disease resistance genes in lettuce. Plant Cell 16, 2870–2894.

Kuang, H., Wei, F., Marano, M.R., Wirtz, U., Wang, X., Liu, J., Shum, W.P., Zaborsky, J., Tallon, L.J., Rensink, W., 2005. The R1 resistance gene cluster contains three groups of independently evolving, type I R1 homologues and shows substantial structural variation among haplotypes of *Solanum demissum*. Plant J. 44, 37–51.

Kunin, V., Goldovsky, L., Darzentas, N., Ouzounis, C.A., 2005. The net of life: reconstructing the microbial phylogenetic network. Genome Res. 15, 954–959.

Lackner, G., Hertweck, C., 2011. Impact of endofungal bacteria on infection biology, food safety, and drug development. PLoS Pathog. 7, e1002096.

Noutoshi, Y., Ito, T., Seki, M., Nakashita, H., Yoshida, S., Marco, Y., Shirasu, K., Shinozaki, K., 2005. A single amino acid insertion in the WRKY domain of the *Arabidopsis* TIR–NBS–LRR–WRKY-type disease resistance protein SLH1 (sensitive to low humidity 1) causes activation of defense responses and hypersensitive cell death. Plant J. 43, 873–888.

Parniske, M., Hammond-Kosack, K.E., Golstein, C., Thomas, C.M., Jones, D.A., Harrison, K., Wulff, B.B., Jones, J.D., 1997. Novel disease resistance specificities result from sequence exchange between tandemly repeated genes at the *Cf-4/9* locus of tomato. Cell 91, 821–832.

Partida-Martinez, L.P., Monajembashi, S., Greulich, K.O., Hertweck, C., 2007. Endosymbiont-dependent host reproduction maintains bacterial–fungal mutualism. Curr. Biol. 17, 773–777.

Rensing, S.A., Lang, D., Zimmer, A.D., Terry, A., Salamov, A., Shapiro, H., Nishiyama, T., Perroud, P.-F., Lindquist, E.A., Kamisugi, Y., Tanahashi, T., Sakakibara, K., Fujita, T., Oishi, K., Kuroki, Shin-I.T., Toyoda, Y., Suzuki, A., Hashimoto S-i, Y., Yamaguchi, K., Sugano, S., Kohara, Y., Fujiyama, A., Anterola, A., Aoki, S., Ashton, N., Barbazuk, W.B., Barker, E., Bennetzen, J.L., Blankenship, R., Cho, S.H., Dutcher, S.K., Estelle, M., Fawcett, J.A., Gundlach, H., Hanada, K., Heyl, A., Hicks, K.A., Hughes, J., Lohr, M., Mayer, K., Melkozernov, A., Murata, T., Nelson, D.R., Pils, B., Prigge, M., Reiss, B., Renner, T., Rombauts, S., Rushton, P.J., Sanderfoot, A., Schween, G., Shiu, S.-H., Stueber, K., Theodoulou, F.L., Tu, H., Van de Peer, Y., Verrier, P.J., Waters, E., Wood, A., Yang, L., Cove, D., Cuming, A.C., Hasebe, M., Lucas, S., Mishler, B.D., Reski, R., Grigoriev, I.V., Quatrano, R.S., Boore, J.L., 2008. The *Physcomitrella* genome reveals evolutionary insights into the conquest of land by plants. Science 319, 64–69.

Richards, T.A., Soanes, D.M., Foster, P.G., Leonard, G., Thornton, C.R., Talbot, N.J., 2009. Phylogenomic analysis demonstrates a pattern of rare and ancient horizontal gene transfer between plants and fungi. Plant Cell 21, 1897–1911.

Rinerson, C.I., Rabara, R.C., Tripathi, P., Shen, Q.J., Rushton, P.J., 2015. The evolution of WRKY transcription factors. BMC Plant Biol. 15, 66.

Rushton, P.J., MacDonald, H., Huttly, A.K., Lazarus, C.M., Hooley, R., 1995. Members of a new family of DNA-binding proteins bind to a conserved *cis*-element in the promoters of alpha-Amy2 genes. Plant Mol. Biol. 29, 691–702.

Rushton, P.J., Somssich, I.E., Ringler, P., Shen, Q.J., 2010. WRKY transcription factors. Trends Plant Sci. 15, 247–258.

Schüßler, A., Schwarzott, D., Walker, C., 2001. A new fungal phylum, the Glomeromycota: phylogeny and evolution. Mycol. Res. 105, 1413–1421.

Simon, L., Bousquet, J., Lévesque, R.C., Lalonde, M., 1993. Origin and diversification of endomycorrhizal fungi and coincidence with vascular land plants. Nature 363, 67–69.

Tisserant, E., Malbreil, M., Kuo, A., Kohler, A., Symeonidi, A., Balestrini, R., Charron, P., Duensing, N., Frei dit Frey, N., Gianinazzi-Pearson, V., Gilbert, L.B., Handa, Y., Herr, J.R., Hijri, M., Koul, R., Kawaguchi, M., Krajinski, F., Lammers, P.J., Masclaux, F.G., Murat, C., Morin, E., Ndikumana, S., Pagni, M., Petitpierre, D., Requena, N., Rosikiewicz, P., Riley, R., Saito, K., San Clemente, H., Shapiro, H., van Tuinen, D., Bécard, G., Bonfante, P., Paszkowski, U., Shachar-Hill, Y.Y., Tuskan, G.A., Young, J.P.W., Sanders, I.R., Henrissat, B., Rensing, S.A., Grigoriev, I.V., Corradi, N., Roux, C., Martin, F., 2013. Genome of an arbuscular mycorrhizal fungus provides insight into the oldest plant symbiosis. Proc. Natl. Acad. Sci. USA 110, 20117–20122.

Tripathi, P., Rabara, R.C., Langum, T.J., Boken, A.K., Rushton, D.L., Boomsma, D.D., Rinerson, C.I., Rabara, J., Reese, R.N., Chen, X., Rohila, J.S., Rushton, P.J., 2012. The WRKY transcription factor family in *Brachypodium distachyon*. BMC Genomics 13, 270.

Yamasaki, K., Kigawa, T., Inoue, M., Tateno, M., Yamasaki, T., Yabuki, T., Aoki, M., Seki, E., Matsuda, T., Tomo, Y., Hayami, N., Terada, T., Shirouzu, M., Tanaka, A., Seki, M., Shinozaki, K., Yokoyama, S., 2005. Solution structure of an *Arabidopsis* WRKY DNA binding domain. Plant Cell 17, 944–956.

Yamasaki, K., Kigawa, T., Watanabe, S., Inoue, M., Yamasaki, T., Seki, M., Shinozaki, K., Yokoyama, S., 2012. Structural basis for sequence-specific DNA recognition by an *Arabidopsis* WRKY transcription factor. J. Biol. Chem. 287, 7683–7691.

Zhang, Y., Wang, L., 2005. The WRKY transcription factor superfamily: its origin in eukaryotes and expansion in plants. BMC Evol. Biol. 5, 1.

CHAPTER 12

Structure, Function, and Evolution of the Dof Transcription Factor Family

Shuichi Yanagisawa

Laboratory of Plant Functional Biotechnology, Biotechnology Research Center, The University of Tokyo, Yayoi 1-1-1, Bunkyo-ku, Tokyo, Japan

OUTLINE

12.1 Discovery and Definition of the Dof Transcription Factor Family 183

12.2 Structure and Molecular Characteristics of Dof Transcription Factors 184
   12.2.1 The Domain Structure of Dof Transcription Factors 184
   12.2.2 The Zinc Finger Motif in the Dof Domain 184
   12.2.3 DNA Binding Mediated by the Dof Domain 185
   12.2.4 Interactions Between Dof Transcription Factors and Other Nuclear Proteins 185
      12.2.4.1 Interactions with bZIP-Type Transcription Factors 186
      12.2.4.2 Interactions with MYB-Type Transcription Factors 186
      12.2.4.3 Interactions with bHLH and Zinc-Finger-Type Transcription Factors 186
      12.2.4.4 Interactions with Chromatin-Associated High Mobility Group (HMG) Proteins 187
   12.2.5 The Dof Domain as a Multifunctional Domain 187
   12.2.6 Transcriptional Activation and Repression Domains of Dof Transcription Factors 187

12.3 Molecular Evolution of the Dof Transcription Factor Family 188

12.4 Physiological Functions of Dof Transcription Factors 190
   12.4.1 Arabidopsis CDF and Members of the CDF-Containing Clade 190
      12.4.1.1 CDFs in Arabidopsis 190
      12.4.1.2 CDF Homologs in Other Angiosperms 191
      12.4.1.3 CDF Homologs in NonVascular Plants 191
   12.4.2 The Roles of Dof Transcription Factors in the Regulation of Development and Differentiation 191
   12.4.3 Regulation of Metabolism 192
   12.4.4 Hormonal Regulation 193
   12.4.5 Response to Light 193

12.5 Perspective 194

References 194

12.1 DISCOVERY AND DEFINITION OF THE DOF TRANSCRIPTION FACTOR FAMILY

Dof transcription factors are defined as transcription factors containing the Dof (DNA-binding one finger) domain. The first Dof transcription factor to be described was Dof1 from *Zea mays* (maize; Yanagisawa and Izui, 1993). This was identified as a DNA-binding protein that interacted with a sequence in the cauliflower mosaic *35S RNA* promoter, which was later confirmed to function as a *cis*-regulatory region (Bhullar et al., 2007). Two other maize proteins were subsequently identified from their homology to Dof1, and comparison of the three amino acid sequences revealed a conserved region sufficient for DNA binding. This region includes a $CX_2CX_{21}CX_2C$ motif that differs from those

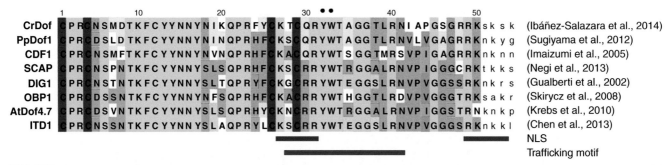

FIGURE 12.1 Amino acid sequences of Dof domains of Dof transcription factors from *C. reinhardtii* (CrDof), the moss *P. patens* (PpDof1), and *Arabidopsis thaliana* (CDF1, SCAP1, DIG1, OBP1, AtDo4.7, and ITD1). As representatives, Dof domains of some Dof transcription factors whose functions have been characterized are indicated. Amino acid residues flanking the C-terminal of the Dof domains are indicated in lower case. CrDof1, PpDof1, and CDF1 form a clade in the phylogenetic tree, while others form different clades (see Section 12.3). Cysteine residues that are probably coordinate to a Zn ion are highlighted in red. The positions of aromatic amino acid residues located outside the Zn finger, but involved in DNA binding, are indicated by black dots. Red bars indicate the position of an NLS identified in AtDof4.7 and the blue bar indicates the region involved in cell-to-cell trafficking of ITD1/AtDof4.1.

known from zinc (Zn) finger proteins from animals and yeast in both the amino acid sequence and the arrangement of the cysteine residues that are predicted to coordinate with a Zn ion. This conserved region (Figure 12.1) was therefore described as a novel Zn finger domain and named the Dof domain (Yanagisawa, 1995, 1996). Although the Dof domain was initially considered to be a sequence of 52 amino acids, the first 2 amino acid residues were later found not to be conserved across all Dof transcription factors (Yanagisawa, 2002, 2004) and thus the core Dof domain contains 50 amino acid residues (Figure 12.1).

Dof transcription factors are found from unicellular algae to angiosperms and are probably ubiquitous among green plants (Moreno-Risueno et al., 2007; Shigyo et al., 2007). However, they appear to be unique to plants, as no genes containing the Dof domain (*Dof* genes) are encoded in the genomes of the yeast *Saccharomyces cerevisiae*, the nematode *Caenorhabditis elegans*, or the insect *Drosophila melanogaster* (Yanagisawa and Sheen, 1998; Yanagisawa, 2002). This is consistent with the failure to identify any *Dof* genes in the genomes of the fish *Danio rerio* or the mammal *Mus musculus* (Moreno-Risueno et al., 2007). Interestingly, *Dof* genes have not been found in the genomes of the red alga *Cyanidioschyzon merolae* or the diatom *Thalassiosira pseudonana* (Shigyo et al., 2007). Accordingly, it is likely that the original gene encoding a Dof transcription factor arose in the common ancestor of green eukaryotic organisms.

12.2 STRUCTURE AND MOLECULAR CHARACTERISTICS OF DOF TRANSCRIPTION FACTORS

12.2.1 The Domain Structure of Dof Transcription Factors

Dof transcription factors are proteins that usually consist of 200–400 amino acid residues. As well as the Dof DNA-binding domain, they usually include a domain involved in transcriptional activation or repression and, consistent with their regulatory roles in nuclei, a nuclear localization signal (NLS). However, beyond the highly conserved Dof domain, the amino acid sequence is not well conserved across members of the family and individual proteins have very divergent sequences. Unlike some transcription factor families, which include both members with a single DNA-binding domain and those with multiple DNA-binding domains, all known Dof transcription factors contain a single copy of the Dof domain in the N-terminal or central region of the protein (Yanagisawa, 2002).

The domain structure of maize Dof1 is shown as a representative example of Dof transcription factors (Figure 12.2). Maize Dof1 consists of 238 amino acid residues and contains the Dof domain (amino acids 49–98 inclusive), a region of basic amino acids, which resembles both the SV40 and bipartite NLSs and confers nuclear localization, and a transcriptional activation domain composed of 44 amino acid residues at the C-terminal of the protein (Yanagisawa, 2001). Since the Dof and the C-terminal transcriptional activation domains of maize Dof1 independently confer DNA-binding and transcriptional activation (Yanagisawa and Sheen, 1998), these are separate domains (Figure 12.2).

12.2.2 The Zinc Finger Motif in the Dof Domain

Several lines of evidence indicate that the Dof domain forms a C_2—C_2 Zn finger. Firstly, divalent metal chelators (1,10-*o*-phenanthroline and EDTA) inhibit DNA binding by maize Dof1 and other Dof transcription factors (Yanagisawa, 2002, 2004), but the DNA-binding activity of maize Dof1 can be restored by the addition of Zn ions (Yanagisawa, 1995). Secondly, mutation of the cysteine residues, which are predicted to coordinate to a Zn ion (Figure 12.2), diminishes the DNA-binding activity *in*

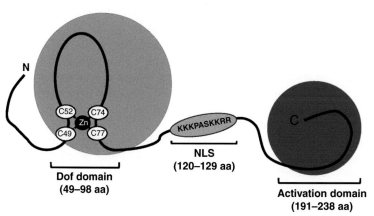

FIGURE 12.2 **Domain structure of a Dof transcription factor.** The structure of maize Dof1 is shown as a representative example.

vitro (Shimofurutani et al., 1998; Yanagisawa, 1995) and *in vivo* (Mena et al., 1998). Finally, inductively coupled argon plasma mass spectrometry indicates that the Dof domain from the ascorbate oxidase gene binding protein (AOBP), a pumpkin Dof protein, contains a Zn ion (Umemura et al., 2004). Conclusive empirical evidence for the Zn finger is still to be obtained, as the three-dimensional structure of the Dof domain has not yet been analyzed by X-ray crystallography or nuclear magnetic resonance (NMR) spectroscopy, but such a structure has been predicted using computer software *in silico* (Kushwaha et al., 2013).

The Dof domain contains a Zn finger motif composed of 29 amino acid residues as well as an additional 21 amino acid residues at its C-terminal region (Figure 12.1). These residues may also be involved in the interaction between the Dof domain and DNA, as mutation of two aromatic amino acids (Tyr and Trp), situated outside the Zn finger (Figure 12.1), considerably reduced the sequence-specific DNA-binding activity of the protein (Shimofurutani et al., 1998; Umemura et al., 2004). The presence of these aromatic amino acid residues has led to the suggestion that similarities exist between the Dof domain, the C_2—C_2 Zn finger domains of the GATA family, and the steroid hormone receptor family (Umemura et al., 2004), although the amino acid sequences of the Zn fingers in these proteins are otherwise completely different from one another.

12.2.3 DNA Binding Mediated by the Dof Domain

The sequence-specific DNA binding by Dof transcription factors or Dof domains has been examined in many *in vitro* and *in vivo* experiments (reviewed in Yanagisawa, 2002, 2004). There is high homology between the Dof domains of different proteins, and all the Dof transcription factors tested bind to the motif AAAG (or its complimentary sequence, CTTT), other than pumpkin AOBP whose recognition sequence is AGTA (Kisu et al., 1998). Experiments in which binding sites were selected from random sequence oligonucleotides clarified the importance of the AAAG motif in DNA recognition by Dof transcription factors, as well as preferential binding to (A/T)AAAG over (G/C)AAAG (Yanagisawa and Schmidt, 1999). Dof transcription factors therefore recognize a relatively short DNA sequence, compared with other plant transcription factors, as, for example, WRKY transcription factors bind to 7 base pair (bp) sequence elements (Ülker and Somssich, 2004) and plant basic leucine zipper (bZIP) transcription factors generally bind to 6 bp sequence elements (bZIP Research Group, 2002).

Due to the recognition of a relatively short sequence by Dof transcription factors, putative Dof-binding sites are very frequently found in promoter sequences and regions for the transcriptional regulation of many genes. However, most are probably nonfunctional sites *in vivo*. Although maize Dof1 can bind the AAAG motif in both naked DNA and at the surface of nucleosomes reconstructed *in vitro*, in the latter case binding is dependent upon the position within the nucleosome of the AAAG motif (Cavalar et al., 2003), suggesting that motif position circumscribes the binding of Dof transcription factors to DNA *in vivo*. Furthermore, as discussed in detail in Section 12.2.4, Dof transcription factors may need to interact with other transcription factors to bind to DNA or modulate transcription at precise sites in the genome.

12.2.4 Interactions Between Dof Transcription Factors and Other Nuclear Proteins

A variety of physiologically relevant interactions of Dof transcription factors with other nuclear proteins have been reported: some are mediated through the Dof domain, while others involve amino acid sequences outside the Dof domain.

12.2.4.1 Interactions with bZIP-Type Transcription Factors

Arabidopsis octopine synthase (*ocs*) element binding factor (OBF) binding protein 1 (OBP1) was the first Dof domain-containing protein found to interact with other transcription factors. It interacts with the *Arabidopsis* bZIP transcription factors OBF4 and OBF5, and this interaction is mediated by the Dof domain of OBP1 (Zhang et al., 1995). OBP1 enhances binding of OBF5 to the *ocs*-like sequence in the promoter of its putative target gene, *Arabidopsis glutathione S-transferase6* (*GST6*), which contains both an *ocs*-element and an OBP1 binding site with only a 13 bp distance between them (Chen et al., 1996). A similar interaction between Dof domain and bZIP transcription factors was identified using an endosperm-specific Dof transcription factor from maize, the prolamin-box (P-box) binding factor (PBF; Vicente-Carbajosa et al., 1997). Endosperm-specific gene expression of storage protein genes in cereals is regulated by a *cis*-regulatory element consisting of a 7 bp sequence (TGTAAAG), termed the prolamin box (P-box), and an adjacent GCN4-like motif (TGA(G/C)TCA). PBF binds to the P-box in a sequence-specific manner, and also specifically interacts with Opaque2, a bZIP transcription factor that binds to the GCN4-like motif. Furthermore, the P-box is necessary for Opaque2-mediated transcriptional activation (Vicente-Carbajosa et al., 1997). Synergistic activation of P-box-dependent transcription has also been reported for a rice PBF homolog (RPBF) and a bZIP transcriptional activator, RISBZ1 (Yamamoto et al., 2006). Thus, Dof domains of some proteins mediate both DNA-binding and protein–protein interactions that are critical for precise interactions with target sites *in vivo* and/or transcriptional regulation.

12.2.4.2 Interactions with MYB-Type Transcription Factors

Some Dof transcription factors interact with an R2R3-type MYB protein involved in gibberellin (GA)-regulated gene expression (GAMYB). This physical interaction plays a critical role in controlling GA-induced expression of hydrolytic enzyme genes in the aleurone layer during the germination of cereal grains. The barley homolog of maize PBF (BPBF; Mena et al., 1998; Mena et al., 2002) interacts with barley GAMYB (HvGAMYB), and HvGAMYB requires the presence of BPBF to transactivate the *B-hordein* gene promoter. Unlike Dof–bZIP interactions, it is the C-terminal region of BPBF outside the Dof domain that interacts with HvGAMYB (Diaz et al., 2002). Another barley Dof transcription factor, SAD, which promotes transcription from the promoters of the endosperm-specific genes encoding B-hordein and trypsin inhibitor BTI–CMe, also interacts with HvGAMYB (Isabel-LaMoneda et al., 2003; Diaz et al., 2005).

A physical interaction between a rice Dof transcription factor (OsDOF3) and GAMYB was detected using yeast two-hybrid assays (Washio, 2003). OsDOF3, the expression of which is induced by GA in the scutellum and aleurone layer, functions as an activator of the promoters of *Type III carboxypeptidase* (which is upregulated by GA in the aleurone layer of germinated cereal grains) and *rice α-amylase* (*RAmy1A*) (a GA-inducible gene that is the predominant GA-responsive gene in the aleurone layer; Washio, 2001, 2003).

Furthermore, activity of the promoter of the GA-inducible α-amylase gene *Amy32b* was modulated by the relative levels of the repressor or activator complexes, including different Dof transcription factors, in barley aleurone cells (Zou et al., 2008). In the model proposed in Figure 12.3, SAD interacts with HvGAMYB and activates the *Amy32b* promoter in the presence of GA, while BPBF interacts with HvWRKY38 and represses it in the absence of GA, illustrating the physiological relevance of the interactions between Dof domain proteins and other transcription factors.

12.2.4.3 Interactions with bHLH and Zinc-Finger-Type Transcription Factors

The *Arabidopsis* Dof transcription factor, DOF6/AtDof3.2, which negatively affects germination in non-after-ripened seeds, interacts with TCP14, a member of

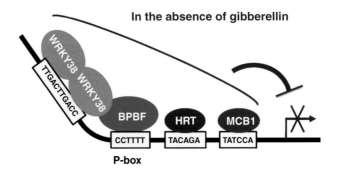

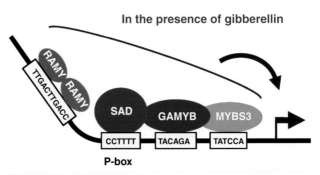

FIGURE 12.3 **A model of transcriptional control mediated by the interactions of Dof transcription factors with other transcription factors.** In the absence of GA, the BPBF–WRKY38 complex, and other negative regulators, bind to the corresponding *cis*-elements in the *Amy32b* promoter, and transcription is switched off. In the presence of GA, the SAD–GAMYB–MYBS3 complex, and other positive regulators, bind to the corresponding *cis*-elements, and transcription is switched on.

the TEOSINTE BRANCHED1, CYCLOIDEA, and PCF (TCP) domain protein family. TCP14 is a basic helix–loop–helix (bHLH)-type transcription factor that positively regulates seed germination (Rueda-Romero et al., 2012). Furthermore, another *Arabidopsis* Dof transcription factor, AtDof4.7, which participates in the control of abscission by directly regulating the expression genes encoding cell wall hydrolysis enzymes, interacts with another abscission-related transcription factor, ZINC FINGER PROTEIN 2 (Wei et al., 2010). Moreover, a physical interaction between Dof domains has been detected by pull-down assays *in vitro* (Yanagisawa, 1997), although the physiological significance of this is obscure.

12.2.4.4 Interactions with Chromatin-Associated High Mobility Group (HMG) Proteins

In addition to their interactions with other transcription factors, Dof transcription factors can also physically interact with the chromatin-associated HMG proteins of the HMG-box (HMGB) family. HMGB proteins contain a single HMGB DNA-binding domain and bind to DNA nonspecifically. The HMGB domain can interact with the Dof domain of Dof transcription factors (Yanagisawa, 1997; Krohn et al., 2002). Five different HMGB proteins of maize were found to enhance the binding of maize Dof2 to DNA with different efficiencies. Furthermore, phosphorylation of HMGB1 by the protein kinase CK2 abolished the interaction between HMGB1 and Dof2 (Krohn et al., 2002), implying that phosphorylation of HMGB proteins is involved in the regulation of Dof2 activity. Although some HMGB proteins have been implicated in transcriptional regulation, they are not authentic transcriptional activators. Their interactions with Dof transcription factors may have a role in the selection of specific binding sites or in enhancing the binding of the transcription factors to DNA *in vivo*.

12.2.5 The Dof Domain as a Multifunctional Domain

Maize Dof1 contains a basic region that acts as an NLS outside the Dof domain (Figure 12.2). However, any such NLS was not found in the sequences of *Arabidopsis* Dof transcription factors. Conversely, examination of *Arabidopsis* Dof4.7 identified a novel type of NLS, which is a bipartite NLS heavily overlapping the Dof domain containing a basic motif (Figure 12.1). The first part of the NLS consists of amino acid residues of the Dof domain (residues 27–31) and the second of the last two residues of the Dof domain (residues 49 and 50) together with subsequent residues outside the Dof domain (Krebs et al., 2010). Although the amino acid sequence outside the Dof domain is not, in general, conserved between Dof transcription factors, there are frequently a pair of basic amino acid residues at these corresponding positions (Figure 12.1)

and, therefore, this region may also function as an NLS in other Dof transcription factors.

Another, novel function of the Dof domain was recently found in *Arabidopsis* INTERCELLULAR TRAFFICKING DOF 1 (ITD1; also known as AtDof4.1). ITD1 is a noncell-autonomous transcription factor, and cell-to-cell trafficking of ITD1 is mediated by the Zn finger motif of the Dof domain and the flanking N-terminal variable region (Chen et al., 2013).

Such results show that, although it was originally identified as a DNA-binding domain, the Dof domain is a multifunctional domain involved in DNA binding, interactions with other transcription factors, nuclear localization, and cell-to-cell trafficking. The Dof domain alone is sufficient to confer DNA-binding on Dof transcription factors, although the possibility that amino acid residues outside the Dof domain modify DNA recognition in some cases cannot be excluded. Amino acid residues both inside and outside the Dof domain are necessary; however, to confer other additional functions including nuclear localization. Thus, although all Dof domains are probably capable of binding to DNA, some may additionally contribute to other molecular processes, such as nuclear localization and cell-to-cell trafficking (Krebs et al., 2010; Chen et al., 2013).

12.2.6 Transcriptional Activation and Repression Domains of Dof Transcription Factors

Domains involved in the transcriptional activation or repression of Dof transcription factors have not been well characterized to date, although a few examples are known. The minimal transcriptional activation domain of maize Dof1 was mapped to a region consisting of 44 amino acid residues at the C-terminal of the protein. This transcriptional activation domain is functional in yeast and human cells as well as in plant cells, and an aromatic amino acid residue in the domain plays a critical role in transcriptional activation (Yanagisawa and Sheen, 1998; Yanagisawa, 2001). Similarly, transcriptional activation domains of barley BPBF and *Arabidopsis* OBP1, OBP2, and OBP3 were mapped to their C-terminal regions by domain-swapping experiments with a yeast transcription factor, GAL4 (Kang and Singh, 2000; Diaz et al., 2002).

Some Dof transcription factors function as transcriptional repressors (Imaizumi et al., 2005; Sugiyama et al., 2012). Because the Dof domain is generally located in the N-terminal regions of the protein, transcriptional repression domains may be located in the C-terminal regions of Dof proteins acting as transcriptional repressors. However, Dof transcription factors with no additional domain beyond the Dof domain might also modulate transcription by functioning as competitors in DNA-binding or interacting with other transcription factors.

In support of this, transactivation by maize Dof1 in protoplasts was hampered by coexpression of the Dof domain alone (Yanagisawa and Sheen, 1998). Furthermore, Dof transcription factors, such as maize Dof2, may function as dual transcription factors that control both transcriptional activation and repression, depending on the content of individual promoters (Yanagisawa, 2000).

12.3 MOLECULAR EVOLUTION OF THE DOF TRANSCRIPTION FACTOR FAMILY

Dof transcription factors are probably present in all green eukaryotic organisms (Moreno-Risueno et al., 2007; Shigyo et al., 2007). The number of *Dof* genes in the genomes of different organisms is highly variable, and ranges from one copy in *Chlamydomonas reinhardtii* to more than 50 copies in maize (Table 12.1). The approximate average copy number of *Dof* genes encoded in the genome of each angiosperm is 30. The genome of the moss *Physcomitrella patens* contains 19 *Dof* genes (including 7 pairs of highly homologous genes that were produced by a genome-wide duplication event in *P. patens*). This number is intermediate between the numbers of *Dof* genes found in *C. reinhardtii* and angiosperms, and thus it is probable that the ancestral *Dof* gene arose before the evolutionary divergence of green algae from other multicellular lineages, and the current *Dof* gene family has evolved *via* a number of duplication events that have occurred since the divergence of algae from land plants and nonvascular plants from vascular plants. In support of this idea, sets of closely related *Dof* genes in the *Arabidopsis* genome appear to be the vestiges of four different large-scale genome duplications that are estimated to have occurred between 100 and 200 million years ago (MYA; Yanagisawa, 2002). However, analysis of the evolution of the *Dof* gene family in rice revealed comparable numbers of *Dof* genes in the *japonica* and *indica* subspecies of *Oryza sativa* from Asia and the African rice *Oryza glaberrima* S. (26–30 genes; Table 12.1), but only 10 *Dof* genes in the wild annual African species, *Oryza brachyantha* (Jacquemin et al., 2014). It is estimated that *O. glaberrima*

TABLE 12.1 Numbers of *Dof* Genes in a Variety of Plant Species

Species	Number of *Dof* genes	References
DICOT		
A. thaliana	36	Yanagisawa (2002)
Solanum lycopersicum L. (tomato)	34	Cai et al. (2013)
Glycine max (soybean)	28	Wang et al. (2007)
Populus trichocarpa (poplar)	41	Yang et al. (2006)
MONOCOT		
O. sativa (Asian rice)	30	Lijavetzky et al. (2003)
O. sativa ssp. *japonica*	30	Jacquemin et al. (2014)
O. sativa ssp. *indica*	28	Jacquemin et al. (2014)
O. glaberrima (African rice)	26	Jacquemin et al. (2014)
O. brachyantha (wild rice)	10	Jacquemin et al. (2014)
Z. mays (maize)	51	Jiang et al. (2012)
Brachypodium distachyon (purple false brome)	27	Hernando-Amado et al. (2012)
Sorghum bicolor (sorghum)	28	Kushwaha et al. (2011)
Hordeum vulgare (barley)	26	Moreno-Risueno et al. (2007)
Triticum aestivum L. (wheat)	28	Romeuf et al. (2010)
T. aestivum L. (wheat)	31	Shaw et al. (2009)
MOSS		
P. patens	19	Shigyo et al. (2007)
GREEN ALGA		
C. reinhardtii	1	Shigyo et al. (2007), Moreno-Risueno et al. (2007)

S. and *O. brachyantha* diverged approximately 15 MYA, and so differential amplification of *Dof* genes may occur in related plant species. Alternatively, it is possible that *O. brachyantha* has lost more than half of its *Dof* genes during the last 15 MYA, in which case species-specific reduction of *Dof* genes may also be important in determining the Dof transcription factor family in a species. The number of *Dof* genes in *O. brachyantha* is much smaller than expected, confusing our understanding of the evolutionary processes of the Dof transcription factor family.

Several phylogenetic trees have been put together using genomic information from different plant species (Cai et al., 2013; Gupta et al., 2014; Hernando-Amado et al., 2012; Kushwaha et al., 2011; Lijavetzky et al., 2003; Moreno-Risueno et al., 2007; Shaw et al., 2009; Shigyo et al., 2007; Wang et al., 2007; Yanagisawa, 2002; Yang et al., 2006). Whole amino acid sequences of proteins were applied to generate sequence alignment in some studies, whereas others used amino acid sequences of the Dof domain, due to limited availability of transcript and insufficiently reliable predictions of splicing site locations. However, although they classify Dof transcription factors into subgroups in somewhat different ways due to the complicated patterns of the branches, the trees produced by these studies are fundamentally similar. In all phylogenetic trees, the clade that contains CYCLING DOF FACTORs (CDFs) is distant from others, while the remaining clusters are contiguous with one another and thereby their classification is somewhat arbitrary (Figure 12.4). The single Dof transcription factor that is present in *C. reinhardtii* (CrDof) is classified into the CDF-containing clade (Moreno-Risueno et al., 2007; Shigyo et al., 2007). This might imply that at least two *Dof* genes were present before the *C. reinhardtii* lineage diverged from the ancestral precursors of land plants, and that one of them is the ancestor of CrDof and others were lost in *C. reinhardtii* (Moreno-Risueno et al., 2007; Shigyo et al., 2007).

Although, overall, the amino acid sequences in the regions outside the Dof domain are very divergent, the Dof transcription factors that group closely together in the

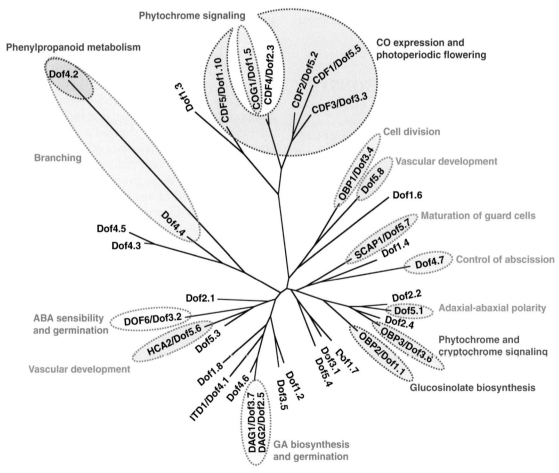

FIGURE 12.4 **The phylogenetic tree produced from the multiple alignments of all Dof domains from *Arabidopsis*.** The physiological function of each Dof transcription factor, classified into photoperiodic flowering (purple), development and differentiation (green), metabolism (red), hormonal regulation (yellow), and light signaling (blue), is denoted.

phylogenetic trees frequently share several short amino acid sequence motifs in addition to the conserved Dof DNA-binding domain (Lijavetzky et al., 2003; Yanagisawa, 2002; Yang et al., 2006). These motifs may be involved in determining the role of a Dof transcription factor, although the effect of these short-sequence motifs has only been characterized so far in CDFs (see Section 12.4.1.1) and some Dof transcription factors do not contain any of these common motifs.

As expected, closely related Dof transcription factors may play redundant or related roles. Conversely, two closely related genes may play different roles. For example, AtDOF4.2 and its close homolog AtDOF4.4 are implicated in the regulation of shoot branching in *Arabidopsis* (Zou et al., 2013). *DOF AFFECTING GERMINATION* (*DAG*) *1* and *2*, which originated in a genome duplication event 100 MYA, have opposing effects on seed germination (Gualberti et al., 2002), whereas *Dof4.7* and *Dof5.7*, which result from an earlier genome duplication 170 MYA, now have very different physiological functions, as they act on abscission (Wei et al., 2010) and guard cell maturation (Negi et al., 2013), respectively (Figure 12.4), indicating that not all closely related Dof transcription factors play similar roles. Hence, it is still difficult to discuss the relationship between functional diversification and gene multiplication.

12.4 PHYSIOLOGICAL FUNCTIONS OF DOF TRANSCRIPTION FACTORS

The physiological functions of a number of Dof transcription factors from several plant species have been characterized to date. However, although Dof transcription factors from the CDF-containing clade have been characterized from several plant species, including vascular and nonvascular plants, there has been no systematic analysis of other Dof transcription factors and, for this reason, it is difficult to discuss the conservation of function of each individual type. Accordingly, a simple summary of our current knowledge of the functions of different Dof transcription factors, other than CDFs, is presented in this section.

12.4.1 *Arabidopsis* CDF and Members of the CDF-Containing Clade

12.4.1.1 CDFs in Arabidopsis

CYCLIC DOF FACTOR 1 (CDF1) was initially identified as an *Arabidopsis* protein that interacted with FLAVIN-BINDING KELCH REPEAT F-BOX 1 (FKF1; Imaizumi et al., 2005), a component of a Skp1 Cullin F-box (SCF) E3 ubiquitin ligase complex governing the daytime expression pattern of *CONSTANS* (*CO*; Imaizumi and Kay, 2006). CDF1 is a transcriptional repressor of *CO*, which encodes a direct regulator of the *FLOWERING LOCUS T* (*FT*) gene whose product, FT, is a mobile signal that moves through the phloem to the meristem where it triggers flowering. CDF also directly interacts with GIGANTEA (GI), a protein that forms a complex with FKF1 (Sawa et al., 2007). Because the FKF1–GI complex directly regulates stability of CDF1 in the afternoon, the FKF1–GI-complex-mediated degradation of CDF1 is part of the molecular mechanism underlying daytime *CO* transcription (Imaizumi and Kay, 2006; Song et al., 2012). Consistent with the fact that FT is a mobile signal in the phloem, it has been shown that CDF1 and its functional homologs, CDF2, CDF3, and CDF5, are expressed in the phloem (Fornara et al., 2009). *Arabidopsis* small ubiquitin-like modifier (SUMO)-targeted ubiquitin ligase (AT-STUbL4) reduces the levels of CDF2, thereby increasing the levels of *CO* mRNA and promoting flowering through the photoperiodic pathway (Elrouby et al., 2012).

In *Arabidopsis*, *CDF1* forms a clade, previously referred to as group II by Yanagisawa (2002), group A by Shigyo et al., (2007) or subfamily A by Moreno-Risueno et al., (2007), together with six other Dof transcription factor genes named *CDF2*–*CDF5*, *Cogwheel 1* (*COG1*), and *AtDof1.3* (Yanagisawa, 2002; Le-Hir and Bellini, 2013), although *CDF4* and *COG1* are separated from the others in a phylogenetic tree that was rooted using *CrDof* as an outgroup (Noguero et al., 2013). All *CDFs*, except for *CDF4*, play redundant roles in repressing *CO* expression and photoperiodic flowering, and thereby the quadruple *cdf1cdf2cdf3cdf5* mutant showed a severe photoperiodic flowering phenotype (Fornara et al., 2009).

Consistent with their redundant roles, in addition to the Dof domain, these proteins share three conserved motifs composed of approximately 10–30 amino acid residues in their C-terminal regions. These CDF-specific motifs may be a characteristic of Dof transcription factors displaying functional redundancy with CDF1. Mutation of a lysine residue within one of these conserved amino acid motifs (amino acid residue 253 of CDF1) markedly reduces the physical interaction between CDF1 and FKF1, indicating a critical role for this residue in CDF1 degradation (Imaizumi et al., 2005).

CDF4, despite its name, does not share any of these conserved motifs and is not involved in the regulation of *CO* expression (Fornara et al., 2009; Corrales et al., 2014). COG1, suggested by phenotypic analysis to be a negative regulator of the phytochrome-signaling pathway (Park et al., 2003), also does not possess these motifs. By contrast, AtDof1.3, whose function has not yet been characterized, contains two of the three CDF-specific motifs. Thus, the CDF-containing clade of *Arabidopsis* may include at least two different classes of Dof transcription factors.

12.4.1.2 CDF Homologs in Other Angiosperms

As in *Arabidopsis* CDFs, CDF homologs from potato (*Solanum tuberosum*), StCDF1.1, StCDF1.2, and StCDF1.3, have been found to play a critical role in controlling the transition from the vegetative to the reproductive phase and tuber development at the stolon termini (Kloosterman et al., 2013). Overexpression of *StCDF1.2* represses the transcription of two *CONSTANS* genes, *StCO1* and *StCO2*, and modifies the expression of potato *FT* homologs *StSP5G* and *StSP6A*, inducing tuberization at an early stage under long-day (LD) conditions. Furthermore, StCDFs, like *Arabidopsis* CDFs, physically interact with the potato homologs of FKF1 and GI1.

Despite this clear demonstration of similar roles for CDFs in *Arabidopsis* and potato, the situation is different in rice. *Dof daily fluctuations 1* (*Rdd1*) is a rice homolog of *Arabidopsis* CDFs and contains CDF-specific amino acid motifs (Iwamoto et al., 2009). Although *Rdd1* is regulated by the circadian clock and phytochromes and shows daily oscillations in expression (Iwamoto et al., 2009), and thus resembles *Arabidopsis CDF* expression, constitutive expression of either the sense or antisense strands of *Rdd1* failed to affect expression levels of the rice *CO* homolog, *Heading date 1* (*Hd1*). Expression of the *Rdd1* antisense strand resulted in late-flowering plants with smaller body sizes, and also decreases in grain length, width, and 1000-grain weight; thus, *Rdd1* is likely to control grain development in rice rather than regulate expression of the *CO* homolog. Furthermore, overexpression of *OsDof12*, another rice *Dof* gene from the CDF-containing clade, promoted flowering, accompanied with upregulation of *Hd3a* expression under LD conditions but not short-day (SD) conditions (Li et al., 2009).

SlCDF1–5, CDF homologs from tomato, have been shown to be transcriptional regulators involved in responses to drought and salt stress as well as in the control of flowering time (Corrales et al., 2014). Unlike *Arabidopsis* CDFs, SlCDF1-5 are transcriptional activators (Corrales et al., 2014). A CDF homolog from *Jatropha curcas*, JcDof3, has been reported to interact with ASK2 (the product of one of the *Arabidopsis Skp1-like* genes), part of the SCF E3 ubiquitin ligase complex, as well as with *Arabidopsis* LOV KELCH REPEAT PROTEIN 2 (LKP2; Yang et al., 2011). These various findings mean the extent to which the roles of CDFs, in the regulation of *CO* expression, are conserved across flowering plants is uncertain, although the current molecular genetic and biochemical evidence adequately demonstrates the involvement of CDFs in the control of phase transition through long-distance signaling in *Arabidopsis* and potato.

12.4.1.3 CDF Homologs in NonVascular Plants

The only functional analyses of Dof transcription factors from nonflowering plants that have been performed to date have involved the unicellular alga *C. reinhardtii* and the moss *P. patens*. Analyzed Dof transcription factors in these nonvascular plants all belong to the CDF-containing clade. The CDF-containing clade of *P. patens* includes six genes, *PpDof1–PpDof6* (Shigyo et al., 2007). The two highly homologous genes, *PpDof3* and *PpDof4*, bear the closest resemblance to the *CDF* genes of *Arabidopsis* and encode two of the three CDF-specific motifs (Sugiyama et al., 2012). They show a diurnal expression pattern and encode transcriptional repressors, similar to *Arabidopsis* CDFs, but disruption of both *PpDof3* and *PpDof4* by gene-targeted mutagenesis did not affect diurnal expression of the three *CO*-like genes of *P. patens* (Ishida et al., 2014). Moreover, targeted disruption of *PpDof1* induced delays in gametophore formation and development of the caulonema from the chloronema, as well as the formation of smaller colonies with a reduced frequency of branching of protonemal filaments. The extent of the change in branching frequency in colonies with disrupted *PpDof1* depended on the ratio or total amount of carbon and nitrogen nutrients and, therefore, PpDof1 is likely to control nutrient-dependent growth of filaments (Sugiyama et al., 2012).

Recently, it was suggested that CrDof, the sole Dof transcription factor present in *C. reinhardtii*, is involved in fatty acid metabolism, because its overexpression increased the amount of lipids around twofold (Ibáñez-Salazara et al., 2014). Therefore, the available evidence concerning the roles of Dof transcription factors from the CDF-containing clade suggests their functions are not well conserved between vascular and nonvascular plants.

12.4.2 The Roles of Dof Transcription Factors in the Regulation of Development and Differentiation

Arabidopsis OBP1 was initially identified as playing a role in the activation of *ocs* elements, which are responsive to auxin and salicylic acid (Zhang et al., 1995). It has since been shown to be involved in the control of cell division, as it regulates expression of the core cell cycle gene *CYCD3;3* and the replication-specific transcription factor gene *AtDOF2;3*. Short-term activation of OBP1 in cell cultures reduces the duration of both the G1 phase and the overall cell cycle length, whereas constitutive overexpression of OBP1 in plants produces a dwarfish phenotype resulting from reductions in cell size and number (Skirycz et al., 2008).

Several studies clearly indicate that several Dof transcription factors are involved in differentiation. Because many Dof transcription factor genes are expressed in vascular tissues (Le-Hir and Bellini, 2013) and some, including *AtDof2.1*, *AtDof2.4*, *AtDof4.6*, *AtDof5.3*, and *AtDof5.8*, are specifically expressed in the early stages of vascular development in *Arabidopsis* (Gardiner et al., 2010; Konishi and Yanagisawa, 2007), these genes are implicated in the control of vascular development. Indeed, *AtDof5.8*, a gene targeted by MONOPTEROS (also known as Auxin Response Factor 5, ARF5), is expressed at the preprocambial cell selection stage, and a mutation in *dof5.8* affects vascular patterning in the *arf5* mutant (Konishi et al., 2015).

HCA2/AtDof5.6 was also identified as a positive regulator promoting initiation of the interfascicular cambium from interfascicular parenchyma cells (Guo et al., 2009). *HCA2* is preferentially expressed in the vasculature of all organs, but particularly in the cambium, phloem, and interfascicular parenchyma cells of inflorescence stems. Elevated expression of *HCA2* increases cambial activity, whereas repression of *HCA2* leads to disruption of interfascicular cambium formation and development in inflorescence stems. The pattern and effect of *HCA2* expression indicate HCA2 is a positive regulator promoting the initiation of the interfascicular cambium from interfascicular parenchyma cells. *HCA2*, *AtDof5.8*, and other *Dof* genes specifically expressed in the early vascular development in *Arabidopsis* are; however, classified into different clades in the phylogenetic tree (Figure 12.4).

AtDOF4.2 and its close homolog AtDOF4.4 are implicated in the regulation of shoot branching and seed coat formation in *Arabidopsis*. Plants overexpressing *AtDOF4.2* exhibit a phenotype of increased branching, probably as a result of upregulation of three branching-related genes, *SHOOT MERISTEMLESS* (*STM*), encoding a member of the KNOTTED transcription factor family, *TERMINAL FLOWER 1* (*TFL1*), and *CYTOCHROME P450 83B1* (*CYP83B1*). They also produce seeds with a collapsed morphology, probably through the direct activation of *EXPANSINA9*, which is involved in loosening the cell wall. Although the *dof4.2* mutant does not exhibit obvious changes in branching and seed coat morphology, overexpression of *AtDof4.4* also increases shoot branching, suggesting that these genes act redundantly (Zou et al., 2013).

Kim et al. (2010) reported that AtDof5.1 regulates adaxial–abaxial polarity by modulating the expression levels of *REVOLUTA* (*REV*), a HD-ZIPIII transcription factor, which is necessary for the establishment of leaf polarity in *Arabidopsis*. An activation-tagged mutant overexpressing AtDof5.1 displayed an upward-curling leaf phenotype due to enhanced expression of *REV*, while constitutive overexpression of the DNA-binding domain of AtDof5.1 produced a downward-curling phenotype in plants, probably due to competitive binding of the truncated AtDof5.1 to DNA. Direct binding of AtDof5.1 to the *REV* promoter was verified experimentally using chromatin immunoprecipitation assays.

STOMATAL CARPENTER 1 (*SCAP1*), a gene expressed in maturing guard cells of *Arabidopsis* but not in guard mother cells, was recently found to be an essential regulator of stomatal guard cell maturation (Negi et al., 2013). The *scap1* mutant develops irregularly shaped guard cells, and this phenotype is accompanied by reductions in expression levels of the genes encoding the K^+ channel protein, the MYB60 transcription factor, and pectin methylesterase. These findings are consistent with the results of several previous studies, which indicated that Dof-binding sites contribute to guard cell-specific activity of the *AtMYB60* promoter (Cominelli et al., 2011) and that Dof-binding motifs are overrepresented in the promoter sequences of *Arabidopsis* genes expressed in stomatal guard cells, including those encoding CYTOCHROME P450 86A2 (CYP86A2) mono-oxygenase, the PLEIOTROPIC DRUG RESISTANCE 3 transporter, and PP2C protein phosphates (Galbiati et al., 2008). *SCAP1* function is probably conserved in a variety of vascular plants, including potato, as the guard-cell-specific gene expression of a potato K^+ channel gene (*KST1*) is regulated by two regions that contain a Dof-binding sequence (Plesch et al., 2001).

12.4.3 Regulation of Metabolism

As maize Dof1 activates the promoters of several genes associated with carbohydrate metabolism, including phosphoenolpyruvate carboxylase genes, in maize protoplasts (Yanagisawa and Sheen, 1998; Yanagisawa, 2000), it is a possible regulator of carbohydrate metabolism and therefore was used in metabolic engineering. Transgenic expression of maize *Dof1* in *Arabidopsis* enhanced nitrogen assimilation, probably *via* activation of the 2-oxogularate synthesis pathway that provides carbon skeletons for nitrogen assimilation (Yanagisawa et al., 2004). As this phenotype is an agriculturally important trait, several studies were examined to determine whether Dof transcription factors produced similar effects in other plant species. Altered carbohydrate metabolism and positive effects on nitrogen assimilation in response to transgenic *Dof* gene expression have been reported from potato (Tanaka et al., 2009), tobacco (Wang et al., 2013), and rice (Kurai et al., 2011), although such effects were not observed in poplar (Lin et al., 2013). Expression of a rice Dof transcription factor, *OsDof25*, a probable homolog of maize *Dof1*, altered carbon and nitrogen metabolism in *Arabidopsis* (Santos et al., 2012). Interestingly, constitutive expression of *SlCDF3*, a tomato CDF homolog, which affected stress responses, also produced positive effects on nitrogen metabolism and caused sucrose and certain amino acids to accumulate to higher levels in *Arabidopsis* (Corrales et al., 2014).

Maize Dof2 is closely related to maize Dof1 and thus it might also be involved in the regulation of primary metabolism. Dof2 is a bifunctional transcription factor that exhibits transcriptional activator or repressor activity depending on the nature of the promoter (Yanagisawa, 2000) and, in protoplasts, it can interfere with transcriptional activation by maize Dof1 (Yanagisawa and Sheen, 1998). Expression analysis of *EcDof1* and *EcDof2*,

Dof1 and *Dof2* homologs from finger millet, indicated that the *EcDof1/EcDof2* ratio is differentially modified in two genotypes of finger millet with different grain protein contents (high and low-protein genotypes). The *Dof1/Dof2* ratio may serve, therefore, as an index of the nitrogen utilization efficiency of crops (Gupta et al., 2014). Although the regulation of primary metabolism by these Dof transcription factors has not yet been conclusively proven, such studies indicate that they have the potential to modify primary metabolism.

On the other hand, PpDof5 from maritime pine (*Pinus pinaster*) is directly involved in regulating primary metabolism, as it differentially regulates the promoters of two glutamine synthetase genes, *GS1a* and *GS1b*, in photosynthetic and nonphotosynthetic tissues. In vascular cells, PpDof5 activates the promoter of *GS1b*, whose product is involved in generating glutamine for nitrogen transport and recycling the ammonium released in lignin biosynthesis, whereas, in the chlorophyllous parenchyma of photosynthetic tissues, PpDof5 acts as a transcriptional repressor of *GS1a*, which is expressed almost exclusively in this cell type and whose product is probably associated with ammonium assimilation in photosynthetic cells (Rueda-López et al., 2008). Wang et al. (2007) showed that overexpression in *Arabidopsis* of soybean *GmDof4* and *GmDof11* increased the total fatty acid content and lipid level in seeds by upregulating genes associated with fatty acid biosynthesis. Similarly, overexpression of *CrDof* increased the amount of lipids in *C. reinhardtii* by around twofold (Ibáñez-Salazara et al., 2014).

A role for Dof proteins in secondary metabolism regulation has been reported by Skirycz et al. (2006). They showed that *Arabidopsis* OBP2 is involved in the control of the biosynthesis of glucosinolates, a group of secondary metabolites that function as defense substances. Overexpression of *OBP2* increases expression of *CYP83B1*, which encodes cytochrome P450, an enzyme involved in the biogenesis of indole glucosinolates, whereas an RNA interference-mediated reduction in *OBP2* expression resulted in decreased *CYP83B1* expression. *AtDOF4.2*, recently implicated in shoot branching and seed coat formation in *Arabidopsis* (discussed in Section 12.4.2; Zou et al., 2013), is also involved in phenylpropanoid metabolism (Skirycz et al., 2007), and overexpression and RNAi-mediated silencing of *AtDOF4.2* have opposing effects upon expression levels of flavonoid biosynthesis-related genes and accumulation of flavonoids. Thus, Dof transcription factors are likely to be involved in the regulation of both primary and secondary metabolism.

12.4.4 Hormonal Regulation

A tobacco Dof transcription factor, NtBBF1, binds to the sequence ACTTTA, which is essential for auxin-regulated expression of the *rolB* oncogene (De Paolis et al., 1996; Baumann et al., 1999). Furthermore, the pumpkin AOBP binds to the silencer region of the pumpkin ascorbate oxidase gene, whose expression is induced by auxin (Kisu et al., 1998). These findings suggest Dof transcription factors may be involved in auxin-regulated gene expression, although further analyses have not yet been reported. By contrast, the involvement of Dof transcription factors in GA biosynthesis and GA-regulated gene expression has been analyzed in detail. As already described (Section 12.2.4.2), several Dof transcription factors are key players in GA-regulated gene expression in cereals. Furthermore, in *Arabidopsis*, DAG1, which is specifically expressed in the phloem, binds to and negatively regulates the promoter of *GA3b-hydroxylase* (*AtGA3ox1*), a gene encoding a key enzyme of GA biosynthesis (Gabriele et al., 2010). Furthermore, DAG1 cooperates with GA INSENSITIVE (GAI) in repressing *AtGA3ox1*; *GAI* is itself activated by PHYTOCHROME INTERACTING FACTOR 3-LIKE 5 (PIL5), the master repressor of light-mediated seed germination (Boccaccini et al., 2014). Consistent with its molecular function, the *dag1* mutant shows an altered response to red and far-red light and a delay in light-mediated seed germination (Papi et al., 2000; Papi et al., 2003). Thus, DAG1 is a negative regulator of phyB- and GA-dependent seed germination. DAG2, a close relative of DAG1 (Figure 12.4), has an opposing role in the regulation of seed germination downstream of PIL5 (Gualberti et al., 2002). It has been suggested that DAG1 also regulates seed germination *via* its effect on the expression of *EARLY LIGHT-INDUCED PROTEIN 1* and *2* (*ELIP1* and *ELIP2*) (Rizza et al., 2011).

12.4.5 Response to Light

The roles of Dof transcription factors in regulating the expression of storage protein genes during seed development (discussed in Section 12.2.4.1) and of GA-regulated gene expression in the aleurone layers of germinating seeds in cereals, including maize, wheat, barley, and rice (discussed in Section 12.2.4.2), have been well characterized in a number of studies (Diaz et al., 2002; Diaz et al., 2005; Dong et al., 2007; Isabel-LaMoneda et al., 2003; Martínez et al., 2005; Marzábal et al., 2008; Mena et al., 1998, 2002; Moreno-Risueno et al., 2007; Vicente-Carbajosa et al., 1997; Washio, 2001, 2003; Yamamoto et al., 2006; Zou et al. 2008). Another well-documented role for Dof transcription factors is the regulation of gene expression in response to light. As already discussed (Section 12.4.1.1), COG1 negatively regulates phytochrome signaling in *Arabidopsis* (Park et al., 2003). Another Dof domain protein, OBP3, whose overexpression results in growth defects (Kang and Singh, 2000; Kang et al., 2003), is also reported to modulate phytochrome and cryptochrome signaling in *Arabidopsis* (Ward et al., 2005). In the gain-of-function mutant, isolated as *suppressor of the phytochrome*

B (phyB) missense allele (sob1-D), overexpression of OBP3 suppressed the long-hypocotyl phenotype of the phyB missense allele. On the other hand, transgenic *RNAi* lines with reduced *OBP3* expression showed larger cotyledons, and the light-dependent cotyledon phenotype of the *OBP –RNAi* lines is most dramatic in blue light. Thus, OBP3 is likely to be a positive regulator of phyB-mediated inhibition of hypocotyl elongation as well as a negative regulator of cryptochrome-1-mediated cotyledon expansion, although the relationship between the OBP3 function and the previously observed growth defects of OBP3 overexpressers is unclear. It has also been suggested that JcDof1 from *J. curcas* is a light-responsive Dof transcription factor involved in the circadian clock (Yang et al., 2010).

12.5 PERSPECTIVE

A number of studies indicate that multiple rounds of gene duplication over the course of green plant evolution have created a plant-specific gene family of diverse Dof transcription factors, which show considerable structural diversification and now play various critical roles in plant differentiation and development. The structural and functional diversification of Dof transcription factors probably proceeded in parallel with the development of highly organized regulatory networks during the evolution of higher plants. In particular, the structural diversification of Dof transcription factors has enabled a great variety of interactions with other classes of transcription factors and thus produced sophisticated regulation of many pathways.

Approximately half of the Dof transcription factors known from *Arabidopsis* have been characterized to date (Figure 12.4) and so our knowledge on Dof transcription factors is still far from complete, largely because of the lack of systematic and cross-sectional analyses of Dof transcription factors from multiple plant species. It is, for example, currently uncertain whether the physiological roles of particular Dof transcription factors are conserved between different groups of higher plants. Thus, it will be necessary to characterize the roles played by Dof transcription factors in various plants systematically, as well as to uncover their novel functions. Such efforts will expand our understanding of how complex and sophisticated regulatory networks have arisen over the course of higher plant evolution. Finally, as some Dof transcription factors have the potential to enhance yield or metabolite production in plants, such understanding may be particularly important for developing novel applications in plant biotechnology.

Acknowledgments

The author acknowledges partial support from the Japan Society for the Promotion of Science (grants no. 25252014 and 26221103).

References

Baumann, K., De Paolis, A., Costantino, P., Gualberti, G., 1999. The DNA binding site of the Dof protein NtBBF1 is essential for tissue-specific and auxin-regulated expression of the *rolB* oncogene in plants. Plant Cell 11, 323–333.

Boccaccini, A., Santopolo, S., Capauto, D., Lorrai, R., Minutello, E., Serino, G., et al., 2014. The DOF protein DAG1 and the DELLA protein GAI cooperate in negatively regulating the AtGA3ox1 gene. Mol. Plant 7, 1486–1489.

Bhullar, S., Datta, S., Advani, S., Chakravarthy, S., Gautam, T., Pental, D., et al., 2007. Functional analysis of cauliflower mosaic virus 35S promoter: re-evaluation of the role of subdomains B5, B4 and B2 in promoter activity. Plant Biotechnol. J. 5, 696–708.

bZIP Research Group, 2002. bZIP transcription factors in *Arabidopsis*. Trends Plant Sci 7, 106–111.

Cai, X., Zhang, Y., Zhang, C., Zhang, T., Hu, T., Ye, J., et al., 2013. Genome-wide analysis of plant-specific Dof transcription factor family in tomato. J. Integr. Plant Biol. 55, 552–566.

Cavalar, M., Möller, C., Offermann, S., Krohn, N.M., Grasser, K.D., Peterhänsel, C., 2003. The interaction of DOF transcription factors with nucleosomes depends on the positioning of the binding site and is facilitated by maize HMGB5. Biochemistry 42, 2149–2157.

Chen, H., Ahmad, M., Rim, Y.J., Lucas, W.J., Kim, J.-Y., 2013. Evolutionary and molecular analysis of Dof transcription factors identified a conserved motif for intercellular protein trafficking. New Phytol. 198, 1250–1260.

Chen, W., Chao, G., Singh, K.B., 1996. The promoter of an H2O2-inducible, *Arabidopsis* glutathione S-transferase gene contains closely linked OBF and OBP1-binding sites. Plant J. 10, 955–966.

Cominelli, E., Galbiati, M., Albertini, A., Fornara, F., Conti, L., Coupland, G., Tonelli, C., 2011. DOF-binding sites additively contribute to guard cell-specificity of AtMYB60 promoter. BMC Plant Biol. 11, 162.

Corrales, A.R., Nebauer, S.G., Carrillo, L., Fernández-Nohales, P., Marqués, J., Renau-Morata, B., et al., 2014. Characterization of tomato Cycling Dof Factors reveals conserved and new functions in the control of flowering time and abiotic stress responses. J. Exp. Bot. 65, 995–1012.

De Paolis, A., Sabatini, S., De Pascalis, L., Costantino, P., Capone, I., 1996. A *rolB* regulatory factor belongs to a new class of single zinc finger plant proteins. Plant J. 10, 215–223.

Diaz, I., Martinez, M., Isabel-LaMoneda, I., Rubio-Somoza, I., Carbonero, P., 2005. The DOF protein, SAD, interacts with GAMYB in plant nuclei and activates transcription of endosperm-specific genes during barley seed development. Plant J. 42, 652–662.

Diaz, I., Vicente-Carbajosa, J., Abraham, Z., Martínez, M., Moneda, I.I.-L., Carbonero, P., 2002. The GAMYB protein from barley interacts with the DOF transcription factor BPBF and activates endosperm-specific genes during seed development. Plant J. 29, 453–464.

Dong, G., Ni, Z., Yao, Y., Nie, X., Sun, Q., 2007. Wheat Dof transcription factor WPBF interacts with TaQM and activates transcription of an alpha-gliadin gene during wheat seed development. Plant Mol. Biol. 63, 73–84.

Elrouby, N., Bonequi, M.V., Porri, A., Coupland, G., 2012. Identification of *Arabidopsis* SUMO-interacting proteins that regulate chromatin activity and developmental transitions. Proc. Natl. Acad. Sci. USA 110, 19956–19961.

Fornara, F., Panigrahi, K.C.S., Gissot, L., Sauerbrunn, N., Rühl, M., Jarillo, J.A., et al., 2009. *Arabidopsis* DOF transcription factors act redundantly to reduce *CONSTANS* expression and are essential for a photoperiodic flowering response. Dev. Cell 17, 75–86.

Gabriele, S., Rizza, A., Martone, J., Circelli, P., Costantino, P., Vittorioso, P., 2010. The Dof protein DAG1 mediates PIL5 activity on seed germination by negatively regulating GA biosynthetic gene *AtGA3ox1*. Plant J. 61, 312–323.

Galbiati, M., Simoni, L., Pavesi, G, Cominelli, E, Francia, P., et al., 2008. Gene trap lines identify *Arabidopsis* genes expressed in stomatal guard cells. Plant J. 53, 750–762.

Gardiner, J., Sherr, I., Scarpella, E., 2010. Expression of DOF genes identifies early stages of vascular development in *Arabidopsis* leaves. Int. J. Dev. Biol. 54, 1389–1396.

Gualberti, G., Papi, M., Bellucci, L., Ricci, I., Bouchez, D., Camilleri, C., et al., 2002. Mutations in the Dof zinc finger genes *DAG2* and *DAG1* influence with opposite effects the germination of *Arabidopsis* seeds. Plant Cell 14, 1253–1263.

Guo, Y., Qin, G., Gu, H., Qua, L.J., 2009. Dof5.6/HCA2, a Dof transcription factor gene, regulates interfascicular cambium formation and vascular tissue development in *Arabidopsis*. Plant Cell 21, 3518–3534.

Gupta, S., Gupta, S.M., Gupta, A.K., Singh Gaur, V.S., Kumar, A., 2014. Fluctuation of Dof1/Dof2 expression ratio under the influence of varying nitrogen and light conditions: involvement in differential regulation of nitrogen metabolism in two genotypes of finger millet (*Eleusine coracana* L.). Gene 546, 327–335.

Hernando-Amado, S., González-Calle, V., Carbonero, P., Barrero-Sicilia, C., 2012. The family of DOF transcription factors in *Brachypodium distachyon*: phylogenetic comparison with rice and barley DOFs and expression profiling. BMC Plant Biol. 12, 202.

Ibáñez-Salazara, A., et al., 2014. Over-expression of Dof-type transcription factor increases lipid production in *Chlamydomonas reinhardtii*. J. Biotechnol. 184, 27–38.

Imaizumi, T., Kay, S.A., 2006. Photoperiodic control of flowering: not only by coincidence. Trends Plant Sci. 11, 550–558.

Imaizumi, T., Schultz, T.F., Harmon, F.G., Ho, L.A., Kay, S.K., 2005. FKF1 F-box protein mediates cyclic degradation of a repressor of *CONSTANS* in *Arabidopsis*. Science 309, 293–297.

Isabel-LaMoneda, I., Diaz, I., Martinez, M., Mena, M., Carbonero, P., 2003. SAD: a new DOF protein from barley that activates transcription of a cathepsin B-like thiol protease gene in the aleurone of germinating seeds. Plant J. 33, 329–340.

Ishida, T., Sugiyama, T., Tabei, N., Yanagisawa, S., 2014. Diurnal expression of *CONSTANS-like genes* is independent of the function of cycling DOF factor (CDF)-like transcriptional repressors in *Physcomitrella patens*. Plant Biotechnol. 31, 293–299.

Iwamoto, M., Higo, K., Takano, M., 2009. Circadian clock- and phytochrome-regulated Dof-like gene, *Rdd1*, is associated with grain size in rice. Plant Cell Environ. 32, 592–603.

Jacquemin, J., Ammiraju, J.S.S., Haberer, G., Billheimer, D.D., Yu, Y., Liu, L.C., et al., 2014. Fifteen million years of evolution in the *Oryza* genus shows extensive gene family expansion. Mol. Plant 7, 642–656.

Jiang, Y., Zeng, B., Zhao, H., Zhang, M., Xie, S., Lai, J., 2012. Genome-wide transcription factor gene prediction and their expressional tissue-specificities in maize. J. Integr. Plant Biol. 54, 616–630.

Kang, H.-G., Singh, K.B., 2000. Characterization of salicylic acid-responsive, *Arabidopsis* Dof domain proteins: overexpression of OBP3 leads to growth defects. Plant J. 21, 329–339.

Kang, H.G., Foley, R.C., Oñate-Sánchez, L., Lin, C., Singh, K.B., 2003. Target genes for OBP3, a Dof transcription factor, include novel basic helix–loop–helix domain proteins inducible by salicylic acid. Plant J. 35, 362–372.

Kim, H.S., Jin Kim, S.J., Abbasi, N., Bressan, R.A., Yun, D.J., Yoo, D.J., et al., 2010. The DOF transcription factor Dof5.1 influences leaf axial patterning by promoting *Revoluta* transcription in *Arabidopsis*. Plant J. 64, 524–535.

Kisu, Y., Ono, T., Shimofurutani, N., Suzuki, M., Esaka, M., 1998. Characterization and expression of a new class of zinc finger protein that binds to silencer region of ascorbate oxidase gene. Plant Cell Physiol. 39, 1054–1064.

Kloosterman, B., Abelenda, J.A., Carretero-Gomez, M.D.M., Oortwijn, M., de Boer, J.M., et al., 2013. Naturally occurring allele diversity allows potato cultivation in northern latitudes. Nature 495, 246–250.

Konishi, M., Yanagisawa, S., 2007. Sequential activation of two Dof transcription factor gene promoters during vascular development in *Arabidopsis thaliana*. Plant Physiol. Biochem. 45, 623–629.

Konishi, M., Donner, T.J., Scarpella, E., Yanagisawa, S.S., 2015. MONOPTEROS directly activates the auxin-inducible promoter of Dof5.8 transcription factor gene in *Arabidopsis thaliana* leaf provascular cells. J. Exp. Bot. 66, 283–291.

Krebs, J., Mueller-Roeber, B., Ruzicic, S., 2010. A novel bipartite nuclear localization signal with anatypically long linker in DOF transcription factors. J. Plant Physiol. 167, 583–586.

Krohn, N.M., Yanagisawa, S., Grasser, K.D., 2002. Specificity of the stimulatory interaction between chromosomal HMGB proteins and the transcription factor Dof2 and its negative regulation by protein kinase CK2-mediated phosphorylation. J. Biol. Chem. 277, 32438–32444.

Kurai, T., Wakayama, M., Abiko, T., Yanagisawa, S., Aoki, N., Ohsugi, R., 2011. Introduction of ZmDof1 gene into rice enhances carbon and nitrogen assimilation under low nitrogen condition. Plant Biotechnol. J. 9, 826–837.

Kushwaha, H., Gupta, S., Singh, V.K., Rastogi, S., Yadav, D., 2011. Genome wide identification of *Dof* transcription factor gene family in sorghum and its comparative phylogenetic analysis with rice and *Arabidopsis*. Mol. Biol. Rep. 38, 5037–5053.

Kushwaha, H., Gupta, S., Singh, V.K., Bisht, N.C., Sarangi, B.K., Yadav, D., 2013. Cloning, *in silico* characterization and prediction of three dimensional structure of SbDof1, SbDof19, SbDof23 and SbDof24 proteins from sorghum [*Sorghum bicolor* (L.) Moench]. Mol. Biotechnol. 54, 1–12.

Le-Hir, R., Bellini, C., 2013. The plant-specific Dof transcription factors family: new players involved in vascular system development and functioning in *Arabidopsis*. Front. Plant Sci. 4, 164.

Li, D., Yang, C., Li, X., Gan, Q., Zhao, X., Zhu, L., 2009. Functional characterization of rice *OsDof12*. Planta 229, 1159–1169.

Lijavetzky, D., Carbonero, P., Vicente-Carbajosa, J., 2003. Genome-wide comparative phylogenetic analysis of the rice and *Arabidopsis* Dof gene families. BMC Evol. Biol. 3, 17.

Lin, W., Hagen, E., Fulcher, A., Hren, M.T., Cheng, Z.-M., 2013. Overexpressing the *ZmDof1* gene in *Populus* does not improve growth and nitrogen assimilation under low-nitrogen conditions. Plant Cell Tissue Organ Cult. 113, 51–61.

Martínez, M., Rubio-Somoza, I., Fuentes, R., Lara, P., Carbonero, P., Díaz, I., 2005. The barley cystatin gene (*Icy*) is regulated by DOF transcription factors in aleurone cells upon germination. J. Exp. Bot. 56, 547–556.

Marzábal, P., Gas, E., Fontanet, P., Vicente-Carbajosa, J., Torrent, M., Ludevid, M.D., 2008. The maize Dof protein PBF activates transcription of gamma-zein during maize seed development. Plant Mol. Biol. 67, 441–454.

Mena, M., Cejudo, F.J., Isabel-Lamoneda, I., Carbonero, P., 2002. A role for the Dof transcription factor BPBF in the regulation of gibberellin-responsive genes in barley aleurone. Plant Physiol. 130, 111–119.

Mena, M., Vicente-Carbajosa, J., Schmidt, R.J., Carbonero, P., 1998. An endosperm-specific DOF protein from barley, highly conserved in wheat, binds to and activates transcription from the prolamin-box of a native B-hordein promoter in barley endosperm. Plant J. 16, 53–62.

Moreno-Risueno, M.A., Martinéz, M., Vicente-Carbajosa, J., Carbonero, P., 2007. The family of DOF transcription factors: from green unicellular algae to vascular plants. Mol. Genet. Genomics 277, 379–390.

Negi, J., Moriwaki, K., Konishi, M., Yokoyama, R., Nakano, T., Kusumi, K., et al., 2013. A Dof transcription factor, SCAP1, is essential for the development of functional stomata in *Arabidopsis*. Curr. Biol. 23, 479–484.

Noguero, M., Atif, R.M., Ochatt, S., Richard, D., Thompson, R.D., 2013. The role of the DNA-binding One Zinc Finger (DOF) transcription factor family in plants. Plant Sci. 209, 32–45.

Papi, M., Sabatini, S., Altamura, M.M., Hennig, L., Schäfer, E., Costantino, P., Vittorioso, P., 2003. Inactivation of the phloem-specific Dof zinc finger gene *dag1* affects response to light and integrity of the testa of *Arabidopsis* seeds. Plant Physiol. 128, 411–417.

Papi, M., Sabatini, S., Bouchez, D., Camilleri, C., Costantino, P., Vittorioso, P., 2000. Identification and disruption of an *Arabidopsis* zinc finger gene controlling seed germination. Genes Dev. 14, 28–33.

Park, D.H., Lim, P.O., Kim, J.S., Cho, D.S., Hong, S.H., Nam, H.G., 2003. The *Arabidopsis COG1* gene encodes a Dof domain transcription factor and negatively regulates phytochrome signaling. Plant J. 34, 161–171.

Plesch, G., Ehrhardt, T., Mueller-Roeber, B., 2001. Involvement of TAAAG elements suggests a role for Dof transcription factors in guard cell-specific gene expression. Plant J. 28, 455–464.

Rizza, A., Boccaccini, A., Lopez-Vidriero, I., Costantino, P., Vittorioso, P., 2011. Inactivation of the *ELIP1* and *ELIP2* genes affects *Arabidopsis* seed germination. New Phytol. 190, 896–905.

Romeuf, I., Tessier, D., Dardevet, M., Branlard, G., Charmet, G., Ravel, C., 2010. wDBTF: an integrated database resource for studying wheat transcription factor families. BMC Genomics 11, 185.

Rueda-López, M., Crespillo, R., Cánovas, F.M., Ávila, C., 2008. Differential regulation of two glutamine synthetase genes by a single Dof transcription factor. Plant J. 56, 73–85.

Rueda-Romero, P., Barrero-Sicilia, C., Gomez-Cadenas, A., Carbonero, P., Onáte-Sanchez, L., 2012. *Arabidopsis thaliana* DOF6 negatively affects germination in non-after-ripened seeds and interacts with TCP14. J. Exp. Bot. 63, 1937–1949.

Santos, L.A., de Souza, S.R., Fernandes, M.S., 2012. OsDof25 expression alters carbon and nitrogen metabolism in *Arabidopsis* under high N-supply. Plant Biotechnol. Rep. 6, 327–337.

Sawa, M., Nusinow, D.A., Kay, S.A., Imaizumi, T., 2007. FKF1 and GIGANTEA complex formation is required for day-length measurement in *Arabidopsis*. Science 318, 261–265.

Shaw, L.M., McIntyre, C.L., Gresshoff, P.M., Xue, G.P., 2009. Members of the Dof transcription factor family in *Triticum aestivum* are associated with light-mediated gene regulation. Funct. Integr. Genomics 9, 485–498.

Shigyo, M., Tabei, N., Yoneyama, T., Yanagisawa, S., 2007. Evolutionary processes during the formation of the plant-specific Dof transcription factor family. Plant Cell Physiol. 48, 179–185.

Shimofurutani, N., Kisu, Y., Suzuki, M., Esaka, M., 1998. Functional analyses of the Dof domain, a zinc finger DNA-binding domain, in a pumpkin DNA-binding protein AOBP. FEBS Lett. 430, 251–256.

Skirycz, A., Jozefczuk, S., Stobiecki, M., Muth, D., Zanor, M.I., Witt, I., et al., 2007. Transcription factor AtDOF4;2 affects phenylpropanoid metabolism in *Arabidopsis thaliana*. New Phytol. 175, 425–438.

Skirycz, A., Radziejwoski, A., Busch, W., Hannah, M.A., Czeszejko, J., Kwasniewski, M., et al., 2008. The DOF transcription factor OBP1 is involved in cell cycle regulation in *Arabidopsis thaliana*. Plant J. 56, 779–792.

Skirycz, A., Reichelt, M., Burow, M., Birkemeyer, C., Rolcik, J., Kopka, J., et al., 2006. DOF transcription factor AtDof1.1 (OBP2) is part of a regulatory network controlling glucosinolate biosynthesis in *Arabidopsis*. Plant J. 47, 10–24.

Song, Y.H., Smith, R.W., To, B.J., Millar, A.J., Imaizumi, T., 2012. FKF1 conveys crucial timing information for CONSTANS stabilization in the photoperiodic flowering. Science 336, 1045–1049.

Sugiyama, T., Ishida, T., Tabei, N., Shigyo, M., Konishi, M., Yoneyama, T., et al., 2012. Involvement of PpDof1 transcriptional repressor in the nutrient condition-dependent growth control of protonemal filaments in *Physcomitrella patens*. J. Exp. Bot. 63, 3185–3197.

Tanaka, M., Yasuhiro Takahata, Y., Nakayama, H., Nakatani, M., Tahara, M., 2009. Altered carbohydrate metabolism in the storage roots of sweet potato plants overexpressing the SRF1 gene, which encodes a Dof zinc finger transcription factor. Planta 230, 737–746.

Ülker, B., Somssich, I.E., 2004. WRKY transcription factors: from DNA binding towards biological function. Curr. Opin. Plant Biol. 7, 491–498.

Umemura, Y., Ishiduka, T., Yamamoto, R., Esaka, M., 2004. The Dof domain, a zinc finger DNA-binding domain conserved only in higher plants, truly functions as a Cys2/Cys2 Zn finger domain. Plant J. 7, 741–749.

Vicente-Carbajosa, J., Moose, S.P., Parsons, R., Schmidt, R.J., 1997. A maize zinc-finger protein binds the prolamin box in zein gene promoters and interacts with the basic leucine zipper transcriptional activator Opaque2. Proc. Natl. Acad. Sci. USA 94, 7685–7690.

Wang, H.-W., Zhang, B., Hao, Y.-J., Huang, J., Tian, A.-G., Liao, Y., et al., 2007. The soybean Dof-type transcription factor genes, *GmDof4* and *GmDof11*, enhance lipid content in the seeds of transgenic *Arabidopsis* plants. Plant J. 52, 716–729.

Wang, Y., Fu, B., Pan, L., Chen, L., Fu, X., Kunzhi Li, K., 2013. Overexpression of *Arabidopsis Dof1, GS1* and *GS2* enhanced nitrogen assimilation in transgenic tobacco grown under low-nitrogen conditions. Plant Mol. Biol. Rep. 31, 886–900.

Ward, J.M., Cufr, C.A., Denzel, M.A., Neff, M.M., 2005. The Dof transcription factor OBP3 modulates phytochrome and cryptochrome signaling in *Arabidopsis*. Plant Cell 217, 475–485.

Washio, K., 2001. Identification of Dof proteins with implication in the gibberellin-regulated expression of a peptidase gene following the germination of rice grains. Biochim. Biophys. Acta 1520, 54–62.

Washio, K., 2003. Functional dissections between GAMYB and Dof transcription factors suggest a role for protein–protein associations in the gibberellin-mediated expression of the *RAmy1A* gene in the rice aleurone. Plant Physiol. 133, 850–863.

Wei, P.-C., Tan, F., Gao, X.-Q., Zhang, X.-Q., Wang, G.-Q., Xu, H., et al., 2010. Overexpression of AtDOF4.7, an *Arabidopsis* DOF family transcription factor, induces floral organ abscission deficiency in *Arabidopsis*. Plant Physiol. 153, 1031–1045.

Yamamoto, M.P., Onodera, Y., Touno, S.M., Takaiwa, F., 2006. Synergism between RPBF Dof and RISBZ1 bZIP activators in the regulation of rice seed expression genes. Plant Physiol. 141, 1694–1707.

Yang, J., Yang, M.-F., Wang, D., Chen, F., Shen, S.-H., 2010. *JcDof1*, a Dof transcription factor gene, is associated with the light-mediated circadian clock in *Jatropha curcas*. Plant Physiol. 139, 324–334.

Yang, J., Yang, M.F., Zhanga, W.P., Chen, F., Shen, S.-H., 2011. A putative flowering-time-related Dof transcription factor gene, *JcDof3*, is controlled by the circadian clock in *Jatropha curcas*. Plant Sci. 181, 667–674.

Yang, X., Tuskan, G.A., Cheng, Z.-M., 2006. Divergence of the Dof gene families in poplar, *Arabidopsis*, and rice suggests multiple modes of gene evolution after duplication. Plant Physiol. 142, 820–830.

Yanagisawa, S., 1995. A novel DNA binding domain that may form a single zinc finger motif. Nucleic Acid Res. 23, 3403–3410.

Yanagisawa, S., 1996. Dof DNA binding domains contain a novel zinc finger motif. Trends Plant Sci. 1, 213–214.

Yanagisawa, S., 1997. Dof DNA-binding domains of plant transcription factors contribute to multiple protein–protein interactions. Eur. J. Biochem. 250, 403–410.

Yanagisawa, S., 2000. Dof1 and Dof2 transcription factors are associated with expression of multiple genes involved in carbon metabolism in maize. Plant J. 21, 281–288.

Yanagisawa, S., 2001. The transcriptional activation domain of the plant-specific Dof1 factor functions in plant, animal, and yeast cells. Plant Cell Physiol. 42, 813–822.

Yanagisawa, S., 2002. The Dof family of plant transcription factors. Trends Plant Sci. 7, 555–560.

Yanagisawa, S., 2004. Dof domain proteins: plant-specific transcription factors associated with diverse phenomena unique to plants. Plant Cell Physiol. 45, 386–391.

Yanagisawa, S., Akiyama, A., Kisaka, H., Uchimiya, H., Miwa, T., 2004. Metabolic engineering with Dof1 transcription factor in plants: improved nitrogen assimilation and growth under low nitrogen conditions. Proc. Natl. Acad. Sci. USA 101, 7833–7838.

Yanagisawa, S., Izui, K., 1993. Molecular cloning of two DNA binding proteins of maize that are structurally different but interact with the same sequence motif. J. Biol. Chem. 268, 16028–16036.

Yanagisawa, S., Schmidt, R.J., 1999. Diversity and similarity among recognition sequences of Dof transcription factors. Plant J. 17, 209–214.

Yanagisawa, S., Sheen, J., 1998. Involvement of maize Dof zinc finger proteins in tissue-specific and light-regulated gene expression. Plant Cell 10, 75–89.

Zhang, B., Chen, W., Foley, R.C., Büttner, M., Singh, K.B., 1995. Interactions between distinct types of DNA binding proteins enhance binding to *ocs* element promoter sequences. Plant Cell 7, 2241–2252.

Zou, H.-F., Zhang, Y.-Q., Wei, W., Chen, H.-W., Song, Q.-X., Liu, Y.-F., et al., 2013. The transcription factor AtDOF4.2 regulates shoot branching and seed coat formation in *Arabidopsis*. Biochem. J. 449, 373–388.

Zou, X., Neuman, D., Shen, Q.J., 2008. Interactions of two transcriptional repressors and two transcriptional activators in modulating gibberellin signaling in aleurone cells. Plant Physiol. 148, 176–186.

CHAPTER

13

NAC Transcription Factors: From Structure to Function in Stress-Associated Networks

Ditte H. Welner*, Farah Deeba**,†, Leila Lo Leggio‡, Karen Skriver**

*Joint BioEnergy Institute, Lawrence Berkeley National Laboratory, Berkeley, CA, USA
**Department of Biology, University of Copenhagen, Copenhagen, Denmark
†Department of Biochemistry, PMAS Arid Agriculture University Rawalpindi, Rawalpindi, Pakistan
‡Department of Chemistry, University of Copenhagen, Copenhagen, Denmark

OUTLINE

13.1 Introduction	199
13.2 NAC Structure	200
13.2.1 The C-Terminal Domain	200
13.2.2 The NAC Domain Fold	200
13.2.3 The Complex with DNA	200
13.2.4 The NAC Dimer	202
13.3 Evolution of NAC Proteins	202
13.3.1 Unraveling NAC Evolution	202
13.3.2 Evolution of the NAC Fold	202
13.4 NAC Proteins: From Structure to Interactions with DNA and Other Proteins	204
13.4.1 DNA Sequence Specificity of the NAC Domain and Target Sequences	204
13.4.2 Structural Basis for DNA Recognition and Specificity of NAC Proteins	204
13.4.3 Direct Interactions of NAC Proteins with Other Proteins	205
13.5 NAC Networks in Abiotic Stress Responses	207
13.5.1 Networks of Stress-Associated ANAC019, ANAC055, and ANAC072	207
13.5.2 Networks of Stress-Associated Rice NAC Proteins	208
13.5.3 Other Transcription Factors in the ANAC019/055/072 Network	209
13.6 Conclusion	209
References	210

13.1 INTRODUCTION

NAM/ATAF1/CUC2 (NAC) proteins constitute a large and widespread transcription factor (TF) family with more than 100 representatives in many plants including the important model organisms *Arabidopsis thaliana* and rice (reviewed in Nakashima et al., 2012; Puranik et al., 2012). As of July 2014, Pfam (protein family; Punta et al., 2014) family 2365 contains 2410 NAC sequences distributed over 114 land plant species. The phylogeny and evolution of NAC proteins will be further described in Section 13.3.

The remarkable diversification and amplification of NAC genes across the plant kingdom reflect their numerous functions. These include plant development, senescence, nutrient distribution, cell wall biosynthesis, and abiotic and biotic stress responses. The diverse roles are thought to be mediated through homo- and heterodimerization within complex networks involving multiple binding partners, miRNA-mediated regulation, alternative splicing and posttranslational modifications such as phosphorylation (reviewed in Nakashima et al., 2012; Puranik et al., 2012; Lindemose et al., 2013). NAC networks in stress responses will be described in Section 13.5 in relation to other TFs discussed in more detail in Section 13.6. The central roles especially in stress responses and wall biosynthesis show great potential for NAC protein tailoring for applications within crop optimization for food, feed, and biofuel production (reviewed in Tran et al., 2010; Puranik et al., 2012; Hussey et al., 2013).

NAC proteins have a modular organization. In the majority, an N-terminal ~150-amino-acid conserved DNA-binding domain (DBD) termed the NAC domain is linked to a longer C-terminal domain (CTD) with transcriptional regulatory activity, although variations from this predominant domain organization have been reported (Puranik et al., 2012). The structural features of NAC proteins as well as of the complex with DNA will be described in detail in Section 13.2, while structure-derived knowledge of DNA-binding specificities and protein–protein interaction properties will be discussed in Section 13.4 together with related biochemical data.

13.2 NAC STRUCTURE

In order to understand and engineer NAC proteins, detailed knowledge on the molecular level is needed. Atomic resolution structural information is a powerful tool to obtain this knowledge.

While the NAC domain has been structurally characterized, the CTD has proven less amenable to such studies. This section will therefore start out by summarizing the sparse existing data on the CTD, followed by a longer subsection on the NAC domain, emphasizing our current structural knowledge base, and the functional aspects that can be learned from it.

13.2.1 The C-Terminal Domain

Canonical NAC proteins contain a highly variable C-terminal transcription regulation region with a high content of low-complexity amino acid repeats (Jensen et al., 2010). In addition to its transcription regulatory activity, it has been shown to be involved in protein–protein interactions (Ren et al., 2005; Kim et al., 2007; Jeong et al., 2009; Kleinow et al., 2009; Kjaersgaard et al., 2011). Although sequence motifs indicating the presence of enzymatic domains such as hydrolases and proteases can be found bioinformatically in some NAC proteins, to the best of our knowledge this has not been shown experimentally. A transmembrane domain in a subset of NAC proteins, the NAC with transmembrane motif 1 (NTM1)-like (NTL) proteins, has; however, been identified, and membrane localization as well as translocation to the nucleus upon stress-induced proteolytic cleavage have been reported (reviewed in Seo et al., 2008). Recent *in silico* analyses indicated intrinsic disorder in the C-termini of *Arabidopsis* NAC proteins (Jensen et al., 2010) and this has been confirmed biophysically for two barley NAC proteins, *Hv*NAC005 and *Hv*NAC013 (Kjaersgaard et al., 2011).

13.2.2 The NAC Domain Fold

The first crystallographic structure of a NAC domain was reported 10 years ago and has, in spite of the extensive scientific interest in the NAC field, only been followed by one other. These structures represent the conserved NAC domains from the *Arabidopsis* ANAC019 protein (Ernst et al., 2004) and the rice SNAC1 protein (Chen et al., 2011). The two domains contain 168 and 174 residues, respectively, and have a sequence similarity of 64%. With a positional root mean square deviation of 1.1 Å over the 485 backbone atoms in a structural alignment, the folds of the two domains are very similar and can be expected to represent a canonical NAC domain fold. Recently, the structure of the AtNAC1 NAC domain has been modeled using the ANAC019 structure as a template (Zhu et al., 2014).

The structure showed no well-known DNA-binding motif at the time. Instead, the fold is a 7-stranded antiparallel twisted β-sheet flanked by an α-helical element on either side of the strand (Figure 13.1 and Figure 13.2). The N-terminus is in an extended conformation and poorly defined in all structures, except where involved in protein–protein contacts.

13.2.3 The Complex with DNA

It was not possible from the first structures to determine the site or mode of DNA binding, although it was noted that one face of the structure was particularly positively charged and therefore well suited for recognizing the overall negatively charged DNA molecule (Ernst et al., 2004). This was recently confirmed to indeed be the interaction surface, when the crystallographic structure of the ANAC019 NAC domain in complex with a double-stranded oligonucleotide was solved to 4.1 Å (Figure 13.1; Welner et al., 2012). It shows the NAC domain inserting the outer β3-strand (numbering from Ernst et al., 2004) of the core β-sheet into the major groove near-perpendicularly to the DNA helical axis. The stress-responsive NAC1 (SNAC1) structure (Chen et al., 2011) is compatible with this and shows an overall positively charged resulting interaction surface area. A recent docking study of an AtNAC1 *in silico* model shows a similar DNA-binding mode (Zhu et al., 2014). In all, it is likely that this DNA recognition mode is common for NAC proteins.

Due to the limited resolution of the ANAC019–DNA complex structure, individual amino acid–nucleotide

13.2 NAC STRUCTURE

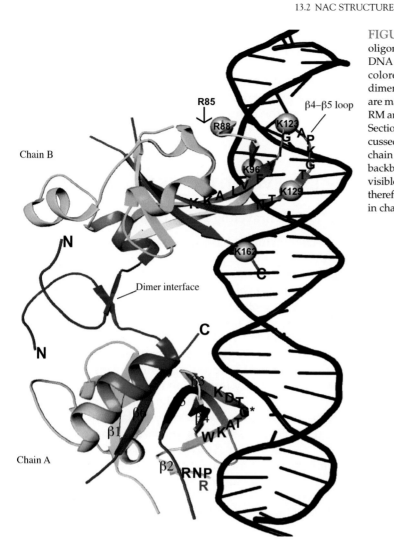

FIGURE 13.1 **The NAC–DNA complex.** The ANAC019–26 bp oligonucleotide complex structure (PDB ID: 3SWP) is shown with DNA (black) bound to one NAC dimer. Each NAC chain is rainbow colored from the N-terminus (blue) to the C-terminus (red); the dimer interface is also indicated. The β-strands of the core β-sheet are marked in chain A using numbering from Ernst et al. (2004). The RM and the four-residue region implicated in DNA binding (RPNR, Section 13.4) are highlighted as a one-letter code in chain A. The discussed hydrophobic and basic loop β4–β5 is similarly highlighted in chain B. In addition, residues that putatively interact with the DNA backbone (see Section 13.4) are shown as gray spheres. Arg85 is not visible in the X-ray data, and the general position of this residue is therefore indicated by a gray one-letter code in chain A, and an arrow in chain B. PDB, Protein Data Bank.

contacts cannot be probed. However, it is a credible interpretation that the β3-strand that protrudes into the major groove (Figure 13.1) and contacts the sugars/bases of the DNA confers specificity, while basic residues found in the DNA-backbone-binding distance in adjacent strands and loops contribute general affinity to the interaction (Welner et al., 2012; Zhu et al., 2014), as will be discussed further in Section 13.4. The residues of the contacting β3-strand, which has the recognition motif (RM) 95-WKATGTDK, are highly conserved except for the second threonine and will be further discussed in Sections 13.3 and 13.4.

A conserved glycine residue in the RM (Gly-99 in ANAC019) has been noted to introduce an unusual curvature in the β3-strand (Yamasaki et al., 2005), and this can be seen in the complex structure bringing it deeper into the major groove (Figure 13.1). An *Arabidopsis* mutant, *sog1-1*, which has an aberrant genomic stress response, has been reported to encode a malfunctioning NAC protein, the suppressor of gamma response 1 (SOG1) where this very glycine is replaced by a valine (Yoshiyama et al., 2009).

This stipulates the importance of Gly-99, which will be further discussed in Section 13.3.

A sequence of conserved basic and hydrophobic residues located in the β4-strand and the β4–β5 loop has received much attention (Figure 13.1). The loop was highlighted early on as a possible DNA-binding site (Duval et al., 2002, Ernst et al., 2004). The ANAC019–DNA complex structure showed that the loop, as well as part of the β4-strand, were indeed within hydrogen-bonding distance of the DNA backbone, and the conserved basic amino acids therefore possibly play a role in increasing the affinity of NAC–DNA recognition (Welner et al., 2012). However, at least two other possible explanations for conservation of these residues have been presented. These residues of ANAC019 (114-KKALVFYIGKAPKGTKTN) have been suggested to constitute a nuclear localization signal (Greve et al., 2003), and NAC proteins have generally been found by *in silico* analyses to carry nuclear localization signals at this position in their NAC domains (reviewed in Olsen et al., 2005a; Le et al., 2011). Putative signal sequences have, to the best of our knowledge, not been probed experimentally. The second functional suggestion for the β4–β5 region is that it contains a stretch of hydrophobic amino acids (LVFY) that could be part of a transcriptional repressor domain (Hao et al., 2010). Curiously, some NAC proteins have been shown to be better transcriptional activators in the absence of the NAC domain, and Hao et al. (2010) suggest this is the result of repressive action of the hydrophobic LVFY sequence, based on deletion mutants of soybean NAC proteins in a yeast transcriptional activation assay. In conclusion, it is not possible with the existing data to dissect the molecular function of disputed amino acids. It is conceivable that the conservation and observed functional effects can be due to a role in both cellular localization, DNA binding, and/or transcriptional repression, or even a complex interplay of these.

13.2.4 The NAC Dimer

Several NAC proteins have been shown to form homo- and heterodimers *in vitro* and this dimerization has been found to be necessary for high-affinity *in vitro* DNA binding (Olsen et al., 2005a; Nakashima et al., 2012). A few of these studies report a dependency on the presence of the CTD (e.g., Hegedus et al., 2003; Jeong et al., 2009), but most of these interactions have been identified with isolated NAC domains (e.g., Xie et al., 2000; Olsen et al., 2004; Takasaki et al., 2010; Chen et al., 2011). The two available crystal structures confirm NAC homodimerization and reveal a very similar dimer interface formed by the highly conserved N-terminal part of the NAC domain (Ernst et al., 2004; Chen et al., 2011). AtNAC1 was also found to be compatible with this dimerization mode by *in silico* analysis (Zhu et al., 2014). The interface consists of a short (3–4 amino acids) antiparallel β-sheet made up of a β-strand from each monomer, two intermolecular salt bridges between amino acid side chains, plus many hydrophobic interactions. In spite of a relatively small interaction area (~800 Å^2 buried surface for each monomer), the extensive contact network between conserved residues and the similar interface in the two structures strongly indicate that this is the functional dimer. Indeed, this dimer was found bound to DNA in the ANAC019–DNA complex structure (Welner et al., 2012), which further corroborates the functional state of NAC domains to be dimeric.

The dimer interface is connected to the core β-sheet through a hinge region, which allows the NAC dimer to be flexible. Dimer angles ranging from ~120° (open) to ~100° (closed) have been reported so far. Based on small-angle X-ray scattering studies, the open conformation seems to be dominant in solution in the absence of DNA (Welner et al., 2012). In crystal, open, semiclosed (108°), and closed conformations have been observed. In the ANAC019–DNA complex structure the DNA-bound dimer has a closed dimer conformation. This trend is also reported for the AtNAC1 model (Zhu et al., 2014), which has angles of 120° and 110° for the apo- and DNA-bound state, respectively. The angle when bound to DNA might be dictated by the spacing between the two halves of the palindromic NAC binding site (NACBS; see Section 13.4).

13.3 EVOLUTION OF NAC PROTEINS

13.3.1 Unraveling NAC Evolution

Several studies have addressed the phylogenetic relationship of NAC proteins. Most of these focus on the model plants *Arabidopsis* and rice (Kikuchi et al., 2000; Nuruzzaman et al., 2010; Jensen et al., 2010) or other specific plants such as poplar, soybean, banana, and grape (Hu et al., 2010; Le et al., 2011; Wang et al., 2013; Cenci et al., 2014). However, the completion of many plant genome sequences has made comparative phylogenetic analysis of NAC proteins possible (Zhu et al., 2012), thereby markedly improving our understanding of the evolution of the NAC family. This reconfirmed that NAC proteins are found only in land plants (Olsen et al., 2005a), and that most plants have large NAC families (>60 members). However, fewer NAC proteins (≤30) were predicted for mosses and lycophytes representing early-diverged land plants, suggesting that expansion of NAC proteins took place after the evolution of vascular plants (Zhu et al., 2012). In general, in accordance with previous phylogenetic analyses (Nuruzzaman et al., 2010; Jensen et al., 2010), 21 NAC subfamilies were defined based on clustering, and several of these are associated with specific functions (Jensen et al., 2010; Nuruzzaman et al., 2010; Jensen and Skriver, 2014). Six of these subfamilies were found in early-diverged land plants, whereas 15 of these include only angiosperm proteins. Based on the analyses by Zhu et al. (2012), the NAC family was hypothesized to date back more than 400 million years.

13.3.2 Evolution of the NAC Fold

Recent progress in structural characterization has revealed similarities between NAC proteins and the WRKY TF family, providing a glimpse of the evolutionary origin of NAC proteins reaching outside the family itself. Thus, the conserved DNA-binding WRKY contains a central twisted β-sheet with the same topology as NAC (Figure 13.2A). The structure of the WRKY–DNA complex (Yamasaki et al., 2012) shows the β-strand corresponding to the DNA-binding β3-strand of NAC inserting deeply into the major groove in a perpendicular manner similar to NAC. The DNA-binding

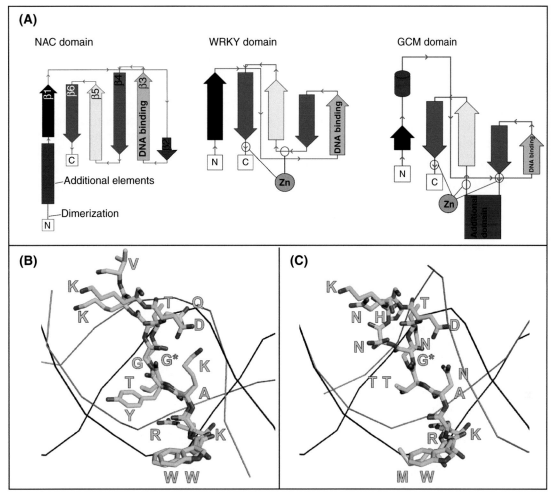

FIGURE 13.2 **Structural comparison of NAC, WRKY, and GCM domains.** The DBDs from ANAC019 (1UT7), WRKY4 (2LEX), and GCM (1ODH) represent the three TF families in this figure. (A) Topology diagrams modified from the PDBsum database (de Beer et al., 2014) with emphasis on the common core β-sheet. The diagrams are further annotated with important features, including the DNA-binding strand, the NAC dimerization domain, and the Zn-binding motif found in WRKY and GCM, where Zn-interacting regions are indicated by circles. Secondary structure elements of the NAC domain have been numbered according to Ernst et al. (2004). (B, C) Crystal structures of complexes with the DNA ID 3SWP, ANAC019, 2LEX (WRKY4), 1ODH (GCM) have been used in structural alignments of the DNA-binding strands of ANAC019 (green) and WRKY4 (cyan, B) or GCM (cyan, C). The ANAC019-bound oligonucleotide is shown as a black ribbon, and the WRKY/GCM-bound oligonucleotide is shown in gray. The conserved glycine discussed is marked by a star.

β-strand contains the family-defining WRKY motif (WRKYGQK), where the conserved glycine residue induces an unusual curvature in the strand, allowing it to penetrate deeply into the major groove, as described for NAC in Section 13.2.3. In spite of the divergence in sequence, the conserved RMs of NAC and WRKY align well, with a resulting similar orientation of the bound DNA (Figure 13.2B). The disputed glycine is a point of spatial divergence, which could be of biological significance.

In a broader context, NAC and WRKY have been found to be related to the mammalian glial cells missing (GCM) TF (Cohen et al., 2003; Olsen et al., 2005a; Yamasaki et al., 2005; Welner et al., 2012; Yamasaki et al., 2012; Chapter 4). GCM has the same core fold, although the RM is highly divergent (MRNTNNHN). Nonetheless, it aligns well with the NAC RM, and the NAC glycine again seems to be a point of spatial divergence (Figure 13.2C). The DNA-binding modes of NAC, WRKY, and GCM are likewise very similar.

WRKY and GCM share a structurally important Zn-binding motif, which is absent in NAC proteins, although Zn finger TFs with sequence similarity to the NAC domain have been identified in *Arabidopsis* (Mitsuda et al., 2004). Another important difference is that NAC domains can dimerize, while the functional state of WRKY and GCM seems to be monomeric. In spite of these differences, the structural similarities and common DNA-binding mode suggest an evolutionary relationship between NAC, WRKY, and GCM.

13.4 NAC PROTEINS: FROM STRUCTURE TO INTERACTIONS WITH DNA AND OTHER PROTEINS

In this section we review the interactions of NAC proteins with focus on studies at the molecular level.

13.4.1 DNA Sequence Specificity of the NAC Domain and Target Sequences

The diverse functionalities of different NAC proteins are believed to some extent to be conferred by DBD specificity and affinity for target DNA sequences, even though other mechanisms for selective activation of proper response pathways exist (e.g., mediated by the CTD which most often contains the transcriptional regulation domains, as described in Jensen et al. (2010). Thus, many research efforts have focused on interactions with DNA, resulting in major progress in defining the DNA sequence preference of NAC proteins.

The most common DNA sequences recognized by *Arabidopsis* NAC proteins have been reviewed in Jensen and Skriver (2014). Target NACBSs have been identified either in the context of natural promoters, or by *in vitro* selection techniques for identification of optimal binding sites. Most of the identified NACBSs contain the core motif **CGT**[AG], both in *Arabidopsis* and other species (e.g., wheat NAC69; Xue et al., 2006, and barley *Hv*NAC005 and *Hv*NAC013, Kjaersgaard et al., 2011). Often this core motif is palindromically repeated, as in the oligonucleotide used for structural studies (Welner et al., 2012). There is some variation in optimal separation between palindromic repeats. For example, wheat NAC69 had a maximal binding activity with a 6 bp (base pair) spacer between CGT and ACG, slightly reduced activity with a 7 bp spacer, and almost no activity with shorter or longer spacers (Xue et al., 2006). In contrast, ANAC019 (Welner et al., 2012; Jensen et al., 2010) has shown good DNA binding with a 8 bp spacer, while ANAC092/ORE1 shows optimal binding with a 7 or 8 bp spacer (Jensen et al., 2010; Matallana-Ramirez et al., 2013). Furthermore, several functional promoters of target NAC genes (such as those reviewed in Jensen and Skriver, 2014) only contain one consensus NACBS and some contain nonpalindromic repeats.

It is clear that the DNA region flanking the core motif is also in contact with the RM on NAC proteins and contributes to the recognition. In a recent study of 12 representative *Arabidopsis* NAC proteins (Lindemose et al., 2014), several clusters of NAC were identified, based on the preferred DNA sequence bound. Clusters 1a and 1b, containing among others the well-characterized ANAC092 and ANAC019, respectively, showed clear preferences for **TTGCGT** and **TTACGT**, which fits well with the general core motif. In contrast Cluster 3 (NTL6/NTL8) prefers the sequence TTXCTT, though binding of sequences containing the core motif has previously been shown at least for NTL6 (Seo et al., 2010). Calmodulin-binding NAC (CBNAC) could have a similar specificity as Cluster 3, since the optimal DNA sequence recognized was reported to be TTGCTTANNNNNNAAG (Kim et al., 2007).

ATAF2, a NAC protein involved in viral defense, also deviates from the general consensus. Although it belongs to the same phylogenetic group (III-3; Jensen et al., 2010) as ANAC019 and other NAC proteins recognizing the core motif, it has been reported instead to recognize AT-rich sequences, such as the imperfect palindrome CAAATNNNATTTG in the nitrilase 2 (NIT2) gene promoter (Huh et al., 2012) and a defensin-like protein At1g68907 promoter TCAGAAGAGCAATCAAATTA-AAAC (Wang and Culver, 2012). Furthermore, an extended DNA-binding site is necessary for efficient binding. Another NAC protein, the turnip crinkle virus interacting protein (TIP), has been reported to bind DNA in a non-sequence-specific manner (Ren et al., 2005). CBNAC, ATAF2, and TIP were not included in the study by Lindemose et al. (2014).

13.4.2 Structural Basis for DNA Recognition and Specificity of NAC Proteins

As already described in Section 13.2, *in vitro* studies on NAC proteins agree that they bind DNA as dimers, as first demonstrated by Olsen et al. (2005b) and shown later by others. Disruption of the dimerization interface (described in Section 13.2) leads to decreased affinity for the DNA, as shown both for ANAC019 in Olsen et al. (2005b) and later studies (e.g., on wheat NAC69; Xue et al., 2006).

Despite considerable information on the DNA sequence preferences of many NAC proteins, we are far from understanding NAC–DNA interactions in detail. Site-directed mutagenesis (sdm) studies on NAC proteins have been very limited and to the best of our knowledge none has addressed specificity. The structure of the complex of ANAC019 with a palindromic NACBS (Welner et al., 2012) elucidated by crystallography, while important, has a resolution of less than 4 Å precluding any detailed conclusions. The only other detailed structural study on DNA binding comes from an *in silico* modeling and docking investigation (Zhu et al., 2014) of AtNAC1, which is somewhat biased as it uses the ANAC019 structure as a template and is not corroborated by experimental studies.

Based on the structural information available ANAC019 conserved or semiconserved residues Arg88, Lys96, Lys123, Lys129, Lys162, and Arg85 potentially interact with backbone phosphates (see Figure 13.1 for the location on the structure), and all but Lys96 have been investigated by sdm. Arg88 seems to be the most important residue, since its substitution alone reduces drastically the affinity for DNA (Welner et al., 2012). Furthermore, substitution of the RPNR motif (Figure 13.1) in wheat NAC69 (Xue et al., 2006), containing the equivalents of ANAC019 Arg85 and Arg88, completely abolished DNA-binding activity. Several of the other residues highlighted in Figure 13.1 were also pinpointed by the computational study on AtNAC1 (Zhu et al., 2014) and may also contribute to binding, but to a lesser extent according to available experimental data.

The crystal structure of the complex (Welner et al., 2012), as already described in Section 13.2, suggests that the RM is positioned appropriately to mediate specific recognition of bases but the structure has insufficient resolution to delineate specific interactions of amino acid residues with bases. An AtNAC1 computational study (Zhu et al., 2014) also identifies residues in the RM as interacting with DNA bases. Somewhat strangely, however, equivalent bases in the two halves of the perfect palindrome are recognized by different residues in the two NAC monomers, while one would expect a symmetrical interaction mode.

Any possible correspondence between the RM sequence and preferred recognized DNA sequences from Lindemose et al. (2014), and other studies, was investigated (see Table 13.1 for selected sequences). Generally, there is very little correlation. For example, NST2 (Cluster 1a) and VND3/VND7 (Cluster 1b) have the same RM sequence despite a different DNA sequence preference. Even more strikingly, ATAF2 and ATAF1 have identical RM sequences even though ATAF2 does not preferentially recognize the CGT core motif. Some trends may be present; for example, NTL8 and NTL6 (Cluster 3) have an unusual terminal R in the RM and ANAC003, which has so far failed to show any DNA binding (Jensen et al., 2010; Lindemose et al., 2014), also diverges from the consensus RM (underlined residues in Table 13.1). The lack of correlation between the RM sequence and recognized base points to the additional importance of other structural regions for specific recognition (e.g., loops). Furthermore, in other TFs (e.g., the helix–turn–helix class), a very important factor in sequence-specific recognition is the indirect readout (i.e., the influence on specificity given by the propensity of specific DNA sequences to deviate from canonical B-DNA conformation; Rohs et al., 2010). The importance of indirect readout in DNA recognition by NAC proteins is almost unexplored. Should this be important, its interplay with the reported flexibility of the NAC dimer (Section 13.2) will also need to be elucidated. It is clear that further studies are necessary to address the structural basis of DNA binding and specificity in NAC proteins.

Another factor to be considered is that heterodimerization and other protein–protein interactions (see Section 13.4.3) could affect both affinity and specificity for DNA. Finally, even though in most experimental reports the DNA-binding ability of NAC has been assigned to the

TABLE 13.1 Sequence of the RM in Selected NAC Proteins

NAC protein	RM sequence	Cluster (from Lindemose et al., 2014) and/or preferred DNA sequence motif
NST2	WKATGRDK	Cluster 1a (TTGCGT)
SND1	WKATGRDK	Cluster 1a (TTGCGT)
ANAC092	WKATGKDK	Cluster 1a (TTGCGT)
ANAC019	WKATGTDK	Cluster 1b (TTACGT)
NAP	WKATGTDK	Cluster 1b (TTACGT)
VND3	WKATGRDK	Cluster 1b (TTACGT)
VND7	WKATGRDK	Cluster 1b (TTACGT)
ANAC055	WKATGTDK	Cluster 1b (TTACGT)
ATAF1	WKATGADK	Cluster 1b (TTACGT)
ATAF2	WKATGADK	AT-rich sequence (Huh et al., 2012; Wang and Culver, 2012)
NTL6	WKATGKDR	Cluster 3 (TTXCTT)
NTL8	WKATGKER	Cluster 3 (TTXCTT)
ANAC003	WKSTGRPK	No DNA binding (Jensen et al., 2010; Lindemose et al., 2014)

N-terminal conserved NAC domain, some studies show the effect of the CTD (Lindemose et al., 2014), where the affinity of full-length ANAC092 for DNA was found to be higher than that of the NAC DBD alone, albeit with only a minor difference in specificity.

13.4.3 Direct Interactions of NAC Proteins with Other Proteins

The most important protein–protein interaction in which NAC proteins are involved is probably dimerization, where two NAC monomers form either a homodimer (same NAC protein) or heterodimer (two different NAC proteins) to generate a functional DNA-binding TF, as described earlier (Section 13.2 and 13.4.2).

In addition, the key role of NAC proteins as regulators of many essential plant cell functions implies a variety of other protein–protein interactions. However, surprisingly few studies show direct physical interactions with other proteins, mostly from *in vivo* yeast two-hybrid studies coupled with *in vitro* pull-down experiments. Only a handful of studies include more thorough characterization, dealing primarily with localization of interacting protein regions by means of experiments with truncated variants.

The first evidence of direct physical interaction between a NAC protein and a non-NAC protein came to the best of our knowledge from the identification of geminivirus RepA-binding (GRAB) proteins specifically interacting with the wheat dwarf geminivirus protein RepA

(Xie et al., 1999). RepA-binding activity was localized in the N-terminal NAC domain. GRAB proteins have been shown to inhibit geminivirus DNA replication in cultured cells. There are several additional examples of interactions with viral proteins. A recent report (Suyal et al., 2014) identified *Arabidopsis* ANAC083 as the interaction partner of mungbean yellow mosaic India virus (another geminivirus) Rep protein, leading to speculation that this interaction might also affect viral DNA replication. Furthermore, an ANAC083-derived peptide, corresponding to the region of β6 in the NAC domain, was shown to interact with Rep by phage display. ATAF2 was shown to interact with a tobacco mosaic virus replicase, resulting in ATAF2 degradation (Wang et al., 2009), presumably as a mechanism by which the pathogen tries to counteract the plant's own defenses. Another NAC protein, TIP, was found to interact with the turnip crinkle virus capsid protein (CP) (Ren et al., 2000) and, initially at least, this interaction was thought to mediate the resistance response of the plant, although this role has since been debated in the literature (Donze et al., 2014). In contrast to the many examples above, the CP-interacting region on TIP consists of 100 C-terminal amino acids (Ren et al., 2005). Two other geminivirus proteins, the tomato leaf curl virus replication enhancer and its homolog from tomato golden mosaic virus, were found to interact with the tomato NAC protein SlNAC1, and through this interaction viral replication was enhanced (Selth et al., 2005). To date, interactions of NAC proteins with viral proteins are some of the most frequently reported NAC interactions. The emerging picture is that these interactions can contribute to either side of the tug of war between plants and viral pathogens, either as elicitors of plant defense responses or as part of the viral mechanisms designed to evade these responses, with the virus often having the upper hand.

TFs sometimes interact with each other to regulate transcription. Heterodimerization of NAC proteins could provide an interesting mechanism of regulation, but its relevance *in vivo* has yet to be established. Additionally, physical interactions between NAC proteins and TFs belonging to other classes have been reported. ANAC096 was found to interact specifically with two basic leucine zipper (bZIP) TFs, ABRE binding factor/ABA-responsive element (ABRE) binding protein (ABF/AREB)2 and ABF4 (but not ABF3), both *in vitro* and *in vivo* (Xu et al., 2013), as part of synergistic transcription regulation of osmotic stress response (Section 13.5.1). Zinc finger homeodomain 1 (ZFHD1) was shown to interact with the DBDs of three NAC proteins ANAC019, ANAC055, and ANAC072 (Tran et al., 2007) to synergistically activate the *EARLY RESPONSIVE TO DEHYDRATION STRESS1 (ERD1)* gene (see Section 13.5.1). ANAC092 (ORE1) interacts with the G2-like TFs, GLK1 and GLK2, and for GLK2 the interaction was confirmed in *Arabidopsis* (Rauf et al., 2013). In contrast to the previous examples, ANAC092 has an inhibitory effect on the transcription of GLK target genes involved in chloroplast maintenance, and thus favors leaf senescence.

NAC proteins also take part in protein–protein interactions where they are covalently posttranslationally modified (e.g., by ubiquitinylation). In the first report of this phenomenon (Xie et al., 2002), the RING protein SINAT5 was shown to ubiquitinate NAC1, involved in auxin signaling. Overexpression of SINAT5 resulted in reduced NAC1 in roots, consistent with the role of ubiquitinylation in targeting NAC1 for degradation. ANAC055 (Bu et al., 2009) and the NAC domain of ANAC019 (Greve et al., 2003) also interact with RING ubiquitin ligases, such as RHA2a. Interaction with RHA2a could suggest a targeting pathway for proteasome-dependent degradation (see Section 13.5.1).

NTL6 is a membrane-associated NAC implicated in cold-induced pathogen resistance and is proteolytically activated (Seo et al., 2010) to a mature nuclear-associated form. Processed NTL6 is phosphorylated primarily at Thr142 (in the NAC domain) by SnRK2.8 (Kim et al., 2012). Thus, proteolysis and phosphorylation are also demonstrated as mechanisms by which chemical modification regulates NAC proteins (see Chapter 25 for the regulation of NAC proteins by proteolysis). Further, the NAC protein ATAF1 has been shown to interact with kinases (Kleinow et al., 2009). The possible importance of a phosphorylation state for the NAC function is underlined by known interactions with phosphatases. For example, ANAC019 is known to be dephosphorylated *in vivo* by the REGULATOR OF C-REPEAT BINDING FACTOR (CBF) GENE EXPRESSION 2 (RCF2)/CTD phosphatase-like 1 (CPL1; see Section 13.5.1; Guan et al., 2014).

A couple of other reports about the direct interactions of NAC proteins are noteworthy. CBNAC has been shown to bind the calcium-binding protein calmodulin through its CTD (Kim et al., 2007), which seems to enhance its transcriptional repression activity. *Hv*NAC013 (Kjaersgaard et al., 2011) has been shown to specifically interact with Radical Induced Cell Death 1 (RCD1). The interaction was localized to the intrinsically disordered last 96 C-terminal residues of *HvNAC013*, which are not needed for transactivation; this is the best characterized NAC interaction with intrinsically disordered domains. The physiological role of the interaction is however not known.

As seen above, NAC proteins have been shown to be involved in a number of protein–protein interactions, and more can be expected to emerge in the future. Their functional importance will be further discussed in Section 13.5. As to the structural elements involved in interaction, no consistent picture has emerged from the small amount of information available. For example, viral proteins have been shown to interact with the NAC domain and/or CTD, and clearly much more work is necessary to understand these interactions at the molecular and atomic level.

13.5 NAC NETWORKS IN ABIOTIC STRESS RESPONSES

One of the best characterized NAC clades is associated with stress responses and senescence and has been named subgroup III-3 in *Arabidopsis* (Jensen et al., 2010) and stress-responsive NAC (SNAC) in rice (Nuruzzaman et al., 2010). The structure, function, and gene regulatory networks (GRNs) of this subgroup has been the focus of several recent reviews (Nakashima et al., 2012; Puranik et al., 2012; Lindemose et al., 2013; Jensen and Skriver, 2014) and knowledge of this subgroup in model plants and crops has accumulated rapidly (Guan et al., 2014) and therefore deserves special attention, as summarized in Figure 13.3.

13.5.1 Networks of Stress-Associated ANAC019, ANAC055, and ANAC072

ANAC019, ANAC055, and ANAC072 (RD26) of subgroup III-3 are associated with different abiotic stress responses, such as drought, high salinity, and the stress hormones abscisic acid (ABA) and jasmonic acid (JA), for about a decade (Greve et al., 2003; Tran et al., 2004; Fujita et al., 2004). Early studies demonstrated the ability of these three NAC proteins to bind the *ERD1* promoter (Tran et al., 2004), thereby contributing the first components to the fast-growing ANAC019/055/072 GRN (Figure 13.3). Curiously, the three NAC proteins were dependent on ZFHD1 for regulation of *ERD1* and thereby for the regulation of drought responses (Tran et al., 2007).

Very recently one of the three, ANAC019, was implicated in novel protein–protein interactions regulating abiotic stress responses. RCF2/CPL1, a positive regulator of heat-stress-responsive gene expression and thermotolerance (Guan et al., 2014), was shown in yeast two-hybrid assays to interact with ANAC019 (Bang et al., 2008; Section 13.4). Overexpression of *RCF2/CPL1* or *ANAC019* in *Arabidopsis* increases thermotolerance. Furthermore, chromatin immunoprecipitation (ChIP) showed that ANAC019 binds to NACBSs in the promoters of several heat stress TFs (HSFs). Altogether, the novel results of Guan et al. (2014) contribute a new regulatory circuit to the growing ANAC019 network. Thus, heat stress induces *ANAC019* and *RCF2/CPL1* expression, RCF2/CPL1 dephosphorylates ANAC019, and both proteins are required for heat induction of *HSF*, and thereby synthesis of HEAT SHOCK PROTEINS (HSPs), which as molecular chaperones are important components of the responses to heat stress (Guan et al., 2014).

Likewise, very recently ANAC096 of subgroup IV-1 (Jensen et al., 2010) was shown to cooperate with ABF2 and ABF4 in regulating dehydration and osmotic stress responses (Xu et al., 2013; Section 13.4.3). ANAC096 and ABF2 synergistically activate *RD29A* transcription. Interestingly, functional interaction was also shown for ANAC019 and ABF2, but in this case the interaction resulted in repression of *RD29* expression thereby repressing stress adaptation (Xu et al., 2013). This demonstrates how delicate interplays between TFs from the same, and different, TF families can regulate network rewiring.

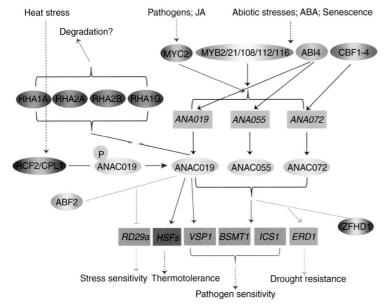

FIGURE 13.3 GRN of ANAC019, ANAC055, and ANAC072/RD29. Direct interactions between the NAC proteins and DNA are shown by arrows (positive effect on expression) or blocked lines (negative effect on expression). Protein–protein interactions are shown by straight lines, and the agonistic effects of these interactions are shown by arrows, whereas the antagonistic effects are shown by blocked lines. Dashed arrows refer to effects that may involve additional components. Connections between components involved in the same processes are shown using the same color. Homo- and heterodimerization is not shown.

As mentioned in Section 13.4.3, the implications of the interactions of ANAC019 with the functional E3 ubiquitin ligase RHA2a, a positive regulator of ABA signaling (Bu et al., 2009), as well as additional small RING-H2 proteins, (Greve et al., 2003) remain elusive, but still deserve attention in the ANAC019 network.

The three NAC TFs directly target additional genes. For example, they mediate coronatine toxin-induced bacterial propagation by inhibiting the accumulation of the key plant immune signal salicylic acid (SA). They do so by activating the expression of the SA metabolic gene *S-ADENOSYLMETHIONINE-DEPENDENT METHYL-TRANSFERASE* (*BSMT1*) and repressing the expression of the SA biosynthesis gene *ISOCHORISMATE SYNTHASE1* (*ICS1*) (Zheng et al., 2012). An association with defense responses was also suggested from the demonstration that the *anac019anac055* double-mutant showed increased resistance to the necrotrophic pathogen *Botrytis cinerea* compared to wild-type plants (Bu et al., 2008). In accordance with this, methyl jasmonate(JA)-induced expression of *ANAC019* and *ANAC055* was dependent on the JA-signaling basic helix–loop–helix TF, *At*MYC2, which directly targets *NAC019/055/072* (Zheng et al., 2012). Together the examples summarized earlier demonstrate how ANAC019/055/072 function as upstream regulators connecting different circuits implicated in biotic stress responses involving JA and SA, as well as abiotic stress responses involving ABA.

The above examples clearly demonstrate how ANAC019/055/72 function as stress regulators through their specific binding of promoter regions in downstream target genes or other proteins. However, upstream regulators of NAC019/055/072 GRNs have also been identified (Hickman et al., 2013). The ABA-associated bZIP TFs, ABF3 and ABF4, bound promoter fragments of all three NAC proteins in a yeast one-hybrid screening, and another mediator of ABA-signaling, ABI4, an APETALA2 (AP2) family TF, bound *ANAC019* and *ANAC055* promoter fragments. Based on the roles of ABF3, ABF4, and ABI4 in ABA-signaling these physical interactions are likely to be of physiological relevance. Several members of the myeloblastosis (MYB) TF family also bound fragments of the three NAC promoters, and MYB2 and MYB108 may mediate senescence-associated regulation of *ANAC019/ANAC055/ANAC072* expression. The analyses also suggested that the four AP2 TFs, C-REPEAT/DEHYDRATION RESPONSIVE ELEMENT BINDING FACTOR1 (CBF1) 1, CBF2, CBF3, and CBF4, which can bind and regulate the expression of *ANAC072*, play different regulatory roles during stress exposure and senescence (Hickman et al., 2013). The ANAC019/055/072 GRN in Figure 13.3, including only components with direct interactions with the three NAC proteins, demonstrates the involvement of these three TFs in many aspects of plant biology.

13.5.2 Networks of Stress-Associated Rice NAC Proteins

Although most studies of stress-associated NAC proteins have been performed in the model plant *Arabidopsis*, interesting stress-associated functions have also been found for NAC proteins in rice. These studies also utilize the tools associated with a model organism and generate results of applicative value for crop plants as well. In rice, the expression of several SNAC genes (*SNAC1*, *OsNAC6/SNAC2*, *OsNAC3*, *OsNAC4*, and *OsNAC5*) is induced by ABA, drought, salinity, and cold stresses (Hu et al., 2008; Takasaki et al., 2010).

SNAC1 and *OsNAC6/SNAC2* are closely related to each other and show similar responses to different abiotic stress conditions (Ooka et al., 2003). Several years ago, overexpression of *SNAC1* was shown to increase drought resistance in transgenic rice as well as rice in the field. This was explained by increased ABA hypersensitivity and decreased water loss due to increased stomatal closure (Hu et al., 2006). Recently, the functional role of SNAC1 was emphasized further by the demonstration that a rice homolog of *SIMILAR TO RCD ONE*, *OsSRO1c*, is regulated by SNAC1 (Figure 13.4). *OsSRO1c* is differentially expressed in guard cells under drought stress, and overexpression of *OsSRO1c* results in decreased stomatal aperture and reduced water loss by regulation of H_2O_2 homeostasis (You et al., 2013). In summary, the work by You et al. (2013) suggests that *Os*SRO1c and SNAC1 play a role in both the drought responses and oxidative stress tolerances of rice.

OsNAC5 and *OsNAC6/SNAC2* both act as transcriptional activators and target the promoter of *OsLATE EMBRYO-GENESIS ABUNDANT3* (*OsLEA3*), but the effects of these two NAC proteins on growth are different (Figure 13.4). Thus, the growth of *OsNAC5*-overexpressing plants was similar to that of control plants, whereas the growth of *OsNAC6/SNAC2*-overexpressing transgenic plants was relatively retarded (Nakashima et al., 2007; Takasaki et al., 2010), as is often seen by overexpression of TFs implicated in stress regulation (Lindemose et al., 2013). *OsNAC6/SNAC2* is also implicated in development through another type of regulatory mechanism – epigenetic regulation by the histone deacetylase (HDAC), *Os*HDAC1 (Figure 13.4). ChIP assays of the *Os*NAC6 promoter suggested that *Os*HDAC1 represses the expression of *OsNAC6/SNAC2* by deacetylating specific lysine residues on histone3 (H3) and histone4 (H4) (Chung et al., 2009).

*Os*NAC4 is another SNAC member with a significant biological role – in this case in plant immune response. Upon induction by a pathogenic signal, *Os*NAC4 is transported into the nucleus in a phosphorylation-dependent manner. Many genes, including *IMMUNE-RELATED ENDONUCLEASE* (*IREN*), encoding a Ca^{2+}-dependent nuclease, and *OsHSP90*, are differentially expressed

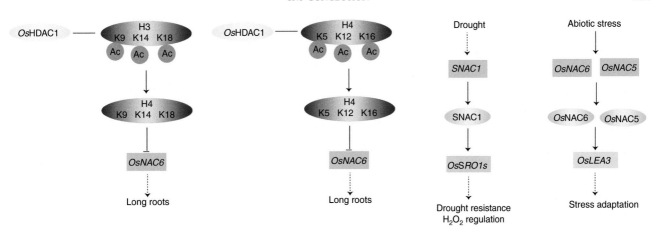

FIGURE 13.4 **Emerging regulatory networks of rice SNAC genes and proteins.** For explanations see the legend of Figure 13.3. In addition, Ac is the acetylation of specific lysine residues (K) in histone proteins H3 and H4.

by OsNAC4 during hypersensitive response (HR) cell death. IREN and OsHSP90 have specific functions in nuclear fragmentation and loss of membrane integrity during induction of HR cell death, respectively (Kaneda et al., 2009). In conclusion, the results summarized show that SNAC proteins have important and diverse functions in a monocot crop plant.

13.5.3 Other Transcription Factors in the ANAC019/055/072 Network

Sections 13.4.3 and 13.5.1 clearly demonstrate (Figures 13.3–13.4) that the strictly defined ANAC019/055/072 GRN involves several TFs from different TF families. MYC2, a basic helix–loop–helix TF, functions as an upstream regulator of *ANAC019/055/072* genes in a regulatory loop connecting the well-known role of MYC2 in pathogen responses with downstream pathogen sensitivity. MYC2, referred to as the master of action, regulates crosstalk between signaling pathways involving several plant hormones, including JA, ABA, and SA (Kazan and Manners, 2012). The GRN also contains TFs from different families, which are key regulators of stress responses and ABA signaling. For example, the AP2 TF, ABI4, and the bZIP TF, ABF2, are implicated in the GRN. ANAC019/055/072 agonistically cooperate with the Zn finger homeodomain TF, ZFHD1, in regulating *ERD1*, whereas cooperation between ANAC019 and ABF2 to regulate *RD29a* is antagonistic. In the downstream part of the GRN, ANAC019 directly targets several different *HSF* genes in the process of regulating thermotolerance. *HSF* genes and additional direct NAC target genes, such as *RD29a*, are also regulated by direct interactions with additional TFs, as reflected by their complex promoter regions containing several different *cis*-acting elements (Lindemose et al., 2013). In addition to being regulated by agonistic and antagonistic interactions between ABFs and ANAC096 or ANA019, *RD29a* (Xu et al., 2013) is directly regulated by AP2 TFs: CBF3 which is also part of the ANAC019/055/072 GRN; DREB2a (Liu et al., 1998) which is a key regulator of abiotic stress responses; the NAC TF VNI2 (Yang et al., 2011) which integrates ABA stress signaling with aging; and the WRKY TF WRKY57 (Jiang et al., 2012). In conclusion, the ANAC019/055/072 GRN integrates multiple signaling pathways and TFs from different pathways to regulate specific physiological processes.

13.6 CONCLUSION

NAC proteins have been found only in land plants and play important roles in development, abiotic and biotic stress responses, and biosynthesis. Their structures are typical of gene-specific TFs. The N-terminal DBD, the NAC domain, forms a 7-stranded antiparallel twisted β-sheet flanked by α-helices. Upon DNA binding the NAC domain inserts an outer β-strand into the major groove of the target DNA. Both structures and DNA-binding modes suggest an evolutionary relationship between NAC, WRKY, and GCM TFs. NAC proteins interact with many different types of proteins, including viral proteins, E3 ubiquitin ligases, phosphatases, and kinases using both their NAC domain and CTD. At the molecular level they regulate the expression of genes involved in a variety of physiological processes, and directly target some of these through binding to their core NACBS, CGT[AG]. The rapidly increasing GRN of the stress-associated *Arabidopsis* NAC proteins ANAC019, ANAC055, and ANAC072 reveal crosstalk between signaling pathways involving several plant hormones, including JA, ABA, and SA, and regulation of both biotic and abiotic stress responses. The emerging networks for corresponding rice NAC proteins reveal epigenetic regulation of growth. Although single NAC genes can markedly regulate plant stress responses, knowledge of

NAC protein structure–function relationships and GRNs is important for future agricultural attempts to generate plants with improved stress tolerance.

Acknowledgments

This work was supported by the Danish Agency for Science Technology and Innovation (grant numbers 274-07-0173 and 10-084503, to K.S. and L.L.L., respectively).

References

Bang, W.Y., Kim, S.W., Jeong, I.S., Koiwa, H., Bahk, J.D., 2008. The C-terminal region (640-967) of *Arabidopsis* CPL1 interacts with the abiotic stress- and ABA-responsive transcription factors. Biochem. Biophys. Res. Commun. 372, 907–912.

Bu, Q., Jiang, H., Li, C.B., Zhai, Q., Zhang, J., Wu, X., Sun, J., Xie, Q., Li, C., 2008. Role of the *Arabidopsis thaliana* NAC transcription factors ANAC019 and ANAC055 in regulating jasmonic acid-signaled defense responses. Cell Res. 18, 756–767.

Bu, Q., Li, H., Zhao, Q., Jiang, H., Zhai, Q., Zhang, J., Wu, X., Sun, J., Xie, Q., Wang, D., Li, C., 2009. The *Arabidopsis* RING finger E3 ligase RHA2a is a novel positive regulator of abscisic acid signaling during seed germination and early seedling development. Plant Phys. 150, 463–481.

Cenci, A., Guignon, V., Roux, N., Rouard, M., 2014. Genomic analysis of NAC transcription factors in banana (*Musa acuminata*) and definition of NAC orthologous groups for monocots and dicots. Plant Mol. Biol. 85 (1–2), 63–80.

Chen, Q., Wang, Q., Xiong, L., Lou, Z., 2011. A structural view of the conserved domain of rice stress-responsive NAC1. Protein Cell 2, 55–63.

Chung, P.J., Kim, Y.S., Jeong, J.S., Park, S.H., Nahm, B.H., Kim, J.K., 2009. The histone deacetylase OsHDAC1 epigenetically regulates the OsNAC6 gene that controls seedling root growth in rice. Plant J. 59, 764–776.

Cohen, S.X., Moulin, M., Hashemolhosseini, S., Kilian, K., Wegner, M., Müller, C.W., 2003. Structure of the GCM domain–DNA complex: a DNA-binding domain with a novel fold and mode of target site recognition. EMBO J. 22, 1835–1845.

de Beer, T.A., Berka, K., Thornton, J.M., Laskowski, R.A., 2014. PDBsum additions. Nucleic Acids Res. 42, D292–D296.

Donze, T., Qu, F., Twigg, P., Morris, T.J., 2014. Turnip crinkle virus coat protein inhibits the basal immune response to virus invasion in *Arabidopsis* by binding to the NAC transcription factor. Virology 449, 207–214.

Duval, M., Hsieh, T.-F., Kim, S.Y., Thomas, T.L., 2002. Molecular characterization of *AtNAM*: a member of the *Arabidopsis* NAC domain superfamily. Plant Mol. Biol. 50, 237–248.

Ernst, H.A., Olsen, A.N., Skriver, K., Larsen, S.Y., Lo Leggio, L., 2004. Structure of the conserved domain of ANAC, a member of the NAC family of transcription factors. EMBO Rep. 5, 297–303.

Fujita, M., Fujita, Y., Maruyama, K., Seki, M., Hiratsu, K., Ohme-Takagi, M., Tran, L.-S.P., Yamaguchi-Shinozaki, K., Shinozaki, K., 2004. A dehydration-induced NAC protein, RD26, is involved in a novel ABA-dependent stress-signaling pathway. Plant J. 39, 863–876.

Greve, K., La Cour, T., Jensen, M.K., Poulsen, F.M., Skriver, K., 2003. Interactions between plant RING-H2 and plant-specific NAC (NAM/ATAF1/2/CUC2) proteins: RING-H2 molecular specificity and cellular localization. Biochem. J. 371, 97–108.

Guan, Q., Yue, X., Zeng, H., Zhu, J., 2014. The protein phosphatase RCF2 and its interacting partner NAC019 are critical for heat stress-responsive gene regulation and thermotolerance in *Arabidopsis*. Plant Cell 26, 438–453.

Hao, Y.J., Song, Q.X., Chen, H.W., Zou, H.F., Wei, W., Kang, W.S., Ma, B., Zhang, W.K., Zhang, J.S., Chen, S.Y., 2010. Plant NAC-type transcription factor proteins contain a NARD domain for repression of transcriptional activation. Planta 232, 1033–1043.

Hegedus, D., Yu, M., Baldwin, D., Gruber, M., Sharpe, A., Parkin, I., Whitwill, S., Lydiate, D., 2003. Molecular characterization of *Brassica napus* NAC domain transcriptional activators induced in response to biotic and abiotic stress. Plant Mol. Biol. 53, 383–397.

Hickman, R., Hill, C., Penfold, C.A., Breeze, E., Bowden, L., Moore, J.D., Zhang, P., Jackson, A., Cooke, E., Bewicke-Copley, F., et al., 2013. A local regulatory network around three NAC transcription factors in stress responses and senescence in *Arabidopsis* leaves. Plant J. 75 (1), 26–39.

Hu, H., Dai, M., Yao, J., Xiao, B., Li, X., Zhang, Q., Xiong, L., 2006. Overexpressing a NAM, ATAF, and CUC (NAC) transcription factor enhances drought resistance and salt tolerance in rice. Proc. Natl. Acad. Sci. USA 103, 12987–12992.

Hu, H., You, J., Fang, Y., Zhu, X., Qi, Z., Xiong, L., 2008. Characterization of transcription factor gene SNAC2 conferring cold and salt tolerance in rice. Plant Mol. Biol. 67, 169–181.

Hu, R., Qi, G., Kong, Y., Kong, D., Gao, Q., Zhou, G., 2010. Comprehensive analysis of NAC domain transcription factor gene family in *Populus trichocarpa*. BMC Plant Biol. 10, 145.

Huh, S.U., Lee, S.B., Kim, H.H., Paek, K.H., 2012. ATAF2, a NAC transcription factor, binds to the promoter and regulates NIT2 gene expression involved in auxin biosynthesis. Mol. Cells 34, 305–313.

Hussey, S.G., Mizrachi, E., Creux, N.M., Myburg, A.A., 2013. Navigating the transcriptional roadmap regulating plant secondary cell wall deposition. Front. Plant Sci. 4, 325.

Jensen, M.K., Skriver, K., 2014. NAC transcription factor gene regulatory and protein–protein interaction networks in plant stress responses and senescence. IUBMB Life 66, 156–166.

Jensen, M.K., Kjaersgaard, T., Nielsen, M.M., Galberg, P., Petersen, K., O'Shea, C., Skriver, K., 2010. The *Arabidopsis thaliana* NAC transcription factor family: structure–function relationships and determinants of ANAC019 stress signalling. Biochem. J. 426, 183–196.

Jeong, J.S., Park, Y.T., Jung, H., Park, S.H., Kim, J.K., 2009. Rice NAC proteins act as homodimers and heterodimers. Plant Biotechnol. Rep. 3, 127–134.

Jiang, Y., Liang, G., Yu, D., 2012. Activated expression of WRKY57 confers drought tolerance in *Arabidopsis*. Mol. Plant 5 (6), 1375–1388.

Kaneda, T., Taga, Y., Takai, R., Iwano, M., Matsui, H., Takayama, S., Isogai, A., Che, F.S., 2009. The transcription factor OsNAC4 is a key positive regulator of plant hypersensitive cell death. EMBO J. 28, 926–936.

Kazan, K., Manners, J.M., 2012. JAZ repressors and the orchestration of phytohormone crosstalk. Trends Plant Sci. 17, 22–31.

Kikuchi, K., Ueguchi-Tanaka, M., Yoshida, K.T., Nagato, Y., Matsusoka, M., Hirano, H.Y., 2000. Molecular analysis of the NAC gene family in rice. Mol. Gen. Genet. 262, 1047–1051.

Kim, H.S., Park, B.O., Yoo, J.H., Jung, M.S., Lee, S.M., Han, H.J., Kim, K.E., Kim, S.H., Lim, C.O., Yun, D.J., Lee, S.Y., Chung, W.S., 2007. Identification of a calmodulin-binding NAC protein (CBNAC) as a transcriptional repressor in *Arabidopsis*. J. Biol. Chem. 282, 36292–36302.

Kim, M.J., Park, M.J., Seo, P.J., Song, J.S., Kim, H.J., Park, C.M., 2012. Controlled nuclear import of NTL6 transcription factor reveals a cytoplasmic role of SnRK2.8 in drought stress response. Biochem. J. 448, 353–363.

Kjaersgaard, T., Jensen, M.K., Christiansen, M.W., Gregersen, P., Kragelund, B.B., Skriver, K., 2011. Senescence-associated barley NAC (NAM, ATAF1,2, CUC) transcription factor interacts with radical-induced cell death 1 through a disordered regulatory domain. J. Biol. Chem. 14, 35418–35429.

Kleinow, T., Himbert, S., Krenz, B., Jeske, H., Koncz, C., 2009. NAC domain transcription factor ATAF1 interacts with SNF1-related kinases and silencing of its subfamily causes severe developmental defects in *Arabidopsis*. Plant Sci. 177, 360–370.

Le, D.T., Nishiyama, R., Watanabe, Y., Mochida, K., Yamaguchi-Shinozaki, K., Shinozaki, K., Tran, L.S.P., 2011. Genome-wide survey and expression analysis of the plant-specific NAC transcription factor family in soybean during development and dehydration stress. DNA Res. 18, 263–276.

Lindemose, S., O'Shea, C., Jensen, M.K., Skriver, K., 2013. Structure, function and networks of transcription factors involved in abiotic stress responses. Int. J. Mol. Sci. 14, 5842–5878.

Lindemose, S., Jensen, M.K., de Velde, J.V., O'Shea, C., Heyndrickx, K.S., Workman, C.T., Vandepoele, K., Skriver, K., De Masi, F., 2014. A DNA-binding-site landscape and regulatory network analysis for NAC transcription factors in *Arabidopsis thaliana*. Nucleic Acids Res. 42, 7681–7693.

Liu, Q., Kasuga, M., Sakuma, Y., Abe, H., Miura, S., Yamaguchi-Shinozaki, K., Shinozaki, K., 1998. Two transcription factors, DREB1 and DREB2, with an EREBP/AP2 DNA binding domain separate two cellular signal transduction pathways in drought- and low-temperature-responsive gene expression, respectively, in *Arabidopsis*. Plant Cell 10, 1391–1406.

Matallana-Ramirez, L.P., Rauf, M., Farage-Barhom, S., Dortay, H., Xue, G.P., Droge-Laser, W., Lers, A., Balazadeh, S., Mueller-Roeber, B., 2013. NAC transcription factor ORE1 and senescence-induced BIFUNCTIONAL NUCLEASE1 (BFN1) constitute a regulatory cascade in *Arabidopsis*. Mol. Plant 6, 1438–1452.

Mitsuda, N., Hisabori, T., Takeyasu, K., Sato, M.H., 2004. VOZ; isolation and characterization of novel vascular plant transcription factors with a one-zinc finger from *Arabidopsis thaliana*. Plant Cell Physiol. 45, 845–854.

Nakashima, K., Tran, .L.S., Van Nguyen, D., Fujita, M., Maruyama, K., Todaka, D., et al., 2007. Functional analysis of a NAC-type transcription factor OsNAC6 involved in abiotic and biotic stress-responsive gene expression in rice. Plant J. 51, 617–630.

Nakashima, K., Takasaki, H., Mizoi, J., Shinozaki, K., Yamaguchi-Shinozaki, K., 2012. NAC transcription factors in plant abiotic stress responses. Biochim. Biophys. Acta 1819, 97–103.

Nuruzzaman, M., Manimekalai, R., Sharoni, A.M., Satoh, K., Kondoh, H., Ooka, H., et al., 2010. Genome-wide analysis of NAC transcription factor family in rice. Gene 465, 30–44.

Olsen, A.N., Ernst, H.A., Lo Leggio, L., Johansson, E., Larsen, S., Skriver, K., 2004. Preliminary crystallographic analysis of the NAC domain of ANAC, a member of the plant-specific NAC transcription factor family. Acta Crystallogr. D60, 112–115.

Olsen, A.N., Ernst, H.A., Lo Leggio, L., Skriver, K., 2005a. NAC transcription factors: structurally distinct, functionally diverse. Trends Plant Sci. 10, 79–87.

Olsen, A.N., Ernst, H.A., Lo Leggio, L., Skriver, K., 2005b. DNA-binding specificity and molecular functions of NAC transcription factors. Plant Sci. 169, 785–797.

Ooka, H., Satoh, K., Doi, K., Nagata, T., Otomo, Y., Murakami, K., Matsubara, K., Osato, N., Kawai, J., Carninci, P., Hayashizaki, Y., Suzuki, K., Kojima, K., Takahara, Y., Yamamoto, K., Kikuchi, S., 2003. Comprehensive analysis of NAC family genes in *Oryza sativa* and *Arabidopsis thaliana*. DNA Res. 20, 239–247.

Punta, M., Coggill, P.C., Eberhardt, R.Y., Mistry, J., Tate, J., Boursnell, C., Pang, N., Forslund, K., Ceric, G., Clements, J., Heger, A., Holm, L., Sonnhammer, E.L.L., Eddy, S.R., Bateman, A., Finn, R.D., 2014. Pfam: the protein families database. Nucleic Acids Res. 42, D222–D230.

Puranik, S., Sahu, P.P., Srivastava, P.S., Prasad, M., 2012. NAC proteins: regulation and role in stress tolerance. Trends Plant Sci. 17, 369–381.

Rauf, M., Arif, M., Dortay, H., Matallana-Ramírez, L.P., Waters, M.T., Nam, H.G., Lim, P.-O., Mueller-Roeber, B., Balazadeh, S., 2013. ORE1 balances leaf senescence against maintenance by antagonizing G2-like-mediated transcription. EMBO Rep. 14 (4), 382–388.

Ren, T., Qu, F., Morris, T.J., 2000. HRT gene function requires interaction between a NAC protein and viral capsid protein to confer resistance to turnip crinkle virus. Plant Cell 12, 1917–1925.

Ren, T., Qu, F., Morris, T.J., 2005. The nuclear localization of the *Arabidopsis* transcription factor TIP is blocked by its interaction with the coat protein of turnip crinkle virus. Virology 331, 316–324.

Rohs, R., Jin, X., West, S.M., Joshi, R., Honig, B., Mann, R.S., 2010. Origins of specificity in protein–DNA recognition. Annu. Rev. Biochem. 79, 233–269.

Selth, L.A., Dogra, S.C., Rasheed, M.S., Healy, H., Randles, J.W., Rezaian, M.A., 2005. A NAC domain protein interacts with *Tomato leaf curl virus* replication accessory protein and enhances viral replication. Plant Cell 17, 311–325.

Seo, P.J., Kim, S.G., Park, C.-M., 2008. Membrane-bound transcription factors in plants. Trends Plant Sci. 13, 550–556.

Seo, P.J., Kim, M.J., Park, J.Y., Kim, S.Y., Jeon, J., Lee, Y.H., Kim, J., Park, C.M., 2010. Cold activation of a plasma membrane-tethered NAC transcription factor induces a pathogen resistance response in *Arabidopsis*. Plant J. 61, 661–671.

Suyal, G., Rana, V.S., Mukherjee, S.K., Wajid, S., Choudhury, N.R., 2014. *Arabidopsis thaliana* NAC083 protein interacts with mungbean yellow mosaic India virus (MYMIV) Rep protein. Virus Genes 48 (3), 486–493.

Takasaki, H., Maruyama, K., Kidokoro, S., Ito, Y., Fujita, Y., Shinozaki, K., Yamaguchi-Shinozaki, K., Nakashima, K., 2010. The abiotic stress-responsive NAC-type transcription factor OsNAC5 regulates stress-inducible genes and stress tolerance in rice. Mol. Genet. Genomics 284, 173–183.

Tran, L.S.P., Nakashima, K., Sakuma, Y., Simpson, S.D., Fujita, Y., Maruyama, K., Fujita, M., Seki, M., Shinozaki, K., Yamaguchi-Shinozaki, K., 2004. Isolation and functional analysis of *Arabidopsis* stress-inducible NAC transcription factors that bind to a drought-responsive *cis*-element in the early responsive to dehydration stress 1 promoter. Plant Cell 16, 2481–2498.

Tran, L.P., Urao, T., Qin, F., Maruyama, K., Kakimoto, T., Shinozaki, K., Yamaguchi-Shinozaki, K., 2007. Functional analysis of AHK1/ATHK1 and cytokinin receptor histidine kinases in response to abscisic acid, drought, and salt stress in *Arabidopsis*. Proc. Natl. Acad. Sci. USA 104, 20623–20628.

Tran, L.-S.P., Nishiyama, R., Yamaguchi-Shinozaki, K., Shinozaki, K., 2010. Potential utilization of NAC transcription factors to enhance abiotic stress tolerance in plants by biotechnological approach. GM Crops 1, 32–39.

Wang, X., Culver, J.N., 2012. DNA binding specificity of ATAF2, a NAC domain transcription factor targeted for degradation by tobacco mosaic virus. BMC Plant Biol. 12, 157.

Wang, X., Goregaoker, S.P., Culver, J.N., 2009. Interaction of the *tobacco mosaic virus* replicase protein with a NAC domain transcription factor is associated with the suppression of systemic host defenses. J. Virol. 83, 9720–9730.

Wang, N., Zheng, Y., Xin, H., Fang, L., Li, S., 2013. Comprehensive analysis of NAC domain transcription factor gene family in *Vitis vinifera*. Plant Cell Rep. 32, 61–75.

Welner, D.H., Lindemose, S., Grossman, G.J., Møllegaard, N.E., Olsen, A.N., Helgstrand, C., Skriver, K., Lo Leggio, L., 2012. DNA binding by plant NAC proteins: a firm link to WRKY and mammalian GCM transcription factors. Biochem. J. 444, 395–404.

Xie, Q., Sanz-Burgos, A.P., Guo, H., García, J.A., Gutierrez, C., 1999. GRAB proteins, novel members of the NAC domain family, isolated by their interaction with a geminivirus protein. Plant Mol. Biol. 39, 647–656.

Xie, Q., Frugis, G., Colgan, D., Chua, N.-H., 2000. *Arabidopsis* NAC1 transduces auxin signal downstream of TIR1 to promote lateral root development. Genes Dev. 14, 3024–3036.

Xie, Q., Guo, H.S., Dallman, G., Fang, S.Y., Weissman, A.M., Chua, N.H., 2002. SINAT5 promotes ubiquitin-related degradation of NAC1 to attenuate auxin signals. Nature 419, 167–170.

Xu, Z.Y., Kim, S.Y., Hyeon Do, Y., Kim, D.H., Dong, T., Park, Y., et al., 2013. The *Arabidopsis* NAC transcription factor ANAC096 cooperates with bZIP-type transcription factors in dehydration and osmotic stress responses. Plant Cell 25, 4708–4724.

Xue, G.P., Bower, N.I., McIntyre, C.L., Riding, G.A., Kazan, K., Shorter, R., 2006. TaNAC69 from the NAC superfamily of transcription factors is up-regulated by abiotic stresses in wheat and recognises two consensus DNA-binding sequences. Funct. Plant Biol. 33, 43–57.

Yamasaki, K., Kigawa, T., Inoue, M., Tateno, M., Yamasaki, T., Yabuki, T., Aoki, M., Seki, E., Matsuda, T., Tomo, Y., Hayami, N., Terada, T., Shirouzu, M., Tanaka, A., Seki, M., Shinozaki, K., Yokoyama, S., 2005. Solution structure of an *Arabidopsis* WRKY DNA binding domain. Plant Cell 17, 944–956.

Yamasaki, K., Kigawa, T., Watanabe, S., Inoue, M., Yamasaki, T., Seki, M., Shinozaki, K., Yokohama, S., 2012. Structural basis for sequence-specific DNA recognition by an *Arabidopsis* WRKY transcription factor. J. Biol. Chem. 287, 7683–7691.

Yang, S.D., Seo, P.J., Yoon, H.K., Park, C.M., 2011. The *Arabidopsis* NAC transcription factor VNI2 integrates abscisic acid signals into leaf senescence via the COR/RD genes. Plant Cell 23, 2155–2168.

Yoshiyama, K., Conklin, P.A., Huefner, N.D., Britt, A.B., 2009. Suppressor of gamma response 1 (SOG1) encodes a putative transcription factor governing multiple responses to DNA damage. Proc. Natl. Acad. Sci. USA 106, 12843–12848.

You, J., Zong, W., Li, X., Ning, J., Hu, H., Xiao, J., Xiong, L., 2013. The SNAC1-targeted gene OsSRO1c modulates stomatal closure and oxidative stress tolerance by regulating hydrogen peroxide in rice. J. Exp. Bot. 64, 569–583.

Zheng, X.Y., Spivey, N.W., Zeng, W., Liu, P.P., Fu, Z.Q., Klessig, D.F., et al., 2012. Coronatine promotes *Pseudomonas syringae* virulence in plants by activating a signaling cascade that inhibits salicylic acid accumulation. Cell Host Microbe 11, 587–596.

Zhu, T., Nevo, E., Sun, D., Peng, J., 2012. Phylogenetic analyses unravel the evolutionary history of NAC proteins in plants. Evolution 66 (6), 1833–1848.

Zhu, Q., Zou, J., Zhu, M., Liu, Z., Feng, P., Fan, G., Wang, W., Liao, H., 2014. *In silico* analysis on structure and DNA binding mode of AtNAC1, a NAC transcription factor from *Arabidopsis thaliana*. J. Mol. Model. 20, 2117.

SECTION C

FUNCTIONAL ASPECTS OF PLANT TRANSCRIPTION FACTOR ACTION

14	*Homeobox transcription factors and the regulation of meristem development and maintenance*	215
15	*CUC transcription factors: to the meristem and beyond*	229
16	*The role of TCP transcription factors in shaping flower structure, leaf morphology and plant architecture*	249
17	*GROWTH-REGULATING FACTORs, a transcription factor family regulating more than just plant growth*	269
18	*The multifaceted roles of miR156-targeted SPL transcription factors in plant developmental transitions*	281
19	*Functional aspects of GRAS family proteins*	295
20	*DELLA proteins, a group of GRAS transcription regulators that mediate gibberellin signaling*	313
21	*bZIP and bHLH family members integrate transcriptional responses to light*	329
22	*What do we know about homeodomain-leucine zipper i transcription factors? Functional and biotechnological considerations*	343

CHAPTER 14

Homeobox Transcription Factors and the Regulation of Meristem Development and Maintenance

Katsutoshi Tsuda, Sarah Hake

U.S. Department of Agriculture-Agricultural Research Service, Plant and Microbial Biology Department, Plant Gene Expression Center, University of California at Berkeley, Berkeley, CA, USA

OUTLINE

14.1 Introduction	215
14.2 KNOX and BELL: TALE Superfamily Homeobox Genes	216
14.2.1 Classification of KNOX and BELL Families	216
14.2.2 Functions of TALE–HD Genes in Plant Lineages	217
14.2.3 Functions of KNOX Genes in Plant Shoot Meristems	218
14.2.4 KNOX Downstream Pathways	219
14.2.4.1 KNOX and Other TFs	219
14.2.4.2 KNOX and Hormones	220
14.2.5 KNOX DNA-Binding Preferences	221
14.2.6 Upstream Regulators of KNOX Genes	221
14.2.7 WUSCHEL Function in Stem Cell Regulation	222
14.2.8 WOX Protein Structure and DNA-Binding Preferences	222
14.2.9 WUS Mutant Phenotype and its Function	223
14.2.10 Functions of Other WOX Family Genes	223
14.2.11 Multiple WUS Regulatory Loops for Shoot Stem Cell Maintenance	223
14.2.11.1 WUSCHEL–CLAVATA Negative Feedback Loop	223
14.2.11.2 WUSCHEL–Cytokinin Positive Feedback Loop	224
14.2.11.3 WUSCHEL–HECATE Negative Feedback Loop	224
14.2.12 Other WUS Targets	224
14.2.13 Termination of WUS Expression in Floral Stem Cells	225
References	225

14.1 INTRODUCTION

Homeobox genes, initially found from studies of *Drosophila* homeotic mutants, encode transcription factors (TFs) that have a characteristic DNA-binding domain called a homeodomain (HD). HDs are typically composed of three α-helices connected by two loops that are approximately 60 amino acids long. HDs are found exclusively in eukaryotes, although they share structural similarities with prokaryotic helix–turn–helix DNA-binding domains (Treisman et al., 1992). Although HDs were lost in several lineages of unicellular parasites and intracellular symbionts, they are present in all major eukaryote lineages, indicating that the last common ancestor of extant eukaryotes had already acquired HDs (Derelle et al., 2007). During eukaryote evolution, HDs were recruited to higher order processes such as cell–cell communication and cell differentiation. In fungi and green alga, HDs are involved in mating-cell-type specification and sexual development (Johnson, 1995; Casselton and Olesnicky, 1998; Lee et al., 2008). In animals and plants, HDs regulate cell differentiation in various developmental processes (Gehring et al., 1994b; Hake et al., 2004). The increase of HDs in these two multicellular lineages implies that HDs contributed to the evolution of multicellular body plans (Derelle et al., 2007).

The shoot apical meristem (SAM) of vascular plants is an indeterminate structure comprised of pluripotent self-renewing stem cells in its center and daughter cells at the periphery. The SAM is responsible for initiation of the aerial parts of land plants. Lateral organs such as leaves and flowers are initiated from the flank of the SAM at the expense of its stem cell population. The two opposing activities, replenishment of the stem cell population and recruitment of daughter cells into lateral organs, must be balanced in the SAM. Thus, the SAM is central to plant organogenesis, and the regulatory mechanisms of this dynamic structure have attracted many biologists. Extensive genetic studies aimed at exploring the regulators of SAM activities have led to the discovery of two classes of homeobox genes, TALE (3-amino-acid loop extension; *KNOX* and *BLH*) and *WUSCHEL*, which are crucial regulators for establishment and maintenance of the SAM. In this chapter we will introduce the function of these homeobox genes in shoot meristems.

14.2 KNOX AND BELL: TALE SUPERFAMILY HOMEOBOX GENES

14.2.1 Classification of *KNOX* and *BELL* Families

The maize *knotted1* (*kn1*) gene, cloned by transposon tagging the dominant *Knotted1* mutant, was the first homeobox gene described in plants (Hake et al., 1989; Vollbrecht et al., 1991). *kn1* belongs to TALE superfamily homeobox genes. TALE proteins have HDs with three extra amino acids between helix1 and helix2 (Bertolino et al., 1995). Although TALE superfamily proteins are usually referred to as "atypical" HDs, both TALE HDs and non-TALE HDs are present in the last common ancestor of extant eukaryotes. Therefore, these two HDs are sister groups (Derelle et al., 2007).

In plants, TALE homeobox genes are classified into two families: *knotted1*-like homeobox (*KNOX*) and *BELL1*-like homeobox (*BELL* or *BLH*) genes. KNOX proteins have the N-terminal KNOX domain (with KNOX1 and KNOX2 subdomains) and the ELK domain followed by the HD at the C-terminus. The KNOX domain shares significant sequence similarities between that of the human TALE HD protein myeloid ectopic viral integration site (MEIS), and therefore these domains were collectively designated MEINOX (MEIS and KNOX). Sequence similarities in the MEINOX domain indicate that this subclass of TALE superfamily was already present prior to the divergence of plant and animal lineages 1.6 billion years ago (Burglin, 1997). BELL proteins are comprised of SKY and BELL domains at the N-terminus and HD in the C-terminus. SKY and BELL domains are required for the formation of heterodimers with KNOX proteins through the interaction with MEINOX domains. This heterodimerization suggests that KNOX and BELL act in the same context (Bellaoui et al., 2001; Byrne et al., 2003; Smith and Hake, 2003).

Both families are found in all plant lineages; however, the number of gene family members varies and has increased with multicellularity (Table 14.1, data from Mukherjee et al., 2009). Species in unicellular lineages, including both green and red algae, have only a single *KNOX* gene and one or two *BELL* genes. In angiosperms, multiple genes are found in both *KNOX* and *BELL* families (Mukherjee et al., 2009). *KNOX* genes further fall into two classes (class I and II) based on their sequence similarities, the position of introns, and expression patterns (Kerstetter et al., 1994). Class I *KNOX* genes are specifically expressed in the SAM and unelongated internodes but there are exceptions in compound-leafed species (Section 14.2.2), whereas class II genes are expressed in various tissues including both shoots and roots. Class I and II genes are found in all terrestrial plants but not in unicellular algae, indicating that the duplication occurred in the last common ancestor of land plant lineages. A third class of

TABLE 14.1 *KNOX*, *BELL*, and *WOX* Family Genes in Plant Lineages

	Eudicots		Monocots		Bryophyta	Lycopodiophyta	Unicellular green algae			Red algae
	Arabidopsis	Poplar	Maize	Rice	*Pp*	*Sm*	*Cr*	*Ol*	*Ot*	*Cm*
KNOX	8	15	13	12	5	5	1	1	1	1
BELL	13	19	17	14	4	2	1	2	2	1
WOX	16	19	16	15	3	6	–	1	1	–

Sm, *Selaginella moellendorffii*; *Cr*, *Chlamydomonas reinhardtii*; *Ol*, *Ostreococcus lucimarinus*; *Ot*, *Ostreococcus tauri*; *Cm*, *Cyanodioschyzon merola*.
Data was adopted from Mukherjee et al., 2009.

KNOX genes, class M, lack the HD and have been found only in dicot species so far (Kimura et al., 2008; Magnani and Hake, 2008). Thus, some *KNOX* subfamilies have coevolved with plant lineages.

14.2.2 Functions of TALE–HD Genes in Plant Lineages

TALE homeobox genes have diverse functions across the different lineages, but so far are specific to the diploid generation. In the unicellular green alga, *Chlamydomonas reinhardtii*, the BELL protein Gamete-specific *plus1* (Gsp1) and KNOX protein Gamete-specific *minus1* (Gsm1) are expressed in plus and minus gametes, respectively. Gsp1 and Gsm1 form a heterodimer upon gamete fusion to initiate the zygotic developmental program (Lee et al., 2008). This function is similar to that of TALE–HD genes in fungi which regulate mating-cell-type differentiation and zygotic gene expression (Johnson, 1995), suggesting that the original function of KNOX and BELL proteins in the plant lineage ancestor was also to regulate gene expression during sexual and zygote development.

In the moss *Physcomitrella patens*, a model system of basal land plant lineages (bryophytes), haploid generation is dominant over diploid generation. Gametophytes, the haploid plant body, produce shoot-like structures with an indeterminate apical cell responsible for repetitive organ production, whereas sporophytes, the diploid body, do not have indeterminate growth potential and terminate with the formation of gametes. Both class I and II *KNOX* genes are exclusively expressed in sporophytes. Disruption of class I genes causes a slight reduction of cell proliferation in zygotes (Sakakibara et al., 2008). On the other hand, disruption of class II genes causes the formation of gametophyte-like structures in sporophytes without meiosis, showing that class II *KNOX* genes are required for repression of the gametophyte developmental program in diploid sporophytes (Sakakibara et al., 2013). Thus, these studies highlight the important function of class II *KNOX* genes to regulate haploid to diploid morphological transition in bryophytes.

In vascular plants, the diploid phase is dominant over the haploid. Upon gamete fusion and fertilization, the zygote does not immediately proceed to meiosis but continues to develop higher orders of shoot and root systems. In angiosperms, class I *KNOX* genes have been intensely studied and shown to play indispensable roles in the establishment and maintenance of the SAM (Section 14.2.3). In addition, class I *KNOX* genes are involved in compound leaf development. In species with compound leaves, class I genes are first downregulated in incipient leaves, but reactivated in developing leaf primordia to facilitate leaflet formation (Hareven et al., 1996; Hay and Tsiantis, 2006; Bharathan et al., 2002).

In contrast, limited knowledge has been accumulated for angiosperm class II genes. Class II genes are expressed in various tissues including leaves, roots, as well as meristems (Kerstetter et al., 1994). In *Arabidopsis* a class II *KNOX* gene *KNAT7* represses secondary cell wall biosynthesis in interfascicular fibers of inflorescence stems by repressing genes in the lignin biosynthetic pathway (Li et al., 2012; Gong et al., 2014).

The TALE class protein, BELL, plays crucial functions in plant development in concert with KNOX proteins. BELL and KNOX proteins form heterodimers and their expression patterns and functions are diverse. For example, *BEL1*, founder member of the *BELL* family, is required for ovule development (Reiser et al., 1995). *SAWTOOTH1* (*SAW1*) and *SAW2* are expressed in lateral organs and are required to repress class I *KNOX* gene expression and leaf margin growth (Kumar et al., 2007). Three others are important for SAM maintenance: the triple mutants of *ARABIDOPSIS THALIANA HOMEOBOX1* (*ATH1*), *PENNYWISE* (*PNY*), and *POUND-FOOLISH* (*PNF*) phenocopy *stm* mutants (Rutjens et al., 2009). Yeast two-hybrid studies showed that each BELL protein has the potential to interact with multiple KNOX proteins including both class I and II (Hackbusch et al., 2005; Bellaoui et al., 2001). However, the precise combination of the KNOX–BELL interaction *in vivo* is still unknown.

The protein–protein interaction between KNOX and BELL is known to regulate subcellular localization. STM alone localizes to the cytosol and translocates to the nucleus upon dimerization with BELL partners (Bhatt et al., 2004; Cole et al., 2006). SKY and BELL domains at the N-terminus of BELL proteins are required to interact with the MEINOX domain of KNOX proteins. This interaction was suggested to hide nuclear export signals harboring in the BELL domain, thus the KNOX–BELL heterodimer is driven to the nucleus (Rutjens et al., 2009). This protein interaction mechanism is also utilized to titrate KNOX–BELL activity to regulate leaf development in dicots. Class M KNOX proteins, which lack the HD, are expressed in developing leaves and interact with BELL proteins in tomato and *Arabidopsis*. Natural variation which led to overexpression of the tomato class M gene *PETROSELINUM* (*PTS*) results in increased leaflet formation (Kimura et al., 2008). PTS interacts with the tomato BELL-type protein BIPINNATA (BIP; a homolog of *Arabidopsis* SAW) and this complex localizes in the cytoplasm. The loss-of-function mutations of *BIP* also increase compound leaf complexity. Furthermore, overexpression of *Arabidopsis* class M gene *KNATM* suppresses the phenotype caused by SAW overexpression (Magnani and Hake, 2008). Thus, these genetic and molecular analyses clearly show that class M *KNOX* proteins have a role in preventing BELL proteins from interacting with other KNOX proteins by sequestering BELL proteins and thereby increasing leaf shape complexity.

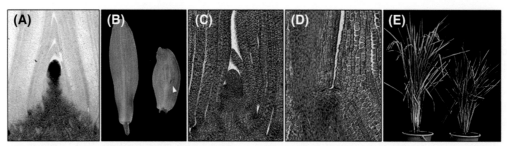

FIGURE 14.1 **Expression pattern of KN1 protein and *KNOX* gain-of and loss-of function phenotypes.** (A) Immunolocalization of the maize vegetative shoot meristem using anti-KN1 antibody. (B) First leaf blades from wild-type (left) and *Kn1-N* dominant mutant (right). An arrowhead indicates a knot formed on the mutant leaf blade. (C, D) Shoot apex sections in wild-type (C) and *kn1* loss-of-function mutant in W23 background. (E) Loss-of-function phenotype of the rice *OSH15* gene. Mature wild-type and *d6* mutant plants are shown on left and right, respectively.

14.2.3 Functions of *KNOX* Genes in Plant Shoot Meristems

Expression patterns and overexpression studies suggest that class I *KNOX* genes play an important function in shoot meristems. The maize *kn1* transcript accumulates in all vegetative and reproductive shoot meristems, but is excluded from leaf primordia. Importantly, *kn1* is downregulated at the site of leaf initiation in the SAM called Plastochron 0 (P0) before any morphological sign appears (Figure 14.1A; Jackson et al., 1994). In other species with simple leaves such as rice and *Arabidopsis*, class I *KNOX* gene expression is similar to *kn1*; it is highly expressed in the SAM with overlapping pattern and downregulated in P0 and older leaf primordia (Lincoln et al., 1994; Long et al., 1996; Sentoku et al., 1999). Therefore, *KNOX* genes are used as markers for the undifferentiated or meristematic state in various species.

In maize *Kn1* dominant mutants, *kn1* is ectopically expressed in early stages of leaf development and outgrowths (or knots), caused by extra cell divisions and growth, which form sporadically along lateral veins in leaf blades (Figure 14.1B; Smith et al., 1992; Freeling and Hake, 1985; Ramirez et al., 2009). Leaf sheath characteristics often intrude into the leaf blade with displaced ligules. Ectopic ligule formation is also found along lateral veins. Thus, distal leaf parts are shifted to more proximal cell fates. Given that *kn1* is expressed in the meristem, it was suggested that KN1 represses or retards developmental programs for leaf differentiation (Muehlbauer et al., 1997). Overexpression studies in many species (Chuck et al., 1996; Lincoln et al., 1994; Sentoku et al., 2000) confirmed that *KNOX* genes have a role in preventing cell differentiation.

The first evidence for the importance of KNOX genes in SAM establishment and maintenance was from the study of *Arabidopsis* loss-of-function mutants of the *SHOOT-MERISTEMLESS* (*STM*) gene (Long et al., 1996). *stm* mutants fail to establish the SAM during embryogenesis. The strong alleles of *stm* mutants completely lack the SAM after germination and are not able to form any shoot organs (Barton and Poethig, 1993; Long and Barton, 1998).

In weak *stm* alleles, although a dome-shaped structure of the SAM is lacking, nucleocytoplasmic cells, which are characteristic of cells in the SAM, are still observed. Weak *stm* mutants frequently recover from adventitious shoots along leaf petioles. These are eventually terminated and reiteration of this process results in bushy plants. This phenotype indicates that *STM* is required for SAM formation and maintenance but not for lateral organ initiation (Clark et al., 1996). The weak allele occasionally bolts and forms flowers with reduced numbers of floral organs especially in inner wholes such as petal, anther, and carpel. Thus, *STM* is required to maintain shoot meristems in all developmental phases.

In grasses, class I *KNOX* genes are also required to establish and maintain the SAM. Loss-of-function mutants of maize *kn1* show defects in shoot meristems; however, phenotypic severity depends on the inbred background (Kerstetter et al., 1997, Vollbrecht et al., 2000). In W23, an inbred with shorter SAMs, about 80% of mutants show a shootless phenotype (Figure 14.1C and D). In the B73 inbred, which has taller SAMs, penetrance of the shootless phenotype is less than 5% and the majority of mutants reach maturity. In this permissive background, mutants often fail to produce tillers and ears, and the tassel branch number is reduced. At low frequency, extra leaves form in the axils of leaves. These results suggest that *kn1* plays a crucial role not only in shoot meristem formation and maintenance, but also in boundary establishment between leaves and meristems. The phenotype dependence on the genetic background suggests there is a modifier(s) of *KNOX* loss-of-function phenotypes in different inbreds, presumably associated with regulation of SAM height.

In rice, loss of function of the rice *kn1* ortholog, *Oryza sativa homeobox1* (*OSH1*), is able to form the SAM during embryogenesis, but fails to maintain the SAM just after germination, resulting in a terminated shoot with a few leaves (Tsuda et al., 2011). *osh1* mutants are able to maintain the SAM and produce panicles when regenerated from a callus; however, double mutants with mutation in another member of class I *KNOX* gene *OSH15* cannot regenerate functional shoots. This suggests that the contribution of other redundant *KNOX* genes to SAM maintenance differs in the conditions that prevail in the early vegetative phase and *in vitro* shoot regeneration.

KNOX gene function in internode development is highlighted in both rice and *Arabidopsis*. *dwarf6* (*d6*), loss-of-function mutants of *OSH15*, show a dwarf phenotype as a result of impaired elongation at lower internodes (Figure 14.1E; Sato et al., 1999). In *d6* internodes, cell size in the longitudinal direction and number of cells are reduced. *Arabidopsis brevipedicellus* (*bp*) mutants show a markedly altered architecture with reduced height owing to irregularly shortened internodes and downward-pointing and very shortened pedicels (Douglas et al., 2002; Venglat et al., 2002). In addition, *bp* internodes show a precocious accumulation of lignin, a secondary cell wall component (Mele et al., 2003). Collectively, these observations of rice and *Arabidopsis* mutants indicate that a subset of *KNOX* gene family has a role in preventing precocious differentiation and elaborating internode structure.

Although single *bp* mutants show no abnormality in SAM function, the combination of *bp* mutants with weak *stm* alleles greatly reduces the recovery of organogenesis, indicating that *BP* has a redundant role with *STM* (Byrne et al., 2002). Similarly, the mutation in *KNAT6* does not cause any developmental abnormalities but, like *bp*, it enhances the defects of a weak *stm* allele (Belles-Boix et al., 2006). The mutation in *KNAT2* shows no such interaction with *STM*, suggesting that *KNAT2* does not contribute to SAM maintenance in a redundant manner with other class I *KNOX* genes. Similarly, enhancement of a SAM defect was observed in higher order mutants in grasses. In rice, *osh1 d6* double mutants fail to establish the SAM during embryogenesis or to regenerate shoots from a callus, suggesting that *OSH15* plays a redundant role with *OSH1* in SAM formation and maintenance even though *d6* single mutants show no defects in the SAM (Tsuda et al., 2011). Similarly, the loss-of-function mutant of the maize *KNOX* gene, *rough sheath1* (*rs1*) has no defect; however, the *kn1–rs1* double-mutant completely lacks the SAM in embryos even in a permissive background such as that of B73 (Bolduc et al., 2014).

These genetic analyses show that class I *KNOX* genes have pivotal roles in shoot meristem formation and its maintenance throughout the plant life cycle. Although their functions are significantly redundant, some of the members acquired specific roles such as *OSH15* and *BP* in internode development. It is hard to verify whether this functional differentiation is attributed to the difference in specific expression patterns or in protein function, or both.

14.2.4 KNOX Downstream Pathways

Recent advances in sequencing technologies have greatly expanded our view of KNOX downstream regulatory networks. Using chromatin immunoprecipitation followed by next-generation sequencing (ChIP-seq)

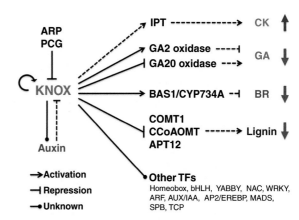

FIGURE 14.2 *KNOX* regulatory network. Solid lines represent direct transcriptional regulation supported by experimental evidence. Red and blue vertical arrows represent positive and negative effects, respectively, on the accumulation of products in corresponding pathways.

combined with transcriptome analyses, a genome-wide view of the KNOX downstream network was uncovered in maize and rice inflorescence meristems (Figure 14.2; Bolduc et al., 2012; Tsuda et al., 2014). In this section, we will summarize the accumulated knowledge on KNOX downstream pathways with the support of these genome-wide studies.

14.2.4.1 KNOX and Other TFs

ChIP-seq studies identified more than 5000 putative direct targets of KN1 in maize inflorescence meristems, and a similar number of OSH1 putative direct targets was identified in rice. The most enriched category is that of the TF. KN1 was found to target various TF families such as homeobox, ARF, AUX-IAA, AP2/EREBP, bHLH, YABBY, MADS, NAC, SPB, TCP, and WRKY. The homeobox category contained TFs important for SAM formation/maintenance such as KNOX and BELL. In rice, it was reported that OSH1 directly activates itself and four other *KNOX* genes and that the positive autoregulation of *OSH1* is essential for SAM maintenance (Tsuda et al., 2011). In addition, a rice *BELL* gene *qSH1*, which encodes a putative KNOX partner protein, was upregulated after induction of *OSH1* overexpression (Tsuda et al., 2014). Therefore, critical *KNOX* gene functions include activating their own expression and that of their partners as well. *Arabidopsis* STM is known to directly activate *CUC1*, which is essential for organ boundary formation (Spinelli et al., 2011). KN1 and OSH1 also target *CUC* genes. These results may explain the boundary-related defects observed in *KNOX* gain and loss-of-function mutants (Kerstetter et al., 1997; Clark et al., 1996; Tsuda et al., 2014). Many KN1 and OSH1 targets are TFs that are known to be involved in leaf and flower development such as HD-ZIPIII, YABBY, and MADS. These findings suggest that KNOX proteins regulate TFs involved not only in SAM maintenance, but also in leaf and floral organ differentiation.

14.2.4.2 KNOX and Hormones

Early overexpression studies suggested hormones were downstream of *KNOX* genes. *kn1*-overexpressing tobacco plants produced ectopic shoots on leaves, something also documented in transgenic tobacco plants that overexpress the cytokinin(CK) biosynthetic gene *isopentenyl transferase* (*ipt*) (Estruch et al., 1991; Li et al., 1992; Sinha et al., 1993). Measurement of hormone levels in *KNOX*-overexpressing plants revealed that multiple hormone levels were altered in these plants (Tamaoki et al., 1997; Kusaba et al., 1998). The most striking difference was observed in gibberellins (GA) and CK levels; bioactive GAs were found to be greatly reduced, whereas CK levels were dramatically increased in *KNOX*-overexpressing tobacco plants.

14.2.4.2.1 GIBBERELLIN

GA acts in differentiated organs to accelerate cell elongation. Detailed quantification of GA and its precursors in *NTH15* (tobacco *STM* ortholog)-overexpressing tobacco revealed that the levels of the final product GA_1 and its precursor GA_{20} were greatly reduced, whereas an upstream intermediate GA_{19} was increased (Tanaka-Ueguchi et al., 1998). Further studies demonstrated that NTH15 directly binds to the intron of *Ntc12*, a gene-encoding GA20 oxidase that produces GA_{20} from GA_{19} and represses its expression in the SAM (Sakamoto et al., 2001). Similar repression of the GA20 oxidase gene by *KNOX* genes was reported in *Arabidopsis* (Hay et al., 2002). In addition, maize KN1 activates the GA catabolism enzyme gene *GA2 oxidase1* by direct binding to its intron (Bolduc and Hake, 2009). *GA2ox1* expression overlaps that of *kn1* in the shoot apex, and its expression level increases in dominant *Kn1* mutant leaves. Thus, KNOX proteins directly regulate GA biosynthetic and catabolism genes and in so doing prevent accumulation of GA in the SAM.

14.2.4.2.2 CYTOKININ

CK is a class of phytohormone that promotes shoot formation and represses root formation. CK is required for SAM maintenance (Kurakawa et al., 2007). An increase of CK in *KNOX*-overexpressing transgenic plants suggested that *KNOX* genes activate the CK biosynthetic pathway. Indeed, inducible overexpression of *STM* leads to a rapid increase in *AtIPT7* expression (Yanai et al., 2005; Jasinski et al., 2005). CK treatment recovers organogenesis potential when a strong *stm* allele fails to form any leaves. In addition, expression of CK-inactivating genes under the *STM* promoter in the mutant background in which GA signaling is constitutively active in phenocopied *stm* mutants, suggesting that the combination of a high level of CK and a low level of GA is important for SAM function. Whether STM directly activates these genes is still unknown, although upregulation of *AtIPT7* occurred a short time after *STM* induction. It is also reported that *OsIPT2* expression increased in transgenic rice overexpressing *OSH15* (Sakamoto et al., 2006). Thus, the *KNOX* function to activate CK biosynthesis is likely conserved in both dicots and monocots.

14.2.4.2.3 AUXIN

The KN1 ChIP-seq study revealed the prevalence of hormone genes as another prominent downstream category (Bolduc et al., 2012). Surprisingly, KN1 directly binds to loci encoding various components of hormonal pathways not only for GA and CK, but also for auxin, brassinosteroid (BR), ethylene, and abscisic acid. Even in the GA and CK pathway, many genes that were previously unlinked to *KNOX* functions, such as receptor and signaling genes, were found to be direct targets. Among these hormones, the auxin pathway is the most differentially expressed in *kn1* loss-of-function and gain-of function mutants. KN1 directly binds auxin-related genes at all levels, including biosynthesis, degradation, export, import, perception, and signaling. In *Kn1-N* dominant mutants carrying *DR5:RFP*, a reporter of auxin-signaling output, stronger RFP expression was observed in the vasculature of leaves compared with that of the wild type. These observations suggest that *KN1* has a positive impact on the auxin-signaling pathway (Bolduc et al., 2012).

14.2.4.2.4 BRASSINOSTEROID

BR, a growth-promoting phytohormone involved in diverse aspects of plant growth and development (Clouse and Sasse, 1998), is also an important downstream target of KNOX proteins. Indeed, inducible overexpression of *OSH1* resulted in BR-insensitive phenotypes and *osh1* mutants showed BR overproduction phenotypes (Tsuda et al., 2014). Of the BR pathway genes that were directly bound by OSH1, three *CYP734A* genes that encode BR catabolism enzymes (Sakamoto et al., 2011) were rapidly upregulated upon *OSH1* induction. RNA interference (RNAi) knockdown of these *CYP734A* genes occasionally arrested shoot growth. Cells in their shoot apices and stems were abnormally enlarged and vacuolated, suggesting that BR inactivation is important for the prevention of premature differentiation of cells in the SAM. Thus, these observations indicate that BR inactivation is an important component of *KNOX* gene function.

14.2.4.2.5 LIGNIN

KNOX genes have been known to repress lignin biosynthesis. The comparison of gene expression profiles using microarrays revealed a significant number of genes involved in cell wall biosynthesis, especially in the lignin biosynthesis pathway, something that was differentially expressed in *bp* mutants compared to wild-type *Arabidopsis* plants. Premature and increased lignin deposition was observed in *bp* mutant stems. In contrast, lignin deposition decreased in the stem of *BP*-overexpressing plants

(Mele et al., 2003). Similarly, a reduction in lignin content was reported in maize leaves of dominant mutants such as *Kn1*, *Rs1*, and *Gn1* (Townsley et al., 2013).

14.2.5 KNOX DNA-Binding Preferences

The third HD helix functions as the DNA-recognition helix (Gehring et al., 1994a), in which the amino acid in position X of WFXN is critical for DNA-binding specificity (Treisman et al., 1989). Human TALE protein MEIS contains WFIN at this position and recognizes the TGACAGG/CT motif (Chang et al., 1997). KNOX and BELL proteins also have WFIN, and *in vitro* binding experiments show that KN1 recognizes TGAC sequences with higher affinity in the presence of its BELL partner protein (Smith et al., 2002; Chen et al., 2004). On the other hand, motif discovery analyses using genomic regions directly bound by KNOX proteins in the ChIP-seq experiment revealed that the most enriched motif in KNOX-binding regions contains two GA cores separated by 3 bp (base pairs) with G enriched in between (Bolduc et al., 2012; Tsuda et al., 2014). The second GA is frequently followed by C/T. The preferences of nucleotides flanking two core GA sequences are not as obvious *in vitro*, although this motif shares similarities with the TGAC motif determined by *in vitro* experiments. This finding may reflect the fact that KNOX proteins form complexes with various BELL proteins *in vivo* and that each combination of the KNOX–BELL complex has different preferences for its flanking nucleotides.

14.2.6 Upstream Regulators of *KNOX* Genes

The downregulation of *KNOX* genes during leaf initiation is maintained throughout leaf development (Jackson et al., 1994). Failure to maintain *KNOX* silencing results in disturbed acquisition of proper cell fate and organ shape (Timmermans et al., 1999; Tsiantis et al., 1999; Byrne et al., 2000; Hay and Tsiantis, 2006). Thus, *KNOX* repression is also crucial for normal development.

Epigenetic histone modification plays important roles in *KNOX* silencing. Polycomb-repressive complex2 (PRC2) catalyzes the trimethylation of histone H3 Lysine 27 (H3K27me3), and this histone modification recruits PRC1 which brings about monoubiquitylation of the histone H2A subunit, resulting in a somatically heritable silencing state over multiple cell cycles during development (Bemer and Grossniklaus, 2012). In loss-of-function mutants of *Arabidopsis* PRC2 proteins CURLY LEAF (CLF), SWINGER (SWN), and FERTILISATION INDEPENDENT ENDOSPERM (FIE) and in double-mutants of PRC1 protein AtRING1a and AtRING1b, *KNOX* genes are derepressed, indicating that both PRC2 and PRC1 components are required for *KNOX* silencing (Katz et al., 2004; Schubert et al., 2006; Xu and Shen, 2008).

ASYMMETRIC LEAVES1 (*AS1*) in *Arabidopsis*, *ROUGH SHEATH2* (*RS2*) in maize, and *PHANTASTICA* in snapdragon constitute the ARP genes that encode Myb-type TFs that maintain the silencing of *KNOX* genes in leaves. In loss-of-function *ARP* mutants, the initial downregulation of *KNOX* genes occurs but reactivation of their expression is observed in later leaf development (Timmermans et al., 1999; Tsiantis et al., 1999; Byrne et al., 2000). In maize *rs2* mutants the ectopic accumulation of KNOX proteins was observed in patches as if the leaves were mosaics of *KNOX*-expressing and nonexpressing sectors (Timmermans et al., 1999). This result suggested that the RS2 protein acts as a cellular memory, similar to chromatin-remodeling factors. *Arabidopsis* AS1 forms heterodimers with ASYMMETRIC LEAVES2 (AS2), a LATERAL ORGAN BOUNDARY (LOB) TF, and functions as a repressive chromatin complex via recruitment of histone chaperone histone regulatory protein A (HIRA; Phelps-Durr et al., 2005). Furthermore, the AS1–AS2 complex directly binds to cis-elements in the promoters of *BP* and *KNAT2*, and is required to recruit PRC2 and accumulate the H3K27me3 repressive mark (Guo et al., 2008; Lodha et al., 2013). The AS1–AS2 complex physically interacts with multiple components of PRC2. Thus, ARP proteins act as epigenetic chromatin regulators and thereby maintain *KNOX* gene repression during leaf development.

Auxin signaling is also required for the repression of *KNOX* genes. Auxin distribution is dynamically regulated by a class of efflux carrier PINFORMED (PIN) family proteins. PIN1 asymmetrically localizes on cell membranes close to future leaf primordia and creates auxin concentration maxima (Krecek et al., 2009). This local auxin maximum is required to trigger leaf initiation, given that *pin1* mutants fail to initiate flowers (lateral organs in the inflorescence meristem), and microapplication of auxin on *pin1* mutant apices restores leaf initiation (Okada et al., 1991; Reinhardt et al., 2000; Reinhardt et al., 2003). The pattern of auxin maxima coincides with the site of initial downregulation of *KNOX* genes, suggesting antagonism between auxin and *KNOX* genes. This notion was supported by several experiments. Scanlon reported that treatment with NPA, an auxin efflux inhibitor, inhibits KN1 downregulation and leaf initiation in shoot apex tissue culture (Scanlon, 2003). In *Arabidopsis*, the double-mutant that is *PIN1* and *PINOID*, encoding a serine/threonine kinase required for the regulation of subcellular localization of PIN proteins, causes *STM* misexpression in cotyledon primordia, and removal of the *STM* function from *pin1–pid* restores cotyledon formation (Furutani et al., 2004). Furthermore, PIN1 and AXR1, which are involved in auxin signaling, are also required for the repression of *BP* in leaves (Hay et al., 2006). Together, these studies strongly suggest important roles for the auxin pathway in *KNOX* downregulation. Nevertheless, how auxin represses *KNOX* transcription at a mechanistic level remains elusive.

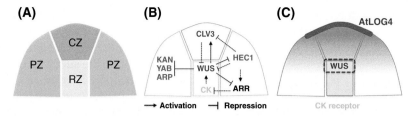

FIGURE 14.3 **Multiple feedback loop through WUS gene function.** (A) Subdomains in the shoot apical meristem. CZ, central zone; PZ, peripheral zone; RZ, rib zone. (B) WUS regulatory networks. Solid lines represent direct transcriptional regulation. (C) Spatial representation of CK and its receptor to define WUS expression domain. Orange background indicates the distribution of CK suggested by the expression of AtLOG4. The yellow region is the domain in which CK receptor AHK4 is expressed.

14.2.7 WUSCHEL Function in Stem Cell Regulation

The SAM is composed of three functionally distinct domains (Figure 14.3). The central zone (CZ) harbors the stem cell population located at the tip of the SAM. Stem cells in the CZ slowly divide and provide daughter cells to the peripheral zone (PZ), which surrounds the CZ. As these daughter cells leave the CZ, they have the potential to differentiate and be incorporated into lateral organs. The rib zone (RZ) is located below the CZ and provides cells for the stem. The replenishment of indeterminate cells in the formation of differentiated tissues must be robustly balanced.

Parallel with the maintenance of indeterminacy of meristematic tissues through the KNOX–BELL pathway are the regulation and maintenance of stem cell populations, which are also critical for SAM function. Genetic studies identified a homeobox gene WUSCHEL (WUS) as a critical regulator of shoot stem cell maintenance in Arabidopsis. wus mutants fail to maintain the shoot stem cell population and the SAM prematurely terminates (Laux et al., 1996).

WUS is the founder member of the WUSCHEL homeobox (WOX) gene family. WOX genes are present only in green alga and land plant lineages, indicating that this family was present in the last common ancestor of green plants (Chlorobionta; Mukherjee et al., 2009). WOX family members can be divided into three clades: WUS, intermediate, and ancient. In green algae and bryophytes such as the moss Physcomitrella patens, only ancient clade genes can be found, indicating that the WUS and intermediate clades diverged after the emergence of vascular plants (Van Der Graaff et al., 2009). Intermediate clade genes are found in vascular plants, including flowering plants and lycophytes. The WUS clade is most recent because genes in this clade are found only in seed plants.

14.2.8 WOX Protein Structure and DNA-Binding Preferences

In contrast to a typical HD of 60 amino acids, the WUS HD has 66 amino acids and other WOX members have 65 amino acids (Haecker et al., 2004). WUS and its orthologs in other angiosperms have a 2-amino-acid extension between helix1 and helix2, whereas all other WOX members have a 1-amino-acid insertion in this position. The HD is at the N-terminus in WOX proteins. Four amino acids are inserted in all WOX proteins between helix 2 and helix 3 (Zhang et al., 2010). Although WOX proteins have been shown to localize to the nucleus, there is no clear nuclear localization signal in this family (Van Der Graaff et al., 2009). WUS homodimerizes through its C-terminus region (Busch et al., 2010; Nagasaki et al., 2005), and directly binds the TAAT DNA motif in cis-acting elements of downstream target loci (Lohmann et al., 2001; Yadav et al., 2011). Busch et al. performed ChIP–chip experiments to determine the chromatin regions directly bound by WUS at the genome-wide level and identified another consensus WUS-binding sequence TCACGTGA as an enriched DNA motif in WUS-bound regions (Busch et al., 2010). Both motifs were experimentally shown to be bound by WUS in vitro; however, the mechanism by which WUS recognizes these different DNA motifs and the transcriptional regulatory effect on downstream targets through these motifs are still elusive.

Based on amino acid sequence comparison between WUS and its orthologs in other species, the WUS box was identified as a highly conserved motif downstream of the HD. The WUS box is also found among other members of the WUS clade subgroup, suggesting the importance of this motif (Haecker et al., 2004). In addition, an ERF-associated amphiphilic repression (EAR)-like motif, which interacts with TOPLESS (TPL) and related proteins that belong to a class of corepressor proteins, was found in some members of WUS-clade proteins (WUS, WOX5, and WOX7). An acidic domain upstream of the WUS box is found in WUS and some members of the WUS clade, although this is not the case in monocots (Haecker et al., 2004; Zhang et al., 2010). In Antirrhinum, a truncated allele that lacks both the C-terminus and the WUS box behaves in a dominant negative manner (Kieffer et al., 2006). In addition, overexpression studies showed that, when the WUS box was mutated, WUS lost the ability to induce ectopic shoot stem cell identity and resulted in a phenotype similar to wus-1 loss-of-function mutants, suggesting that WUS proteins lacking the WUS box function compete with native WUS proteins for their DNA-binding sites and disturb downstream gene regulation (Ikeda et al., 2009). The EAR-like motif is not essential for WUS repression activity. Mutation in the EAR-like motif in WUS does not abolish repression activity *in vitro* or *in vivo*. In addition, WOX11 and WOX3, which lack the EAR-like motif, act as repressors in rice (Dai et al., 2007; Zhao et al., 2009). Although these studies suggest that WUS mainly functions as a repressor, other examples suggest it may act as an activator (Sections 14.2.11.1, 14.2.12 and 14.2.13).

14.2.9 WUS Mutant Phenotype and its Function

In the embryo, *wus* mutants fail to form the SAM. However, they initiate new shoot meristems repetitively after termination of the former shoot meristem, resulting in a bushy phenotype. These secondary meristems also prematurely terminate. Flowers formed in *wus* mutants are abnormal with fewer anthers and missing carpels, indicating that the *WUS* function is required for both vegetative and reproductive meristems (Laux et al., 1996). Cells in the CZ of *wus* mutants are larger and more vacuolated than those of the wild type, indicating these cells are differentiated or fail to maintain stem cell identity. Consistent with this result, WUS has the ability to confer shoot stem cell identity when ectopically expressed (Gallois et al., 2004). Thus, the *WUS* function is to specify stem cell fate. *WUS* is expressed in cells beneath stem cells (Mayer et al., 1998). This discrepancy between the *WUS* expression domain and the site of its action suggests that the WUS protein or its downstream unknown signals act non-cell-autonomously to promote stem cell identity in the overlying CZ cells (Mayer et al., 1998).

14.2.10 Functions of Other WOX Family Genes

In *Arabidopsis* the function of other *WOX* gene family members is highlighted in embryo patterning (Breuninger et al., 2008; Haecker et al., 2004). *WOX2* and *WOX8* are coexpressed in the zygote, but become restricted to apical and basal lineages, respectively, after first cell division. *WOX2*, together with other partially redundant *WOX* genes expressed in the apical half such as *WOX1*, *PRESSED FLOWER* (*PRS*)/*WOX3*, and *WOX5*, play important roles in proper cell division patterning and shoot formation. *WOX8* and its closest homolog *WOX9*, which is also expressed in the basal lineage, are required not only for basal lineage development but also for apical lineages by promoting *WOX2* and *PIN* expression to form a localized auxin response in the embryo (Breuninger et al., 2008).

WOX genes also play important roles in various aspects of postembryonic development. *PRS/WOX3* and their orthologs in rice and maize are required for lamina growth of leaf blade and floral organs (Matsumoto and Okada, 2001; Ishiwata et al., 2013; Nardmann et al., 2004). *Arabidopsis WOX1* and its orthologs in some other dicot species also promote leaf blade lamina outgrowth (Nakata et al., 2012; Vandenbussche et al., 2009; Tadege et al., 2011). *WOX4* promotes the proliferation of procambium/cambium stem cells in *Arabidopsis*, and its ortholog in rice is required for SAM maintenance (Hirakawa et al., 2010; Ohmori et al., 2013). *PRETTY FEW SEEDS2/WOX6* is involved in ovule development (Park et al., 2005). *WOX5* is required for root columella stem cell maintenance (Sarkar et al., 2007). Interestingly, *WOX5* and *WOX3* can complement *wus* mutants when driven under the *WUS* promoter, indicating that these *WUS*-clade genes have similar protein functions and their functional differentiation is mainly due to different expression patterns (Sarkar et al., 2007, Shimizu et al., 2009).

14.2.11 Multiple WUS Regulatory Loops for Shoot Stem Cell Maintenance

One of the fundamental questions in biology is how the homeostasis of stem cell populations is maintained. To elucidate this question, extensive studies have been conducted on the regulation of *WUS* and its downstream targets, revealing that *WUS* participates in multiple regulatory networks to maintain the stem cell population (Figure 14.3).

14.2.11.1 WUSCHEL–CLAVATA Negative Feedback Loop

The first regulatory network described was a negative feedback loop between *WUS* and the *CLAVATA* (*CLV*) pathway. *WUS* promotes expression of *CLV3*, a negative regulator of *WUS* that is expressed in stem cells, in a non-cell-autonomous manner. *WUS* transcript expression is confined to a few cells just beneath the stem cells, but its protein moves to overlaying stem cells and directly binds to *cis*-elements at the *CLV3* locus to activate expression (Yadav et al., 2011). WUS protein movement is critical to its function because immobile WUS proteins fail to rescue *wus* mutants. The *CLV* pathway, which is composed of a short-signal peptide CLV3 and its receptors CLV1, CLV2, CORYNE, and RECEPTOR-LIKE PROTEIN KINASE 2 (RPK2), provides a signal cascade to limit the stem cell population (Jeong et al., 1999; Fletcher et al., 1999; Clark et al., 1997; Muller et al., 2008; Kinoshita et al., 2010). Mutations in the *CLV* pathway cause expansion of the *WUS*-expressing domain and overproliferation of stem cells, whereas overexpression of *CLV3* reduces *WUS* expression and results in termination of the SAM, indicating that this pathway confines *WUS* expression to where

WUS is normally expressed (Schoof et al., 2000; Brand et al., 2000). Thus, *WUS* and *CLV3*, transcribed in adjacent compartments of the SAM, serve as mobile signals to form a negative feedback regulatory loop to balance the stem cell population.

14.2.11.2 WUSCHEL–Cytokinin Positive Feedback Loop

In the SAM, cell divisions push daughter cells from the stem cell population into lateral organ recruitment. During this dynamic process, each cell recognizes its relative position and in so doing maintains the organized structure of the SAM. For example, the *WUS*-expressing domain is maintained at a constant distance from the L1 layer of the SAM. Although the *WUS–CLV* negative feedback loop explains how stem cell populations are maintained, it does not explain how the spatial domain of *WUS* expression is stably maintained. Studies on the crosstalk between *WUS* and the phytohormone CK have revealed another layer of communication between the *WUS*-expressing domain and stem cells, which could answer this question.

WUS directly represses expression of CK-signaling inhibitors such as *ARR7* and *ARR15*; on the other hand, overexpression of *ARR7* decreases *WUS* expression, suggesting *WUS* and *ARR* functions antagonize each other. In addition, overexpression of the constitutively active form of *ARR7* causes the SAM to stop, in much the same way as done by *wus* mutants. Therefore, the repression of *ARR* genes by WUS is essential for maintenance of the SAM (Leibfried et al., 2005). Repression of CK signaling via *ARR* genes also plays an important role in controlling SAM size, because loss of function of the maize *ARR* gene, *abphyl1*, and reduced expression of *Arabidopsis ARR7* and *ARR15* result in enlargement of the SAM (Giulini et al., 2004; Zhao et al., 2010). Thus, WUS-mediated increases in CK sensitivity are important for stem cell maintenance.

Data suggest that CK also activates *WUS* expression. Treatment with a high concentration of CK results in increased *WUS* expression (Lindsay et al., 2006), and this upregulation of *WUS* requires CK receptors (Gordon et al., 2009). The CK receptor gene *ARABIDOPSIS HISTIDINE KINASE 4* (*AHK4*) is strongly expressed in the RZ, supporting the idea that sensitivity to CK is high in the *WUS* domain. Thus, *WUS* increases CK sensitivity via direct repression of *ARR* genes, and in turn, CK signaling activates *WUS*, forming a positive feedback loop to maintain the *WUS* expression level.

Interestingly, *WUS* is not expressed in the entire RZ where the CK receptor *AHK4* is expressed; instead, *WUS* is expressed only in the apical part of the RZ. The *WUS* expression domain expands toward the proximal direction upon CK treatment (Chickarmane et al., 2012). These observations suggest there is a higher concentration of CK in the distal part of the RZ than in the proximal region. Consistent with this notion, *AtLOG4* is expressed only in the L1 layer of the SAM. Based on these results in tandem with support from computational modeling, Chickarmane et al. proposed that CK acts as a stem-cell-derived positional cue to define *WUS* expression (Chickarmane et al., 2012). *WUS* expression is allowed in the overlapping region of the active CK derived from the stem cell epidermis and the CK-sensitive RZ defined by *AHK4*. Furthermore, their computational modeling suggested the presence of negative regulation of CK biosynthesis by *WUS* to strengthen the robustness of stem cell homeostasis. Thus, regulation of CK distribution and sensitivity is likely to play an important role in maintaining the *WUS* expression domain.

14.2.11.3 WUSCHEL–HECATE Negative Feedback Loop

Another negative feedback loop has been identified (Schuster et al., 2014). A basic helix–loop–helix (bHLH) TF HECATE1 (HEC1) antagonizes WUS activity and stimulates stem cell proliferation at the same time. WUS, in turn, directly represses *HEC1*. This repression is important for WUS activity because ectopic *HEC1* expression from the *WUS* promoter results in stem cell termination in much the same way as does the *wus* mutant. Interestingly, a triple mutant of *hec1* and its closely related paralogs, *hec2* and *hec3*, has a smaller SAM with reduced *WUS* and increased *CLV3* expression. This deviation from the *WUS–CLV* negative feedback loop model suggests that *HEC1* is involved in additional feedback regulation between *WUS* and stem cells. *HEC1* and *WUS* share highly overlapping downstream genes but regulate in opposite ways, supporting the antagonistic function of *HEC1* against *WUS* (Busch et al., 2010; Schuster et al., 2014). In addition, *HEC1* activates cell-cycle regulators to stimulate stem cell proliferation, and activates CK-signaling inhibitors *ARR5*, *ARR7*, and *ARR15* to repress *WUS* expression as well. Thus, *HEC1* and *WUS* form a negative feedback loop as a result of CK signaling (Schuster et al., 2014).

14.2.12 Other WUS Targets

Transcriptome analyses combined with ChIP–chip experiments identified direct targets of WUS that are responsive to its function (Busch et al., 2010). Upon *WUS* induction, auxin biosynthesis and responsive genes were repressed whereas reduction of *WUS* by inducible RNAi resulted in elevation of auxin response in the PZ and enlarged organ primordia (Yadav et al., 2010). Furthermore, it was proposed that the CZ has lower sensitivity to auxin, whereas the PZ has higher auxin sensitivity based on observation of the expression pattern of auxin-signaling components (Vernoux et al., 2011). These observations suggest that *WUS* has the potential to decrease responsiveness to auxin in addition to increasing CK sensitivity. This study also revealed that WUS directly represses *CLV1* (Busch et al., 2010). Because *CLV1* expression overlaps that of *WUS*, this regulatory loop is thought to maintain the expression of *WUS* by reducing CLV signaling in the *WUS* domain. In addition, WUS directly activates a transcriptional corepressor TPL which is known to interact physically with WUS protein (Busch et al., 2010; Kieffer et al., 2006). Therefore, WUS activates its own protein partner as well. Another transcriptome analysis of the *WUS*-responsive gene network combined with high spatial resolution of gene expression revealed that WUS directly represses expression of TFs that promote lateral organ differentiation such as *KANADI1*, *KANADI2*, *AS2*, and *YAB3*, revealing a function in preventing premature differentiation of stem cell progenitors (Yadav et al., 2013).

14.2.13 Termination of WUS Expression in Floral Stem Cells

Floral meristems eventually differentiate into carpels (female reproductive organs), and hence *WUS* must be repressed. In this process, *WUS* activates a MADS box gene *AGAMOUS* (*AG*), which specifies the central whorl identity for carpel differentiation, by direct binding to *cis*-elements in the intron of *AG* (Lohmann et al., 2001; Lenhard et al., 2001). In turn, *AG* represses *WUS* in later stages to terminate floral stem cell activity. In *ag* mutants, floral stem cells are not terminated and persistent *WUS* expression is observed in the center of the floral meristem (Lenhard et al., 2001; Lohmann et al., 2001). *AG* is known to repress *WUS* in two ways. First, AG indirectly represses *WUS* via direct activation of *KNUCKLES*, which terminates *WUS* expression and stem cell fate in floral meristems (Sun et al., 2009). In parallel, AG also directly represses *WUS* and promotes deposition of H3K27me3 on the *WUS* locus (Liu et al., 2011). Collectively, these studies establish a negative feedback loop between *WUS* and *AG*, which brings floral stem cell activity to an end and specifies central whorl identity.

Acknowledgment

The work was funded by NIFA grant 2012-67014-19429 (to S.H.) and JSPS Postdoctoral Fellowships for Research Abroad (to K.T.).

References

Barton, M.K., Poethig, R.S., 1993. Formation of the shoot apical meristem in *Arabidopsis thaliana* - an analysis of development in the wild type and in the *shoot meristemless* mutant. Development 119, 823–831.

Bellaoui, M., Pidkowich, M.S., Samach, A., Kushalappa, K., Kohalmi, S.E., Modrusan, Z., Crosby, W.L., Haughn, G.W., 2001. The *Arabidopsis* BELL1 and KNOX TALE homeodomain proteins interact through a domain conserved between plants and animals. Plant Cell 11, 2455–2470.

Belles-Boix, E., Hamant, O., Witiak, S.M., Morin, H., Traas, J., Pautot, V., 2006. KNAT6: an *Arabidopsis* homeobox gene involved in meristem activity and organ separation. Plant Cell 18, 1900–1907.

Bemer, M., Grossniklaus, U., 2012. Dynamic regulation of Polycomb group activity during plant development. Curr. Opin. Plant Biol. 15, 523–529.

Bertolino, E., Reimund, B., Wildt-Perinic, D., Clerc, R.G., 1995. A novel homeobox protein which recognizes a TGT core and functionally interferes with a retinoid-responsive motif. J. Biol. Chem. 270, 31178–31188.

Bharathan, G., Goliber, T.E., Moore, C., Kessler, S., Pham, T., Sinha, N.R., 2002. Homologies in leaf form inferred from *KNOX1* gene expression during development. Science 296, 1858–1860.

Bhatt, A.M., Etchells, J.P., Canales, C., Lagodienko, A., Dickinson, H., 2004. VAAMANA – a BEL1-like homeodomain protein, interacts with KNOX proteins BP and STM and regulates inflorescence stem growth in *Arabidopsis*. Gene 328, 103–111.

Bolduc, N., Hake, S., 2009. The maize transcription factor KNOTTED1 directly regulates the gibberellin catabolism gene *ga2ox1*. Plant Cell 21, 1647–1658.

Bolduc, N., Yilmaz, A., Mejia-Guerra, M.K., Morohashi, K., O'Connor, D., Grotewold, E., Hake, S., 2012. Unraveling the KNOTTED1 regulatory network in maize meristems. Genes Dev. 26, 1685–1690.

Bolduc, N., Tyers, R., Freeling, M., Hake, S., 2014. Unequal redundancy in maize knox genes. Plant Physiol. 164, 229–238.

Brand, U., Fletcher, J.C., Hobe, M., Meyerowitz, E.M., Simon, R., 2000. Dependence of stem cell fate in *Arabidopsis* on a feedback loop regulated by CLV3 activity. Science 289, 617–619.

Breuninger, H., Rikirsch, E., Hermann, M., Ueda, M., Laux, T., 2008. Differential expression of WOX genes mediates apical–basal axis formation in the *Arabidopsis* embryo. Dev. Cell 14, 867–876.

Burglin, T.R., 1997. Analysis of TALE superclass homeobox genes (MEIS, PBC, KNOX, Iroquois, TGIF) reveals a novel domain conserved between plants and animals. Nucleic Acids Res. 25, 4173–4180.

Busch, W., Miotk, A., Ariel, F.D., Zhao, Z., Forner, J., Daum, G., Suzaki, T., Schuster, C., Schultheiss, S.J., Leibfried, A., Haubeiss, S., Ha, N., Chan, R.L., Lohmann, J.U., 2010. Transcriptional control of a plant stem cell niche. Dev. Cell 18, 849–861.

Byrne, M.E., Barley, R., Curtis, M., Arroyo, J.M., Dunham, M., Hudson, A., Martienssen, R.A., 2000. *Asymmetric leaves 1* mediates leaf patterning and stem cell function in *Arabidopsis*. Nature 408, 967–971.

Byrne, M.E., Simorowski, J., Martienssen, R.A., 2002. *ASYMMETRIC LEAVES1* reveals knox gene redundancy in *Arabidopsis*. Development 129, 1957–1965.

Byrne, M.E., Groover, A.T., Fontana, J.R., Martienssen, R.A., 2003. Phyllotactic pattern and stem cell fate are determined by the *Arabidopsis* homeobox gene *BELLRINGER*. Development 130, 3941–3950.

Casselton, L.A., Olesnicky, N.S., 1998. Molecular genetics of mating recognition in basidiomycete fungi. Microbiol. Mol. Biol. Rev. 62, 55–70.

Chang, C.-P., Jacobs, Y., Nakamura, T., Jenkins, N.A., Copeland, N.G., Cleary, M.L., 1997. Meis proteins are major *in vivo* DNA binding partners for wild-type but not chimeric Pbx proteins. Mol. Cell. Biol. 17, 5679–5687.

Chen, H., Banerjee, A.K., Hannapel, D.J., 2004. The tandem complex of BEL and KNOX partners is required for transcriptional repression of ga20ox1. Plant J. 38, 276–284.

Chickarmane, V.S., Gordon, S.P., Tarr, P.T., Heisler, M.G., Meyerowitz, E.M., 2012. Cytokinin signaling as a positional cue for patterning the apical–basal axis of the growing *Arabidopsis* shoot meristem. Proc. Natl. Acad. Sci. USA 109, 4002–4007.

Chuck, G., Lincoln, C., Hake, S., 1996. KNAT1 induces lobed leaves with ectopic meristems when overexpressed in *Arabidopsis*. Plant Cell 8, 1277–1289.

Clark, S.E., Jacobsen, S.E., Levin, J.Z., Meyerowitz, E.M., 1996. The *CLAVATA* and *SHOOT MERISTEMLESS* loci competitively regulate meristem activity in *Arabidopsis*. Development 122, 1567–1575.

Clark, S.E., Williams, R.W., Meyerowitz, E.M., 1997. The *CLAVATA1* gene encodes a putative receptor kinase that controls shoot and floral meristem size in *Arabidopsis*. Cell 89, 575–585.

Clouse, S.D., Sasse, J.M., 1998. BRASSINOSTEROIDS: essential regulators of plant growth and development. Annu. Rev. Plant Physiol. Plant Mol. Biol. 49, 427–451.

Cole, M., Nolte, C., Werr, W., 2006. Nuclear import of the transcription factor SHOOT MERISTEMLESS depends on heterodimerization with BLH proteins expressed in discrete sub-domains of the shoot apical meristem of *Arabidopsis thaliana*. Nucleic Acids Res. 34, 1281–1292.

Dai, M., Hu, Y., Zhao, Y., Liu, H., Zhou, D.X., 2007. A WUSCHEL-LIKE HOMEOBOX gene represses a YABBY gene expression required for rice leaf development. Plant Physiol. 144, 380–390.

Derelle, R., Lopez, P., Le Guyader, H., Manuel, M., 2007. Homeodomain proteins belong to the ancestral molecular toolkit of eukaryotes. Evol. Dev. 9, 212–219.

Douglas, S.J., Chuck, G., Dengler, R.E., Pelecanda, L., Riggs, C.D., 2002. *KNAT1* and *ERECTA* regulate inflorescence architecture in *Arabidopsis*. Plant Cell 14, 547–558.

Estruch, J.J., Prinsen, E., Van Onckelen, H., Schell, J., Spena, A., 1991. Viviparous leaves produced by somatic activation of an inactive cytokinin-synthesizing gene. Science 254, 1364–1367.

Fletcher, J.C., Brand, U., Running, M.P., Simon, R., Meyerowitz, E.M., 1999. Signaling of cell fate decisions by CLAVATA3 in *Arabidopsis* meristems. Science 183, 1911–1914.

Freeling, M., Hake, S., 1985. Developmental genetics of mutants that specify Knotted leaves in maize. Genetics 111, 617–634.

Furutani, M., Vernoux, T., Traas, J., Kato, T., Tasaka, M., Aida, M., 2004. PIN-FORMED1 and PINOID regulate boundary formation and cotyledon development in *Arabidopsis* embryogenesis. Development 131, 5021–5030.

Gallois, J.L., Nora, F.R., Mizukami, Y., Sablowski, R., 2004. WUSCHEL induces shoot stem cell activity and developmental plasticity in the root meristem. Genes Dev. 18, 375–380.

Gehring, W., Qian, Y.Q., Billeter, M., Furukubo-Tokunaga, K., Schier, A., Resendez-Perez, D., Affolter, M., Otting, G., Wuthrich, K., 1994a. Homeodomain–DNA recognition. Cell 78, 211–223.

Gehring, W.J., Affolter, M., Burglin, T., 1994b. Homeodomain proteins. Annu. Rev. Biochem. 63, 487–526.

Giulini, A., Wang, J., Jackson, D., 2004. Control of phyllotaxy by the cytokinin-inducible response regulator homologue ABPHYL1. Nature 430, 1031–1034.

Gong, S.Y., Huang, G.Q., Sun, X., Qin, L.X., Li, Y., Zhou, L., Li, X.B., 2014. Cotton KNL1, encoding a class II KNOX transcription factor, is involved in regulation of fibre development. J. Exp. Bot. 65, 4133–4147.

Gordon, S.P., Chickarmane, V.S., Ohno, C., Meyerowitz, E.M., 2009. Multiple feedback loops through cytokinin signaling control stem cell number within the *Arabidopsis* shoot meristem. Proc. Natl. Acad. Sci. USA 106, 16529–16534.

Guo, M., Thomas, J., Collins, G., Timmermans, M.C., 2008. Direct repression of KNOX loci by the ASYMMETRIC LEAVES1 complex of *Arabidopsis*. Plant Cell 20, 48–58.

Hackbusch, J., Richter, K., Muller, J., Salamini, F., Uhrig, J.F., 2005. A central role of *Arabidopsis thaliana* ovate family proteins in networking and subcellular localization of 3-aa loop extension homeodomain proteins. Proc. Natl. Acad. Sci. USA 102, 4908–4912.

Haecker, A., Gross-Hardt, R., Geiges, B., Sarkar, A., Breuninger, H., Herrmann, M., Laux, T., 2004. Expression dynamics of WOX genes mark cell fate decisions during early embryonic patterning in *Arabidopsis thaliana*. Development 131, 657–668.

Hake, S., Vollbrecht, E., Freeling, M., 1989. Cloning *Knotted*, the dominant morphological mutant in maize using *Ds2* as a transposon tag. EMBO J. 8, 15–22.

Hake, S., Smith, H.M.S., Holtan, H., Magnani, E., Mele, G., Ramirez, J., 2004. The role of *KNOX* genes in plant development. Annu. Rev. Cell Dev. Biol. 20, 125–151.

Hareven, D., Gutfinger, T., Parnis, A., Eshed, Y., Lifschitz, E., 1996. The making of a compound leaf: genetic manipulation of leaf architecture in tomato. Cell 84, 735–744.

Hay, A., Tsiantis, M., 2006. The genetic basis for differences in leaf form between *Arabidopsis thaliana* and its wild relative *Cardamine hirsuta*. Nat. Genet. 38, 942–947.

Hay, A., Kaur, H., Phillips, A., Hedden, P., Hake, S., Tsiantis, M., 2002. The gibberellin pathway mediates KNOTTED1-type homeobox function in plants with different body plans. Curr. Biol. 12, 1557–1565.

Hay, A., Barkoulas, M., Tsiantis, M., 2006. ASYMMETRIC LEAVES1 and auxin activities converge to repress BREVIPEDICELLUS expression and promote leaf development in *Arabidopsis*. Development 133, 3955–3961.

Hirakawa, Y., Kondo, Y., Fukuda, H., 2010. TDIF peptide signaling regulates vascular stem cell proliferation via the WOX4 homeobox gene in *Arabidopsis*. Plant Cell 22, 2618–2629.

Ikeda, M., Mitsuda, N., Ohme-Takagi, M., 2009. *Arabidopsis* WUSCHEL is a bifunctional transcription factor that acts as a repressor in stem cell regulation and as an activator in floral patterning. Plant Cell 21, 3493–3505.

Ishiwata, A., Ozawa, M., Nagasaki, H., Kato, M., Noda, Y., Yamaguchi, T., Nosaka, M., Shimizu-Sato, S., Nagasaki, A., Maekawa, M., Hirano, H.Y., Sato, Y., 2013. Two WUSCHEL-related homeobox genes, narrow leaf2 and narrow leaf3, control leaf width in rice. Plant Cell Physiol. 54, 779–792.

Jackson, D., Veit, B., Hake, S., 1994. Expression of maize *KNOTTED1* related homeobox genes in the shoot apical meristem predicts patterns of morphogenesis in the vegetative shoot. Development 120, 405–413.

Jasinski, S., Piazza, P., Craft, J., Hay, A., Woolley, L., Rieu, I., Phillips, A., Hedden, P., Tsiantis, M., 2005. KNOX action in *Arabidopsis* is mediated by coordinate regulation of cytokinin and gibberellin activities. Curr. Biol. 15, 1560–1565.

Jeong, S., Trotochaud, A.E., Clark, S.E., 1999. The *Arabidopsis CLAVATA2* gene encodes a receptor-like protein required for the stability of the CLAVATA1 receptor-like kinase. Plant Cell 11, 1925–1933.

Johnson, A.D., 1995. Molecular mechanisms of cell-type determination in budding yeast. Curr. Opin. Genet. Dev. 5, 552–558.

Katz, A., Oliva, M., Mosquna, A., Hakim, O., Ohad, N., 2004. FIE and CURLY LEAF polycomb proteins interact in the regulation of homeobox gene expression during sporophyte development. Plant J. 37, 707–719.

Kerstetter, R., Vollbrecht, E., Lowe, B., Veit, B., Yamaguchi, J., Hake, S., 1994. Sequence analysis and expression patterns divide the maize *knotted1*-like homeobox genes into two classes. Plant Cell 6, 1877–1887.

Kerstetter, R.A., Laudencia-Chingcuanco, D., Smith, L.G., Hake, S., 1997. Loss of function mutations in the maize homeobox gene, *knotted1*, are defective in shoot meristem maintenance. Development 124, 3045–3054.

Kieffer, M., Stern, Y., Cook, H., Clerici, E., Maulbetsch, C., Laux, T., Davies, B., 2006. Analysis of the transcription factor WUSCHEL and its functional homologue in *Antirrhinum* reveals a potential mechanism for their roles in meristem maintenance. Plant Cell 18, 560–573.

Kimura, S., Koenig, D., Kang, J., Yoong, F.Y., Sinha, N., 2008. Natural variation in leaf morphology results from mutation of a novel KNOX gene. Curr. Biol. 18, 672–677.

Kinoshita, A., Betsuyaku, S., Osakabe, Y., Mizuno, S., Nagawa, S., Stahl, Y., Simon, R., Yamaguchi-Shinozaki, K., Fukuda, H., Sawa, S., 2010. RPK2 is an essential receptor-like kinase that transmits the CLV3 signal in *Arabidopsis*. Development 137, 3911–3920.

Krecek, P., Skupa, P., Libus, J., Naramoto, S., Tejos, R., Friml, J., Zazimalova, E., 2009. The PIN-FORMED (PIN) protein family of auxin transporters. Genome Biol. 10, 249.

Kumar, R., Kushalappa, K., Godt, D., Pidkowich, M.S., Pastorelli, S., Hepworth, S.R., Haughn, G.W., 2007. The *Arabidopsis* BEL1-LIKE HOMEODOMAIN proteins SAW1 and SAW2 act redundantly to regulate KNOX expression spatially in leaf margins. Plant Cell 19, 2719–2735.

Kurakawa, T., Ueda, N., Maekawa, M., Kobayashi, K., Kojima, M., Nagato, Y., Sakakibara, H., Kyozuka, J., 2007. Direct control of shoot meristem activity by a cytokinin-activating enzyme. Nature 445, 652–655.

Kusaba, S., Kano-Murakami, Y., Matsuoka, M., Tamaoki, M., Sakamoto, T., Yamaguchi, I., Fukumoto, M., 1998. Alteration of hormone levels in transgenic tobacco plants overexpressing the rice homeobox gene OSH1. Plant Physiol. 116, 471–476.

Laux, T., Mayer, K.F.X., Berger, J., Jurgens, G., 1996. The wuschel gene is required for shoot and floral meristem integrity in *Arabidopsis*. Development 122, 87–96.

Lee, J.H., Lin, H., Joo, S., Goodenough, U., 2008. Early sexual origins of homeoprotein heterodimerization and evolution of the plant KNOX/BELL family. Cell 133, 829–840.

Leibfried, A., To, J.P., Busch, W., Stehling, S., Kehle, A., Demar, M., Kieber, J.J., Lohmann, J.U., 2005. WUSCHEL controls meristem function by direct regulation of cytokinin-inducible response regulators. Nature 438, 1172–1175.

Lenhard, M., Bohnert, A., Jurgens, G., Laux, T., 2001. Termination of stem cell maintenance in *Arabidopsis* floral meristems by interactions between WUSCHEL and AGAMOUS. Cell 105, 805–814.

Li, E., Bhargava, A., Qiang, W., Friedmann, M.C., Forneris, N., Savidge, R.A., Johnson, L.A., Mansfield, S.D., Ellis, B.E., Douglas, C.J., 2012. The class II KNOX gene KNAT7 negatively regulates secondary wall formation in *Arabidopsis* and is functionally conserved in *Populus*. New Phytol. 194, 102–115.

Li, Y., Hagen, G., Guilfoyle, T.J., 1992. Altered morphology in transgenic tobacco plants that overproduce cytokinins in specific tissues and organs. Dev. Biol. 153, 386–395.

Lincoln, C., Long, J., Yamaguchi, J., Serikawa, K., Hake, S., 1994. A Knotted1-like homeobox gene in *Arabidopsis* is expressed in the vegetative meristem and dramatically alters leaf morphology when overexpressed in transgenic plants. Plant Cell 6, 1859–1876.

Lindsay, D., Sawhney, V., Bonham-Smith, P., 2006. Cytokinin-induced changes in CLAVATA1 and WUSCHEL expression temporally coincide with altered floral development in *Arabidopsis*. Plant Sci. 170, 1111–1117.

Liu, X., Kim, Y.J., Muller, R., Yumul, R.E., Liu, C., Pan, Y., Cao, X., Goodrich, J., Chen, X., 2011. AGAMOUS terminates floral stem cell maintenance in *Arabidopsis* by directly repressing WUSCHEL through recruitment of Polycomb group proteins. Plant Cell 23, 3654–3670.

Lodha, M., Marco, C.F., Timmermans, M.C., 2013. The ASYMMETRIC LEAVES complex maintains repression of KNOX homeobox genes *via* direct recruitment of Polycomb-repressive complex2. Genes Dev. 27, 596–601.

Lohmann, J.U., Hong, R.L., Hobe, M., Busch, M.A., Parcy, F., Simon, R., Weigel, D., 2001. A molecular link between stem cell regulation and floral patterning in *Arabidopsis*. Cell 105, 793–803.

Long, J.A., Barton, M.K., 1998. The development of apical embryonic pattern in *Arabidopsis*. Development 125, 3027–3035.

Long, J.A., Moan, E.I., Medford, J.I., Barton, M.K., 1996. A member of the KNOTTED class of homeodomain proteins encoded by the *SHOOTMERISTEMLESS* gene of *Arabidopsis*. Nature 379, 66–69.

Magnani, E., Hake, S., 2008. KNOX lost the OX: the *Arabidopsis* KNATM gene defines a novel class of KNOX transcriptional regulators missing the homeodomain. Plant Cell 20, 875–887.

Matsumoto, N., Okada, K., 2001. A homeobox gene, PRESSED FLOWER, regulates lateral axis-dependent development of *Arabidopsis* flowers. Genes Dev. 15, 3355–3364.

Mayer, K.F., Schoof, H., Haecker, A., Lenhard, M., Jurgens, G., Laux, T., 1998. Role of WUSCHEL in regulating stem cell fate in the *Arabidopsis* shoot meristem. Cell 95, 805–815.

Mele, G., Ori, N., Sato, Y., Hake, S., 2003. The *knotted1*-like homeobox gene *BREVIPEDICELLUS* regulates cell differentiation by modulating metabolic pathways. Genes Dev. 17, 2088–2093.

Muehlbauer, G.J., Fowler, J.E., Freeling, M., 1997. Sectors expressing the homeobox gene *liguleless3* implicate a time-dependent mechanism for cell fate acquisition along the proximal–distal axis of the maize leaf. Development 124, 5097–5106.

Mukherjee, K., Brocchieri, L., Burglin, T.R., 2009. A comprehensive classification and evolutionary analysis of plant homeobox genes. Mol. Biol. Evol. 26, 2775–2794.

Muller, R., Bleckmann, A., Simon, R., 2008. The receptor kinase CORYNE of *Arabidopsis* transmits the stem cell-limiting signal CLAVATA3 independently of CLAVATA1. Plant Cell 20, 934–946.

Nagasaki, H., Matsuoka, M., Sato, Y., 2005. Members of TALE and WUS subfamilies of homeodomain proteins with potentially important functions in development form dimers within each subfamily in rice. Genes Genet. Syst. 80, 261–267.

Nakata, M., Matsumoto, N., Tsugeki, R., Rikirsch, E., Laux, T., Okada, K., 2012. Roles of the middle domain-specific WUSCHEL-RELATED HOMEOBOX genes in early development of leaves in *Arabidopsis*. Plant Cell 24, 519–535.

Nardmann, J., Ji, J., Werr, W., Scanlon, M.J., 2004. The maize duplicate genes narrow sheath1 and narrow sheath2 encode a conserved homeobox gene function in a lateral domain of shoot apical meristems. Development 131, 2827–2839.

Ohmori, Y., Tanaka, W., Kojima, M., Sakakibara, H., Hirano, H.Y., 2013. WUSCHEL-RELATED HOMEOBOX4 is involved in meristem maintenance and is negatively regulated by the CLE gene FCP1 in rice. Plant Cell 25, 229–241.

Okada, K., Ueda, J., Komaki, M.K., Bell, C.J., Shimura, Y., 1991. Requirement of the auxin polar transport system in early stages of *Arabidopsis* floral bud formation. Plant Cell 3, 677–684.

Park, S.O., Zheng, Z., Oppenheimer, D.G., Hauser, B.A., 2005. The PRETTY FEW SEEDS2 gene encodes an *Arabidopsis* homeodomain protein that regulates ovule development. Development 132, 841–849.

Phelps-Durr, T.L., Thomas, J., Vahab, P., Timmermans, M.C., 2005. Maize rough sheath2 and its *Arabidopsis* orthologue ASYMMETRIC LEAVES1 interact with HIRA, a predicted histone chaperone, to maintain knox gene silencing and determinacy during organogenesis. Plant Cell 17, 2886–2898.

Ramirez, J., Bolduc, N., Lisch, D., Hake, S., 2009. Distal expression of knotted1 in maize leaves leads to re-establishment of proximal/distal patterning and leaf dissection. Plant Physiol. 151, 1878–1888.

Reinhardt, D., Mandel, T., Kuhlemeier, C., 2000. Auxin regulates the initiation and radial position of plant lateral organs. Plant Cell 12, 507–518.

Reinhardt, D., Pesce, E.R., Stieger, P., Mandel, T., Baltensperger, K., Bennett, M., Traas, J., Friml, J., Kuhlemeier, C., 2003. Regulation of phyllotaxis by polar auxin transport. Nature 426, 255–260.

Reiser, L., Modrusan, Z., Margossian, L., Samach, A., Ohad, N., Haughn, G.W., Fischer, R.L., 1995. The *BELL1* gene encodes a homeodomain protein involved in pattern formation in the *Arabidopsis* ovule primordium. Cell 83, 735–742.

Rutjens, B., Bao, D., Van Eck-Stouten, E., Brand, M., Smeekens, S., Proveniers, M., 2009. Shoot apical meristem function in *Arabidopsis* requires the combined activities of three BEL1-like homeodomain proteins. Plant J. 58, 641–654.

Sakakibara, K., Nishiyama, T., Deguchi, H., Hasebe, M., 2008. Class 1 KNOX genes are not involved in shoot development in the moss *Physcomitrella patens* but do function in sporophyte development. Evol. Dev. 10, 555–566.

Sakakibara, K., Ando, S., Yip, H.K., Tamada, Y., Hiwatashi, Y., Murata, T., Deguchi, H., Hasebe, M., Bowman, J.L., 2013. KNOX2 genes regulate the haploid-to-diploid morphological transition in land plants. Science 339, 1067–1070.

Sakamoto, T., Kamiya, N., Ueguchi-Tanaka, M., Iwahori, S., Matsuoka, M., 2001. KNOX homeodomain protein directly suppresses the expression of a gibberellin biosynthetic gene in the tobacco shoot apical meristem. Genes Dev. 15, 581–590.

Sakamoto, T., Sakakibara, H., Kojima, M., Yamamoto, Y., Nagasaki, H., Inukai, Y., Sato, Y., Matsuoka, M., 2006. Ectopic expression of KNOTTED1-like homeobox protein induces expression of cytokinin biosynthesis genes in rice. Plant Physiol. 142, 54–62.

Sakamoto, T., Kawabe, A., Tokida-Segawa, A., Shimizu, B., Takatsuto, S., Shimada, Y., Fujioka, S., Mizutani, M., 2011. Rice CYP734As function as multisubstrate and multifunctional enzymes in brassinosteroid catabolism. Plant J. 67, 1–12.

Sarkar, A.K., Luijten, M., Miyashima, S., Lenhard, M., Hashimoto, T., Nakajima, K., Scheres, B., Heidstra, R., Laux, T., 2007. Conserved factors regulate signalling in *Arabidopsis thaliana* shoot and root stem cell organizers. Nature 446, 811–814.

Sato, Y., Sentoku, N., Miura, Y., Hirochika, H., Kitano, H., Matsuoka, M., 1999. Loss-of-function mutations in the rice homeobox gene OSH15 affect the architecture of internodes resulting in dwarf plants. EMBO J. 18, 992–1002.

Scanlon, M.J., 2003. The polar auxin transport inhibitor N-1-naphthylphthalamic acid disrupts leaf initiation, KNOX protein regulation, and formation of leaf margins in maize. Plant Physiol. 133, 597–605.

Schoof, H., Lenhard, M., Haecker, A., Mayer, K.F., Jurgens, G., Laux, T., 2000. The stem cell population of *Arabidopsis* shoot meristems is maintained by a regulatory loop between the CLAVATA and WUSCHEL genes. Cell 100, 635–644.

Schubert, D., Primavesi, L., Bishopp, A., Roberts, G., Doonan, J., Jenuwein, T., Goodrich, J., 2006. Silencing by plant Polycomb-group genes requires dispersed trimethylation of histone H3 at lysine 27. EMBO J. 25, 4638–4649.

Schuster, C., Gaillochet, C., Medzihradszky, A., Busch, W., Daum, G., Krebs, M., Kehle, A., Lohmann, J.U., 2014. A regulatory framework for shoot stem cell control integrating metabolic, transcriptional, and phytohormone signals. Dev. Cell 28, 438–449.

Sentoku, N., Sato, Y., Kurata, N., Ito, Y., Kitano, H., Matsuoka, M., 1999. Regional expression of the rice KN1-type homeobox gene family during embryo, shoot, and flower development. Plant Cell 11, 1651–1664.

Sentoku, N., Sato, Y., Matsuoka, M., 2000. Overexpression of rice OSH genes induces ectopic shoots on leaf sheaths of transgenic rice plants. Dev. Biol. 220, 358–364.

Shimizu, R., Ji, J., Kelsey, E., Ohtsu, K., Schnable, P.S., Scanlon, M.J., 2009. Tissue specificity and evolution of meristematic WOX3 function. Plant Physiol. 149, 841–850.

Sinha, N.R., Williams, R.E., Hake, S., 1993. Overexpression of the maize homeo box gene, KNOTTED-1, causes a switch from determinate to indeterminate cell fates. Genes Dev. 7, 787–795.

Smith, H.M.S., Hake, S., 2003. The interaction of two homeobox genes, BREVIPEDICELLUS and PENNYWISE, regulates internode patterning in the *Arabidopsis* inflorescence. Plant Cell 15, 1–12.

Smith, L.G., Greene, B., Veit, B., Hake, S., 1992. A dominant mutation in the maize homeobox gene, Knotted-1, causes its ectopic expression in leaf cells with altered fates. Development 116, 21–30.

Smith, H.M.S., Boschke, I., Hake, S., 2002. Selective interaction of plant homeodomain proteins mediates high DNA-binding affinity. Proc. Natl. Acad. Sci. USA 99, 9579–9584.

Spinelli, S.V., Martin, A.P., Viola, I.L., Gonzalez, D.H., Palatnik, J.F., 2011. A mechanistic link between STM and CUC1 during *Arabidopsis* development. Plant Physiol. 156, 1894–1904.

Sun, B., Xu, Y., Ng, K.H., Ito, T., 2009. A timing mechanism for stem cell maintenance and differentiation in the *Arabidopsis* floral meristem. Genes Dev. 23, 1791–1804.

Tadege, M., Lin, H., Bedair, M., Berbel, A., Wen, J., Rojas, C.M., Niu, L., Tang, Y., Sumner, L., Ratet, P., Mchale, N.A., Madueno, F., Mysore, K.S., 2011. STENOFOLIA regulates blade outgrowth and leaf vascular patterning in *Medicago truncatula* and *Nicotiana sylvestris*. Plant Cell 23, 2125–2142.

Tamaoki, M., Kusaba, S., Kano-Murakami, Y., Matsuoka, M., 1997. Ectopic expression of a tobacco homeobox gene, NTH15, dramatically alters leaf morphology and hormone levels in transgenic tobacco. Plant Cell Physiol. 38, 917–927.

Tanaka-Ueguchi, M., Itoh, H., Oyama, N., Koshioka, M., Matsuoka, M., 1998. Over-expression of a tobacco homeobox gene, *NTH15*, decreases the expression of a gibberellin biosynthetic gene encoding GA 20-oxidase. Plant J. 15, 391–400.

Timmermans, M.C., Hudson, A., Becraft, P.W., Nelson, T., 1999. ROUGH SHEATH2: a Myb protein that represses knox homeobox genes in maize lateral organ primordia. Science 284, 151–153.

Townsley, B.T., Sinha, N.R., Kang, J., 2013. KNOX1 genes regulate lignin deposition and composition in monocots and dicots. Front. Plant Sci. 4, 121.

Treisman, J., Goncyz, P., Vashishtha, M., Harris, E., Desplan, C., 1989. A single amino acid can determine the DNA binding specificity of homeodomain proteins. Cell 59, 553–562.

Treisman, J., Harris, E., Wilson, D., Desplan, C., 1992. The homeodomain: a new face for the helix–turn–helix? BioEssays 14, 145–150.

Tsiantis, M., Schneeberger, R., Golz, J.F., Freeling, M., Langdale, J.A., 1999. The maize rough sheath2 gene and leaf development programs in monocot and dicot plants. Science 284, 154–156.

Tsuda, K., Ito, Y., Sato, Y., Kurata, N., 2011. Positive autoregulation of a KNOX gene is essential for shoot apical meristem maintenance in rice. Plant Cell 23, 4368–4381.

Tsuda, K., Kurata, N., Ohyanagi, H., Hake, S., 2014. Genome-wide study of KNOX regulatory network reveals brassinosteroid catabolic genes important for shoot meristem function in rice. Plant Cell 26, 3488–3500.

Van Der Graaff, E., Laux, T., Rensing, S.A., 2009. The WUS homeobox-containing (WOX) protein family. Genome Biol. 10, 248.

Vandenbussche, M., Horstman, A., Zethof, J., Koes, R., Rijpkema, A.S., Gerats, T., 2009. Differential recruitment of WOX transcription factors for lateral development and organ fusion in *Petunia* and *Arabidopsis*. Plant Cell 21, 2269–2283.

Venglat, S.P., Dumonceaux, T., Rozwadowski, K., Parnell, L., Babic, V., Keller, W., Martienssen, R., Selvaraj, G., Datla, R., 2002. The homeobox gene BREVIPEDICELLUS is a key regulator of inflorescence architecture in *Arabidopsis*. Proc. Natl. Acad. Sci. USA 99, 4730–4735.

Vernoux, T., Brunoud, G., Farcot, E., Morin, V., Van Den Daele, H., Legrand, J., Oliva, M., Das, P., Larrieu, A., Wells, D., Guedon, Y., Armitage, L., Picard, F., Guyomarc'h, S., Cellier, C., Parry, G., Koumproglou, R., Doonan, J.H., Estelle, M., Godin, C., Kepinski, S., Bennett, M., De Veylder, L., Traas, J., 2011. The auxin signalling network translates dynamic input into robust patterning at the shoot apex. Mol. Syst. Biol. 7, 508.

Vollbrecht, E., Veit, B., Sinha, N., Hake, S., 1991. The developmental gene Knotted-1 is a member of a maize homeobox gene family. Nature (London) 350, 241–243.

Vollbrecht, E., Reiser, L., Hake, S., 2000. Shoot meristem size is dependent on inbred background and presence of the maize homeobox gene, *knotted1*. Development 127, 3161–3172.

Xu, L., Shen, W.-H., 2008. Polycomb silencing of *KNOX* genes confines shoot stem cell niches in *Arabidopsis*. Curr. Biol. 18, 1966–1971.

Yadav, R.K., Tavakkoli, M., Reddy, G.V., 2010. WUSCHEL mediates stem cell homeostasis by regulating stem cell number and patterns of cell division and differentiation of stem cell progenitors. Development 137, 3581–3589.

Yadav, R.K., Perales, M., Gruel, J., Girke, T., Jonsson, H., Reddy, G.V., 2011. WUSCHEL protein movement mediates stem cell homeostasis in the *Arabidopsis* shoot apex. Genes Dev. 25, 2025–2030.

Yadav, R.K., Perales, M., Gruel, J., Ohno, C., Heisler, M., Girke, T., Jonsson, H., Reddy, G.V., 2013. Plant stem cell maintenance involves direct transcriptional repression of differentiation program. Mol. Syst. Biol. 9, 654.

Yanai, O., Shani, E., Dolezal, K., Tarkowski, P., Sablowski, R., Sandberg, G., Samach, A., Ori, N., 2005. *Arabidopsis* KNOXI proteins activate cytokinin biosynthesis. Curr. Biol. 15, 1566–1571.

Zhang, X., Zong, J., Liu, J., Yin, J., Zhang, D., 2010. Genome-wide analysis of WOX gene family in rice, sorghum, maize, *Arabidopsis* and poplar. J. Integr. Plant Biol. 52, 1016–1026.

Zhao, Y., Hu, Y., Dai, M., Huang, L., Zhou, D.X., 2009. The WUSCHEL-related homeobox gene WOX11 is required to activate shoot-borne crown root development in rice. Plant Cell 21, 736–748.

Zhao, Z., Andersen, S.U., Ljung, K., Dolezal, K., Miotk, A., Schultheiss, S.J., Lohmann, J.U., 2010. Hormonal control of the shoot stem-cell niche. Nature 465, 1089–1092.

CHAPTER

15

CUC Transcription Factors: To the Meristem and Beyond

Aude Maugarny, Beatriz Gonçalves, Nicolas Arnaud, Patrick Laufs

INRA, UMR1318, AgroParisTech, Institut Jean-Pierre Bourgin, Versailles, France

OUTLINE

15.1 Introduction 230

15.2 Evolution and Structure of NAM/CUC3 Proteins 230
  15.2.1 The NAM/CUC3 Proteins are Part of the Large Plant-Specific Family of NAC Transcription Factors 230
    15.2.1.1 New Insights into the Origin of the NAC Family 230
    15.2.1.2 Origin and Early Evolution of the NAM/CUC3 Family 230
    15.2.1.3 Recent Evolution Within the NAM/CUC3 Family 230
  15.2.2 NAM/CUC3 Protein Organization and Specific Domains 231
    15.2.2.1 The Amino-Terminal NAC Domain 231
    15.2.2.2 The Carboxy-Terminal Domain 232

15.3 NAM/CUC3 Genes Define Boundaries in Meristems and Beyond 234
  15.3.1 Identification of the NAM/CUC3 Genes: Role in Boundary and Meristem Formation 234
    15.3.1.1 Identification of NAM/CUC3 in Petunia and *Arabidopsis* 234
    15.3.1.2 Role in Other Dicots and Monocots 235
  15.3.2 Role of CUC Genes in Other Meristematic Territories 235
    15.3.2.1 Axillary Meristems 235
    15.3.2.2 Floral Organ Boundaries 235
    15.3.2.3 Gynecium 236
    15.3.2.4 Organ Abscission 236
  15.3.3 Role of CUC Genes in Leaf Development 236

15.4 Multiple Regulatory Pathways Contribute to the Fine Regulation of NAM/CUC3 Genes 236
  15.4.1 Hormonal Regulation of NAM/CUC3 Gene Expression 237
    15.4.1.1 The Interplay between NAM/CUC3 Genes and Auxin 237
    15.4.1.2 Brassinosteroids, New Regulators of CUC Expression 238
  15.4.2 miR164 FineTunes NAM Gene Expression 238
    15.4.2.1 miR164 Regulation is Essential for Shoot Development 238
    15.4.2.2 Evolution and Specialization of the *MIR164* Genes 239
    15.4.2.3 Transcriptional Control of *miR164* Expression 240
  15.4.3 Transcriptional Regulation of NAM/CUC3 Expression 240
    15.4.3.1 Transcription Factors Regulating CUC Expression During Embryogenesis 240
    15.4.3.2 Transcription Factors Regulating CUC Expression During Axillary Meristem Formation 241
    15.4.3.3 Transcription Factors Regulating NAM/CUC3 Expression During Leaf Development 241
    15.4.3.4 Transcription Factors Regulating GOB Expression During Abscission 241
    15.4.3.5 Regulation of CUC Expression by Chromatin Modifications 241

15.5 NAM/CUC3 Control Plant Development via Modifications of the Cellular Behavior 242
  15.5.1 CUC-Dependent Cellular Effects 242
  15.5.2 How Does CUC Impact Cell Proliferation? 242
  15.5.3 CUC Direct Targets 243
  15.5.4 Other Regulators: KNOX, LFY, LAS 243

15.6 Conclusion 243

References 244

15.1 INTRODUCTION

In 1996, Souer et al. reported the phenotypic characterization of a petunia mutant that fails to develop a shoot apical meristem (SAM) called *"nam"* (*no apical meristem*). The *NAM* gene is expressed in the boundaries of meristems and primordia, and the NAM protein shares a conserved N-terminal domain with other proteins, suggesting that it is part of a novel class of proteins. Indeed, the following year, Aida et al. (1997) identified an *Arabidopsis* mutant with no apical meristem but with cotyledons fused along their edge resulting in a cup-like structure, hence named *"cuc"* (*cup-shaped cotyledon*). This phenotype results from the combination of two mutations, one of which affects *CUC2*, a gene showing strong homology with the petunia *NAM* gene. These two papers began the story of the *NAM/CUC3* genes and founded the basis for the NAM, ATAF1, ATAF2, and CUC (NAC) family of plant-specific transcription factors. Here, we retrace the phylogenetic and evolutionary context of *NAM/CUC3* genes and review the important roles they play as boundary-defining actors during plant development. In particular, we discuss the mechanisms that regulate their expression patterns and how they affect plant development via their effects on cellular behavior.

15.2 EVOLUTION AND STRUCTURE OF NAM/CUC3 PROTEINS

15.2.1 The NAM/CUC3 Proteins are Part of the Large Plant-Specific Family of NAC Transcription Factors

15.2.1.1 New Insights into the Origin of the NAC Family

Together with ATAF1 and ATAF2, the petunia NAM and *Arabidopsis* CUC proteins are the founding members of the NAC family of plant-specific transcription factors (Aida et al., 1997). In an effort to trace the evolutionary origin of NAC proteins, Zhu et al. (2012) searched the full genome or expressed sequence tag (EST) data of 16 different species including eudicots, monocots, a lycophyte and a moss, chlorophytes, a red algae, and glaucophytes. Whereas a large number of NAC proteins have been identified in flowering plants (66–44, depending on the species) only 30 and 20 have been identified in *Physcomitrella patens* and *Selaginella moellendorffii*, respectively. This suggests that NAC proteins expanded as land plants evolved. Interestingly, no NAC proteins could be identified in the aquatic species analyzed, which suggests that NAC proteins may be specific to land plants. However, the analysis of Zhu and coworkers did not include any representatives of the charophytes which are thought to contain the sister group to land plants (Finet et al., 2010). Using Basic Local Alignment Search Tool (BLAST) searches we identified transcriptome shotgun assembly (TSA) sequences from the charophytes *Coleochaete* sp. (loci JO249122 and JO249294), *Penium margaritaceum* (locus JO233410), *Chaetosphaeridium globosum* (locus JO158096), and *Nitella mirabilis* (locus JV748667) whose putative translation yields proteins showing a conserved NAC domain (Figure 15.1). This observation strongly suggests that NAC proteins appeared before the transition from water to land, about 450 million years ago.

15.2.1.2 Origin and Early Evolution of the NAM/CUC3 Family

The phylogeny of NAC proteins has been analyzed by several groups who often determined the position of NACs from a particular species in relation to *Arabidopsis* and rice NACs (e.g., Fang et al., 2008; Hu et al., 2010; Ooka et al., 2003; Pinheiro et al., 2009; Shen et al., 2009; Zhu et al., 2012). Results from these phylogenetic analyses show some variability: for instance, the number of subfamilies varies from 5 (Fang et al., 2008) to 21 (Zhu et al., 2012). Despite these variations, NAM/CUC3 proteins are often associated with the same group of proteins, although with a variable topology, forming an entire or part of a subfamily (Figure 15.2A).

Proteins that belong to the NAM/CUC3 family can be clearly divided into two clades: the NAM clade that includes the petunia NAM and *Arabidopsis* CUC1 and CUC2 proteins, and the CUC3 clade (Blein et al., 2008, Zimmermann and Werr, 2005). These two clades are based on the sequence of the NAC domain (Section 15.2.2.1), but also overlap with the presence/absence of a microRNA-binding site. Indeed, all *NAM* genes possess a binding site for the *microRNA164* (*miR164*), whereas *CUC3* genes do not. Members of these two clades can be found in eudicots, monocots, and early-diverging angiosperms such as *Amborella trichopoda* (Adam et al., 2011; Blein et al., 2008; Zimmermann and Werr, 2005; Vialette-Guiraud et al., 2011). On the other hand, gymnosperm genes possessing a *miR164*-binding site are not grouped within angiosperm NAM or CUC3 clades, rather they occupy a sister position to the combined NAM + CUC3 clade. This suggests that a unique NAM + CUC3 lineage regulated by *miR164* was present in the last common ancestor of extant seed plants, and that a duplication event generated the NAM and CUC3 clades in the angiosperm lineage after its divergence from gymnosperms. In this scenario, loss of *miR164* regulation would have occurred later in the CUC3 lineage (Vialette-Guiraud et al., 2011; Figure 15.2B).

15.2.1.3 Recent Evolution Within the NAM/CUC3 Family

Additional duplication events further complicated the phylogeny of NAM/CUC3 proteins in angiosperms. Such duplication events can either be recent, resulting in two closely related paralogs (such as the pea proteins

15.2 EVOLUTION AND STRUCTURE OF NAM/CUC3 PROTEINS

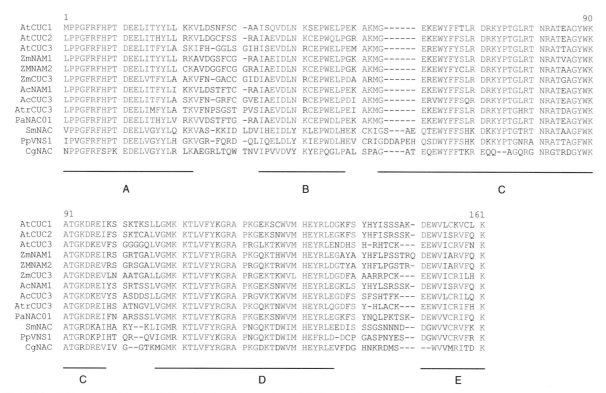

FIGURE 15.1 Alignment of the NAC domain of representatives of different plant groups showing five conserved subdomains named "A to E." At, *Arabidopsis thaliana*; Zm, *Zea mays*; Ac, *Aquilegia coerulea*; Atr, *Amborella trichopoda*; Pa, *Picea abies*; Sm, *Selaginella moellendorffii*; Pp, *Physcomitrella patens*; Cg, *Chaetosphaeridium globosum*. The alignment was produced using MultAlin (Corpet, 1988).

PsNAM1 and PsNAM2 that share 95% identity) or more ancient leading to more divergent genes like the CUC1 and CUC2 proteins found in the Brassicaceae lineage (the *Arabidopsis* CUC1 and CUC2 proteins share only 50% identity mostly concentrated in the NAC domain). Indeed, phylogenetic analysis and reconstruction of genome duplication events suggest that two rounds of gene duplication followed by gene loss led to the distinct CUC1 and CUC2 lineages in Brassicaceae, which have partially divergent functions (Hasson et al., 2011, Vialette-Guiraud et al., 2011). Because the CUC1 and CUC2 lineages are specific to Brassicaceae and possibly other closely related Brassicales, the names "CUC1" and "CUC2" should be exclusively used for genes identified in these groups, while for other species "NAM" should be used. Here, we use "*NAM/CUC3*" when referring to genes belonging to either of the two clades, and "*CUC*" when referring specifically to *Arabidopsis* genes.

15.2.2 NAM/CUC3 Protein Organization and Specific Domains

NAM/CUC3 proteins, like other NAC transcription factors, can be subdivided into two main functional domains: an amino-terminal domain including the conserved NAC domain, and a more divergent carboxy-terminal domain (CTD; Duval et al., 2002; Taoka et al., 2004). Domain-swapping experiments between the NAC or CTD domains of CUC1, CUC2 and the more distantly related ATAF1 protein showed that the ability of the CUC1/2 proteins to promote *in vitro* adventitious shoot formation lies in their NAC domain, suggesting that this part of the CUC1 and CUC2 proteins determines their specific functions (Taoka et al., 2004).

15.2.2.1 The Amino-Terminal NAC Domain

This domain can be subdivided into five highly conserved regions (Figure 15.1) and has been implicated in the DNA-binding properties of several NAC proteins (e.g., Duval et al., 2002; Jensen et al., 2010). Most mutations disrupting CUC1 function fall into its NAC domain, thus highlighting its importance (Figure 15.3).

The DNA-binding mechanisms of NAC proteins have begun to be elucidated. ANAC019 and ANAC092 bind to a CGT[A/G] consensus site (Olsen et al., 2005; Tran et al., 2004; Xu et al., 2013). Binding affinity to this motif varies between NAC proteins (Jensen et al., 2010). Lindemose et al. (2014) showed that 12 NACs can be divided into three groups with different binding specificities. Two groups recognize variants of the previously identified CGT[A/G] target sequence while the third recognizes an unrelated motif. These groups largely match the phylogenetic differences between NAC proteins.

NAC proteins can form both homo- and heterodimers via interaction of their NAC domains, for which both the

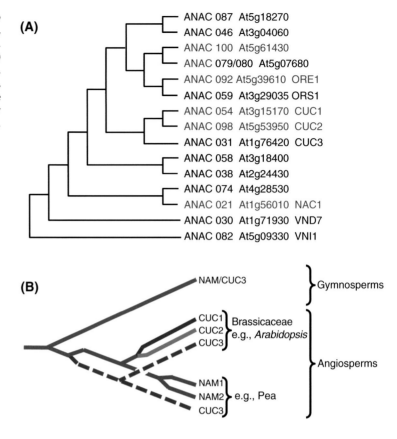

FIGURE 15.2 (A) Schematic phylogeny of the subgroup of *Arabidopsis thaliana* NAC proteins containing the CUC1, CUC2 and CUC3 proteins (adapted from Zhu et al., 2012). The genes targeted by the *miR164* are indicated in red. (B) Schematic evolution of the *NAM/CUC3* genes in seed plants. Members of the NAM + CUC3 clade are indicated in violet, members of the NAM clade are in blue, and members of the CUC3 clade are in red. Solid lines indicate lineages targeted by *miR164* while dotted lines are lineages not targeted by *miR164*. (Vialette-Guiraud et al., 2011).

interacting surface and two essential salt-bridge-forming residues have been identified (Ernst et al., 2004; Olsen et al., 2005). NACs stably bind DNA as dimers by recognizing two palindromic binding sites, but a single binding site is also sufficient for NAC binding, both *in vitro* (Olsen et al., 2005) and *in vivo* (Tran et al., 2004; Xu et al., 2013). ANAC019 dimers can exist either in an open or a closed conformation (Welner et al., 2012). While the open conformation is predominant in solution, dimers mostly adopt a closed conformation when bound to DNA. Variation in dimer conformation may account for the recognition of DNA stretches with either single binding sites or a variable number of base pairs separating two binding sites.

The NAC domain of ANAC019 contains a central twisted antiparallel ß-sheet, which is packed between two α-helices on both extremities (Ernst et al., 2004; Chapter 4). Part of this ß-sheet formed by conserved WKATGTD amino acids protrudes into the major groove of DNA and interacts with the sugar/base region of DNA providing specificity to the recognition, while other parts interact with the DNA backbone potentially increasing affinity (Welner et al., 2012; Chapter 13). This mode of interaction shows similarities with those of plant WRKY and mammalian glial cells missing (GCM) transcription factors.

To date, no structure data on any of the NAM/CUC3 proteins nor their DNA binding sites have been determined. A recent study suggests that, like ANAC019 and other phylogenetically related proteins, NAM/CUC3 may recognize a TT[A/G]CGT[A/G] motif (Lindemose et al., 2014). However, because the WKATG<u>TD</u> residues that contribute to ANAC019 specificity are replaced by WKATG<u>KD</u> in NAM/CUC3 proteins, it is not clear how conserved the core binding site can be. It is therefore essential to determine the binding specificity of NAM/CUC3 proteins experimentally.

15.2.2.2 The Carboxy-Terminal Domain

The CTD of NAM/CUC3 proteins is more variable than the NAC domain, but several small domains can be recognized. However, these domains are not all found in all NAM/CUC3 proteins and neither are they specific to these proteins, as they can be found in other NACs. Initially, Taoka et al. (2004) identified three domains called the V (TEHVSCFS), L (SLPPL), and W motifs (WNY) as well as a serine-rich domain, but further analyses identified additional domains (Adam et al., 2011; Larsson et al., 2012; Zimmermann and Werr, 2005; Figure 15.3). When fused to the GAL4 DNA-binding domain, the CTD of CUC1 and CUC2 proteins, like that of other NAC proteins, shows transcription activation in yeast cells and tobacco BY-2 cells (Taoka et al., 2004). Its serine-rich and W domains are necessary for transcriptional activity in yeast, while the V and L motifs are dispensable. The W motif is also important *in planta* as the strong *cuc1–3* and weak *cuc1–6* alleles affect this domain (Hibara et al., 2006; Takada et al., 2001; Figure 15.3). In contrast to the serine-rich and

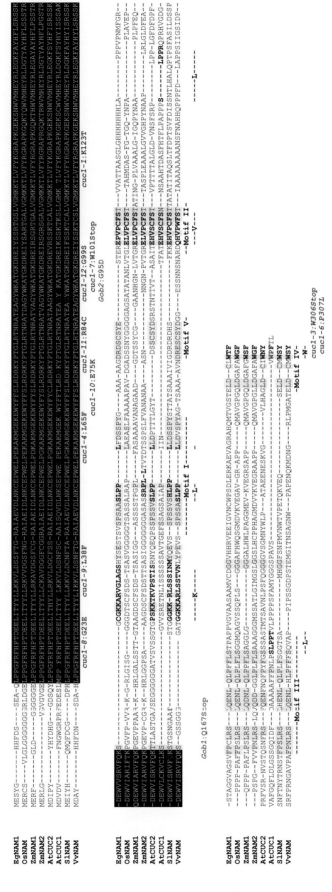

FIGURE 15.3 Alignment of the NAM proteins (adapted from Adam et al., 2011), are on a gray background while those defined by Taoka et al. (2004) and Larsson et al. (2012); (domains K, L, V, and W) are in boldface. Mutations affecting either AtCUC1 or SINAM (also known as GOB) are indicated in red and blue, respectively. Eg, *Elaeis guineensis*; Os, *Oriza sativa*; Zm, *Zea mays*; At, *Arabidopsis thaliana*; Sl, *Solanum lycopersicum*; Vv, *Vitis vinifera*.

W domains that positively contribute to the transcriptional activity of CUC1/2 proteins, a hydrophobic region that contains the K domain (described by Larsson et al., 2012) represses CUC2 activity: deletion of this domain leads to a 14-fold increase in transcriptional activity in yeast (Taoka et al., 2004). Interestingly, the K domain is not only absent from monocot NAM proteins but also from CUC1, which correlates with a higher activity of CUC1 compared to CUC2 (Hasson et al., 2011).

15.3 NAM/CUC3 GENES DEFINE BOUNDARIES IN MERISTEMS AND BEYOND

As mentioned in the introduction, *NAM/CUC3* genes were identified in genetic screens in petunia and *Arabidopsis* as arrested-development mutants showing seedlings with fused cotyledons. Here we present the mutant phenotypes, genetic studies, and expression pattern analysis that led to the characterization of the *NAM/CUC3* functions during plant development.

15.3.1 Identification of the NAM/CUC3 Genes: Role in Boundary and Meristem Formation

15.3.1.1 Identification of NAM/CUC3 in Petunia and Arabidopsis

Petunia *nam* and *Arabidopsis cuc1–cuc2* seedlings share similar phenotypes characterized by fused cotyledons and no SAM (Souer et al., 1996, Aida et al., 1997). This phenotype appears early on during embryonic development with an ectopic bulging at the central apical part of heart-shaped embryos. Simultaneous bulging within this region and at cotyledon primordia effectively leads to fusion of the two cotyledons (Figure 15.4A, B). Therefore, the role of *CUC1/2* and *NAM* genes in cotyledon separation has been ascribed to inhibition of growth in the boundary region. Cotyledon fusion in these mutants is accompanied by a lack of embryonic SAM development. Indeed, presumptive SAM cells in *cuc1–cuc2* double-mutants do not express the meristem marker *SHOOT MERISTEMLESS* (*STM*; Aida et al., 1999). Together, these observations suggest that, in addition to their role in cotyledon separation, *CUC1*, *CUC2*, and *NAM* genes are also implicated in SAM initiation. Accordingly, these genes are expressed during embryogenesis in a region encompassing the presumptive SAM (Aida et al., 1999; Takada et al., 2001). In later stages, this expression disappears from the initiating SAM and becomes restricted to the boundaries between the developing cotyledons and the SAM (Figure 15.4G). This observation suggests that *CUC1*, *CUC2*, and *NAM* have an early role in separating cotyledons and specifying SAM initiation, and a later one separating the undifferentiated SAM from the differentiating cotyledons.

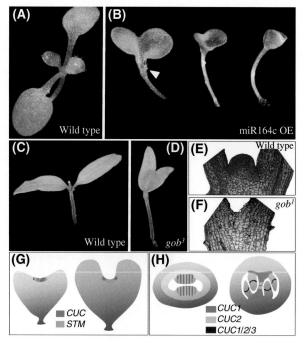

FIGURE 15.4 **Mutant phenotypes and expression patterns of *NAM/CUC3* genes in *Arabidopsis* and tomato.** (A) Wild-type *Arabidopsis* seedlings and (B) transgenic seedlings showing various degrees of cotyledon fusion phenotypes resulting from reduced expression of *CUC1/2* due to overexpression of its regulator *miR164*. Weak phenotypes show partial cotyledon fusion and reduced meristematic activity (arrowhead points to young leaves) and strongly silenced lines show complete cotyledon fusion and no meristematic activity. (C) Tomato wild-type seedlings and (D) *gob* mutant seedlings showing partial cotyledon fusion. (E, F) The cotyledon fusion phenotype of *gob* mutants is accompanied by an absence of meristem that can be identified in the wild type as a bulge between cotyledon primordia. (G) The expression domain of *CUC* genes at the central apical region of heart-stage embryos (left) overlaps that of shoot apical meristem marker *STM*. At later stages (right) *CUC* expression is restricted to the boundary between cotyledon primordia and the meristematic zone. (H) *CUC* genes are expressed during gynecium development at the adaxial side (lighter gray at the left) of the medial region in the presumptive septum in a region encompassing the future placenta (in orange at the left). This expression is sustained at the medial ridge tips during septum development and eventual fusion. *CUC* genes are also expressed in the boundaries and at the base of ovule primordia and in a ring at the boundary between the nucellus and chalaza.

Much like in the embryo, *CUC* genes are expressed at a variety of frontier regions in the mature plant, such as the boundary between the apical meristem and leaf primordia, between the inflorescence and floral meristems, or even between different floral organs. Accordingly, regenerated shoots of *cuc* double-mutants show organ fusions at all these levels (Aida et al., 1997). Therefore, these genes have been classified as general regulators of organ separation, or, simply put, boundary genes.

While *cuc1–cuc2* double-mutants show strong fusion phenotypes with no SAM initiation, single *cuc1* or *cuc2* mutants are phenotypically normal for the most part, with few showing incomplete cotyledon fusions that produce heart-shaped seedlings. The incomplete penetrance of single mutations and the overlap of *CUC1* and *CUC2* expression domains suggest that a certain degree of functional redundancy exists between these genes (Takada et al., 2001). The characterization of the *CUC3* gene increased the degree of functional redundancy between *CUC* genes. Indeed, this paralog has overall similar expression patterns to *CUC1* and *CUC2* as well as additive phenotypic effects (Vroemen et al., 2003).

15.3.1.2 Role in Other Dicots and Monocots

Since the characterization of *nam* and *cuc* mutants in petunia and *Arabidopsis*, similar roles in boundary specification and organ separation have been revealed for *NAM/CUC3* genes in other species. For example, *nam* mutants in *Medicago truncatula* have fused cotyledons and lack primary apical meristems (Cheng et al., 2012). In *Antirrhinum majus*, the *CUPULIFORMIS* (*CUP*) gene has been identified through its mutant phenotype, which presents strong organ fusion both at the embryonic and vegetative level (Weir et al., 2004). Despite strong defects in meristem initiation, *cup* mutants can produce secondary meristems at the hypocotyl. These develop severe fusions of leaves and floral organs as well as phyllotaxis perturbations. Overall, organ fusion defects in *cup* mutants are more severe than observed in *Arabidopsis cuc1–cuc2*, suggesting that the redundancy level between *NAM/CUC3* paralogs can vary. Tomato (*Solanum lycopersicum*) *goblet* (*gob*) mutants show similar phenotypes of cotyledon fusion and SAM absence (Figure 15.4C–F; Berger et al., 2009; Blein et al., 2008; Brand et al., 2007). The role of *NAM/CUC3* genes in monocots has not yet been functionally tested, but characterization of the maize *ZmNAM1*, *ZmNAM2*, and *ZmCUC3* genes as well as the oil palm *EgNAM1* and *EgCUC3* genes showed that they have similar expression patterns to *Arabidopsis* homologs, with transcripts being found in meristematic tissues and in cells separating adjacent organs. Interestingly, small differences may exist between monocots and dicots: for instance, *ZmCUC3* is expressed later than *ZmNAM1/2* during maize embryo development whereas *CUC3* is activated earlier than *CUC1/2* in *Arabidopsis* (Zimmermann and Werr, 2005). Nevertheless, these proteins appear to have a conserved function, as oil palm homologs are capable of increasing leaf serration when ectopically expressed in *Arabidopsis* (for the role *NAM/CUC3* plays in leaf development see Section 15.3.3) and restore organ fusion defects in *cuc* mutants (Adam et al., 2011).

15.3.2 Role of CUC Genes in Other Meristematic Territories

As full cup-shaped mutants usually lack a SAM, the study of their effect at the postembryonic level is dependent on the formation of escape or regenerated shoots. Studies of such regenerated shoots have allowed for additional roles of *NAM/CUC3* genes during later vegetative and flowering stages to be characterized.

15.3.2.1 Axillary Meristems

Axillary meristems form near the shoot apex during vegetative and reproductive development at the axils of developing rosette and cauline leaves (Grbic and Bleecker, 2000). *CUC1/2/3* transcripts have been detected in axillary meristems at the boundary between leaf primordia and the shoot apex, and *cuc3* mutants fail to initiate axillary meristems in rosette leaf axils (Aida et al., 1999; Hibara et al., 2006; Ishida et al., 2000; Raman et al., 2008; Takada et al., 2001). This phenotype is greatly enhanced by the *cuc2* mutation but is not observed in other single-mutants or combination of mutants, suggesting that, although both *CUC2* and *CUC3* are required for axillary meristem specification, *CUC3* contribution is greater (Hibara et al., 2006; Raman et al., 2008). Alternatively, plants expressing *miR164*-resistant variants of *CUC1/2* genes form accessory axillary meristems (Raman et al., 2008). Collectively, these results show that *CUC* genes redundantly promote shoot meristem formation both during embryonic and postembryonic development.

15.3.2.2 Floral Organ Boundaries

Flowers of *Arabidopsis cuc1–cuc2* double-mutants show strong organ fusions between sepals and stamens and also have fewer petals and stamens (Aida et al., 1997). Floral phenotypes in single-mutants are much less severe, suggesting once again a certain degree of functional redundancy between *CUC* genes (Hibara et al., 2006). Accordingly, *CUC1/2/3* have mostly overlapping expression patterns in the boundaries between floral organs, both between organs of the same whorl and between different whorls (Hibara et al., 2006; Ishida et al., 2000; Takada et al., 2001; Vroemen et al., 2003). Similar to the way they function in the SAM, CUCs act at the boundaries of organ primordia suppressing cell proliferation and bulging, which allows for clean organ separation. The roles *CUC1/2* play in floral organ number and separation are also dependent upon their regulation by *miR164*. Indeed, *eep1*, a mutant allele of *MIR164C*, leads to the production of supernumerary petals in regions adjacent to normal organs, which is associated with an increase in *CUC1* and *CUC2* expression (Baker et al., 2005).

The role of *NAM/CUC3* genes in floral organ patterning and separation also appears to be conserved across angiosperms. Floral organ fusions are observed in *Medicago truncatula nam*, tomato *gob*, and *Antirrhinum cup* mutants, with the corresponding genes being expressed at floral organ boundaries (Berger et al., 2009; Cheng et al., 2012; Weir et al., 2004). Interestingly, the expression of a *miR164*-resistant variant of *SlGOB* leads to the production of accessory organs mostly in the petal and carpel whorls, suggesting conservation of the *miR164/CUC* module during tomato flower development (Berger et al., 2009).

15.3.2.3 Gynecium

Arabidopsis CUCs are also expressed within carpel tissues and around developing ovules, suggesting they play a role in gynecium and ovule development (Galbiati et al., 2013; Ishida et al., 2000; Kamiuchi et al., 2014; Nahar et al., 2012; Takada et al., 2001; Vroemen et al., 2003). In *Arabidopsis*, gynecia are composed of two carpels that fuse along two opposing longitudinal medial ridges. The two medial ridge meristems form the placenta, a tissue with meristematic properties, which develops ovules and central outgrowths that fuse to form the septum. Gynecia that lack *CUC* activity fail to initiate medial ridge meristems resulting in severe septum fusion defects and fewer ovules (Ishida et al., 2000; Kamiuchi et al., 2014; Nahar et al., 2012). The early expression of *CUC1/2* at presumptive medial ridges, the absence of meristem marker *STM* expression in the double mutant, and the enlargement of carpel margins in plants expressing *miR164*-resistant forms of *CUC1* or *CUC2* indicate that these genes act both to initiate medial ridge meristems and to maintain their meristematic state (Figure 15.4H; Kamiuchi et al., 2014). In some mutants, incipient medial ridge meristems are formed in an asymmetric fashion suggesting that *CUC1/2* are also required for proper positioning of meristems. In a more extreme case, *miR164*-dependent *CUC2* misregulation leads to incomplete carpel fusion, as medial ridges are incompletely formed (Larue et al., 2009; Nikovics et al., 2006; Sieber et al., 2007). Several lines of evidence suggest *CUC* genes play a role in ovule development, notably the reduced number of ovules in *cuc* double-mutants and the expression of *CUC1/3* between ovule primordia and *CUC1/2/3* between nucellus and chalaza (Aida et al., 1999; Ishida et al., 2000; Vroemen et al., 2003). Although, the exact mechanisms through which *CUC* genes regulate ovule development are still unknown, a recent model involving the integration of auxin signaling has been proposed (Section 15.4.1.1; Galbiati et al., 2013).

Other results are also suggestive of conservation of *NAM/CUC3* roles in gynecium and ovule development across angiosperms. In *Medicago truncatula*, *nam* mutant carpel margins are incompletely fused and fewer ovules with altered embryo sac development are formed, leading to female sterility (Cheng et al., 2012). *Antirrhinum cup* mutants not only produce fewer ovules and/or fused ovules, they are also female sterile (Weir et al., 2004).

15.3.2.4 Organ Abscission

Abscission – the detachment of aged, mature, or diseased organs such as leaves and seeds – occurs in specific regions that display a set of characteristics reminiscent of meristematic tissues such as small cells with dense cytoplasms (Nakano et al., 2013). These abscission zones situated at key hinge regions share characteristics with boundaries. Indeed, in tomato, the *GOB* gene and other genes known to promote meristematic identity in axillary meristems are expressed in the abscission zone.

15.3.3 Role of CUC Genes in Leaf Development

Arabidopsis leaves are simple with small serrations on their margins. While *cuc2* mutants produce leaves with smooth margins, plants with increased *CUC2* expression as a result of defective *miR164* regulation show deeper and larger serrations than the wild type (Nikovics et al., 2006). *cuc3* mutants also show reduced serrations, while *CUC1*, which is not expressed in leaves, plays no role in *Arabidopsis* leaf development. Whereas *CUC2* acts early on with the onset of teeth, *CUC3* is thought to act only at later stages to sustain teeth outgrowth (Hasson et al., 2011). Interestingly, chimeric constructs, where the *CUC2* promoter drives the expression of CUC1 rescue normal leaf serration in *cuc2* mutants, also induce leaflet formation in genetic backgrounds lacking *miR164*. These results show that, even though *CUC1* is not expressed in developing leaves, the CUC1 protein is partially functionally interchangeable with CUC2.

In species with compound leaves the role of *NAM/CUC3* genes is extended to specify the boundaries between leaflets. Indeed, these genes are expressed at the boundaries of leaflet primordia, and their inactivation results in fused and fewer leaflets (Berger et al., 2009; Blein et al., 2008; Cheng et al., 2012; Wang et al., 2013). Alternatively, tomato plants expressing the gain-of-function *miR164*-resistant allele *Gob4-d* produce deeply lobed leaflets (Berger et al., 2009). Altogether, these observations are suggestive of a conservation of the mechanisms controlling boundary specification between the apex and leaf primordia with different architectures.

15.4 MULTIPLE REGULATORY PATHWAYS CONTRIBUTE TO THE FINE REGULATION OF NAM/CUC3 GENES

Section 15.3 focused on *nam/cuc3* mutant phenotypes and highlighted the precise expression patterns of these genes during development. *NAM/CUC3* genes are expressed in narrow and discontinuous domains, often restricted to a few cells at the boundary between two outgrowing structures. Regulation of this expression pattern is essential for proper organ development as *CUC* overexpression leads to severe phenotypes (Hibara et al., 2006; Laufs et al., 2004). When *CUC2* is uniformly expressed across the leaf margin instead of its discrete expression pattern at the teeth sinuses, a smooth leaf margin is formed in place of the typical serrated form (Bilsborough et al., 2011).

15.4 MULTIPLE REGULATORY PATHWAYS CONTRIBUTE TO THE FINE REGULATION OF NAM/CUC3 GENES

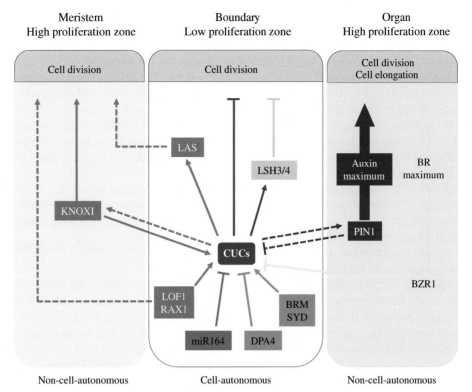

FIGURE 15.5 **CUC regulatory network.** CUC transcription factors are regulated by chromatin regulators such as BRM, SYD, and DPA4. CUC1 and CUC2 are negatively regulated by *miR164*. LOF1 and RAX1 induce CUCs during axillary meristem formation. BRs and auxin, which promote lateral organ proliferation, participate in the CUC discrete boundary expression pattern. BR modify BZR1 activity, which in turn negatively regulates CUCs. Auxin maxima, formed via a PIN1-dependent mechanism in growing primordia, restrict CUC expression to boundaries. In turn, CUCs act via a nonidentified mechanism dependent on PIN1 to modify auxin levels. KNOX genes, which are essential for meristem maintenance, induce CUC expression and activate KNOX expression in a feedforward regulatory loop as well. KNOX, LOF, RAX1, and LAS all contribute to maintaining the high division rate in meristematic zones. CUCs directly activate LSH3 and LSH4 in boundary cells. This complex regulatory network allows both the definition of the boundary by locally repressing growth and organ outgrowth, and meristem initiation and maintenance by promoting growth in a non-cell-autonomous manner. Solid arrows represent direct interaction; dashed arrows represent nonlocal genetic interactions.

This section discusses the factors that contribute to establishing the precise expression patterns of *NAM/CUC3* genes throughout plant development (Figure 15.5). First, we discuss how hormonal regulation shapes *NAM/CUC3* expression. Then, we consider the role of *miR164* in the posttranscriptional regulation of *NAM* genes. Finally, we describe *NAM/CUC3* transcriptional regulation.

15.4.1 Hormonal Regulation of NAM/CUC3 Gene Expression

15.4.1.1 *The Interplay between* **NAM/CUC3** *Genes and Auxin*

Numerous works suggest that *CUC2* expression is repressed by PIN1-generated auxin maxima. The *PIN-FORMED1* (*PIN1*) gene encodes an auxin efflux carrier that has a polar distribution within the cell thus contributing to differential auxin accumulation in *Arabidopsis*. In developing embryos, PIN1-dependent auxin maxima induce cotyledon formation (Friml et al., 2003). In *pin1* mutants, the *CUC1* expression domain is extended to the entire apical region whereas *CUC2* is expressed in patches restricted to the center and sides of the embryo (Aida et al., 2002). The *PINOID* (*PID*) gene encodes a serine/threonine kinase that acts as a positive regulator of PIN1-mediated polar auxin transport. *pin1–pid* double-mutant embryos completely lack cotyledons and show broad expression of *CUC1* and slight enlargement of the *CUC2* expression domain. Additionally, *pin1–pid–cuc1* triple-mutants form small cotyledons which suggests that ectopic expression of *CUC1* in *pin1–pid* embryos is responsible for the absence of cotyledons (Furutani et al., 2004). *pasticcino1* (*pas1*) mutants show defective cotyledon development and associate altered membrane localization of PIN1 with an enlargement of the domain expressing *CUC2* (Roudier et al., 2010). Overall, these results indicate that PIN1-mediated auxin transport is necessary to regulate *CUC1/2* expression in the embryo.

During postembryonic development, *pin1* mutants produce a naked inflorescence (Okada et al., 1991). In similarly to what as happens in the embryo, primordia positioning in the SAM is determined by PIN1-driven auxin maxima (Reinhardt et al., 2003). In *pin1* mutants, *CUC2* expression is enlarged forming a circle around

C. FUNCTIONAL ASPECTS OF PLANT TRANSCRIPTION FACTOR ACTION

the inflorescence SAM (Vernoux et al., 2000). Moreover, live imaging experiments suggest that *CUC2* expression is downregulated in tissues where convergent PIN1 polarities are expected to accumulate high auxin levels (Heisler et al., 2005). Together, these results suggest that *CUC2* expression in the SAM is inhibited by PIN1-generated auxin activity maxima. As explained in Section 15.3.2.1, *CUC2* genes redundantly promote axillary meristem formation (Raman et al., 2008). Two articles suggest that an auxin minimum is required for axillary meristem formation in *Arabidopsis* and tomato (Wang et al., 2014a; Wang et al., 2014b). Although this has not been tested, this auxin minimum could allow *CUC* expression thus inducing axillary meristem formation.

pin1 Arabidopsis mutants form leaves that lack serrations. PIN1-mediated auxin response foci at the leaf margin are interspaced with regions showing high *CUC2* and *CUC3* expression (Hasson et al., 2011; Hay et al., 2006; Nikovics et al., 2006). Auxin treatments are able to abolish expression of a *CUC2* reporter in leaf primordia, suggesting that auxin negatively regulates *CUC2* expression during simple leaf development (Bilsborough et al., 2011).

As mentioned in Section 15.3.2.3, *CUC1* and *CUC2* are involved in carpel margin meristem initiation required for ovule initiation. In this context, *MONOPTEROS* (*MP*) is expressed in a similar pattern to *CUC1* and *CUC2*. Moreover, in *mp* mutants, *CUC1* and *CUC2* expression is reduced in inflorescences and leaves. Chromatin immunoprecipitation (ChIP) experiments have established that MP directly binds *CUC1* and *CUC2* genomic regions (Galbiati et al., 2013). These results strongly suggest that MP positively regulates *CUC1* and *CUC2*, providing a molecular link between auxin signaling and *CUC* genes. However, these results are difficult to reconcile with data obtained in the embryo where *CUC1* expression domain is enlarged in *mp* mutant embryos (Aida et al., 2002), indicating that *MP* negatively regulates *CUC1* expression in the embryo. Moreover, it is surprising that an auxin response factor would positively regulate *CUC* expression when in most organs auxin maxima negatively regulate *CUC* expression.

To date, it is not clear to what extent the relationship between auxin and *CUC* genes identified in *Arabidopsis* is conserved. In *Cardamine hirsuta*, a close relative of *Arabidopsis* with compound leaves, it is not known whether *ChCUC* expression is controlled by polar auxin transport as in *Arabidopsis*. Interestingly, the expression of *GOB* in tomato is not modified upon auxin treatment. Moreover, *ENTIRE*, an auxin-response repressor, acts on leaf dissection in a parallel pathway independent of *GOB* (Ben-Gera et al., 2012; Berger et al., 2009). Alternatively, auxin-induced downregulation of *NAM* genes seems to be a general feature during embryonic and postembryonic development, and some evidence points to conservation of this role. Indeed, in the gymnosperm *Picea abies* a *NAM/CUC3* ortholog is also regulated by polar auxin transport (Larsson et al., 2012).

A recent work reported a link between cytokinins and *CUCs* (Li et al., 2010). A line overproducing cytokinins produces more flowers, a phenotype that is dependent on *CUC2* and *CUC3* overexpression. Moreover, in the cytokinin receptor *ahk2–ahk3* double-mutant, *CUC1* and *CUC2* expression is strongly reduced, suggesting that cytokinin signaling promotes *CUC* expression. Interestingly, there is increasing evidence that cytokinin signaling controls polar auxin transport (Marhavy et al., 2014). Therefore, further investigations are required to determine whether regulation of *CUC* genes by cytokinins is mediated by auxin.

15.4.1.2 Brassinosteroids, New Regulators of CUC Expression

Brassinosteroids (BRs) are plant steroid hormones that regulate cell proliferation and other developmental processes (Kim and Wang, 2010). They act through a complex signaling pathway that leads to activation of two transcription factors, BZR1 and BES1, which in turn modify the expression of over 1000 genes.

Recent findings suggest a link between BRs and *CUC* genes. Plants with increased BR content or signaling show axillary shoot, stamen, and cotyledon fusions, reflecting abnormal boundary establishment (Gendron et al., 2012). Alternatively, mutants with reduced biosynthesis or sensitivity to BRs have deeper axillary separations and form ectopic boundaries. This suggests that low BR signaling is sufficient and necessary for proper boundary formation. Genetic and pharmacological experiments show that low BR signaling induces *CUC* expression in the SAM, whereas high BR signaling inhibits it. Additionally, ChIP experiments indicate that BZR1 strongly binds the *CUC3* promoter suggesting direct regulation. Overall, these results indicate that BR signaling negatively regulates *CUC* gene expression.

15.4.2 *miR164* FineTunes NAM Gene Expression

miR164 was among the first identified plant miRNAs. In *Arabidopsis*, it is encoded by three loci, *MIR164A*, *B*, and *C*. It regulates the expression of six transcription factors of the NAC family: *CUC1* and *CUC2*, *NAC1* which is known to regulate lateral root induction, *ORESARA1* (*ORE1*) which controls leaf senescence, and two uncharacterized *NACs* (At5g61430 and At5g07680; Schwab et al., 2005).

15.4.2.1 miR164 Regulation is Essential for Shoot Development

miR164 controls inflorescence and floral development. Plants expressing *CUC1* or *CUC2 miR164*-resistant variants show extra petals and enlarged sepal boundaries (Laufs et al., 2004; Mallory et al., 2004). Another *CUC2 miR164*-resistant allele has shorter and wider siliques with tissue projections along the valve margins (Larue et al., 2009). Accordingly, *early extra petals1* (*eep1*), a *mir164c* mutant, also presents extra petals and defects in carpel fusion (Baker et al., 2005). This indicates *MIR164C* plays a role in regulating *CUC1* and *CUC2* during flower development. A similar role has been proposed for *SlmiR164* in tomato, in which expression of a *SlmiR164*-resistant *GOB* variant results in extra petals and ectopic carpels (Berger et al., 2009). Expression of *miR164*-resistant *CUC2* leads to modified phyllotaxy compared to wild type (Peaucelle et al., 2007), as also observed in the *mir164abc* triple-mutant (Sieber et al., 2007). Strikingly, in both genotypes, the formation of primordia in the meristem appears to be normal. Taken together, this reveals that phyllotaxy is postmeristematically maintained via *miR164*-dependent negative regulation of *CUC2*.

miR164 also regulates leaf development. Both *mir164a* Arabidopsis mutants and *CUC2 miR164*-resistant lines present leaves with deeper serrations than the wild type. Moreover, *MIR164A* is expressed in leaf margin sinuses in a pattern overlapping *CUC2* expression (Nikovics et al., 2006). Thus, in *Arabidopsis*, *MIR164A* regulates the level of *CUC2* expression, which in turn governs the level of leaf serration. Interestingly, quantitative trait locus (QTL) mapping has revealed a single nucleotide polymorphism in *MIR164A* miRNA* which modifies *MIR164A* biogenesis and drastically reduces its accumulation (Todesco et al., 2012). This indicates that natural variation in *MIR164A* maturation can contribute to leaf serration polymorphism. In the compound leaves of tomato, the *GOB* miRNA-resistant allele, *Gob-4d*, harbors leaflets with deeper and wider lobes than the wild type, whereas *gob* mutants show smooth fused leaflets (Berger et al., 2009).

In contrast to *Arabidopsis*, *GOB* and *SlMIR164* show complementary expression profiles and the *GOB* expression pattern becomes wider in the *Gob-4d* allele. This suggests, that *SlmiR164* defines the sharp domain of the *GOB* expression pattern rather than controlling its expression level as in *Arabidopsis*.

miR164 controls axillary meristem development. Expression of *CUC1* or *CUC2 miR164*-resistant variants leads to the formation of accessory buds in leaf axils, a phenotype also observed in *mir164abc* mutants. Concurrently, *MIR164A* and *MIR164C* are expressed in the boundary between the leaf primordium and the SAM, from where the axillary meristem subsequently emerges (Raman et al., 2008). Overall, *CUC1* and *CUC2* mRNA cleavage by *miR164* is required to negatively regulate the formation of accessory buds in leaf axils.

These results establish *CUC1/2–miR164* as a conserved genetic module that is recruited multiple times during the evolution of aerial organs (Figure 15.6). Moreover, *miR164* plays a crucial role in regulating *NAC1* during lateral root induction (Guo et al., 2005) and inhibiting *ORE1* expression during leaf senescence (Kim et al., 2009). *miR164* is therefore an important regulator of plant development (Pulido and Laufs, 2010). Interestingly, *miR164* does not seem to regulate its targets always in the same manner: while it regulates the timing of *ORE1* expression, it regulates the *GOB* expression pattern spatially during tomato leaf development and controls the level of *CUC1* and *CUC2* expression during flower and leaf development in *Arabidopsis*.

15.4.2.2 Evolution and Specialization of the MIR164 Genes

miR164 is found in dicots, monocots, and gymnosperms, indicating that, much like its target *NAC* genes, it was likely present in the last common ancestor of

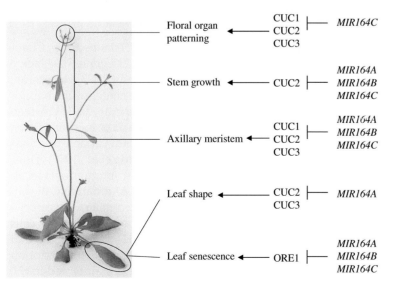

FIGURE 15.6 *CUC/miR164*, a genetic module essential for plant shoot development. The roles played by each *CUC* (and the related NAC gene, *ORE1*) and *MIR164* genes during *Arabidopsis* development are indicated.

gymnosperms and angiosperms (Section 15.2.1.2; Axtell and Bartel, 2005). Multiple genes code for *miR164*: 3 in *Arabidopsis* but up to 12 in soybean. Mature *miR164* genes encoded by different members can be identical, such as in soybean, or show small sequence variations as in *Arabidopsis* in which mature *miR164c* differs by one nucleotide from mature *mir164a* or *mir164b*. An evo-devo study of the *MIR164* family suggests that two lineages were present in the last common ancestor of extant angiosperms: a B clade containing the *Arabidopsis MIR164B* gene whose members tend to be highly expressed in roots; and another clade whose members, such as the *Arabidopsis MIR164A* and *C* genes, tend to be less expressed in roots than in other tissues (Jasinski et al., 2010).

In some developmental processes, the three *Arabidopsis MIR164* genes seem to be functionally redundant. For instance, *miR164a, miR164b*, and *miR164c* act redundantly to downregulate *CUC2* expression during the postmeristematic maintenance of phyllotaxy (Sieber et al., 2007). In contrast, *MIR164C* plays a more important role during floral development than the two others (Sieber et al., 2007), while *MIR164A* is the negative regulator of *CUC2* that controls leaf shape (Nikovics et al., 2006). The extent to which each *MIR164* gene regulates different developmental processes varies and is likely to result from differences in their expression patterns.

15.4.2.3 Transcriptional Control of miR164 Expression

Transcription factors of the plant-specific TCP family (TEOSINTE BRANCHED/CYCLOIDEA/PROLIFERATING CELL FACTOR) are well known for the role they play in regulating developmental processes. They are divided into two classes (Martin-Trillo and Cubas, 2009; Chapter 16). In particular, class II *CINCINNATA-like* (*CIN-like*) genes redundantly regulate cell proliferation and promote differentiation during leaf development. Plants expressing TCP3-EAR, a fusion with the EAR repression domain, have rosette leaves with exaggerated serrations and lobed cotyledons with ectopic shoot meristems (Koyama et al., 2007). CUC genes are overexpressed in these plants and *miR164* accumulation is decreased. Moreover, *cuc1* and *cuc2* mutations suppress the *TCP3-EAR* phenotype, indicating that it mainly results from increased *CUC* expression. *TCP3* has been found to directly activate *MIR164A* transcription (Koyama et al., 2010). Interestingly, some class I TCP members also regulate *CUC* expression (Uberti-Manassero et al., 2012). Thus, TCPs appear to be positive regulators of *miR164* and indirect inhibitors of *CUC* expression.

Auxin regulates *CUC* expression both directly (Section 15.4.1.1) and indirectly via modification of *miR164* levels. Indeed, auxin treatments can induce *miR164* expression in roots (Guo et al., 2005), and some mutants with disrupted auxin signaling show enhanced leaf serration due to reduced *MIR164A* expression (Bilsborough et al., 2011). Thus, auxin regulates *miR164* levels during root and leaf development.

miR164 expression is regulated by ethylene during leaf aging (Kim et al., 2009). In the *ethylene insensitive 2* (*ein2*) mutant, no reduction of *miR164* levels is observed, which suggests that *EIN2* is required for *MIR164* downregulation. EIN3 acts downstream of EIN2 and binds to *MIR164A*, *MIR164B*, and *MIR164C* promoters to repress their activity (Li et al., 2013). Overall, ethylene negatively regulates *miR164* levels by activating EIN2 which in turn promotes EIN3 transcriptional repression activity.

Interestingly, all three *MIR164* genes are marked with H3K27me3 repressive histone modification during leaf development (Lafos et al., 2011), suggesting that modification of chromatin dynamics also contributes to their regulation.

During floral development, a C_2H_2 zinc finger transcriptional repressor named RABBIT EARS (RBE) is specifically required for proper formation of second-whorl boundaries (Krizek et al., 2006). *rbe* mutants show fused sepals and aberrant petals, which recapitulate the floral phenotype of *cuc1–cuc2* double-mutants. RBE directly binds to the *MIR164C* promoter, negatively regulating its activity in floral boundaries. Moreover, genetic analyses reveal that RBE negatively regulates *MIR164B* expression while activating *MIR164A* expression in floral buds (Huang et al., 2012). Thus, RBE would differentially regulate the expression of *MIR164* genes during floral development, promoting their functional differentiation.

15.4.3 Transcriptional Regulation of NAM/CUC3 Expression

15.4.3.1 Transcription Factors Regulating CUC Expression During Embryogenesis

Class I KNOTTED-like homeobox genes (KNOXI) that code for homeodomain transcription factors are essential for SAM initiation and maintenance (for reviews see Hamant and Pautot, 2010; Hay and Tsiantis, 2009; Chapter 14). Plants mutated in the *KNOXI* gene *STM* lack a SAM and show reduced *CUC1* and *CUC3* expression, which is restricted to a stripe in the center of the boundary between two cotyledons (Takada et al., 2001; Vroemen et al., 2003). *CUC2* expression is even more modified in *stm* embryos, being limited to small spots at variable positions between developing cotyledons (Aida et al., 1999). Thus, *STM* regulates *CUC* expression contributing to its localization at the center of the embryo. Alternatively, the absence of *CUC* downregulation in the center of the embryo could also be attributed to the lack of meristematic cells in the *stm* mutant. Spinelli et al. (2011) demonstrated that inducing *STM* expression activates *CUC1* transcription in a direct manner since induction is maintained in the presence of a translational inhibitor. A binding site for STM in the *CUC1* promoter has been identified and validated *in vitro*, in yeast, and *in planta*. Overall, this indicates that *STM* directly induces *CUC1* expression.

Besides *STM*, there are other *KNOXI* genes that contribute to the establishment and maintenance of the SAM. Notably, their inactivation aggravates the phenotype of a weak *stm* allele, *stm-2*. For example, *knat6-1–stm-2* double-mutants show no SAM and strong cotyledon fusion compared with *stm-2*. Although *CUC3* expression is not altered in *knat6* mutants, it is completely lost in *knat6-1–stm-2* double-mutants indicating that *KNAT6* and *STM* redundantly contribute to *CUC3* activation (Belles-Boix et al., 2006).

In addition to KNOXI proteins the homeobox transcription factor WUSCHEL (WUS) also contributes to stem cell maintenance (Chapter 14). Although there is no evidence to suggest that WUS regulates *CUC* genes, other members of the same *WUSCHEL-RELATED HOMEOBOX* (*WOX*) clade control *CUC* expression in the cotyledon boundary. *wox2 stimpy-like* (*stpl/wox8*) double-mutants show partial cotyledon fusion, which correlates with asymmetrical *CUC2* and *CUC3* expression at one side of the embryo (Lie et al., 2012). Conversely, *CUC1* expression in some embryos is expanded in the protodermal layer. Overall, *WOX2* and *STPL* differentially regulate *CUC* genes, restricting *CUC1* expression and allowing symmetrical expression of *CUC2* and *CUC3*.

15.4.3.2 Transcription Factors Regulating CUC Expression During Axillary Meristem Formation

Two independent groups identified three genes coding for MYB domain transcription factors –*REGULATOR OF AXILLARY MERISTEM 1* (*RAX1*), *RAX2*, and *RAX3* – which are redundantly required for early induction of axillary meristems in *Arabidopsis* (Keller et al., 2006; Muller et al., 2006). Like the CUC genes, *RAX1* and *RAX3* are expressed in the axils of leaf primordia. Interestingly, *in situ* hybridizations show that *CUC2* expression in *rax1* is missing at the exact position of a future axillary meristem, indicating that *RAX1* induces local *CUC2* expression to promote axillary meristem formation (Keller et al., 2006).

Like *RAX1-3* genes, *LATERAL ORGAN FUSION1* (*LOF1*) also encodes a MYB domain transcription factor involved in axillary meristem formation and expressed in leaf axils. *lof1* mutants show reduced expression levels of *CUC1/2/3* and *RAX1* that could be indirectly mediated by changes in *RAX1* activity (Lee et al., 2009). Overall, LOF1 and RAX1 are transcription factors that act upstream of *CUC* genes during axillary meristem formation.

15.4.3.3 Transcription Factors Regulating NAM/CUC3 Expression During Leaf Development

Besides their central role in meristem formation, *KNOXI* genes are also involved in the development of most compound leaves (Blein et al., 2010; Chapter 14). *KNOXI* expression is initially downregulated both in simple and compound incipient leaf primordia. Such downregulation is permanent in species with simple leaves, whereas it is transient in primordia of compound leaves, being reactivated later during primordia development (Bharathan et al., 2002). In *Cardamine hirsuta*, the expression of *KNOXI* genes is required for leaflet formation and their overexpression leads to more leaflets and deeper serrations (Hay and Tsiantis, 2006). *KNOXI* overexpression increases *CUC* expression, while silencing *CUC* genes in plants overexpressing *KNOXI* suppresses their phenotype, indicating that *KNOXI* genes promote leaflet formation by activating *CUC* expression (Blein et al., 2008).

Although observed in many species, the reactivation of *KNOXI* genes during compound leaf development is not a general mechanism. Instead, some *Fabacea* show activation of *UNIFOLIATA*, an ortholog of the *Arabidopsis LEAFY* (*LFY*) gene, which controls leaflet formation in these species (Hofer et al., 1997). In *Pisum sativum*, the *uni* mutant forms simple smooth leaves where neither *NAM* nor *CUC3* expression could be detected (Blein et al., 2008). Interestingly, *CUC2* has been shown to be a possible target of LFY in the *Arabidopsis* inflorescence (Winter et al., 2011). All in all, *UNI/LFY* could be a positive regulator of *NAM/CUC3* expression.

15.4.3.4 Transcription Factors Regulating GOB Expression During Abscission

Two MADS box domain transcription factors, JOINTLESS and MACROCALYX (MC), promote abscission zone formation during tomato fruit development (Nakano et al., 2012). Transcriptional studies on plants misexpressing *JOINTLESS* or *MC* show that *GOB* expression is probably positively regulated by the *JOINTLESS/MC* heterodimer (Nakano et al., 2012). Another gene induced by the JOINTLESS/MC heterodimer is the AP2/ERF transcription factor *ETHYLENE RESPONSE FACTOR 52* (*SlERF52*). Plants with reduced *SlERF52* levels are impaired when pedicel abscission is activated and present reduced *GOB* expression, indicating that SlERF52 is also a positive regulator of *GOB* expression (Nakano et al., 2014). Overall, this designates JOINTLESS/MC as early activators of *GOB* expression and SlERF52 as a late *GOB* activator during fruit abscission.

15.4.3.5 Regulation of CUC Expression by Chromatin Modifications

Gene expression regulation depends not only on the presence of transcription factors that bind to specific promoter domains but also on chromatin availability to transcription factors. The chromatin dynamic is regulated by nucleosome-modifying enzymes that catalyze histone and DNA-covalent modifications as well as chromatin-remodeling complexes that remodel histone octamers/DNA interactions. Switch/sucrose nonfermentable (SWI/SNF) complexes are chromatin-remodeling factors conserved between yeast, mammals, and plants. Remodeling SWI/SNF complexes are recruited to promoters and regulate the accessibility of binding sites to transcription factors (Jerzmanowski, 2007).

In an enhancer screen of the *cuc2* cotyledon fusion phenotype, three mutations in the *BRAHMA* (*BRM*) gene were identified. BRM is an adenosine triphosphatase (ATPase) of the SWI2/SNF2 family (Kwon et al., 2006). A mutation in another SWI2/SNF2 member, *splayed* (*syd*), also enhances the cotyledon fusion phenotype of *cuc1* and *cuc3* mutants. Real time polymerase chain reaction (RT-PCR) and β-glucuronidase gene (GUS) reporter analyses established that *BRM* positively regulates the expression of the three *CUC* genes, and that *SYD* induces *CUC2* expression. This result indicates that general regulators of gene expression are also required for proper *CUC* expression.

Among factors regulating the chromatin dynamic are the modifying enzymes of histone octamers. These enzymes catalyze posttranslational modifications of histones, thus changing their interaction with DNA. One of the best-characterized histone modifications is the trimethylation of histone 3 on lysine 27 (H3K27me3), which leads to chromatin compaction and transcriptional repression. This mark is deposited by Polycomb group (PcG) proteins assembled in the Polycomb repressive complex 2 (PRC2). H3K27me3 is subsequently recognized by PRC1, which mediates locus repression (Schatlowski et al., 2008). Interestingly, *CUC2* and *CUC3* carry the H3K27me3 repressive mark. *CUC2* shows this mark in the meristem and leaves, whereas *CUC3* specifically carries the H3K27me3 mark in the leaves (Lafos et al., 2011). Thus, developmentally regulated deposition of repressive histone marks is likely to contribute to proper *CUC2* and *CUC3* expression.

Engelhorn et al. (2012) screened for genes expressed in the plant apex which were regulated by PRC1. They characterized the *DEVELOPMENT-RELATED PCG TARGET IN THE APEX 4* (*DPA4*) gene, which encodes a transcriptional repressor containing a B3 DNA-binding domain. *DPA4*, like *CUC* genes, is expressed in the boundary domains of the meristem and leaf primordia. *DPA4* negatively regulates *CUC2* expression and, accordingly, *dpa4* mutants show increased leaf serration, whereas a *DPA4* overexpressor presents smooth leaves. Thus, DPA4 appears to be an upstream negative regulator of *CUC2* expression.

15.5 NAM/CUC3 CONTROL PLANT DEVELOPMENT VIA MODIFICATIONS OF THE CELLULAR BEHAVIOR

Organ boundaries act both as frontiers and growth organizer centers (Aida and Tasaka, 2006). Boundaries cells display typically reduced growth activity, delimiting the frontier between different cell types. Besides this role in organ/tissue separation, boundaries participate in organ initiation and meristematic activity maintenance. Therefore, CUCs are likely to play different roles in controlling multiple aspects of plant growth and morphogenesis. Here, we focus on the effects downstream of CUCs, exploring the cellular effects dependent on CUCs, how they are achieved, and what molecular actors are involved (Figure 15.5).

15.5.1 CUC-Dependent Cellular Effects

Genetic analysis of *cuc* mutant combinations, coupled with morphologic analysis, suggest that *CUC1/2/3* repress growth in boundaries thus allowing organ separation (Aida et al., 1997; Aida et al., 1999; Takada et al., 2001; Vroemen et al., 2003). Growth integrates cell division and cell expansion parameters; therefore, reduced growth activity from cells localized at boundaries can be due to decreased cell division rate, reduced cell expansion, or both. Several pieces of work investigating various species report that cells located at boundaries display reduced cell division (Breuil-Broyer et al., 2004; Gaudin et al., 2000). However, experimental work linking cell proliferation and CUC transcription factors is scarce. In the wild-type *Arabidopsis* inflorescence meristem, floral primordia are formed 5–6 cells apart from each other (Heisler et al., 2005; Reddy et al., 2004). In the *mir164abc* triple-mutant, mature flowers are separated by roughly the same number of cells indicating that plant cell division in *mir164abc* is repressed between flowers during stem development. This correlates with local increase of *CUC1* expression suggesting a function for CUC1 in controlling cell division (Sieber et al., 2007). To test this hypothesis, Sieber and coworkers ectopically expressed CUC1 and examined sepal cells. Sepal length was dramatically reduced in plants overexpressing *CUC1*, but the cell number per area unit was not different from the wild type suggesting that CUC1 plays a role in cell division regulation. Taken together these results indicate that CUCs act as growth antagonists through local repression of cell division.

Leaf development constitutes an excellent model to study cellular parameters controlled by *CUC* genes. By analogy with cellular mechanisms occurring at lateral organ primordia boundaries, *CUC2* has been suggested to restrict growth of sinuses at the leaf margin (Nikovics et al., 2006). In contrast, *CUC2* promotes tooth outgrowth via a non-cell-autonomous pathway involving auxin (Bilsborough et al., 2011; Kawamura et al., 2010). These opposing results highlight the fact that CUCs control cell proliferation in different ways to allow differential growth.

15.5.2 How Does CUC Impact Cell Proliferation?

In plants, cell proliferation depends on the action of phytohormones. BRs, for example, constitute a major class of polyhydroxysteroid hormones, structurally similar to steroid hormones in animals, promoting growth in various developmental processes (Mussig, 2005). BRs promote growth by controlling both cell elongation and cell division. BR-insensitive mutants display dwarf phenotypes, partially as a result of impaired mitotic activity (Gonzalez-Garcia et al., 2011; Zhiponova et al., 2013). The *LATERAL ORGAN BOUNDARIES* (*LOB*) gene negatively controls BR accumulation in boundaries, while BRs repress other boundary identity genes, such as *CUC* genes, in a feedback loop to control boundary formation (Section 15.3; Bell et al., 2012; Gendron et al., 2012). These studies reveal the fundamental role BRs play in boundary delimitation and link BR signaling to boundary identity genes.

Another hormone playing a key role in boundary formation is auxin. Spatiotemporal auxin accumulation relies on controlled expression and subcellular localization of auxin efflux transporters PIN1 (Friml et al., 2004; Okada et al., 1991). JAGGED LATERAL ORGAN (JLO), a boundary identity gene and member of the LATERAL ORGAN BOUNDARY DOMAIN (LBD) transcription factor family – to which LOB belongs – controls *PIN* expression (Bureau et al., 2010; Rast and Simon, 2012). CUC2 promotes auxin accumulation via an unknown PIN1-dependent mechanism in leaves and, in turn, auxin represses *CUC2* expression forming a regulatory feedback loop. *In silico* models accounting for such a regulatory loop recapitulate wild-type leaf margin development and teeth formation patterns (Bilsborough et al., 2011).

Interestingly, BRs and auxins act synergistically to regulate photomorphogenesis by modulating AUXIN RESPONSE FACTOR2 (ARF2) activity of BR signaling components (Vert et al., 2008). Therefore, it is probable that the integrated action of these two hormones regulates boundary domain formation as well.

Although it is clear from the work described above that the *CUC* genes, auxins, and BRs play important roles in boundary delimitation, the underlying molecular mechanisms still need to be elucidated.

15.5.3 CUC Direct Targets

cuc mutant boundary phenotypes can be enhanced by mutations in several other genes, including transcription factors (Gomez-Mena and Sablowski, 2008; Lee et al., 2009; Lie et al., 2012), chromatin-remodeling factors (Kwon et al., 2006), and auxin flux regulators (Furutani et al., 2004). Taken together these studies show that CUC transcription factors cooperate with various biological processes to regulate boundary formation. Despite efforts to identify the molecular factors responsible for boundary delimitation, little is known about the regulatory network involved in this developmental process.

So far, only two CUC direct targets have been identified. Using rat glucocorticoid-receptor-inducible cell lines overexpressing *CUC1*, Takeda et al. (2011) showed that *LIGHT-DEPENDENT SHORT HYPOCOTYL 4* (*LSH4*) and its homolog *LSH3* are directly activated by CUC1 in boundary cells. These genes encode proteins belonging to the ALOG family (*Arabidopsis* LSH1 and *Oryza* G1) which are predicted to bind DNA and modulate transcriptional activity (Iyer and Aravind, 2012). LSH3 (also known as OBO1; Cho and Zambryski, 2011) and LSH4 are located in the nuclei of boundary cells and, therefore, may play a role in boundary formation (Takeda et al., 2011). Constitutive *LSH4* expression results in developmental defects such as inhibition of leaf growth and formation of ectopic meristems highlighting its potential role during plant development. Conversely, constitutive LSH4 expression cannot rescue the developmental defects of *cuc1–cuc2* mutants, suggesting that other regulators act downstream of *CUC1* to delimit boundaries.

15.5.4 Other Regulators: KNOX, LFY, LAS

Other regulators are known to act downstream of CUC transcription factors, but their molecular links are still missing. This is the case for *KNOXI* genes. Hibara et al. (2003) have shown that *KNOXI* genes such as *STM* and *BREVIPEDICELLUS* (*BP*) are ectopically expressed in *Arabidopsis* plants overexpressing *CUC1*. More generally, the accumulation of *KNOXI/LFY*-like transcripts is reduced in leaves when *NAM/CUC3* genes are silenced (Blein et al., 2008). These results, together with the data presented in Section 15.4.3.1, reveal the existence of a feedforward regulatory loop between *KNOXI/LFY*-like genes and *NAM/CUC3* genes during leaf development, which is likely to be conserved widely across eudicots. The *LATERAL SUPPRESSOR* (*LAS*) gene encodes a member of the GAI, RGA, SCR (GRAS) family of putative transcription factors, which is expressed at the SAM boundary (Greb et al., 2003). *LAS* expression decreases when *CUC* activity is reduced suggesting that LAS acts downstream of CUC (Raman et al., 2008). Accordingly, the higher level of *CUC* mRNA accumulation in *mir164abc* mutants correlates with an increase in *LAS* expression.

15.6 CONCLUSION

Since their first identification almost 20 years ago, a wealth of data have been accumulated on *NAM/CUC3* genes, establishing their central role in plant boundary formation. Fine analyses have shown a strong conservation of their function from species to species and in different organs of aerial parts, but have also underlined variations within this general trend. However, these conclusions are mostly based on genetic analyses. The challenge for the next years will be to reveal the molecular links, in particular, the genetic regulatory network between NAM/CUC3 transcription factors and boundary biology.

References

Adam, H., Marguerettaz, M., Qadri, R., Adroher, B., Richaud, F., Collin, M., Thuillet, A.C., Vigouroux, Y., Laufs, P., Tregear, J.W., Jouannic, S., 2011. Divergent expression patterns of miR164 and CUP-SHAPED COTYLEDON genes in palms and other monocots: implication for the evolution of meristem function in angiosperms. Mol. Biol. Evol. 28, 1439–1454.

Aida, M., Tasaka, M., 2006. Morphogenesis and patterning at the organ boundaries in the higher plant shoot apex. Plant Mol. Biol. 60, 915–928.

Aida, M., Ishida, T., Fukaki, H., Fujisawa, H., Tasaka, M., 1997. Genes involved in organ separation in *Arabidopsis*: an analysis of the cup-shaped cotyledon mutant. Plant Cell 9, 841–857.

Aida, M., Ishida, T., Tasaka, M., 1999. Shoot apical meristem and cotyledon formation during *Arabidopsis* embryogenesis: interaction among the CUP-SHAPED COTYLEDON and SHOOT MERISTEMLESS genes. Development 126, 1563–1570.

Aida, M., Vernoux, T., Furutani, M., Traas, J., Tasaka, M., 2002. Roles of PIN-FORMED1 and MONOPTEROS in pattern formation of the apical region of the *Arabidopsis* embryo. Development 129, 3965–3974.

Axtell, M.J., Bartel, D.P., 2005. Antiquity of microRNAs and their targets in land plants. Plant Cell 17, 1658–1673.

Baker, C.C., Sieber, P., Wellmer, F., Meyerowitz, E.M., 2005. The early extra petals1 mutant uncovers a role for microRNA miR164c in regulating petal number in *Arabidopsis*. Curr. Biol. 15, 303–315.

Bell, E.M., Lin, W.C., Husbands, A.Y., Yu, L., Jaganatha, V., Jablonska, B., Mangeon, A., Neff, M.M., Girke, T., Springer, P.S., 2012. *Arabidopsis* LATERAL ORGAN BOUNDARIES negatively regulates brassinosteroid accumulation to limit growth in organ boundaries. Proc. Natl. Acad. Sci. USA 109, 21146–21151.

Belles-Boix, E., Hamant, O., Witiak, S.M., Morin, H., Traas, J., Pautot, V., 2006. KNAT6: an *Arabidopsis* homeobox gene involved in meristem activity and organ separation. Plant Cell 18, 1900–1907.

Ben-Gera, H., Shwartz, I., Shao, M.R., Shani, E., Estelle, M., Ori, N., 2012. ENTIRE and GOBLET promote leaflet development in tomato by modulating auxin response. Plant J. 70, 903–915.

Berger, Y., Harpaz-Saad, S., Brand, A., Melnik, H., Sirding, N., Alvarez, J.P., Zinder, M., Samach, A., Eshed, Y., Ori, N., 2009. The NAC-domain transcription factor GOBLET specifies leaflet boundaries in compound tomato leaves. Development 136, 823–832.

Bharathan, G., Goliber, T.E., Moore, C., Kessler, S., Pham, T., Sinha, N.R., 2002. Homologies in leaf form inferred from KNOXI gene expression during development. Science 296, 1858–1860.

Bilsborough, G.D., Runions, A., Barkoulas, M., Jenkins, H.W., Hasson, A., Galinha, C., Laufs, P., Hay, A., Prusinkiewicz, P., Tsiantis, M., 2011. Model for the regulation of *Arabidopsis thaliana* leaf margin development. Proc. Natl. Acad. Sci. USA 108, 3424–3429.

Blein, T., Pulido, A., Vialette-Guiraud, A., Nikovics, K., Morin, H., Hay, A., Johansen, I.E., Tsiantis, M., Laufs, P., 2008. A conserved molecular framework for compound leaf development. Science 322, 1835–1839.

Blein, T., Hasson, A., Laufs, P., 2010. Leaf development: what it needs to be complex. Curr. Opin. Plant Biol. 13, 75–82.

Brand, A., Shirding, N., Shleizer, S., Ori, N., 2007. Meristem maintenance and compound-leaf patterning utilize common genetic mechanisms in tomato. Planta 226, 941–951.

Breuil-Broyer, S., Morel, P., De Almeida-Engler, J., Coustham, V., Negrutiu, I., Trehin, C., 2004. High-resolution boundary analysis during *Arabidopsis thaliana* flower development. Plant J. 38, 182–192.

Bureau, M., Rast, M.I., Illmer, J., Simon, R., 2010. JAGGED LATERAL ORGAN (JLO) controls auxin dependent patterning during development of the *Arabidopsis* embryo and root. Plant Mol. Biol. 74, 479–491.

Cheng, X., Peng, J., Ma, J., Tang, Y., Chen, R., Mysore, K.S., Wen, J., 2012. NO APICAL MERISTEM (MtNAM) regulates floral organ identity and lateral organ separation in *Medicago truncatula*. New Phytol. 195, 71–84.

Cho, E., Zambryski, P.C., 2011. ORGAN BOUNDARY1 defines a gene expressed at the junction between the shoot apical meristem and lateral organs. Proc. Natl. Acad. Sci. USA 108, 2154–2159.

Corpet, F., 1988. Multiple sequence alignment with hierarchical clustering. Nucleic Acids Res. 16, 10881–10890.

Duval, M., Hsieh, T.F., Kim, S.Y., Thomas, T.L., 2002. Molecular characterization of AtNAM: a member of the *Arabidopsis* NAC domain superfamily. Plant Mol. Biol. 50, 237–248.

Engelhorn, J., Reimer, J.J., Leuz, I., Gobel, U., Huettel, B., Farrona, S., Turck, F., 2012. Development-related PcG target in the apex 4 controls leaf margin architecture in *Arabidopsis thaliana*. Development 139, 2566–2575.

Ernst, H.A., Nina Olsen, A., Skriver, K., Larsen, S., Lo Leggio, L., 2004. Structure of the conserved domain of ANAC, a member of the NAC family of transcription factors. EMBO Rep. 5, 297–303.

Fang, Y., You, J., Xie, K., Xie, W., Xiong, L., 2008. Systematic sequence analysis and identification of tissue-specific or stress-responsive genes of NAC transcription factor family in rice. Mol. Genet. Genomics 280, 547–563.

Finet, C., Timme, R.E., Delwiche, C.F., Marletaz, F., 2010. Multigene phylogeny of the green lineage reveals the origin and diversification of land plants. Curr. Biol. 20, 2217–2222.

Friml, J., Vieten, A., Sauer, M., Weijers, D., Schwarz, H., Hamann, T., Offringa, R., Jurgens, G., 2003. Efflux-dependent auxin gradients establish the apical-basal axis of *Arabidopsis*. Nature 426, 147–153.

Friml, J., Yang, X., Michniewicz, M., Weijers, D., Quint, A., Tietz, O., Benjamins, R., Ouwerkerk, P.B., Ljung, K., Sandberg, G., Hooykaas, P.J., Palme, K., Offringa, R., 2004. A PINOID-dependent binary switch in apical-basal PIN polar targeting directs auxin efflux. Science 306, 862–865.

Furutani, M., Vernoux, T., Traas, J., Kato, T., Tasaka, M., Aida, M., 2004. PIN-FORMED1 and PINOID regulate boundary formation and cotyledon development in *Arabidopsis* embryogenesis. Development 131, 5021–5030.

Galbiati, F., Sinha Roy, D., Simonini, S., Cucinotta, M., Ceccato, L., Cuesta, C., Simaskova, M., Benkova, E., Kamiuchi, Y., Aida, M., Weijers, D., Simon, R., Masiero, S., Colombo, L., 2013. An integrative model of the control of ovule primordia formation. Plant J. 76, 446–455.

Gaudin, V., Lunness, P.A., Fobert, P.R., Towers, M., Riou-Khamlichi, C., Murray, J.A., Coen, E., Doonan, J.H., 2000. The expression of D-cyclin genes defines distinct developmental zones in snapdragon apical meristems and is locally regulated by the cycloidea gene. Plant Physiol. 122, 1137–1148.

Gendron, J.M., Liu, J.S., Fan, M., Bai, M.Y., Wenkel, S., Springer, P.S., Barton, M.K., Wang, Z.Y., 2012. Brassinosteroids regulate organ boundary formation in the shoot apical meristem of *Arabidopsis*. Proc. Natl. Acad. Sci. USA 109, 21152–21157.

Gomez-Mena, C., Sablowski, R., 2008. ARABIDOPSIS THALIANA HOMEOBOX GENE1 establishes the basal boundaries of shoot organs and controls stem growth. Plant Cell 20, 2059–2072.

Gonzalez-Garcia, M.P., Vilarrasa-Blasi, J., Zhiponova, M., Divol, F., Mora-Garcia, S., Russinova, E., Cano-Delgado, A.I., 2011. Brassinosteroids control meristem size by promoting cell cycle progression in *Arabidopsis* roots. Development 138, 849–859.

Grbic, V., Bleecker, A.B., 2000. Axillary meristem development in *Arabidopsis thaliana*. Plant J. 21, 215–223.

Greb, T., Clarenz, O., Schafer, E., Muller, D., Herrero, R., Schmitz, G., Theres, K., 2003. Molecular analysis of the LATERAL SUPPRESSOR gene in *Arabidopsis* reveals a conserved control mechanism for axillary meristem formation. Genes Dev. 17, 1175–1187.

Guo, H.S., Xie, Q., Fei, J.F., Chua, N.H., 2005. MicroRNA directs mRNA cleavage of the transcription factor NAC1 to downregulate auxin signals for *Arabidopsis* lateral root development. Plant Cell 17, 1376–1386.

Hamant, O., Pautot, V., 2010. Plant development: a TALE story. C.R. Biol. 333, 371–381.

Hasson, A., Plessis, A., Blein, T., Adroher, B., Grigg, S., Tsiantis, M., Boudaoud, A., Damerval, C., Laufs, P., 2011. Evolution and diverse roles of the CUP-SHAPED COTYLEDON genes in *Arabidopsis* leaf development. Plant Cell 23, 54–68.

Hay, A., Tsiantis, M., 2006. The genetic basis for differences in leaf form between *Arabidopsis thaliana* and its wild relative *Cardamine hirsuta*. Nat. Genet. 38, 942–947.

Hay, A., Tsiantis, M., 2009. A KNOX family TALE. Curr. Opin. Plant Biol. 12, 593–598.

Hay, A., Barkoulas, M., Tsiantis, M., 2006. ASYMMETRIC LEAVES1 and auxin activities converge to repress BREVIPEDICELLUS expression and promote leaf development in *Arabidopsis*. Development 133, 3955–3961.

Heisler, M.G., Ohno, C., Das, P., Sieber, P., Reddy, G.V., Long, J.A., Meyerowitz, E.M., 2005. Patterns of auxin transport and gene expression during primordium development revealed by live imaging of the *Arabidopsis* inflorescence meristem. Curr. Biol. 15, 1899–1911.

Hibara, K., Takada, S., Tasaka, M., 2003. CUC1 gene activates the expression of SAM-related genes to induce adventitious shoot formation. Plant J. 36, 687–696.

Hibara, K., Karim, M.R., Takada, S., Taoka, K., Furutani, M., Aida, M., Tasaka, M., 2006. *Arabidopsis* CUP-SHAPED COTYLEDON3 regulates postembryonic shoot meristem and organ boundary formation. Plant Cell 18, 2946–2957.

Hofer, J., Turner, L., Hellens, R., Ambrose, M., Matthews, P., Michael, A., Ellis, N., 1997. UNIFOLIATA regulates leaf and flower morphogenesis in pea. Curr. Biol. 7, 581–587.

Hu, R., Qi, G., Kong, Y., Kong, D., Gao, Q., Zhou, G., 2010. Comprehensive analysis of NAC domain transcription factor gene family in *Populus trichocarpa*. BMC Plant Biol. 10, 145.

Huang, T., Lopez-Giraldez, F., Townsend, J.P., Irish, V.F., 2012. RBE controls microRNA164 expression to effect floral organogenesis. Development 139, 2161–2169.

Ishida, T., Aida, M., Takada, S., Tasaka, M., 2000. Involvement of CUP-SHAPED COTYLEDON genes in gynoecium and ovule development in *Arabidopsis thaliana*. Plant Cell Physiol. 41, 60–67.

Iyer, L.M., Aravind, L., 2012. ALOG domains: provenance of plant homeotic and developmental regulators from the DNA-binding domain of a novel class of DIRS1-type retroposons. Biol. Direct 7, 39.

Jasinski, S., Vialette-Guiraud, A.C., Scutt, C.P., 2010. The evolutionary-developmental analysis of plant microRNAs. Philos. Trans. R. Soc. Lond. B Biol. Sci. 365, 469–476.

Jensen, M.K., Kjaersgaard, T., Nielsen, M.M., Galberg, P., Petersen, K., O'Shea, C., Skriver, K., 2010. The *Arabidopsis thaliana* NAC transcription factor family: structure–function relationships and determinants of ANAC019 stress signalling. Biochem. J. 426, 183–196.

Jerzmanowski, A., 2007. SWI/SNF chromatin remodeling and linker histones in plants. Biochim. Biophys. Acta 1769, 330–345.

Kamiuchi, Y., Yamamoto, K., Furutani, M., Tasaka, M., Aida, M., 2014. The CUC1 and CUC2 genes promote carpel margin meristem formation during *Arabidopsis* gynoecium development. Front. Plant Sci. 5, 165.

Kawamura, E., Horiguchi, G., Tsukaya, H., 2010. Mechanisms of leaf tooth formation in *Arabidopsis*. Plant J. 62, 429–441.

Keller, T., Abbott, J., Moritz, T., Doerner, P., 2006. *Arabidopsis* REGULATOR OF AXILLARY MERISTEMS1 controls a leaf axil stem cell niche and modulates vegetative development. Plant Cell 18, 598–611.

Kim, T.W., Wang, Z.Y., 2010. Brassinosteroid signal transduction from receptor kinases to transcription factors. Annu. Rev. Plant Biol. 61, 681–704.

Kim, J.H., Woo, H.R., Kim, J., Lim, P.O., Lee, I.C., Choi, S.H., Hwang, D., Nam, H.G., 2009. Trifurcate feed-forward regulation of age-dependent cell death involving miR164 in *Arabidopsis*. Science 323, 1053–1057.

Koyama, T., Furutani, M., Tasaka, M., Ohme-Takagi, M., 2007. TCP transcription factors control the morphology of shoot lateral organs via negative regulation of the expression of boundary-specific genes in *Arabidopsis*. Plant Cell 19, 473–484.

Koyama, T., Mitsuda, N., Seki, M., Shinozaki, K., Ohme-Takagi, M., 2010. TCP transcription factors regulate the activities of ASYMMETRIC LEAVES1 and miR164, as well as the auxin response, during differentiation of leaves in *Arabidopsis*. Plant Cell 22, 3574–3588.

Krizek, B.A., Lewis, M.W., Fletcher, J.C., 2006. RABBIT EARS is a second-whorl repressor of AGAMOUS that maintains spatial boundaries in *Arabidopsis* flowers. Plant J. 45, 369–383.

Kwon, C.S., Hibara, K., Pfluger, J., Bezhani, S., Metha, H., Aida, M., Tasaka, M., Wagner, D., 2006. A role for chromatin remodeling in regulation of CUC gene expression in the *Arabidopsis* cotyledon boundary. Development 133, 3223–3230.

Lafos, M., Kroll, P., Hohenstatt, M.L., Thorpe, F.L., Clarenz, O., Schubert, D., 2011. Dynamic regulation of H3K27 trimethylation during *Arabidopsis* differentiation. PLoS Genet. 7, e1002040.

Larsson, E., Sundstrom, J.F., Sitbon, F., Von Arnold, S., 2012. Expression of PaNAC01, a *Picea abies* CUP-SHAPED COTYLEDON orthologue, is regulated by polar auxin transport and associated with differentiation of the shoot apical meristem and formation of separated cotyledons. Ann. Bot. 110, 923–934.

Larue, C.T., Wen, J., Walker, J.C., 2009. A microRNA–transcription factor module regulates lateral organ size and patterning in *Arabidopsis*. Plant J. 58, 450–463.

Laufs, P., Peaucelle, A., Morin, H., Traas, J., 2004. MicroRNA regulation of the CUC genes is required for boundary size control in *Arabidopsis* meristems. Development 131, 4311–4322.

Lee, D.K., Geisler, M., Springer, P.S., 2009. LATERAL ORGAN FUSION1 and LATERAL ORGAN FUSION2 function in lateral organ separation and axillary meristem formation in *Arabidopsis*. Development 136, 2423–2432.

Li, X.G., Su, Y.H., Zhao, X.Y., Li, W., Gao, X.Q., Zhang, X.S., 2010. Cytokinin overproduction-caused alteration of flower development is partially mediated by CUC2 and CUC3 in *Arabidopsis*. Gene 450, 109–120.

Li, Z., Peng, J., Wen, X., Guo, H., 2013. Ethylene-insensitive3 is a senescence-associated gene that accelerates age-dependent leaf senescence by directly repressing miR164 transcription in *Arabidopsis*. Plant Cell 25, 3311–3328.

Lie, C., Kelsom, C., Wu, X., 2012. WOX2 and STIMPY-LIKE/WOX8 promote cotyledon boundary formation in *Arabidopsis*. Plant J. 72, 674–682.

Lindemose, S., Jensen, M.K., De Velde, J.V., O'Shea, C., Heyndrickx, K.S., Workman, C.T., Vandepoele, K., Skriver, K., Masi, F.D., 2014. A DNA-binding-site landscape and regulatory network analysis for NAC transcription factors in *Arabidopsis thaliana*. Nucleic Acids Res. 42, 7681–7693.

Mallory, A.C., Dugas, D.V., Bartel, D.P., Bartel, B., 2004. MicroRNA regulation of NAC-domain targets is required for proper formation and separation of adjacent embryonic, vegetative, and floral organs. Curr. Biol. 14, 1035–1046.

Marhavy, P., Duclercq, J., Weller, B., Feraru, E., Bielach, A., Offringa, R., Friml, J., Schwechheimer, C., Murphy, A., Benkova, E., 2014. Cytokinin controls polarity of PIN1-dependent auxin transport during lateral root organogenesis. Curr. Biol. 24, 1031–1037.

Martin-Trillo, M., Cubas, P., 2009. TCP genes: a family snapshot ten years later. Trends Plant Sci. 15, 31–39.

Muller, D., Schmitz, G., Theres, K., 2006. Blind homologous R2R3 Myb genes control the pattern of lateral meristem initiation in *Arabidopsis*. Plant Cell 18, 586–597.

Mussig, C., 2005. Brassinosteroid-promoted growth. Plant Biol. 7, 110–117.

Nahar, M.A., Ishida, T., Smyth, D.R., Tasaka, M., Aida, M., 2012. Interactions of CUP-SHAPED COTYLEDON and SPATULA genes control carpel margin development in *Arabidopsis thaliana*. Plant Cell Physiol. 53, 1134–1143.

Nakano, T., Kimbara, J., Fujisawa, M., Kitagawa, M., Ihashi, N., Maeda, H., Kasumi, T., Ito, Y., 2012. MACROCALYX and JOINTLESS interact in the transcriptional regulation of tomato fruit abscission zone development. Plant Physiol. 158, 439–450.

Nakano, T., Fujisawa, M., Shima, Y., Ito, Y., 2013. Expression profiling of tomato pre-abscission pedicels provides insights into abscission zone properties including competence to respond to abscission signals. BMC Plant Biol. 13, 40.

Nakano, T., Fujisawa, M., Shima, Y., Ito, Y., 2014. The AP2/ERF transcription factor SlERF52 functions in flower pedicel abscission in tomato. J. Exp. Bot. 65, 3111–3119.

Nikovics, K., Blein, T., Peaucelle, A., Ishida, T., Morin, H., Aida, M., Laufs, P., 2006. The balance between the MIR164A and CUC2 genes controls leaf margin serration in *Arabidopsis*. Plant Cell 18, 2929–2945.

Okada, K., Ueda, J., Komaki, M.K., Bell, C.J., Shimura, Y., 1991. Requirement of the auxin polar transport system in early stages of *Arabidopsis* floral bud formation. Plant Cell 3, 677–684.

Olsen, A.N., Ernst, H.A., Lo Leggio, L., Skriver, K., 2005. DNA-binding specificity and molecular functions of NAC transcription factors. Plant Sci. 169, 785–797.

Ooka, H., Satoh, K., Doi, K., Nagata, T., Otomo, Y., Murakami, K., Matsubara, K., Osato, N., Kawai, J., Carninci, P., Hayashizaki, Y., Suzuki, K., Kojima, K., Takahara, Y., Yamamoto, K., Kikuchi, S., 2003. Comprehensive analysis of NAC family genes in *Oryza sativa* and *Arabidopsis thaliana*. DNA Res. 10, 239–247.

Peaucelle, A., Morin, H., Traas, J., Laufs, P., 2007. Plants expressing a miR164-resistant CUC2 gene reveal the importance of post-meristematic maintenance of phyllotaxy in *Arabidopsis*. Development 134, 1045–1050.

Pinheiro, G.L., Marques, C.S., Costa, M.D., Reis, P.A., Alves, M.S., Carvalho, C.M., Fietto, L.G., Fontes, E.P., 2009. Complete inventory of soybean NAC transcription factors: sequence conservation and expression analysis uncover their distinct roles in stress response. Gene 444, 10–23.

Pulido, A., Laufs, P., 2010. Co-ordination of developmental processes by small RNAs during leaf development. J. Exp. Bot. 61, 1277–1291.

Raman, S., Greb, T., Peaucelle, A., Blein, T., Laufs, P., Theres, K., 2008. Interplay of miR164, CUP-SHAPED COTYLEDON genes and LATERAL SUPPRESSOR controls axillary meristem formation in *Arabidopsis thaliana*. Plant J. 55, 65–76.

Rast, M.I., Simon, R., 2012. *Arabidopsis* JAGGED LATERAL ORGANS acts with ASYMMETRIC LEAVES2 to coordinate KNOX and PIN expression in shoot and root meristems. Plant Cell 24, 2917–2933.

Reddy, G.V., Heisler, M.G., Ehrhardt, D.W., Meyerowitz, E.M., 2004. Real-time lineage analysis reveals oriented cell divisions associated with morphogenesis at the shoot apex of *Arabidopsis thaliana*. Development 131, 4225–4237.

Reinhardt, D., Pesce, E.R., Stieger, P., Mandel, T., Baltensperger, K., Bennett, M., Traas, J., Friml, J., Kuhlemeier, C., 2003. Regulation of phyllotaxis by polar auxin transport. Nature 426, 255–260.

Roudier, F., Gissot, L., Beaudoin, F., Haslam, R., Michaelson, L., Marion, J., Molino, D., Lima, A., Bach, L., Morin, H., Tellier, F., Palauqui, J.C., Bellec, Y., Renne, C., Miquel, M., Dacosta, M., Vignard, J., Rochat, C., Markham, J.E., Moreau, P., Napier, J., Faure, J.D., 2010. Very-long-chain fatty acids are involved in polar auxin transport and developmental patterning in *Arabidopsis*. Plant Cell 22, 364–375.

Schatlowski, N., Creasey, K., Goodrich, J., Schubert, D., 2008. Keeping plants in shape: polycomb-group genes and histone methylation. Semin. Cell Dev. Biol. 19, 547–553.

Schwab, R., Palatnik, J.F., Riester, M., Schommer, C., Schmid, M., Weigel, D., 2005. Specific effects of microRNAs on the plant transcriptome. Dev. Cell 8, 517–527.

Shen, H., Yin, Y., Chen, F., Xu, Y., Dixon, R.A., 2009. A bioinformatic analysis of NAC genes for plant cell wall development in relation to lignocellulosic bioenergy production. Bioenerg. Res. 2, 217–232.

Sieber, P., Wellmer, F., Gheyselinck, J., Riechmann, J.L., Meyerowitz, E.M., 2007. Redundancy and specialization among plant microRNAs: role of the MIR164 family in developmental robustness. Development 134, 1051–1060.

Souer, E., Van Houwelingen, A., Kloos, D., Mol, J., Koes, R., 1996. The *No Apical Meristem* gene of petunia is required for pattern formation in embryos and flower and is expressed at meristem and primordia boundaries. Cell 85, 159–170.

Spinelli, S.V., Martin, A.P., Viola, I.L., Gonzalez, D.H., Palatnik, J.F., 2011. A mechanistic link between STM and CUC1 during *Arabidopsis* development. Plant Physiol. 156, 1894–1904.

Takada, S., Hibara, K., Ishida, T., Tasaka, M., 2001. The CUP-SHAPED COTYLEDON1 gene of *Arabidopsis* regulates shoot apical meristem formation. Development 128, 1127–1135.

Takeda, S., Hanano, K., Kariya, A., Shimizu, S., Zhao, L., Matsui, M., Tasaka, M., Aida, M., 2011. CUP-SHAPED COTYLEDON1 transcription factor activates the expression of LSH4 and LSH3, two members of the ALOG gene family, in shoot organ boundary cells. Plant J. 66, 1066–1077.

Taoka, K., Yanagimoto, Y., Daimon, Y., Hibara, K., Aida, M., Tasaka, M., 2004. The NAC domain mediates functional specificity of CUP-SHAPED COTYLEDON proteins. Plant J. 40, 462–473.

Todesco, M., Balasubramanian, S., Cao, J., Ott, F., Sureshkumar, S., Schneeberger, K., Meyer, R.C., Altmann, T., Weigel, D., 2012. Natural variation in biogenesis efficiency of individual *Arabidopsis thaliana* microRNAs. Curr. Biol. 22, 166–170.

Tran, L.S., Nakashima, K., Sakuma, Y., Simpson, S.D., Fujita, Y., Maruyama, K., Fujita, M., Seki, M., Shinozaki, K., Yamaguchi-Shinozaki, K., 2004. Isolation and functional analysis of *Arabidopsis* stress-inducible NAC transcription factors that bind to a drought-responsive *cis*-element in the early responsive to dehydration stress 1 promoter. Plant Cell 16, 2481–2498.

Uberti-Manassero, N.G., Lucero, L.E., Viola, I.L., Vegetti, A.C., Gonzalez, D.H., 2012. The class I protein AtTCP15 modulates plant development through a pathway that overlaps with the one affected by CIN-like TCP proteins. J. Exp. Bot. 63, 809–823.

Vernoux, T., Kronenberger, J., Grandjean, O., Laufs, P., Traas, J., 2000. PIN-FORMED 1 regulates cell fate at the periphery of the shoot apical meristem. Development 127, 5157–5165.

Vert, G., Walcher, C.L., Chory, J., Nemhauser, J.L., 2008. Integration of auxin and brassinosteroid pathways by Auxin Response Factor 2. Proc. Natl. Acad. Sci. USA 105, 9829–9834.

Vialette-Guiraud, A.C., Adam, H., Finet, C., Jasinski, S., Jouannic, S., Scutt, C.P., 2011. Insights from ANA-grade angiosperms into the early evolution of CUP-SHAPED COTYLEDON genes. Ann. Bot. 107, 1511–1519.

Vroemen, C.W., Mordhorst, A.P., Albrecht, C., Kwaaitaal, M.A., De Vries, S.C., 2003. The CUP-SHAPED COTYLEDON3 gene is required for boundary and shoot meristem formation in *Arabidopsis*. Plant Cell 15, 1563–1577.

Wang, Z., Chen, J., Weng, L., Li, X., Cao, X., Hu, X., Luo, D., Yang, J., 2013. Multiple components are integrated to determine leaf complexity in *Lotus japonicus*. J. Integr. Plant Biol. 55, 419–433.

Wang, Q., Kohlen, W., Rossmann, S., Vernoux, T., Theres, K., 2014a. Auxin depletion from the leaf axil conditions competence for axillary meristem formation in *Arabidopsis* and tomato. Plant Cell 26, 2068–2079.

Wang, Y., Wang, J., Shi, B., Yu, T., Qi, J., Meyerowitz, E.M., Jiao, Y., 2014b. The stem cell niche in leaf axils is established by auxin and cytokinin in *Arabidopsis*. Plant Cell 26, 2055–2067.

Weir, I., Lu, J., Cook, H., Causier, B., Schwarz-Sommer, Z., Davies, B., 2004. CUPULIFORMIS establishes lateral organ boundaries in *Antirrhinum*. Development 131, 915–922.

Welner, D.H., Lindemose, S., Grossmann, J.G., Mollegaard, N.E., Olsen, A.N., Helgstrand, C., Skriver, K., Lo Leggio, L., 2012. DNA binding by the plant-specific NAC transcription factors in crystal and solution: a firm link to WRKY and GCM transcription factors. Biochem. J. 444, 395–404.

Winter, C.M., Austin, R.S., Blanvillain-Baufume, S., Reback, M.A., Monniaux, M., Wu, M.F., Sang, Y., Yamaguchi, A., Yamaguchi, N., Parker, J.E., Parcy, F., Jensen, S.T., Li, H., Wagner, D., 2011. LEAFY target genes reveal floral regulatory logic, cis motifs, and a link to biotic stimulus response. Dev. Cell 20, 430–443.

Xu, Z.Y., Kim, S.Y., Hyeon Do, Y., Kim, D.H., Dong, T., Park, Y., Jin, J.B., Joo, S.H., Kim, S.K., Hong, J.C., Hwang, D., Hwang, I., 2013. The *Arabidopsis* NAC transcription factor ANAC096 cooperates with bZIP-type transcription factors in dehydration and osmotic stress responses. Plant Cell 25, 4708–4724.

Zhiponova, M.K., Vanhoutte, I., Boudolf, V., Betti, C., Dhondt, S., Coppens, F., Mylle, E., Maes, S., Gonzalez-Garcia, M.P., Cano-Delgado, A.I., Inze, D., Beemster, G.T., De Veylder, L., Russinova, E., 2013. Brassinosteroid production and signaling differentially control cell division and expansion in the leaf. New Phytol. 197, 490–502.

Zhu, T., Nevo, E., Sun, D., Peng, J., 2012. Phylogenetic analyses unravel the evolutionary history of NAC proteins in plants. Evolution 66, 1833–1848.

Zimmermann, R., Werr, W., 2005. Pattern formation in the monocot embryo as revealed by NAM and CUC3 orthologues from *Zea mays* L. Plant Mol. Biol. 58, 669–685.

CHAPTER 16

The Role of TCP Transcription Factors in Shaping Flower Structure, Leaf Morphology, and Plant Architecture

Michael Nicolas, Pilar Cubas

Department of Plant Molecular Genetics, Centro Nacional de Biotecnología (CNB-CSIC), Madrid, Spain

OUTLINE

16.1 Introduction	250
16.2 TCP Genes and the Control of Leaf Development	**250**
16.2.1 TCP Genes and the Development of Simple Leaves	250
16.2.1.1 CIN Genes and Simple Leaf Development	250
16.2.1.2 CYC/TB1 Genes and Simple Leaf Development	252
16.2.1.3 Class I TCP and Simple Leaf Development	252
16.2.2 Role of TCP Genes During Compound Leaf Development	253
16.2.3 TCP-Regulated Networks in the Control of Leaf Development	253
16.2.3.1 TCP Regulation of the CUC/KNOX1 Gene Pathway	253
16.2.3.2 TCP Regulation of Growth-Regulating Factors	254
16.2.3.3 TCP Regulation of Auxin Signaling During Leaf Development	255
16.2.3.4 TCP Regulation of CK Signaling During Leaf Development	255
16.2.3.5 TCP Genes and Gibberellin Signaling	255
16.2.3.6 TCP Genes and Jasmonate Signaling	256
16.2.4 Posttranslational Regulation of TCP Genes During Leaf Development	256
16.3 TCP Genes and the Control of Shoot Branching	**256**
16.3.1 TB1/BRC1 Genes Prevent Lateral Branch Outgrowth	256
16.3.1.1 Expression of TB1/BRC1 Genes	257
16.3.1.2 Regulation of TB1/BRC1 Genes by Light Quality	258
16.3.1.3 Hormonal Regulation of TB1/BRC1 Genes	258
16.3.1.4 Sucrose Regulation of TB1/BRC1 Genes	258
16.3.1.5 TB1/BRC1 Genetic Networks: Up- and Downstream Genes	258
16.4 TCP Genes and the Control of Flower Shape	**259**
16.4.1 CYC2 Genes Control Floral Zygomorphy in Dicots	259
16.4.1.1 CYC2 Genes Control Floral Identity of Capitulum Inflorescence	260
16.4.2 CYC/TB1 Genes Control Flower Asymmetry in Monocots	260
16.4.3 Role of CIN Genes in Flower Development	261
16.4.4 Role of Class I Genes in Flower Development	261
16.4.5 TCP Hormone Regulation of Flower Development	261
16.5 TCP Genes Affect Flowering Time	**262**
16.6 Concluding Remarks	**262**
References	**263**

Plant Transcription Factors. http://dx.doi.org/10.1016/B978-0-12-800854-6.00016-6
Copyright © 2016 Elsevier Inc. All rights reserved.

16.1 INTRODUCTION

The TEOSINTE BRANCHED1/CYCLOIDEA/PROLIFERATING CELL FACTOR (TCP) family is a plant-specific group of genes that encode transcription factors. Many members of this family influence greatly the growth patterns of tissues and organs during plant development and, therefore, are key determinants of plant form. TCP genes fall into two classes, class I and II, distinguishable by specific amino acids in the TCP domain. Within class II, two further lineages are found in angiosperms, *CINCINNATA* (*CIN*) and *CYCLOIDEA /TEOSINTE BRANCHED 1* (*CYC/TB1*) genes (Martín-Trillo and Cubas, 2010; Chapter 9). The most prominent functions associated with TCP genes are the control of leaf and flower size and shape as well as the suppression of shoot branching. These traits are essential for evolutionary fitness and affect important aspects of plant ecophysiology such as light interception efficiency, adaptation to resource availability and pollination success.

Probably because of their ability to modulate growth patterns, some TCP genes were recruited during angiosperm evolution to generate new morphological traits and, in crop domestication, to stabilize and improve architectural features. For instance, class II *CYC* played an essential role in the emergence of floral bilateral symmetry (zygomorphy), a pivotal innovation that evolved independently several times as an adaptation to fertilization by specialized pollinators. Class II *Tb1* was a main target for selection during maize domestication, which resulted in strong apical dominance and improved plant architecture.

In recent years, additional TCP genes have been implicated in increasingly large numbers of central biological processes such as hormone synthesis and signaling, flowering time, and regulation of circadian rhythms. Due to the breadth of this topic, we will focus here on recent discoveries related to the best-characterized TCP gene functions, namely, control of leaf and flower shape as well as suppression of shoot branching. We also summarize how TCP genes are integrated in the gene regulatory networks and hormone-signaling pathways that control these processes.

16.2 TCP GENES AND THE CONTROL OF LEAF DEVELOPMENT

16.2.1 TCP Genes and the Development of Simple Leaves

There is now compelling evidence that TCP genes play a critical role during leaf development, specifically in the control of maturation timing and differentiation. Leaf maturation occurs after the young leaf primordium has established its abaxial–adaxial asymmetry, the petiole has undergone elongation, and the lamina (leaf blade) has initiated expansion but is still formed of histologically uniform dividing cells (Efroni et al., 2010). In species such as *Nicotiana tabacum*, *Antirrhinum majus*, and *Arabidopsis thaliana*, leaf maturation follows a basiplastic (apical-to-basal) pattern, forming a cell cycle arrest front that can be monitored by sequential appearance of morphogenetic markers such as trichomes, provascular strands, and stomata (Efroni et al., 2010), and a decline in molecular cell cycle markers such as *HISTONE 4* (*H4*) and *CYCLIN D3b* (*CYC D3b*; Nath et al., 2003). Class II *CIN* genes appear to have a prominent role in the regulation of leaf maturation, although recent evidence indicates that other TCPs, especially those of class I, might also act in a highly redundant manner to regulate leaf growth and shape.

16.2.1.1 CIN *Genes and Simple Leaf Development*

In *Antirrhinum*, mutations in the *CIN* locus were the first to show that this gene affects the gradual process of cell division arrest in developing leaves (Nath et al., 2003; Crawford et al., 2004). *CIN* is expressed in the lamina, proximal to the arrest front, and its expression is stronger in the marginal than in the medial region (Nath et al., 2003). Whereas in wild-type plants the cell division arrest front is weakly convex, in *cin* mutants it is strongly concave. Moreover, in *cin* leaves, the progression front, leaf blade cell expansion, and decline in histone H4 (*H4*) and CYCLIND3b (*CYCD3b*) expression proceed more slowly through the middle region than in the wild type (Nath et al., 2003). This results in greater growth of marginal regions, both in width and length, which has an impact on final leaf morphology: *cin* leaves are strongly concave and curled, and cannot be flattened, unlike weakly convex wild-type leaves (Figure 16.1A).

FIGURE 16.1 **Leaf phenotype of plants with altered TCP function.** In each panel, wild-type (WT, left) and mutant(s) (right) phenotypes are shown. (A) *A. majus cin* mutant (Nath et al., 2003). (B) Single, double and triple-mutants of *CIN* genes *TCP2*, *TCP4*, and *TCP10* in *A. thaliana* (Schommer et al., 2008). (C) Triple, quadruple, and quintuple *cin* mutants in *A. thaliana* (Koyama et al., 2010a). (D) Plants overexpressing *miR319a* (equivalent of *jaw-D* mutants or a *tcp2/3/4/10/24* quintuple mutant, center) and plants overexpressing *miR319a* plus the artificial *miR-3TCP* (which targets *TCP5*, *TCP13*, and *TCP17*, equivalent to an octuple *cin* mutant, right; Efroni et al., 2008). (E) Plants expressing a dominant loss-of-function *TCP14* (*TCP14:SRDX*), in *A. thaliana* (Kieffer et al., 2011). *A. thaliana* (F) and *T. fournieri* (G) overexpressing a dominant loss-of-function form of *TCP3* (*TCP3:SRDX*; Koyama et al., 2007; Narumi et al., 2011). (H) *Cyclamen persicum* plant overexpressing a dominant loss-of-function form of the *CIN* gene *CpTCP1* (*CpTCP1:SRDX*; Tanaka et al., 2011). (I) Overexpressors of a *miR319a*-resistant form of *TCP3* in *A. thaliana* (Koyama et al., 2010a). (J) Heterozygote (center) and homozygote (right) form of the *miR319*-resistant *LA* allele *La-2* in tomato (Ori et al., 2007). (K) Leaf and leaflet (insets) of loss-of-function mutant *La-6* in tomato (Ori et al., 2007). (L) *Arabidopsis* plants expressing a hyperactive form of *TCP4* (*TCP4:VP16*; Sarvepalli and Nath, 2011). (M) Leaf and leaflet (insets) of *miR319* overexpressors in tomato (Shleizer-Burko et al., 2011). (N) *Rosa hybrida* overexpressing *AtTCP3:SRDX* (Gion et al., 2011). *Reproduced with permission from Nath et al. (2003), Schommer et al. (2008), Efroni et al. (2008), Kieffer et al. (2011), Koyama et al. (2007, 2010a), Narumi et al. (2011), Tanaka et al. (2011), Ori et al. (2007), Sarvepalli and Nath (2011), Shleizer-Burko et al. (2011), Gion et al. (2011).*

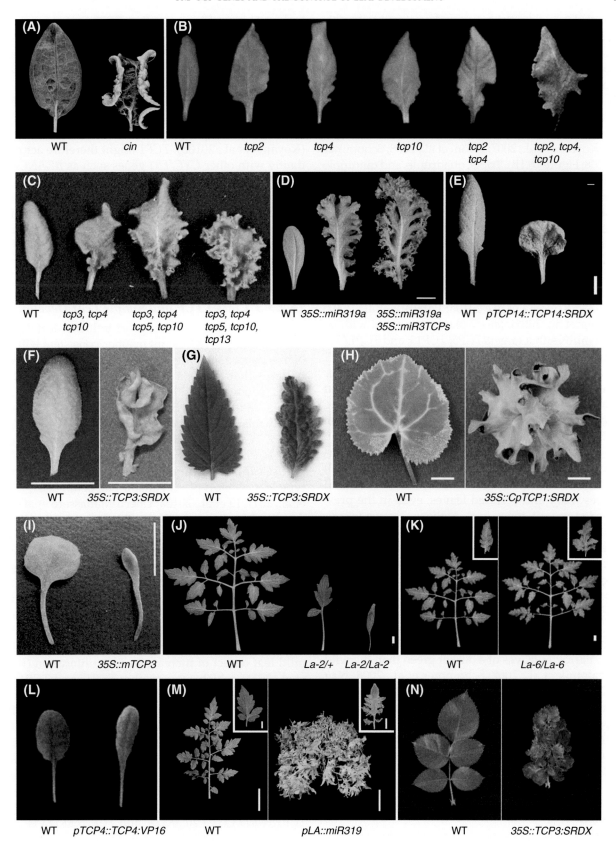

In *Arabidopsis*, a similar role was described for the *CIN* ortholog genes. In this species, eight *CIN* genes (*TCP2, TCP3, TCP4, TCP5, TCP10, TCP13, TCP17,* and *TCP24*) have partially redundant roles (Palatnik et al., 2003; Koyama et al., 2007; Koyama et al., 2010a; Schommer et al., 2008; Efroni et al., 2008). A dose-dependent phenotypic gradient is observed in their mutants. Single *tcp2, tcp3, tcp4,* or *tcp10* mutants have weak phenotypes of slightly enlarged leaves (Koyama et al., 2007; Efroni et al., 2008; Schommer et al., 2008; Figure 16.1B). Triple, quadruple, and quintuple mutants of these genes show much stronger phenotypes of serrated crinkly leaves that cannot be flattened due to delayed leaf maturation (Figure 16.1B, C; Schommer et al., 2008; Efroni et al., 2008; Koyama et al., 2010a). These phenotypes resemble those of *jagged and wavy-D* (*jaw-D*) mutants, which overexpress the micro-RNA *miR319A*, which targets *TCP2, TCP3, TCP4, TCP10,* and *TCP24* simultaneously (Figure 16.1D; Palatnik et al., 2003; Koyama et al., 2007; Koyama et al., 2010a; Palatnik et al., 2007; Schommer et al., 2008). *Jaw-D* plants combined with an artificial miRNA that targets *TCP5, TCP13,* and *TCP17* behave as an octuple *tcp2–tcp3–tcp4–tcp5–tcp10–tcp13–tcp17–tcp24* mutant and display even stronger leaf phenotypes (Figure 16.1D; Efroni et al., 2008). Similarly, crinkly leaf phenotypes are observed in *miR319*-overexpressing *Brassica rapa* plants (Mao et al., 2014). Overexpression of a *TCP3* chimeric gene that encodes a TCP3 protein fused to the transcriptional repressor domain superman repression domain X (SRDX; Hiratsu et al., 2003) causes dominant repression of this gene function and generates wavy and serrated leaves, even in the presence of redundant genes in *A. thaliana, Torenia fournieri, Chrysanthemum morifolium,* and *Cyclamen persicum* (Figure 16.1F–H; Koyama et al., 2007; Koyama et al., 2010a; Narumi et al., 2011; Tanaka et al., 2011). In contrast, *CIN* gene gain-of-function lines show the opposite phenotype; plants that overexpress *miR319*-resistant versions of *TCP3* (Figure 16.1I), *TCP4* (*soj8* and *soj6* mutants) or bear a *miR319A* mutation (*mir319a^{129}*) show accelerated leaf maturation and reduced proliferation. This results in smaller, narrower leaves of a darker green than the wild type (Palatnik et al., 2007; Schommer et al., 2008; Nag et al., 2009; Rodriguez et al., 2010; Li and Zachgo, 2013; Koyama et al., 2013). Cell lines that express a hyperactive form of *TCP4* (*pTCP4::TCP4:VP16*, viral protein 16) also show premature onset of maturation and decreased cell proliferation in leaves (Figure 16.1L; Sarvepalli and Nath, 2011).

16.2.1.2 CYC/TB1 *Genes and Simple Leaf Development*

Limited evidence indicates a role for *CYC/TB1*-like genes in leaf development. Koyama et al. (2010b) found that the promoter of *Arabidopsis CYC/TB1* gene *TCP1* is active in the midrib region, petiole, and distal part of expanding leaves. This could suggest that *TCP1* is also involved in regulating the cell cycle arrest front during leaf development in *Arabidopsis*. Although *TCP1* loss-of-function mutants do not display abnormal leaf phenotypes (Cubas et al., 2001; Koyama et al., 2010b), overexpression of a dominant negative form of *TCP1* (*35S::TCP1:SRDX*) generates plants with reduced elongation of the leaf petioles and blades, possibly due to limited cell elongation (Koyama et al., 2010b). *TCP1* might therefore act redundantly with other *CYC/TB1* genes such as *BRANCHED1* (*BRC1*) and/or *BRC2* during leaf development. Indeed, *Arabidopsis BRC1* and maize *TB1* are also expressed in axillary bud leaves (husk leaves in maize), and their loss of function results in faster growth of these lateral organs (Hubbard et al., 2002; Aguilar-Martínez et al., 2007). In the case of *tb1* mutants, this results in very long husk leaves (Hubbard et al., 2002).

16.2.1.3 Class I TCP *and Simple Leaf Development*

Several *Arabidopsis* class I genes influence leaf development in a largely redundant manner, consistent with their high sequence similarity and their overlapping expression patterns that follow a basipetal wave, like that of *CIN* genes (Kieffer et al., 2011; Uberti-Manassero et al., 2012; Aguilar Martínez and Sinha, 2013). *Arabidopsis* has seven closely related class I genes, *TCP7, TCP8, TCP14, TCP15, TCP21, TCP22,* and *TCP23,* some of which are close paralogs (*TCP7* and *TCP21; TCP8, TCP22,* and *TCP23; TCP14,* and *TCP15*). Single mutants of most of them have wild-type leaf phenotypes (Pruneda-Paz et al., 2009; Kieffer et al., 2011; Li et al., 2012a; Uberti-Manassero et al., 2012; Aguilar-Martínez and Sinha, 2013). In *tcp14* and *tcp15* single-mutants, LeafAnalyser quantitative imaging (Weight et al., 2008) nonetheless allowed detection of small, almost imperceptible morphological changes in leaves, which are slightly broader toward the base (Kieffer et al., 2011). Aguilar Martínez and Sinha (2013) found that mutants in *TCP15* also had short petioles and in *TCP23* increased blade length, width, perimeter, and area. The *tcp14–tcp15* double-mutants displayed an additional mild leaf defect of increased cell proliferation (Kieffer et al., 2011), which suggests that *TCP14* and *TCP15* prevent proliferation during leaf development. The quintuple-mutant *tcp8–tcp15–tcp21–tcp22–tcp23* has leaf blades broader than the wild type, a phenotype associated with increased expression of *CYCLINA1;1* (*CYCA1;1*) and *CYCA2;3* (Aguilar-Martínez and Sinha, 2013). As in the case of *CIN* genes, mutant phenotypes are more pronounced in cell lines that overexpress dominant negative forms of these genes. *TCP7:SRDX, TCP11:SRDX, TCP14:SRDX, TCP15:SRDX, TCP20:SRDX,* and *TCP23:SRDX* cell lines produce plants with shorter leaf petioles, upward-curved leaves with increased cell density, and groups of small undifferentiated cells (Figure 16.1E; Hervé et al., 2009; Kieffer et al., 2011; Viola et al., 2011; Li et al., 2012a;

Uberti-Manassero et al., 2012). These findings indicate that at least some class I genes prevent leaf proliferation and/or promote leaf maturation during development.

Notably, other studies suggest the opposite function, at least for *TCP14* and *TCP15*; in *Arabidopsis* and tomato, overexpression of these genes induces cytokinin (CK) responses and delayed leaf maturation (Steiner et al., 2012a; Steiner et al., 2012b; Section 16.2.3.4). In other tissues, *TCP14*, *TCP15*, and *TCP20* promote or repress cell proliferation, depending on the timing and cell types involved (Hervé et al., 2009; Kieffer et al., 2011). *TCP15* also negatively controls other cell-cycle-related processes such as endoreduplication of leaf trichomes and rosette leaf cells (Li et al., 2012a). These and other observations indicate that class I TCPs modulate leaf cell division, growth, and maturation in a complex and largely context-dependent manner. This contradicts the classical proposal that class I TCPs promote growth and proliferation, whereas class II TCPs promote growth arrest (Li et al., 2005; Hervé et al., 2009; Koyama et al., 2010a; Martín-Trillo and Cubas, 2010).

16.2.2 Role of TCP Genes During Compound Leaf Development

Compound leaves are formed by lateral appendages, the leaflets, whose margins can give rise to secondary leaflets in a reiterative manner. *CIN*-like genes are involved in the elaboration of compound leaves in species such as tomato (*Solanum lycopersicum*). In tomato, it was proposed that the *CIN* ortholog (*LANCEOLATE, LA*) regulates the timing and location of leaf maturation, thus determining leaf shape (Ori et al., 2007; Shleizer-Burko et al., 2011). *LA* gain-of-function point mutations (*La-2*) that confer resistance to *miR319* lead to increased *LA* activity in young leaf primordia and precocious differentiation of leaf margins. As a consequence, large compound tomato leaves are converted into small, simpler ones with fewer leaflets (Figure 16.1J; Ori et al., 2007). In contrast, loss-of-function point mutations (*La-6*) or overexpression of *miR319*, which downregulates *LA*-like genes, enhance compound leaf patterns and promote the formation of larger leaflets with continuous growth of leaf margins (Figures 16.1K, M; Ori et al., 2007). The precisely regulated spatial and temporal activity of *LA* therefore appears to be crucial for leaf maturation in tomato. It has been speculated that evolutionary changes in *LA* expression patterns could be responsible for the evolution of leaves with different morphologies in Solanaceae, due to variations in leaf and leaflet maturation time (Shleizer-Burko et al., 2011). This could now be tested in *Chelidomium majus* (Papaveraceae) a species with compound leaves in which *CIN* genes have been isolated (Ikeuchi et al., 2013). In *Rosa hybrida* (Rosaceae), another species with compound leaves, overexpression of an *Arabidopsis TCP3:SDRX* construct produces leaves more serrated and with increased number of leaflets (Figure 16.1N; Gion et al., 2011) suggesting that an endogenous *CIN* function plays a role in the elaboration of this compound leaf.

16.2.3 TCP-Regulated Networks in the Control of Leaf Development

A general conclusion of studies on how TCPs influence leaf maturation and development is that several TCP-dependent pathways could converge (directly or indirectly) in the negative regulation of the *CUP-SHAPED COTYLEDON/Class I KNOTTED*-like homeobox (*CUC/KNOX1*) module in leaves. In addition, some TCPs prevent cell proliferation by negatively regulating *GROWTH-REGULATING FACTORs* (GRFs), which have a positive effect on cell proliferation of lateral organs (Kim et al., 2003; Jones-Rhoades and Bartel, 2004; Rodriguez et al., 2010; Chapter 17). Finally TCPs influence auxin and CK signaling, genetic pathways that have great impact on the control of leaf shape.

16.2.3.1 TCP Regulation of the CUC/KNOX1 Gene Pathway

CUC genes are positive regulators of the *KNOX1* genes *SHOOTMERISTEMLESS* (*STM*), *BREVIPEDICELLUS* (*BP*), *KNOTTED-LIKE FROM A. THALIANA* (*KNAT1*), *KNAT2*, and *KNAT6* (Aida et al., 1999; Takada et al., 2001; Hibara et al., 2003; Spinelli et al., 2011; Chapter 15). *KNOX1* genes maintain the undifferentiated state of leaf cells and promote cell proliferation by activating CK biogenesis (Figure 16.2; Yanai et al., 2005; Chapter 14). Their overexpression can severely affect leaf morphogenesis, leading to the formation of lobes and ectopic meristems in *Arabidopsis* and *Nicotiana* (Sinha et al., 1993; Lincoln et al., 1994; Chuck et al., 1996; Pautot et al., 2001; Gallois et al., 2002).

In *Arabidopsis*, *CIN* genes negatively regulate *CUC1*, *CUC2*, and *CUC3* (Koyama et al., 2007, 2010a; Sarvepalli and Nath, 2011). *TCP3* dominant loss-of-function plants (*TCP3:SRDX*) show ectopic expression of *CUC* genes (Koyama et al., 2007) and form ectopic shoots on cotyledons, a *cuc* gain-of-function phenotype (Hibara et al., 2003); *cuc1* and *cuc2* mutations suppress this phenotype (Koyama et al., 2007). *TCP3* gain-of-function cell lines (*TCP3 miR319a*-resistant overexpressors; *35S::mTCP3*) display *cuc* mutant phenotypes of fused cotyledons (Li and Zachgo, 2013). A hyperactive form of *TCP4* (*pTCP4::TCP4:VP16*) also has cup-shaped leaves (Sarvepalli and Nath, 2011).

It is proposed that *TCP3* (and probably other *CIN* genes) directly activate the microRNA *miR164A*, a negative regulator of *CUC1* and *CUC2* (Figure 16.2; Laufs et al., 2004; Mallory et al., 2004; Nikovics et al., 2006; Sieber et al., 2007). *TCP3* also binds the promoter and activates the expression of *ASYMMETRIC LEAVES 1* (*AS1*), which

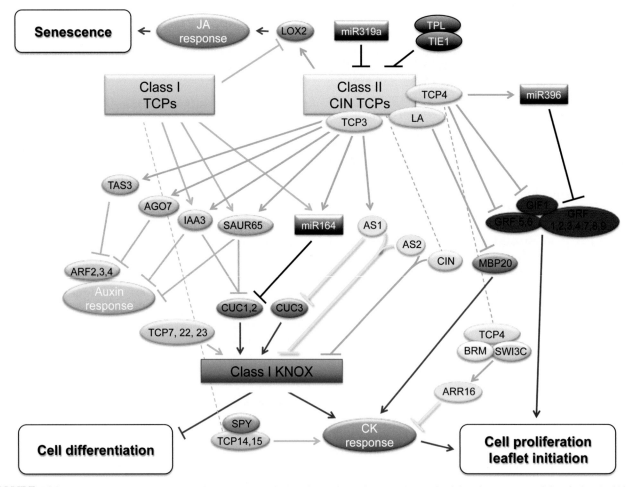

FIGURE 16.2 Gene regulatory networks of class I (turquoise) and CIN (green) genes during leaf development. Model includes *Arabidopsis* and tomato data (see text).

encodes a myeloblastosis (MYB) transcriptional repressor of *CUC3* (Figure 16.2; Byrne et al., 2000; Sieber et al., 2007; Koyama et al., 2010a). AS1 also forms complexes with AS2 to repress transcription of *BP*, *KNAT2*, and *KNAT6* (Ori et al., 2000; Semiarti et al., 2001; Guo et al., 2008). TCP3, TCP4, TCP10, and TCP24, in turn, interact physically with AS2 to directly downregulate *BP* and *KNAT2* expression (Figure 16.2; Li et al., 2012b).

Class I TCPs also negatively regulate the *CUC/KNOX1* module in *Arabidopsis* leaves, although their effect is not identical to that of *CIN* genes. Dominant *TCP15* loss of function (*pTCP15::TCP15:EAR*) generates plants with abnormal leaves, in which *miR164* is downregulated and *CUC1* and *CUC2* upregulated (Uberti-Manassero et al., 2012). In contrast, *AS1*, its target *CUC3*, *STM*, and *BP* are unaffected in *pTCP15::TCP15:EAR* plants (Uberti-Manassero et al., 2012); nonetheless, the quintuple class I mutant *tcp8–tcp15–tcp21–tcp22–tcp23* does show upregulation of *STM* and *BP* as well as of *AS2*. In addition, TCP7 homodimers and TCP7–TCP23, TCP7–TCP22, and TCP22–TCP23 heterodimers are able to bind the K-box of the *STM* promoter (Aguilar-Martínez and Sinha, 2013), which indicates that class I proteins probably regulate *KNOX1* genes directly (Figure 16.2).

In tomato, the molecular link between *LA* and *KNOX1* activity is still unclear. Nevertheless, *LA* can directly inactivate the expression of the *MADS–BOX PROTEIN20* (*MBP20*) gene, which maintains the CK response downstream of *KNOX1* genes (Figure 16.2; Burko et al., 2013).

16.2.3.2 TCP Regulation of Growth-Regulating Factors

TCP4 negatively regulates genes that encode several GRFs, which have a positive effect on cell proliferation of lateral organs (Kim et al., 2003; Horiguchi et al., 2005; Rodriguez et al., 2010). This control occurs at both transcriptional and posttranscriptional levels. First, TCP4 activates *miR396*, which targets seven *GRFs* (Jones-Rhoades and Bartel, 2004; Liu et al., 2009; Rodriguez et al., 2010). TCP4 subsequently downregulates expression of *GRF5* and *GRF6*, which are not under the control of *miR396*, and that of the *GRF-INTERACTING FACTOR1* (*GIF1*) gene, which encodes a GRF transcriptional coactivator (Figure 16.2; Kim and Kende, 2004; Lee et al., 2009; Chapter 17).

16.2.3.3 TCP Regulation of Auxin Signaling During Leaf Development

During leaf development auxin has a central role in determining the final size and shape of simple and compound leaves by promoting expression of the *CUC/KNOX1* gene module (Vernoux et al., 2000; Aida et al., 2002; Hu et al., 2003; Furutani et al., 2004; Cheng et al., 2006; Scarpella et al., 2006; Hay et al., 2006; Koyama et al., 2007; Koyama et al., 2010a; Ori et al., 2007; Efroni et al., 2008; Uberti-Manassero et al., 2012; Li and Zachgo, 2013). *CIN* genes, as well as probably class I *TCP14* and *TCP15*, reduce auxin transport and sensitivity, and thus contribute to *CUC/KNOX1* downregulation. Several *CIN* genes, as well as *TCP14* and *TCP15*, are direct transcriptional activators of the *INDOLE-3-ACETIC ACID INDUCIBLE3* (*IAA3*) and *SMALL AUXIN UP RNA 65* (*SAUR65*) genes (Figure 16.2; Koyama et al., 2010a; Uberti-Manassero et al., 2012). Direct activation of *Antirrhinum IAA3* by *CIN* was recently demonstrated (Das Gupta et al., 2014). Many other auxin-inducible genes are downstream of *TCP3* or *TCP4* including a large group of genes encoding SAUR proteins, which are negative regulators of auxin synthesis and transport (Hagen and Guilfoyle, 2002; Kant et al., 2009) and IAA factors, which are auxin response factor (ARF) repressors (Liscum and Reed, 2002; Weijers et al., 2005; Koyama et al., 2010a; Sarvepalli and Nath, 2011). However, direct transcriptional control of these genes by TCPs has not been established.

In addition, *TCP3* promotes expression of *TRANS-ACTING SHORT-INTERFERING RNA3* (*TAS3*) and *ARGONAUTE7* (*AGO7*), both negative posttranscriptional regulators of *ARF2*, *ARF3*, and *ARF4* through a trans-acting short-interfering RNA mechanism (Figure 16.2; Adenot et al., 2006; Fahlgren et al., 2006; Garcia et al., 2006; Hunter et al., 2006; Koyama et al., 2010a). *TCP3* also influences the expression of *PIN-FORMED* (*PIN*) genes involved in auxin transport; in *TCP3:SRDX* cell lines *PIN1*, *PIN5*, and *PIN6* are upregulated and *PIN3*, *PIN4*, and *PIN7* downregulated (Koyama et al., 2010a; Li and Zachgo, 2013). Finally, transcriptomic analyses show that genes that encode the polar auxin transport protein AUXIN TRANSPORTER PROTEIN1 (AUX1; Swarup et al., 2001), the auxin receptor TRANSPORT INHIBITOR RESPONSE1 (Ruegger et al., 1998), the INDOLE-3-ACETIC ACID AMIDO SYNTHASE (GH3 17; Staswick et al., 2005), and IAA29 (Riechmann, 2000) are downregulated in *TCP3* gain-of-function cell lines (*mTCP3*) and upregulated in loss-of-function cell lines (*TCP3:SRDX* overexpressing plants; Li and Zachgo, 2013).

16.2.3.4 TCP Regulation of CK Signaling During Leaf Development

In developing leaves, *CIN* genes and class I TCPs also cause indirect downregulation of CK signaling via downregulation of *KNOX1* genes (Section 16.2.3.1; Yanai et al., 2005; Koyama et al., 2010a; Uberti-Manassero et al., 2012). In addition, the CIN factor TCP4 might dampen CK response and sensitivity through a second, *KNOX1*-independent mechanism (Efroni et al., 2013). TCP4 interacts physically with switch/sucrose nonfermentable (SWI/SNF; Peterson et al., 1994) chromatin-remodeling adenosine triphosphate (ATPase) BRAHMA (BRM); BRM-TCP4 could then bind the promoter and activate expression of the CK response inhibitor *ARABIDOPSIS RESPONSE REGULATOR16* (*ARR16*; Figure 16.2; To et al., 2004; Ren et al., 2009; Efroni et al., 2013). In accordance, the leaf phenotype of *miR319* overexpressors (equivalent to quintuple *cin* mutants) can be compensated by *ARR16* overexpression in leaves (Efroni et al., 2013). Other *CIN* genes could have similar roles; for example, TCP3 and TCP5 proteins also interact with BRM and the chromatin-remodeling subunit SWI/SNF complex subunit 3C in yeast two-hybrid (YTH) assays (Archacki et al., 2009; Hurtado et al., 2006). In contrast, the *Antirrhinum CIN* gene promotes CK signaling at the transition zone between dividing and differentiating leaf cells. CIN directly activates *HISTIDINE KINASE 4* (*HK4*), the ortholog of the *Arabidopsis* CK receptor, *CYTOKININ RESPONSE 1* (*CRE1*, Yamada et al., 2001; Das Gupta et al., 2014). Probably as a consequence, several CK response genes such as *ARR* are upregulated.

Class I TCP14 and TCP15 promote CK responsiveness. Their overexpression in *Arabidopsis* leads to typical CK responses, such as upregulation of the *RESPONSE REGULATOR 5* (*ARR5*) gene and the mitotic factor *CYCLIN B1;2* (Steiner et al., 2012a). The phenotype of tomato plants overexpressing these genes supports this proposal (Steiner et al., 2012b). In *Arabidopsis*, *tcp14–tcp15* double-mutants are hyposensitive and *TCP14*-overexpressing cell lines are hypersensitive to CK (Steiner et al., 2012a). This effect on CK response requires a functional *SPINDLY* (*SPY*) gene, which encodes an O-GLCNAC TRANSFERASE (OGT). *SPY* promotes similar CK-related responses (Maymon et al., 2009; Steiner et al., 2012a, 2012b). TCP14 and TCP15 interact with SPY in YTH and *in vitro* pull-down assays. It is thus proposed that TCP14 and TCP15 require an O-GlcNAc modification by SPY to promote CK response in leaves (Figure 16.2; Steiner et al., 2012a). Class I TCP8, TCP19, TCP23, and class II TCP2 and TCP24 can also be O-GlcNAc-modified by another *Arabidopsis* OGT, SECRET AGENT (SEC; Steiner et al., 2012a), which suggests that both TCP classes act in collaboration with SPY/SEC to modulate CK signaling.

16.2.3.5 TCP Genes and Gibberellin Signaling

In tomato, *LA* activity in leaf development is mediated in part by positive regulation of the gibberellin (GA) response. *LA* and *mir319* gain-of-function cell lines have, respectively, increased and decreased mRNA levels of the GA biosynthesis gene *SlGA20-OXIDASE1*. Increased GA levels or response lead to leaf phenotypes similar to those of *LA* gain-of-function mutants. Exogenous application of GA suppresses the *miR319* overexpression leaf phenotype, whereas *LA* gain-of-function lines partially suppress the phenotype due to overexpression of the GA deactivation gene *GIBBERELLIN 2-OXIDASE 4* (Yanai et al., 2011).

16.2.3.6 TCP Genes and Jasmonate Signaling

Class I and class II TCP genes act antagonistically in leaves to control expression of *LIPOXYGENASE2* (*LOX2*), a gene involved in jasmonate (JA) biosynthesis (Vick and Zimmerman, 1983; Schommer et al., 2008; Danisman et al., 2012). Methyl-JA represses cell proliferation and promotes senescence in leaves (Pauwels et al., 2008). In young leaves, the class I TCP dimers TCP20–TCP8 and TCP20–TCP22 bind the *LOX2* promoter and repress its transcription (Figure 16.2). It is likely that TCP9, which is also transcriptionally activated by TCP20, binds to the *LOX2* promoter, but experimental evidence is lacking (Danisman et al., 2012). In mature leaves, *TCP4* is expressed strongly and antagonizes TCP20-mediated *LOX2* repression (Figure 16.2); it promotes *LOX2* transcription, probably by binding to DNA motifs different from those bound by TCP20 (Schommer et al., 2008; Aggarwal et al., 2010).

16.2.4 Posttranslational Regulation of TCP Genes During Leaf Development

Several *CIN* genes (*TCP2*, *TCP3*, *TCP4*, *TCP10*, and *TCP24*) are posttranscriptionally regulated by *miR319* (Section 16.2.1.1; Palatnik et al., 2003). However, very few TCP posttranslational regulators have been identified. The *TCP INTERACTOR-CONTAINING EAR MOTIF PROTEIN1* (*TIE1*) gene, which codes for a transcriptional repressor, has been shown to be a key modulator of TCP protein activity during leaf development (Tao et al., 2013). Like *CIN* genes, *TIE* is expressed at the differentiation front of developing leaves. The TIE1 protein not only interacts with TOPLESS (TPL) corepressors through the ethylene-responsive factor (ERF) amphiphilic repressor (EAR) domain, but also with CIN proteins. This led to a model in which TIE1 mediates the interaction of CIN factors with TPL transcriptional corepressors, thus transforming TCPs into transcriptional repressors. As predicted, *TIE1* overexpressors (*tie1-D*) have curly-leaved small-cell phenotypes similar to those of *cin* multiple knockout mutants. Moreover, expression of *CIN* target genes (*AS1*, *IAA3*, *SAUR65*, and *LOX2*) is greatly reduced in these cell lines (Tao et al., 2013).

16.3 TCP GENES AND THE CONTROL OF SHOOT BRANCHING

Branches are generated from lateral shoot meristems (axillary meristems, AMs) formed at the base of leaves. Axillary meristems form leaf primordia at their periphery and develop into axillary buds. Axillary buds can remain dormant for long time periods or elongate to give rise to a branch. Several *CYC/TB1* genes have functions related to the development of AMs and buds, which give rise to lateral shoots.

16.3.1 TB1/BRC1 Genes Prevent Lateral Branch Outgrowth

In grasses, the *CYC/TB1* clade underwent at least two duplications, giving rise to three lineages termed groups 18, 19, and 20 (Mondragón-Palomino and Trontin, 2011; Chapter 9). Group 18 includes all the genes reported to control lateral shoot development in monocots: maize *Tb1* (Doebley et al.,1997), rice *OsTB1* (*FINE CULM1*, *FC1*; Takeda et al., 2003), and sorghum *SbTB1* (Kebrom et al., 2006). In core eudicots, there are also three *CYC/TB1* subclades (CYC1, CYC2, and CYC3; Howarth and Donoghue, 2006), of which CYC1 has the most prominent role in the control of lateral branch outgrowth, and includes *Arabidopsis BRC1* (Aguilar-Martínez et al., 2007; Finlayson, 2007) and its orthologs. The CYC2 subclade includes *Arabidopsis TCP1*, expressed in the dorsal region of vegetative axillary buds (Cubas et al., 2001). Nonetheless, no discernible branching phenotypes have been identified in *tcp1* mutants. CYC3 genes such as *Arabidopsis BRC2* could play a minor role in the control of shoot branching, particularly in short-day photoperiods and under poor light conditions (Aguilar-Martínez et al., 2007; Finlayson, 2007; Finlayson et al., 2010).

Mutant analyses indicate that *Tb1* prevents axillary bud outgrowth in maize at lower nodes, promotes the shortening of female inflorescence internodes, and probably promotes female floret stamen abortion. Maize *Tb1* mutants are highly branched (Figure 16.3A) and resemble its wild ancestor teosinte (*Zea mays* ssp. *parviglumis*). In lower nodes they have lateral branches; in upper nodes female inflorescences are replaced by lateral branches (Doebley et al., 1997; Hubbard et al., 2002; Wang et al., 1999). Rice *OsTB1* (*FC1*; Takeda et al., 2003) has a similar role, as *fc1* mutants have a significantly increased number of branches (tillers) at the base of the main shoot (Figure 16.3B; Takeda et al., 2003). The *Tb1* orthologs sorghum *SbTb1* (Kebrom et al., 2006), barley *INTERMEDIUM-C* (Ramsay et al., 2011), and wheat *TaTB1* (Kebrom et al., 2012) are also probably involved in the negative control of axillary bud outgrowth, suggesting conservation of this function in the monocot lineage.

In dicots, the *BRC1* function also appears to be conserved. *Arabidopsis* loss-of-function *brc1* mutants have increased rosette leaf branching compared to wild-type plants (Figure 16.3C; Aguilar-Martínez et al., 2007; Finlayson, 2007). Pea *Psbrc1* mutants have more basal branches than the wild type, especially in cotyledonary axils and in nodes 1 and 2, which indicates that *PsBRC1* is more

FIGURE 16.3 Shoot-branching phenotypes of mutants in **BRC1/TB1** genes or downstream genes in different plant species. In each panel, wild-type (WT, left) and mutant (right) phenotypes are shown. (A) Maize *Tb1* mutant (Hubbard et al., 2002). (B) Mutant in the rice *FINE CULM1* gene (Minakuchi et al., 2010). (C) *Arabidopsis brc1* mutant (Aguilar-Martínez et al., 2007). (D) Pea *Psbrc1* mutant (Braun et al., 2012). (E) Tomato *Slbrc1b* mutant (Martín-Trillo et al., 2011). (F) Maize *gt1* mutant (Whipple et al., 2011). *Reproduced with permission from Hubbard et al. (2002), Minakuchi et al. (2010), Aguilar-Martínez et al. (2007), Martín-Trillo et al. (2011), Whipple et al. (2011), Catherine Rameau, Braun et al. (2012).*

active in the lowest nodes of the plant (Figure 16.3D, Braun et al., 2012). *Psbrc1* mutants also have thinner stems, a phenotype not observed in *Arabidopsis brc1* mutants. In Solanaceae, two *BRC1*-like genes have been identified, *BRC1a* and *BRC1b*, whose functions are still not fully clear. Tomato *SlBRC1b* loss-of-function (*SlBRC1b–RNAi*) plants have increased basal branching (Figure 16.3E), which supports a more important role for these genes at the basal nodes of the plant (Martín-Trillo et al., 2011). In contrast, *Slbrc1a* point mutants do not have abnormal phenotypes, which could indicate that *SlBRC1a* has a minor function in tomato; in accordance with this, *SlBRC1a* is expressed at very low levels in this species. Nonetheless, *BRC1a* might have a more active role in other *Solanum* species, since tomato introgression lines bearing a *Solanum pennellii BRC1a* (*SpBRC1a*) gene instead of a *SlBRC1a* gene, express fourfold higher *BRC1a* transcript levels and have fewer branches than wild-type tomato plants (Martín-Trillo et al., 2011).

16.3.1.1 Expression of TB1/BRC1 Genes

A common feature of *TB1/BRC1* genes is that their expression throughout bud development is spatially restricted to axillary buds, with few exceptions. Transcripts are detected in young axillary meristems, bud leaf primordia, as well as the provascular and vascular tissue subtending the buds (Hubbard et al., 2002; Arite et al., 2007; Aguilar-Martínez et al., 2007; Martín-Trillo et al., 2011; Braun et al., 2012). It is notable that they are not expressed in the shoot apical meristem, which is developmentally equivalent to an axillary meristem, where *TB1/BRC1* genes would arrest plant growth. *TB1/BRC1* gene expression is closely associated to axillary bud dormancy: there is a negative correlation between *TB1/BRC1* mRNA levels and the degree of axillary bud outgrowth in most species analyzed (Hubbard et al., 2002; Takeda et al., 2003; Kebrom et al., 2006; Kebrom et al., 2010; Aguilar-Martínez et al., 2007; Finlayson, 2007; Arite et al., 2007; Mason et al., 2014). Indeed, *BRC1* expression is the earliest marker of bud dormancy identified insofar as it is upregulated earlier than dormancy markers such as *DORMANCY1* (*DRM1*) and *DRM2* (Stafstrom et al., 1998; Tatematsu et al., 2005). However, exceptions to this correlation of *TB1/BRC1* expression with bud dormancy were found in monocots; *SbTb1* is not upregulated when bud arrest is caused by defoliation (Kebrom et al., 2010), and rice *FC1* remains upregulated in growing basal axillary buds of *d10* mutants (Arite et al., 2007).

16.3.1.2 Regulation of TB1/BRC1 Genes by Light Quality

The expression of *TB1/BRC1* gene transcript levels often responds to environmental and endogenous stimuli that control bud dormancy. This indicates that *BRC1* genes might act as integrators of at least some of the signals that regulate bud outgrowth, and translate them into a growth arrest response (Aguilar-Martínez et al., 2007). For instance, branching (or tillering) is greatly affected by planting density, as plants grown at low density are bushier than genotypically identical plants grown under crowded conditions (Casal et al., 1986). Increased bud dormancy is one of the developmental responses of the shade avoidance syndrome (SAS) triggered in plants exposed to a low red:far-red (R:FR) light ratio. Changes in R:FR are perceived by the *PHYTOCHROMEB* (*PHYB*) photoreceptor and are interpreted as signals of impending shading by nearby competitors (Casal, 2012). *TB1/BRC1* genes in both monocots and dicots are regulated during the SAS response: sorghum *SbTB1* and *Arabidopsis BRC1* mRNA levels are increased in low R:FR, and their expression is negatively regulated by *PHYB* (Kebrom et al., 2006; Finlayson et al., 2010; González-Grandío et al., 2013). Nevertheless, *Arabidopsis brc1* mutants and rice *fc1* mutants show some reduction in lateral branching at high planting density, which shows that additional signals regulate this response (Takeda et al., 2003; Aguilar-Martínez et al., 2007; Finlayson et al., 2010; González-Grandío et al., 2013).

16.3.1.3 Hormonal Regulation of TB1/BRC1 Genes

Hormones that have central roles in the control of shoot branching, such as strigolactones (SLs) and CKs, regulate *TB1/BRC1* transcriptional levels, although their function during branch suppression is probably not limited to this role. CKs have been classically associated with the promotion of branch outgrowth, and they act antagonistically with *TB1/BRC1* genes. In rice and pea, CKs negatively regulate *FC1* and *PsBRC1* mRNA levels, respectively (Minakuchi et al., 2010; Braun et al., 2012).

SLs are recently discovered hormones that prevent plant shoot branching in response to environmental stresses such as phosphate starvation (López-Ráez et al., 2008; Mayzlish-Gati and De-Cuyper, 2012; Foo et al., 2013). *Arabidopsis* and pea SL mutants have very low *BRC1* and *PsBRC1* transcript levels, respectively, even in tissues other than axillary buds, such as cauline leaves (Aguilar-Martínez et al., 2007; Finlayson, 2007; Braun et al., 2012; Dun et al., 2012, 2013; Chevalier et al., 2014). Moreover, exogenous application of SLs to pea axillary buds enhances *PsBRC1* expression (Braun et al., 2012). However, SL-dependent regulation of *TB1*-like genes in monocots is not well established. Although *FC1* is necessary for SL-mediated inhibition of bud outgrowth, *FC1* expression is not reduced in rice *d* mutants and there is no evidence for SL regulation of *FC1* transcription (Arite et al., 2007; Minakuchi et al., 2010). In maize, SL signaling appears to be uncoupled from *Tb1* expression: mutant plants for the SL-synthesis gene *CAROTENOID CLEAVAGE DIOXYGENASE8* have *Tb1* mRNA levels higher than wild-type plants, and no *Tb1* transcriptional response is detectable after SL treatment (Guan et al., 2012). This is not unexpected as *Tb1* alleles selected during maize domestication constitutively express this gene at high levels. Enhanced *Tb1* expression is attributed to the presence of a HOPSCOTCH retrotransposon in a region 70 kb (kilobase) upstream of the transcription start site of the gene (Wang et al., 1999; Clark et al., 2006; Studer et al., 2011).

16.3.1.4 Sucrose Regulation of TB1/BRC1 Genes

A classical hypothesis proposed that access to plant nutrients is the major factor that regulates branching (Phillips, 1975). Sugar availability has been shown to be necessary and sufficient for axillary bud outgrowth; application of exogenous sugars to intact plants resulted in rapid bud release (Mason et al., 2014). In wheat, Kebrom et al. (2012) found an association between low sucrose levels in axillary buds and upregulation of *TB1*. Conversely, an artificial increase in sucrose levels led to transcriptional repression of *BRC1* in *Arabidopsis* (Mason et al., 2014). These findings seem to indicate that *TB1/BRC1* genes mediate the branching response to changes in sucrose content. Whether sugar content triggers hormone signaling (i.e., auxin signaling) or acts directly as a signal that controls *TB1/BRC1* expression remains to be elucidated.

16.3.1.5 TB1/BRC1 Genetic Networks: Up- and Downstream Genes

Environmental, hormonal, and nutritional signals that control *TB1/BRC1* function must be translated into the activity of specific *TB1/BRC1* upstream regulators. One of these is the rice gene *IDEAL PLANT ARCHITECTURE 1* (*IPA1*), which encodes a SQUAMOSA PROMOTER-BINDING PROTEIN-LIKE 14 gene (Jiao et al., 2010; Miura et al., 2010). IPA1 binds directly to the *FC1* promoter to activate its transcription and suppress tillering (Lu et al., 2013). IPA1 also interacts with class I proliferating cell factor (PCF1) and PCF2 factors (and weakly with FC1) to bind TGGGCC/T motifs (Lu et al., 2013; Chapter 9), which suggests that TCPs could enable IPA1 binding to TCP motifs. IPA1 activity on branching appears to be additive to that of SL signaling, which supports the partially independent function of SL and *TB1*-like genes in branch suppression in monocots (Luo et al., 2012).

TB1/BRC1 downstream genes are also being studied. One putative target of maize *Tb1* is *GRASSY TILLERS1* (*GT1*), which codes for a homeodomain–leucine zipper (HD-ZIP) I protein whose expression is greatly reduced in *tb1* mutants (Whipple et al., 2011). *GT1* maps to a quantitative trait locus that regulates lateral branching. Therefore, it is proposed that, like *TB1* and *IPA1*, *GT1* could have been a target for selection during maize domestication. Indeed, *gt1* mutants have more tillers than the wild type, like *tb1* mutants (Figure 16.3F).

Transcriptomic analyses of wild-type and *brc1* axillary buds revealed putative *BRC1* downstream genes. Two coregulated gene networks of cell-cycle-related and ribosome-related genes are significantly downregulated in response to *BRC1*. Genes in these networks have promoters enriched in TCP-binding sites, which indicates that they could be controlled directly by TCP factors (González-Grandío et al., 2013). Genes in these networks have been tested for their response to transient expression of *BRC1*, and they respond rapidly after *BRC1* induction (González-Grandío et al., 2013; González-Grandío and Cubas, 2014). This indicates that *BRC1* could promote bud dormancy by causing generalized downregulation of cell cycle and protein synthesis genes.

In addition, a network of abscisic acid (ABA) responding genes is significantly upregulated in response to *BRC1*. ABA has been classically associated with bud dormancy in many species. Moreover, ABA levels in axillary buds directly correlate with the degree of bud dormancy in *Arabidopsis* (Reddy et al., 2013). González-Grandío and Cubas (2014) showed that two central ABA response genes, *ABSCISIC ACID RESPONSIVE ELEMENTS-BINDING FACTOR3* (*ABF3*) and *ABA INSENSITIVE5* (*ABI5*), which contain TCP-binding sites in their promoters, are rapidly activated after transient *BRC1* induction in axillary buds (González-Grandío and Cubas, 2014). Therefore, *BRC1* could be responsible for enhancing or maintaining ABA signaling in buds.

16.4 TCP GENES AND THE CONTROL OF FLOWER SHAPE

Based on their symmetry, flowers can either be radially symmetrical (actinomorphic) if they have several symmetry planes or bilaterally symmetrical (zygomorphic) if they have a single symmetry plane. In zygomorphic flowers, floral organs of the same whorl have different shapes along the dorsiventral axis. Floral zygomorphy evolved independently many times from the radially symmetric condition, as a specialized adaptation to animal pollinators (Stebbins, 1974). During angiosperm evolution, *CYC/TB1* genes were repeatedly coopted to control flower asymmetry in several groups of monocots and dicots.

16.4.1 CYC2 Genes Control Floral Zygomorphy in Dicots

In dicots, the CYC2 subclass of *CYC/TB1* genes plays a major role in the control of flower zygomorphy. In species with bilateral flowers, these genes are expressed asymmetrically, usually in the dorsal part of developing flowers, from meristem initiation to petal and stamen differentiation. Exceptionally, CYC2 genes in Asteraceae are expressed ventrally (Broholm et al., 2008; Juntheikki-Palovaara et al., 2014). In some species, CYC2 genes also promote stamen abortion and control floral organ number.

CYC2 genes were originally characterized in *A. majus* (Scrophulariaceae, Lamiales), a species with bilaterally symmetrical flowers with morphologically different dorsal, lateral, and ventral petals and a dorsalmost aborted staminode. In *Antirrhinum*, two CYC2 paralogs, *CYC* (Luo et al., 1996) and *DICHOTOMA* (*DICH*; Luo et al., 1999), establish dorsal floral organ identity and number. *cyc–dich* double mutants thus have radially symmetric flowers, with six (instead of five) sepals, petals, and stamens, identical to the ventral ones of the wild type. In the closely related *Linaria vulgaris*, silencing of *LvCYC* is sufficient to generate fully radially symmetrical flowers although organ number is not altered (Figure 16.4A; Cubas et al., 1999). Many additional studies have analyzed the role of CYC2 genes in other rosids and asterids (e.g., Gesneriaceae, Fabaceae, Brassicaceae, Dipsacales). Throughout these groups, bilateral symmetry is associated with dorsal petal expression of CYC2 genes. Functional studies also support the implication of these genes in the control of dorsal petal shape (Feng et al., 2006; Busch and Zachgo, 2007; Wang et al., 2008; Wang et al., 2010; Xu et al., 2013, reviewed in Hileman, 2014a and Hileman, 2014b). In pea (*Pisum sativum*) and *Lotus japonicus*, a double-mutant for two CYC2 genes (*CYC2* and *CYC3*) has flowers with ventralized corollas, although petals retain internal asymmetry (Figure 16.4B; Feng et al., 2006; Wang et al., 2008). A third pea CYC2 gene, *PsCYC3*, also expressed in dorsal and lateral petals, has a lateralizing function (Wang et al., 2008). In basal eudicot species with bilaterally symmetrical flowers (e.g., *Consolida regalis*, Ranunculaceae and *Capnoides sempervirens*, Papaveraceae), CYC2 gene expression is also asymmetric (Damerval et al., 2007; Jabbour et al., 2014), which is consistent with

FIGURE 16.4 **Flower phenotype of class II TCP gene mutants.** (A) Inflorescences of wild type (WT) and *Lvcyc* mutant *L. vulgaris* (Cubas et al., 1999). (B) Dissected petals of a wild-type (left) and a *k lts1* double-mutant (right) pea flower. Note the morphological differences between dorsal, lateral, and ventral organs in the wild-type and the ventralized petals of the mutant. The dorsiventral (D/V) axis of symmetry as well as the petal symmetry axes are indicated in red (Wang et al., 2008). Sunflower inflorescences (top) and florets (bottom) from wild-type (WT, C), *double-flowered* (*dbl*, D) and *tubular-rayed* (*tub*, E) individuals. Florets found in different positions of the inflorescence are arranged so that outer florets are in the left and inner florets in the right of each panel (Chapman et al., 2012). (F) *Arabidopsis* flowers of wild type and quintuple *tcp3-tcp4-tcp5-tcp10-tcp13* mutant (Koyama et al., 2011). (G) *Arabidopsis* flowers of wild-type plants and of plants in which *miR319* activity is impaired by constitutive expression of target mimics (Rubio-Somoza and Weigel, 2013). (H) *Arabidopsis* wild-type flower (left) and flower of a plant expressing a miR319-resistant *TCP4* gene under the control of APETALA3 (Nag et al., 2009). *Reproduced with permission from Cubas et al. (1999), Wang et al. (2008), Chapman et al. (2012), Koyama et al. (2011), Rubio-Somoza and Weigel (2013), Nag et al. (2009).*

a role for these genes in determining the zygomorphy of their flowers.

16.4.1.1 CYC2 Genes Control Floral Identity of Capitulum Inflorescence

The capitulum is a complex, highly compressed inflorescence formed by small flowers or florets characteristic of dicot families such as Asteraceae and Dipsaceae (Figure 16.4C; Broholm et al., 2008; Kim et al., 2008; Fambrini et al., 2011; Chapman et al., 2012; Tähtiharju et al., 2012; Juntheikki-Palovaara et al., 2014; Carlson et al., 2011). In heterogamous capitula, the outer ray florets are usually zygomorphic, have elongated petals, and are sterile or female. The inner disk florets are frequently actinomorphic, small, and have fertile stamens and carpels (Figure 16.4C). In Asteraceae and Dipsaceae, CYC2 genes have undergone a remarkable amplification in association with the increased complexity of the capitulum inflorescence, and they play an important role in the specification of ray vs. disk floral morphologies (reviewed in Broholm et al., 2014; Chapter 9). In *Senecio vulgaris*, the *RAY* locus, which controls formation of ray florets, corresponds to two linked CYC2 genes that are specifically upregulated in ray flower primordia (Kim et al., 2008). In *Gerbera hybrida*, at least three CYC2 genes (*GhCYC2*, *GhCYC3*, *GhCYC4*) could have redundant functions in the regulation of ray petal development. These genes are upregulated during early stages of ray flower development and, when ectopically expressed in transgenic plants, lead to the transformation of disk florets into ray-like florets; they promote ligule growth by enhanced cell proliferation and suppress stamen development (Broholm et al., 2008; Juntheikki-Palovaara et al., 2014). The *Gerbera* cultivar crested, in which all the florets are of the ray type, shows strong upregulation of *GhCYC3* in the centralmost flower primordia (Juntheikki-Palovaara et al., 2014). In sunflowers, a mutant with radially symmetrical ray florets (*tubular ray flower*, *tub*; Figure 16.4E) bears a transposon insertion in the *HaCYC2c* gene which results in the production of a premature stop codon (Fambrini et al., 2011; Chapman et al., 2012) and another mutant (*double-flowered*, *dbl*) with disk florets converted into ray-like florets shows ectopic expression of *HaCYC2c* throughout the capitulum (Figure 16.4D; Chapman et al., 2012).

16.4.2 CYC/TB1 Genes Control Flower Asymmetry in Monocots

The monocot sister orders Zingiberales and Commelinales contain species with radially symmetrical flowers and others with bilateral flowers. In species with zygomorphic flowers such as *Heliconia stricta* (Zingiberaceae) and *Commelina communis* (Commelinaceae), group 20 TCP genes (according to Mondragón-Palomino and Trontin, 2011) *ZinTBL2/TB1a* are expressed ventrally at midstages of flower development. In contrast, in species with radially symmetrical corollas, such as *Costus spicatus* (Zingiberaceae) and *Tradescantia pallida* (Commelinaceae), these genes are expressed homogeneously or are undetectable (Bartlett and Specht, 2011; Preston and Hileman, 2012). Another group 20 gene, *RETARDED PALEA1* (*REP1*), controls the formation of bilaterally symmetrical flowers in the Gramineae (grass) family. The grass floret consists of 3 types of organs: a pistil, 1 or 2 whorls of stamens, and 2–3 lodicules subtended by 2 bracts, the palea and lemma (Rudall and Bateman, 2004). The rice *REP1* gene, a Zin*TBL2/TB1a* ortholog (Preston and Hileman, 2012), regulates the identity and development of the palea (Yuan et al., 2009). Early in flower development, *REP1* is expressed in the palea primordium and, later, in the stamens and lemma and palea vascular bundles. In the *rep1* mutant, palea development is delayed significantly and has morphological traits of the wild-type lemma (Yuan et al., 2009).

In addition to its role in the control of axillary bud outgrowth, maize *TB1* (group 18; Mondragón-Palomino and Trontin, 2011) might promote stamen abortion in the female florets as *TB1* mRNA accumulation is detected in the female floret stamen primordia, which later develop into arrested staminodes (Hubbard et al., 2002).

16.4.3 Role of CIN Genes in Flower Development

Probably because floral organs are evolutionarily modified leaves, the *CIN* genes, which are involved in leaf development, also have a role in flower development. In *Arabidopsis*, quadruple *tcp3–tcp4–tcp5–tcp10* and quintuple *tcp3–tcp4–tcp5–tcp10–tcp13* mutants have flowers with wavy, serrated petals (Figure 16.4F; Koyama et al., 2011). *miR319A*-overexpressing cell lines and *CIN*-dominant loss-of-function cell lines (*CIN:SRDX* fusions) also show strong flower phenotypes (Koyama et al., 2007; Koyama et al., 2010a). Conversely, reduction of *miR319* activity results in flowers with short, underdeveloped floral organs (Figure 16.4G), and expression of *miR319*-resistant versions of *CIN* genes lead to petal and stamen reduction in tomato or loss in *Arabidopsis* (Figure 16.4H; Ori et al., 2007; Nag et al., 2009; Rubio-Somoza and Weigel, 2013). These phenotypes are attributed to indirect negative control of *ARF6* and *ARF8* by *TCP4* (Section 16.4.3). In contrast, *Antirrhinum CIN* genes seem to have a divergent role relative to *Arabidopsis CIN* genes during flower development. Mutants in *CIN* genes have flowers with reduced petal lobes and flat (instead of conical) cells, a phenotype associated with downregulation of *CYCLIN D3b* and *HISTONE H4* (Nath et al., 2003; Crawford et al., 2004) indicating reduced cell proliferation in petals of *cin* mutants.

In species in which variations in flower morphology could be of commercial interest, *CIN* gene function has been manipulated to create novel floral forms. In *Cyclamen persicum* and *Torenia fournieri*, *TCP3:SRDX* domain fusions led to flowers with curly petals (Narumi et al., 2011; Tanaka et al., 2011). In *Rosa hybrida*, transgenic plants bearing these constructs yielded flowers with wavy petals and compound leafy sepals (Gion et al., 2011). In contrast, a *CIN:SRDX* construct in *Chrysanthemum morifolium* suppresses floral organ development as described for *Antirrhinum* (Narumi et al., 2011).

16.4.4 Role of Class I Genes in Flower Development

Plants carrying SRDX fusions of some class I TCPs (*TCP7*, *TCP15*, *TCP23*) have flowers with shorter petals and, in the case of *TCP15:SDRX* cell lines, shorter sepals and stamens and altered carpels (Aguilar-Martínez and Sinha, 2013; Uberti-Manassero et al., 2012). In addition, *TCP20* could be involved in petal initiation as it is expressed in petal primordia (Danisman et al., 2012). However, its function in flowers has not yet been analyzed.

16.4.5 TCP Hormone Regulation of Flower Development

During flower development, TCP4 indirectly antagonizes auxin and jasmonate signaling; it interacts with MYB33 to promote the transcription of *miR167a*, which in turn targets *ARF6* and *ARF8* transcripts for degradation (Rubio-Somoza and Weigel, 2013). ARF6/ARF8 promote auxin transport, enhance auxin sensitivity, downregulate class I *KNOX* genes thus inactivating CK signaling, and modulate JA biosynthesis (Ori et al., 2000; Scofield et al., 2007; Tabata et al., 2010). As a consequence, *ARF6* and *ARF8* promote petal expansion, stamen filament elongation, and gynecium maturation (Reeves et al., 2012). Thus, reduced *ARF6/8* (or increased *TCP4*) function delays sepal, petal, and anther maturation and results in underdeveloped floral organs as well as male and female sterility (Figure 16.4H; Nag et al., 2009; Rubio-Somoza and Weigel, 2013).

Class I *TCP14* and *TCP15* also promote CK responsiveness. Their overexpression in *Arabidopsis* leads to typical CK responses, such as trichome development on sepals (Steiner et al., 2012a). The relationship between *TCP14* and *TCP15* and CK signaling has already been reviewed in relation to leaf development.

The only evidence of interaction between TCP and brassinosteroids (BRs) is derived from a study of the *Arabidopsis TCP1* gene, an ortholog of the *CYC* gene involved in generation of floral dorsiventral asymmetry in species with zygomorphic flowers (Section 16.4.1). In *Arabidopsis*, TCP1 binds to and activates the promoter of *DWARF4* (*DWF4*), a hydroxylase that catalyzes a rate-limiting step during BR biosynthesis (Choe et al., 1998; Kim et al., 2006; Guo et al., 2010). A *TCP1* gain-of-function allele (*tcp1-1D*) suppresses the phenotype of *bri1-5*, a BR receptor mutant (Guo et al., 2010; An et al., 2011). *TCP1* expression can in turn be activated by brassinolide, the final product of the BR biosynthetic pathway and the most active BR form (Guo et al., 2010; An et al., 2011); this could result in a positive feedback loop between BR synthesis and *TCP1* activity. *TCP1* might thus counteract the negative feedback loop of the BR pathway operated by BRI1-EMS SUPPRESSOR1 (BES1) and BRASSINAZOLE RESISTANT1 (BZR1; Wang et al., 2002; Yin et al., 2002; Tanaka et al., 2005). Unequal *TCP1* distribution in developing floral organs could thus lead to asymmetric BR distribution, which in turn might contribute to asymmetric growth patterns. *DWF4* is also expressed in leaf margins, where it is required during leaf differentiation (Reinhardt et al., 2007); this might be consistent with *CIN* genes regulating *DWF4* expression to control leaf shape.

TCP4 negatively modulates JA signaling in the last steps of floral organ maturation (Schommer et al., 2008). In developing floral organs, *LOX2* promotes sepal and petal maturation while another JA biosynthetic enzyme, *DEFECTIVE IN ANTHER DEHISCENCE1*, could promote anther maturation (Nagpal et al., 2005; Tabata et al., 2010). These genes are activated by ARF6 and ARF8, which are in turn negatively regulated by the *TCP4/miR167* module, as just described (Rubio-Somoza and Weigel, 2013).

16.5 TCP GENES AFFECT FLOWERING TIME

TCPs have been implicated in an additional role related to reproductive development: the control of flowering time in shoot apical and axillary meristems. Timing of the flowering transition is determined by the activity of the florigen, a systemic signal that moves from leaves to the shoot apex. In *Arabidopsis* this function is carried out by the genes *FLOWERING LOCUS T* (*FT*) and *TWIN SISTER OF FT* (*TSF*). The BRC1 protein can interact with FT and TSF in YTH assays, and *brc1* mutants show early floral transition in lateral shoots. Therefore, it has been proposed that BRC1 could prevent premature flowering in axillary buds by interfering with FT and TSF (Niwa et al., 2013). Interestingly, an independent study showed that a *brc1* mutant is late flowering in the main shoot (Ho and Weigel, 2014), although other putative knockout *brc1* alleles do not display this phenotype (Aguilar-Martínez et al., 2007). In addition, several *Arabidopsis* and apple CIN proteins also interact with FT proteins in YTH and/or transient plant assays (Mimida et al., 2011; Ho and Weigel, 2014), and *cin* loss-of-function mutant combinations are late flowering (Palatnik et al., 2003; Efroni et al., 2008; Efroni et al., 2013; Schommer et al., 2008; Koyama et al., 2010a; Ho and Weigel, 2014). Accordingly, *Arabidopsis TCP4* and tomato *LA* gain-of-function lines are early flowering (Sarvepalli and Nath, 2011; Burko et al., 2013). As *CIN* genes are expressed in leaves, their positive effect on flowering could arise through interactions with FT/TSF in leaves, even before FT is transported to the shoot apex. Finally, several class I TCP proteins (TCP7, TCP8, TCP9, TCP14, TCP15, TCP21, TCP22, and TCP23) have also been found to interact with FT in bimolecular fluorescence complementation (BiFC) assays; however, their effect on flowering has not yet been analyzed except for TCP11 whose mutants display late flowering (Ho and Weigel, 2014).

16.6 CONCLUDING REMARKS

Until now, particular CIN, CYC-like, and TB1-like genes had been shown to play key roles in the development of leaves, flowers, and lateral branches, respectively. Now, class I TCPs have been found to participate in the control of leaf and flower shape in a highly redundant manner, and CIN genes have been reported to affect flower shape. This indicates that the function of many TCP genes is more overlapping than previously suspected. Future research should tell us whether CIN and class I TCPs also regulate shoot branching and if CYC/TB1 genes play relevant roles during leaf development. Moreover, the role of TCPs in the control of leaf and flower form in monocots remains to be elucidated.

New studies also allow us to refine and qualify previously postulated antagonisms between class I and class II genes. Reports that identify TCP protein interactors as well as TCP direct target genes indicate that this is true in some cases (e.g., the control of *LOX2* expression during leaf development and senescence). However, in some other developmental processes (e.g., leaf primordia maturation), both classes cooperate and share gene targets. Multiple mutants affecting both class I and class II genes will help us elucidate additional synergistic or antagonistic gene interactions and shed more light on the global impact of TCP genes in plant development.

Finally, it is becoming clear that the TCP control of cell growth and development does not exclusively rely on direct transcriptional regulation of cell division genes (e.g., cyclins or proliferating cell nuclear antigens, *PCNAs*) by these proteins. Perhaps more importantly, TCPs act by influencing key gene regulatory networks comprising microRNA-controlled pathways (miR164, miRNA167, miR396), transcription factors involved in growth control (KNOX, CUCs, GRFs), and hormone (auxin, GA, CK, JA, and ABA) signaling and responses.

Systematic chromatin immunoprecipitation of TCP-bound DNA followed by massive sequencing (ChIP-seq) combined with massive transcriptomic analyses (RNA-seq) will reveal further direct TCP targets activated or repressed by TCPs. High-throughput protein–protein assays and system biology approaches will help complement the knowledge about the gene regulatory networks in which TCPs participate. These studies would provide a significant step toward an integrated view of how TCP genes control plant architecture.

Acknowledgments

The authors acknowledge Catherine Mark for editorial assistance. This work was supported by the Spanish Ministerio de Educación y Ciencia (grants BIO2008-00581 and CSD2007- 00057), Ministerio de Ciencia y Tecnología (grant BIO2011-25687), and Marie Curie Action (FP7-PEOPLE-2010-IEF, grant agreement 272481).

References

Adenot, X., Elmayan, T., Lauressergues, D., Boutet, S., Bouché, N., Gasciolli, V., Vaucheret, H., 2006. DRB4-dependent TAS3 trans-acting siRNAs control leaf morphology through AGO7. Curr. Biol. 16, 927–932.

Aggarwal, P., Das Gupta, M., Joseph, A.P., Chatterjee, N., Srinivasan, N., Nath, U., 2010. Identification of specific DNA binding residues in the TCP family of transcription factors in *Arabidopsis*. Plant Cell 22, 1174–1189.

Aguilar-Martínez, J.A., Sinha, N., 2013. Analysis of the role of *Arabidopsis* class I TCP genes AtTCP7, AtTCP8, AtTCP22, and AtTCP23 in leaf development. Front. Plant Sci. 4, 406.

Aguilar-Martínez, J.A., Poza-Carrión, C., Cubas, P., 2007. *Arabidopsis* BRANCHED1 acts as an integrator of branching signals within axillary buds. Plant Cell 19, 458–472.

Aida, M., Ishida, T., Tasaka, M., 1999. Shoot apical meristem and cotyledon formation during *Arabidopsis* embryogenesis: interaction among the CUP-SHAPED COTYLEDON and SHOOT MERISTEMLESS genes. Development 126, 1563–1570.

Aida, M., Vernoux, T., Furutani, M., Traas, J., Tasaka, M., 2002. Roles of PIN-FORMED1 and MONOPTEROS in pattern formation of the apical region of the *Arabidopsis* embryo roles of PIN-FORMED1 and MONOPTEROS in pattern formation of the apical region of the *Arabidopsis* embryo. Development 129, 3965–3974.

An, J., Guo, Z., Gou, X., Li, J., 2011. TCP1 positively regulates the expression of DWF4 in *Arabidopsis thaliana*. Plant Signal. Behav. 6, 1117–1118.

Archacki, R., Sarnowski, T.J., Halibart-Puzio, J., Brzeska, K., Buszewicz, D., Prymakowska-Bosak, M., Koncz, C., Jerzmanowski, A., 2009. Genetic analysis of functional redundancy of BRM ATPase and ATSWI3C subunits of *Arabidopsis* SWI/SNF chromatin remodelling complexes. Planta 229, 1281–1292.

Arite, T., Iwata, H., Ohshima, K., 2007. DWARF10, an RMS1/MAX4/DAD1 ortholog, controls lateral bud outgrowth in rice. Plant J. 51, 1019–1029.

Bartlett, M.E., Specht, C.D., 2011. Changes in expression pattern of the teosinte branched1-like genes in the Zingiberales provide a mechanism for evolutionary shifts in symmetry across the order. Am. J. Bot. 98, 227–243.

Braun, N., de Saint Germain, A., Pillot, J.-P., Boutet-Mercey, S., Dalmais, M., Antoniadi, I., Li, X., Maia-Grondard, A., Le Signor, C., Bouteiller, N., Luo, D., Bendahmane, A., Turnbull, C., Rameau, C., 2012. The pea TCP transcription factor PsBRC1 acts downstream of strigolactones to control shoot branching. Plant Physiol. 158, 225–238.

Broholm, S.K., Tähtiharju, S., Laitinen, R., Albert, V.A., Teeri, T.H., Elomaa, P., 2008. A TCP domain transcription factor controls flower type specification along the radial axis of the *Gerbera* (Asteraceae) inflorescence. Proc. Natl. Acad. Sci. USA 105, 9117–9122.

Broholm, S.K., Teeri, T.H., Elomaa, P., 2014. The molecular genetics of floral transition and flower development. Advances in Botanical Research. Elsevier, Oxford, pp. 297–333.

Burko, Y., Shleizer-Burko, S., Yanai, O., Shwartz, I., Zelnik, I.D., Jacob-Hirsch, J., Kela, I., Eshed-Williams, L., Ori, N., 2013. A role for APETALA1/fruitfull transcription factors in tomato leaf development. Plant Cell 25, 2070–2083.

Busch, A., Zachgo, S., 2007. Control of corolla monosymmetry in the Brassicaceae *Iberis amara*. Proc. Natl. Acad. Sci. USA 104, 16714–16719.

Byrne, M.E., Barley, R., Curtis, M., Arroyo, J.M., Dunham, M., Hudson, A., Martienssen, R.A., 2000. Asymmetric leaves1 mediates leaf patterning and stem cell function in *Arabidopsis*. Nature 408, 967–971.

Carlson, S.E., Howarth, D.G., Donoghue, M.J., 2011. Diversification of CYCLOIDEA-like genes in Dipsacaceae (Dipsacales): implications for the evolution of capitulum inflorescences. BMC Evol. Biol., 11, 325.

Casal, J.J., 2012. Shade avoidance. Arabidopsis Book 10, e0157.

Casal, J.J., Sanchez, R.A., Deregibus, V.A., 1986. The effect of plant density on tillering: the involvement of R/FR ratio and the proportion of radiation intercepted per plant. Environ. Exp. Bot. 26, 365–371.

Chapman, M.A., Tang, S., Draeger, D., Nambeesan, S., Shaffer, H., Barb, J.G., Knapp, S.J., Burke, J.M., 2012. Genetic analysis of floral symmetry in Van Gogh's sunflowers reveals independent recruitment of CYCLOIDEA genes in the Asteraceae. PLoS Genet. 8, e1002628.

Cheng, Y., Dai, X., Zhao, Y., 2006. Auxin biosynthesis by the YUCCA flavin monooxygenases controls the formation of floral organs and vascular tissues in *Arabidopsis*. Genes Dev. 20, 1790–1799.

Chevalier, F., Nieminen, K., Sánchez-Ferrero, J.C., Rodríguez, M.L., Chagoyen, M., Hardtke, C.S., Cubas, P., 2014. Strigolactone promotes degradation of DWARF14, an α/β hydrolase essential for strigolactone signaling in *Arabidopsis*. Plant Cell 26, 1134–1150.

Choe, S., Dilkes, B.P., Fujioka, S., Takatsuto, S., Sakurai, A., Feldmann, K.A., 1998. The DWF4 gene of *Arabidopsis* encodes a cytochrome P450 that mediates multiple 22alpha-hydroxylation steps in brassinosteroid biosynthesis. Plant Cell 10, 231–243.

Chuck, G., Lincoln, C., Hake, S., 1996. KNAT1 induces lobed leaves with ectopic meristems when overexpressed in *Arabidopsis*. Plant Cell 8, 1277–1289.

Clark, R.M., Wagler, T.N., Quijada, P., Doebley, J., 2006. A distant upstream enhancer at the maize domestication gene tb1 has pleiotropic effects on plant and inflorescent architecture. Nat. Genet. 38, 594–597.

Crawford, B.C.W., Nath, U., Carpenter, R., Coen, E.S., 2004. CINCINNATA controls both cell differentiation and growth in petal lobes and leaves of *Antirrhinum*. Plant Physiol. 135, 244–253.

Cubas, P., Vincent, C., Coen, E., 1999. An epigenetic mutation responsible for natural variation in floral symmetry. Nature 401, 157–161.

Cubas, P., Coen, E., Martinez Zapater, J.M., 2001. Ancient asymmetries in the evolution of flowers. Curr. Biol. 11, 1050–1052.

Damerval, C., Le Guilloux, M., Jager, M., Charon, C., 2007. Diversity and evolution of CYCLOIDEA-like TCP genes in relation to flower development in Papaveraceae. Plant Physiol. 143, 759–772.

Danisman, S., van der Wal, F., Dhondt, S., Waites, R., de Folter, S., Bimbo, A., van Dijk, A.D.J., Muino, J.M., Cutri, L., Dornelas, M.C., Angenent, G.C., Immink, R.G.H., 2012. *Arabidopsis* class I and class II TCP transcription factors regulate jasmonic acid metabolism and leaf development antagonistically. Plant Physiol. 159, 1511–1523.

Das Gupta, M., Aggarwal, P., Nath, U., 2014. CINCINNATA in *Antirrhinum majus* directly modulates genes involved in cytokinin and auxin signaling. New Phytol. 204, 901–912.

Doebley, J., Stec, A., Hubbard, L., 1997. The evolution of apical dominance in maize. Nature 386, 485–488.

Dun, E.A., de Saint Germain, A., Rameau, C., Beveridge, C.A., 2012. Antagonistic action of strigolactone and cytokinin in bud outgrowth control. Plant Physiol. 158, 487–498.

Dun, E.A., de Saint Germain, A., Rameau, C., Beveridge, C.A., 2013. Dynamics of strigolactone function and shoot branching responses in Pisum sativum. Mol. Plant 6, 128–140.

Efroni, I., Blum, E., Goldshmidt, A., Eshed, Y., 2008. A protracted and dynamic maturation schedule underlies Arabidopsis leaf development. Plant Cell 20, 2293–2306.

Efroni, I., Eshed, Y., Lifschitz, E., 2010. Morphogenesis of simple and compound leaves: a critical review. Plant Cell 22, 1019–1032.

Efroni, I., Han, S.K., Kim, H.J., Wu, M.F., Steiner, E., Birnbaum, K.D., Hong, J.C., Eshed, Y., Wagner, D., 2013. Regulation of leaf maturation by chromatin-mediated modulation of cytokinin responses. Dev. Cell 24, 438–445.

Fahlgren, N., Montgomery, T.A., Howell, M.D., Allen, E., Dvorak, S.K., Alexander, A.L., Carrington, J.C., 2006. Regulation of AUXIN RESPONSE FACTOR3 by TAS3 ta-siRNA affects developmental timing and patterning in Arabidopsis. Curr. Biol. 16, 939–944.

Fambrini, M., Salvini, M., Pugliesi, C., 2011. A transposon-mediated inactivation of a CYCLOIDEA-like gene originates polysymmetric and androgynous ray flowers in Helianthus annuus. Genetica 139, 1521–1529.

Feng, X., Zhao, Z., Tian, Z., Xu, S., Luo, Y., Cai, Z., Wang, Y., Yang, J., Wang, Z., Weng, L., Chen, J., Zheng, L., Guo, X., Luo, J., Sato, S., Tabata, S., Ma, W., Cao, X., Hu, X., Sun, C., Luo, D., 2006. Control of petal shape and floral zygomorphy in Lotus japonicus. Proc. Natl. Acad. Sci. USA 103, 4970–4975.

Finlayson, S., 2007. Arabidopsis TEOSINTE BRANCHED1-LIKE 1 regulates axillary bud outgrowth and is homologous to monocot TEOSINTE BRANCHED1. Plant Cell Physiol. 48, 667–677.

Finlayson, S., Krishnareddy, S., Kebrom, T.H., Casal, J.J., 2010. Phytochrome regulation of branching in Arabidopsis. Plant Physiol. 152, 1914–1927.

Foo, E., Yoneyama, K., Hugill, C.J., Quittenden, L.J., Reid, J.B., 2013. Strigolactones and the regulation of pea symbioses in response to nitrate and phosphate deficiency. Mol. Plant 6, 76–87.

Furutani, M., Vernoux, T., Traas, J., Kato, T., Tasaka, M., Aida, M., 2004. PIN-FORMED1 and PINOID regulate boundary formation and cotyledon development in Arabidopsis embryogenesis. Development 131, 5021–5030.

Gallois, J., Woodward, C., Reddy, G., Sablowski, R., 2002. Combined SHOOT MERISTEMLESS and WUSCHEL trigger ectopic organogenesis in Arabidopsis. Development, 3207–3217, July 1st.

Garcia, D., Collier, S., Byrne, M.E., Martienssen, R., 2006. Specification of leaf polarity in Arabidopsis via the trans-acting siRNA pathway. Curr. Biol. 16, 933–938.

Gion, K., Suzuri, R., Shikata, M., Mitsuda, N., Oshima, Y., Koyama, T., Ohme-Takagi, M., Ohtsubo, N., Tanaka, Y., 2011. Morphological changes of Rosa × hybrida by a chimeric repressor of Arabidopsis TCP3. Plant Biotechnol. 28, 149–152.

González-Grandío, E., Cubas, P., 2014. Identification of gene functions associated to active and dormant buds in Arabidopsis. Plant Signal. Behav. 9, e27994.

González-Grandío, E., Poza-Carrión, C., Sorzano, C.O.S., Cubas, P., 2013. BRANCHED1 promotes axillary bud dormancy in response to shade in Arabidopsis. Plant Cell 25, 834–850.

Guan, J.C., Koch, K.E., Suzuki, M., Wu, S., Latshaw, S., Petruff, T., Goulet, C., Klee, H.J., McCarty, D.R., 2012. Diverse roles of strigolactone signaling in maize architecture and the uncoupling of a branching-specific subnetwork. Plant Physiol. 160, 1303–1317.

Guo, M., Thomas, J., Collins, G., Timmermans, M.C.P., 2008. Direct repression of KNOX loci by the ASYMMETRIC LEAVES1 complex of Arabidopsis. Plant Cell 20, 48–58.

Guo, Z., Fujioka, S., Blancaflor, E.B., Miao, S., Gou, X., Li, J., 2010. TCP1 modulates brassinosteroid biosynthesis by regulating the expression of the key biosynthetic gene DWARF4 in Arabidopsis thaliana. Plant Cell 22, 1161–1173.

Hagen, G., Guilfoyle, T., 2002. Auxin-responsive gene expression: genes, promoters and regulatory factors. Plant Mol. Biol. 49, 373–385.

Hay, A., Barkoulas, M., Tsiantis, M., 2006. ASYMMETRIC LEAVES1 and auxin activities converge to repress BREVIPEDICELLUS expression and promote leaf development in Arabidopsis. Development 133, 3955–3961.

Hervé, C., Dabos, P., Bardet, C., Jauneau, A., Auriac, M.C., Ramboer, A., Lacout, F., Tremousaygue, D., 2009. In vivo interference with AtTCP20 function induces severe plant growth alterations and deregulates the expression of many genes important for development. Plant Physiol. 149, 1462–1477.

Hibara, K., Takada, S., Tasaka, M., 2003. CUC1 gene activates the expression of SAM-related genes to induce adventitious shoot formation. Plant J. 36, 687–696.

Hileman, L.C., 2014a. Trends in flower symmetry evolution revealed through phylogenetic and developmental genetic advances. Philos. Trans. R. Soc. Lond. B Biol. Sci. 369, 20130348.

Hileman, L.C., 2014b. Bilateral flower symmetry – how, when and why? Curr. Opin. Plant Biol. 17, 146–152.

Hiratsu, K., Matsui, K., Koyama, T., Ohme-Takagi, M., 2003. Dominant repression of target genes by chimeric repressors that include the EAR motif, a repression domain, in Arabidopsis. Plant J. 34, 733–739.

Ho, W.W.H., Weigel, D., 2014. Structural features determining flower-promoting activity of Arabidopsis FLOWERING LOCUS T. Plant Cell 26, 552–564.

Horiguchi, G., Kim, G.-T., Tsukaya, H., 2005. The transcription factor AtGRF5 and the transcription coactivator AN3 regulate cell proliferation in leaf primordia of Arabidopsis thaliana. Plant J. 43, 68–78.

Howarth, D.G., Donoghue, M.J., 2006. Phylogenetic analysis of the "ECE" (CYC/TB1) clade reveals duplications predating the core eudicots. Proc. Natl. Acad. Sci. USA 103, 9101–9106.

Hu, Y., Xie, Q., Chua, N.H., 2003. The Arabidopsis auxin-inducible gene ARGOS controls lateral organ size. Plant Cell 15, 1951–1961.

Hubbard, L., McSteen, P., Doebley, J., Hake, S., 2002. Expression patterns and mutant phenotype of teosinte branched1 correlate with growth suppression in maize and teosinte. Genetics 162, 1927.

Hunter, C., Willmann, M.R., Wu, G., Yoshikawa, M., de la Luz Gutiérrez-Nava, M., Poethig, S.R., 2006. Trans-acting siRNA-mediated repression of ETTIN and ARF4 regulates heteroblasty in Arabidopsis. Development 133, 2973–2981.

Hurtado, L., Farrona, S., Reyes, J.C., 2006. The putative SWI/SNF complex subunit BRAHMA activates flower homeotic genes in Arabidopsis thaliana. Plant Mol. Biol. 62, 291–304.

Ikeuchi, M., Tatematsu, K., Yamaguchi, T., Okada, K., Tsukaya, H., 2013. Precocious progression of tissue maturation instructs basipetal initiation of leaflets in Chelidonium majus subsp. asiaticum (Papaveraceae). Am. J. Bot. 100, 1116–1126.

Jabbour, F., Cossard, G., Le Guilloux, M., Sannier, J., Nadot, S., Damerval, C., 2014. Specific duplication and dorsoventrally asymmetric expression patterns of cycloidea-like genes in zygomorphic species of Ranunculaceae. PLoS ONE 9, e95727.

Jiao, Y., Wang, Y., Xue, D., Wang, J., Yan, M., Liu, G., Dong, G., Zeng, D., Lu, Z., Zhu, X., Qian, Q., Li, J., 2010. Regulation of OsSPL14 by OsmiR156 defines ideal plant architecture in rice. Nat. Genet. 42, 541–544.

Jones-Rhoades, M.W., Bartel, D.P., 2004. Computational identification of plant microRNAs and their targets, including a stress-induced miRNA. Mol. Cell 14, 787–799.

Juntheikki-Palovaara, I., Tähtiharju, S., Lan, T., Broholm, S.K., Rijpkema, A.S., Ruonala, R., Kale, L., Albert, V.A., Teeri, T.H., Elomaa, P., 2014. Functional diversification of duplicated CYC2 clade genes in regulation of inflorescence development in Gerbera hybrida (Asteraceae). Plant J. 79, 783–796.

REFERENCES

Kant, S., Bi, Y.-M., Zhu, T., Rothstein, S.J., 2009. SAUR39, a small auxin-up RNA gene, acts as a negative regulator of auxin synthesis and transport in rice. Plant Physiol. 151, 691–701.

Kebrom, T., Burson, B., Finlayson, S., 2006. Phytochrome B represses Teosinte Branched1 expression and induces sorghum axillary bud outgrowth in response to light signals. Plant Physiol. 140, 1109–1117.

Kebrom, T., Brutnell, T.P., Finlayson, S.A., 2010. Suppression of sorghum axillary bud outgrowth by shade, phyB and defoliation signalling pathways. Plant. Cell Environ., 48–58.

Kebrom, T.H., Chandler, P.M., Swain, S.M., King, R.W., Richards, R.A., Spielmeyer, W., 2012. Inhibition of tiller bud outgrowth in the tin mutant of wheat is associated with precocious internode development. Plant Physiol. 160, 308–318.

Kieffer, M., Master, V., Waites, R., Davies, B., 2011. TCP14 and TCP15 affect internode length and leaf shape in *Arabidopsis*. Plant J. 68, 147–158.

Kim, J.H., Kende, H., 2004. A transcriptional coactivator, AtGIF1, is involved in regulating leaf growth and morphology in *Arabidopsis*. Proc. Natl. Acad. Sci. USA 101, 13374–13379.

Kim, J.H., Choi, D., Kende, H., 2003. The AtGRF family of putative transcription factors is involved in leaf and cotyledon growth in *Arabidopsis*. Plant J. 36, 94–104.

Kim, H., Park, P.J., Hwang, H.J., Lee, S.Y., Oh, M.H., Kim, S.G., 2006. Brassinosteroid signals control expression of the AXR3/IAA17 gene in the cross-talk point with auxin in root development. Biosci. Biotechnol. Biochem. 70, 768–773.

Kim, M., Cui, M.L., Cubas, P., Gillies, A., Lee, K., Chapman, M.A., Abbott, R.J., Coen, E., 2008. Regulatory genes control a key morphological and ecological trait transferred between species. Science 322, 1116–1119.

Koyama, T., Furutani, M., Tasaka, M., Ohme-Takagi, M., 2007. TCP transcription factors control the morphology of shoot lateral organs via negative regulation of the expression of boundary-specific genes in *Arabidopsis*. Plant Cell 19, 473–484.

Koyama, T., Mitsuda, N., Seki, M., Shinozaki, K., Ohme-Takagi, M., 2010a. TCP transcription factors regulate the activities of ASYMMETRIC LEAVES1 and miR164, as well as the auxin response, during differentiation of leaves in *Arabidopsis*. Plant Cell 22, 3574–3588.

Koyama, T., Sato, F., Ohme-Takagi, M., 2010b. A role of TCP1 in the longitudinal elongation of leaves in *Arabidopsis*. Biosci. Biotechnol. Biochem. 74, 2145–2147.

Koyama, T., Ohme-Takagi, M., Sato, F., 2011. Generation of serrated and wavy petals by inhibition of the activity of TCP transcription factors in *Arabidopsis thaliana*. Plant Signal. Behav. 6, 697–699.

Koyama, T., Nii, H., Mitsuda, N., Ohta, M., Kitajima, S., Ohme-Takagi, M., Sato, F., 2013. A regulatory cascade involving class II ETHYLENE RESPONSE FACTOR transcriptional repressors operates in the progression of leaf senescence. Plant Physiol. 162, 991–1005.

Laufs, P., Peaucelle, A., Morin, H., Traas, J., 2004. MicroRNA regulation of the CUC genes is required for boundary size control in *Arabidopsis* meristems. Development 131, 4311–4322.

Lee, B.H., Ko, J.H., Lee, S., Lee, Y., Pak, J.H., Kim, J.H., 2009. The *Arabidopsis* GRF-INTERACTING FACTOR gene family performs an overlapping function in determining organ size as well as multiple developmental properties. Plant Physiol. 151, 655–668.

Li, S., Zachgo, S., 2013. TCP3 interacts with R2R3-MYB proteins, promotes flavonoid biosynthesis and negatively regulates the auxin response in *Arabidopsis thaliana*. Plant J. 76, 901–913.

Li, C., Potuschak, T., Colón-Carmona, A., Gutiérrez, R.A., Doerner, P., 2005. *Arabidopsis* TCP20 links regulation of growth and cell division control pathways. Proc. Natl. Acad. Sci. USA 102, 12978–12983.

Li, Z.Y., Li, B., Dong, A.W., 2012a. The *Arabidopsis* transcription factor AtTCP15 regulates endoreduplication by modulating expression of key cell-cycle genes. Mol. Plant 5, 270–280.

Li, Z.Y., Li, B., Shen, W.H., Huang, H., Dong, A., 2012b. TCP transcription factors interact with AS2 in the repression of class-I KNOX genes in *Arabidopsis thaliana*. Plant J. 71, 99–107.

Lincoln, C., Long, J., Yamaguchi, J., Serikawa, K., Hake, S., 1994. A knotted1-like homeobox gene in *Arabidopsis* is expressed in the vegetative meristem and dramatically alters leaf morphology when overexpressed in transgenic plants. Plant Cell 6, 1859–1876.

Liscum, E., Reed, J.W., 2002. Genetics of Aux/IAA and ARF action in plant growth and development. Auxin Mol. Biol., 387–400.

Liu, D., Song, Y., Chen, Z., Yu, D., 2009. Ectopic expression of miR396 suppresses GRF target gene expression and alters leaf growth in *Arabidopsis*. Plant Physiol. 136, 223–236.

López-Ráez, J.A., Charnikhova, T., Gomez-Roldan, V., Matusova, R., Kohlen, W., De Vos, R., Verstappen, F., Puech-Pagès, V., Bécard, G., Mulder, P., Bouwmeester, H., 2008. Tomato strigolactones are derived from carotenoids and their biosynthesis is promoted by phosphate starvation. New Phytol. 178, 863–874.

Lu, Z., Yu, H., Xiong, G., Wang, J., Jiao, Y., Liu, G., Jing, Y., Meng, X., Hu, X., Qian, Q., Fu, X., Wang, Y., Li, J., 2013. Genome-wide binding analysis of the transcription activator IDEAL PLANT ARCHITECTURE1 reveals a complex network regulating rice plant architecture. Plant Cell 25, 3743–3759.

Luo, D., Carpenter, R., Vincent, C., Copsey, L., Coen, E., 1996. Origin of floral asymmetry in *Antirrhinum*. Nature 383, 794–799.

Luo, D., Carpenter, R., Copsey, L., Vincent, C., Clark, J., Coen, E., 1999. Control of organ asymmetry in flowers of *Antirrhinum*. Cell 99, 367–376.

Luo, L., Li, W., Miura, K., Ashikari, M., Kyozuka, J., 2012. Control of tiller growth of rice by OsSPL14 and strigolactones, which work in two independent pathways. Plant Cell Physiol. 1, 1793–1801.

Mallory, A.C., Dugas, D.V., Bartel, D.P., Bartel, B., 2004. MicroRNA regulation of NAC-domain targets is required for proper formation and separation of adjacent embryonic, vegetative, and floral organs. Curr. Biol. 14, 1035–1046.

Mao, Y., Wu, F., Yu, X., Bai, J., He, Y., 2014. miR319a-targeted BrpTCP genes modulate head shape in *Brassica rapa* by differential cell division arrest in leaf regions. Plant Physiol. 164, 710–720.

Martín-Trillo, M., Cubas, P., 2010. TCP genes: a family snapshot ten years later. Trends Plant Sci. 15, 31–39.

Martín-Trillo, M., Grandío, E.G., Serra, F., Marcel, F., Rodríguez-Buey, M.L., Schmitz, G., Theres, K., Bendahmane, A., Dopazo, H., Cubas, P., 2011. Role of tomato BRANCHED1-like genes in the control of shoot branching. Plant J. 67, 701–714.

Mason, M.G., Ross, J.J., Babst, B.A., Wienclaw, B.N., Beveridge, C.A., 2014. Sugar demand, not auxin, is the initial regulator of apical dominance. Proc. Natl. Acad. Sci. USA 111, 6092–6097.

Maymon, I., Greenboim-Wainberg, Y., Sagiv, S., Kieber, J.J., Moshelion, M., Olszewski, N., Weiss, D., 2009. Cytosolic activity of SPINDLY implies the existence of a DELLA-independent gibberellin-response pathway. Plant J. 58, 979–988.

Mayzlish-Gati, E., De-Cuyper, C., 2012. Strigolactones are involved in root response to low phosphate conditions in *Arabidopsis*. Plant Physiol. 160, 1329–1341.

Mimida, N., Kidou, S.I., Iwanami, H., Moriya, S., Abe, K., Voogd, C., Varkonyi-Gasic, E., Kotoda, N., 2011. Apple FLOWERING LOCUS T proteins interact with transcription factors implicated in cell growth and organ development. Tree Physiol. 31, 555–566.

Minakuchi, K., Kameoka, H., Yasuno, N., Umehara, M., Luo, L., Kobayashi, K., Hanada, A., Ueno, K., Asami, T., Yamaguchi, S., Kyozuka, J., 2010. FINE CULM1 (FC1) works downstream of strigolactones to inhibit the outgrowth of axillary buds in rice. Plant Cell Physiol. 51, 1127–1135.

Miura, K., Ikeda, M., Matsubara, A., Song, X.J., Ito, M., Asano, K., Matsuoka, M., Kitano, H., Ashikari, M., 2010. OsSPL14 promotes panicle branching and higher grain productivity in rice. Nat. Genet. 42, 545–549.

Mondragón-Palomino, M., Trontin, C., 2011. High time for a roll call: gene duplication and phylogenetic relationships of TCP-like genes in monocots. Ann. Bot. 107, 1533–1544.

Nag, A., King, S., Jack, T., 2009. miR319a targeting of TCP4 is critical for petal growth and development in *Arabidopsis*. Proc. Natl. Acad. Sci. USA 106, 106.

Nagpal, P., Ellis, C.M., Weber, H., Ploense, S.E., Barkawi, L.S., Guilfoyle, T.J., Hagen, G., Alonso, J.M., Cohen, J.D., Farmer, E.E., Ecker, J.R., Reed, J.W., 2005. Auxin response factors ARF6 and ARF8 promote jasmonic acid production and flower maturation. Development 132, 4107–4118.

Narumi, T., Aida, R., Koyama, T., Yamaguchi, H., Sasaki, K., Shikata, M., Nakayama, M., Ohme-Takagi, M., Ohtsubo, N., 2011. *Arabidopsis* chimeric TCP3 repressor produces novel floral traits in *Torenia fournieri* and *Chrysanthemum morifolium*. Plant Biotechnol. 28, 131–140.

Nath, U., Crawford, B.C.W., Carpenter, R., Coen, E., 2003. Genetic control of surface curvature. Science 299, 1404–1407.

Nikovics, K., Blein, T., Peaucelle, A., Ishida, T., Morin, H., Aida, M., Laufs, P., 2006. The balance between the MIR164A and CUC2 genes controls leaf margin serration in *Arabidopsis*. Plant Cell 18, 2929–2945.

Niwa, M., Daimon, Y., Kurotani, K., Higo, A., Pruneda-Paz, J.L., Breton, G., Mitsuda, N., Kay, S.A., Ohme-Takagi, M., Endo, M., Araki, T., 2013. BRANCHED1 interacts with FLOWERING LOCUS T to repress the floral transition of the axillary meristems in *Arabidopsis*. Plant Cell 25, 1228–1242.

Ori, N., Eshed, Y., Chuck, G., Bowman, J.L., Hake, S., 2000. Mechanisms that control knox gene expression in the *Arabidopsis* shoot. Development 127, 5523–5532.

Ori, N., Cohen, A.R., Etzioni, A., Brand, A., Yanai, O., Shleizer, S., Menda, N., Amsellem, Z., Efroni, I., Pekker, I., Alvarez, J.P., Blum, E., Zamir, D., Eshed, Y., 2007. Regulation of LANCEOLATE by miR319 is required for compound-leaf development in tomato. Nat. Genet. 39, 787–791.

Palatnik, J.F., Allen, E., Wu, X., Schommer, C., Schwab, R., Carrington, J.C., Weigel, D., 2003. Control of leaf morphogenesis by microRNAs. Nature 425, 257–263.

Palatnik, J.F., Wollmann, H., Schommer, C., Schwab, R., Rodriguez, R., Warthmann, N., Allen, E., Dezulian, T., Huson, D., Carrington, J.C., Weigel, D., 2007. Sequence and expression differences underlie functional specialization of *Arabidopsis* microRNAs miR159 and miR319. Dev. Cell 13, 115–125.

Pautot, V., Dockx, J., Hamant, O., Kronenberger, J., Grandjean, O., Jublot, D., Traas, J., 2001. KNAT2: evidence for a link between knotted-like genes and carpel development. Plant Cell 13, 1719–1734.

Pauwels, L., Morreel, K., De Witte, E., Lammertyn, F., Van Montagu, M., Boerjan, W., Inzé, D., Goossens, A., 2008. Mapping methyl jasmonate-mediated transcriptional reprogramming of metabolism and cell cycle progression in cultured *Arabidopsis* cells. Proc. Natl. Acad. Sci. USA 105, 1380–1385.

Peterson, C., Dingwall, A., Scott, M.P., 1994. Five SWI/SNF gene products are components of a large multisubunit complex required for transcriptional enhancement. Proc. Natl. Acad. Sci. USA 91, 2905–2908.

Phillips, I., 1975. Apical dominance. Annu. Rev. Plant Physiol. 26, 341–367.

Preston, J.C., Hileman, L.C., 2012. Parallel evolution of TCP and B-class genes in Commelinaceae flower bilateral symmetry. EvoDevo 3, 6.

Pruneda-Paz, J.L., Breton, G., Para, A., Kay, S.A., 2009. A functional genomics approach reveals CHE as a component of the *Arabidopsis* circadian clock. Science 323, 1481–1485.

Ramsay, L., Comadran, J., Druka, A., Marshall, D.F., Thomas, W.T.B., Macaulay, M., MacKenzie, K., Simpson, C., Fuller, J., Bonar, N., Hayes, P.M., Lundqvist, U., Franckowiak, J.D., Close, T.J., Muehlbauer, G.J., Waugh, R., 2011. INTERMEDIUM-C, a modifier of lateral spikelet fertility in barley, is an ortholog of the maize domestication gene TEOSINTE BRANCHED 1. Nat. Genet. 43, 169–172.

Reddy, S.K., Holalu, S.V., Casal, J.J., Finlayson, S.A., 2013. Abscisic acid regulates axillary bud outgrowth responses to the ratio of red to far-red light. Plant Physiol. 163, 1047–1058.

Reeves, P.H., Ellis, C.M., Ploense, S.E., Wu, M.F., Yadav, V., Tholl, D., Chételat, A., Haupt, I., Kennerley, B.J., Hodgens, C., Farmer, E.E., Nagpal, P., Reed, J.W., 2012. A regulatory network for coordinated flower maturation. PLoS Genet. 8, e1002506.

Reinhardt, B., Hänggi, E., Müller, S., Bauch, M., Wyrzykowska, J., Kerstetter, R., Poethig, S., Fleming, A.J., 2007. Restoration of DWF4 expression to the leaf margin of a dwf4 mutant is sufficient to restore leaf shape but not size: the role of the margin in leaf development. Plant J. 52, 1094–1104.

Ren, B., Liang, Y., Deng, Y., Chen, Q., Zhang, J., Yang, X., Zuo, J., 2009. Genome-wide comparative analysis of type-A *Arabidopsis* response regulator genes by overexpression studies reveals their diverse roles and regulatory mechanisms in cytokinin signaling. Cell Res. 19, 1178–1190.

Riechmann, J.L., 2000. *Arabidopsis* transcription factors: genome-wide comparative analysis among eukaryotes. Science 290, 2105–2110.

Rodriguez, R.E., Mecchia, M.A., Debernardi, J.M., Schommer, C., Weigel, D., Palatnik, J.F., 2010. Control of cell proliferation in *Arabidopsis thaliana* by microRNA miR396. Development 137, 103–112.

Rubio-Somoza, I., Weigel, D., 2013. Coordination of flower maturation by a regulatory circuit of three microRNAs. PLoS Genet. 9, e1003374.

Rudall, P.J., Bateman, R.M., 2004. Evolution of zygomorphy in monocot flowers: iterative patterns and developmental constraints. New Phytol. 162, 25–44.

Ruegger, M., Dewey, E., Gray, W.M., Hobbie, L., Turner, J., Estelle, M., 1998. The TIR1 protein of *Arabidopsis* functions in auxin response and is related to human SKP2 and yeast Grr1p. Genes Dev. 12, 198–207.

Sarvepalli, K., Nath, U., 2011. Hyper-activation of the TCP4 transcription factor in *Arabidopsis thaliana* accelerates multiple aspects of plant maturation. Plant J. 67, 595–607.

Scarpella, E., Marcos, D., Friml, J., Berleth, T., 2006. Control of leaf vascular patterning by polar auxin transport. Genes Dev. 20, 1015–1027.

Schommer, C., Palatnik, J.F., Aggarwal, P., Chételat, A., Cubas, P., Farmer, E.E., Nath, U., Weigel, D., 2008. Control of jasmonate biosynthesis and senescence by miR319 targets. PLoS Biol. 6, e230.

Scofield, S., Dewitte, W., Murray, J.A.H., 2007. The KNOX gene SHOOT MERISTEMLESS is required for the development of reproductive meristematic tissues in *Arabidopsis*. Plant J. 50, 767–781.

Semiarti, E., Ueno, Y., Tsukaya, H., Iwakawa, H., Machida, C., Machida, Y., 2001. The ASYMMETRIC LEAVES2 gene of *Arabidopsis thaliana* regulates formation of a symmetric lamina, establishment of venation and repression of meristem-related homeobox genes in leaves. Development 128, 1771–1783.

Shleizer-Burko, S., Burko, Y., Ben-Herzel, O., Ori, N., 2011. Dynamic growth program regulated by LANCEOLATE enables flexible leaf patterning. Development 138, 695–704.

Sieber, P., Wellmer, F., Gheyselinck, J., Riechmann, J.L., Meyerowitz, E.M., 2007. Redundancy and specialization among plant microRNAs: role of the MIR164 family in developmental robustness. Development 134, 1051–1060.

Sinha, N.R., Williams, R.E., Hake, S., 1993. Overexpression of the maize homeo box gene, KNOTTED-1, causes a switch from determinate to indeterminate cell fates. Genes Dev. 7, 787–795.

Spinelli, S.V., Martin, A.P., Viola, I.L., Gonzalez, D.H., Palatnik, J.F., 2011. A mechanistic link between STM and CUC1 during *Arabidopsis* development. Plant Physiol. 156, 1894–1904.

Stafstrom, J.P., Ripley, B.D., Devitt, M.L., Drake, B., 1998. Dormancy-associated gene expression in pea axillary buds. Planta 205, 547–552.

Staswick, P., Serban, B., Rowe, M., Tiyaki, I., Maldonado, M.T., Maldonado, M.C., Suza, W., 2005. Characterization of an *Arabidopsis* enzyme family that conjugates amino acids to indole-3-acetic acid. Plant Cell 17, 616–627.

Stebbins, G., 1974. Flowering Plants: Evolution Above the Species Level. Cambridge Harvard University Press, Massachusetts, 399 p.

Steiner, E., Efroni, I., Gopalraj, M., Saathoff, K., Tseng, T.-S., Kieffer, M., Eshed, Y., Olszewski, N., Weiss, D., 2012a. The *Arabidopsis* O-linked

N-acetylglucosamine transferase SPINDLY interacts with class I TCPs to facilitate cytokinin responses in leaves and flowers. Plant Cell 24, 96–108.

Steiner, E., Yanai, O., Efroni, I., Ori, N., Eshed, Y., Weiss, D., 2012b. Class I TCPs modulate cytokinin-induced branching and meristematic activity in tomato. Plant Signal. Behav. 7, 807–810.

Studer, A., Zhao, Q., Ross-Ibarra, J., Doebley, J., 2011. Identification of a functional transposon insertion in the maize domestication gene tb1. Nat. Genet. 43, 1160–1163.

Swarup, R., Friml, J., Marchant, A., Ljung, K., Sandberg, G., Palme, K., Bennett, M., 2001. Localization of the auxin permease AUX1 suggests two functionally distinct hormone transport pathways operate in the Arabidopsis root apex. Genes Dev. 15, 2648–2653.

Tabata, R., Ikezaki, M., Fujibe, T., Aida, M., Tian, C.E., Ueno, Y., Yamamoto, K.T., Machida, Y., Nakamura, K., Ishiguro, S., 2010. Arabidopsis AUXIN RESPONSE FACTOR6 and 8 regulate jasmonic acid biosynthesis and floral organ development via repression of class 1 KNOX genes. Plant Cell Physiol. 51, 164–175.

Tähtiharju, S., Rijpkema, A.S., Vetterli, A., Albert, V.A., Teeri, T.H., Elomaa, P., 2012. Evolution and diversification of the CYC/TB1 gene family in Asteraceae – a comparative study in Gerbera (Mutisieae) and sunflower (Heliantheae). Mol. Biol. Evol. 29, 1155–1166.

Takada, S., Hibara, K., Ishida, T., Tasaka, M., 2001. The CUP-SHAPED COTYLEDON1 gene of Arabidopsis regulates shoot apical meristem formation. Development 128, 1127–1135.

Takeda, T., Suwa, Y., Suzuki, M., Kitano, H., Ueguchi-Tanaka, M., Ashikari, M., Matsuoka, M., Ueguchi, C., 2003. The OsTB1 gene negatively regulates lateral branching in rice. Plant J. 33, 513–520.

Tanaka, K., Asami, T., Yoshida, S., Nakamura, Y., Matsuo, T., Okamoto, S., 2005. Brassinosteroid homeostasis in Arabidopsis is ensured by feedback expressions of multiple genes involved in its metabolism. Plant Physiol. 138, 1117–1125.

Tanaka, Y., Yamamura, T., Oshima, Y., Mitsuda, N., Koyama, T., Ohme-Takagi, M., Terakawa, T., 2011. Creating ruffled flower petals in Cyclamen persicum by expression of the chimeric cyclamen TCP repressor. Plant Biotechnol. 28, 141–147.

Tao, Q., Guo, D., Wei, B., Zhang, F., Pang, C., Jiang, H., Zhang, J., Wei, T., Gu, H., Qu, L.-J., Qin, G., 2013. The TIE1 transcriptional repressor links TCP transcription factors with TOPLESS/TOPLESS-RELATED corepressors and modulates leaf development in Arabidopsis. Plant Cell 25, 421–437.

Tatematsu, K., Ward, S., Leyser, O., Kamiya, Y., Nambara, E., 2005. Identification of cis-elements that regulate gene expression during initiation of axillary bud outgrowth in Arabidopsis. Plant Physiol. 138, 757–766.

To, J.P.C., Haberer, G., Ferreira, F.J., Deruère, J., Mason, M.G., Schaller, G.E., Alonso, J.M., Ecker, J.R., Kieber, J.J., 2004. Type-A Arabidopsis response regulators are partially redundant negative regulators of cytokinin signaling. Plant Cell 16, 658–671.

Uberti-Manassero, N.G., Lucero, L.E., Viola, I.L., Vegetti, A.C., Gonzalez, D.H., 2012. The class I protein AtTCP15 modulates plant development through a pathway that overlaps with the one affected by CIN-like TCP proteins. J. Exp. Bot. 63, 809–823.

Vernoux, T., Kronenberger, J., Grandjean, O., Laufs, P., Traas, J., 2000. PIN-FORMED 1 regulates cell fate at the periphery of the shoot apical meristem. Development 127, 5157–5165.

Vick, B.A., Zimmerman, D.C., 1983. The biosynthesis of jasmonic acid: a physiological role for plant lipoxygenase. Biochem. Biophys. Res. Commun., 470–477.

Viola, I.L., Uberti Manassero, N.G., Ripoll, R., Gonzalez, D.H., 2011. The Arabidopsis class I TCP transcription factor AtTCP11 is a developmental regulator with distinct DNA-binding properties due to the presence of a threonine residue at position 15 of the TCP domain. Plant J. 435, 143–155.

Wang, R.L., Stec, A., Hey, J., Lukens, L., Doebley, J., 1999. The limits of selection during maize domestication. Nature 398, 236–239.

Wang, Z.Y., Nakano, T., Gendron, J., He, J., Chen, M., Vafeados, D., Yang, Y., Fujioka, S., Yoshida, S., Asami, T., Chory, J., 2002. Nuclear-localized BZR1 mediates brassinosteroid-induced growth and feedback suppression of brassinosteroid biosynthesis. Dev. Cell 2, 505–513.

Wang, Z., Luo, Y., Li, X., Wang, L., Xu, S., Yang, J., Weng, L., Sato, S., Tabata, S., Ambrose, M., Rameau, C., Feng, X., Hu, X., Luo, D., 2008. Genetic control of floral zygomorphy in pea (Pisum sativum L.). Proc. Natl. Acad. Sci. USA 105, 10414–10419.

Wang, J., Wang, Y., Luo, D., 2010. LjCYC genes constitute floral dorsoventral asymmetry in Lotus japonicus. J. Integr. Plant Biol. 52, 959–970.

Weight, C., Parnham, D., Waites, R., 2008. LeafAnalyser: a computational method for rapid and large-scale analyses of leaf shape variation. Plant J. 53, 578–586.

Weijers, D., Benkova, E., Jäger, K.E., Schlereth, A., Hamann, T., Kientz, M., Wilmoth, J.C., Reed, J.W., Jürgens, G., 2005. Developmental specificity of auxin response by pairs of ARF and Aux/IAA transcriptional regulators. EMBO J. 24, 1874–1885.

Whipple, C.J., Kebrom, T.H., Weber, A.L., Yang, F., Hall, D., Meeley, R., Schmidt, R., Doebley, J., Brutnell, T.P., Jackson, D.P., 2011. Grassy Tillers1 promotes apical dominance in maize and responds to shade signals in the grasses. Proc. Natl. Acad. Sci. USA 108, E506–E512.

Xu, S., Luo, Y., Cai, Z., Cao, X., Hu, X., Yang, J., Luo, D., 2013. Functional diversity of CYCLOIDEA-like TCP genes in the control of zygomorphic flower development in Lotus japonicus. J. Integr. Plant Biol. 55, 221–231.

Yamada, H., Suzuki, T., Terada, K., Takei, K., Ishikawa, K., Miwa, K., Yamashino, T., Mizuno, T., 2001. The Arabidopsis AHK4 histidine kinase is a cytokinin-binding receptor that transduces cytokinin signals across the membrane. Plant Cell Physiol. 42, 1017–1023.

Yanai, O., Shani, E., Dolezal, K., Tarkowski, P., Sablowski, R., Sandberg, G., Samach, A., Ori, N., 2005. Arabidopsis KNOXI proteins activate cytokinin biosynthesis. Curr. Biol. 15, 1566–1571.

Yanai, O., Shani, E., Russ, D., Ori, N., 2011. Gibberellin partly mediates LANCEOLATE activity in tomato. Plant J. 68, 571–582.

Yin, Y., Wang, Z.Y., Mora-Garcia, S., Li, J., Yoshida, S., Asami, T., Chory, J., 2002. BES1 accumulates in the nucleus in response to brassinosteroids to regulate gene expression and promote stem elongation. Cell 109, 181–191.

Yuan, Z., Gao, S., Xue, D.W., Luo, D., Li, L.T., Ding, S.Y., Yao, X., Wilson, Z.A., Qian, Q., Zhang, D.B., 2009. RETARDED PALEA1 controls palea development and floral zygomorphy in rice. Plant Physiol. 149, 235–244.

CHAPTER 17

Growth-Regulating Factors, A Transcription Factor Family Regulating More than Just Plant Growth

Ramiro E. Rodriguez, María Florencia Ercoli, Juan Manuel Debernardi, Javier F. Palatnik

Instituto de Biología Molecular y Celular de Rosario (IBR), CONICET, Facultad de Ciencias Bioquímicas y Farmacéuticas, Universidad Nacional de Rosario, Rosario, Argentina

OUTLINE

17.1 GROWTH-REGULATING FACTORs, a Plant-specific Family of Transcription Factors	269
17.1.1 Seminal Identification of the GROWTH-REGULATING FACTORs	270
17.1.2 Functional Protein Domains	270
17.1.2.1 DNA and Protein–Protein Interaction Domains in the N-terminal Region	270
17.1.2.2 A C-terminal Region with Transcriptional Activation Activity	271
17.2 Control of GRF Activity	272
17.2.1 Transcriptional Regulation of GRF Expression	272
17.2.2 GRF-INTERACTING FACTORs	273
17.2.3 Repression of GRF Expression by MicroRNAs	273
17.3 Role of GRFs in Organ Growth and Other Developmental Processes	274
17.3.1 Control of Leaf Development by the GRFs	274
17.3.2 GRFs in the Development and Growth of Other Plant Structures	275
17.3.2.1 Embryo and Shoot Apical Meristem	275
17.3.2.2 Root Development	276
17.3.2.3 Reproductive Development	276
17.3.3 Beyond Organ Growth	277
17.3.3.1 GRFs are Involved in Developmental Plasticity and Reprogramming in Response to External Cues	277
17.3.3.2 GRFs and Leaf Senescence	277
17.4 Conclusion and Perspectives	277
References	278

17.1 GROWTH-REGULATING FACTORs, A PLANT-SPECIFIC FAMILY OF TRANSCRIPTION FACTORS

Plant organs develop from meristems, which are regions with limited cell differentiation and active cell proliferation. Many different types of meristems exist in plants. For example, the apical meristems are responsible for the primary growth of plants (Heidstra and Sabatini, 2014), while the vascular cambium, a lateral secondary meristem, generates the secondary phloem and xylem (wood) (Zhang et al., 2014).

Groups of cells often exit meristems to originate at their flanks the primordia of lateral organs, such as the leaves and flowers that develop from the vegetative and reproductive shoot apical meristem (SAM), respectively.

Importantly, once the primordia are initiated, extensive cell proliferation increases the size and determines the shape of the organ. Finally, cell proliferation ceases, and the final size, shape, and function are achieved by cell expansion and differentiation.

Numerous regulatory networks have been implicated in the establishment and maintenance of meristems and in promoting cell proliferation of developing organ primordia, including the GROWTH-REGULATING FACTOR (GRF) family of transcription factors (TFs) described in this chapter.

17.1.1 Seminal Identification of the GROWTH-REGULATING FACTORs

Oryza sativa GROWTH-REGULATING FACTOR1 (OsGRF1) is the founding member of the GROWTH-REGULATING FACTOR family of TFs. It was identified as one of the genes induced in deep-water rice (*O. sativa* L.) internodes in response to gibberellins (GA; Van Der Knaap et al., 2000). This plant hormone promotes the growth of rice stems after flooding by activating cell proliferation in the intercalary meristem (Kende et al., 1998), a region at the base of internodes where new cells are produced. OsGRF1 was detected in regions of the rice plant that contain meristematic tissues and was induced just before cell proliferation genes in response to GA, suggesting that this gene stimulates growth and mitosis, hence prompting this gene family's name.

Sequence analysis of OsGRF1 homologs highlighted two highly conserved protein domains: the QLQ (Gln, Leu, Gln), involved in protein–protein interactions, and the WRC (Trp, Arg, Cys), containing a putative nuclear localization signal and a DNA-binding domain (Figure 17.1A and C). It was also shown that *Arabidopsis thaliana* GRFs (AtGRFs) have transcriptional transactivation activity in yeast, further confirming their role as TFs (Kim and Kende, 2004).

Characterization of the GRF gene family in *Arabidopsis* revealed nine members (Figure 17.1A) that contain the same characteristic regions of OsGRF1: the QLQ and WRC domains (Kim et al., 2003). AtGRF genes were detected in actively growing and developing tissues, such as vegetative shoot apices, inflorescences, and roots tips. Triple insertional null mutants of AtGRF1, AtGRF2, and AtGRF3 had smaller leaves (Figure 17.1B) confirming that GRFs promote organ growth. Further analysis of *grf* mutants indicated that their smaller leaves contain fewer cells, which implicated a link between GRFs and the control of cell proliferation (Horiguchi et al., 2005; Kim and Kende, 2004; Kim et al., 2003).

The GRF family of TFs has not only been found in all land plant genomes, including representatives of angiosperms and gymnosperms, but also in the moss *Physcomitrella patens* and the lycophyte *Selaginella moellendorffii*. So far, no GRF has been identified in the genomes of chlorophytes or charophytes, or in species from other kingdoms, indicating that this family of TFs is specific of embryophytes, the more complex multicellular members of the Plantae kingdom.

17.1.2 Functional Protein Domains

The GRF family of TFs is defined by the presence of two conserved regions (Figure 17.1A and C) in the N-terminal region: the WRC and the QLQ domains. Moreover, a C-terminal region with limited sequence conservation among GRFs plays a role in transcriptional modulation.

17.1.2.1 DNA and Protein–Protein Interaction Domains in the N-terminal Region

The WRC domain is named after a conserved Trp-Arg-Cys motif (Figure 17.1C). It has two distinctive features: a putative nuclear localization signal and a zinc (Zn) finger motif involved in DNA binding. Nuclear localization has been experimentally confirmed for several GRFs (Kim et al., 2012; Van Der Knaap et al., 2000; Liang et al., 2014; Osnato et al., 2010), and a bipartite nuclear localization signal has been predicted inside the WRC domain (Van Der Knaap et al., 2000).

The DNA-binding activity of the WRC domain was initially proposed based on the presence of a putative C3H Zn finger motif consisting in the conserved spacing of three Cys and one His with the sequence $CX_9CX_{10}CX_2H$ (Figure 17.1C). The amino acid sequence of the WRC domain and the spacing of the Cys and His residues do not fit into any of the more common classes of Zn fingers. However, this unusual motif was found in three repeats in the C-terminal region of HORDEUM REPRESSOR OF TRANSCRIPTION (HRT; Figure 17.1C), a transcriptional repressor from barley which binds to GA-responsive elements (GAREs; Raventos et al., 1998). The biochemical characterization of the C3H motif from HRT showed that the conserved Cys and His are essential for Zn^{2+} and DNA binding (Raventos et al., 1998).

No systematic analysis to determine the DNA-binding sites of GRFs has been performed yet, neither the biochemical nor the structural properties of the DNA-binding activity was analyzed. However, *in vivo* and *in vitro* DNA-binding activities have been detected for particular GRFs, which in all cases were mediated by the WRC domain (Liu et al., 2014; Kim et al., 2012; Kuijt et al., 2014; Osnato et al., 2010).

The QLQ domain is named for a conserved Gln-Leu-Gln motif in the N-terminal region of GRFs (Figure 17.1). It is similar to the N-terminal region of the yeast SWITCH2/SUCROSE NONFERMENTING2 (SWI2/SNF2) adenosine triphosphatases (ATPases) involved in chromatin-remodeling processes and mediates the interaction

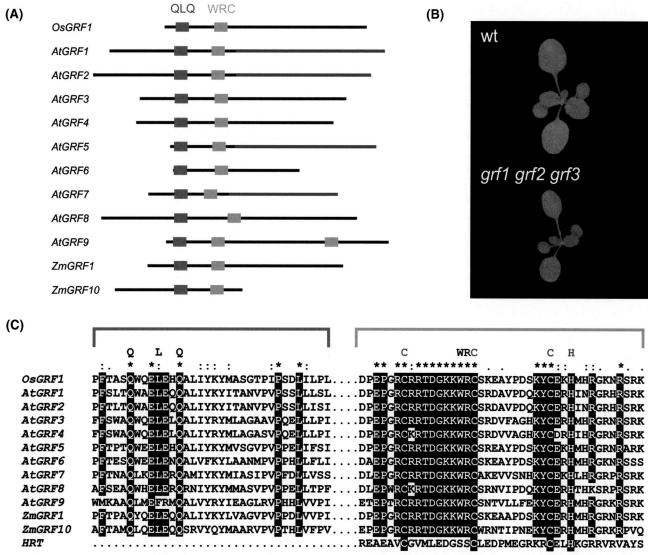

FIGURE 17.1 **The plant-specific GRF family of TFs.** (A) QLQ and WRC domain organization in representative GRFs from *A. thaliana* (AtGRF), *O. sativa* (OsGRF), and *Zea mays* (ZmGRF). Red indicates the C-terminal regions from AtGRF1, AtGRF2, AtGRF5, and AtGRF7, which display transcriptional transactivation activity in yeast (Kim et al., 2003; Kim et al., 2012). (B) The *Arabidopsis* triple mutant grf1–grf2–grf3 has small leaves. (C) Multiple sequence alignments of the QLQ and WRC domains. Red indicates the conserved residues of the C3H Zn finger motif present in the WRC and in the barley *HRT* transcriptional repressor.

between GRFs and a family of transcriptional coactivators described in Section 17.2.2.

17.1.2.2 A C-terminal Region with Transcriptional Activation Activity

The C-terminal region of *GRFs* is more divergent in sequence and size among the members of this TF family. It is rich in Pro, Gln, His, Ala/Gly, and Ser/Thr, amino acids frequently found in the transcriptional activation domains of other TFs (Triezenberg, 1995). Indeed, deletion studies with several *GRFs* from *Arabidopsis* and rice indicated that the C–terminal region (Figure 17.1A) is responsible for transcriptional transactivation activity in yeast or *Arabidopsis* protoplasts (Choi et al., 2004; Kim and Kende, 2004; Liu et al., 2014).

Maize *ZmGRF10* is a naturally occurring version of *GRF* that lacks the C-terminal region (Figure 17.1A) but has functional QLQ and WRC domains. It has no transcriptional transactivation activity in yeast, and its overexpression in maize restricted leaf size by decreasing cell proliferation (Wu et al., 2014).

GRFs are usually regarded as transcriptional activators, although several examples of GRFs acting as repressors have been described (Osnato et al., 2010; Kuijt et al., 2014; Kim et al., 2012; Wu et al., 2014). For example, of the nine *AtGRFs*, only GRF7 has the capacity to bind *in vitro* and *in vivo* to the promoter of the stress-responsive gene *DREB2A* to repress its expression, and this activity depends on the WRC and QLQ domains, together with unidentified sequences present in the C-terminal region (Kim et al., 2012).

17.2 CONTROL OF GRF ACTIVITY

17.2.1 Transcriptional Regulation of GRF Expression

Analysis of *GRF* gene expression in various species has consistently shown that the highest levels occur in developing tissues with active cell proliferation (Bazin et al., 2013; Choi et al., 2004; Debernardi et al., 2012, 2014; Horiguchi et al., 2005; Kim et al., 2003; Van Der Knaap et al., 2000; Zhang et al., 2008). Analysis of several plant species in the gene expression atlas, such as *Arabidopsis* (Schmid et al., 2005) and rice (Wang et al., 2010), also shows that the highest levels of *GRF* transcripts are detected in tissues with the greatest expression levels of mitotic specific genes (Figure 17.2A).

The development of techniques involving chromatin immunoprecipitation followed by next-generation sequencing (ChIP-Seq) has allowed identification of the binding sites of several plant TFs at the genome level. Through these approaches, several putative upstream regulators of *GRFs* have been identified.

The promoters of *Arabidopsis GRF8* and *GRF1* were identified by ChIP–Chip from inflorescence tissues as genomic *loci*-containing regions bound by *APETALA2* (*AP2*) and *SCHLAFMUTZE* (*SMZ*), two TFs involved in the transition to flowering and in the specification of floral organ identity (Yant et al., 2010).

Furthermore, *GRF5* was identified as a target gene of the plant-specific transcription factor *LEAFY* (*LFY*), which is a master regulator of flowering and directs floral organ patterning (Winter et al., 2011).

JAGGED is a C2H2 Zn finger transcription factor that stimulates vegetative and reproductive organ growth in *Arabidopsis* (Ohno et al., 2004; Schiessl et al., 2012). ChIP-Seq analyses in *Arabidopsis* inflorescences detected *JAGGED* binding sites in *GRF4*, -5, -6, -8, and -9, but only *GRF8* seems to be a direct target of *JAGGED* according to transcriptome studies (Schiessl et al., 2014).

Finally, *APETALA1* (*AP1*) and *SEPALLATA3* (*SEP3*), two MADS-box genes that regulate floral meristem identity and behave as homeotic regulators of sepal and petal identity, were shown to bind regulatory sequences of several *Arabidopsis GRFs* in different stages of flower

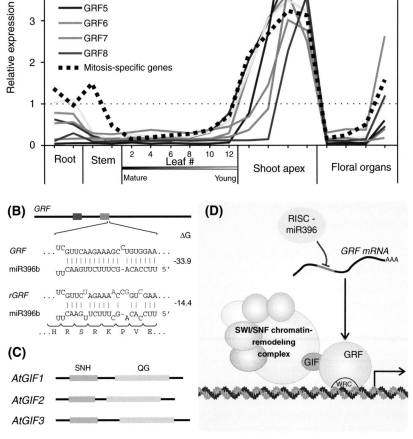

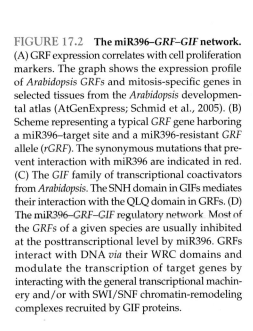

FIGURE 17.2 **The miR396–GRF–GIF network.** (A) GRF expression correlates with cell proliferation markers. The graph shows the expression profile of *Arabidopsis GRFs* and mitosis-specific genes in selected tissues from the *Arabidopsis* developmental atlas (AtGenExpress; Schmid et al., 2005). (B) Scheme representing a typical *GRF* gene harboring a miR396–target site and a miR396-resistant *GRF* allele (*rGRF*). The synonymous mutations that prevent interaction with miR396 are indicated in red. (C) The *GIF* family of transcriptional coactivators from *Arabidopsis*. The SNH domain in GIFs mediates their interaction with the QLQ domain in GRFs. (D) The miR396–*GRF*–*GIF* regulatory network. Most of the *GRFs* of a given species are usually inhibited at the posttranscriptional level by miR396. GRFs interact with DNA *via* their WRC domains and modulate the transcription of target genes by interacting with the general transcriptional machinery and/or with SWI/SNF chromatin-remodeling complexes recruited by GIF proteins.

development. The binding was further supported by chromatin accessibility and transcriptome studies (Pajoro et al., 2014). The observation of similar flower phenotypes in plants defective in *AP1* or *GRFs* supported the biological relevance of this regulation, suggesting a role played by *GRF* downstream of the floral modulators *AP1* and *SEP3*.

It is important to take into account that these approaches identify transcriptional target candidates for the particular TF analyzed, but the biological relevance of the putative regulation needs to be further assessed using biochemical and genetic approaches. ChIP-Seq experiments have not been reported for GRFs themselves, something that would be interesting to see in the future.

17.2.2 GRF-INTERACTING FACTORs

TFs with DNA-binding domains often cooperate with transcriptional coactivators to promote transcription by stimulating the assembly of the basal transcription apparatus or by recruiting chromatin remodelers (Lemon and Tjian, 2000).

In a search for partner proteins of AtGRF1, a small gene family of transcriptional coactivators, named the *GRF–INTERACTING FACTORs* (*GIFs*), was identified. The *Arabidopsis GIF* family has three members. *AtGIF1*, also known as *ANGUSTIFOLIA3* (*AN3*), along with *AtGIF2* and *AtGIF3* play key roles in *Arabidopsis* development (Horiguchi et al., 2005; Ferjani et al., 2007; Kim and Kende, 2004; Lee et al., 2009, 2014). The classical *an3/gif1* mutant has similar reductions in leaf size as those found in *grf* mutants, and combinations of *gif* and *grf* mutations show cooperative effects (Kim and Kende, 2004; Horiguchi et al., 2005).

The N-terminal region of GIF proteins is homologous to the synovial translocation N-terminal (SNH) domain of the human SYNOVIAL TRANSLOCATION (SYT) coactivator (Figure 17.2C; Kim et al., 2003; Horiguchi et al., 2005), which was shown to interact with human SWI/SNF chromatin-remodeling complexes. In addition, regions rich in Gln and Gly are found in the C-terminal portion of GIF proteins (Figure 17.2C).

AtGIF1 has been demonstrated to physically interact with AtGRF1, AtGRF2, AtGRF4, AtGRF5, AtGRF7, and AtGRF9 *in vitro* and in yeast two-hybrid (Y2H) studies (Kim and Kende, 2004; Kim et al., 2012). In particular, the interaction between AtGIF1 and AtGRF5 and AtGRF3 was confirmed *in vivo* by coimmunoprecipitation (Co-IP) followed by mass spectrometry analysis in *Arabidopsis* inflorescences (Debernardi et al., 2014). Biochemical and truncation analyses indicated that the formation of the GRF–GIF complex relies on the QLQ and SNH domains (Kim et al., 2003), and similar GRF–GIF complexes have been described in rice (Liu et al., 2014) and maize (Wu et al., 2014).

GIF proteins are involved in transcriptional transactivation activity in yeast *per se* (Kim and Kende, 2004; Liu et al., 2014) but, most interestingly, they have the capacity to enhance the transcriptional modulating capacities of *Arabidopsis* and rice GRFs (Liu et al., 2014; Kim et al., 2012). Insights into the molecular mechanisms by which GIFs exert this action came from Co–IP experiments in which plant SWI/SNF complexes associated with AtGIF1 could be reconstructed (Vercruyssen et al., 2014; Debernardi et al., 2014). Altogether, this indicates that GIF1 forms a bridge between SWI/SNF complexes and GRFs which helps modify the chromatin structure for efficient transcription (Figure 17.2D).

17.2.3 Repression of GRF Expression by MicroRNAs

Small RNAs are key regulators of gene expression in animals and plants (Bologna and Voinnet, 2014). MicroRNAs (miRNAs) are a class of small RNAs defined by their biogenesis pathway, which requires cleavage of a noncoding RNA harboring a fold-back precursor by the type III ribonuclease called DICER-LIKE1 (DCL1; Meyers et al., 2008). MiRNAs are around 21 nucleotides (nts) long and function in the context of an RNA-induced silencing complex (RISC) containing an ARGONAUTE (AGO) protein, generally AGO1 (Mallory et al., 2008). The small RNA guides these complexes to their binding sites in target RNAs by means of base complementarity and represses their expression at the posttranscriptional level (Bologna and Voinnet, 2014).

Arabidopsis contains more than 200 *MIRNA* genes in its genome, including members of the 21 families of miRNAs deeply conserved in angiosperms. MiRNA miR396 is one of these conserved miRNAs and has been detected in angiosperms and gymnosperms (Debernardi et al., 2012). miR396 has binding sites in seven of the nine *Arabidopsis GRFs* (Jones-Rhoades and Bartel, 2004; Figure 17.2B). miR396 binding sites are located in the coding sequence of GRFs (Figure 17.2B), in the C-terminal end of the WRC domain, and can be found in *GRFs* from angiosperms and gymnosperms (Debernardi et al., 2012). The interaction between miR396 and *GRFs* has a bulge between positions 7 and 8 from the 5′ end of the miRNA (Figure 17.2B). This bulge dampens the activity of the miRNA suggesting that miR396 is a quantitative regulator of *GRF* expression (Debernardi et al., 2012). Interestingly, monocots have an additional miR396 variant that represses GRFs more efficiently thanks to an insertion between positions 7 and 8 (Debernardi et al., 2012).

The miR396-mediated repression of *GRFs* provides an excellent means of manipulating their levels. For example, miR396 over- or missexpression reduces the expression levels of target *GRFs* (Debernardi et al., 2012; Liang et al., 2014; Rodriguez et al., 2010; Liu et al., 2009; Pajoro et al., 2014). In this way, weak 35S:*miR396* plants are similar to the triple *grf1–grf2–grf3* mutant or *grf5*, circumventing the difficulties posed by the functional redundancy of these TFs in their functional analyses. In any event, phenotypes observed after overexpression of miR396 need to be taken cautiously due to the existence of additional miR396 targets that are different from *GRFs* (Debernardi et al., 2012; Chorostecki et al., 2012).

On the other hand, *GRF* activity can be increased either by overexpression from strong constitutive promoters (Horiguchi et al., 2005; Kim and Kende, 2004), or by decreasing the activity of miR396. The repression caused by a miRNA can be eliminated by introducing silent mutations in the target sequence which abolish interaction with the small RNA, generating miRNA-resistant alleles (Figure 17.2B). Introducing miRNA–resistant *GRFs* into plants increases leaf size causing, as expected, opposite effects to the overexpression of miR396 (Section 17.3).

In summary, the activity of *GRFs* is determined by a combination of different regulatory layers: transcriptional regulation, posttranscriptional repression by miR396, and stimulation of GRF activity by GIF coregulators. These elements define a miR396–*GRF*–*GIF* network (Figure 17.2D; Debernardi et al., 2014; Rubio-Somoza et al., 2009), which controls the several developmental processes described in Section 17.3.

17.3 ROLE OF *GRFs* IN ORGAN GROWTH AND OTHER DEVELOPMENTAL PROCESSES

17.3.1 Control of Leaf Development by the GRFs

In dicotyledonous plants, leaf primordia are initiated at the peripheral zone of the vegetative SAM (Figure 17.3A). Initially, they have a rod-like structure that soon takes on dorsoventral polarity to constitute a flat lamina consisting of two anatomically distinct surfaces: the adaxial and abaxial sides (reviewed in Rodriguez et al., 2014). In the next step, the leaf lamina expands to acquire its final size and shape. First, cell proliferation occurs throughout the small-leaf primordium, which can be visualized using reporters of mitotic cyclins such as CYCLINB1;1-GUS (Donnelly et al., 1999; Figure 17.3B). As the organ grows, the region containing proliferative cells becomes restricted to the base of the organ, and the cells located in the distal part of the leaf begin their expansion (Figure 17.3B). Finally, cell proliferation ceases and the leaf lamina continues to grow only by cell expansion. A similar spatial organization of the processes contributing to organ growth occurs in monocotyledonous plants. In a growing maize leaf, cells divide only in the division zone at the base of the leaf, while they expand at the more distally located expansion zone. In both dicots and monocots, the size and persistence of the division zone is one of the main

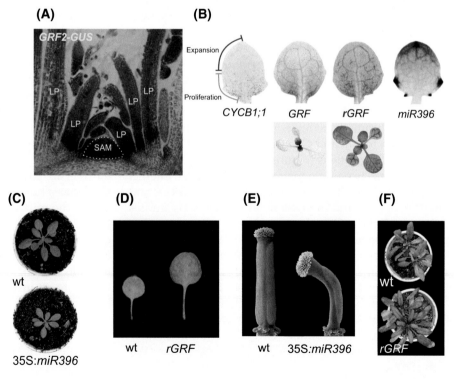

FIGURE 17.3 **Role of GRFs in plant development.** (A) Architecture of the vegetative SAM in longitudinal cross-sections of plants expressing a GUS reporter of *AtGRF2*. Leaf primordia (LP) originate at the flanks of the SAM. (B) Expression pattern of a mitotic cyclin (*CYCB1;1*), *GRF2*, miR396-resistant *GRF2* (*rGRF2*), and miR396 in developing leaves. Note that cell proliferation and *GRF2* expression are restricted to the proximal part of the leaf, while cell expansion and miR396 expression occur in the distal part. Adapted with permission from Debernardi et al. (2012). Copyright 2012 PLOS. (C) Rosette phenotype of 18-day-old plants overexpressing miR396. 35S:*miR396* plants have reduced *GRF* levels. (D) Phenotypes of plants expressing a miR396-resistant *GRF* (*rGRF*) transgene. Note the large leaves without major changes in shape. (E) Pistil defects in plants overexpressing miR396. The carpels are shown in green. (F) Delayed senescence observed in *rGRF* plants.

factors determining final leaf size (Nelissen et al., 2012; Rodriguez et al., 2010; Ferjani et al., 2007; Vercruyssen et al., 2014).

Arabidopsis mutants in *AtGRFs* have small leaves with fewer cells (Kim et al., 2003, 2012; Kim and Kende, 2004; Figure 17.1B). Additionally, similar phenotypes are found in plants overexpressing miR396 (Rodriguez et al., 2010; Liu et al., 2009; Figure 17.3C) or in *gif1/an3* mutants. The leaf size seen in *gif1* is significantly reduced as a result of combination with *gif2* or *gif3* mutants (Lee et al., 2009). Detailed analysis of these mutants indicated that cell proliferation is prematurely terminated (Rodriguez et al., 2010; Ferjani et al., 2007).

On the other hand, increasing *GRF* levels by overexpression of *GRF1*, *GRF2*, or *GRF5* under strong promoters (Horiguchi et al., 2005; Kim et al., 2003; Gonzalez et al., 2010), or by introducing miR396-resistant alleles of *GRF2* or *GRF3* (Rodriguez et al., 2010; Debernardi et al., 2014), produced bigger leaves with more cells (Figure 17.3D). Overexpression of *GIF1* also increased leaf size (Kim and Kende, 2004) and, in agreement with its function as a coactivator, synergistically boosted the effect of *rGRF3* and *GRF5* on leaf size (Debernardi et al., 2014).

In young developing leaves, miR396 and *GRFs* have opposing gradients of expression. MiR396 is expressed in a gradient along the longitudinal axis of the organ, with higher expression at the distal part (Figure 17.3B). In turn, miR396 represses *GRFs*, generating an opposing gradient of expression with higher levels in the proximal part of the organ (Figure 17.3B). Posttranscriptional regulation of *GRFs* by miR396 results in the expression of *GRFs* in cells undergoing mitosis (Debernardi et al., 2012; Rodriguez et al., 2010). Expression of *AtGIFs* is also restricted to the proximal part of leaf primordia, as expected from their function as GRFs coactivators (Ha Lee and Hoe Kim, 2014).

Similar defects in leaves and comparable *GRF*, *GIF*, and miR396 expression patterns in relation to cell proliferation were obtained in monocotyledonous plants (Liu et al., 2014; Candaele et al., 2014), indicating a conserved function for *GRFs* in leaf development across plant species.

Altogether these results indicate that the miR396–*GRF*–*GIF* network controls leaf organ growth after primordia initiation. In developing leaves, GRF–GIF protein complexes promote cell proliferation in the proximal part of the leaf, while miR396 restricts the activity of the complexes by posttranscriptionally repressing *GRF* expression in the distal part, where cell expansion and differentiation occur.

Overexpression of miR396 in sensitized backgrounds such as *rdr6* or *as1/as2* brought about plants with lotus-like or needle-like leaves with loss of their adaxial side (Mecchia et al., 2013; Wang et al., 2011a). These phenotypes were partially rescued by a miR396-insensitive allele of *AtGRF9* (Wang et al., 2011a). Moreover, *AtGRF2* expression was preferentially detected in the adaxial side of the leaves (Mecchia et al., 2013). The results suggested that the miR396–*GRF*–*GIF* network might also play a role in dorsoventral polarity of the organ after primordia initiation.

17.3.2 GRFs in the Development and Growth of Other Plant Structures

17.3.2.1 Embryo and Shoot Apical Meristem

Several lines of evidence indicate that *GRFs* are required not only for leaf primordium growth, but also for the development and maintenance of the SAM from which they originate.

Together with miR396, *Arabidopsis GRF2* is detected at the RNA and protein levels in the vegetative SAM of *Arabidopsis* plants (Figure 17.3A). MiR396 overexpression caused a decrease in the size of the SAM, while the opposite effect was observed in plants with a miR396-insensitive allele of *GRF2* (Rodriguez et al., 2010). These changes correlated with the cell proliferation activity of the SAM, suggesting that GRFs promote cell proliferation not only in organ primordia, but also in meristems. Plants with a further decrease in *GRF* activity, achieved by overexpression of miR396 in *gif1* mutants (Rodriguez et al., 2010) or in quadruple *grf1*–*grf2*–*grf3*–*grf4* mutants (Kim and Lee, 2006), lacked a functional SAM and displayed fused cotyledons that resulted from defects in embryo development. Finally, the size of the SAM in seedlings correlated with the dosage of GIFs (Lee et al., 2009). Altogether, these results indicate that the miR396–*GRF*–*GIF* network regulates embryo development, SAM establishment and/or maintenance, and determines the size of the SAM by controlling cell proliferation.

Class I Knotted1-like homeobox (KNOXI) genes are essential to SAM formation and maintenance and important for vegetative and reproductive morphogenetic processes as well (Hake et al., 2004; Chapter 14). A small number of these genes have been shown to be regulated by GRFs.

For example, barley *GROWTH-REGULATING FACTOR1* (*BGRF1*) binds *in vitro* to an unidentified sequence in a regulatory intron of *BKN3*, a KNOXI gene involved in barley floral patterning (Osnato et al., 2010). The binding of *BGRF1* to the regulatory intron was confirmed *in vivo* in rice protoplasts, and a repressing activity for GRFs on KNOXI genes has been proposed (Kuijt et al., 2014).

The rice KNOXI genes *OsKN2* and *OsKN3* are expressed in the SAM and downregulated in developing organ primordia (Postma-Haarsma et al., 2002). Using the promoter of *OsKN2* as bait in yeast one-hybrid (Y1H) screening, *OSGRF3* and *OSGRF10* were identified (Kuijt et al., 2014). When *GRF* levels were manipulated in transgenic rice plants, an inverse relationship between *OSGRF* and KNOXI gene expression was found, indicating once again that *GRFs* may act as direct repressors of

KNOXI genes in rice. The exact sequences of barley or rice GRF-binding sites to these KNOXI genes was not identified, although it was proposed that GRF binding is associated with the presence of CAG or CTG repeats (Kuijt et al., 2014).

17.3.2.2 Root Development

In plants, root growth is sustained via coordinated cell division and expansion at the root tip. Cell proliferation occurs in the root apical meristem (RAM), located in the more distal part of the organ, while cell elongation and differentiation take place above this meristematic zone (Petricka et al., 2012). Six *GRFs* were identified in the model legume *Medicago truncatula*, all of which possessed a miR396-binding site (Bazin et al., 2013). *MtGRFs* and miR396 are expressed in opposite gradients in root tips, suggesting that miR396 may restrict GRF expression to regions that have active cell proliferation of the RAM, as has been observed in *Arabidopsis* leaves (Rodriguez et al., 2010). Repression of *MtGRFs* by overexpression of miR396 or by GRF-specific RNAi caused a reduction in primary root length due to a reduction in the size and activity of the meristem, together with a reduction in the expression of cell proliferation markers (Bazin et al., 2013).

Conflicting results were obtained when the function of the miR396–GRF–GIF network was analyzed in *Arabidopsis* (Hewezi et al., 2012). Both, downregulation of *GRFs* in plants overexpressing miR396 and overexpression of *AtGRF1* or *AtGRF3* resulted in short roots, indicating that a precise balance between miR396 and *GRF* expression in Brassicaceae is needed for root development.

17.3.2.3 Reproductive Development

The reproductive organs in the Poaceae (grasses) family are the basic units determining grain yield in cereal crops. Spikelets are the basic inflorescence units in rice and typically consist of a flower of interlocked lemma and palea forming a husk, two lodicules, six stamens, and one pistil. Both overexpression of miR396, with the resulting downregulation of its target genes, and *osgrf6–osgrf10* double-mutants, exhibited abnormal spikelets with open husks, long sterile lemmas, and anomalous numbers of pistils and stamens (Liu et al., 2014). In particular, the open husks were suggested to be a result of lemma and palea not growing normally enough to reach each other.

Part of the function of *OsGRF6* and *OsGRF10* in rice flower development could be explained by direct transcriptional regulation of the rice jumonji domain 2 (JM-JD2) family *jmjC* gene 706, which encodes an H3K9 demethylase (*OsJMJ706*; Sun and Zhou, 2008) and *O. sativa* crinkly4 receptor-like kinase (*OsCR4*; Pu et al., 2012), two genes needed for husk integrity and flower organ identity and number. This transcriptional regulation seems to be mediated by the direct binding of OsGRF6 or OsGRF10 to GA-responsive elements (TAACARA, R = G or A) present in *OsJMJ706* and *OsCR4* promoters, as demonstrated *in vitro* by electrophoretic mobility shift assay (EMSA) and *in vivo* by ChIP or reporter assays in *Arabidopsis* protoplasts. Interestingly, the authors also showed that the interaction with OsGIFs enhances the transcriptional activity of both GRF6 and GRF10 on *OsJMJ706* and *OsCR4* promoters.

The rice mutant *rdh1* with a modified heading date (equivalent to flowering time) was shown to have reduced levels of *OsGRF1* (Luo et al., 2005). Targeted downregulation of *OsGRF1* by RNA interference resulted in small leaves and a delay in flowering, indicating that *OsGRF* is involved in regulating not only organ growth and development at the vegetative and reproductive phase, but also may be involved in flowering time regulation in rice (Luo et al., 2005).

It is worth noting that some of the TFs identified so far, which appear to regulate *GRF* transcription in *Arabidopsis* are involved in floral meristem identity and flower patterning (Pajoro et al., 2014; Schiessl et al., 2014; Winter et al., 2011; Yant et al., 2010). Surprisingly, no flower patterning phenotypes have been observed in any of the single or multiple *GRF* mutants analyzed so far in this plant (Liang et al., 2014; Kim et al., 2003). In any event, two lines of evidence support a role for *GRFs* in dicotyledonous flower development.

First, the *Arabidopsis gif1–gif2–gif3* triple-mutant has defects in flower development, including reduced numbers of organs in each whorl, small sepals and petals, unfused or missing carpels, short ovule integuments, defective gametogenesis and organs with mosaic identity, among others (Lee et al., 2009, 2014; Liang et al., 2014). Second, the overexpression of miR396 produced similar phenotypes in *Arabidopsis* (Pajoro et al., 2014; Liang et al., 2014; Figure 17.3E) and tobacco (Yang et al., 2009), which were fully complemented by overexpression of a miR396-insensitive *GRF* (Liang et al., 2014).

Interestingly, some of the phenotypes observed in monocot and dicot plants that had defects in GRF–GIF complexes, such as open husks in rice or small petals and short integuments in *Arabidopsis*, have been interpreted as a result of a defect in cell proliferation on already established organ primordia. Instead, other phenotypes, such as the mosaic organs and defective organ number in each whorl, highlight a role for the miR396–GRF–GIF network in flower organ patterning and specification.

The final output of reproductive development is the seed, and its size is a major determinant of crop yield. Gene expression analyses in different lines of rapeseed (*Brassica napus*) identified a positive correlation between oil content and *BnGRF2* expression (Liu et al., 2012). Heterologous overexpression of *BnGRF2* in *Arabidopsis* from a seed-specific promoter increased seed size and oil content ca. 30% due to a higher number of cells in the embryo without affecting its structure or cell size (Liu et al., 2012).

17.3.3 Beyond Organ Growth

17.3.3.1 GRFs are Involved in Developmental Plasticity and Reprogramming in Response to External Cues

Owing to the challenges imposed by their sessile lifestyle, plants regulate their development to adapt to external conditions. The resulting developmental plasticity is exemplified by modifications in leaf size and shape in response to a variety of environmental factors, such as light quality and quantity (Kim et al., 2005), water availability (Skirycz et al., 2011), and wounding (Zhang and Turner, 2008).

In particular, inhibition of leaf growth is a typical response to solar UV-B radiation in many plant species, a reduction that can be caused by modifications in cell expansion and/or proliferation. It has been shown that miR396 is prematurely induced by UV-B in developing leaves with active cell proliferation (Casadevall et al., 2013). This causes repression of *GRF* expression, which decreases cell production, leading to smaller leaves with fewer cells.

miR396 has been reported to be regulated by other abiotic stresses, such as drought (Zhou et al., 2010; Kantar et al., 2011; Li et al., 2011; Wang et al., 2011b), aluminum (Chen et al., 2012), cadmium (Ding et al., 2011), salt and alkali stress (Ding et al., 2009; Gao et al., 2010). As already demonstrated for UV-B stress, it would be interesting to confirm whether miR396-mediated repression of *GRFs* underlies the response of plants to these stresses.

Pathogens modify their host's biology to succeed during infections. One example is the developmental reprogramming of differentiated root vascular cells into a feeding syncytium brought about by cyst nematodes, obligated parasitic roundworms that cause great losses in crop production. The initial phase of syncytium development requires miR396 downregulation and *AtGRF1* and *AtGRF3* induction in differentiated nonproliferating tissues, suggesting that modulation of the miR396–GRF–GIF network by the parasite is essential for successful infection (Hewezi et al., 2012).

DEHYDRATION-RESPONSIVE ELEMENT-BINDING PROTEIN2A (DREB2A) is a TF that increases tolerance to abiotic stress by functioning as a transcriptional activator of a number of genes involved in osmotic and heat stress tolerance. Concomitantly, it affects plant growth and reproduction, suggesting that, under normal conditions, a mechanism repressing its activity must exist. GRF7 was demonstrated to interact *in vivo* with the promoter of *DREB2A* and repress its expression under normal conditions (Kim et al., 2012). *Arabidopsis GRF7* mutants have increased resistance to osmotic and drought stress and higher expression levels of stress-responsive genes, including *DREB2A*. With this in mind, it has been proposed that *AtGRF7* acts, under nonstress conditions, as a suppressor of *DREB2A* and other stress-responsive genes that might inhibit plant growth. This transcriptional repression was suggested to be mediated by direct binding to a GRF7-targeting *cis*-element (TGTCAGG) present in the promoter of *DREB2A* and other stress-responsive genes.

Of the nine GRFs present in the *Arabidopsis* genome, only GRF7 has the capacity to bind to the promoter of the stress-responsive gene *DREB2A* (Kim et al., 2012). The WRC domain was found to be essential for the binding in Y1H experiments of GRF7 to the target sequence TGTCAGG present in this promoter. Moreover, an unidentified protein sequence in the C-terminal region of GRF7 is also needed. This interaction was confirmed *in vivo* using transient expression in *Arabidopsis* protoplasts and ChIP-qPCR (quantitative polymerase chain reaction) in stable GRF7-overexpressing transgenic plants.

17.3.3.2 GRFs and Leaf Senescence

The last developmental stage of a leaf is senescence, a complex and active program that recycles its resources and relocates nutrients to seeds (Lim et al., 2007; Breeze et al., 2011). Interestingly, Debernardi and coworkers have demonstrated that the miR396–GRF–GIF network regulates leaf longevity. Analysis of *Arabidopsis* plants with increased levels of *GRF3* or *GRF5* revealed a delay in leaf senescence (Figure 17.3F) and in the expression of specific markers of leaf senescence, such as *SEN1*, *SEN4*, and *SAG12* (Debernardi et al., 2014). Similar to the effects on leaf size, the delay in senescence was boosted by cooverexpression of *GIF1*. In agreement with these results, plants overexpressing miR396 or single *grf3*, *grf5*, and *gif1* mutants senesced faster than wild-type plants. Therefore, GRFs not only stimulate cell proliferation during the early stages of leaf development, they also prevent them from entering the senescence program prematurely.

17.4 CONCLUSION AND PERSPECTIVES

GRFs have been shown to be positive regulators of plant growth. Although they have been primarily studied in leaves, GRFs also have functions in SAMs and RAMs as well as during the reproductive phases of the plant. Later results have not only implicated GRFs in the response of the plant to UV and drought, but also during senescence. In *Arabidopsis*, at least, distinct GRF family members have largely overlapping functions in plant development. The levels and activity of GRFs are under strong control by miRNA miR396 and GIF coactivators, which can in turn recruit chromatin remodelers. Although the mechanisms regulating GRF activity are beginning to be unraveled, the genes directly regulated by GRFs are largely unknown. Experiments designed to identify the targets of GRFs will certainly provide insights into the processes regulated by this family of TFs.

Acknowledgments

The authors acknowledge Dr Nicolás Blanco for critical reading of the manuscript. M.F.E. and J.M.D. acknowledge support provided by fellowships from CONICET and the Josefina Prats Foundation. J.P. and R.E.R. are members of CONICET. J.F.P. acknowledges the support of grants from Agencia Nacional de Promoción Científica y Tecnológica and the Howard Hughes Medical Institute. R.E.R. acknowledges the support of grants from Agencia Nacional de Promoción Científica y Tecnológica.

References

Bazin, J., Khan, G.A., Combier, J.P., Bustos-Sanmamed, P., Debernardi, J.M., Rodriguez, R., Sorin, C., Palatnik, J., Hartmann, C., Crespi, M., Lelandais-Briere, C., 2013. miR396 affects mycorrhization and root meristem activity in the legume Medicago truncatula. Plant J. 74, 920–934.

Bologna, N.G., Voinnet, O., 2014. The diversity, biogenesis, and activities of endogenous silencing small RNAs in Arabidopsis. Annu. Rev. Plant Biol. 65, 473–503.

Breeze, E., Harrison, E., McHattie, S., Hughes, L., Hickman, R., Hill, C., Kiddle, S., Kim, Y.S., Penfold, C.A., Jenkins, D., Zhang, C., Morris, K., Jenner, C., Jackson, S., Thomas, B., Tabrett, A., Legaie, R., Moore, J.D., Wild, D.L., Ott, S., Rand, D., Beynon, J., Denby, K., Mead, A., Buchanan-Wollaston, V., 2011. High-resolution temporal profiling of transcripts during Arabidopsis leaf senescence reveals a distinct chronology of processes and regulation. Plant Cell 23, 873–894.

Candaele, J., Demuynck, K., Mosoti, D., Beemster, G.T., Inze, D., Nelissen, H., 2014. Differential methylation during maize leaf growth targets developmentally regulated genes. Plant Physiol. 164, 1350–1364.

Casadevall, R., Rodriguez, R.E., Debernardi, J.M., Palatnik, J.F., Casati, P., 2013. Repression of growth regulating factors by the microRNA396 inhibits cell proliferation by UV-B radiation in Arabidopsis leaves. Plant Cell 25, 3570–3583.

Chen, L., Wang, T., Zhao, M., Tian, Q., Zhang, W.H., 2012. Identification of aluminum-responsive microRNAs in Medicago truncatula by genome-wide high-throughput sequencing. Planta 235, 375–386.

Choi, D., Kim, J.H., Kende, H., 2004. Whole genome analysis of the OsGRF gene family encoding plant-specific putative transcription activators in rice (Oryza sativa L.). Plant Cell Physiol. 45, 897–904.

Chorostecki, U., Crosa, V.A., Lodeyro, A.F., Bologna, N.G., Martin, A.P., Carrillo, N., Schommer, C., Palatnik, J.F., 2012. Identification of new microRNA-regulated genes by conserved targeting in plant species. Nucleic Acids Res. 40, 8893–8904.

Debernardi, J.M., Rodriguez, R.E., Mecchia, M.A., Palatnik, J.F., 2012. Functional specialization of the plant miR396 regulatory network through distinct microRNA-target interactions. PLoS Genet. 8, e1002419.

Debernardi, J.M., Mecchia, M.A., Vercruyssen, L., Smaczniak, C., Kaufmann, K., Inze, D., Rodriguez, R.E., Palatnik, J.F., 2014. Post-transcriptional control of GRF transcription factors by microRNA miR396 and GIF co-activator affects leaf size and longevity. Plant J. 79, 413–426.

Ding, D., Zhang, L., Wang, H., Liu, Z., Zhang, Z., Zheng, Y., 2009. Differential expression of miRNAs in response to salt stress in maize roots. Ann. Bot. 103, 29–38.

Ding, Y., Chen, Z., Zhu, C., 2011. Microarray-based analysis of cadmium-responsive microRNAs in rice (Oryza sativa). J. Exp. Bot. 62, 3563–3573.

Donnelly, P.M., Bonetta, D., Tsukaya, H., Dengler, R.E., Dengler, N.G., 1999. Cell cycling and cell enlargement in developing leaves of Arabidopsis. Dev. Biol. 215, 407–419.

Ferjani, A., Horiguchi, G., Yano, S., Tsukaya, H., 2007. Analysis of leaf development in fugu mutants of Arabidopsis reveals three compensation modes that modulate cell expansion in determinate organs. Plant Physiol. 144, 988–999.

Gao, P., Bai, X., Yang, L., Lv, D., Li, Y., Cai, H., Ji, W., Guo, D., Zhu, Y., 2010. Over-expression of osa-MIR396c decreases salt and alkali stress tolerance. Planta 231, 991–1001.

Gonzalez, N., De Bodt, S., Sulpice, R., Jikumaru, Y., Chae, E., Dhondt, S., Van Daele, T., De Milde, L., Weigel, D., Kamiya, Y., Stitt, M., Beemster, G.T., Inze, D., 2010. Increased leaf size: different means to an end. Plant Physiol. 153, 1261–1279.

Ha Lee, B., Hoe Kim, J., 2014. Spatio-temporal distribution patterns of GRF-INTERACTING FACTOR expression, leaf size control. Plant Signal. Behav. 9, e29697.

Hake, S., Smith, H.M., Holtan, H., Magnani, E., Mele, G., Ramirez, J., 2004. The role of knox genes in plant development. Annu. Rev. Cell Dev. Biol. 20, 125–151.

Heidstra, R., Sabatini, S., 2014. Plant and animal stem cells: similar yet different. Nat. Rev. Mol. Cell Biol. 15, 301–312.

Hewezi, T., Maier, T.R., Nettleton, D., Baum, T.J., 2012. The Arabidopsis microRNA396–GRF1/GRF3 regulatory module acts as a developmental regulator in the reprogramming of root cells during cyst nematode infection. Plant Physiol. 159, 321–335.

Horiguchi, G., Kim, G.T., Tsukaya, H., 2005. The transcription factor AtGRF5 and the transcription coactivator AN3 regulate cell proliferation in leaf primordia of Arabidopsis thaliana. Plant J. 43, 68–78.

Jones-Rhoades, M.W., Bartel, D.P., 2004. Computational identification of plant microRNAs and their targets, including a stress-induced miRNA. Mol. Cell 14, 787–799.

Kantar, M., Lucas, S.J., Budak, H., 2011. miRNA expression patterns of Triticum dicoccoides in response to shock drought stress. Planta 233, 471–484.

Kende, H., Van Der Knaap, E., Cho, H.T., 1998. Deepwater rice: a model plant to study stem elongation. Plant Physiol. 118, 1105–1110.

Kim, J.H., Kende, H., 2004. A transcriptional coactivator, AtGIF1, is involved in regulating leaf growth and morphology in Arabidopsis. Proc. Natl. Acad. Sci. USA 101, 13374–13379.

Kim, J., Lee, B., 2006. GROWTH-REGULATING FACTOR4 of Arabidopsis thaliana is required for development of leaves, cotyledons, and shoot apical meristem. J. Plant Biol. 49, 463–468.

Kim, J.H., Choi, D., Kende, H., 2003. The AtGRF family of putative transcription factors is involved in leaf and cotyledon growth in Arabidopsis. Plant J. 36, 94–104.

Kim, G.T., Yano, S., Kozuka, T., Tsukaya, H., 2005. Photomorphogenesis of leaves: shade-avoidance and differentiation of sun and shade leaves. Photochem. Photobiol. Sci. 4, 770–774.

Kim, J.S., Mizoi, J., Kidokoro, S., Maruyama, K., Nakajima, J., Nakashima, K., Mitsuda, N., Takiguchi, Y., Ohme-Takagi, M., Kondou, Y., Yoshizumi, T., Matsui, M., Shinozaki, K., Yamaguchi-Shinozaki, K., 2012. Arabidopsis growth-regulating factor7 functions as a transcriptional repressor of abscisic acid- and osmotic stress-responsive genes, including DREB2A. Plant Cell 24, 3393–3405.

Kuijt, S.J., Greco, R., Agalou, A., Shao, J., 't Hoen, C.C, Overnas, E., Osnato, M., Curiale, S., Meynard, D., Van Gulik, R., De Faria Maraschin, S., Atallah, M., De Kam, R.J., Lamers, G.E., Guiderdoni, E., Rossini, L., Meijer, A.H., Ouwerkerk, P.B., 2014. Interaction between the GROWTH-REGULATING FACTOR and KNOTTED1-LIKE HOMEOBOX families of transcription factors. Plant Physiol. 164, 1952–1966.

Lee, B.H., Ko, J.H., Lee, S., Lee, Y., Pak, J.H., Kim, J.H., 2009. The Arabidopsis GRF-INTERACTING FACTOR gene family performs an overlapping function in determining organ size as well as multiple developmental properties. Plant Physiol. 151, 655–668.

Lee, B.H., Wynn, A.N., Franks, R.G., Hwang, Y.S., Lim, J., Kim, J.H., 2014. The Arabidopsis thaliana GRF-INTERACTING FACTOR gene family plays an essential role in control of male and female reproductive development. Dev. Biol. 386, 12–24.

Lemon, B., Tjian, R., 2000. Orchestrated response: a symphony of transcription factors for gene control. Genes Dev. 14, 2551–2569.

Li, B., Qin, Y., Duan, H., Yin, W., Xia, X., 2011. Genome-wide characterization of new and drought stress responsive microRNAs in *Populus euphratica*. J. Exp. Bot. 62, 3765–3779.

Liang, G., He, H., Li, Y., Wang, F., Yu, D., 2014. Molecular mechanism of microRNA396 mediating pistil development in *Arabidopsis*. Plant Physiol. 164, 249–258.

Lim, P.O., Kim, H.J., Nam, H.G., 2007. Leaf senescence. Annu. Rev. Plant Biol. 58, 115–136.

Liu, D., Song, Y., Chen, Z., Yu, D., 2009. Ectopic expression of miR396 suppresses GRF target gene expression and alters leaf growth in *Arabidopsis*. Plant Physiol. 136, 223–236.

Liu, J., Hua, W., Yang, H.L., Zhan, G.M., Li, R.J., Deng, L.B., Wang, X.F., Liu, G.H., Wang, H.Z., 2012. The BnGRF2 gene (GRF2-like gene from *Brassica napus*) enhances seed oil production through regulating cell number and plant photosynthesis. J. Exp. Bot. 63, 3727–3740.

Liu, H., Guo, S., Xu, Y., Li, C., Zhang, Z., Zhang, D., Xu, S., Zhang, C., Chong, K., 2014. OsmiR396d-regulated OsGRFs function in floral organogenesis in rice through binding to their targets OsJMJ706 and OsCR4. Plant Physiol. 165, 160–174.

Luo, A.D., Liu, L., Tang, Z.S., Bai, X.Q., Cao, S.Y., Chu, C.C., 2005. Downregulation of OsGRF1 gene in rice rhd1 mutant results in reduced heading date. J. Integr. Plant Biol. 47, 745–752.

Mallory, A.C., Elmayan, T., Vaucheret, H., 2008. MicroRNA maturation and action – the expanding roles of ARGONAUTEs. Curr. Opin. Plant Biol. 11, 560–566.

Mecchia, M.A., Debernardi, J.M., Rodriguez, R.E., Schommer, C., Palatnik, J.F., 2013. MicroRNA miR396 and RDR6 synergistically regulate leaf development. Mech. Dev. 130, 2–13.

Meyers, B.C., Axtell, M.J., Bartel, B., Bartel, D.P., Baulcombe, D., Bowman, J.L., Cao, X., Carrington, J.C., Chen, X., Green, P.J., Griffiths-Jones, S., Jacobsen, S.E., Mallory, A.C., Martienssen, R.A., Poethig, R.S., Qi, Y., Vaucheret, H., Voinnet, O., Watanabe, Y., Weigel, D., Zhu, J.K., 2008. Criteria for annotation of plant MicroRNAs. Plant Cell 20, 3186–3190.

Nelissen, H., Rymen, B., Jikumaru, Y., Demuynck, K., Van Lijsebettens, M., Kamiya, Y., Inze, D., Beemster, G.T., 2012. A local maximum in gibberellin levels regulates maize leaf growth by spatial control of cell division. Curr. Biol. 22, 1183–1187.

Ohno, C.K., Reddy, G.V., Heisler, M.G., Meyerowitz, E.M., 2004. The *Arabidopsis* JAGGED gene encodes a zinc finger protein that promotes leaf tissue development. Development 131, 1111–1122.

Osnato, M., Stile, M.R., Wang, Y., Meynard, D., Curiale, S., Guiderdoni, E., Liu, Y., Horner, D.S., Ouwerkerk, P.B., Pozzi, C., Muller, K.J., Salamini, F., Rossini, L., 2010. Cross talk between the KNOX and ethylene pathways is mediated by intron-binding transcription factors in barley. Plant Physiol. 154, 1616–1632.

Pajoro, A., Madrigal, P., Muino, J.M., Matus, J.T., Jin, J., Mecchia, M.A., Debernardi, J.M., Palatnik, J.F., Balazadeh, S., Arif, M., O'Maoileidigh, D.S., Wellmer, F., Krajewski, P., Riechmann, J.L., Angenent, G.C., Kaufmann, K., 2014. Dynamics of chromatin accessibility and gene regulation by MADS-domain transcription factors in flower development. Genome Biol. 15, R41.

Petricka, J.J., Winter, C.M., Benfey, P.N., 2012. Control of *Arabidopsis* root development. Annu. Rev. Plant Biol. 63, 563–590.

Postma-Haarsma, A.D., Rueb, S., Scarpella, E., Den Besten, W., Hoge, J.H., Meijer, A.H., 2002. Developmental regulation and downstream effects of the knox class homeobox genes Oskn2 and Oskn3 from rice. Plant Mol. Biol. 48, 423–441.

Pu, C.X., Ma, Y., Wang, J., Zhang, Y.C., Jiao, X.W., Hu, Y.H., Wang, L.L., Zhu, Z.G., Sun, D., Sun, Y., 2012. Crinkly4 receptor-like kinase is required to maintain the interlocking of the palea and lemma, and fertility in rice, by promoting epidermal cell differentiation. Plant J. 70, 940–953.

Raventos, D., Skriver, K., Schlein, M., Karnahl, K., Rogers, S.W., Rogers, J.C., Mundy, J., 1998. HRT, a novel zinc finger, transcriptional repressor from barley. J. Biol. Chem. 273, 23313–23320.

Rodriguez, R.E., Mecchia, M.A., Debernardi, J.M., Schommer, C., Weigel, D., Palatnik, J.F., 2010. Control of cell proliferation in *Arabidopsis thaliana* by microRNA miR396. Development 137, 103–112.

Rodriguez, R.E., Debernardi, J.M., Palatnik, J.F., 2014. Morphogenesis of simple leaves: regulation of leaf size and shape. Interdiscip. Rev. Dev. Biol. 3, 41–57.

Rubio-Somoza, I., Cuperus, J.T., Weigel, D., Carrington, J.C., 2009. Regulation and functional specialization of small RNA-target nodes during plant development. Curr. Opin. Plant Biol. 12, 622–627.

Schiessl, K., Kausika, S., Southam, P., Bush, M., Sablowski, R., 2012. JAGGED controls growth anisotropy and coordination between cell size and cell cycle during plant organogenesis. Curr. Biol. 22, 1739–1746.

Schiessl, K., Muino, J.M., Sablowski, R., 2014. *Arabidopsis* JAGGED links floral organ patterning to tissue growth by repressing Kip-related cell cycle inhibitors. Proc. Natl. Acad. Sci. USA 111, 2830–2835.

Schmid, M., Davison, T.S., Henz, S.R., Pape, U.J., Demar, M., Vingron, M., Scholkopf, B., Weigel, D., Lohmann, J.U., 2005. A gene expression map of *Arabidopsis thaliana* development. Nat. Genet. 37, 501–506.

Skirycz, A., Claeys, H., De Bodt, S., Oikawa, A., Shinoda, S., Andriankaja, M., Maleux, K., Eloy, N.B., Coppens, F., Yoo, S.D., Saito, K., Inze, D., 2011. Pause-and-stop: the effects of osmotic stress on cell proliferation during early leaf development in *Arabidopsis* and a role for ethylene signaling in cell cycle arrest. Plant Cell 23, 1876–1888.

Sun, Q., Zhou, D.X., 2008. Rice jmjC domain-containing gene JMJ706 encodes H3K9 demethylase required for floral organ development. Proc. Natl. Acad. Sci. USA 105, 13679–13684.

Triezenberg, S.J., 1995. Structure and function of transcriptional activation domains. Curr. Opin. Genet. Dev. 5, 190–196.

Van Der Knaap, E., Kim, J.H., Kende, H., 2000. A novel gibberellin-induced gene from rice and its potential regulatory role in stem growth. Plant Physiol. 122, 695–704.

Vercruyssen, L., Verkest, A., Gonzalez, N., Heyndrickx, K.S., Eeckhout, D., Han, S.K., Jegu, T., Archacki, R., Van Leene, J., Andriankaja, M., De Bodt, S., Abeel, T., Coppens, F., Dhondt, S., De Milde, L., Vermeersch, M., Maleux, K., Gevaert, K., Jerzmanowski, A., Benhamed, M., Wagner, D., Vandepoele, K., De Jaeger, G., Inze, D., 2014. ANGUSTIFOLIA3 binds to SWI/SNF chromatin remodeling complexes to regulate transcription during *Arabidopsis* leaf development. Plant Cell 26, 210–229.

Wang, L., Xie, W., Chen, Y., Tang, W., Yang, J., Ye, R., Liu, L., Lin, Y., Xu, C., Xiao, J., Zhang, Q., 2010. A dynamic gene expression atlas covering the entire life cycle of rice. Plant J. 61, 752–766.

Wang, L., Gu, X., Xu, D., Wang, W., Wang, H., Zeng, M., Chang, Z., Huang, H., Cui, X., 2011a. miR396-targeted AtGRF transcription factors are required for coordination of cell division and differentiation during leaf development in *Arabidopsis*. J. Exp. Bot. 62, 761–773.

Wang, T., Chen, L., Zhao, M., Tian, Q., Zhang, W.H., 2011b. Identification of drought-responsive microRNAs in *Medicago truncatula* by genome-wide high-throughput sequencing. BMC Genomics 12, 367.

Winter, C.M., Austin, R.S., Blanvillain-Baufume, S., Reback, M.A., Monniaux, M., Wu, M.F., Sang, Y., Yamaguchi, A., Yamaguchi, N., Parker, J.E., Parcy, F., Jensen, S.T., Li, H., Wagner, D., 2011. LEAFY target genes reveal floral regulatory logic, cis motifs, and a link to biotic stimulus response. Dev. Cell 20, 430–443.

Wu, L., Zhang, D., Xue, M., Qian, J., He, Y., Wang, S., 2014. Overexpression of the maize GRF10, an endogenous truncated growth-regulating factor protein, leads to reduction in leaf size and plant height. J. Integr. Plant Biol. 56, 1053–1063.

Yang, F., Liang, G., Liu, D., Yu, D., 2009. *Arabidopsis* MiR396 mediates the development of leaves and flowers in transgenic tobacco. J. Plant Biol. 52, 475–481.

Yant, L., Mathieu, J., Dinh, T.T., Ott, F., Lanz, C., Wollmann, H., Chen, X., Schmid, M., 2010. Orchestration of the floral transition and floral development in *Arabidopsis* by the bifunctional transcription factor APETALA2. Plant Cell 22, 2156–2170.

Zhang, Y., Turner, J.G., 2008. Wound-induced endogenous jasmonates stunt plant growth by inhibiting mitosis. PLoS ONE 3, e3699.

Zhang, D.F., Li, B., Jia, G.Q., Zhang, T.F., Dai, J.R., Li, J.S., Wang, S.C., 2008. Isolation and characterization of genes encoding GRF transcription factors and GIF transcriptional coactivators in maize (*Zea mays* L.). Plant Sci. 175, 809–817.

Zhang, J., Nieminen, K., Serra, J.A., Helariutta, Y., 2014. The formation of wood and its control. Curr. Opin. Plant Biol. 17, 56–63.

Zhou, L., Liu, Y., Liu, Z., Kong, D., Duan, M., Luo, L., 2010. Genome-wide identification and analysis of drought-responsive microRNAs in *Oryza sativa*. J. Exp. Bot. 61, 4157–4168.

CHAPTER 18

The Multifaceted Roles of miR156-targeted SPL Transcription Factors in Plant Developmental Transitions

Jia-Wei Wang

National Key Laboratory of Plant Molecular Genetics, Institute of Plant Physiology and Ecology, Shanghai Institutes for Biological Sciences, Shanghai, China

OUTLINE

- 18.1 Introduction to Developmental Transitions — 281
- 18.2 miR156 and its Targets — 282
- 18.3 miR156-SPL Module in Timing Embryonic Development — 283
- 18.4 miR156-SPL Module in Juvenile-to-Adult Phase Transition in Higher Plants — 283
- 18.5 The miR156-SPL Module Regulates Flowering Time in Higher Plants — 285
- 18.6 The miR156-SPL Module in Developmental Transitions in Moss — 286
- 18.7 The miR156-SPL Module in Other Developmental Processes — 287
  - 18.7.1 The Role of the miR156-SPL Module in Age-Related Resistance — 287
  - 18.7.2 The Role of the miR156-SPL Module in Stress Response — 287
  - 18.7.3 The Role of the miR156-SPL Module in Leaf Initiation Rate (Plastochron) — 287
  - 18.7.4 The Role of the miR156-SPL Module in Ligule Differentiation — 287
  - 18.7.5 The Role of the miR156-SPL Module in Plant Architecture — 289
  - 18.7.6 Temporal Regulation of Anthocyanin Biosynthesis by the miR156-SPL Module — 289
  - 18.7.7 Temporal Regulation of Trichome Development on Stem and Floral Organs by the miR156-SPL Module — 289
  - 18.7.8 The Role of the miR156-SPL Module in Floral Organ Development — 289
- 18.8 Perspectives — 290
- References — 291

18.1 INTRODUCTION TO DEVELOPMENTAL TRANSITIONS

Each scale of physiology and development – the division of cells, the production of metabolites, the emergence of patterns, and the formation of organs – requires proper timing. Instead of inventing new gene functions, the differences in developmental timing, even when subtle, cause physiological defects or a novel morphology that confers an evolutionary advantage.

Despite the most obvious morphological changes in plants being the establishment of seedlings (seed germination and photomorphogenesis) and the production of reproductive structures (flowering), plants also undergo dramatic changes in vegetative morphology during the juvenile-to-adult transition, also referred to as the vegetative phase transition (Poethig, 2010, 2013; Huijser and Schmid, 2011; Bäurle and Dean, 2006). These changes include the shape and size of leaves, the pattern of epidermal cell differentiation, the number of lateral shoots, resistance against disease and stress, the capacity for adventitious root production, and reproductive competence for external stimuli (Brink, 1962; Kerstetter and Poethig, 1998).

The changes underlying the vegetative phase transition vary in different plants. First, the juvenile and adult phases are defined on the basis of species-specific traits. A specific trait might mark the juvenile phase in one species but the adult phase in another species. For example, leaf complexity (the number of leaflets) is increased during development in *Cardamine hirsuta* (Canales et al., 2010), but decreased in *Acacia confusa* (Kaplan, 1980). Second, the juvenile-to-adult transition involves quantitative changes as well as rapid qualitative changes. For instance, *Eucalyptus* species exhibit a striking and abrupt change from juvenile to adult phase (James and Bell, 2001), whereas a similar transition in *Arabidopsis thaliana* is gradual (Telfer et al., 1997). Therefore, none of the specific morphological changes can be used as a universal marker for the juvenile-to-adult transition in plants.

The timing of developmental transitions is regulated by diverse cues. The vegetative phase transition is influenced by exogenous cues such as day length and temperature, as well as endogenous cues such as plant hormones and age. In *A. thaliana*, the appearance of abaxial trichomes, an adult trait, is delayed when day length is shortened, but accelerated by treatment with gibberellin (GA), a plant diterpene hormone (Telfer et al., 1997). Similarly, floral transition – namely, the switch from vegetative-to-reproductive phase – is coordinately controlled by multiple genetic pathways in response to various developmental and environmental cues (Bäurle and Dean, 2006; Srikanth and Schmid, 2011; Andres and Coupland, 2012; Romera-Branchat et al., 2014). Age, day length (photoperiod), GA, and temperature are the most important of these cues. Thus, developmental timing is aligned with both internal and external conditions, thereby maximizing environmental adaptation and reproductive success.

Developmental transition is subject to strong genetic control and varies markedly both within and between species. Genetic and population genomics studies of precocious and normal ecotypes of *Eucalyptus globules* revealed several quantitative trait loci (QTLs) responsible for vegetative phase transition and flowering (Hudson et al., 2014). Similarly, the vegetative phase transition in *A. thaliana* is more accelerated in the Landsberg erecta accession than the Columbia accession (Telfer et al., 1997). Moreover, there is extensive natural variation of flowering time in response to high temperature (27°C) or cold treatment (Balasubramanian et al., 2006; Coustham et al., 2012; Dittmar et al., 2014; Shindo et al., 2006).

Although the definition of developmental transitions – timing and morphological changes – is species-specific, recent studies have revealed that these transitions are coordinately regulated by a decrease in expression of microRNA156 (miR156), which represses the expression of *SQUAMOSA PROMOTER BINDING PROTEIN-LIKE* (*SPL*) transcription factors. This chapter begins with a brief introduction to the discovery of miR156 as the master regulator of plant developmental transitions. Then, the discussion turns to the role played by miR156 and its targets in other aspects of plant development and physiology, which require proper timing.

18.2 miR156 AND ITS TARGETS

MicroRNAs (miRNAs) are a class of 20-nucleotide to 24-nucleotide endogenous small RNAs that repress gene expression through transcript cleavage or translational inhibition (Rogers and Chen, 2013). miR156 was initially discovered by genome-wide small RNA sequencing in *A. thaliana* (Reinhart et al., 2002), and subsequent studies revealed that this miRNA is evolutionarily conserved and present in all land plants including mosses. miR157 is one nucleotide longer than miR156, and differs from miR156 by having three nucleotides in sequence. miR157 is likely to share the same targets as miR156 because miR157 overexpression lines show a similar phenotype to that of miR156 (Shikata et al., 2009, 2012). However, its expression pattern and the corresponding loss-of-function mutants have yet to be characterized.

The targets of miR156 belong to a group of plant-specific transcription factors, called SQUAMOSA PROMOTER BINDING PROTEIN (SBP box)-like genes (SPL; Cardon et al., 1999; Klein et al., 1996). In the Columbia accession of *A. thaliana*, 11 of 17 *SPL* genes are targeted by miR156 (Xing et al., 2010; Figure 18.1A). These miR156-targeted *SPLs* can be structurally divided into two groups, represented by *SPL3* and *SPL9* (Xing et al., 2010). SPL9 protein is larger than SPL3 because it harbors a C-terminal domain with unknown function (Figure 18.1B). It is suggested that this domain is involved in protein–protein interaction because the removal of this domain abolishes the interaction between SPL9 and DELLA proteins (Yu et al., 2012). A second difference between *SPL3* and *SPL9* is the mechanism by which miR156 regulates their expression. miR156 represses *SPL9* expression through transcript cleavage, but inhibits *SPL3* expression mainly through translational inhibition (Gandikota et al., 2007). This difference is probably caused by the different location of miR156-binding sites in the mRNAs: the miR156-binding site is located in the 3'UTR (untranslated region) of *SPL3* mRNA, but within the coding region of *SPL9* mRNA. The biological roles played by *SPL3* and *SPL9* group genes in *A. thaliana* are divergent as well. For example, *SPL9* and *SPL15* redundantly regulate leaf initiation rate, leaf size, and shoot maturation (Wu et al., 2009; Wang et al., 2009a; Usami et al., 2009; Schwarz et al., 2008), whereas *SPL3*

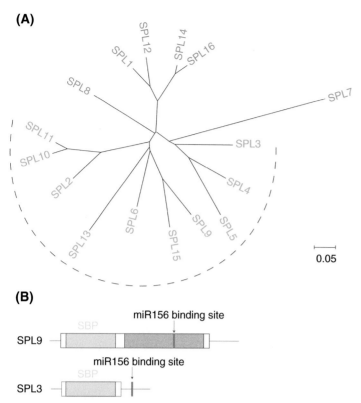

FIGURE 18.1 **The SPL gene family in *A. thaliana*.** (A) Phylogenetic analysis of *SPL* genes in *A. thaliana*. The neighbor-joining algorithm based on the conserved SBP domain was used. Dashed line indicates miR156/7-targeted *SPL* genes (green). Nontargeted *SPL* genes are in pink. The alignment used to generate this tree is according to Xing et al. (2010). (B) Structural analysis of SPL3 and SPL9. Box and black line indicate coding and untranslated regions, respectively. The red box in SPL9 represents the C-terminal domain with unknown function.

plays a specific role in floral transition (Wang et al., 2009a; Wu and Poethig, 2006; Jung et al., 2011).

18.3 miR156-SPL MODULE IN TIMING EMBRYONIC DEVELOPMENT

In angiosperms, sporophytic generation is initiated by double-fertilization, resulting in the formation of seeds. After fertilization the zygote divides asymmetrically and generates two daughter cells with different properties and developmental programs. Subsequent divisions change the small apical daughter cells into the proembryo cell (Capron et al., 2009). At the 8-cell stage, the proembryo is divided into an upper tier of cells giving rise to the shoot and a lower tier of cells giving rise to the hypocotyl and embryonic root. An *A. thaliana* mutant defective in *DICER-LIKE1* (*DCL1*), a key factor for miRNA biogenesis, exhibits growth arrest as early as the 8-cell stage (Nodine and Bartel, 2010). Microarray analyses have revealed that, in the 8-cell embryos of *dcl1*, the expression of two miR156 targets, *SPL10* and *SPL11*, is elevated. In agreement, overexpression of *SPL10* causes a weak but similar phenotype to the *dcl1* mutation. Interestingly, genes normally induced during the embryonic maturation phase are precociously expressed in the *SPL10* overexpressor and the *dcl1* mutant. Thus, these observations indicate that plant miRNAs enable proper embryonic patterning by preventing precocious expression of differentiation promoting transcription factors. Because the *SPL11* overexpressor does not fully phenocopy *dcl1*, these results further suggest that other miRNAs might also be involved in this process.

18.4 miR156-SPL MODULE IN JUVENILE-TO-ADULT PHASE TRANSITION IN HIGHER PLANTS

A. thaliana and maize have been used as models to investigate the genetic basis for vegetative phase transition for over three decades. In *A. thaliana*, the juvenile and adult phase can be clearly defined (Telfer et al., 1997; Figure 18.2). The most obvious difference between the juvenile and adult phase is leaf shape. The shape of juvenile leaves is round, whereas that of adult leaves is oval. As mentioned earlier, the second difference lies in the spatial distribution of trichomes, the hairy cells on the epidermis of leaves. Only the adult leaf bears trichomes on the abaxial (lower) side. The formation of leaf serrations also marks the transition from juvenile to adult phase. Adult leaves produce serrations on the leaf margin, whereas that of juvenile leaves is smooth. It should be noted that the transition from juvenile to adult phase in *A. thaliana* is gradual. The numbers of serrations and abaxial trichomes are progressively increased during development.

FIGURE 18.2 **Leaf morphology of *A. thaliana* and *C. flexuosa*.** The successive leaves of *A. thaliana* and *C. flexuosa* were detached and scanned.

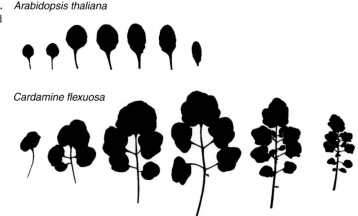

The past five years have seen a great advance in our understanding of the molecular basis of vegetative phase transition. Mutant characterizations and genetic analyses in *A. thaliana* revealed that vegetative phase change is controlled by miR156, the master regulator for juvenility (Wu et al., 2009). The decline in miR156 with time is responsible for the morphological difference between juvenile and adult phases. Notably, manipulation of miR156 suffices to regulate the length of the juvenile phase. Elevated miR156 levels lead to a prolonged juvenile phase, whereas lower levels cause loss of the juvenile phase (Figure 18.3).

In *A. thaliana*, mature miR156 is highly enriched in seedlings and its expression is subsequently decreased (Wu and Poethig, 2006; Wang et al., 2009a). This age-dependent expression pattern is observed in all the annual plants examined, including Chinese cabbage (*Brassica rapa*), tomato (*Solanum lycopersicum*), rice (*Oryza sativa*), and maize (*Zea mays*; Chuck et al., 2007; Wang et al., 2014; Zhang et al., 2011; Xie et al., 2012). In perennial woody plants such as *Acacia confusa*, *Acacia colei*, *Eucalyptus globulus*, *Hedera helix*, *Quercus acutissima*, and *Populus × canadensis* (poplar), the expression of miR156 is decreased with time as well, although at a lower rate (Wang et al., 2011). Thus, these observations suggest that the developmental decline of miR156 is regulated by an endogenous timing cue rather than chronological age.

Defoliation studies in *A. thaliana* and *Nicotiana benthamiana* indicated that the developmental decline in miR156 is mediated by a signal produced by leaf primordia (Yang et al., 2011). Removal of either root or cotyledons

FIGURE 18.3 **miR156 regulates vegetative phase transition and flowering in *A. thaliana*.** (A) Vegetative phase transition in *A. thaliana*. Orange marks the region that bears abaxial trichomes. Compared to wild type, *35S::MIR156* plants, in which miR156 is overexpressed, have more juvenile leaves, while *35S::MIM156* plants, in which miR156 activity is attenuated, accelerate the appearance of adult traits. (B) miR156 delays flowering. *35S::MIR156* plants flower later than the wild type under short-day conditions. *Photo courtesy Dr Rebecca Schwab.*

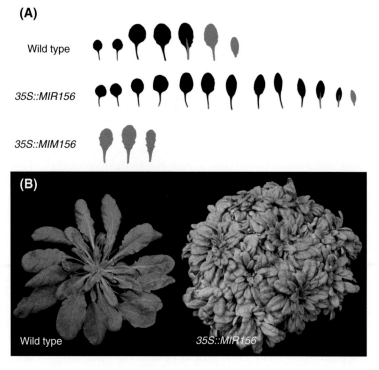

does not appear to be important for this event. This conclusion is supported by previous studies showing that defoliation and severe pruning increases the production of juvenile leaves in both the herbaceous plant *Ipomoea caerulea* (Njoku, 1956) and woody plants such as *Pinus radiata* (Libby and Hood, 1976; Schaffalitzky de Muckadell, 1954).

In light of these findings, it has been noted that sugar, the product of photosynthesis in leaves, is responsible for the temporal expression pattern of miR156 (Yang et al., 2013; Yu et al., 2013). Sugar treatment leads to rapid decrease in miR156 levels, accompanied by acceleration of vegetative phase transition. In agreement with this, the *A. thaliana cao/ch1* mutant, which shows reduced level of chlorophyll b and low efficiency of photosynthesis, exhibits a prolonged juvenile phase that accompanies elevated levels of miR156.

The identification of sugar as the upstream regulator of miR156 provides an explanation for previous observations that carbohydrates regulate the length of the juvenile phase. Using the water fern *Marsilea drummondii*, Allsopp showed that plants continue producing juvenile leaves in the absence of exogenous sugar, whereas supplementing the growth medium with sugar accelerates the production of adult leaves (Allsopp, 1952, 1953). Similarly, reducing the photosynthesis rate by suppression of the *RUBISCO SMALL SUBUNIT* (*RBCS*) results in prolonging the juvenile phase in tobacco (Tsai et al., 1997).

How sugar represses miR156 expression is largely unknown. It is suggested that this process occurs at two different levels. At the transcriptional level, sugar represses miR156 expression partially through the glucose sensor HEXOKINASE1 (HXK1; Yang et al., 2013; Moore et al., 2003), whereas, at the posttranscriptional level, glucose is able to promote degradation of the primary transcript of miR156 (Yu et al., 2013). However, the precise underlying molecular mechanism awaits further investigation.

The molecular basis by which miR156 regulates morphological changes during vegetative phase transition is not well understood. It has been shown that the effect of miR156 on phase-dependent epidermal cell differentiation relies on miR172 (Poethig, 2013). miR172 acts downstream of miR156, exhibiting an opposite expression pattern to that of miR156. One of the miR172 coding loci, *MIR172B*, was identified as a direct downstream target of miR156-targeted SPL9 (Wu et al., 2009). Consistently with this, overexpression of miR172 compared with the wild type causes the production of abaxial trichomes on the first rosette leaf and is sufficient to suppress the late abaxial trichome phenotype of a miR156 overexpressor (Wu et al., 2009).

As is the case in dicots, miR156 also plays an important role in maintenance of the juvenile phase in monocots, such as rice and maize. The dominant *Corngrass1* (*Cg1*) mutant of maize displays phenotypes that may be present in the grass-like ancestors of maize (Chuck et al., 2007). Map-based cloning revealed that *Cg1* encodes two tandem miR156 genes that are overexpressed in the meristem and lateral organs. In the maize genome there are at least 13 potential target genes of miR156, including *teosinte glume architecture 1* (*tga1*), a gene responsible for a major QTL in the evolution of maize from teosinte. In agreement with the sequential role played by miR156 and miR172 in vegetative phase transition in *A. thaliana*, *Cg1* mutants have lower levels of miR172, suggesting that monocots and dicots use the same miRNA cascade for juvenile-to-adult phase transition. Transgenic rice overproducing miR156 shows a similar phenotype to that seen in the maize *cg1* mutant (Xie et al., 2006).

In herbaceous perennials the role of miR156 in leaf morphology is also evident. Young *Cardamine flexuosa* plants produce simple leaves, but they give rise to compound leaves with 4–6 leaflets at late stages. This temporal regulation of leaf complexity is controlled by miR156. Inactivation of miR156 by target mimicry causes an acceleration of leaflet formation, whereas overexpression of miR156 leads to leaves with reduced leaflet number (Zhou et al., 2013).

Vegetative phase transition in woody perennials occurs after months or years and involves major changes in shoot architecture and leaf morphology. For example, *A. confusa* produces horizontally oriented bipinnately compound leaves in the juvenile phase, but vertically oriented simple leaves in the adult phase. Similarly, the vegetative phase change in *Populus × canadensis*is is marked by changes in leaf shape and internode length. Overexpression of miR156 in transgenic *P. × canadensis* reduced the expression of miR156-targeted SPL genes and miR172, and drastically prolonged the juvenile phase (Wang et al., 2011). Therefore, miR156 is an evolutionarily conserved regulator of vegetative phase change in both annual herbaceous plants and perennial trees.

18.5 THE miR156-SPL MODULE REGULATES FLOWERING TIME IN HIGHER PLANTS

Studies of the annual models *A. thaliana* and rice identified several flowering time pathways named: autonomous, GA, photoperiod, and vernalization (exposure to prolonged cold temperature; Amasino and Michaels, 2010). These multiple floral inductive cues are integrated into several flowering time integrator genes, including MADS-box genes such as *APETALA 1* (*AP1*), *FRUITFULL* (*FUL*), *SUPPRESSOR OF OVEREXPRESSION OF CO1* (*SOC1*), *FLOWERING LOCUS T* (*FT*), and the plant-specific transcription factor *LEAFY* (*LFY*; Srikanth and Schmid, 2011; Lee and Lee, 2010; Amasino and Michaels, 2010). The role of miR156 in flowering was revealed by the Detlef

Weigel Laboratory in 2005. In *A. thaliana*, overexpression of miR156 delays flowering under both long-day and short-day conditions (Schwab et al., 2005). Because the expression of miR156 is regulated by age but not affected by exogenous cues, such as temperature, day length and plant hormones, the miR156-mediated flowering pathway is called the "age pathway" (Wang et al., 2009a; Wang, 2014; Zhou and Wang, 2013).

Genetic studies and misexpression experiments revealed that miR156 regulates flowering time in both leaf and shoot apical meristem via distinct mechanisms. In leaves the miR156-SPL module regulates flowering mainly through miR172, which targets AP2-like transcription factors. This connection is fulfilled by the activation of *MIR172B* by SPL9-group transcription factors, the same regulatory circuit for juvenile-to-adult phase transition, thereby revealing the basis for the association between vegetative phase change and reproductive competence. In the *A. thaliana* genome, miR172 targets five AP2-like genes, which act as flowering repressors by inhibiting the expression of *FT*, which encodes a mobile floral promoter in the photoperiod pathway (Andres and Coupland, 2012). Overexpression of miR172 causes early flowering whereas increased levels of AP2-like proteins lead to late flowering, especially under noninductive conditions. In addition, a recent study revealed that miR156-targeted *SPL3*-group genes could directly regulate *FT* expression in the leaf to control ambient temperature-responsive flowering (Kim et al., 2012).

At the shoot apex, miR156-targeted *SPLs* (both *SPL3* and *SPL9*-group genes) and *FT* promote flowering by activating an overlapping set of targets such as *AP1*, *FUL*, *LFY*, and *SOC1* (Yamaguchi et al., 2009; Wang et al., 2009a). Additionally, SOC1 and FT enhance the expression of *SPL3*-group genes (*SPL3*, *SPL4*, and *SPL5*) by directly binding to their gene promoters (Jung et al., 2011). Thus, these results demonstrate the redundant activities and the feedforward action of the miR156-SPL and FT modules in flowering control, and reveal a crosstalk between photoperiod and age pathways (Wang et al., 2009a).

How does the age pathway integrate with other flowering time pathways? Studies have revealed that this integration is exerted via different mechanisms. GA promotes flowering in many plant species. This action is largely mediated by a group of transcriptional repressors, called "DELLAs" (Chapter 20). The *Arabidopsis* genome encodes five DELLA genes: *REPRESSOR OF GA1-3* (*RGA*), *GA INSENSITIVE* (*GAI*), *RGA-LIKE 1* (*RGL1*), *RGL2*, and *RGL3* (Murase et al., 2008). DELLA proteins are eliminated in response to GA by ubiquitin–proteasome degradation (Dill et al., 2001). Elevated GA levels bring about a reduction in the number of DELLAs, thereby leading to activation of flowering time genes. Integration of the GA and miR156-SPL modules is mediated by the physical interaction between SPL9-group SPLs and DELLAs. In leaves, DELLAs bind to miR156-targeted SPL9 and inhibit its activation toward miR172, which subsequently results in a reduced level of FT. At the shoot apex, such binding leads to inactivation of *SOC1* and *FUL* (Yu et al., 2012).

The winter annual type *A. thaliana* is late flowering. Such a late-flowering phenotype can be suppressed by vernalization. Genetic studies have revealed that *FLOWERING LOCUS C* (*FLC*), a MADS-box gene, acts as the master regulator in the vernalization pathway. FLC represses flowering by inhibiting *FT* in leaves and *SOC1* at the shoot apex (Searle et al., 2006). Transcription of *FLC* rapidly decreases in response to vernalization by complex mechanisms involving long-noncoding RNAs (lncRNAs), histone modification, and higher-order chromatin assembly (Crevillen et al., 2013; Rosa et al., 2013; Sun et al., 2013; Zografos and Sung, 2012; Song et al., 2012). Although both age and vernalization pathways exist in *A. thaliana*, integration of these two pathways does not occur. However, two recent studies have found that such a crosstalk does operate in *Arabis alpina* and *C. flexuosa*, two herbaceous perennials closely related to *A. thaliana* (Zhou et al., 2013; Bergonzi et al., 2013). In both species, the levels of miR156 and miR172 are correlated with sensitivity to vernalization. Overexpression of miR156 prevents flowering in response to vernalization, whereas reduced activity of miR156 or miR172-targeted *AP2*-like genes results in accelerated acquisition of floral competence in response to vernalization. *FLC* expression in *C. flexuosa* is not reduced when miR172-targeted *AP2*-group genes are suppressed by miR172 overexpression, whereas *PEP1*, the *A. alpina* *FLC* ortholog (Wang et al., 2009b), is decreased when *PEP2* is inactivated. Interestingly, the expression of miR172 is coupled with miR156 in *C. flexuosa* but not in *A. alpina*, where a rise in miR172 abundance is observed in developing floral primordia (Bergonzi et al., 2013).

As is the case in vegetative phase transition, the role of miR156 in flowering time is conserved. Overexpression of miR156 delays flowering, whereas overexpression of a miR156-nontargetable version of *SPL* results in early flowering (Preston and Hileman, 2010; Zhang et al., 2011; Xie et al., 2006; Wang et al., 2014; Bhogale et al., 2014; Fu et al., 2012).

18.6 THE miR156-SPL MODULE IN DEVELOPMENTAL TRANSITIONS IN MOSS

miR156 is present in all major plant taxa, including bryophytes. Unlike land plants (embryophytes), bryophytes are gametophyte dominant: the predominant longer lived plant is the haploid gametophyte. Diploid sporophytes appear only occasionally and remain attached to and nutritionally dependent on the gametophyte.

miR156 accumulation is temporally regulated in *Physcomitrella patens* gametophytes, as seen in other higher plants. After spore germination, miR156 levels are gradually increased and reach the maximum level when young protonemata switch to adult leafy gametophores. Afterward, miR156 is decreased as the leafy gametophores mature (Cho et al., 2012).

Characterization of transgenic *P. patens* (moss) plants revealed that miR156 promotes a developmental change from young filamentous protonemata to leafy gametophores, opposite to its role as an inhibitor of development in flowering plants. The role of miR156 in developmental transitions in *P. patens* is in part mediated by miR390, which acts as a trigger for the accumulation of trans-acting small interfering RNAs (tasiRNAs). Overexpression of miR390 causes a slower formation of gametophores. Inactivation of miR156 results in an increased level of miR390-triggered tasiRNA accumulation, decreased accumulation of tasiRNA targets, and thereby delayed sporophyte-to-gametophyte transition.

18.7 THE miR156-SPL MODULE IN OTHER DEVELOPMENTAL PROCESSES

The role of miR156-targeted *SPL* is not restricted to vegetative phase transition and flowering. Growing evidence indicates that these transcription factors also play critical roles in other developmental processes that require proper timing (Table 18.1).

18.7.1 The Role of the miR156-SPL Module in Age-Related Resistance

Age-related resistance (ARR) has been observed in a number of plant species. For example, *A. thaliana* plants become more resistant, or less susceptible, to virulent *Pseudomonas syringae* as plants mature. However, little is known about the biochemical or molecular mechanisms involved in this response.

Plant TIR-NB-LRR (toll-like/interleukin-1 receptor-nucleotide -binding-leucine-rich repeat) immune receptors induce defense signaling in response to pathogen-encoded effectors. Padmanabhan et al. investigated the role of SPL in defense against tobacco mosaic virus (TMV) in *N. benthamiana* and against *P. syringae* in *A. thaliana*. During active immune response, the *N. benthamiana* TIR-NB-LRR N immune receptor binds to NbSPL6, a miR156-targeted SBP-box transcription factor in *N. benthamiana*, within distinct nuclear compartments (Padmanabhan et al., 2013). *NbSPL6*-silenced plants exhibit a loss-of-resistance phenotype, indicating that *NbSPL6* encodes an essential factor for N-mediated resistance to TMV. Similarly, the *Arabidopsis* ortholog *AtSPL6* was found to be required for resistance mediated by the TIR-NB-LRR RPS4 against *P. syringae* harboring the avrRps4 effector. AtSPL6 positively regulates a subset of defense genes. Because SPL6 is negatively regulated by miR156, these results suggest that an age-dependent increase in *SPL6* level might underlie age-related resistance.

18.7.2 The Role of the miR156-SPL Module in Stress Response

As sessile organisms, plants have to cope with recurring stress conditions to ensure survival and reproductive success. Three miR156-encoding genes, *MIR156C*, *MIR156D* and *MIR156H*, are highly induced after heat shock (HS). Despite not being the only miRNA induced after HS, miR156 overexpression is sufficient to boost HS memory by promoting sustained expression of HS-responsive genes. These findings indicate that the miR156-SPL module serves as an integrator of HS responses and development in plants, and suggest that the induction of miR156 by stress might provide an avenue for plants to avoid flowering under stress conditions by prolonging the juvenile phase (Stief et al., 2014).

18.7.3 The Role of the miR156-SPL Module in Leaf Initiation Rate (Plastochron)

Overexpression of miR156 causes the leaf initiation rate in *A. thaliana*, to increase. The same occurs with Chinese cabbage, maize, tomato, poplar, potato, rice, switchgrass, and *Torenia fournieri*, indicating that the role of miR156 in leaf initiation rate is conserved across plant species (Wang et al., 2011, 2014; Bhogale et al., 2014; Fu et al., 2012; Chuck et al., 2007, 2011; Schwab et al., 2005; Xie et al., 2006; Shikata et al., 2012; Zhang et al., 2011; Martin et al., 2010). *In situ* hybridization studies in *A. thaliana* revealed that *SPL9* is expressed in leaf primordia but absent from shoot apical meristems, demonstrating that SPL9 nonautonomously inhibits initiation of new leaves at the shoot apical meristem. This conclusion is further supported by misexpression experiments, where specific expression of miR156 in leaves causes the same effect on the leaf initiation rate. Intriguingly, the effect of miR156-targeted SPL genes on the leaf initiation rate is correlated with changes in leaf size, suggesting a potential compensatory mechanism that links the rate at which leaves are produced to final leaf size (Wang et al., 2008; Usami et al., 2009).

18.7.4 The Role of the miR156-SPL Module in Ligule Differentiation

The maize ligule and auricle are structures on the maize leaf that develop at the boundary of the sheath and the blade. The maize *liguleless1* (*lg1*) mutant does not form a ligule and auricle and the blade–sheath boundary does not develop as an exact line between sheath and blade (Moreno et al., 1997). Map-based cloning demonstrated

TABLE 18.1 Known Functions of miR156-Targeted *SPL* Genes

Developmental process	Organisms	miR156-SPL genes	Direct downstream target	References
Embryonic development	*A. thaliana*	*SPL10*, *SPL11*	?	Nodine and Bartel (2010)
Vegetative phase transition	*A. thaliana*	*SPL9*, *SPL10*, *SPL11*, *SPL15* miR156	miR172	Wu and Poethig (2006); Wu et al. (2009)
	B. rapa			Wang et al. (2014)
	Z. mays			Chuck et al. (2007)
	T. fournieri			Shikata et al. (2012)
Flowering time	*A. majus*	*SPL3*, *SPL4* *SPL9*, *SPL15* miR156	*MIR172B*, *LFY*, *SOC1*, *FUL*, *AP1*, *FT*	Preston and Hileman (2010)
	A. thaliana			Wang et al. (2009a); Kim et al. (2012); Wu et al. (2009)
	B. rapa			Wang et al. (2014)
	O. sativa			Xie et al. (2006)
	P. virgatum			Fu et al. (2012)
	S. lycopersicum			Zhang et al. (2011)
	S. tuberosum			Bhogale et al. (2014)
	Z. mays			Chuck et al. (2007)
Leaf initiation rate	*A. thaliana*	*SPL9*, *SPL15* miR156	?	Wang et al. (2008); Martin et al. (2010)
	B. rapa			Wang et al. (2014)
	O. sativa			Xie et al. (2006)
	P. × canadensis			Wang et al. (2011)
	P. virgatum			Fu et al. (2012)
	S. lycopersicum			Zhang et al. (2011)
	S. tuberosum			Bhogale et al. (2014)
	T. fournieri			Shikata et al. (2012)
	Z. mays			Chuck et al. (2007)
Anthocyanin biosynthesis	*A. thaliana*	*SPL9*, *SPL15*	*DFR*	Gou et al. (2011)
Trichome distribution on stem	*A. thaliana*	*SPL9*, *SPL15*	*TCL1*, *TRY*	Yu et al. (2010)
Heat stress	*A. thaliana*	*SPL2*	?	Stief et al. (2014)
Pathogen resistance	*A. thaliana*	*SPL6*	?	Padmanabhan et al. (2013)
	N. benthamiana	*NbSPL6*		
Branching	*A. thaliana*	*SPL9*, *SPL15* miR156	?	Schwab et al. (2005); Schwarz et al. (2008)
	O. sativa			Jiao et al. (2010); Miura et al. (2010); Xie et al. (2006)
	P. × canadensis			Wang et al. (2011)
Stamen development	*A. thaliana*	SPL9-group genes	?	Xing et al. (2010)
Gynoecium development	*A. thaliana*	SPL9-group genes	?	Xing et al. (2013)
Fruit development	*S. lycopersicum*	miR156 *CNR*	?	Silva et al. (2014); Manning et al. (2006)
Bract development	*Z. mays*	*TSH4* (*ZmSBP6*)	?	Chuck et al. (2010)
Ligule differentiation	*Z. mays*	*lg1*	?	Moreno et al. (1997)
Tuberization	*S. tuberosum*	miR156	?	Bhogale et al. (2014); Eviatar-Ribak et al. (2013)

that *LG1* encodes a putative transcription factor, similar to SBP1 and SBP2 in *Antirrhinum majus*. *LG1* is expressed at very low levels in the ligular region of developing maize leaf primordia. The molecular mechanism by which *LG1* regulates ligule differentiation is still unknown.

18.7.5 The Role of the miR156-SPL Module in Plant Architecture

Plant architectures are defined by the degree of branching, the length of internodes, and shoot determinancy (Wang and Li, 2008). Overexpression of miR156 in *A. thaliana* reduces apical dominance, revealing a role played by miR156 and its targets in shoot branching (Schwab et al., 2005). A similar effect was observed in rice. The ideal plant architecture (*IPA1*)/*OsSPL14* point mutation in rice, which disturbs miR156 recognition, results in a rice plant with a reduced tiller number (the number of branches), increased lodging resistance, and enhanced grain yield (Jiao et al., 2010; Miura et al., 2010). Genome-wide chromatin immunoprecipitation on chip (ChIP-chip) and ChIP followed by sequencing (ChIP-seq) experiments revealed that IPA1 directly binds to the promoter of rice TEOSINTE BRANCHED1, a negative regulator of tiller bud outgrowth (Chapter 16), to suppress rice tillering. IPA1 directly and positively regulates DENSE AND ERECT PANICLE1, an important gene-regulating panicle architecture, to influence plant height and panicle length (Lu et al., 2013).

The gradual suppression of bracts during floral transition is another feature shaping inflorescence architecture. In maize, *TASSEL SHEATH 4* (*TSH4*, *ZmSBP6*) encodes an SBP-box transcription factor targeted by miR156. Expression analyses showed that *TSH4* is expressed in a broad domain near the top of the inflorescence in the primordia anlagen. As the primordia grow, *TSH4* transcripts are localized in a group of cells adaxial to spikelet pair meristems, where the bract is suppressed. Consistent with this expression pattern, *TSH4* is found to play a major role in the establishment of lateral meristems and the repression of bract initiation. The *tsh4* mutant displays altered phyllotaxy and ectopic bract formation at the expense of the lateral meristem (Chuck et al., 2010).

In potato (*Solanum tuberosum*), overexpression of miR156 also causes altered plant architecture and extends tuber-forming potential to distal axillary buds (Eviatar-Ribak et al., 2013). Notably, miR156 is present in phloem cells, and mobility assays in heterografts suggest that miR156 is a graft-transmissible and phloem-mobile signal regulating leaf morphology and tuberization (Bhogale et al., 2014).

18.7.6 Temporal Regulation of Anthocyanin Biosynthesis by the miR156-SPL Module

Anthocyanins and flavonols are derived from phenylalanine and share common precursors, dihydroflavonols, which are substrates for both flavonol synthase and dihydroflavonol 4-reductase. In the stems of *A. thaliana*, anthocyanins are accumulated in an acropetal manner, with the highest level at the junction between rosette and stem. This accumulation pattern is generated by the gradual change of miR156-targeted SPL9-group genes along the stem. The miR156 level is correlated with anthocyanin content. Increased miR156 activity promotes accumulation of anthocyanins, whereas reduced miR156 activity results in high levels of flavonols. Mechanistically, SPL9 negatively regulates anthocyanin accumulation by directly preventing expression of anthocyanin biosynthetic genes such as dihydroflavonol reductase (*DFR*) by destabilizing a myeloblastosis (MYB), basic helix–loop–helix (bHLH), WD40 transcriptional activation complex (Gou et al., 2011).

18.7.7 Temporal Regulation of Trichome Development on Stem and Floral Organs by the miR156-SPL Module

Gradual increase in the level of miR156-targeted *SPL*s not only directs metabolic flux in the flavonoid biosynthetic pathway, but also affects the temporal production of trichomes on the stem. After bolting, the number of trichomes is progressively reduced on the inflorescence stem in *A. thaliana*. Plants overexpressing miR156 develop ectopic trichomes on the stem and floral organs, whereas elevated levels of SPLs give rise to an opposite phenotype. The effect of miR156 on trichome development is mediated by SPL9-related genes but not SPL3-group genes. Expression analyses of an *SPL9*-inducible line and ChIP assays demonstrated that SPL9 directly activates *TRICHOMELESS1* (*TCL1*) and *TRIPTYCHON* (*TRY*), two MYB transcription factor genes negatively regulating trichome initiation (Yu et al., 2010).

Another axis regulating trichome production on the stem is mediated by miR171-targeted LOST MERISTEMS 1 (LOM1), LOM2, and LOM3, encoding GRAS family members previously known to maintain shoot apical meristems. Reduced *LOM* abundance leads to decreased trichome density on stems and floral organs; conversely, elevated expression of *LOM* promotes trichome production. The role of LOMs in shaping trichome distribution is dependent on SPLs. LOM binds to SPL and suppresses its transcriptional activity, thereby blunting the expression of trichome repressor genes *TCL1* and *TRY* (Xue et al., 2014).

18.7.8 The Role of the miR156-SPL Module in Floral Organ Development

In the *A. thaliana* genome, not all *SPL* genes are targeted by miR156 (Xing et al., 2010; Chen et al., 2010). Studies have revealed that *SPL8*, a non-miR156-targeted *SPL* gene, shares functional redundancy with those targeted by miR156. An *SPL8* loss-of-function mutant exhibits a semisterile phenotype (Xing et al., 2010). Overexpression of miR156 enhances this defect, resulting in fully sterile plants. Histological analyses demonstrated that the sterility of this plant is caused by an almost complete absence of sporogenous and anther wall tissue differentiation, a phenotype similar to that reported for *nozzle* (*nzz*) mutant anthers. Expression studies further revealed that, together with *NZZ* or independently, these *SPL* genes orchestrate the expression of genes mediating cell division, differentiation, and specification early in anther development.

Interestingly, *SPL8* and miR156-targeted *SPL* genes are also expressed within the developing gynoecium, where they redundantly control development of the female reproductive tract. Although the gynoecium of *spl8* mutants is largely normal, the additional inactivation of miR156-targeted *SPL* genes by miR156 overexpression results in a shortened style and an apically swollen ovary narrowing onto an elongated gynophore. Auxin sensitivity and expression analyses suggested that these SPLs regulate gynoecium patterning through interference with auxin homeostasis and signaling (Xing et al., 2013).

Tomato (*S. lycopersicum*) was also used as a model to study the role of miR156 and its targets in ovary and fruit development. miR156-targeted *SPLs* are dynamically expressed in developing flowers and ovaries, while miR156 is highly enriched in the meristematic tissues of the ovary, including the placenta and ovules. Transgenic tomato overexpressing miR156 exhibits abnormal fruit morphology with extra carpels and ectopic structures. In agreement with this, expression analyses found that the genes associated with meristem maintenance and initiation are induced in the developing ovaries of transgenic plants. Therefore, the miR156-SPL module is involved in the maintenance of the meristematic state of ovary tissues, thereby controlling initial steps of fleshy fruit development and determinacy (Silva et al., 2014). A more direct role of *SPL* in tomato fruit development was revealed by analyzing a fruit-ripening mutant, *colorless non-ripening* (*cnr*). The *cnr* mutation causes colorless fruit with a substantial loss of cell-to-cell adhesion. Positional cloning found the mutant phenotype is due to a spontaneous epigenetic change in *CNR*, which encodes an SBP-box transcription factor, similar to SPL3 in *A. thaliana* (Manning et al., 2006).

The role played by *SPL* is not restricted to fruit development. It has been reported that *OsSPL16* (*GW8*), another miR156-targeted SPL, regulates grain size and shape in rice by promoting cell proliferation. Elevated expression of *OsSPL16* promotes cell division and grain filling, whereas inactivation of *OsSPL16* results in the formation of a more slender grain (Wang et al., 2012). In maize, *tga1* is responsible for ear formation, one of the key events in the domestication of maize (*Z. mays* ssp. *mays*) from its progenitor teosinte (*Z. mays* ssp. *parviglumis*). Phenotypic and genetic studies showed that *tga1* encodes an SPL transcriptional regulator, determining the ease with which the fruit can be separated from surrounding inflorescence structures (Wang et al., 2005).

18.8 PERSPECTIVES

Previous efforts using *A. thaliana*, rice, and maize have revealed that miR156 plays a central role in developmental timing in plants. The precise timing of miR156 abundance and the diversity of SPL downstream targets constitute a powerful and flexible pathway regulating many aspects of plant developmental and physiological processes that require proper timing. However, our understanding of this pathway has just begun.

First, the regulatory basis of the developmental decline of miR156 remains unclear. One major question is how sugar represses miR156 expression. As a result of mutations in *HXK1*, which encodes a glucose sensor, not blocking the decline of miR156 (Yu et al., 2013; Yang et al., 2013), there must be an unidentified factor participating in sugar sensing and subsequently repressing miR156 expression. Despite the identification of sugar as the upstream regulator of miR156 explaining the irreversible nature of developmental transitions, another question is whether miR156 can be regulated by other factors. Indeed, ChIP–chip data revealed that FUSCA3 (FUS3), a B3 domain transcription factor regulating seed development, is associated with regulatory regions of *MIR156A* and *MIR156C* in *A. thaliana* (Wang and Perry, 2013). Furthermore, miR156 is encoded by eight loci in *A. thaliana*. It remains to be determined whether individual loci are functional and whether they exhibit different expression patterns.

Second, downstream events of miR156-targeted *SPLs* are also poorly understood. What causes the functional divergence between SPL3-group and SPL9-group genes? Both group genes are able to activate *FUL* expression (Wang et al., 2009a; Kim et al., 2012), whereas *MIR172B* is only activated by SPL9-group genes (Wu et al., 2009). A plausible explanation for this is that SPL9 differs from SPL3 in that its C-terminal domain is involved in protein–protein interaction. Therefore, the identification of SPL9-binding proteins and genome-wide identification of SPL3/SPL9 targets might help us to answer this question.

Third, how does SPL concurrently coordinate the diverse developmental and physiological processes? It has been shown that SPL9 acts as a bifunctional transcription factor and regulates downstream gene expression *via* distinct mechanisms. It regulates the trichome distribution pattern by activating *TCL1* and *TRY* (Yu et al., 2010), but controls anthocyanin biosynthesis by repressing dihydroflavonol 4-reductase (*DFR*) genes (Gou et al., 2011). Thus, illustration of the precise molecular mechanism by which SPL9 regulates specific downstream gene expression in each developmental context will be of particular interest.

Fourth, despite the miR156-SPL module regulating morphological changes during vegetative phase transition, it is still unknown whether it also plays a role in physiological change, such as the biosynthesis of primary or secondary metabolites, during vegetative phase transition and flowering.

Finally, the multiple roles of miR156-targeted *SPLs* in plant development and physiology offer promising candidate genes for biotechnology. For instance, overexpression of miR156 markedly increases biomass (Chuck et al., 2011; Fu et al., 2012), which has a major impact on biofuel production. In addition, by altering the levels of miR156 or its downstream factor miR172, it is possible to either prolong or shorten juvenile development in fruit trees. Therefore, identification of the species-specific role of the miR156-SPL module in agricultural traits of crops and other commercial plants will broaden the application of this miRNA.

Acknowledgments

The author acknowledges the staff of the Wang Laboratory for helpful comments on the manuscript. The work done in the Wang Laboratory is supported by grants from the National Natural Science Foundation of China (31430013, 31222029, 91217306), State Key Basic Research Program of China (2013CB127000), Shanghai Pujiang Program (12PJ1409900), Recruitment Program of Global Expects (China), and Key Research Program from NKLPMG (SIPPE, SIBS).

References

Allsopp, A., 1952. Experimental and analytical studies of Pteridophytes XVII. The effect of various physiologically active substances on the development of *Marsilea* in sterile culture. Ann. Bot. 16, 165–183.

Allsopp, A., 1953. Experimental and analytical studies of Pteridophytes XXI. Investigations on *Marsilea*. 3. The effect of various sugars on development and morphology. Ann. Bot. 17, 447–463.

Amasino, R.M., Michaels, S.D., 2010. The timing of flowering. Plant Physiol. 154, 516–520.

Andres, F., Coupland, G., 2012. The genetic basis of flowering responses to seasonal cues. Nat. Rev. Genet. 13, 627–639.

Balasubramanian, S., Sureshkumar, S., Lempe, J., Weigel, D., 2006. Potent induction of *Arabidopsis thaliana* flowering by elevated growth temperature. PLoS Genet. 2, e106.

Bäurle, I., Dean, C., 2006. The timing of developmental transitions in plants. Cell 125, 655–664.

Bergonzi, S., Albani, M.C., Ver Loren Van Themaat, E., Nordstrom, K.J., Wang, R., Schneeberger, K., Moerland, P.D., Coupland, G., 2013. Mechanisms of age-dependent response to winter temperature in perennial flowering of *Arabis alpina*. Science 340, 1094–1097.

Bhogale, S., Mahajan, A.S., Natarajan, B., Rajabhoj, M., Thulasiram, H.V., Banerjee, A.K., 2014. MicroRNA156: a potential graft-transmissible microRNA that modulates plant architecture and tuberization in *Solanum tuberosum* ssp. *andigena*. Plant Physiol. 164, 1011–1027.

Brink, R.A., 1962. Phase change in higher plants and somatic cell heredity. Q. Rev. Biol. 37, 1–22.

Canales, C., Barkoulas, M., Galinha, C., Tsiantis, M., 2010. Weeds of change: *Cardamine hirsuta* as a new model system for studying dissected leaf development. J. Plant Res. 123, 25–33.

Capron, A., Chatfield, S., Provart, N., Berleth, T., 2009. Embryogenesis: pattern formation from a single cell. Arabidopsis Book 7, e0126.

Cardon, G., Hohmann, S., Klein, J., Nettesheim, K., Saedler, H., Huijser, P., 1999. Molecular characterisation of the *Arabidopsis* SBP-box genes. Gene 237, 91–104.

Chen, X., Zhang, Z., Liu, D., Zhang, K., Li, A., Mao, L., 2010. SQUAMOSA promoter-binding protein-like transcription factors: star players for plant growth and development. J. Integr. Plant Biol. 52, 946–951.

Cho, S.H., Coruh, C., Axtell, M.J., 2012. miR156 and miR390 regulate tasiRNA accumulation and developmental timing in *Physcomitrella patens*. Plant Cell 24, 4837–4849.

Chuck, G., Cigan, A.M., Saeteurn, K., Hake, S., 2007. The heterochronic maize mutant *Corngrass1* results from overexpression of a tandem microRNA. Nat. Genet. 39, 544–549.

Chuck, G., Whipple, C., Jackson, D., Hake, S., 2010. The maize SBP-box transcription factor encoded by *tasselsheath4* regulates bract development and the establishment of meristem boundaries. Development 137, 1243–1250.

Chuck, G.S., Tobias, C., Sun, L., Kraemer, F., Li, C., Dibble, D., Arora, R., Bragg, J.N., Vogel, J.P., Singh, S., Simmons, B.A., Pauly, M., Hake, S., 2011. Overexpression of the maize *Corngrass1* microRNA prevents flowering, improves digestibility, and increases starch content of switchgrass. Proc. Natl. Acad. Sci. USA 108, 17550–17555.

Coustham, V., Li, P., Strange, A., Lister, C., Song, J., Dean, C., 2012. Quantitative modulation of polycomb silencing underlies natural variation in vernalization. Science 337, 584–587.

Crevillen, P., Sonmez, C., Wu, Z., Dean, C., 2013. A gene loop containing the floral repressor FLC is disrupted in the early phase of vernalization. EMBO J. 32, 140–148.

Dill, A., Jung, H.S., Sun, T.P., 2001. The DELLA motif is essential for gibberellin-induced degradation of RGA. Proc. Natl. Acad. Sci. USA 98, 14162–14167.

Dittmar, E.L., Oakley, C.G., Agren, J., Schemske, D.W., 2014. Flowering time QTL in natural populations of *Arabidopsis thaliana* and implications for their adaptive value. Mol. Ecol. 23, 4291–4303.

Eviatar-Ribak, T., Shalit-Kaneh, A., Chappell-Maor, L., Amsellem, Z., Eshed, Y., Lifschitz, E., 2013. A cytokinin-activating enzyme promotes tuber formation in tomato. Curr. Biol. 23, 1057–1064.

Fu, C., Sunkar, R., Zhou, C., Shen, H., Zhang, J.Y., Matts, J., Wolf, J., Mann, D.G., Stewart, Jr., C.N., Tang, Y., Wang, Z.Y., 2012. Overexpression of miR156 in switchgrass (*Panicum virgatum* L.) results in various morphological alterations and leads to improved biomass production. Plant Biotechnol. J. 10, 443–452.

Gandikota, M., Birkenbihl, R.P., Hohmann, S., Cardon, G.H., Saedler, H., Huijser, P., 2007. The miRNA156/157 recognition element in the 3′ UTR of the *Arabidopsis* SBP box gene SPL3 prevents early flowering by translational inhibition in seedlings. Plant J. 49, 683–693.

Gou, J.Y., Felippes, F.F., Liu, C.J., Weigel, D., Wang, J.W., 2011. Negative regulation of anthocyanin biosynthesis in *Arabidopsis* by a miR156-targeted SPL transcription factor. Plant Cell 23, 1512–1522.

Hudson, C.J., Freeman, J.S., Jones, R.C., Potts, B.M., Wong, M.M., Weller, J.L., Hecht, V.F., Poethig, R.S., Vaillancourt, R.E., 2014. Genetic control of heterochrony in *Eucalyptus globulus*. G3 (Bethesda) 4, 1235–1245.

Huijser, P., Schmid, M., 2011. The control of developmental phase transitions in plants. Development 138, 4117–4129.

James, S.A., Bell, D.T., 2001. Leaf morphological and anatomical characteristics of heteroblastic *Eucalyptus globulus* ssp. *globulus* (Myrtaceae). Aust. J. Bot. 49, 259–269.

Jiao, Y., Wang, Y., Xue, D., Wang, J., Yan, M., Liu, G., Dong, G., Zeng, D., Lu, Z., Zhu, X., Qian, Q., Li, J., 2010. Regulation of *OsSPL14* by OsmiR156 defines ideal plant architecture in rice. Nat. Genet. 42, 541–544.

Jung, J.H., Ju, Y., Seo, P.J., Lee, J.H., Park, C.M., 2011. The SOC1–SPL module integrates photoperiod and gibberellic acid signals to control flowering time in *Arabidopsis*. Plant J. 69, 577–588.

Kaplan, D.R., 1980. Heteroblastic leaf development in *Acacia*. Morphological and morphogenetic implications. Cellule 73, 137–203.

Kerstetter, R.A., Poethig, R.S., 1998. The specification of leaf identity during shoot development. Annu. Rev. Cell Dev. Biol. 14, 373–398.

Kim, J.J., Lee, J.H., Kim, W., Jung, H.S., Huijser, P., Ahn, J.H., 2012. The microRNA156-SQUAMOSA PROMOTER BINDING PROTEIN-LIKE3 module regulates ambient temperature-responsive flowering via FLOWERING LOCUS T in Arabidopsis. Plant Physiol. 159, 461–478.

Klein, J., Saedler, H., Huijser, P., 1996. A new family of DNA binding proteins includes putative transcriptional regulators of the Antirrhinum majus floral meristem identity gene SQUAMOSA. Mol. Gen. Genet. 250, 7–16.

Lee, J., Lee, I., 2010. Regulation and function of SOC1, a flowering pathway integrator. J. Exp. Bot. 61, 2247–2254.

Libby, W.J., Hood, J.V., 1976. Juvenility in hedged radiata pine. Acta Horticult. 56, 91–98.

Lu, Z., Yu, H., Xiong, G., Wang, J., Jiao, Y., Liu, G., Jing, Y., Meng, X., Hu, X., Qian, Q., Fu, X., Wang, Y., Li, J., 2013. Genome-wide binding analysis of the transcription activator IDEAL PLANT ARCHITECTURE1 reveals a complex network regulating rice plant architecture. Plant Cell 25, 3743–3759.

Manning, K., Tor, M., Poole, M., Hong, Y., Thompson, A.J., King, G.J., Giovannoni, J.J., Seymour, G.B., 2006. A naturally occurring epigenetic mutation in a gene encoding an SBP-box transcription factor inhibits tomato fruit ripening. Nat. Genet. 38, 948–952.

Martin, R.C., Asahina, M., Liu, P.P., Kristof, J.R., Coppersmith, J.L., Pluskota, W.E., Bassel, G.W., Goloviznina, N.A., Nguyen, T.T., Martinez-Andujar, C., Arun Kumar, M.B., Pupel, P., Nonogaki, H., 2010. The regulation of post-germinative transition from the cotyledon- to vegetative-leaf stages by microRNA-targeted SQUAMOSA PROMOTER-BINDING PROTEIN LIKE13 in Arabidopsis. Seed Sci. Res. 20, 89–96.

Miura, K., Ikeda, M., Matsubara, A., Song, X.J., Ito, M., Asano, K., Matsuoka, M., Kitano, H., Ashikari, M., 2010. OsSPL14 promotes panicle branching and higher grain productivity in rice. Nat. Genet. 42, 545–549.

Moore, B., Zhou, L., Rolland, F., Hall, Q., Cheng, W.H., Liu, Y.X., Hwang, I., Jones, T., Sheen, J., 2003. Role of the Arabidopsis glucose sensor HXK1 in nutrient, light, and hormonal signaling. Science 300, 332–336.

Moreno, M.A., Harper, L.C., Krueger, R.W., Dellaporta, S.L., Freeling, M., 1997. liguleless1 encodes a nuclear-localized protein required for induction of ligules and auricles during maize leaf organogenesis. Genes Dev. 11, 616–628.

Murase, K., Hirano, Y., Sun, T.P., Hakoshima, T., 2008. Gibberellin-induced DELLA recognition by the gibberellin receptor GID1. Nature 456, 459–463.

Njoku, E., 1956. The effect of defoliation on leaf shape in Ipomoea caerulea. New Phytol. 55, 213–228.

Nodine, M.D., Bartel, D.P., 2010. MicroRNAs prevent precocious gene expression and enable pattern formation during plant embryogenesis. Genes Dev. 24, 2678–2692.

Padmanabhan, M.S., Ma, S., Burch-Smith, T.M., Czymmek, K, Huijser, P., Dinesh-Kumar, S.P., 2013. Novel positive regulatory role for the SPL6 transcription factor in the N TI –NB–LRR receptor-mediated plant innate immunity. PLoS Pathog. 9, e1003235.

Poethig, R.S., 2010. The past, present, and future of vegetative phase change. Plant Physiol. 154, 541–544.

Poethig, R.S., 2013. Vegetative phase change and shoot maturation in plants. Curr. Top. Dev. Biol. 105, 125–152.

Preston, J.C., Hileman, L.C., 2010. SQUAMOSA-PROMOTER BINDING PROTEIN 1 initiates flowering in Antirrhinum majus through the activation of meristem identity genes. Plant J. 62, 704–712.

Reinhart, B.J., Weinstein, E.G., Rhoades, M.W., Bartel, B., Bartel, D.P., 2002. MicroRNAs in plants. Genes Dev. 16, 1616–1626.

Rogers, K., Chen, X., 2013. Biogenesis, turnover, and mode of action of plant microRNAs. Plant Cell 25, 2383–2399.

Romera-Branchat, M., Andres, F., Coupland, G., 2014. Flowering responses to seasonal cues: what's new? Curr. Opin. Plant Biol. 21C, 120–127.

Rosa, S., De Lucia, F., Mylne, J.S., Zhu, D., Ohmido, N., Pendle, A., Kato, N., Shaw, P., Dean, C., 2013. Physical clustering of FLC alleles during polycomb-mediated epigenetic silencing in vernalization. Genes Dev. 27, 1845–1850.

Schaffalitzky de Muckadell, M., 1954. Juvenile stages in woody plants. Physiol. Plant. 7, 782–796.

Schwab, R., Palatnik, J.F., Riester, M., Schommer, C., Schmid, M., Weigel, D., 2005. Specific effects of microRNAs on the plant transcriptome. Dev. Cell 8, 517–527.

Schwarz, S., Grande, A.V., Bujdoso, N., Saedler, H., Huijser, P., 2008. The microRNA regulated SBP-box genes SPL9 and SPL15 control shoot maturation in Arabidopsis. Plant Mol. Biol. 67, 183–195.

Searle, I., He, Y., Turck, F., Vincent, C., Fornara, F., Krober, S., Amasino, R.A., Coupland, G., 2006. The transcription factor FLC confers a flowering response to vernalization by repressing meristem competence and systemic signaling in Arabidopsis. Genes Dev. 20, 898–912.

Shikata, M., Koyama, T., Mitsuda, N., Ohme-Takagi, M., 2009. Arabidopsis SBP-box genes SPL10, SPL11 and SPL2 control morphological change in association with shoot maturation in the reproductive phase. Plant Cell Physiol. 50, 2133–2145.

Shikata, M., Yamaguchi, H., Sasaki, K., Ohtsubo, N., 2012. Overexpression of Arabidopsis miR157b induces bushy architecture and delayed phase transition in Torenia fournieri. Planta 236, 1027–1035.

Shindo, C., Lister, C., Crevillen, P., Nordborg, M., Dean, C., 2006. Variation in the epigenetic silencing of FLC contributes to natural variation in Arabidopsis vernalization response. Genes Dev. 20, 3079–3083.

Silva, G.F., Silva, E.M., Azevedo, M.D., Guivin, M.A., Ramiro, D.A., Figueiredo, C.R., Carrer, H., Peres, L.E., Nogueira, F.T., 2014. microRNA156-targeted SPL/SBP-box transcription factors regulate tomato ovary and fruit development. Plant J. 78, 604–618.

Song, J., Angel, A., Howard, M., Dean, C., 2012. Vernalization – a cold-induced epigenetic switch. J. Cell Sci. 125, 3723–3731.

Srikanth, A., Schmid, M., 2011. Regulation of flowering time: all roads lead to Rome. Cell Mol. Life Sci. 68, 2013–2037.

Stief, A., Altmann, S., Hoffmann, K., Pant, B.D., Scheible, W.R., Baurle, I., 2014. Arabidopsis miR156 regulates tolerance to recurring environmental stress through SPL transcription factors. Plant Cell 26, 1792–1807.

Sun, Q., Csorba, T., Skourti-Stathaki, K., Proudfoot, N.J., Dean, C., 2013. R-loop stabilization represses antisense transcription at the Arabidopsis FLC locus. Science 340, 619–621.

Telfer, A., Bollman, K.M., Poethig, R.S., 1997. Phase change and the regulation of trichome distribution in Arabidopsis thaliana. Development 124, 645–654.

Tsai, C.H., Miller, A., Spalding, M., Rodermel, S., 1997. Source strength regulates an early phase transition of tobacco shoot morphogenesis. Plant Physiol. 115, 907–914.

Usami, T., Horiguchi, G., Yano, S., Tsukaya, H., 2009. The more and smaller cells mutants of Arabidopsis thaliana identify novel roles for SQUAMOSA PROMOTER BINDING PROTEIN-LIKE genes in the control of heteroblasty. Development 136, 955–964.

Wang, J.W., 2014. Regulation of flowering time by the miR156-mediated age pathway. J. Exp. Bot. 65, 4723–4730.

Wang, Y., Li, J., 2008. Molecular basis of plant architecture. Annu. Rev. Plant Biol. 59, 253–279.

Wang, F., Perry, S.E., 2013. Identification of direct targets of FUSCA3, a key regulator of Arabidopsis seed development. Plant Physiol. 161, 1251–1264.

Wang, H., Nussbaum-Wagler, T., Li, B., Zhao, Q., Vigouroux, Y., Faller, M., Bomblies, K., Lukens, L., Doebley, J.F., 2005. The origin of the naked grains of maize. Nature 436, 714–719.

Wang, J.W., Schwab, R., Czech, B., Mica, E., Weigel, D., 2008. Dual effects of miR156-targeted SPL genes and CYP78A5/KLUH on plastochron length and organ size in Arabidopsis thaliana. Plant Cell 20, 1231–1243.

Wang, J.W., Czech, B., Weigel, D., 2009a. miR156-regulated SPL transcription factors define an endogenous flowering pathway in *Arabidopsis thaliana*. Cell 138, 738–749.

Wang, R., Farrona, S., Vincent, C., Joecker, A., Schoof, H., Turck, F., Alonso-Blanco, C., Coupland, G., Albani, M.C., 2009b. *PEP1* regulates perennial flowering in *Arabis alpina*. Nature 459, 423–427.

Wang, J.W., Park, M.Y., Wang, L.J., Koo, Y., Chen, X.Y., Weigel, D., Poethig, R.S., 2011. miRNA control of vegetative phase change in trees. PLoS Genet. 7, e1002012.

Wang, S., Wu, K., Yuan, Q., Liu, X., Liu, Z., Lin, X., Zeng, R., Zhu, H., Dong, G., Qian, Q., Zhang, G., Fu, X., 2012. Control of grain size, shape and quality by *OsSPL16* in rice. Nat. Genet. 44, 950–954.

Wang, Y., Wu, F., Bai, J., He, Y., 2014. *BrpSPL9* (*Brassica rapa* ssp. *pekinensis SPL9*) controls the earliness of heading time in Chinese cabbage. Plant Biotechnol. J. 12, 312–321.

Wu, G., Poethig, R.S., 2006. Temporal regulation of shoot development in *Arabidopsis thaliana* by miR156 and its target SPL3. Development 133, 3539–3547.

Wu, G., Park, M.Y., Conway, S.R., Wang, J.W., Weigel, D., Poethig, R.S., 2009. The sequential action of miR156 and miR172 regulates developmental timing in *Arabidopsis*. Cell 138, 750–759.

Xie, K., Wu, C., Xiong, L., 2006. Genomic organization, differential expression, and interaction of SQUAMOSA promoter-binding-like transcription factors and microRNA156 in rice. Plant Physiol. 142, 280–293.

Xie, K., Shen, J., Hou, X., Yao, J., Li, X., Xiao, J., Xiong, L., 2012. Gradual increase of miR156 regulates temporal expression changes of numerous genes during leaf development in rice. Plant Physiol. 158, 1382–1394.

Xing, S., Salinas, M., Hohmann, S., Berndtgen, R., Huijser, P., 2010. miR156-targeted and nontargeted SBP-box transcription factors act in concert to secure male fertility in *Arabidopsis*. Plant Cell 22, 3935–3950.

Xing, S., Salinas, M., Garcia-Molina, A., Hohmann, S., Berndtgen, R., Huijser, P., 2013. SPL8 and miR156-targeted SPL genes redundantly regulate *Arabidopsis* gynoecium differential patterning. Plant J. 75, 566–577.

Xue, X.Y., Zhao, B., Chao, L.M., Chen, D.Y., Cui, W.R., Mao, Y.B., Wang, L.J., Chen, X.Y., 2014. Interaction between two timing microRNAs controls trichome distribution in *Arabidopsis*. PLoS Genet. 10, e1004266.

Yamaguchi, A., Wu, M.F., Yang, L., Wu, G., Poethig, R.S., Wagner, D., 2009. The microRNA-regulated SBP-box transcription factor SPL3 is a direct upstream activator of *LEAFY, FRUITFULL,* and *APETALA1*. Dev. Cell 17, 268–278.

Yang, L., Conway, S.R., Poethig, R.S., 2011. Vegetative phase change is mediated by a leaf-derived signal that represses the transcription of miR156. Development 138, 245–249.

Yang, L., Xu, M., Koo, Y., He, J., Poethig, R.S., 2013. Sugar promotes vegetative phase change in *Arabidopsis thaliana* by repressing the expression of *MIR156A* and *MIR156C*. eLife 2, e00260.

Yu, N., Cai, W.J., Wang, S., Shan, C.M., Wang, L.J., Chen, X.Y., 2010. Temporal control of trichome distribution by microRNA156-targeted SPL genes in *Arabidopsis thaliana*. Plant Cell 22, 2322–2335.

Yu, S., Galvao, V.C., Zhang, Y.C., Horrer, D., Zhang, T.Q., Hao, Y.H., Feng, Y.Q., Wang, S., Schmid, M., Wang, J.W., 2012. Gibberellin regulates the *Arabidopsis* floral transition through miR156-targeted SQUAMOSA PROMOTER BINDING-LIKE transcription factors. Plant Cell 24, 3320–3332.

Yu, S., Cao, L., Zhou, C.M., Zhang, T.Q., Lian, H., Sun, Y., Wu, J.Q., Huang, J.R., Wang, G.D., Wang, J.W., 2013. Sugar is an endogenous cue for juvenile-to-adult phase transition in plants. eLife 2, e00269.

Zhang, X., Zou, Z., Zhang, J., Zhang, Y., Han, Q., Hu, T., Xu, X., Liu, H., Li, H., Ye, Z., 2011. Over-expression of sly-miR156a in tomato results in multiple vegetative and reproductive trait alterations and partial phenocopy of the sft mutant. FEBS Lett. 585, 435–439.

Zhou, C.M., Wang, J.W., 2013. Regulation of flowering time by microRNAs. J. Genet. Genomics 40, 211–215.

Zhou, C.M., Zhang, T.Q., Wang, X., Yu, S., Lian, H., Tang, H., Feng, Z.Y., Zozomova-Lihova, J., Wang, J.W., 2013. Molecular basis of age-dependent vernalization in *Cardamine flexuosa*. Science 340, 1097–1100.

Zografos, B.R., Sung, S., 2012. Vernalization-mediated chromatin changes. J. Exp. Bot. 63, 4343–4348.

CHAPTER 19

Functional Aspects of GRAS Family Proteins

Cordelia Bolle

Department Biologie I, Lehrstuhl für Molekularbiologie der Pflanzen (Botanik),
Biozentrum der LMU München, Planegg-Martinsried, Germany

OUTLINE

19.1 The Role of GRAS Proteins in Development 296	**19.2 The Role of GRAS Proteins in Signaling** 301
19.1.1 GRAS Proteins in Root Development 297	19.2.1 Gibberellin Signaling 301
19.1.1.1 Ground Tissue 297	19.2.2 Light Signaling 302
19.1.1.2 Middle-Cortex Formation 298	19.2.3 Stress Signaling 302
19.1.1.3 The Root Vascular Tissue 298	**19.3 General Principles of GRAS Function** 303
19.1.1.4 Stem Cell Niche Maintenance 298	19.3.1 GRAS Protein–DNA Interaction 303
19.1.1.5 Elongation Zone and Lateral Root Formation 299	19.3.2 Protein–protein Interactions of GRAS Proteins 306
19.1.1.6 Root Modification (Nodulation, Mycorrhiza) 299	19.3.2.1 Interaction Between GRAS Proteins (Homo- and Heterodimers) 306
19.1.2 GRAS Proteins in Shoot Development 300	19.3.2.2 Interaction of GRAS Proteins with Other Factors 306
19.1.3 GRAS Proteins in Leaf Development 300	19.3.3 Protein Movement of GRAS Proteins 307
19.1.4 GRAS Proteins in Shoot Apical Meristem Development 300	19.3.4 Regulation of GRAS Gene Expression by miRNA 307
19.1.5 GRAS Proteins in Axillary Meristem Development 301	**19.4 Conclusion** 307
19.1.6 GRAS Proteins in Flower, Embryo, and Seed Development 301	**References** 307

The GRAS proteins are an important, mainly plant-specific, protein family whose name derives from its first three identified members in *Arabidopsis thaliana*: gibberellic acid insensitive (GAI), repressor of GA1 (RGA), and scarecrow (SCR; Di Laurenzio et al., 1996; Peng et al., 1997; Silverstone et al., 1998). The gene family comprises at least 30 members in plants but has sometimes also expanded into more than 100 members (Chapter 10). About 8–12 distinct clades can be discriminated, which are usually named after the most prominent member of the clade and which in most cases can be found in all plants including mosses (Figure 19.1, for details see Chapter 10). The conserved GRAS domain can usually be found in the C-terminal part of the proteins, whereas the N-terminal part is very variant in sequence and length. Especially in *Arabidopsis* several GRAS proteins have been characterized according to their biological function, which has demonstrated that this protein family plays an important role in a wide variety of biological processes affecting many different aspects of development and signaling in plants. Changes in gene expression are often observed in loss-of-function mutants suggesting that GRAS

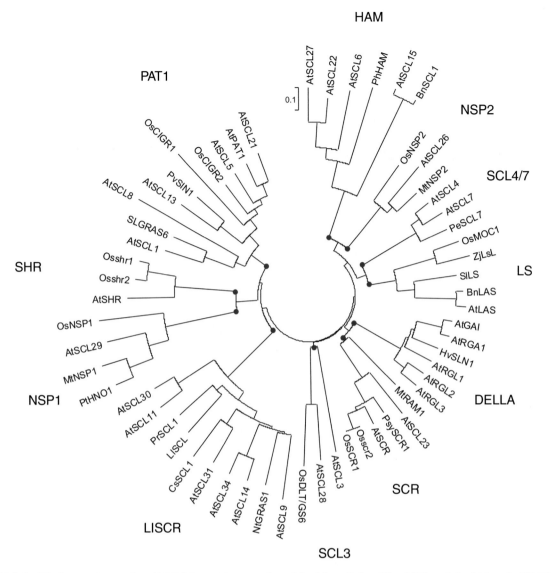

FIGURE 19.1 Phylogenetic tree for all GRAS proteins mentioned in this article with additional Arabidopsis GRAS proteins to show homologs from other species. The eleven main clades are indicated by name and red dot. At, Arabidopsis thaliana; Bn, Brassica napus; Hv, Hordeum vulgare; Ll, Lilium longiflorum; Mt, Medicago trunculata; Nt, Nicotiana tabacum; Os, Oryza sativa; Pe, Populus euphratica; Ph, Petunia × hybrida; Pr, Pinus radiata; Psy, Pinus sylvestris; Pt, Populus trichocarpa; Pv, Phaseolus vulgare; Sl, Solanum lycopersicum; Zj, Zoysia japonica. Alignment was performed with DNA Star/Lasergene bootstrapping; and tree generation with Mega (Neighbor-joining).

proteins are somehow involved in transcriptional control. Nevertheless, many loss-of-function mutants of at least *Arabidopsis* GRAS proteins do not show any obvious phenotype compared to the wild type (Lee et al., 2008; Bolle, unpublished results), either because these phenotypes are subtle or because of the enlargement of the protein family resulting in functional redundancy (Chapter 10). Whereas the biological processes some of them are involved in have been studied in great detail, the biochemical function of GRAS proteins is still not very clear.

19.1 THE ROLE OF GRAS PROTEINS IN DEVELOPMENT

Development is based on cell differentiation and elongation. To determine cell fate this process has to be coordinated very precisely both in time and space. In recent years the involvement of several GRAS factors in these processes has been demonstrated. They seem to be especially important for cell differentiation and organ boundary formation.

19.1.1 GRAS Proteins in Root Development

Root development occurs in a very precise and controlled fashion. In the root apex the quiescent center (QC), a set of mostly mitotically inactive cells, can be found, protected by the root cap and columella cells. Stem cells or initial cells surround the QC. They undergo asymmetric divisions with one cell remaining an initial cell while the other divides again periclinically. These initial cells differentiate leading to the different tissue layers found. The outer layers of a root, epidermis, and ground tissue (consisting of cortex and endodermis) are arranged as concentric cylinders that surround the central stele (consisting of pericycle and vascular tissue). In later stages of root development cells elongate and differentiate with lateral roots formed from the pericycle (Benfey and Scheres, 2000). Phytohormones such as auxin and gibberellin have been shown to also influence root development in addition to nutrient signals. The two GRAS proteins SCARECROW (SCR) and SHORTROOT (SHR) were identified in a screen for *Arabidopsis* mutants with shorter roots (Benfey et al., 1993; Di Laurenzio et al., 1996; Pysh et al., 1999) and since then it has been shown that they are involved in many different stages of (root) development.

19.1.1.1 Ground Tissue

The cortex/endodermis initial cell (CEI) generates the two ground tissues, cortex and endodermis, via sequential asymmetric divisions. Especially endodermis development has to be under tight control with an evolutionary conserved process, as in nearly all plants it consists of only one cell layer (Cui et al., 2007; Engstrom, 2011). Both *scr* and *shr* mutants lack one of these cell layers (Di Laurenzio et al., 1996; Helariutta et al., 2000; Scheres, 1997). Whereas the single cell layer in *scr* mutants contains markers of both endodermis and cortex, *shr* mutants lack the endodermis completely. Therefore, SCR is involved in the asymmetric division of the cortex-endodermis initial daughters (CEID), whereas SHR is also needed for the formation of endodermal characteristics (Di Laurenzio et al., 1996; Helariutta et al., 2000; Nakajima et al., 2001; Sena et al., 2004). This can also be observed when looking at the functionality of the endodermis, which is characterized by targeted suberin depositions, the casparian band. In an *scr* mutant, the suberin is deposited between the cortical/endodermis cells and pericycle cells, whereas *shr* mutant roots deposit suberin ectopically in the middle lamellae of cells of the stele (Martinka et al., 2012).

SCR is expressed in the same cells where its function can be observed, which is the QC, endodermis, cortex/endodermis initial (CEI), and daughter cells (CEID). *SHR*, on the other hand, is expressed in the stele (Di Laurenzio et al., 1996; Helariutta et al., 2000). The SHR protein, though, moves from the stele to the neighboring cell layers, namely the endodermis, CEI, and QC (Nakajima et al., 2001). For this protein movement a motif in the leucine repeat II domain of the SHR protein (LNELDV) has been shown to be important (Gallagher and Benfey, 2009). Furthermore, SHR has to interact with SHORT-ROOT INTERACTING EMBRYONIC LETHAL (SIEL), a protein associated with endosomes, to gain mobility (Koizumi et al., 2011). SCR increases the expression of *SIEL*, suggesting that SCR at least partly controls SHR movement. Upon entering the endodermis, CEI, or QC, SHR becomes nuclear localized and can stimulate the expression of target genes, among them *SCR* (Levesque et al., 2006). SCR protein in turn binds to its own promoter in the presence of SHR, increasing its own expression (Levesque et al., 2006; Cui et al., 2007). Additionally, SCR interacts with SHR in the nucleus, thereby sequestering SHR to the nucleus and inhibiting its further movement (Cui et al., 2007). Consequently, in the *scr* mutant, SHR stays cytoplasmic in the single ground tissue layer and can even move beyond this layer (Nakajima et al., 2001; Sena et al., 2004). The SCR/SHR complex is rapidly inactivated following the asymmetrical division of the CEI. Moreover, SHR and SCR in the cortex are degraded, while the SCR remains present in the endodermis to sequester SHR (Heidstra et al., 2004). The function of the SCR/SHR complex can be explained by its ability to induce changes in gene expression patterns. Among SHR and SCR target genes, several genes important for cell cycle progression and linked to cyclin-dependent kinase (CDK) activity, have been identified (Sozzani et al., 2010). *CYCD6;1* in particular was found to be directly induced by SHR and SCR, specifically at times of asymmetric division in the CEI and its daughter cell to trigger division. Therefore, SCR and SHR are important for the optimal timing of this asymmetric cell division.

The transcription of three C2H2 zinc (Zn) finger transcription factors or INDETERMINATE DOMAIN (IDD) family proteins, MAGPIE (MGP), JACKDAW (JKD), and NUTCRACKER (NUC), is also under the control of SCR and SHR. SHR promotes JKD and NUC expression, whereas MGP is a direct target of SHR/SCR (Welch et al., 2007; Levesque et al., 2006). JKD also promotes *SCR* expression thereby promoting sequestration of SHR in the nucleus. To add another layer, MGP and JKD act antagonistically and have been shown to interact with SCR (MGP) and SHR (MGP, JKD; Levesque et al., 2006; Cui et al., 2007; Welch et al., 2007). This interplay between the GRAS proteins and IDD proteins is important in setting the boundaries between the endodermis and cortex (Ogasawara et al., 2011).

SCR and SHR homologs can also be found in all other plant species usually with one or two copies. In rice it has been established that OsSHR1 and OsSCR1 have similar functions in endodermis formation with OsSHR1 also interacting with AtSCR and OsSCR1 (Cui et al., 2007). Duplication events of OsSHR led to some diversification of the function as *OsSHR1* expression is still restricted to the stele, while *OsSHR2* seems to expand in the endodermis and some cortex cell layers and has probably gained a more divergent function in rice root development (Kamiya et al., 2003). Instead of the closed meristem formation found in *Arabidopsis* and rice, in conifers such as *Pinus sylvestris* an open root meristem organization can be found. In this case the initial cells are not as quiescent and the boundaries between the peripheral part of the cap and the cortex are unstable with common initials for the stele and columella. *SCR* gene (*PsySCR*) expression occurred also in young root tissue and is present in the initials of the stele and root cap column, but restricted to the endodermis in more developed cells (Laajanen et al., 2007). This suggests a regulatory role being undertaken by SCR, maintaining the endodermal characteristics and radial patterning of roots with open meristem organization.

Hormonal control of ground tissue formation has also been demonstrated to act via GRAS proteins. A vacuolar sorting protein, SHRUBBY, controls root growth and, as hypomorphic mutants cause poor root growth, decreased meristematic activity and extra cells in the cortex and endodermis (Koizumi and Gallagher, 2013). A similar phenotype can be seen after treating plants with the proteasome inhibitor MG132 or the gibberellic acid (GA) synthesis inhibitor paclobutrazol (PAC), suggesting that SHRUBBY acts downstream of GA signaling. SHRUBBY ties into the SCR/SHR pathway by interacting with SHR. Furthermore, the expression of a nondegradable form of the DELLA protein GA insensitive (GAI), specifically in the endodermis, resulted in decrease in root growth. Epidermal and cortex cells of these lines expanded radially, causing severe bulging in the epidermis (Cederholm et al., 2012). This suggests a link between GA, DELLA proteins, SCR, SHR, and SHRUBBY.

19.1.1.2 Middle-Cortex Formation

An additional cell layer, the middle cortex (MC), is formed some days later in root development by asymmetric periclinal division in the endodermis. This division occurs earlier in *scr* mutants but is impaired in *shr* mutants (Paquette and Benfey, 2005; Wysocka-Diller et al., 2000). Therefore, SCR is not directly required for MC formation, but regulates the timing of its formation, whereas SHR is directly required. The expression of CYCD6;1 is needed for this division and *shr* mutants fail to activate this.

Early MC appearance has also been observed in mutants of the SCR interacting protein LIKE HETEROCHROMATIN PROTEIN 1(LHP1; Cui and Benfey, 2009). LHP1 is considered similar to polycomb and can bind to H3K-27me3 (Exner et al., 2009). Therefore, it has been shown to be important in the maintenance of gene silencing in euchromatin.

In *scarecrow-like* (*scl*)3 mutants early MC appearance can also be observed, whereas GA suppresses it. *SCL3*, another member of the GRAS gene family, is a direct target of both SHR and SCR (Cui et al., 2007; Levesque et al., 2006) and has the same expression pattern as *SCR* (Pysh et al., 1999; Heo et al., 2011; Zhang et al., 2011). It has been shown that *SCL3* is downregulated by GA and upregulated by the SHR/SCR complex (Heo et al., 2011; Zhang et al., 2011). Precocious MC formation in *scl3* mutants is enhanced under GA-deficient conditions. Therefore, SCL3 integrates the GA and the SCR/SHR pathways.

19.1.1.3 The Root Vascular Tissue

As mentioned above, *SHR* is expressed in the stele but the protein moves out of this tissue. Nevertheless, *scr* and *shr* mutants exhibit ectopic metaxylem in place of protoxylem in the stele, suggesting that SHR (and SCR) can also indirectly regulate vacuolar tissue formation via gene expression. SHR is especially good at directly binding to regions upstream of the transcription start site for miR165b and miR166a, which repress PHB transcript expression within the vacuolar tissue (Carlsbecker et al., 2010). SHR also controls the transcription of a gene important for cytokinin catabolism (*CYTOKININ OXIDASE* 3), which is mainly expressed in the stele, thereby influencing cytokinin homeostasis and vascular development.

19.1.1.4 Stem Cell Niche Maintenance

The maintenance of cell division in the initial cells is dependent on the presence of a functional QC. Loss of the QC results in premature termination of root growth (van den Berg et al., 1997; Sabatini et al., 2003; Xu et al., 2006). Loss of either SHR or SCR results in the formation of a short root that fails to maintain the QC and meristem, suggesting that both proteins are needed for its maintenance. The SHR protein moves from the stele, not only to the initial cells but also to the QC, where it switches on the expression of *SCR* (Helariutta et al., 2000). Moreover, SCR is required for its own transcription in the QC, which is enhanced by the accumulation of JKD in the QC (Sabatini et al., 2003; Welch et al., 2007).

It has been shown that a RETINOBLASTOMA-RELATED protein (RBR) is necessary to enforce mitotic quiescence within the QC (Wildwater et al., 2005; Cruz-Ramirez et al., 2013). Retinoblastoma proteins are negative regulators of the G1 to S cell cycle transition and reduction of RBR increases the amount of stem cells in roots (Desvoyes et al., 2014; Kuwabara and Gruissem, 2014). RBR interacts with its interactors *via* an LxCxE motif. Such an LxCxE motif can be found at the N-terminal of the GRAS domain in the SCR protein and is highly conserved in all plant SCR homologs (Cruz-Ramirez et al., 2013). Indeed, RBR can interact with SCR and disruption of this interaction promotes asymmetric divisions in the QC. SCR, on the other hand, regulates the RBR pathway by downregulating Kip-related proteins or upregulating genes encoding for D cyclins, such as CYCD6;1 (Wildwater et al., 2005).

CYCD6;1 mediates RBR phosphorylation, which then counteracts SCR activity leading to a feedback control (Sozzani et al., 2010). A genetic link between RBR and SCR–SHR transcriptional activity was suggested by Wildwater et al. (2005). The expression domain of NUC expands as a consequence of RBR silencing and both SHR and SCR have been shown to bind to the NUC promoter (Cui et al., 2007; Cruz-Ramirez et al., 2012), although no direct interaction of the IDD protein NUC on SCR/SHR similar to JKD and MGP has yet been demonstrated.

It is not only SCR/SHR that are important for root meristem development, but also proteins of the HAM clade of the GRAS family. The root of a triple-mutant with the loss of all three members of this branch (SCL6, 22, and 27 or Atham1–3) was significantly smaller than that found in the wild type, a consequence of reduced rates of meristem cell division leading (among other phenotypes) to root meristem arrest (Engstrom et al., 2011).

19.1.1.5 Elongation Zone and Lateral Root Formation

So far, no role has been demonstrated for GRAS proteins in the expansion of the cells in the elongation zone or lateral roots in *Arabidopsis*. Nevertheless, SHR function is required for the initiation and patterning of lateral root primordia. The loss of primary root growth in *shr* is compensated for by the activation of anchor root primordia, whose tissues do not show any growth alterations (Lucas et al., 2011). A *Phaseolus vulgaris* GRAS protein from the PHYTOCHROME A SIGNAL TRANSDUCTION1 (PAT1) clade was shown to be important for lateral root development, as mutants with reduction of SIN1 showed reduced lateral root elongation (Battaglia et al., 2014). However, no effects on root cell organization, density of lateral roots, and length of root hairs could be observed in the mutant.

19.1.1.6 Root Modification (Nodulation, Mycorrhiza)

Roots of many species of plants can associate with microorganisms or fungi to form a symbiosis. These interactions are associated with modifications in the developmental pattern and differentiation of the root. The best studied symbiotic interactions are those between plants and arbuscular mycorrhizal fungi (AMF), and between legumes and nitrogen-fixing rhizobial bacteria (Oldroyd et al., 2011; Gutjahr and Parniske, 2013; Oldroyd, 2013). To form interactions with AMF, strigolactones are released by the plant to attract the fungus which in turn produces mycorrhizal factors and signals that activate the symbiosis signaling pathway in the root, leading to calcium oscillations. The fungus attaches to the root with a hyphopodium and the plant guides and allows the fungus to colonize the inner root cortex intercellularly.

In the case of rhizobia, flavonoids are released by the plant root as signals to rhizobia, which in turn produce nodulation (Nod) factors that are recognized by the plant. Nod factor perception activates the symbiosis signaling pathway, leading also to calcium oscillations. Nodules initiate below the site of bacterial infection and are formed by cortical cells that proliferate with undifferentiated cells. Root hair cells grow around the bacteria attached at the root surface, trapping the bacteria inside a root hair curl. Invaginations of the plant cell allow the invasion of the rhizobia into the root tissue.

The calcium oscillations of both events are perceived by calcium- and calmodulin-dependent protein kinase, which activates the GRAS domain factor nodulation signaling pathway2 (NSP2), which is important for the early stages of infection (Kalo et al., 2005). NSP2 expressed from a constitutive promoter is localized to the endoplasmatic reticulum/nuclear envelope and relocalizes to the nucleus after Nod-factor elicitation.

The GRAS protein NSP1 is important only for nodulation and directly binds the *early nodulation gene (ENOD) 11* promoter through the *cis*-element "AATTT" (Murakami et al., 2006; Hirsch et al., 2009). This binding requires a direct interaction between NSP1 and NSP2. Nevertheless, whether or not this NSP1–NSP2–DNA interaction is directly relevant for inducing *ENOD11* expression specifically is discussed by Cerri et al. (2012).

For AMF, another GRAS protein, RAM1, which is important in the early stages of the colonization of plants, has been identified. RAM1 also binds to NSP2 (Gobbato et al., 2012). Mutants in RAM1 have defects in the formation of hyphopodia. Both NSP1 and RAM1 can form heterocomplexes with NSP2 thereby promoting specific gene expression, but NSP2 is not able to bind DNA directly (Hirsch et al., 2009). Therefore, in legumes the decision to form nodules or AMF depends on the complex that is formed.

Although rice and *A. thaliana* are not able to form nodules or AMF, homologs of NSP1 and NSP2 can be found and can substitute for their *Lotus japonicus* orthologs suggesting they shared a common function before they became important for root modification (Heckmann et al., 2006; Yokota et al., 2010). Indeed, it can be shown that NSP1 and NSP2 are important for the production of strigolactones (SL), which are important for the modulation of root differentiation and the attraction of AMF. The regulation of SL biosynthesis by NSP1 and NSP2 seems to be the ancestral function conserved in higher plants (Liu et al., 2011).

In *P. vulgaris*, silencing of *SIN1* showed that the product of this gene is involved in lateral root elongation. In addition, the number and size of nodules formed upon inoculation with *Rhizobium etli* and the progression of infection threads toward the nodule primordia were affected (Battaglia et al., 2014). SIN1 interacts with a C subunit of the heterotrimeric nuclear factor Y (NF-YC1), also involved in nodule organogenesis and bacterial infection.

In a similar manner to SCR/SHR, SIN1 also influences the expression of gene-encoding proteins important for cell cycles such as the G2/M transition cell cycle genes *CYCLIN B* and *Cell Division Cycle2* which were reduced in *SIN1* RNAi roots.

19.1.2 GRAS Proteins in Shoot Development

Mutants of *scr* and *shr* have not only been identified in root specific screens, but also in a screen for mutants with agravitropic shoots (Fukaki et al., 1996). Both mutants lack the starch sheath (endodermal layers) in the inflorescence stem, which results in loss of gravitropic sensing (Fukaki et al., 1998). Besides the expression of SCR in the starch sheath of the inflorescence stem, SHR and SCR are also expressed in the L1 layer of the shoot apical meristem (SAM; Wysocka-Diller et al., 2000).

19.1.3 GRAS Proteins in Leaf Development

Development of leaves from leaf primordia involves a determinate growth pattern with leaves lacking an equivalent to the QC. Nevertheless, *SHR* and *SCR* are also expressed in leaves, in young leaf primordia, in developing leaf vascular tissue, and bundle sheath cells (Wysocka-Diller et al., 2000; Lim et al., 2005; Dhondt et al., 2010; Gardiner et al., 2011). In *scr* and *shr* mutants a reduced cell division rate, and an early cessation of proliferation, can be observed leading to smaller leaves, most probably acting *via* cell cycle inhibitors (Dhondt et al., 2010). The expression of *SHR* and *SCR* genes in leaves is closely associated with cell division activity in most cell types.

Bundle sheath (BS) cells are a leaf cell type that forms a single cell layer between mesophyll cells and the central vascular tissue. In C3 plants, both mesophyll cells and BS cells are photosynthetic, but the BS cells are elongated with fewer chloroplasts. In contrast, BS cells are the major sites of photosynthesis in most C4 plants, whereas mesophyll cells are involved in CO_2 fixation only. The Kranz anatomy of C4 plants is defined by the ratio between mesophyll and BS cells. Besides *SCR*, *SCL23* is also expressed in *Arabidopsis* BS cells but *SHR* is exclusively expressed in the xylem (Cui et al., 2014). The SHR protein, analogous to its role in roots, moves into the BS cells, where it directly regulates *SCR* and *SCL23* expression. Target genes of SCR seem to be important for sugar transport whereas those of SCL23 are involved in mineral transport (Cui et al., 2014).

Mutations in the two maize *SCR* genes cause an abnormal number of BS cell layers, abnormal differentiation of bundle sheath chloroplasts, vein disorientation, loss of minor veins, and reduction of vein density (Cui et al., 2014; Slewinski et al., 2012). The maize SHR1 protein is also involved in forming the Kranz anatomy, suggesting that both SCR and SHR have a role in establishing the Kranz anatomy in C4 plants, and that similar processes to those in the root might also regulate tissue formation around the veins in leaves (Slewinski et al., 2012, 2014). Therefore, these proteins are of great interest because of how they might play a role in efforts to change a C3 plant into a C4 plant (Slewinski et al., 2012; Slewinski, 2013).

The GRAS proteins from the HAM branch (SCL6, 22, and 27) are also involved in leaf development, but this phenotype is only visible in a triple mutant, which besides a reduced root growth shows increased chlorophyll accumulation and abnormal leaf patterning (Wang et al., 2010; Ma et al., 2014). SCL6, 22, and 27 have been shown to negatively regulate chlorophyll biosynthesis by inhibiting *PROTOCHLOROPHYLLIDE OXIREDUCTASE C (PORC)* gene expression in light-grown plants (Ma et al., 2014). SCL27, in particular, accumulates at high levels and suppresses chlorophyll biosynthesis at the leaf basal proliferation region during leaf development. SCL27 also interacts with DELLA proteins, and their interaction reduces the binding activity of SCL27 to the *PORC* promoter, whereas DELLAs promote protochlorophyllide (Pchlide) biosynthesis (Cheminant et al., 2011).

19.1.4 GRAS Proteins in Shoot Apical Meristem Development

The SAM is organized differently from the root apical meristem, nevertheless stem cells can be found in the central zone with lateral organ primordia at the periphery. In a longitudinal section of the apical meristem the corpus can be discriminated consisting of randomly oriented cells (L3) above a tunica comprised of two cell layers (L1 and L2) giving rise to different tissue types. A vegetative meristem gives rise to leaves or other organs (e.g., axillary buds and internodes), whereas the reproductive meristem gives rise to flowers and internodes.

In petunia, a mutant called *hairy meristem* (*ham*) was identified because its SAM arrests and differentiation takes place, which manifests itself in the enlargement and vacuolization of internal meristem cells and the formation of trichomes ("hairy"; Stuurman et al., 2002). A reduction in PhSHOOTMERISTEMLESS expression, a marker of meristem identity, is also observed in the mutant. *PhHAM* mRNA is confined to basal and peripheral regions of the meristem (L3 layer), most clearly in the provasculature, extending into initiating lateral organs. The absence of detectable *HAM* expression in the meristem apex suggests that PhHAM initiates a signal that moves to the apical meristematic regions. Whereas in petunia only one gene belongs to this subgroup of GRAS proteins, in Arabidopsis three orthologous proteins can be found (SCL6/22/27, alternatively named AtHAM1-3, LOST MERISTEMS [LOM1-3], or Scl6-II, scl6-III and scl6-IV; Llave et al., 2002; Bolle, 2004; Schulze et al., 2010; Engstrom et al., 2011). A triple-mutant confirms a similar phenotype as *phham*, and the role played by these three proteins in maintaining shoot stem cells in an undifferentiated state and in limiting the number of clonally derived meristem layers. Although the functions of all three proteins overlap, some differences can also be observed. SCL6 is expressed at the boundary between the axillary meristem and the SAM, whereas SCL22 and 27 are expressed in the peripheral zone and vascular tissues of the SAM (Schulze et al., 2010; Engstrom et al., 2011). The latter are more important for the maintenance of the meristematic cells and the differentiation of the axillary meristem, whereas SCL6 is more related to branching.

Consistent with the expression pattern of the HAM genes, the triple mutant also exhibits aberrant shoot phyllotaxis, lateral organ abnormalities, and relatively flat shoot apical meristems (Schulze et al., 2010; Engstrom et al., 2011). Reduced expression of these genes in *Hordeum vulgare* resulted in a plant with an increased number of short vegetative phytomers, branching defects, and late flowering as a consequence of changes in the organization of the shoot meristem (Curaba et al., 2013).

The phase transition between vegetative and reproductive meristems is an important step and its timing is highly regulated. This transition not only affects the organs produced by the meristem but also other factors such as trichome density. The expression of the *Squamosa promotor binding protein-like transcription factors (SPLs)* was reduced in the triple *ham* mutant. Furthermore, AtHAM proteins can interact with its N-terminus, with SPL2 and SPL9, which modulate plant phase transition, flowering, and trichome formation (Xue et al., 2014; Chapter 18). Trichome density on stems and floral organs is dependent on the presence of the HAM/SPL complex. In *H. vulgare* the reduction of HAM expression alters the vegetative to reproductive phase transition by activating the miR156 pathway, which targets SPLs (Curaba et al., 2013).

19.1.5 GRAS Proteins in Axillary Meristem Development

Aerial architecture in higher plants occurs postembryonically by the establishment and activation of new meristems in the axils of leaves. Axillary meristems recapitulate the function of the SAM by initiating several leaf primordia, resulting in the formation of axillary buds, which either grow out or remain dormant (Bennett and Leyser, 2006). As a consequence, in mutants of the HAM protein clade, reduced shoot branching phenotypes can be observed (Schulze et al., 2010; Engstrom et al., 2011; Curaba et al., 2013).

In the tomato lateral suppressor (*Ls*) mutant, axillary meristem formation is blocked during the vegetative growth phase and flowers fail to develop petals and display reduced fertility. Characterization of the *Ls* gene revealed that the encoded protein belongs to the GRAS family (Schumacher et al., 1999). The *Arabidopsis* ortholog LATERAL SUPPRESSOR (LAS), which is expressed at the adaxial boundary of leaf primordia, has a similar function with the mutant not able to form lateral shoots during vegetative development. However, there is no impairment in the reproductive phase (LAS; Greb et al., 2003). In addition, in monocots the protein has comparable functions as a mutation in MONOCULM1 (MOC1), the rice ortholog, causes defective tiller formation, altered rachis branches, and modified spikelets (Li et al., 2003). Overexpression of the *Zoysia* grass ortholog, ZjLsL, promoted axillary bud formation (Yang et al., 2012).

MOC1 degradation by the ubiquitin-26S proteasome pathway is induced by the anaphase promoting complex/cyclosome (E3 ligase complex) activated by the tiller enhancer (TE). Tillering and Dwarf 1 (TAD1), a coactivator of the complex, interacts with MOC1 by targeting MOC1 for degradation in a cell-cycle-dependent manner. This leads to the downregulation of the expression of the meristem identity gene *Oryza sativa homeobox 1*, thereby repressing axillary meristem formation (Lin et al., 2012; Xu et al., 2012).

19.1.6 GRAS Proteins in Flower, Embryo, and Seed Development

Not much is known about GRAS proteins in the stages from flower development to seed formation. One GRAS protein in lily, LlSCL, has been identified as being important during the microsporogenesis of the anther inducing the meiosis-associated promoter *lim10* (Morohashi et al., 2003). SCR and SHR expression can already be detected as early as the heart stage during embryo development and are probably responsible for meristem formation at this stage. In rice, a GRAS protein was identified as being responsible for reducing grain size, Grain Size 6 (GS6) otherwise named DWARF AND LOW-TILLERING (DLT; Sun et al., 2013).

19.2 THE ROLE OF GRAS PROTEINS IN SIGNALING

Similar to developmental processes, signal transduction in most cases leads to changes in transcriptional activity. The amplification of the initial signal, and also the sensation of the cessation of a signal, needs a high level of control over the components of a signaling pathway. Therefore, it is no surprise that GRAS proteins can be found in a multitude of signaling cascades.

19.2.1 Gibberellin Signaling

The DELLA proteins are a very conserved subbranch of GRAS proteins and are named on the basis of a short stretch of amino acids (D-E-L-L-A) in their N-terminal region, which is highly conserved among all plant species (Sun and Gubler, 2004). In *Arabidopsis*, five proteins are present, whereas in other species, especially monocots, only one representative can be found. All DELLA proteins known to date are involved in gibberellin signaling. GA binds to the GID1 receptor which then interacts with DELLA proteins; in some cases this happens without GA (for review see Sun, 2011 and Chapter 20). This interaction promotes the ubiquitination of the DELLA proteins by an E3 ubiquitin ligase, and leads to their proteolysis via the 26S proteasome pathway. This destabilization requires the N-terminal region of the protein, containing the DELLA domain (Dill et al., 2001). DELLA proteins interact with different transcription factors, especially from the helix–loop–helix (HLH) group such as the PHYTOCHROME-INTERACTING FACTOR3 (PIF3) and other PIF paralogs (Daviere et al., 2008, Feng et al., 2008; Gallego-Bartolome et al., 2010; Chapter 21). DELLA degradation releases the transcription factors, relieving the suppression of GA-responsive genes (Daviere et al., 2008 and Chapter 20). The localization of DELLA proteins in the nucleus is in agreement with their role in the regulation of gene expression (Silverstone et al., 1998; Dill et al., 2001).

DELLA proteins integrate not only GA-signaling pathways but also jasmonate, auxin, brassinosteroid, and ethylene pathways, thereby constituting a main hub in signaling (Bai et al., 2012; Wild et al., 2012). *SCL3* has been identified as a direct target gene of DELLAs in *Arabidopsis* seedlings (Cui et al., 2007). *SCL3* expression is induced by DELLAs and repressed by GA, but SCL3 functions as a positive regulator of GA signaling. SCL3 can interact with DELLA proteins and this complex regulates *SCL3* transcription. Therefore, SCL3 seems to be the antagonist of the DELLA proteins in controlling GA-mediated development in plants (Fukazawa et al., 2014; Yoshida et al., 2014).

A rice dwarf mutant with reduced content of endogenous GA was identified as a mutation in DLT, which has been previously identified to play a role in brassinosteroid (BR) signaling and was recently identified as GS 6 (Li et al., 2010; Tong and Chu, 2009; Sun et al., 2013). This suggests that GRAS proteins from branches outside of the DELLA clade could also be involved in both GA signaling and crosstalk with other hormonal signaling pathways.

19.2.2 Light Signaling

To sense light intensity, quality, and direction plants have developed several specific photoreceptors, among them is the family of phytochromes (Chen and Chory, 2011; Chapter 21). Phytochromes perceive mainly red and far-red light, thereby sensing the environmental condition important for growth and development. They are therefore especially important for germination efficiency, deetiolation, and avoidance of shading by neighboring plants. The members of one subbranch especially, that of *A. thaliana* GRAS proteins, namely PAT1, SCL21, and SCL13, have been shown to be downstream of the phytochrome signal transduction pathway (Bolle et al., 2000; Torres-Galea et al., 2006, 2013).

PAT1 and SCL21 are most closely related and both are specifically involved in phytochrome A (phyA) signal transduction since both mutants have an elongated hypocotyl specifically under far-red light, a phyA-dependent trait (Bolle et al., 2000; Torres-Galea et al., 2013). Both proteins can interact with each other and a double-mutant shows no additive phenotype, also suggesting genetic interaction. Therefore, both SCL21 and PAT1 are positive regulators of phyA signal transduction for several high-irradiance responses. Interestingly, *pat1-1*, a semidominant mutant, which still expresses the protein albeit without its C-terminal part, shows a stronger phenotype than the loss-of-function mutant. This suggests that the C-terminus is needed either to confer different stability to the protein itself or to establish an interaction, thereby interfering with signal transduction. PAT1 is furthermore needed to regulate *SCL21* gene expression creating a feedback loop.

SCL13, on the other hand, serves as a positive regulator of continuous red light signals downstream of phytochrome B (phyB). Only a distinct subset of phyB-mediated responses is affected in RNAi lines, indicating that SCL13 executes its major role in hypocotyl elongation during deetiolation (Torres-Galea et al., 2006). Genetic evidence suggests that SCL13 is also needed to modulate phyA signal transduction in a phyB-independent way. All three proteins can be localized to the cytoplasm and the nucleus. For the SCL13 protein, it could be shown that overexpression of both a mainly nuclear or cytoplasmic localized SCL13 protein leads to a hypersensitive phenotype under red light, indicating that SCL13 could be biologically active in both compartments (Torres-Galea et al., 2006).

19.2.3 Stress Signaling

Several GRAS proteins have been associated with a role in stress signaling by changing the transcriptional activity of their target genes. For SCR and SHR several direct targets have been identified that are associated with stress responses. Consequently, both *scr* and *shr* mutants were found to be hypersensitive to abscisic acid (ABA) and high levels of glucose (Glc) but not affected by high salinity and osmotic stress (Cui, 2012). One of the major responses seems to be due to SCR-dependent repression of *ABI4* and via this pathway SCR also regulates the sugar response in the root apical meristem, important for normal root growth. Moreover, low phosphate levels result in a stress reaction in roots. SCR, SHR, and DELLA proteins are involved in the signaling pathway of this stress and the resulting adaptive responses (Sato and Miura, 2011).

Osmotic stress on dividing leaf cells in *Arabidopsis* leads to an arrest of cell cycle progression and a stabilization of DELLA proteins could be observed. Subsequently, downregulation of the anaphase-promoting complex/cyclosome activity led to reduced levels of inhibitors of the developmental transition from mitosis to endoreduplication (Claeys et al., 2012).

SCL14 is essential for the activation of stress-inducible promoters, especially SA- and 2,4-D-inducible promoters (Fode et al., 2008). This activation is mediated by the interaction of SCL14 with class II TGA transcription factors. The TGA/SCL14 complex seems therefore to be involved in the activation of a general broad-spectrum detoxification network when a plant is challenged by xenobiotics.

On the other hand, in many cases it has been reported that expression of certain GRAS proteins is highly upregulated under different kinds of stress. Examples are the *Brassica oleracea* SCL13 ortholog, BoGRAS, under heat stress (Park et al., 2013), and the salt- and drought-inducible poplar GRAS protein SCL7, which in turn confers salt and drought tolerance in *A. thaliana* upon overexpression (Ma et al., 2010). Moreover, constitutive overexpression of BnLAS in *Arabidopsis* results in a reduction of water loss in leaves and an enhanced drought tolerance (Yang et al., 2011).

GRAS proteins in particular seem to react in expression to biotic factors, but their role in the defense system is not yet clear. Six tomato SlGRAS transcripts accumulate during the onset of disease resistance to *Pseudomonas syringae* pv. Tomato, and mechanical stress is dependent on jasmonic acid. Suppression of *SlGRAS6* gene expression could impair the resistance to *P. syringae* (Mayrose et al., 2006). Moreover, *NtGRAS1* expression was induced in leaf tissue upon antimycin A treatment or following *P. syringae* infection (Czikkel and Maxwell, 2007). Reminiscent of the role of GRAS proteins in nodulation and AMF interaction, the elicitor *N*-acetyl chitooligosaccharide could induce two rice GRAS proteins from the PAT1 branch, chitin-inducible gibberellin-responsive (CIGR)1 and 2 (Day et al., 2003). In a study of *Arabidopsis*, four GRAS genes were also affected by chitooctaose but not by fungal or jasmonic acid treatment (Libault et al., 2007). Also, two GRAS proteins from tomato were identified as putative targets for peptides, which stimulate root growth, that are secreted by root-knot nematodes in order to transform the recipient cells into enlarged multinucleate feeding cells called "giant cells" (Huang et al., 2006).

19.3 GENERAL PRINCIPLES OF GRAS FUNCTION

As described previously, a lot is known about the biological roles GRAS proteins are involved in. Nevertheless, it seems less clear what their biochemical function is. In the literature GRAS proteins are usually described as transcription factors. This is due to their (mostly) nuclear localization, similarities with animal-specific signal transducers and activators of transcription (STAT) which have been described to bind DNA (Richards et al., 2000), stretches of polyhomomeric amino acids which sometimes occur in transcription factors, and the fact that the leucine-rich repeats sometimes have a spacing of seven amino acids between leucines therefore resembling leucine zippers. However, the homology to STAT factors has been put in doubt since they would adopt a β-barrel structure which is different from the predicted secondary structure of the GRAS domain (Zhang et al., 2012; Chapter 10), and although their role in transcriptional regulation seems clear direct DNA binding has been shown in only a few cases. Furthermore, not all GRAS proteins localize strictly to the nucleus. For many GRAS proteins, such as LLSCR, PAT1, SCL21, NSP1, NSP2, and DELLAs, transcriptional activity could be detected in the yeast two-hybrid system (Bolle, unpublished results; Morohashi et al., 2003; Hirsch et al., 2009; Fukazawa et al., 2014). This activity could in several cases be mapped to the N-terminal domains of the GRAS proteins, suggesting that these proteins have an intrinsic transcription-promoting function. Nevertheless, this activity could also be an artifact of the Y2H system and therefore has to be evaluated with caution.

19.3.1 GRAS Protein–DNA Interaction

To detect DNA interaction, in most cases two methods are employed: chromatin immunoprecipitation (ChIP) and electromobility shift assays. In only two GRAS proteins has a *cis*-element on the DNA been identified to which these proteins bind. An electrophoretic mobility shift assay revealed that SCL27 directly interacted with the G(A/G)(A/T)AA(A/T)GT *cis*-elements of the *PORC* promoter (Ma et al., 2014), whereas NSP1 interacted with the *ENOD11* promoter which had a shorter motif, AATTT (Hirsch et al., 2009). Both isolated LRI and LRII domains of NSP1 could interact with this motif. Whether or not this motif is the decisive *cis*-element necessary for nodulation-dependent activation is still under discussion (Cerri et al., 2012).

Nevertheless, several GRAS proteins have been identified to associate *in vivo* with promoter regions. These ChIP assays do not actually show if a protein is directly bound to DNA or is part of a complex that is binding to the chromatin. For example, SCR, SHR, NSP1, and DELLA proteins have been shown to bind to target genes with these assays (see Table 19.1). Several of them, such as SCR, SHR, and SCL3, have also been shown to interact with their own promoters to form a feedback loop (Heidstra et al., 2004; Cui et al., 2007; Cui et al., 2014). Additionally, GRAS proteins interact with the promoters of several genes that encode proteins that directly interact with the

TABLE 19.1 Known GRAS Protein Interactions with Proteins and DNA

GRAS protein	Interactor	Category	References
INTERACTION WITH OTHER GRAS PROTEINS			
SLR1	SLR1	GRAS	Itoh et al. (2005)
DELLAs	SCL3	GRAS	Zhang et al. (2011); Yoshida et al. (2014)
SCR	SHR	GRAS	Levesque et al. (2006); Cui et al. (2007)
PAT1	SCL21	GRAS	Torres-Galea et al. (2013)
NSP2	NSP1	GRAS	Hirsch et al. (2009)
NSP2	RAM1	GRAS	Gobbato et al. (2012)
DELLA	SCL27	GRAS	Ma et al. (2014)
PROTEINS INVOLVED IN TRANSCRIPTION			
DELLAs	PIFs (PIF3, PIF4)	TF, bHLH	Feng et al. (2008); Gallego-Bartolome et al. (2010); de Lucas and Prat (2014)
DELLAs	SPATULA, SPT	TF, bHLH	Josse et al. (2011)
DELLAs	MYC2	TF, bHLH	Hong et al. (2012)
DELLAs	ALCATRAZ (ALC)	TF, bHLH	Arnaud et al. (2010)
DELLAs	INDETERMINATE DOMAIN(IDD)1, ENY	TF C2H2 Zn finger, IDD	Feurtado et al. (2011)
DELLA	GAI-ASSOCIATED FACTOR1 (GAF1)	TF C2H2 Zn finger, IDD	Fukazawa et al. (2014)
SCR	MAGPIE (MGP)	TF C2H2 Zn finger, IDD	Cui et al. (2007)
SHR	JACKDAW (JKD), MGP	TF C2H2 Zn finger, IDD	Welch et al. (2007)
SCL3	IDDs	TF, C2H2 Zn finger, IDD	Yoshida et al. (2014)
SHR	Zn finger domain	TF	Koizumi and Gallagher (2013)
DELLAs	JASMONATE-ZINCDOMAIN PROTEIN1, JAZ	TF	Josse et al. (2011); Wild et al. (2012)
DELLAs	EIN3	TF, EIL	An et al. (2012)
DELLAs	BES1, BZR1	TF	Bai et al. (2012); Gallego-Bartolome et al. (2012)
DELLAs	SPLs	TF	Yu et al. (2012)
HAM/LOM	SPLs	TF	Xue et al. (2014)
SCR	C2 domain-containing protein	TF	Cui and Benfey (2009)
SCR	MADS box (AGL14, 20)	TF	Cui and Benfey (2009)
SCR	B3 family protein	TF	Cui and Benfey (2009)
SCR	Homeobox–leucine zipper	TF	Cui and Benfey (2009)
SCL14	Class II TGA factors	TF	Fode et al. (2008)
SIN1	NUCLEAR-FACTOR Y, C1	TF	Battaglia et al. (2014)
CHROMATIN			
BNSCL1	AtHDA19	Chromatin	Gao et al. (2004)
SCR	RBR	Chromatin	Cruz-Ramirez et al. (2012)
DELLA	SWI3C	Chromatin	Sarnowska et al. (2013)
SCR	LIKE HETEROCHROMATIN PROTEIN 1 (LHP1)	Chromatin	Cui and Benfey (2009)

19.3 GENERAL PRINCIPLES OF GRAS FUNCTION

TABLE 19.1 Known GRAS Protein Interactions with Proteins and DNA (cont.)

GRAS protein	Interactor	Category	References
MODIFYING ENZYMES, RECEPTORS			
DELLA PROTEIN SLR1	EL1	Casein kinase I	Dai and Xue (2010)
RGA	TOPP4	Phosphatase	Qin et al. (2014)
DELLA	GID1	Receptor	Willige et al. (2007)
DELLA	BOIs	RING domain proteins	Park et al. (2013)
DELLA	SLEEPY1, SLEEZY	F-box protein	Ariizumi et al. (2011)
MOCI	TAD1	E3 ligase	Xu et al. (2012)
Tomato SCL	Bioactive nematode peptide RKN 16D10	Peptide	Huang et al. (2006)
SCR	Transcription elongation factor 1a		Cui and Benfey (2009)
CYTOPLASMIC INTERACTORS			
SHR	SHRUBBY	Vacuolar sorting protein	Koizumi and Gallagher (2013)
DELLAs	Prefoldin	Microtubule orientation	Locascio et al. (2013)
SHR	SHORT-ROOT INTERACTING EMBRYONIC LETHAL (SIEL)	Associated with endosomes	Koizumi et al. (2011)
SCR	SEC14 family protein	Membrane trafficking	Cui and Benfey (2009)
SCR	GASA1	Cysteine rich	Cui and Benfey (2009)
SCR	PER32	Peroxidase	Cui and Benfey (2009)
SCR	NAD-dependent epimerase		Cui and Benfey (2009)
DNA			
SCR/SHR	Stress-responsive genes	DNA	Iyer-Pascuzzi et al. (2011); Cui (2012)
SCR (+SHR)	*SCR* promoter	DNA	Levesque et al. (2006); Cui et al. (2007)
SCR SHR	*SCL3*	DNA	Cui et al. (2007)
SHR	*miRNA165b/166a*	DNA	Carlsbecker et al. (2010)
SHR, SCR	*MGP, NUC*	DNA	Cui et al. (2007); Welch et al. (2007)
SHR	*SCR, SCL23*	DNA	Cui et al. (2014)
SCR	*ABI4, ABI5*	DNA	Cui (2012)
SHR, SCR	*CYCD6;1*	DNA	Sozzani et al. (2010)
SCR	Genes encoding D-cyclins, kip-related proteins	DNA	Sozzani et al. (2010)
SHR	*JKD*	DNA	Welch et al. (2007)
LLSCR	*lim10* promoter	DNA	Morohashi et al. (2003)
DELLAs	*SOMNUS* promoter	DNA	Lim et al. (2013)
DELLA	*SCR3*	DNA	Zentella et al. (2007)
DELLA	*XERICO*	DNA	Zentella et al. (2007)
DELLA	*MYB, bHLH (BHLH137)*	DNA	Zentella et al. (2007)
DELLA	*LBD40*	DNA	Zentella et al. (2007)
DELLA	*GID1a, 1b*	DNA	Zentella et al. (2007)
NSP1 (+NSP2)	*ENOD11* promoter	DNA	Hirsch et al. (2009)
SCL27 (HAM)	*PORC*	DNA	Ma et al. (2014)
HAM	*miR171*	DNA	Xue et al. (2014)

Category defines the nature of the interactor. TF, transcription factor.

C. FUNCTIONAL ASPECTS OF PLANT TRANSCRIPTION FACTOR ACTION

GRAS protein itself, such as DELLAs with the *SCL3* and *GID1* promoter, SHR with *JKD*, and SCR with *MGP*, adding another layer of complexity to the regulation (see Table 19.1 and references therein).

19.3.2 Protein–protein Interactions of GRAS Proteins

19.3.2.1 Interaction Between GRAS Proteins (Homo- and Heterodimers)

Interactions between GRAS proteins have been demonstrated several times. Only one GRAS protein has been reported to be able to homodimerize, that is the rice DELLA protein SLENDER RICE 1 (Itoh et al., 2005). Several GRAS proteins, even from different subbranches, can heterodimerize, such as SHR–SCR, MtNSP1–MtNSP2, MtNSP2–MtRAM1, SCL27–DELLAs, AtSCL3–DELLAs, and PAT1–SCL21, and this heterodimerization is often important for their functionality (see Table 19.1 and references therein). NSP2, for example, would only be able to bind DNA after the formation of a complex with NSP1 or RAM1, and SCL3 can only bind to its own promoter after binding to DELLAs (Hirsch et al., 2009; Yoshida et al., 2014).

19.3.2.2 Interaction of GRAS Proteins with Other Factors

Identification of GRAS interaction factors using a yeast two-hybrid screen is very often hampered due to the fact that GRAS proteins have an intrinsic transactivation activity as mentioned earlier. Several studies circumvented this problem by either deleting the transactivating domain or adding a repressor (e.g., Fukazawa et al., 2014) to find new interaction partners. Therefore, new interactors have been found, but for many of them the biological function has not yet been shown (e.g., Cui and Benfey, 2009).

Specific interaction with proteins is either due to the variable N-terminal regions of the GRAS proteins or the conserved GRAS domain. Therefore, it is not surprising that several GRAS proteins can interact with the same protein family (see Table 19.1 and citations therein). One example are the basic helix–loop–helix (bHLH) proteins, which DELLA proteins and other GRAS proteins interact with. The interaction of PIF and DELLAs has been studied in more detail (Chapter 20), but in addition bHLHs such as SPATULA (SPT), MYC2, and ALCATRAZ (ALC) can be found in interaction studies with DELLAs (see Table 19.1 and citations therein). SPT and ALC are important for organ size control, fruit development, and seed dormancy (Groszmann et al., 2011; Makkena and Lamb, 2013; Vaistij et al., 2013), whereas MYC2 has been implicated in many responses, among them ethylene and jasmonate (Song et al., 2014). This underlines the variation in processes regulated by DELLA proteins. Another emerging family is the plant-specific IDD transcription factors which can interact with DELLAs, SCR, and SHR (see Table 19.1 and citations therein) and regulate a wide array of developmental phenotypes. Moreover, SPLs, especially important for phase transitions (Chapter 18), can be bound by at least two different GRAS proteins, DELLAs and HAM (Yu et al., 2012; Xue et al., 2014)

It could be speculated that the increasing amount of transcription factors GRAS proteins can interact with (see Table 19.1) point to a general way of functioning for GRAS proteins. As for the DELLA proteins, it has been shown that the interaction with a transcription factor inactivates that transcription factor and only upon degradation of the GRAS protein does this factor get released. In this scenario, the GRAS protein is meant to inactivate transcription factors and make them available as soon as they are needed. On the other hand, as it seems that most GRAS proteins cannot bind to DNA directly, the binding to a transcription factor might be their way of attaching to DNA and regulating transcription directly. Additionally, GRAS proteins also seem to interact with several factors that regulate chromatin. In 2004 a *Brassica napus* SCARECROW-like protein, BnSCL1, was shown to interact with the histone deacetylase AtHDA19 in a yeast two-hybrid screen (Gao et al., 2004). Interactions between SCR and LHP1 and RBR, and between DELLA proteins and SWI3C, have shown a link between GRAS proteins and chromatin-remodeling factors (see Table 19.1 and citations therein). LHP1 is considered similar to polycomb, and retinoblastoma proteins have in other systems been shown to interact with polycomb complexes. The SWI3C subunit of the switch (SWI)/sucrose nonfermenting (SNF)-type chromatin-remodeling complex DELLA proteins can interact with, is also known to regulate transcription and cell cycle, and mutants can severely impair growth and development (Sarnowska et al., 2013). Future studies will show to what degree GRAS proteins are involved in these complexes.

Other interaction partners of GRAS are important for their functionality, such as DELLA with GID1, the GA receptor and part of an E3 ligase, and the two F-box proteins SLEEPY1 and SNEEZY (Ueguchi-Tanaka et al., 2007; Ariizumi et al., 2011). In addition, MOC1 binds to an E3 ligase, the anaphase-promoting complex (APC/C), and DELLA proteins are involved in this complex which regulates the cell cycle (Claeys et al., 2012; Xu et al., 2012). This shows that degradation and cell cycle control are important mechanisms of GRAS proteins. Furthermore, modifying enzymes such as the phosphatase TOPP4 or a casein kinase I can interact with GRAS proteins (Qin et al., 2014; Dai and Xue, 2010). Another example is SIEL, which interacts with SHR for the migration from one cell to the nucleus of another (Koizumi et al., 2011). Although GRAS proteins are considered primarily to be nuclear localized, other cytosolic proteins have been identified to interact with GRAS proteins, although for many their function is not yet clear (see Table 19.1).

19.3.3 Protein Movement of GRAS Proteins

Radial patterning of *Arabidopsis* roots necessitates SHR movement from the stele to the endodermis. So far this is the only GRAS protein identified for which movement from cell to cell could be demonstrated. Multiple regions within the GRAS domain of SHR are essential for its movement and at least two interacting proteins have been identified that are necessary for this (Koizumi et al., 2011; Koizumi and Gallagher, 2013). The interaction of SHR with SCR in the nucleus sequesters the protein and prevents further movement (Cui et al., 2007). However, this also shows that a movement between the cytoplasm and nucleus is possible and such a movement has been shown for several GRAS proteins (PAT1, SCL13, SCL21; Torres-Galea et al., 2006, 2013). An SCR protein without the N-terminus also has the capacity to move between the cytoplasm and nucleus which shows that the N-terminus can determine the subcellular localization (Gallagher and Benfey, 2009). NSP2 can actually move from the ER to the nucleus (Kalo et al., 2005). Two GRAS proteins have been identified as membrane-bound transcription factors in *Arabidopsis* by Kim et al. (2010). As many cytoplasmic factors have been identified to interact with GRAS proteins (Table 19.1), a role of at least some GRAS proteins outside the nucleus cannot be excluded at the moment.

19.3.4 Regulation of GRAS Gene Expression by miRNA

MicroRNAs (miRNAs) are ≈21-nucleotide noncoding RNAs that direct the degradation of target mRNAs. It has been shown that the miRNA171 and 170 target several GRAS proteins (Llave et al., 2002; Bari et al., 2013). In *Arabidopsis*, three *MIR171* genes (*a*, *b*, and *c*) are predicted to regulate SCL6, 22, and 27 (HAM clade). The interaction sites between the miRNAs and the mRNAs of *HAM* genes in *Arabidopsis* have been identified. Binding sites for ath-miR171a–c and ath-miR170 are located in the CDSs of the HAM genes. The conserved miRNA-binding sequence, GAUAUUGGCGCGGCUCAAUCA, encodes the ILARLN hexapeptide and is highly conserved between plant species (Bari et al., 2013). These miRNAs are conserved throughout the plant species (e.g., in rice, four *GRAS* genes are also complementary to miRNA171). miR171a is most highly expressed in inflorescence where it regulates *SCL22 (SCL6-III)* and *SCL6 (SCL6-IV)* expression through mRNA cleavage (Llave et al., 2002). *Arabidopsis* plants overexpressing miR171c, and the triple *ham* mutant in *Arabidopsis*, show similar pleiotropic phenotypes such as changes in branching, height, chlorophyll accumulation, and primary root elongation, suggesting that these three GRAS proteins are the major targets of miRNA (Wang et al., 2010; Engstrom et al., 2011). *MIR171* gene expression itself is regulated by its targets, as HAM proteins bind to their promoters, activating expression and thereby forming a regulatory loop (Xue et al., 2014).

In legumes, an additional miR171 isoform, miR171h, is present, which negatively regulates mycorrhizal or bacterial colonization (De Luis et al., 2012). *MtNSP2* is expressed in the root elongation zone, which usually remains uncolonized, but overexpression of a miR171h-resistant NSP2 version allows colonization of this zone (Lauressergues et al., 2012).

19.4 CONCLUSION

GRAS protein accumulation seems carefully regulated by plants and in many cases GRAS proteins are very low in abundance. Regulation can be either on the level of transcription, sometimes even by feedback loops, in some cases also involving miRNAs, or in later stages. One feature observed in several GRAS proteins is degradation by the ubiquitin system. The involvement of GRAS proteins in so many divergent processes can be explained by their role as transcriptional regulators, not necessarily binding directly to DNA. All GRAS proteins identified so far are very specific to certain processes and regulate only a specific subset of genes. It is not yet clear how the specificity for a certain process is determined, as the GRAS domains are highly conserved. Either specific changes in the GRAS domain or the more variant N-termini are needed for specificity and functionality. The more interacting proteins and complexes that GRAS proteins associate with are identified, the more our understanding of how GRAS proteins regulate transcription will emerge.

References

An, F., Zhang, X., Zhu, Z., Ji, Y., He, W., Jiang, Z., Li, M., Guo, H., 2012. Coordinated regulation of apical hook development by gibberellins and ethylene in etiolated *Arabidopsis* seedlings. Cell Res. 22, 915–927.

Ariizumi, T., Lawrence, P.K., Steber, C.M., 2011. The role of two f-box proteins, SLEEPY1 and SNEEZY, in *Arabidopsis* gibberellin signaling. Plant Physiol. 155, 765–775.

Arnaud, N., Girin, T., Sorefan, K., Fuentes, S., Wood, T.A., Lawrenson, T., Sablowski, R., Ostergaard, L., 2010. Gibberellins control fruit patterning in *Arabidopsis thaliana*. Genes Dev. 24, 2127–2132.

Bai, M.Y., Shang, J.X., Oh, E., Fan, M., Bai, Y., Zentella, R., Sun, T.P., Wang, Z.Y., 2012. Brassinosteroid, gibberellin and phytochrome impinge on a common transcription module in *Arabidopsis*. Nat. Cell Biol. 14, 810–817.

Bari, A., Orazova, S., Ivashchenko, A., 2013. miR156- and miR171-binding sites in the protein-coding sequences of several plant genes. BioMed Res. Int. 2013, 307145.

Battaglia, M., Ripodas, C., Clua, J., Baudin, M., Aguilar, O.M., Niebel, A., Zanetti, M.E., Blanco, F.A., 2014. A nuclear factor Y interacting protein of the GRAS family is required for nodule organogenesis, infection thread progression, and lateral root growth. Plant Physiol. 164, 1430–1442.

Benfey, P.N., Scheres, B., 2000. Root development. Curr. Biol. 10, R813–R815.

Benfey, P.N., Linstead, P.J., Roberts, K., Schiefelbein, J.W., Hauser, M.T., Aeschbacher, R.A., 1993. Root development in *Arabidopsis*: four mutants with dramatically altered root morphogenesis. Development 119, 57–70.

Bennett, T., Leyser, O., 2006. Something on the side: axillary meristems and plant development. Plant Mol. Biol. 60, 843–854.

Bolle, C., 2004. The role of GRAS proteins in plant signal transduction and development. Planta 218, 683–692.

Bolle, C., Koncz, C., Chua, N.H., 2000. PAT1, a new member of the GRAS family, is involved in phytochrome A signal transduction. Genes Dev. 14, 1269–1278.

Carlsbecker, A., Lee, J.Y., Roberts, C.J., Dettmer, J., Lehesranta, S., Zhou, J., Lindgren, O., Moreno-Risueno, M.A., Vaten, A., Thitamadee, S., Campilho, A., Sebastian, J., Bowman, J.L., Helariutta, Y., Benfey, P.N., 2010. Cell signalling by microRNA165/6 directs gene dose-dependent root cell fate. Nature 465, 316–321.

Cederholm, H.M., Iyer-Pascuzzi, A.S., Benfey, P.N., 2012. Patterning the primary root in *Arabidopsis*. Wiley interdisciplinary reviews. Dev. Biol. 1, 675–691.

Cerri, M.R., Frances, L., Laloum, T., Auriac, M.C., Niebel, A., Oldroyd, G.E., Barker, D.G., Fournier, J., de Carvalho-Niebel, F., 2012. *Medicago truncatula* ERN transcription factors: regulatory interplay with NSP1/NSP2 GRAS factors and expression dynamics throughout rhizobial infection. Plant Physiol. 160, 2155–2172.

Cheminant, S., Wild, M., Bouvier, F., Pelletier, S., Renou, J.P., Erhardt, M., Hayes, S., Terry, M.J., Genschik, P., Achard, P., 2011. DELLAs regulate chlorophyll and carotenoid biosynthesis to prevent photo-oxidative damage during seedling deetiolation in *Arabidopsis*. Plant Cell 23, 1849–1860.

Chen, M., Chory, J., 2011. Phytochrome signaling mechanisms and the control of plant development. Trends Cell. Biol. 21, 664–671.

Claeys, H., Skirycz, A., Maleux, K., Inze, D., 2012. DELLA signaling mediates stress-induced cell differentiation in *Arabidopsis* leaves through modulation of anaphase-promoting complex/cyclosome activity. Plant Physiol. 159, 739–747.

Cruz-Ramirez, A., Diaz-Trivino, S., Blilou, I., Grieneisen, V.A., Sozzani, R., Zamioudis, C., Miskolczi, P., Nieuwland, J., Benjamins, R., Dhonukshe, P., Caballero-Perez, J., Horvath, B., Long, Y., Mahonen, A.P., Zhang, H., Xu, J., Murray, J.A., Benfey, P.N., Bako, L., Maree, A.F., Scheres, B., 2012. A bistable circuit involving SCARECROW-RETINOBLASTOMA integrates cues to inform asymmetric stem cell division. Cell 150, 1002–1015.

Cruz-Ramirez, A., Diaz-Trivino, S., Wachsman, G., Du, Y., Arteaga-Vazquez, M., Zhang, H., Benjamins, R., Blilou, I., Neef, A.B., Chandler, V., Scheres, B., 2013. A SCARECROW-RETINOBLASTOMA protein network controls protective quiescence in the *Arabidopsis* root stem cell organizer. PLoS Biol. 11, e1001724.

Cui, H., 2012. Killing two birds with one stone: transcriptional regulators coordinate development and stress responses in plants. Plant Signal. Behav. 7, 701–703.

Cui, H., Benfey, P.N., 2009. Interplay between SCARECROW, GA and LIKE HETEROCHROMATIN PROTEIN 1 in ground tissue patterning in the *Arabidopsis* root. Plant J. 58, 1016–1027.

Cui, H., Levesque, M.P., Vernoux, T., Jung, J.W., Paquette, A.J., Gallagher, K.L., Wang, J.Y., Blilou, I., Scheres, B., Benfey, P.N., 2007. An evolutionarily conserved mechanism delimiting SHR movement defines a single layer of endodermis in plants. Science 316, 421–425.

Cui, H., Kong, D., Liu, X., Hao, Y., 2014. SCARECROW, SCR-LIKE 23 and SHORT-ROOT control bundle sheath cell fate and function in *Arabidopsis thaliana*. Plant J. 78, 319–327.

Curaba, J., Talbot, M., Li, Z., Helliwell, C., 2013. Over-expression of microRNA171 affects phase transitions and floral meristem determinancy in barley. BMC Plant Biol. 13, 6.

Czikkel, B.E., Maxwell, D.P., 2007. NtGRAS1, a novel stress-induced member of the GRAS family in tobacco, localizes to the nucleus. J. Plant Physiol. 164, 1220–1230.

Dai, C., Xue, H.W., 2010. Rice early flowering1, a CKI, phosphorylates DELLA protein SLR1 to negatively regulate gibberellin signalling. EMBO J. 29, 1916–1927.

Daviere, J.M., de Lucas, M., Prat, S., 2008. Transcriptional factor interaction: a central step in DELLA function. Curr. Opin. Genet. Dev. 18, 295–303.

Day, R.B., Shibuya, N., Minami, E., 2003. Identification and characterization of two new members of the GRAS gene family in rice responsive to N-acetylchitooligosaccharide elicitor. Biochim. Biophys. Acta 1625, 261–268.

de Lucas, M., Prat, S., 2014. PIFs get BRright: PHYTOCHROME INTERACTING FACTORs as integrators of light and hormonal signals. New Phytol. 202, 1126–1141.

De Luis, A., Markmann, K., Cognat, V., Holt, D.B., Charpentier, M., Parniske, M., Stougaard, J., Voinnet, O., 2012. Two microRNAs linked to nodule infection and nitrogen-fixing ability in the legume *Lotus japonicus*. Plant Physiol. 160, 2137–2154.

Desvoyes, B., de Mendoza, A., Ruiz-Trillo, I., Gutierrez, C., 2014. Novel roles of plant RETINOBLASTOMA-RELATED (RBR) protein in cell proliferation and asymmetric cell division. J. Exp. Bot. 65, 2657–2666.

Dhondt, S., Coppens, F., De Winter, F., Swarup, K., Merks, R.M., Inze, D., Bennett, M.J., Beemster, G.T., 2010. SHORT-ROOT and SCARECROW regulate leaf growth in *Arabidopsis* by stimulating S-phase progression of the cell cycle. Plant Physiol. 154, 1183–1195.

Di Laurenzio, L., Wysocka-Diller, J., Malamy, J.E., Pysh, L., Helariutta, Y., Freshour, G., Hahn, M.G., Feldmann, K.A., Benfey, P.N., 1996. The SCARECROW gene regulates an asymmetric cell division that is essential for generating the radial organization of the *Arabidopsis* root. Cell 86, 423–433.

Dill, A., Jung, H.S., Sun, T.P., 2001. The DELLA motif is essential for gibberellin-induced degradation of RGA. Proc. Natl. Acad. Sci. USA 98, 14162–14167.

Engstrom, E.M., 2011. Phylogenetic analysis of GRAS proteins from moss, lycophyte and vascular plant lineages reveals that GRAS genes arose and underwent substantial diversification in the ancestral lineage common to bryophytes and vascular plants. Plant Signal. Behav. 6, 850–854.

Engstrom, E.M., Andersen, C.M., Gumulak-Smith, J., Hu, J., Orlova, E., Sozzani, R., Bowman, J.L., 2011. *Arabidopsis* homologs of the petunia hairy meristem gene are required for maintenance of shoot and root indeterminacy. Plant Physiol. 155, 735–750.

Exner, V., Aichinger, E., Shu, H., Wildhaber, T., Alfarano, P., Caflisch, A., Gruissem, W., Kohler, C., Hennig, L., 2009. The chromodomain of LIKE HETEROCHROMATIN PROTEIN 1 is essential for H3K27me3 binding and function during *Arabidopsis* development. PLoS ONE 4, e5335.

Feng, S., Martinez, C., Gusmaroli, G., Wang, Y., Zhou, J., Wang, F., Chen, L., Yu, L., Iglesias-Pedraz, J.M., Kircher, S., Schafer, E., Fu, X., Fan, L.M., Deng, X.W., 2008. Coordinated regulation of *Arabidopsis thaliana* development by light and gibberellins. Nature 451, 475–479.

Feurtado, J.A., Huang, D., Wicki-Stordeur, L., Hemstock, L.E., Potentier, M.S., Tsang, E.W., Cutler, A.J., 2011. The *Arabidopsis* C2H2 zinc finger INDETERMINATE DOMAIN1/ENHYDROUS promotes the transition to germination by regulating light and hormonal signaling during seed maturation. Plant Cell 23, 1772–1794.

Fode, B., Siemsen, T., Thurow, C., Weigel, R., Gatz, C., 2008. The *Arabidopsis* GRAS protein SCL14 interacts with class II TGA transcription factors and is essential for the activation of stress-inducible promoters. Plant Cell 20, 3122–3135.

Fukaki, H., Fujisawa, H., Tasaka, M., 1996. SGR1, SGR2, SGR3: novel genetic loci involved in shoot gravitropism in *Arabidopsis thaliana*. Plant Physiol. 110, 945–955.

Fukaki, H., Wysocka-Diller, J., Kato, T., Fujisawa, H., Benfey, P.N., Tasaka, M., 1998. Genetic evidence that the endodermis is essential for shoot gravitropism in *Arabidopsis thaliana*. Plant J. 14, 425–430.

Fukazawa, J., Teramura, H., Murakoshi, S., Nasuno, K., Nishida, N., Ito, T., Yoshida, M., Kamiya, Y., Yamaguchi, S., Takahashi, Y., 2014. DELLAs function as coactivators of GAI-ASSOCIATED FACTOR1 in regulation of gibberellin homeostasis and signaling in *Arabidopsis*. Plant Cell 26, 2920–2938.

Gallagher, K.L., Benfey, P.N., 2009. Both the conserved GRAS domain and nuclear localization are required for SHORT-ROOT movement. Plant J. 57, 785–797.

Gallego-Bartolome, J., Minguet, E.G., Marin, J.A., Prat, S., Blazquez, M.A., Alabadi, D., 2010. Transcriptional diversification and functional conservation between DELLA proteins in *Arabidopsis*. Mol. Biol. Evol. 27, 1247–1256.

Gallego-Bartolome, J., Minguet, E.G., Grau-Enguix, F., Abbas, M., Locascio, A., Thomas, S.G., Alabadi, D., Blazquez, M.A., 2012. Molecular mechanism for the interaction between gibberellin and brassinosteroid signaling pathways in *Arabidopsis*. Proc. Natl. Acad. Sci. USA 109, 13446–13451.

Gao, M.J., Parkin, I., Lydiate, D., Hannoufa, A., 2004. An auxin-responsive SCARECROW-like transcriptional activator interacts with histone deacetylase. Plant Mol. Biol. 55, 417–431.

Gardiner, J., Donner, T.J., Scarpella, E., 2011. Simultaneous activation of SHR and ATHB8 expression defines switch to preprocambial cell state in *Arabidopsis* leaf development. Dev. Dyn. 240, 261–270.

Gobbato, E., Marsh, J.F., Vernie, T., Wang, E., Maillet, F., Kim, J., Miller, J.B., Sun, J., Bano, S.A., Ratet, P., Mysore, K.S., Denarie, J., Schultze, M., Oldroyd, G.E., 2012. A GRAS-type transcription factor with a specific function in mycorrhizal signaling. Curr. Biol. 22, 2236–2241.

Greb, T., Clarenz, O., Schafer, E., Muller, D., Herrero, R., Schmitz, G., Theres, K., 2003. Molecular analysis of the LATERAL SUPPRESSOR gene in *Arabidopsis* reveals a conserved control mechanism for axillary meristem formation. Genes Dev. 17, 1175–1187.

Groszmann, M., Paicu, T., Alvarez, J.P., Swain, S.M., Smyth, D.R., 2011. SPATULA and ALCATRAZ are partially redundant, functionally diverging bHLH genes required for *Arabidopsis* gynoecium and fruit development. Plant J. 68, 816–829.

Gutjahr, C., Parniske, M., 2013. Cell and developmental biology of arbuscular mycorrhiza symbiosis. Annu. Rev. Cell Dev. Biol. 29, 593–617.

Heckmann, A.B., Lombardo, F., Miwa, H., Perry, J.A., Bunnewell, S., Parniske, M., Wang, T.L., Downie, J.A., 2006. *Lotus japonicus* nodulation requires two GRAS domain regulators, one of which is functionally conserved in a non-legume. Plant Physiol. 142, 1739–1750.

Heidstra, R., Welch, D., Scheres, B., 2004. Mosaic analyses using marked activation and deletion clones dissect *Arabidopsis* SCARECROW action in asymmetric cell division. Genes Dev. 18, 1964–1969.

Helariutta, Y., Fukaki, H., Wysocka-Diller, J., Nakajima, K., Jung, J., Sena, G., Hauser, M.T., Benfey, P.N., 2000. The SHORT-ROOT gene controls radial patterning of the *Arabidopsis* root through radial signaling. Cell 101, 555–567.

Heo, J.O., Chang, K.S., Kim, I.A., Lee, M.H., Lee, S.A., Song, S.K., Lee, M.M., Lim, J., 2011. Funneling of gibberellin signaling by the GRAS transcription regulator scarecrow-like 3 in the *Arabidopsis* root. Proc. Natl. Acad. Sci. USA 108, 2166–2171.

Hirsch, S., Kim, J., Munoz, A., Heckmann, A.B., Downie, J.A., Oldroyd, G.E., 2009. GRAS proteins form a DNA binding complex to induce gene expression during nodulation signaling in *Medicago truncatula*. Plant Cell 21, 545–557.

Hong, G.J., Xue, X.Y., Mao, Y.B., Wang, L.J., Chen, X.Y., 2012. *Arabidopsis* MYC2 interacts with DELLA proteins in regulating sesquiterpene synthase gene expression. Plant Cell 24, 2635–2648.

Huang, G., Dong, R., Allen, R., Davis, E.L., Baum, T.J., Hussey, R.S., 2006. A root-knot nematode secretory peptide functions as a ligand for a plant transcription factor. Mol. Plant Microbe Interact. 19, 463–470.

Itoh, H., Sasaki, A., Ueguchi-Tanaka, M., Ishiyama, K., Kobayashi, M., Hasegawa, Y., Minami, E., Ashikari, M., Matsuoka, M., 2005. Dissection of the phosphorylation of rice DELLA protein, SLENDER RICE1. Plant Cell Physiol. 46, 1392–1399.

Iyer-Pascuzzi, A.S., Jackson, T., Cui, H., Petricka, J.J., Busch, W., Tsukagoshi, H., Benfey, P.N., 2011. Cell identity regulators link development and stress responses in the *Arabidopsis* root. Dev. Cell 21, 770–782.

Josse, E.M., Gan, Y., Bou-Torrent, J., Stewart, K.L., Gilday, A.D., Jeffree, C.E., Vaistij, F.E., Martinez-Garcia, J.F., Nagy, F., Graham, I.A., Halliday, K.J., 2011. A DELLA in disguise: SPATULA restrains the growth of the developing *Arabidopsis* seedling. Plant Cell 23, 1337–1351.

Kalo, P., Gleason, C., Edwards, A., Marsh, J., Mitra, R.M., Hirsch, S., Jakab, J., Sims, S., Long, S.R., Rogers, J., Kiss, G.B., Downie, J.A., Oldroyd, G.E., 2005. Nodulation signaling in legumes requires NSP2, a member of the GRAS family of transcriptional regulators. Science 308, 1786–1789.

Kamiya, N., Itoh, J., Morikami, A., Nagato, Y., Matsuoka, M., 2003. The SCARECROW gene's role in asymmetric cell divisions in rice plants. Plant J. 36, 45–54.

Kim, S.G., Lee, S., Seo, P.J., Kim, S.K., Kim, J.K., Park, C.M., 2010. Genome-scale screening and molecular characterization of membrane-bound transcription factors in *Arabidopsis* and rice. Genomics 95, 56–65.

Koizumi, K., Gallagher, K.L., 2013. Identification of SHRUBBY, a SHORT-ROOT and SCARECROW interacting protein that controls root growth and radial patterning. Development 140, 1292–1300.

Koizumi, K., Wu, S., MacRae-Crerar, A., Gallagher, K.L., 2011. An essential protein that interacts with endosomes and promotes movement of the SHORT-ROOT transcription factor. Curr. Biol. 21, 1559–1564.

Kuwabara, A., Gruissem, W., 2014. *Arabidopsis* RETINOBLASTOMA-RELATED and polycomb group proteins: cooperation during plant cell differentiation and development. J. Exp. Bot. 65, 2667–2676.

Laajanen, K., Vuorinen, I., Salo, V., Juuti, J., Raudaskoski, M., 2007. Cloning of *Pinus sylvestris* SCARECROW gene and its expression pattern in the pine root system, mycorrhiza and NPA-treated short roots. New Phytol. 175, 230–243.

Lauressergues, D., Delaux, P.M., Formey, D., Lelandais-Briere, C., Fort, S., Cottaz, S., Becard, G., Niebel, A., Roux, C., Combier, J.P., 2012. The microRNA miR171h modulates arbuscular mycorrhizal colonization of *Medicago truncatula* by targeting NSP2. Plant J. 72, 512–522.

Lee, M.H., Kim, B., Song, S.K., Heo, J.O., Yu, N.I., Lee, S.A., Kim, M., Kim, D.G., Sohn, S.O., Lim, C.E., Chang, K.S., Lee, M.M., Lim, J., 2008. Large-scale analysis of the GRAS gene family in *Arabidopsis thaliana*. Plant Mol. Biol. 67, 659–670.

Levesque, M.P., Vernoux, T., Busch, W., Cui, H., Wang, J.Y., Blilou, I., Hassan, H., Nakajima, K., Matsumoto, N., Lohmann, J.U., Scheres, B., Benfey, P.N., 2006. Whole-genome analysis of the SHORT-ROOT developmental pathway in *Arabidopsis*. PLoS Biol. 4, e143.

Li, X., Qian, Q., Fu, Z., Wang, Y., Xiong, G., Zeng, D., Wang, X., Liu, X., Teng, S., Hiroshi, F., Yuan, M., Luo, D., Han, B., Li, J., 2003. Control of tillering in rice. Nature 422, 618–621.

Li, W., Wu, J., Weng, S., Zhang, Y., Zhang, D., Shi, C., 2010. Identification and characterization of dwarf 62, a loss-of-function mutation in DLT/OsGRAS-32 affecting gibberellin metabolism in rice. Planta 232, 1383–1396.

Libault, M., Wan, J., Czechowski, T., Udvardi, M., Stacey, G., 2007. Identification of 118 *Arabidopsis* transcription factor and 30 ubiquitin-ligase genes responding to chitin, a plant-defense elicitor. Mol. Plant Microbe Interact. 20, 900–911.

Lim, J., Jung, J.W., Lim, C.E., Lee, M.H., Kim, B.J., Kim, M., Bruce, W.B., Benfey, P.N., 2005. Conservation and diversification of SCARECROW in maize. Plant Mol. Biol. 59, 619–630.

Lim, S., Park, J., Lee, N., Jeong, J., Toh, S., Watanabe, A., Kim, J., Kang, H., Kim, D.H., Kawakami, N., Choi, G., 2013. ABA-insensitive3, ABA-insensitive5, and DELLAs interact to activate the expression of SOMNUS and other high-temperature-inducible genes in imbibed seeds in *Arabidopsis*. Plant Cell 25, 4863–4878.

Lin, Q., Wang, D., Dong, H., Gu, S., Cheng, Z., Gong, J., Qin, R., Jiang, L., Li, G., Wang, J.L., Wu, F., Guo, X., Zhang, X., Lei, C., Wang, H., Wan, J., 2012. Rice APC/C(TE) controls tillering by mediating the degradation of MONOCULM 1. Nat. Commun. 3, 752.

Liu, W., Kohlen, W., Lillo, A., Op den Camp, R., Ivanov, S., Hartog, M., Limpens, E., Jamil, M., Smaczniak, C., Kaufmann, K., Yang, W.C., Hooiveld, G.J., Charnikhova, T., Bouwmeester, H.J., Bisseling, T., Geurts, R., 2011. Strigolactone biosynthesis in *Medicago truncatula* and rice requires the symbiotic GRAS-type transcription factors NSP1 and NSP2. Plant Cell 23, 3853–3865.

Llave, C., Xie, Z., Kasschau, K.D., Carrington, J.C., 2002. Cleavage of scarecrow-like mRNA targets directed by a class of *Arabidopsis* miRNA. Science 297, 2053–2056.

Locascio, A., Blazquez, M.A., Alabadi, D., 2013. Genomic analysis of DELLA protein activity. Plant Cell Physiol. 54, 1229–1237.

Lucas, M., Swarup, R., Paponov, I.A., Swarup, K., Casimiro, I., Lake, D., Peret, B., Zappala, S., Mairhofer, S., Whitworth, M., Wang, J., Ljung, K., Marchant, A., Sandberg, G., Holdsworth, M.J., Palme, K., Pridmore, T., Mooney, S., Bennett, M.J., 2011. Short-root regulates primary, lateral, and adventitious root development in *Arabidopsis*. Plant Physiol. 155, 384–398.

Ma, H.S., Liang, D., Shuai, P., Xia, X.L., Yin, W.L., 2010. The salt- and drought-inducible poplar GRAS protein SCL7 confers salt and drought tolerance in *Arabidopsis thaliana*. J. Exp. Bot. 61, 4011–4019.

Ma, Z., Hu, X., Cai, W., Huang, W., Zhou, X., Luo, Q., Yang, H., Wang, J., Huang, J., 2014. *Arabidopsis* miR171-targeted scarecrow-like proteins bind to GT cis-elements and mediate gibberellin-regulated chlorophyll biosynthesis under light conditions. PLoS Genet. 10, e1004519.

Makkena, S., Lamb, R.S., 2013. The bHLH transcription factor SPATULA is a key regulator of organ size in *Arabidopsis thaliana*. Plant Signal. Behav. 8, e24140.

Martinka, M., Dolan, L., Pernas, M., Abe, J., Lux, A., 2012. Endodermal cell–cell contact is required for the spatial control of Casparian band development in *Arabidopsis thaliana*. Ann. Bot. 110, 361–371.

Mayrose, M., Ekengren, S.K., Melech-Bonfil, S., Martin, G.B., Sessa, G., 2006. A novel link between tomato GRAS genes, plant disease resistance and mechanical stress response. Mol. Plant Pathol. 7, 593–604.

Morohashi, K., Minami, M., Takase, H., Hotta, Y., Hiratsuka, K., 2003. Isolation and characterization of a novel GRAS gene that regulates meiosis-associated gene expression. J. Biol. Chem. 278, 20865–20873.

Murakami, Y., Miwa, H., Imaizumi-Anraku, H., Kouchi, H., Downie, J.A., Kawaguchi, M., Kawasaki, S., 2006. Positional cloning identifies *Lotus japonicus* NSP2, a putative transcription factor of the GRAS family, required for NIN and ENOD40 gene expression in nodule initiation. DNA Res. 13, 255–265.

Nakajima, K., Sena, G., Nawy, T., Benfey, P.N., 2001. Intercellular movement of the putative transcription factor SHR in root patterning. Nature 413, 307–311.

Ogasawara, H., Kaimi, R., Colasanti, J., Kozaki, A., 2011. Activity of transcription factor JACKDAW is essential for SHR/SCR-dependent activation of SCARECROW and MAGPIE and is modulated by reciprocal interactions with MAGPIE, SCARECROW and SHORT ROOT. Plant Mol. Biol. 77, 489–499.

Oldroyd, G.E., 2013. Speak, friend, and enter: signalling systems that promote beneficial symbiotic associations in plants. Nature reviews. Microbiology 11, 252–263.

Oldroyd, G.E., Murray, J.D., Poole, P.S., Downie, J.A., 2011. The rules of engagement in the legume–rhizobial symbiosis. Annu. Rev. Genet. 45, 119–144.

Paquette, A.J., Benfey, P.N., 2005. Maturation of the ground tissue of the root is regulated by gibberellin and SCARECROW and requires SHORT-ROOT. Plant Physiol. 138, 636–640.

Park, J., Nguyen, K.T., Park, E., Jeon, J.S., Choi, G., 2013. DELLA proteins and their interacting RING finger proteins repress gibberellin responses by binding to the promoters of a subset of gibberellin-responsive genes in *Arabidopsis*. Plant Cell 25, 927–943.

Peng, J., Carol, P., Richards, D.E., King, K.E., Cowling, R.J., Murphy, G.P., Harberd, N.P., 1997. The *Arabidopsis* GAI gene defines a signaling pathway that negatively regulates gibberellin responses. Genes Dev. 11, 3194–3205.

Pysh, L.D., Wysocka-Diller, J.W., Camilleri, C., Bouchez, D., Benfey, P.N., 1999. The GRAS gene family in *Arabidopsis*: sequence characterization and basic expression analysis of the SCARECROW-LIKE genes. Plant J. 18, 111–119.

Qin, Q., Wang, W., Guo, X., Yue, J., Huang, Y., Xu, X., Li, J., Hou, S., 2014. *Arabidopsis* DELLA protein degradation is controlled by a type-one protein phosphatase TOPP4. PLoS Genet. 10, e1004464.

Richards, D.E., Peng, J., Harberd, N.P., 2000. Plant GRAS and metazoan STATs: one family? BioEssays 22, 573–577.

Sabatini, S., Heidstra, R., Wildwater, M., Scheres, B., 2003. SCARECROW is involved in positioning the stem cell niche in the *Arabidopsis* root meristem. Genes Dev. 17, 354–358.

Sarnowska, E.A., Rolicka, A.T., Bucior, E., Cwiek, P., Tohge, T., Fernie, A.R., Jikumaru, Y., Kamiya, Y., Franzen, R., Schmelzer, E., Porri, A., Sacharowski, S., Gratkowska, D.M., Zugaj, D.L., Taff, A., Zalewska, A., Archacki, R., Davis, S.J., Copuland, G., Koncz, C., Jerzmanowski, A., Sarnowski, T.J., 2013. DELLA-interacting SWI3C core subunit of switch/sucrose nonfermenting chromatin remodeling complex modulates gibberellin responses and hormonal cross talk in *Arabidopsis*. Plant Physiol. 163, 305–317.

Sato, A., Miura, K., 2011. Root architecture remodeling induced by phosphate starvation. Plant Signal. Behav. 6, 1122–1126.

Scheres, B., 1997. Cell signaling in root development. Curr. Opin. Genet. Dev. 7, 501–506.

Schulze, S., Schafer, B.N., Parizotto, E.A., Voinnet, O., Theres, K., 2010. LOST MERISTEMS genes regulate cell differentiation of central zone descendants in *Arabidopsis* shoot meristems. Plant J. 64, 668–678.

Schumacher, K., Schmitt, T., Rossberg, M., Schmitz, G., Theres, K., 1999. The lateral suppressor (Ls) gene of tomato encodes a new member of the VHIID protein family. Proc. Natl. Acad. Sci. USA 96, 290–295.

Sena, G., Jung, J.W., Benfey, P.N., 2004. A broad competence to respond to SHORT ROOT revealed by tissue-specific ectopic expression. Development 131, 2817–2826.

Silverstone, A.L., Ciampaglio, C.N., Sun, T., 1998. The *Arabidopsis* RGA gene encodes a transcriptional regulator repressing the gibberellin signal transduction pathway. Plant Cell 10, 155–169.

Slewinski, T.L., 2013. Using evolution as a guide to engineer kranz-type c4 photosynthesis. Front. Plant Sci. 4, 212.

Slewinski, T.L., Anderson, A.A., Zhang, C., Turgeon, R., 2012. Scarecrow plays a role in establishing Kranz anatomy in maize leaves. Plant Cell Physiol. 53, 2030–2037.

Slewinski, T.L., Anderson, A.A., Price, S., Withee, J.R., Gallagher, K., Turgeon, R., 2014. Short-root1 plays a role in the development of vascular tissue and kranz anatomy in maize leaves. Mol. Plant 7, 1388–1392.

Song, S., Huang, H., Gao, H., Wang, J., Wu, D., Liu, X., Yang, S., Zhai, Q., Li, C., Qi, T., Xie, D., 2014. Interaction between MYC2 and ETHYLENE INSENSITIVE3 modulates antagonism between jasmonate and ethylene signaling in *Arabidopsis*. Plant Cell 26, 263–279.

Sozzani, R., Cui, H., Moreno-Risueno, M.A., Busch, W., Van Norman, J.M., Vernoux, T., Brady, S.M., Dewitte, W., Murray, J.A., Benfey, P.N., 2010. Spatiotemporal regulation of cell-cycle genes by SHORTROOT links patterning and growth. Nature 466, 128–132.

Stuurman, J., Jaggi, F., Kuhlemeier, C., 2002. Shoot meristem maintenance is controlled by a GRAS-gene mediated signal from differentiating cells. Genes Dev. 16, 2213–2218.

Sun, T.P., 2011. The molecular mechanism and evolution of the GA–GID1–DELLA signaling module in plants. Curr. Biol. 21, R338–R345.

Sun, T.P., Gubler, F., 2004. Molecular mechanism of gibberellin signaling in plants. Annu. Rev. Plant Biol. 55, 197–223.

Sun, L., Li, X., Fu, Y., Zhu, Z., Tan, L., Liu, F., Sun, X., Sun, X., Sun, C., 2013. GS6, a member of the GRAS gene family, negatively regulates grain size in rice. J. Integr. Plant Biol. 55, 938–949.

Tong, H., Chu, C., 2009. Roles of DLT in fine modulation on brassinosteroid response in rice. Plant Signal. Behav. 4, 438–439.

Torres-Galea, P., Huang, L.F., Chua, N.H., Bolle, C., 2006. The GRAS protein SCL13 is a positive regulator of phytochrome-dependent red light signaling, but can also modulate phytochrome A responses. Mol. Genet. Genomics 276, 13–30.

Torres-Galea, P., Hirtreiter, B., Bolle, C., 2013. Two GRAS proteins, SCARECROW-LIKE21 and PHYTOCHROME A SIGNAL TRANSDUCTION1, function cooperatively in phytochrome A signal transduction. Plant Physiol. 161, 291–304.

Ueguchi-Tanaka, M., Nakajima, M., Katoh, E., Ohmiya, H., Asano, K., Saji, S., Hongyu, X., Ashikari, M., Kitano, H., Yamaguchi, I., Matsuoka, M., 2007. Molecular interactions of a soluble gibberellin receptor, GID1, with a rice DELLA protein, SLR1, and gibberellin. Plant Cell 19, 2140–2155.

Vaistij, F.E., Gan, Y., Penfield, S., Gilday, A.D., Dave, A., He, Z., Josse, E.M., Choi, G., Halliday, K.J., Graham, I.A., 2013. Differential control of seed primary dormancy in Arabidopsis ecotypes by the transcription factor SPATULA. Proc. Natl. Acad. Sci. USA 110, 10866–10871.

van den Berg, C., Willemsen, V., Hendriks, G., Weisbeek, P., Scheres, B., 1997. Short-range control of cell differentiation in the Arabidopsis root meristem. Nature 390, 287–289.

Wang, L., Mai, Y.X., Zhang, Y.C., Luo, Q., Yang, H.Q., 2010. MicroRNA171c-targeted SCL6-II, SCL6-III, and SCL6-IV genes regulate shoot branching in Arabidopsis. Mol. Plant 3, 794–806.

Welch, D., Hassan, H., Blilou, I., Immink, R., Heidstra, R., Scheres, B., 2007. Arabidopsis JACKDAW and MAGPIE zinc finger proteins delimit asymmetric cell division and stabilize tissue boundaries by restricting SHORT-ROOT action. Genes Dev. 21, 2196–2204.

Wild, M., Daviere, J.M., Cheminant, S., Regnault, T., Baumberger, N., Heintz, D., Baltz, R., Genschik, P., Achard, P., 2012. The Arabidopsis DELLA RGA-LIKE3 is a direct target of MYC2 and modulates jasmonate signaling responses. Plant Cell 24, 3307–3319.

Wildwater, M., Campilho, A., Perez-Perez, J.M., Heidstra, R., Blilou, I., Korthout, H., Chatterjee, J., Mariconti, L., Gruissem, W., Scheres, B., 2005. The RETINOBLASTOMA-RELATED gene regulates stem cell maintenance in Arabidopsis roots. Cell 123, 1337–1349.

Willige, B.C., Ghosh, S., Nill, C., Zourelidou, M., Dohmann, E.M., Maier, A., Schwechheimer, C., 2007. The DELLA domain of GA INSENSITIVE mediates the interaction with the GA INSENSITIVE DWARF1A gibberellin receptor of Arabidopsis. Plant Cell 19, 1209–1220.

Wysocka-Diller, J.W., Helariutta, Y., Fukaki, H., Malamy, J.E., Benfey, P.N., 2000. Molecular analysis of SCARECROW function reveals a radial patterning mechanism common to root and shoot. Development 127, 595–603.

Xu, J., Hofhuis, H., Heidstra, R., Sauer, M., Friml, J., Scheres, B., 2006. A molecular framework for plant regeneration. Science 311, 385–388.

Xu, C., Wang, Y., Yu, Y., Duan, J., Liao, Z., Xiong, G., Meng, X., Liu, G., Qian, Q., Li, J., 2012. Degradation of MONOCULM 1 by APC/C(TAD1) regulates rice tillering. Nat. Commun. 3, 750.

Xue, X.Y., Zhao, B., Chao, L.M., Chen, D.Y., Cui, W.R., Mao, Y.B., Wang, L.J., Chen, X.Y., 2014. Interaction between two timing microRNAs controls trichome distribution in Arabidopsis. PLoS Genet. 10, e1004266.

Yang, M., Yang, Q., Fu, T., Zhou, Y., 2011. Overexpression of the Brassica napus BnLAS gene in Arabidopsis affects plant development and increases drought tolerance. Plant Cell Rep. 30, 373–388.

Yang, D.H., Sun, H.J., Goh, C.H., Song, P.S., Bae, T.W., Song, I.J., Lim, Y.P., Lim, P.O., Lee, H.Y., 2012. Cloning of a Zoysia ZjLsL and its overexpression to induce axillary meristem initiation and tiller formation in Arabidopsis and bentgrass. Plant Biol. 14, 411–419.

Yokota, K., Soyano, T., Kouchi, H., Hayashi, M., 2010. Function of GRAS proteins in root nodule symbiosis is retained in homologs of a non-legume, rice. Plant Cell Physiol. 51, 1436–1442.

Yoshida, H., Hirano, K., Sato, T., Mitsuda, N., Nomoto, M., Maeo, K., Koketsu, E., Mitani, R., Kawamura, M., Ishiguro, S., Tada, Y., Ohme-Takagi, M., Matsuoka, M., Ueguchi-Tanaka, M., 2014. DELLA protein functions as a transcriptional activator through the DNA binding of the indeterminate domain family proteins. Proc. Natl. Acad. Sci. USA 111, 7861–7866.

Yu, S., Galvao, V.C., Zhang, Y.C., Horrer, D., Zhang, T.Q., Hao, Y.H., Feng, Y.Q., Wang, S., Schmid, M., Wang, J.W., 2012. Gibberellin regulates the Arabidopsis floral transition through miR156-targeted SQUAMOSA promoter binding-like transcription factors. Plant Cell 24, 3320–3332.

Zentella, R., Zhang, Z.L., Park, M., Thomas, S.G., Endo, A., Murase, K., Fleet, C.M., Jikumaru, Y., Nambara, E., Kamiya, Y., Sun, T.P., 2007. Global analysis of della direct targets in early gibberellin signaling in Arabidopsis. Plant Cell 19, 3037–3057.

Zhang, D., Iyer, L.M., Aravind, L., 2012. Bacterial GRAS domain proteins throw new light on gibberellic acid response mechanisms. Bioinformatics 28, 2407–2411.

Zhang, Z.L., Ogawa, M., Fleet, C.M., Zentella, R., Hu, J., Heo, J.O., Lim, J., Kamiya, Y., Yamaguchi, S., Sun, T.P., 2011. Scarecrow-like 3 promotes gibberellin signaling by antagonizing master growth repressor DELLA in Arabidopsis. Proc. Natl. Acad. Sci. USA 108, 2160–2165.

CHAPTER 20

DELLA Proteins, a Group of GRAS Transcription Regulators that Mediate Gibberellin Signaling

Francisco Vera-Sirera, Maria Dolores Gomez, Miguel A. Perez-Amador

Instituto de Biología Molecular y Celular de Plantas (IBMCP), Universidad Politécnica de Valencia-Consejo Superior de Investigaciones Científicas, Ciudad Politécnica de la Innovación (CPI), Valencia, Spain

OUTLINE

20.1 About DELLAs and Gibberellins — 313	20.2.5 GA-Independent DELLA Proteolytic Signaling — 318
20.1.1 Introduction to Gibberellins — 314	20.2.6 Posttranslational Modifications of DELLA Proteins — 318
20.1.2 Discovery of DELLAs in GA-Mediated Growth and Development — 314	20.3 The Molecular Mechanism of DELLA Action: DELLA–Protein Interactions and Target Genes — 319
20.1.3 DELLA Protein Structure — 315	20.3.1 Interfering Mechanism — 320
20.1.4 Distribution and Evolution of DELLA Proteins — 316	20.3.2 The Transactivation Mechanism — 321
20.1.5 DELLAs and the Green Revolution — 316	20.3.3 Nontranscriptional DELLA Interaction — 323
20.2 GA Signaling through DELLAs — 317	20.3.4 DELLAs Regulate the Expression of Target Genes — 323
20.2.1 GA Receptor GID1 — 317	20.4 Conclusion and Future Perspectives — 324
20.2.2 The GA–GID1–DELLA Module in GA Signaling — 317	References — 324
20.2.3 DELLA Nonproteolytic Signaling — 318	
20.2.4 DELLA-independent GA Response — 318	

20.1 ABOUT DELLAs AND GIBBERELLINS

DELLAs (aspartic acid–glutamic acid–leucine–leucine–alanine) are a subfamily of the plant-specific GRAS (GIB-BERELIC ACID INSENSITIVE REPRESSOR OF *ga1-3* SCARECROW) family of transcriptional regulators that mediate gibberellin (GA) signaling. The first DELLA member was discovered by the characterization of the *GA-insensitive-1* (*gai-1*) mutation of *Arabidopsis thaliana*, which confers a dwarf phenotype with darker green foliage, reduced apical dominance, and late flowering (Koorneef et al., 1985; Peng et al., 1997). Unlike GA biosynthesis mutants, *gai-1* plants fail to respond to exogenous GA and accumulate high levels of GAs, which suggests that the corresponding gene participates in GA signaling and response. The defective *gai-1* allele showed a deletion of 17 amino acids within the N-terminal domain, which included the conserved Asp-Glu-Leu-Leu-Ala (DELLA) amino acid motif (Peng et al., 1997; Peng et al., 1999; Pysh et al., 1999). DELLAs were named after this domain. Later it was demonstrated that the complete removal of all DELLA activity leads to the constitutive activation of all GA responses (Cheng et al., 2004; Ikeda et al., 2001; Jasinski et al., 2008), which indicated DELLAs to be repressors of GA signaling. GAs control most growth and development processes through the GA-dependent degradation of DELLA proteins, which is mediated mainly by the DELLA domain (Sun, 2011).

In this chapter we review advances in the biological and molecular function of DELLAs in the control of GA responses. Further evidence indicates a central role of DELLAs as key integrators of both internal (hormonal factors and gene regulatory networks) and external (environmental and stress responses) signaling to coordinate plant growth, differentiation, and stress responses.

20.1.1 Introduction to Gibberellins

GAs are plant hormones that regulate many growth and development aspects throughout the plants life cycle, including promotion of cell division and elongation, stem and root elongation, bolting, promotion of flower and fruit development, and control of seed dormancy and germination (Fleet and Sun, 2005; Sun, 2011; Swain and Singh, 2005). Some evidence has also revealed a role of GAs in the response to both biotic and abiotic stress (reviewed in Colebrook et al., 2014). In chemical terms, GAs are a large group of tetracyclic diterpenoid carboxylic acids synthesized from geranylgeranyl diphosphate (GGPP). GA biosynthesis is a complex process that has been extensively reviewed (Hedden and Thomas, 2012; Yamaguchi, 2008; Zi et al., 2014). From GGPP, a series of enzymatic activities in the plastid, endoplasmic reticulum, and cytosol lead to the formation of GA_{12}, which is considered the common precursor of all GAs in plants. The activity 2-oxoglutarate-dependent dioxygenases of GA20 (GA20ox, GA 20-OXIDASE) and GA3 (GA3ox, GA 3-OXIDASE) oxidases mediates the synthesis of bioactive gibberellins GA_1, GA_3 (gibberellic acid), and GA_4 from precursor GA_{12} throughout a complex pathway, which involves the synthesis of multiple intermediates. Finally, bioactive GA can be deactivated by the further action of GA2 oxidases.

The discovery of GAs stems from a hunt for the causal agent of the "bakanae" (foolish seedling) disease in rice, which caused huge crop yield losses in Asia during the nineteenth century and the early twentieth century. Bakanae is produced by an infection with the fungus *Gibberella fujikuroi*, and causes yellowing and the development of elongated seedlings, slender leaves, and stunted roots. In the mid-1950s the causal agent of the disease produced by the fungus, gibberellic acid or GA_3, was identified and characterized (Curtis and Cross, 1954; Stodola et al., 1955; Takahashi et al., 1955). Since then, almost 200 GA-related compounds have been described, although only a few have biological activity.

20.1.2 Discovery of DELLAs in GA-Mediated Growth and Development

GAI was the first DELLA protein characterized by the molecular cloning of the *gai-1* mutation of *Arabidopsis* (Peng and Harberd, 1997). *gai-1* is a semidominant mutation that resulted in dwarf dark green plants, much like GA biosynthesis mutants such as *ga1-3*, which cannot be rescued by application of exogenous GAs (Figure 20.1; Koorneef et al., 1985; Peng and Harberd, 1993; Wilson and Somerville, 1995). Molecular cloning revealed that GAI is a 532-amino-acid residue protein with no introns, has a high sequence similarity to SCARECROW (SCR), and is a member of a novel family of putative transcription factors currently known as the GRAS family (reviewed in Chapter 10). The phenotype of *gai-1* resulted from the deletion of the DELLA domain (DELLAVLGYKVRSSEMA) in the N-terminal region of GAI. A second DELLA gene in *Arabidopsis*, RGA (for REPRESSOR OF *ga1-3*), was isolated by a genetic screen for suppressors of *ga1-3* (Silverstone et al., 1998; Silverstone et al., 1997). GAI and RGA are extremely similar, and the 17-amino-acid sequence of the DELLA domain deleted in *gai-1* is highly conserved in RGA, while the elimination of this domain in the pRGA:GFP-*rgaΔ17* transgenic line generates a similar semidominant phenotype to *gai-1* (Dill et al., 2001). Later, it was demonstrated that the DELLA domain is sufficient to promote GA-dependent DELLA protein degradation; the *rgaΔ17* protein in *pRGA:GFP-rgaΔ17* plants is not degraded upon GA treatment. Interestingly, germination and flower development in *ga1-3* are not rescued by the removal of GAI and RGA, which suggests the presence of other DELLA proteins in *Arabidopsis*. Indeed, a

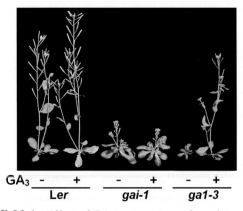

FIGURE 20.1 Effect of GA treatments on the gai-1 mutant of *Arabidopsis*. The gai-1 shows insensitivity to GA_3 treatment, unlike the wild-type plants (Ler) and GA synthesis mutant *ga1-3*. *Reused with permission from Dill et al. (2001). Copyright (2001) National Academy of Sciences, USA.*

sequence analysis identified three other DELLA genes in the *Arabidopsis* genome, *RGA-LIKE 1, 2,* and *3* (*RGL1, RGL2,* and *RGL3*).

In general terms, RGA is the DELLA that shows broader expression patterns, which is coincident with its predominant effect in the control of GA growth and development (Silverstone et al., 1997; Tyler et al., 2004). Other DELLAs have particular expression patterns, and consequently show more specific functions. As stated previously, strong GA-deficient mutant *ga1-3* displays marked reduction in bioactive GAs, which results in a strong GA-deficient phenotype. As DELLAs act mostly as repressors of the GA response, analysis of the phenotypes of the single and multiple loss-of-function DELLA mutant alleles in the *ga1-3* background allows the dissection of the specific and overlapping roles of each DELLA in GA-mediated growth and development, as well as in response to stress. For example, the null mutation of *RGA* partially alleviates several GA deficiency defects, such as stem growth, trichome initiation, flowering time, and apical dominance, which indicates that RGA represses these GA-mediated processes (Silverstone et al., 1997). Another simple DELLA mutant, *rgl2-1*, has been reported to germinate in the presence of paclobutrazol (PAC), a GA biosynthesis inhibitor, which supports a key role of RGL2 in seed germination (Lee et al., 2002; Tyler et al., 2004). The almost complete rescue of vegetative growth and floral initiation has been observed in the double *rga–gai* null mutant, which indicates that both GAI and RGA act synergistically as major repressors for these processes (Dill and Sun, 2001; King et al., 2001). RGA, along with RGL1 and RGL2, control petal and sepal development, stamen filament length, and microsporogenesis in pollen and male fertility (Cheng et al., 2004; Tyler et al., 2004). Fruit set and development are regulated negatively by four DELLAs, GAI, RGA, RGL1, and RGL2, as the *quadruple-DELLA* mutant shows facultative parthenocarpy (fruit growth without seeds; Dorcey et al., 2009; Fuentes et al., 2012). Finally, the *quadruple* and the *quintuple* or *global DELLA* mutants (the last one lacks all five DELLAs) have been reported to give similar constitutive GA responses to a GA-treated wild-type plant, while treatments with PAC have no significant effect (Feng et al., 2008). Interestingly, the *quadruple* and *global DELLA* mutants show no differences in growth and development, which suggests that RGL3 performs a minor function in such processes. So, exactly what is the role of RGL3? It has been reported that RGL3 acts as a positive regulator in the plant defense response by modulating jasmonate (JA) and ethylene signaling (Wild and Achard, 2013; Wild et al., 2012). Finally, DELLAs do not always act as repressors of plant growth; it has been extensively reported that DELLAs also act positively in plant responses to adverse growth conditions, such as salt (Achard et al., 2006; Achard et al., 2008b), deetiolation (Cheminant et al., 2011; de Lucas et al., 2008; Feng et al., 2008), or cold (Achard et al., 2008a). Therefore, DELLAs favor the plasticity and adaptation of plants to their surrounding growth conditions by adjusting proper balance between growth and response to adverse conditions.

Although the role of DELLAs has been assigned to different GA-mediated plant processes for a number of years, we did not have a clear view of the molecular mechanism underlying the DELLA function. How do DELLAs exert their function at the molecular level? We have assumed that DELLAs act as putative transcriptional regulators because they are nuclear proteins that lack a canonical DNA binding domain. Luckily in the past 8 years, tremendous progress has been made thanks to our understanding of GA perception and to the identification of the proteins that act downstream of DELLAs. We now understand how DELLAs perform their function, and most importantly, how they integrate endogenous and external signals to coordinate growth, development, and stress responses. In this chapter we will describe several aspects of the DELLA protein structure, distribution among plants, evolution, and their role in the Green Revolution. We will then focus on the molecular mechanism of the DELLA function, and the identification of DELLA-interacting proteins and target genes.

20.1.3 DELLA Protein Structure

DELLA proteins have a unique structure divided in two major domains that have been extensively studied (Daviere and Achard, 2013; Hauvermale et al., 2012), the C-terminal GRAS functional domain, shared by all the GRAS proteins, and the specific N-terminal GA perception domain, which differentiates the DELLA protein from the rest of the GRAS family (Figure 20.2). The GRAS functional domain is characterized by several motifs: two leucine heptad repeats (LHR), a nuclear localization signal (NLS) responsible for nuclear localization (Dill et al., 2001; Silverstone et al., 2001), two motifs, PFYRE and SAW, which mediate the secondary interactions with

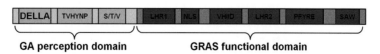

FIGURE 20.2 **DELLA protein structure.** The GA perception domain at the N-terminal of the protein contains the DELLA, and TVHYNP domains involved in GID1 binding, and the S/T/V motif involved in DELLA phosphorylation. The GRAS functional domain at the C-terminal of the protein containing a motif implicated in target binding (LHR1), a nuclear localization signal (NLS), two motifs that interact with the F-box proteins (VHIID and LHR2), and two domains involved in GID1 binding (PFYRE and SAW).

the GA receptor GIBBERELLIN INSENSITIVE DWARF1 (GID1), and a motif called VHIID that interacts, along with one of the LHR motifs, with F-box proteins (Hirano et al., 2010). The N-terminal GA perception domain has three specific motifs: the DELLA and TVHYNP domains, responsible for binding with GID1 (Asano et al., 2009; Dill and Sun, 2001; Itoh et al., 2002; Peng et al., 1997; Silverstone et al., 2007), and the S/T/V motif involved in DELLA phosphorylation.

20.1.4 Distribution and Evolution of DELLA Proteins

The GRAS superfamily suddenly appeared in the land plant lineage, which suggests a horizontal transfer from bacteria close to the origin of land plants (Zhang et al., 2012). Very early after the emergence of land plants, a major functional diversification of the GRAS family occurred and resulted in the appearance of the main GRAS subfamilies, including DELLAs (Zhang et al., 2012). In evolutionary terms, the most antique DELLA-like proteins are those in mosses, *Sphagnum palustre* and *Physcomitrella patens*, although their role as DELLA cannot be demonstrated (Yasumura et al., 2007). Neither bioactive GAs nor GID1 receptor orthologs have been found in mosses. The first functional DELLAs appeared in the lycophytes *Selaginella moellendorffii* and *Selaginella kraussiana* (Yasumura et al., 2007), where the whole and functional GA–GID1–DELLA module has been found (Hirano et al., 2007). Nonetheless, the GA-signaling pathway, as it is known in higher plants, is not completely conserved in lycophytes since GA_3 does not promote *Selaginella* growth (Yasumura et al., 2007) and GA receptor SmGID1 interacts with GA_{51}, which is an inactive GA in higher plants (Hirano et al., 2007). *Bona fide* DELLAs have been identified in gymnosperms and angiosperms (Vandenbussche et al., 2007; Table 20.1). The complexity of the DELLA gene family in seeded plants is very diverse, and the number of DELLAs is not stable in the major families of plants. For example, some monocots have unique DELLA-like *SLR1* (SLENDER RICE1) in rice (Ikeda et al., 2001) or *SLN1* (SLENDER1) in barley (Chandler and Robertson, 1999; Chandler et al., 2002), while others have two DELLAs, like *d8* and *d9*, in maize (Lawit et al., 2010). Among dicots, tomato, in the Solanaceae family, only encodes one DELLA gene, *PROCERA* (Bassel et al., 2008; Jasinski et al., 2008; Martí et al., 2007). In the family Fabaceae (legumes), the pea has two DELLAs, *LA* and *CRY* (Weston et al., 2008), but *Medicago* has three. Finally, citrus, in the Rutaceae family, has at least three (Tadeo and Perez-Amador, unpublished). So, how do GAs regulate growth and development in different plant species with the complexity and variety of DELLA gene families from a single gene to multiple genes? *Arabidopsis*, with five proteins, is a puzzling case. By promoter-swapping experiments, it

TABLE 20.1 DELLA Proteins Identified in Different Angiosperm Species

Plant species	DELLA proteins	References
Arabidopsis	GAI	Peng et al. (1997)
	RGA	Silverstone et al. (1997)
	RGL1	Wen and Chang (2002)
	RGL2	Lee et al. (2002)
	RGL3	Tyler et al. (2004)
Rice	SLR1	Ikeda et al. (2001)
Barley	SLN1	Chandler et al. (2002)
Maize	d8	Lawit et al. (2010); Peng et al. (1999)
	d9	
Grape	VvGAI	Boss and Thomas (2002)
Tomato	PROCERA	Bassel et al. (2008); Martí et al. (2007)
Pea	LA	Weston et al. (2008)
	CRY	
Medicago	MtDELLA1	Floss et al. (2013)
	MtDELLA2	
	MtDELLA3	

has been proposed that the different DELLA proteins in *Arabidopsis* have evolved toward subfunctionalization (Gallego-Bartolome et al., 2010), probably due to changes in their regulatory sequences. This means that they perform a similar molecular function, but are specialized in the timing and spatial pattern of expression.

20.1.5 DELLAs and the Green Revolution

The "Green Revolution" is a phenomenon that occurred between 1940 and the late 1960s due to the combination of improved agronomical techniques that resulted in huge increases in cereal yields worldwide (Evenson and Gollin, 2003). Apart from enhanced agricultural practices based on better irrigation and increased usage of fertilizers and pesticides, one key factor was the development and implementation of high-yielding semidwarf crop varieties, especially in rice and wheat. Yield increase in dwarf varieties is due to an enhanced partitioning of dry matter to growing spikes thanks to less competition from shorter stems, which results in a higher harvest index and better grain yields. Most of these dwarf varieties are caused by mutations in the GA biosynthetic or signaling pathways (Hedden, 2003). For example, IR8, a high-yielding semidwarf rice variety possessing the *semidwarf1* mutant gene, has defects in GA biosynthesis. Another semidwarf variety of wheat, Norin 10, was also crucial for increased yield during the Green Revolution. It was later demonstrated that it contained two homologous dwarfing genes, *Reduced Height1* and *2* (*RHT1* and *RHT2*), which encoded two DELLA genes in wheat (Peng et al., 1999). The mutations in *RHT1* and *RHT2* were caused by the complete or partial deletion of missense mutations in the N-terminal DELLA or adjacent TVHYNP domains, similarly to *gai-1* in *Arabidopsis*, which results in a protein resistant to GA-induced degradation that confers their semidominant phenotype. Other DELLA genes that contributed to the Green Revolution were maize *dwarf8* (Harberd and Freeling, 1989; Winkler and Freeling, 1994), barley *SLN1* (Chandler and Robertson, 1999; Chandler et al., 2002), and rice *SLR1* (Asano et al., 2009; Ikeda et al., 2001; Ogawa et al., 2000).

20.2 GA SIGNALING THROUGH DELLAs

20.2.1 GA Receptor GID1

GAs and DELLA proteins were discovered and characterized in the last century. However, the identification of the GA receptor remained elusive for many years. The discovery of the first GA receptor, GID1, from rice did not occur until 2005 (Ueguchi-Tanaka et al., 2005). Shortly after, the three GID1 genes in *Arabidopsis* (*GID1A*, *GID1B*, and *GID1C*) were described and characterized (Griffiths et al., 2006; Nakajima et al., 2006). Until this discovery, a complete understanding of the GA-signaling cascade and how bioactive GAs promote the degradation of DELLAs remained unclear (Sun, 2011).

GID1s are soluble and predominantly nuclear localized proteins similar to the hormone-sensitive lipase family, with no hydrolase activity, which reversibly bind bioactive GAs with high affinity (Ueguchi-Tanaka et al., 2005; Willige et al., 2007). Bioactive GAs bind to GID1 in a deep binding pocket that is covered by its N-terminal helical switch region, acting as a lid of the binding site. Upon binding, the N-terminal region undergoes a conformational change that blocks the GA molecule in its position (Murase et al., 2008; Shimada et al., 2008). Consequently, the GA–GID1 complex is able to bind to the DELLA and TVHYNP motifs of the DELLA protein, which results in a new conformational modification that further stabilizes the GA–GID1–DELLA complex (Murase et al., 2008; Ueguchi-Tanaka et al., 2007). Upon complex formation, the GRAS C-terminal domain of the DELLA also interacts with the GID1 (Hirano et al., 2010). The GID1–DELLA complex binds GA with more affinity than the sole GID1, so the DELLA indirectly stabilizes the GA–GID1 interaction (Hauvermale et al., 2012; Nakajima et al., 2006; Ueguchi-Tanaka et al., 2007). The formation of the complex is also affected by the differential affinity of GID1 toward DELLAs. *In vitro* assays have shown that the five DELLA proteins of *Arabidopsis* show differential affinities toward the three GID1s, which result in complexes that are more stable than others, such as GA–GID1B–RGA, GA–GID1A–RGL2, and GA–GID1B–GAI (Suzuki et al., 2009). Interestingly, recent evidence has indicated that the proteins that form the stable complexes in pistils of mature flowers are coexpressed, and that those with low affinity, like GID1B-RGL1 or GID1C-RGL2, are not coexpressed in the same tissues (Gallego-Giraldo et al., 2014), which suggests possible coevolution of the expression patterns and binding affinities of GID1s and DELLAs.

20.2.2 The GA–GID1–DELLA Module in GA Signaling

As already mentioned, DELLAs act as key repressors of GA signaling. So, how can the perception of GA by GID1 overcome the repression of plant growth and development imposed by the activity of the DELLA proteins? The formation of the GA–GID1–DELLA complex leads to structural changes in the GRAS domain of the DELLA (Hirano et al., 2010). This change stabilizes the interaction of the DELLA with GID1, but also allows recognition and binding with F-box proteins, GID2 in rice (Gomi et al., 2004; Sasaki et al., 2003) and SLEEPY1 (SLY1) or SNEEZY1 in *Arabidopsis* (Ariizumi et al., 2011; Dill et al., 2004; Dohmann et al., 2010; Fu et al., 2004; McGinnis et al., 2003). These F-box proteins are responsible for the recruitment of DELLAs to the SCF (SKP1, CULLIN, F-box) E3 ubiquitin–ligase complex to be polyubiquitinated and subsequently degraded by the 26S proteasome (Figure 20.3; Lechner et al., 2006). Therefore, GAs control

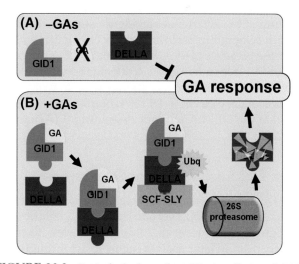

FIGURE 20.3 **Proteolysis-dependent GA-signaling model.** This model illustrates the GA-dependent GID1–DELLA complex formation that results in DELLA recognition, polyubiquitination by SCF^{SLY1} E3, and degradation by the 26S proteasome. (A) In the absence of GA, the GID1–DELLA interaction does not occur and DELLA represses the GA response. (B) In the presence of GA, the GA–GID1–DELLA complex is formed, which triggers DELLA proteolysis to enable GA responses.

the growth and development by the rapid degradation of DELLA repressor proteins through formation of the GA–GID1–DELLA complex, the polyubiquitination of DELLAs, and their degradation by the 26S proteasome. DELLA degradation upon GA treatment is very fast; GA treatment of seedlings of *ga1-3*, the GA biosynthesis mutant that accumulates high levels of DELLAs, results in almost total DELLA disappearance within 10 min (Ariizumi and Steber, 2007; Tyler et al., 2004; Zhang et al., 2011).

Although the proteolytic-dependent degradation of DELLAs has been extensively studied and characterized, three other alternative pathways for GA signaling may also occur, and each one is described below.

20.2.3 DELLA Nonproteolytic Signaling

Evidently, if the whole GA signaling passes through the classical DELLA degradation pathway, the DELLA protein levels should correlate with GA phenotypes. However, there are some cases in which this correlation does not properly fit. Extreme GA mutants, like *ga1-3* or the triple *gid1abc*, show more severe GA phenotypes than the *sly1* F-box mutant, although this mutant accumulates more DELLA proteins (McGinnis et al., 2003; Willige et al., 2007). Different seed lots or mutant alleles of *sly1* can germinate almost normally, while others cannot despite them all accumulating high levels of the RGL2 protein, as they impair the degradation of RGL2, the DELLA responsible for seed germination inhibition (Ariizumi and Steber, 2007). In addition, overexpression of GID1 in the *sly1* mutant can also rescue most GA phenotypes without DELLA degradation (Ariizumi et al., 2008). Finally, the double *ga1-3–sly1* mutant responds to GA treatment (Ariizumi et al., 2008). Such evidence suggests that an alternative signaling pathway may occur, which would require the presence of the GA hormone and GA receptor, but not SLY1-dependent DELLA degradation. Therefore, a DELLA inactivation system should exist, in which the formation of the GA–GID1–DELLA complex can partially inactivate the DELLA without degrading it (Ariizumi et al., 2008).

20.2.4 DELLA-independent GA Response

The DELLA-independent GA-signaling pathway is implicated in fruit development and significantly contributes to fruit growth (Fuentes et al., 2012). Nevertheless, the major mechanism in fruit growth seems to act through the canonical model of GA–GID1–DELLA complex formation and GA-dependent DELLA proteolytic degradation, as previously described. It has been described that the pistils of the *global-DELLA* mutant of *Arabidopsis*, which already show facultative parthenocarpy, can respond to exogenous GA treatment. This response is possible only in the presence of GID1 receptors and the 26S proteasome machinery. It also requires the activity of SPATULA (SPT), a basic helix–loop–helix (bHLH) transcription factor that seems to act as a repressor of fruit growth. Therefore, there is a GA response that still occurs when DELLA activity is absolutely absent (Fuentes et al., 2012).

20.2.5 GA-Independent DELLA Proteolytic Signaling

During the characterization of the GID1 receptor from *Arabidopsis*, GID–DELLA interactions that occur independently of the presence of GAs have been described. In the absence of GAs, GID1b can interact with not only RGL1 (Nakajima et al., 2006), but also with GAI and RGA (Griffiths et al., 2006). Furthermore, a more recent study with a rice *gid1* mutant suppressor has found that a mutation which transforms a proline residue into alanine or serine of rice GID1 allows the modified receptor to interact with SLR1, the DELLA in rice, in a GA-independent manner (Yamamoto et al., 2010). An equivalent proline residue is conserved in *Arabidopsis* GID1a and GID1c, but not GID1b, where proline is substituted for a histidine. It has been reported that the substitution of this histidine back to a proline in the GID1b protein suffices to abolish GA independent GID–DELLA interaction, which demonstrates that this residue, and the loop region where it is located, are key for the interaction (Yamamoto et al., 2010). Other species, like soybean and *Brassica*, also have GID1 receptors that bind DELLAs in a GA-independent way (Yamamoto et al., 2010). The presence of three GID1 genes in *Arabidopsis* allows the existence of different mechanisms for GA perception, and provides great plasticity in the regulation of growth and development mediated by GID1s and DELLAs.

20.2.6 Posttranslational Modifications of DELLA Proteins

One of the ways to modulate the activity of transcription factors is through different posttranslational modifications, and DELLAs are no exception to the rule. DELLA proteins can undergo phosphorylation and O-linked *N*-acetylglucosamine (O-GlcNAc) modification, but the role of these processes remains unclear.

It was initially hypothesized that upon phosphorylation, DELLAs SLR1 from rice and GAI from *Arabidopsis* increase the binding to the F-box proteins (Fu et al., 2004; Gomi et al., 2004). Later, no data were confirmed in rice (Itoh et al., 2005). Unlike the first hypothesis, it has also been shown that phosphorylated DELLAs, SLN1 from barley and RGL2 and RGA from *Arabidopsis*, are more resistant to degradation (Fu et al., 2002; Hussain et al., 2005; Wang et al., 2009a). Finally, rice EARLY FLOWERING1, a serine/threonine protein kinase, can phosphorylate

SLR1 *in vitro* and *in vivo* in two serine residues, one in the GRAS functional domain and the other in the GA perception domain. These phosphorylations are important for maintaining DELLA activity and stability (Dai and Xue, 2010).

SPINDLY (SPY) is an O-GlcNAc transferase that seems to regulate GA signaling negatively in *Arabidopsis*, rice, and barley by the glycosylation of DELLAs (Filardo et al., 2009; Robertson et al., 1998; Shimada et al., 2006; Swain et al., 2001). The effect of such a modification does not seem to be related to DELLA stability (Silverstone et al., 2007), but regulates the activity of the DELLA through competition between the phosphorylation and glycosylation of the same residues (Shimada et al., 2006; Silverstone et al., 2007; Wells et al., 2004). A more recent study has found that SPY can interact with SWI3C, a component of a chromatin remodeler complex, that can also bind to DELLAs (Sarnowska et al., 2013), and suggests a potential role for chromatin structural modifications in the functioning of both DELLAs and SPY proteins.

20.3 THE MOLECULAR MECHANISM OF DELLA ACTION: DELLA–PROTEIN INTERACTIONS AND TARGET GENES

The mechanism by which DELLAs block GA responses has remained unclear until quite recently. DELLAs have been reported to localize in the nucleus (Dill et al., 2001; Silverstone et al., 1998), which was expected for proteins that regulate gene expression. However, they lack a clear DNA-binding domain and there is no evidence for direct binding to DNA. So, how can DELLAs regulate the transcription of target genes? The answer came 7 years ago, when it was discovered that DELLAs control gene transcription through protein interaction with other transcription factors (Daviere et al., 2008) which, in most cases, can interact directly with DNA (Table 20.2). It has been recently described that GAI from *Arabidopsis* can interact with at least 57 different transcription factors (Marin-de la Rosa et al., 2014) to reflect multiple connections of DELLAs with different signaling pathways. But how

TABLE 20.2 Proteins that Interact with DELLAs in *Arabidopsis*

DELLAs	Interacting protein	Functions	References
All DELLAs	PIFs	Hypocotyl elongation, apical hook development, chlorophyll biosynthesis	Cheminant et al. (2011); de Lucas et al. (2008); Feng et al. (2008); Gallego-Bartolome et al. (2010)
GAI, RGA, RGL1, RGL3	BES/BZR1	GA–BR crosstalk, light signaling	Bai et al. (2012); Gallego-Bartolomé et al. (2012); Li et al. (2012)
GAI, RGA	EIN3, EIL1, EIL2	GA–ET crosstalk, apical hook development	An et al. (2012)
GAI, RGA	RAP2.3	Apical hook opening	Marin-de la Rosa et al. (2014)
RGA	SPL9	Floral induction	Yu et al. (2012)
GAI, RGA, RGL2	ALC	Fruit development	Arnaud et al. (2010)
RGA, RGL2	SPT	Seed germination, cotyledon expansion	Gallego-Bartolome et al. (2010); Josse et al. (2011)
RGA, RGL2	GL1, EGL3, GL3	Trichome initiation	Qi et al. (2014)
All DELLAs	JAZs	GA–JA crosstalk, biotic stress, root, and hypocotyl elongation	Hou et al. (2010) Wild et al. (2012) Yang et al. (2012)
All DELLAs	MYC	GA–JA crosstalk, biotic stress, root, and hypocotyl elongation	Hong et al. (2012)
GAI, RGA, RGL1	SCL3	Seed germination and root and hypocotyl elongation	Zhang et al. (2011)
All DELLAs	IDDs	Seed maturation and germination, root development	Feurtado et al. (2011); Yoshida et al. (2014)
All DELLAs	GAF1	Floral induction	Fukazawa et al. (2014)
GAI, RGA	ABI3, ABI5	Seed germination	Lim et al. (2013)
All DELLAs	BOIs	Seed germination, floral induction	Park et al., (2013)
RGA, RGL1, RGL2	SWI3C	GA biosynthesis, root and hypocotyl elongation	Sarnowska et al. (2013)
All DELLAs	PKL	Light signaling	Zhang et al. (2014)
GAI	PFD3, PFD5	Tubulin folding	Locascio et al. (2013b)

can DELLAs interact with so many partners? DELLAs bind other proteins through different regions of the GRAS domain, like the LHR1 domain, which is important for the binding of PHYTOCHROME-INTERACTING FACTOR 4 (PIF4), BRASSINAZOLE-RESISTANT1 (BZR1), or JAZ1 (jasmonate ZIM domain; Arnaud et al., 2010; Gallego-Bartolomé et al., 2012; Hauvermale et al., 2012). Furthermore, DELLA partners interact with DELLAs according to different motifs; that is, DELLAs bind directly to the DNA-binding domain of several transcription factors such as PIFs, BZR1, EIN3 (ETHYLENE INSENSITIVE3), or RELATED TO APETALA2.3 (RAP2.3), to block their activity. Multiple DELLA–protein interactions can also be facilitated by the N-terminal region of the DELLA protein, a disordered region with high structural flexibility that can facilitate specific, but low-affinity, interactions (Hauvermale et al., 2012; Sun et al., 2010).

Several mechanisms that allow DELLAs to control gene expression can be differentiated depending on the partner and how the interaction with DELLAs occurs. DELLA–protein interactions can be divided into two main groups: DELLAs can act as a kidnapper of other transcription factors that interfere with its activity; DELLAs can cooperate with other transcription factors to work as a transactivator to modulate binding to the promoters of target genes (Daviere and Achard, 2013; Yang et al., 2012). A recent study has shown that DELLAs can also interact with PREFOLDIN3 and 5, two proteins with no apparent transcriptional activity (Locascio et al., 2013b). Therefore, DELLAs may control several aspects of plant growth and development by binding to other proteins that are not transcriptional factors/regulators, and by modulating their biochemical functions in the cell. We next describe the different mechanisms involved in the DELLA interaction with other proteins in detail, and review the different proteins that bind to the DELLAs described so far.

20.3.1 Interfering Mechanism

In this mechanism, DELLAs interact with a protein that has transcriptional activity by binding to it to block its function. The DELLA never joins genomic DNA but, depending on whether the interacting protein directly binds with DNA or not, two subtypes can be differentiated. In the first one, which is apparently the commonest, the DELLA binds a transcription factor, which directly regulates gene expression by a physical interaction with DNA. It has been reported that DELLAs bind directly to the DNA-binding motif of the transcriptional factor to block transcriptional activity. Examples of the first mechanism are PIFs, BZR1/BES (BRI1-EMS-SUPPRESSOR), EIN3, RAP2.3, SPL9 (SQUAMOSA PROMOTER BINDING-LIKE 9), ALC (ALCATRAZ), SPT, MYC, GL1 (GLABRA1), GL3 (GLABRA3), and EGL3 (ENHANCER of GLABRA3). In the second one, DELLAs interact with a transcription regulator that does not bind directly to DNA, but can interact with another transcription factor that does bind to it. The JAZ protein can exemplify this mechanism.

Discovery of the interaction between DELLAs and PIF proteins was the first evidence for the downstream mechanism of DELLAs in the control of GA response, and it has been a key factor in understanding how GAs integrate both hormonal and environmental signals to control growth and development (Cheminant et al., 2011; de Lucas et al., 2008; Feng et al., 2008; Gallego-Bartolome et al., 2010). PIFs (including PIF1, PIF3, PIF4, and PIF5) are a subset class of the bHLH family of transcriptional factors that repress phytochrome-mediated light responses by interacting with light receptor phytochromes (Leivar and Quail, 2011; Chapter 21). PIFs function as negative regulators of photomorphogenic seedling development in darkness, and light reverses this repression by stimulating the proteasome-mediated degradation of PIFs, thus releasing the repressive effect of PIFs on photosynthetic gene expression (Cheminant et al., 2011; Leivar and Quail, 2011). Seedling deetiolation is also subject to hormonal regulation (Alabadí et al., 2004). GAs have the opposite effect to light on seedling development; light induces photomorphogenesis, whereas GAs promote etiolated growth in the dark. The mechanism underlying this antagonistic effect relies on the interaction of DELLAs with at least PIF3 and PIF4 factors (de Lucas et al., 2008; Feng et al., 2008). DELLAs block PIF transcriptional activity by binding to the DNA recognition domain of these factors. GAs alleviate this repression by promoting DELLA destabilization, and therefore inducing an accumulation of free PIFs in the nucleus. The interaction of DELLAs with PIF4 and PIF3 provides evidence that these factors play a crucial role in the integration of both GA and light signals to modulate photomorphogenesis.

BRASSINAZOLE-RESISTANT1 (BZR1) and BRI1-EMS-SUPPRESSOR1 (BES1) are transcriptional factors that mediate brassinosteroid (BR)-dependent gene expression (Kim and Wang, 2010; Wang et al., 2012). BRs promote growth by inducing the dephosphorylation of BZR1 by the PROTEIN PHOSPHATASE2A. BRs and GAs promote many similar developmental responses in plants, and it has been shown that this is mediated by direct interaction between the BR-activated BZR1 and DELLAs (Bai et al., 2012; Gallego-Bartolomé et al., 2012; Li et al., 2012). DELLAs directly bind to BZR1/BES and block their ability to bind to the promoters of BR-regulated genes. On the other hand BZR1 can bind to PIF4 to promote PIF activity (Bernardo-García et al., 2014; Oh et al., 2007). All this evidence indicates that BRs and GAs play a similar role in controlling growth in response to light through the promotion of PIF activity by both the GA-dependent degradation of DELLAs and the BR-mediated activation of BZR1 and BES1. The DELLA–BZR1–PIF4 interaction defines a core transcription module that mediates coordinated growth regulation by GA, BR, and light signals (Bai et al., 2012).

ETHYLENE INSENSITIVE3 (EIN3) and EIN3-LIKE1 (EIL1) are positive regulators of the ethylene-signaling pathway that act as master transcription factors to generate the primary output of ethylene responses, and both are necessary and sufficient for the regulation of expression of ethylene-responsive genes (Merchante et al., 2013; Solano et al., 1998). It is known that DELLAs can act in the ethylene response, especially in apical hook development. For example, a treatment with PAC, which stabilizes DELLAs, inhibits hook formation (Achard et al., 2003), while a *global-DELLA* mutant forms an exaggerated hook in an ethylene-dependent way (An et al., 2012). DELLAs participate in ethylene responses by directly binding to the DNA-binding domains of EIN3/EIL1 (An et al., 2012), which represses its DNA-binding function and inhibits hook formation (An et al., 2012). RELATED TO APETALA2.3 (RAP2.3) is a transcriptional factor of the ETHYLENE RESPONSE FACTOR-AP2/ERBP superfamily, whose expression is directly regulated by the transcriptional activity of EIN3. Interestingly, RAP2.3 physically interacts with the DELLA GAI of *Arabidopsis* and blocks its transcriptional activity on target promoters (Marin-de la Rosa et al., 2014).

SPL9 is a transcription factor that belongs to the SQUAMOSA PROMOTER BINDING–LIKE (SPL) family. SPLs promote juvenile-to-adult phase transition and flowering through the activation of miR172, MADS box genes and *LEAFY* (Wang et al., 2009b; Chapter 18). The direct interaction between SPL9 and DELLAs inhibits the transcriptional activation of MADS box genes and miR172, and causes a delay in floral transition (Yu et al., 2012).

ALCATRAZ (ALC) is a bHLH protein, like PIF3 and PIF4, which controls the differentiation of valve margins in *Arabidopsis* fruits (Arnaud et al., 2010). The valve margin is a specialized structure consisting of a lignification layer and a separation layer (SL), which facilitates fruit opening and efficient seed release. ALC is required for the regulation of the genes responsible for SL specification. Prior to valve margin differentiation, ALC is bound to DELLA proteins to repress the expression of its target genes. Local GA synthesis leads to the degradation of DELLAs, and enables ALC to direct the differentiation of the SL. INDEHISCENT is a positive upstream regulator of ALC that controls GA biosynthesis, and contributes to proper valve margin development. Therefore, the DELLA pathway plays a major role in seed dispersal and plant domestication through the interaction and regulation of ALC.

SPATULA (SPT), a PIF homolog, regulates seed dormancy and restrains cotyledon expansion and fruit growth (Josse et al., 2011; Fuentes et al., 2012). SPT and DELLA proteins directly interact (Gallego-Bartolome et al., 2010), while SPT also represses GA biosynthesis and is subjected to negative transcriptional regulation by DELLAs during cotyledon expansion (Josse et al., 2011). The complex cross-regulation of SPT by DELLAs provides a means of preventing detrimental effects on growth, which would result from either an excess or deficiency in these potent growth regulators.

GLABRA1 (GL1), GLABRA3 (GL3), and ENHANCER of GLABRA3 (EGL3) participate in the WD-repeat/bHLH/MYB complex, which mediates trichome development (Grebe, 2012). GL1 is an R2R3 MYB transcription factor, while GL3 and EGL3 are bHLH factors. DELLAs physically interact with these proteins to attenuate trichome initiation, while GAs cause the degradation of DELLAs to promote trichome formation (Qi et al., 2014).

JA ZIM domain (JAZ) proteins are a family of key repressors of JA signaling. This pathway regulates multiple plant growth responses, including defense against pathogens and insects, adaptation to abiotic stresses, and root and pollen development (Wasternack, 2007). JAZs inhibit the transcription of JA-responsive genes through a physical interaction with the bHLHs MYC2, MYC3, and MYC4. The crosstalk between GA and JA signaling in pathogen interaction and plant development can be explained by a direct interaction between JAZ and DELLAs (Hou et al., 2010; Wild et al., 2012; Yang et al., 2012). DELLA proteins bind to JAZ1 and release MYC2 to promote JA signaling, while GA triggers the degradation of DELLAs, which allows JAZ1 to bind to MYC2 and to repress JA signaling. It has also been found that JAZ9 interrupts the RGA–PIF3 interaction, which suggests that, in the absence of JA signaling, some DELLAs can be kidnapped by JAZ proteins, which release PIFs to activate growth programs (Yang et al., 2012). DELLAs also bind to MYC2 under specific and physiological conditions, which not only adds further complexity to JA–GA signaling crosstalk, but also suggests that DELLAs interfere with the MYC2-mediated JA signaling output by competitively binding to MYC2 (Hong et al., 2012).

20.3.2 The Transactivation Mechanism

DELLAs can bind to DNA by the interaction with other proteins with DNA-binding motifs (Daviere and Achard, 2013). It was first shown that DELLAs can bind with the promoters of several GA response genes (Zentella et al., 2007). The DELLA and TVHYNP domains of rice SLR1 also display transactivation activity in yeast one-hybrid assays, while the fusion of SLR1 to the VP16 activation domain reduces stem elongation (Hirano et al., 2012). This activity of DELLAs seems to be regulated by the nonproteolytic DELLA-signaling pathway (Daviere and Achard, 2013; Hirano et al., 2012). Some examples of this interaction are SCL3, IDDs (INDETERMINATE DOMAINs), ABIs (ABA INSENSITIVEs), and

BOIs (BOTRYTIS SUSCEPTIBLE1 INTERACTORs). In this section, we have also included the interaction between DELLAs and chromatin-remodeling machinery components, like SWI3C and PICKLE, which greatly influences gene expression (Ho and Crabtree, 2010). Although we do not yet know how DELLAs interact with these proteins, it is plausible that DELLAs come in close proximity to genomic DNA through these interactions and, thus, regulate gene expression.

SCARECROW-LIKE3 (SCL3) is a GRAS protein with no DNA-binding domain, which is upregulated by DELLAs (Zentella et al., 2007). SCL3 is a positive regulator of GA signaling (Heo et al., 2011; Zhang et al., 2011), and SCL3 binds to DELLAs by attenuating the gene expression of some DELLA target genes, including those of GA biosynthesis like *GA20ox2* or *GA3ox*. SCL3–DELLA binding also disrupts the SCL3-negative regulation of its own promoter (Zhang et al., 2011). Therefore, SCL3 and DELLAs antagonize each other in controlling both downstream GA responses and upstream GA biosynthetic genes by regulating GA homeostasis and controlling GA-mediated growth and development (Zhang et al., 2011).

INDETERMINATE DOMAIN (IDD) is a family of 16 members in *Arabidopsis* characterized by a distinct arrangement of zinc (Zn) finger motifs (Colasanti et al., 2006). Several IDDs have been implicated in many developmental processes, such as seed germination, floral transition, gravitropism, and root patterning (Yoshida et al., 2014). Several members of this family, like IDD1, 3, 4, 5, 9, and 10, interact with DELLAs (Feurtado et al., 2011; Yoshida et al., 2014). GAI-ASSOCIATED FACTOR1 (GAF1), a gene with high homology with IDD1, also binds DELLAs (Fukazawa et al., 2014). The analysis of the IDD1–DELLA interaction has suggested that IDD1 acts as an antagonist of the DELLA negative feedback loop, which controls the expression of GA metabolism and perception genes like *GA20ox2* or *GID1B*, similarly to SCL3 (Feurtado et al., 2011). An alternative mechanism of action for IDD–DELLA binding has been provided in a more recent study which has shown that IDD3, 4, 5, 9, and 10 can interact with not only DELLAs, but also SCL3. IDD would act as transcriptional scaffolds, used by DELLAs and SCL3 to target DNA and to control the expression of GA-regulated genes (Yoshida et al., 2014). IDD–DELLA activates the expression of SCL3 and GA downstream genes, and the subsequent increase in SCL3 results in an increase in the SCL3/IDD complex, which competes with IDD–DELLA (Yoshida et al., 2014). The difference between the IDD1–DELLA interaction and the remaining IDD–DELLA interactions could be due to the presence of a putative repressor domain in IDD1, which is not present in the other IDDs (Mitsuda et al., 2011; Yoshida et al., 2014)

ABA-INSENSITIVE3 (ABI3) and ABI5 are B3 domain and basic leucine zipper transcription factors, respectively, with DNA-binding capacity, which mediate abscisic acid (ABA) response (Nakamura et al., 2001). High temperature inhibits seed germination by activating ABA biosynthesis and repressing GA biosynthesis. Under these conditions ABIs and DELLAs accumulate and interact. The ABI–DELLA complex binds to the *SOMNUS* promoter, a seed inhibitor gene, and induces its expression (Lim et al., 2013).

BOTRYTIS SUSCEPTIBLE1 INTERACTOR (BOI), BOI-RELATED GENE1 (BRG1), BRG2, and BRG3 (collectively referred to as BOIs) belong to the RING domain protein family that regulates various plant developmental processes (Hotton and Callis, 2008). BOIs interact with DELLAs and the complexes that form are targeted to the promoters of a subset of GA-responsive genes, such as *EXPANSIN8* and *PACLOBUTRAZOL RESISTANCE1* and *5*, to regulate their expression. In this way, the BOI–DELLA interaction represses several GA responses like seed germination, juvenile-to-adult transition, and flowering (Park et al., 2013).

Other types of interactions that can be included in this group are those between DELLAs and SWI3C, the central component of the SWI/SNF chromatin-remodeling complex (Sarnowska et al., 2013), or PICKLE (PKL; Zhang et al., 2014). Although these proteins are not transcriptional regulators, chromatin remodeling is a crucial mechanism to control gene expression in eukaryotes by allowing regulatory transcription machinery proteins access to condensed genomic DNA (Ho and Crabtree, 2010). The *swi3c* mutant has GA development defects that are caused by the lowering of the levels of bioactive GAs due to the downregulation of genes *GA3ox2* and *GA3ox3*, and of other genes involved in the GA negative feedback loop, like *GID1A*. These genes are transcriptionally regulated by DELLAs, which suggests that SWI3C controls the transcriptional activity of DELLAs. This idea has also been supported by the interaction between SWI3C and SPY, which are involved in the glycosylation of DELLAs, which can modulate DELLA activity (Sarnowska et al., 2013). PKL, a negative regulator of the light-signaling pathway (Jing and Lin, 2013), encodes an ATP-dependent CHROMODOMAIN HELICASE-DNA BINDING3-type chromatin-remodeling factor. This factor modifies the interactions between DNA and histones and allows the transcriptional complex to contact DNA (Ogas et al., 1999). PKL binds directly with PIF3 to promote hypocotyl growth in *Arabidopsis* as this interaction inhibits the recruitment of repressive histone mark H3K27me3 on the corresponding chromatin, which leads to the activation of cell elongation-related genes. Furthermore, DELLA proteins physically interact with PKL by blocking the PKL–PIF3 interaction and, therefore, the activation of these genes. Therefore, PKL, PIF3, and DELLAs act cooperatively to regulate skotomorphogenic growth in darkness (Zhang et al., 2014).

20.3.3 Nontranscriptional DELLA Interaction

A recent study has proved that DELLAs can interact with PREFOLDIN3 (PFD3) and PFD5 (Locascio et al., 2013b), which are proteins with no transcriptional activity. PFD3 and PFD5 form part of the cochaperone complex responsible for tubulin folding that is normally located in the cytoplasm (Hartl and Hayer-Hartl, 2002). Tubulin is the main component of microtubules, which play a key role in the cell. When the DELLA is present in the complex, it is retained in the nucleus and affects microtubule organization. DELLAs seem to control the growth direction of cells by interfering with the microtubule system. However, a role for PFDs in the regulation of transcription in the nucleus cannot be completely ruled out (Locascio et al., 2013b).

20.3.4 DELLAs Regulate the Expression of Target Genes

DELLA proteins integrate many development processes and stress responses, most probably by regulating basic cellular processes like cell division and expansion. For example, in the root meristem, there is a balance between cytokinins, which promote cell differentiation, and auxin, which promotes cell division (Blilou et al., 2005; Dello Ioio et al., 2008). The role of DELLAs in cytokinin–auxin crosstalk is to control auxin signaling by reducing the abundance and stability of PIN1 auxin transporters, and by balancing the equilibrium toward cell differentiation (Moubayidin et al., 2010; Willige et al., 2011). DELLAs can also control the cell cycle by repressing some core cell cycle genes in rice and *Arabidopsis* and they also promote the expression of cell cycle inhibitors (Achard et al., 2003; Fabian et al., 2000; Ogawa et al., 2003). DELLAs also repress cell expansion in several ways: by inhibiting the expression of the *EXPANSINS* and *XYLOGLUCAN ENDOTRANSGLUSOSYLASE/HYDROLASE* genes responsible for cell wall relaxation (Jan et al., 2004; Lee and Kende, 2002), or by stimulating the detoxification of reactive oxygen species that limit cell expansion and root growth (Achard et al., 2008b). But which genes mediate all these DELLA functions in cell growth and development? A highly successful way of finding DELLA target genes has been to use the transcriptomic methods recently reviewed in Locascio et al. (2013a) and Claeys et al. (2014). We can distinguish three transcriptomic approaches:

- Application of GAs or PAC to Col-0 (Goda et al., 2008) and L*er* (Bari, MacLean, Jones, unpublished) ecotypes of *Arabidopsis*, or application of GAs to *ga1* mutants (Goda et al., 2008; Willige et al., 2007; Zentella et al., 2007).
- DELLA mutants, like *quadruple DELLA gai–rga–rgl1–rgl2* in the *ga1-3* background (Cao et al., 2006), *global DELLA* (Arana et al., 2011), or triple *ga1-3–rga-t2–rgl2-1* compared with the double *ga1-3–rga-t2* (Stamm et al., 2012).
- Transgenic lines in which DELLAs are induced by several treatments, like the dominant *gai-1* induced by heat shock (Gallego-Bartolomé et al., 2011), or the dominant allele *rgaΔ17* induced by dexamethasone (Zentella et al., 2007).

Most of these studies have allowed the identification of many genes regulated by GAs or DELLAs, but do not distinguish between the primary target genes directly regulated by DELLAs and those secondary targets affected as a result of primary changes (Locascio et al., 2013a). Primary targets were only discovered with the last approaches, like those by Zentella et al. (2007) and Gallego-Bartolomé et al. (2011), and by looking for changes in gene expression that occur shortly after DELLA induction. The lists of target genes identified from each approach have very little in common, most probably due to differences in the tissue, and the developmental and environmental factors of each analysis, which strongly influence GA responses. The technical differences among the different methods and platforms used also greatly influence the outcome of the analysis. In spite of all this, a set of common genes has been identified (Claeys et al., 2014), like those involved in GA homeostasis, *GA3ox1*, *GA20ox2*, and *GA20ox1*, which are known to be controlled through DELLA feedback loops (Cao et al., 2006; Hedden and Phillips, 2000; Hou et al., 2008; Zentella et al., 2007), or the GA receptors *GID1A* and *GID1B*, which seem to be controlled by this feedback loop (Griffiths et al., 2006).

Another common DELLA target gene is *SCL3*. As mentioned earlier, SCL3 is an attenuator of GA signaling, which physically interacts with DELLAs and is GA repressed and DELLA induced (Zentella et al., 2007), which suggests that SCL3 is a negative regulator of GA signaling. In contrast with this hypothesis, an *scl3* null mutant has reduced GA response and elevated expression of GA biosynthetic genes. This finding indicates that SCL3 may also act as a positive regulator of GA signaling (Zhang et al., 2011). SCL3 is also able to repress its expression by binding to its own promoter (Zhang et al., 2011).

Another common target gene is *XERICO* (*XER*). XER is a RING-H2 E3 ubiquitin ligase protein that promotes ABA synthesis in response to salt and osmotic stress to increase drought tolerance (Ko et al., 2006). GAs and ABA play antagonistic roles in some development processes, like germination, seedling growth, or floral initiation (Gazzarrini and McCourt, 2003; Koornneef et al., 1991; Razem et al., 2006). DELLAs can induce *XER* expression and, hence, ABA accumulation (Zentella et al., 2007). Chromatin immunoprecipitation assays have shown that RGA can bind to the *XER* promoter, which indicates that *XER* is a primary target of DELLAs (Zentella et al., 2007). More recently, it has been shown that *XER* expression may be mediated by the DELLA nonproteolytic pathway (Ariizumi et al., 2013).

20.4 CONCLUSION AND FUTURE PERSPECTIVES

DELLAs are central components of GA signaling. Recent evidence indicates that DELLAs repress GA signaling by binding to a wide variety of transcriptional regulators, as well as to proteins with no transcriptional activity. DELLA-interacting proteins are also key regulatory elements of different signal cascades and they participate in the molecular complexes responsible for basic cellular processes. This suggests that DELLAs act as the central signaling hubs that integrate both endogenous and external signals to coordinate growth, development, and stress responses. One striking piece of evidence is that DELLAs not only interact with a single component of a signal cascade, but do so at different levels. For example, DELLAs participate in GA–JA crosstalk by a three-way interaction of JAZ, MYC2, and DELLA, or in GA–ethylene crosstalk through the interaction between EIN3/EIL1–DELLA and RAP2.3–DELLA, where *RAP2.3* is a target gene of EIN3. This suggests that DELLAs participate in complex molecular mechanisms that allow the fine adjustment of signaling by not only plant hormones, but also by other environmental cues and stress responses. DELLAs also participate in chromatin-remodeling complexes by means of the interaction with the central core component, SWI3C, or with the chromatin-remodeling factor PICKLE. The dynamic remodeling of chromatin by the chromatin-remodeling complex confers it an epigenetic regulatory role in several key biological processes, such as DNA replication and repair, cell cycle, apoptosis, and development. Therefore, DELLAs have the potential to actively regulate gene expression and to modulate multiple cellular responses in a broader sense. In the near future, we will witness further evidence for the function of DELLAs by describing new DELLA-interacting proteins, as well as target genes that mediate multiple plant responses. It should be pointed out that the wide variety of biological functions that DELLAs perform is apparently achieved by merely a singular molecular mechanism, protein–protein interaction, although other protein functions should not be excluded.

Nearly all existing evidence for the role of DELLAs stems from the work done in the reference plant *Arabidopsis thaliana*. We hypothesize that most roles described in *Arabidopsis* can be extended to other species since we assume an equivalent regulation mechanism will occur in different species. Nevertheless, we need to broaden our knowledge on the role of DELLAs in other species, especially those of agronomical interest.

Acknowledgments

The authors acknowledge Dr J. Carbonell and Dr Francisco Tadeo for their critical reading of the manuscript. They further acknowledge funding provided by grants BIO2008-01039 and BIO2011-26302 from the Spanish Ministry of Science and Innovation, and grants ACOMP/2010/079 and ACOMP/2011/287 from the Generalitat Valenciana.

References

Achard, P., Cheng, H., De Grauwe, L., Decat, J., Schoutteten, H., Moritz, T., Van Der Straeten, D., Peng, J., Harberd, N.P., 2006. Integration of plant responses to environmentally activated phytohormonal signals. Science 311, 91–94.

Achard, P., Gong, F., Cheminant, S., Alioua, M., Hedden, P., Genschik, P., 2008a. The cold-inducible CBF1 factor-dependent signaling pathway modulates the accumulation of the growth-repressing DELLA proteins via its effect on gibberellin metabolism. Plant Cell 20, 2117–2129.

Achard, P., Renou, J.P., Berthome, R., Harberd, N.P., Genschik, P., 2008b. Plant DELLAs restrain growth and promote survival of adversity by reducing the levels of reactive oxygen species. Curr. Biol. 18, 656–660.

Achard, P., Vriezen, W.H., Van Der Straeten, D., Harberd, N.P., 2003. Ethylene regulates *Arabidopsis* development via the modulation of DELLA protein growth repressor function. Plant Cell 15, 2816–2825.

Alabadí, D., Gil, J., Blázquez, M.A., García-Martínez, J.L., 2004. Gibberellins repress photomorphogenesis in darkness. Plant Physiol. 134, 1050–1057.

An, F., Zhang, X., Zhu, Z., Ji, Y., He, W., Jiang, Z., Li, M., Guo, H., 2012. Coordinated regulation of apical hook development by gibberellins and ethylene in etiolated *Arabidopsis* seedlings. Cell Res. 22, 915–927.

Arana, M.V., Marín-de la Rosa, N., Maloof, J.N., Blázquez, M.A., Alabadí, D., 2011. Circadian oscillation of gibberellin signaling in *Arabidopsis*. Proc. Natl. Acad. Sci. USA 108, 9292–9297.

Ariizumi, T., Steber, C.M., 2007. Seed germination of GA-insensitive *sleepy1* mutants does not require RGL2 protein disappearance in *Arabidopsis*. Plant Cell 19, 791–804.

Ariizumi, T., Murase, K., Sun, T.P., Steber, C.M., 2008. Proteolysis-independent downregulation of DELLA repression in *Arabidopsis* by the gibberellin receptor GIBBERELLIN INSENSITIVE DWARF1. Plant Cell 20, 2447–2459.

Ariizumi, T., Lawrence, P.K., Steber, C.M., 2011. The role of two f-box proteins, SLEEPY1 and SNEEZY, in *Arabidopsis* gibberellin signaling. Plant Physiol. 155, 765–775.

Ariizumi, T., Hauvermale, A.L., Nelson, S.K., Hanada, A., Yamaguchi, S., Steber, C.M., 2013. Lifting DELLA repression of *Arabidopsis* seed germination by nonproteolytic gibberellin signaling. Plant Physiol. 162, 2125–2139.

Arnaud, N., Girin, T., Sorefan, K., Fuentes, S., Wood, T.A., Lawrenson, T., Sablowski, R., Østergaard, L., 2010. Gibberellins control fruit patterning in *Arabidopsis thaliana*. Genes Dev. 24, 2127–2132.

Asano, K., Hirano, K., Ueguchi-Tanaka, M., Angeles-Shim, R.B., Komura, T., Satoh, H., Kitano, H., Matsuoka, M., Ashikari, M., 2009. Isolation and characterization of dominant dwarf mutants, *Slr1-d*, in rice. Mol. Genet. Genomics 281, 223–231.

Bai, M.Y., Shang, J.X., Oh, E., Fan, M., Bai, Y., Zentella, R., Sun, T.P., Wang, Z.Y., 2012. Brassinosteroid, gibberellin and phytochrome impinge on a common transcription module in *Arabidopsis*. Nat. Cell Biol. 14, 810–817.

Bassel, G.W., Mullen, R.T., Bewley, J.D., 2008. Procera is a putative DELLA mutant in tomato (*Solanum lycopersicum*): effects on the seed and vegetative plant. J. Exp. Bot. 59, 585–593.

Bernardo-García, S., de Lucas, M., Martínez, C., Espinosa-Ruiz, A., Davière, J.-M., Prat, S., 2014. BR-dependent phosphorylation modulates PIF4 transcriptional activity and shapes diurnal hypocotyl growth. Genes Dev. 28, 1681–1694.

Blilou, I., Xu, J., Wildwater, M., Willemsen, V., Paponov, I., Friml, J., Heidstra, R., Aida, M., Palme, K., Scheres, B., 2005. The PIN auxin efflux facilitator network controls growth and patterning in *Arabidopsis* roots. Nature 433, 39–44.

Boss, P.K., Thomas, M.R., 2002. Association of dwarfism and floral induction with a grape 'green revolution' mutation. Nature 416, 847–850.

Cao, D., Cheng, H., Wu, W., Soo, H.M., Peng, J., 2006. Gibberellin mobilizes distinct DELLA-dependent transcriptomes to regulate seed germination and floral development in *Arabidopsis*. Plant Physiol. 142, 509–525.

Chandler, P., Robertson, M., 1999. Gibberellin dose-response curves and the characterization of dwarf mutants of barley. Plant Physiol. 120, 623–632.

Chandler, P.M., Marion-Poll, A., Ellis, M., Gubler, F., 2002. Mutants at the *Slender1* locus of barley cv Himalaya. Molecular and physiological characterization. Plant Physiol. 129, 181–190.

Cheminant, S., Wild, M., Bouvier, F., Pelletier, S., Renou, J.P., Erhardt, M., Hayes, S., Terry, M.J., Genschik, P., Achard, P., 2011. DELLAs regulate chlorophyll and carotenoid biosynthesis to prevent photooxidative damage during seedling deetiolation in *Arabidopsis*. Plant Cell 23, 1849–1860.

Cheng, H., Qin, L., Lee, S., Fu, X., Richards, D.E., Cao, D., Luo, D., Harberd, N.P., Peng, J., 2004. Gibberellin regulates *Arabidopsis* floral development via suppression of DELLA protein function. Development 131, 1055–1064.

Claeys, H., De Bodt, S., Inze, D., 2014. Gibberellins and DELLAs: central nodes in growth regulatory networks. Trends Plant Sci. 19, 231–239.

Colasanti, J., Tremblay, R., Wong, A.Y.M., Coneva, V., Kozaki, A., Mable, B.K., 2006. The maize *INDETERMINATE1* flowering time regulator defines a highly conserved zinc finger protein family in higher plants. BMC Genomics 7, 158.

Colebrook, E.H., Thomas, S.G., Phillips, A.L., Hedden, P., 2014. The role of gibberellin signalling in plant responses to abiotic stress. J. Exp. Biol. 217, 67–75.

Curtis, P.J., Cross, B.E., 1954. Gibberellic acid – a new metabolite from the culture filtrates of *Gibberella fujikuroi*. Chem. Ind. Lond. 35, 1066.

Dai, C., Xue, H.W., 2010. Rice early flowering1, a CKI, phosphorylates DELLA protein SLR1 to negatively regulate gibberellin signalling. EMBO J. 29, 1916–1927.

Daviere, J.M., Achard, P., 2013. Gibberellin signaling in plants. Development 140, 1147–1151.

Daviere, J.M., de Lucas, M., Prat, S., 2008. Transcriptional factor interaction: a central step in DELLA function. Curr. Opin. Genet. Dev. 18, 295–303.

de Lucas, M., Daviere, J.M., Rodriguez-Falcon, M., Pontin, M., Iglesias-Pedraz, J.M., Lorrain, S., Fankhauser, C., Blazquez, M.A., Titarenko, E., Prat, S., 2008. A molecular framework for light and gibberellin control of cell elongation. Nature 451, 480–484.

Dello Ioio, R., Nakamura, K., Moubayidin, L., Perilli, S., Taniguchi, M., Morita, M.T., Aoyama, T., Costantino, P., Sabatini, S., 2008. A genetic framework for the control of cell division and differentiation in the root meristem. Science 322, 1380–1384.

Dill, A., Sun, T.P., 2001. Synergistic derepression of gibberellin signaling by removing RGA and GAI function in *Arabidopsis thaliana*. Genetics 159, 777–785.

Dill, A., Jung, H.S., Sun, T.P., 2001. The DELLA motif is essential for gibberellin-induced degradation of RGA. Proc. Natl. Acad. Sci. USA 98, 14162–14167.

Dill, A., Thomas, S.G., Hu, J., Steber, C.M., Sun, T.P., 2004. The *Arabidopsis* F-box protein SLEEPY1 targets gibberellin signaling repressors for gibberellin-induced degradation. Plant Cell 16, 1392–1405.

Dohmann, E.M.N., Nill, C., Schwechheimer, C., 2010. DELLA proteins restrain germination and elongation growth in *Arabidopsis thaliana* COP9 signalosome mutants. Eur. J. Cell Biol. 89, 163–168.

Dorcey, E., Urbez, C., Blázquez, M.A., Carbonell, J., Perez-Amador, M.A., 2009. Fertilization-dependent auxin response in ovules triggers fruit development through the modulation of gibberellin metabolism in *Arabidopsis*. Plant J. 58, 318–332.

Evenson, R.E., Gollin, D., 2003. Assessing the impact of the green revolution, 1960 to 2000. Science 300, 758–762.

Fabian, T., Lorbiecke, R., Umeda, M., Sauter, M., 2000. The cell cycle genes *cycA1;1* and *cdc2Os-3* are coordinately regulated by gibberellin *in planta*. Planta 211, 376–383.

Feng, S., Martinez, C., Gusmaroli, G., Wang, Y., Zhou, J., Wang, F., Chen, L., Yu, L., Iglesias-Pedraz, J.M., Kircher, S., Schäfer, E., Fu, X., Fan, L.-M., Deng, X.W., 2008. Coordinated regulation of *Arabidopsis thaliana* development by light and gibberellins. Nature 451, 475–479.

Feurtado, J.A., Huang, D., Wicki-Stordeur, L., Hemstock, L.E., Potentier, M.S., Tsang, E.W.T., Cutler, A.J., 2011. The *Arabidopsis* C2H2 zinc finger INDETERMINATE DOMAIN1/ENHYDROUS promotes the transition to germination by regulating light and hormonal signaling during seed maturation. Plant Cell 23, 1772–1794.

Filardo, F., Robertson, M., Singh, D.P., Parish, R.W., Swain, S.M., 2009. Functional analysis of HvSPY, a negative regulator of GA response, in barley aleurone cells and *Arabidopsis*. Planta 229, 523–537.

Fleet, C.M., Sun, T.P., 2005. A DELLAcate balance: the role of gibberellin in plant morphogenesis. Curr. Opin. Plant Biol. 8, 77–85.

Floss, D.S., Levy, J.G., Lévesque-Tremblay, V., Pumplin, N., Harrison, M.J., 2013. DELLA proteins regulate arbuscule formation in arbuscular mycorrhizal symbiosis. Proc. Natl. Acad. Sci. USA 110, E5025–E5034.

Fu, X., Richards, D.E., Ait-Ali, T., Hynes, L.W., Ougham, H., Peng, J., Harberd, N.P., 2002. Gibberellin-mediated proteasome-dependent degradation of the barley DELLA protein SLN1 repressor. Plant Cell 14, 3191–3200.

Fu, X., Richards, D.E., Fleck, B., Xie, D., Burton, N., Harberd, N.P., 2004. The *Arabidopsis* mutant sleepy1^{gar2-1} protein promotes plant growth by increasing the affinity of the SCF^{SLY1} E3 ubiquitin ligase for DELLA protein substrates. Plant Cell 16, 1406–1418.

Fuentes, S., Ljung, K., Sorefan, K., Alvey, E., Harberd, N.P., Østergaard, L., 2012. Fruit growth in *Arabidopsis* occurs via DELLA-dependent and DELLA-independent gibberellin responses. Plant Cell 24, 3982–3996.

Fukazawa, J., Teramura, H., Murakoshi, S., Nasuno, K., Nishida, N., Ito, T., Yoshida, M., Kamiya, Y., Yamaguchi, S., Takahashi, Y., 2014. DELLAs function as coactivators of GAI-ASSOCIATED FACTOR1 in regulation of gibberellin homeostasis and signaling in *Arabidopsis*. Plant Cell 26, 2920–2938.

Gallego-Bartolome, J., Minguet, E.G., Marin, J.A., Prat, S., Blazquez, M.A., Alabadi, D., 2010. Transcriptional diversification and functional conservation between DELLA proteins in *Arabidopsis*. Mol. Biol. Evol. 27, 1247–1256.

Gallego-Bartolomé, J., Alabadí, D., Blázquez, M.A., 2011. DELLA-induced early transcriptional changes during etiolated development in *Arabidopsis thaliana*. PLoS ONE 6, e23918.

Gallego-Bartolomé, J., Minguet, E.G., Grau-Enguix, F., Abbas, M., Locascio, A., Thomas, S.G., Alabadí, D., Blázquez, M.A., 2012. Molecular mechanism for the interaction between gibberellin and brassinosteroid signaling pathways in *Arabidopsis*. Proc. Natl. Acad. Sci. USA 109, 13446–13451.

Gallego-Giraldo, C., Hu, J., Urbez, C., Gomez, M.D., Sun, T.-p., Perez-Amador, M.A., 2014. Role of the gibberellin receptors GID1 during fruit-set in *Arabidopsis*. Plant J. 79, 1020–1032.

Gazzarrini, S., McCourt, P., 2003. Cross-talk in plant hormone signalling: what *Arabidopsis* mutants are telling us. Ann. Bot. 91, 605–612.

Goda, H., Sasaki, E., Akiyama, K., Maruyama-Nakashita, A., Nakabayashi, K., Li, W., Ogawa, M., Yamauchi, Y., Preston, J., Aoki, K., Kiba, T., Takatsuto, S., Fujioka, S., Asami, T., Nakano, T., Kato, H., Mizuno, T., Sakakibara, H., Yamaguchi, S., Nambara, E., Kamiya, Y., Takahashi, H., Hirai, M.Y., Sakurai, T., Shinozaki, K., Saito, K., Yoshida, S., Shimada, Y., 2008. The AtGenExpress hormone and chemical treatment data set: experimental design, data evaluation, model data analysis and data access. Plant J. 55, 526–542.

Gomi, K., Sasaki, A., Itoh, H., Ueguchi-Tanaka, M., Ashikari, M., Kitano, H., Matsuoka, M., 2004. GID2, an F-box subunit of the SCF E3 complex, specifically interacts with phosphorylated SLR1 protein and regulates the gibberellin-dependent degradation of SLR1 in rice. Plant J. 37, 626–634.

Grebe, M., 2012. The patterning of epidermal hairs in Arabidopsis – updated. Curr. Opin. Plant Biol. 15, 31–37.

Griffiths, J., Murase, K., Rieu, I., Zentella, R., Zhang, Z.L., Powers, S.J., Gong, F., Phillips, A.L., Hedden, P., Sun, T.P., Thomas, S.G., 2006. Genetic characterization and functional analysis of the GID1 gibberellin receptors in Arabidopsis. Plant Cell 18, 3399–3414.

Harberd, N.P., Freeling, M., 1989. Genetics of dominant gibberellin-insensitive dwarfism in maize. Genetics 121, 827–838.

Hartl, F.U., Hayer-Hartl, M., 2002. Molecular chaperones in the cytosol: from nascent chain to folded protein. Science 295, 1852–1858.

Hauvermale, A.L., Ariizumi, T., Steber, C.M., 2012. Gibberellin signaling: a theme and variations on DELLA repression. Plant Physiol. 160, 83–92.

Hedden, P., 2003. The genes of the Green Revolution. Trends Genet. 19, 5–9.

Hedden, P., Phillips, A.L., 2000. Gibberellin metabolism: new insights revealed by the genes. Trends Plant Sci. 5, 523–530.

Hedden, P., Thomas, S.G., 2012. Gibberellin biosynthesis and its regulation. Biochem. J. 444, 11–25.

Heo, J.O., Chang, K.S., Kim, I.A., Lee, M.H., Lee, S.A., Song, S.K., Lee, M.M., Lim, J., 2011. Funneling of gibberellin signaling by the GRAS transcription regulator SCARECROW-LIKE 3 in the Arabidopsis root. Proc. Natl. Acad. Sci. USA 108, 2166–2171.

Hirano, K., Nakajima, M., Asano, K., Nishiyama, T., Sakakibara, H., Kojima, M., Katoh, E., Xiang, H., Tanahashi, T., Hasebe, M., Banks, J.A., Ashikari, M., Kitano, H., Ueguchi-Tanaka, M., Matsuoka, M., 2007. The GID1-mediated gibberellin perception mechanism is conserved in the lycophyte Selaginella moellendorffii but not in the bryophyte Physcomitrella patens. Plant Cell 19, 3058–3079.

Hirano, K., Asano, K., Tsuji, H., Kawamura, M., Mori, H., Kitano, H., Ueguchi-Tanaka, M., Matsuoka, M., 2010. Characterization of the molecular mechanism underlying gibberellin perception complex formation in rice. Plant Cell 22, 2680–2696.

Hirano, K., Kouketu, E., Katoh, H., Aya, K., Ueguchi-Tanaka, M., Matsuoka, M., 2012. The suppressive function of the rice DELLA protein SLR1 is dependent on its transcriptional activation activity. Plant J. 71, 443–453.

Ho, L., Crabtree, G.R., 2010. Chromatin remodelling during development. Nature 463, 474–484.

Hong, G.J., Xue, X.Y., Mao, Y.B., Wang, L.J., Chen, X.Y., 2012. Arabidopsis MYC2 interacts with DELLA proteins in regulating sesquiterpene synthase gene expression. Plant Cell 24, 2635–2648.

Hotton, S.K., Callis, J., 2008. Regulation of cullin RING ligases. Annu. Rev. Plant Biol. 59, 467–489.

Hou, X., Hu, W.W., Shen, L., Lee, L.Y., Tao, Z., Han, J.H., Yu, H., 2008. Global identification of DELLA target genes during Arabidopsis flower development. Plant Physiol. 147, 1126–1142.

Hou, X., Lee, L.Y., Xia, K., Yan, Y., Yu, H., 2010. DELLAs modulate jasmonate signaling via competitive binding to JAZs. Dev. Cell 19, 884–894.

Hussain, A., Cao, D., Cheng, H., Wen, Z., Peng, J., 2005. Identification of the conserved serine/threonine residues important for gibberellin-sensitivity of Arabidopsis RGL2 protein. Plant J. 44, 88–99.

Ikeda, A., Ueguchi-Tanaka, M., Sonoda, Y., Kitano, H., Koshioka, M., Futsuhara, Y., Matsuoka, M., Yamaguchi, J., 2001. slender rice, a constitutive gibberellin response mutant, is caused by a null mutation of the SLR1 gene, an ortholog of the height-regulating gene GAI/RGA/RHT/D8. Plant Cell 13, 999–1010.

Itoh, H., Ueguchi-Tanaka, M., Sato, Y., Ashikari, M., Matsuoka, M., 2002. The gibberellin signaling pathway is regulated by the appearance and disappearance of SLENDER RICE1 in nuclei. Plant Cell 14, 57–70.

Itoh, H., Shimada, A., Ueguchi-Tanaka, M., Kamiya, N., Hasegawa, Y., Ashikari, M., Matsuoka, M., 2005. Overexpression of a GRAS protein lacking the DELLA domain confers altered gibberellin responses in rice. Plant J. 44, 669–679.

Jan, A., Yang, G., Nakamura, H., Ichikawa, H., Kitano, H., Matsuoka, M., Matsumoto, H., Komatsu, S., 2004. Characterization of a xyloglucan endotransglucosylase gene that is up-regulated by gibberellin in rice. Plant Physiol. 136, 3670–3681.

Jasinski, S., Tattersall, A., Piazza, P., Hay, A., Martinez-Garcia, J.F., Schmitz, G., Theres, K., McCormick, S., Tsiantis, M., 2008. PROCERA encodes a DELLA protein that mediates control of dissected leaf form in tomato. Plant J. 56, 603–612.

Jing, Y., Lin, R., 2013. PICKLE is a repressor in seedling de-etiolation pathway. Plant Signal. Behav. 8, e25026.

Josse, E.M., Gan, Y., Bou-Torrent, J., Stewart, K.L., Gilday, A.D., Jeffree, C.E., Vaistij, F.E., Martínez-García, J.F., Nagy, F., Graham, I.A., Halliday, K.J., 2011. A DELLA in disguise: SPATULA restrains the growth of the developing Arabidopsis seedling. Plant Cell 23, 1337–1351.

Kim, T.W., Wang, Z.Y., 2010. Brassinosteroid signal transduction from receptor kinases to transcription factors. Annu. Rev. Plant Biol. 61, 681–704.

King, K.E., Moritz, T., Harberd, N.P., 2001. Gibberellins are not required for normal stem growth in Arabidopsis thaliana in the absence of GAI and RGA. Genetics 159, 767–776.

Ko, J.H., Yang, S.H., Han, K.H., 2006. Upregulation of an Arabidopsis RING-H2 gene, XERICO, confers drought tolerance through increased abscisic acid biosynthesis. Plant J. 47, 343–355.

Koorneef, M., Elgersma, A., Hanhart, C.J., van Loenen-Martinet, E.P., van Rijn, L., Zeevaart, J.A.D., 1985. A gibberellin insensitive mutant of Arabidopsis thaliana. Physiol. Plant. 65, 33–39.

Koornneef, M., Hanhart, C.J., van der Veen, J.H., 1991. A genetic and physiological analysis of late flowering mutants in Arabidopsis thaliana. Mol. Gen. Genet. 229, 57–66.

Lawit, S.J., Wych, H.M., Xu, D., Kundu, S., Tomes, D.T., 2010. Maize DELLA proteins dwarf plant8 and dwarf plant9 as modulators of plant development. Plant Cell Physiol. 51, 1854–1868.

Lechner, E., Achard, P., Vansiri, A., Potuschak, T., Genschik, P., 2006. F-box proteins everywhere. Curr. Opin. Plant Biol. 9, 631–638.

Lee, Y., Kende, H., 2002. Expression of alpha-expansin and expansin-like genes in deepwater rice. Plant Physiol. 130, 1396–1405.

Lee, S., Cheng, H., King, K.E., Wang, W., He, Y., Hussain, A., Lo, J., Harberd, N.P., Peng, J., 2002. Gibberellin regulates Arabidopsis seed germination via RGL2, a GAI/RGA-like gene whose expression is up-regulated following imbibition. Genes Dev. 16, 646–658.

Leivar, P., Quail, P.H., 2011. PIFs: pivotal components in a cellular signaling hub. Trends Plant Sci. 16, 19–28.

Li, Q.F., Wang, C., Jiang, L., Li, S., Sun, S.S.M., He, J.X., 2012. An interaction between BZR1 and DELLAs mediates direct signaling crosstalk between brassinosteroids and gibberellins in Arabidopsis. Sci. Signal. 5, ra72.

Lim, S., Park, J., Lee, N., Jeong, J., Toh, S., Watanabe, A., Kim, J., Kang, H., Kim, D.H., Kawakami, N., Choi, G., 2013. ABA-insensitive3, ABA-insensitive5, and DELLAs interact to activate the expression of SOMNUS and other high-temperature-inducible genes in imbibed seeds in Arabidopsis. Plant Cell 25, 4863–4878.

Locascio, A., Blazquez, M.A., Alabadi, D., 2013a. Genomic analysis of DELLA protein activity. Plant Cell Physiol. 54, 1229–1237.

Locascio, A., Blázquez, M.A., Alabadí, D., 2013b. Dynamic regulation of cortical microtubule organization through prefoldin–DELLA interaction. Curr. Biol. 23, 804–809.

Marin-de la Rosa, N., Sotillo, B., Miskolczi, P., Gibbs, D.J., Vicente, J., Carbonero, P., Onate-Sanchez, L., Holdsworth, M.J., Bhalerao, R., Alabadi, D., Blazquez, M.A., 2014. Large-scale identification of gibberellin-related transcription factors defines group VII ERFs as functional DELLA partners. Plant Physiol. 166, 1022–1032.

Martí, C., Orzáez, D., Ellul, P., Moreno, V., Carbonell, J., Granell, A., 2007. Silencing of DELLA induces facultative parthenocarpy in tomato fruits. Plant J. 52, 865–876.

McGinnis, K.M., Thomas, S.G., Soule, J.D., Strader, L.C., Zale, J.M., Sun, T.-p., Steber, C.M., 2003. The *Arabidopsis SLEEPY1* gene encodes a putative F-box subunit of an SCF E3 ubiquitin ligase. Plant Cell 15, 1120–1130.

Merchante, C., Alonso, J.M., Stepanova, A.N., 2013. Ethylene signaling: simple ligand, complex regulation. Curr. Opin. Plant Biol. 16, 554–560.

Mitsuda, N., Takiguchi, Y., Shikata, M., Sage-Ono, K., Ono, M., Sasaki, K., Yamaguchi, H., Narumi, T., Tanaka, Y., Sugiyama, M., Yamamura, T., Terakawa, T., Gion, K., Suzuri, R., Tanaka, Y., Nakatsuka, T., Kimura, S., Nishihara, M., Sakai, T., Endo-Onodera, R., Saitoh, K., Isuzugawa, K., Oshima, Y., Koyama, T., Ikeda, M., Narukawa, M., Matsui, K., Nakata, M., Ohtsubo, N., Ohme-Takagi, M., 2011. The new FioreDB database provides comprehensive information on plant transcription factors and phenotypes induced by CRES-T in ornamental and model plants. Plant Biotechnol. 28, 123–130.

Moubayidin, L., Perilli, S., Dello Ioio, R., Di Mambro, R., Costantino, P., Sabatini, S., 2010. The rate of cell differentiation controls the *Arabidopsis* root meristem growth phase. Curr. Biol. 20, 1138–1143.

Murase, K., Hirano, Y., Sun, T.P., Hakoshima, T., 2008. Gibberellin-induced DELLA recognition by the gibberellin receptor GID1. Nature 456, 459–463.

Nakajima, M., Shimada, A., Takashi, Y., Kim, Y.C., Park, S.H., Ueguchi-Tanaka, M., Suzuki, H., Katoh, E., Iuchi, S., Kobayashi, M., Maeda, T., Matsuoka, M., Yamaguchi, I., 2006. Identification and characterization of *Arabidopsis* gibberellin receptors. Plant J. 46, 880–889.

Nakamura, S., Lynch, T.J., Finkelstein, R.R., 2001. Physical interactions between ABA response loci of *Arabidopsis*. Plant J. 26, 627–635.

Ogas, J., Kaufmann, S., Henderson, J., Somerville, C., 1999. PICKLE is a CHD3 chromatin-remodeling factor that regulates the transition from embryonic to vegetative development in *Arabidopsis*. Proc. Natl. Acad. Sci. USA 96, 13839–13844.

Ogawa, M., Kusano, T., Katsumi, M., Sano, H., 2000. Rice gibberellin-insensitive gene homolog, *OsGAI*, encodes a nuclear-localized protein capable of gene activation at transcriptional level. Gene 245, 21–29.

Ogawa, M., Hanada, A., Yamauchi, Y., Kuwahara, A., Kamiya, Y., Yamaguchi, S., 2003. Gibberellin biosynthesis and response during *Arabidopsis* seed germination. Plant Cell 15, 1591–1604.

Oh, E., Yamaguchi, S., Hu, J., Yusuke, J., Jung, B., Paik, I., Lee, H.-S., Sun, T.P., Kamiya, Y., Choi, G., 2007. PIL5, a phytochrome-interacting bHLH protein, regulates gibberellin responsiveness by binding directly to the GAI and RGA promoters in *Arabidopsis* seeds. Plant Cell 19, 1192–1208.

Park, J., Nguyen, K.T., Park, E., Jeon, J.S., Choi, G., 2013. DELLA proteins and their interacting RING Finger proteins repress gibberellin responses by binding to the promoters of a subset of gibberellin-responsive genes in *Arabidopsis*. Plant Cell 25, 927–943.

Peng, J., Harberd, N.P., 1993. Derivative alleles of the *Arabidopsis* gibberellin-insensitive (*gai*) mutation confer a wild-type phenotype. Plant Cell 5, 351–360.

Peng, J., Harberd, N.P., 1997. Gibberellin deficiency and response mutations suppress the stem elongation phenotype of phytochrome-deficient mutants of *Arabidopsis*. Plant Physiol. 113, 1051–1058.

Peng, J., Carol, P., Richards, D.E., King, K.E., Cowling, R.J., Murphy, G.P., Harberd, N.P., 1997. The *Arabidopsis GAI* gene defines a signaling pathway that negatively regulates gibberellin responses. Genes Dev. 11, 3194–3205.

Peng, J., Richards, D.E., Hartley, N.M., Murphy, G.P., Devos, K.M., Flintham, J.E., Beales, J., Fish, L.J., Worland, A.J., Pelica, F., Sudhakar, D., Christou, P., Snape, J.W., Gale, M.D., Harberd, N.P., 1999. 'Green revolution' genes encode mutant gibberellin response modulators. Nature 400, 256–261.

Pysh, L.D., Wysocka-Diller, J.W., Camilleri, C., Bouchez, D., Benfey, P.N., 1999. The GRAS gene family in *Arabidopsis*: sequence characterization and basic expression analysis of the *SCARECROW-LIKE* genes. Plant J. 18, 111–119.

Qi, T., Huang, H., Wu, D., Yan, J., Qi, Y., Song, S., Xie, D., 2014. *Arabidopsis* DELLA and JAZ proteins bind the WD-repeat/bHLH/MYB complex to modulate gibberellin and jasmonate signaling synergy. Plant Cell 26, 1118–1133.

Razem, F.A., El-Kereamy, A., Abrams, S.R., Hill, R.D., 2006. The RNA-binding protein FCA is an abscisic acid receptor. Nature 439, 290–294.

Robertson, M., Swain, S.M., Chandler, P.M., Olszewski, N.E., 1998. Identification of a negative regulator of gibberellin action, HvSPY, in barley. Plant Cell 10, 995–1007.

Sarnowska, E.A., Rolicka, A.T., Bucior, E., Cwiek, P., Tohge, T., Fernie, A.R., Jikumaru, Y., Kamiya, Y., Franzen, R., Schmelzer, E., Porri, A., Sacharowski, S., Gratkowska, D.M., Zugaj, D.L., Taff, A., Zalewska, A., Archacki, R., Davis, S.J., Coupland, G., Koncz, C., Jerzmanowski, A., Sarnowski, T.J., 2013. DELLA-interacting SWI3C core subunit of switch/sucrose nonfermenting chromatin remodeling complex modulates gibberellin responses and hormonal cross talk in *Arabidopsis*. Plant Physiol. 163, 305–317.

Sasaki, A., Itoh, H., Gomi, K., Ueguchi-Tanaka, M., Ishiyama, K., Kobayashi, M., Jeong, D.H., An, G., Kitano, H., Ashikari, M., Matsuoka, M., 2003. Accumulation of phosphorylated repressor for gibberellin signaling in an F-box mutant. Science 299, 1896–1898.

Shimada, A., Ueguchi-Tanaka, M., Sakamoto, T., Fujioka, S., Takatsuto, S., Yoshida, S., Sazuka, T., Ashikari, M., Matsuoka, M., 2006. The rice SPINDLY gene functions as a negative regulator of gibberellin signaling by controlling the suppressive function of the DELLA protein, SLR1, and modulating brassinosteroid synthesis. Plant J. 48, 390–402.

Shimada, A., Ueguchi-Tanaka, M., Nakatsu, T., Nakajima, M., Naoe, Y., Ohmiya, H., Kato, H., Matsuoka, M., 2008. Structural basis for gibberellin recognition by its receptor GID1. Nature 456, 520–523.

Silverstone, A.L., Mak, P.Y., Martínez, E.C., Sun, T.-p., 1997. The new *RGA* locus encodes a negative regulator of gibberellin response in *Arabidopsis thaliana*. Genetics 146, 1087–1099.

Silverstone, A.L., Ciampaglio, C.N., Sun, T.-p., 1998. The *Arabidopsis RGA* gene encodes a transcriptional regulator repressing the gibberellin signal transduction pathway. Plant Cell 10, 155–169.

Silverstone, A.L., Jung, H.S., Dill, A., Kawaide, H., Kamiya, Y., Sun, T.-p., 2001. Repressing a repressor: gibberellin-induced rapid reduction of the RGA protein in *Arabidopsis*. Plant Cell 13, 1555–1566.

Silverstone, A.L., Tseng, T.S., Swain, S.M., Dill, A., Jeong, S.Y., Olszewski, N.E., Sun, T.P., 2007. Functional analysis of SPINDLY in gibberellin signaling in *Arabidopsis*. Plant Physiol. 143, 987–1000.

Solano, R., Stepanova, A., Chao, Q., Ecker, J.R., 1998. Nuclear events in ethylene signaling: a transcriptional cascade mediated by ETHYLENE-INSENSITIVE3 and ETHYLENE-RESPONSE-FACTOR1. Genes Dev. 12, 3703–3714.

Stamm, P., Ravindran, P., Mohanty, B., Tan, E.L., Yu, H., Kumar, P.P., 2012. Insights into the molecular mechanism of RGL2-mediated inhibition of seed germination in *Arabidopsis thaliana*. BMC Plant Biol. 12, 179.

Stodola, F.H., Raper, K.B., Fennell, D.I., Conway, H.F., Johns, V.E., Langford, C.T., Jackson, R.W., 1955. The microbial production of gibberellins A and X. Arch. Biochem. Biophys. 54, 240–245.

Sun, T.-p., 2011. The molecular mechanism and evolution of the GA – GID1–DELLA signaling module in plants. Curr. Biol. 21, R338–R345.

Sun, X., Jones, W.T., Harvey, D., Edwards, P.J., Pascal, S.M., Kirk, C., Considine, T., Sheerin, D.J., Rakonjac, J., Oldfield, C.J., Xue, B., Dunker, A.K., Uversky, V.N., 2010. N-terminal domains of DELLA proteins are intrinsically unstructured in the absence of interaction with GID1/gibberellic acid receptors. J. Biol. Chem. 285, 11557–11571.

Suzuki, H., Park, S.H., Okubo, K., Kitamura, J., Ueguchi-Tanaka, M., Iuchi, S., Katoh, E., Kobayashi, M., Yamaguchi, I., Matsuoka, M., Asami, T., Nakajima, M., 2009. Differential expression and affinities of *Arabidopsis* gibberellin receptors can explain variation in phenotypes of multiple knock-out mutants. Plant J. 60, 48–55.

Swain, S.M., Singh, D.P., 2005. Tall tales from sly dwarves: novel functions of gibberellins in plant development. Trends Plant Sci. 10, 123–129.

Swain, S.M., Tseng, T.S., Olszewski, N.E., 2001. Altered expression of *SPINDLY* affects gibberellin response and plant development. Plant Physiol. 126, 1174–1185.

Takahashi, N., Kitamura, H., Kawarada, A., Seta, Y., Takai, M., Tamura, S., Sumiki, Y., 1955. Biochemical studies on bakanae fungus. Isolation of gibberellins and their properties. Bull. Agr. Chem. Soc. Japan 19, 267–277.

Tyler, L., Thomas, S.G., Hu, J., Dill, A., Alonso, J.M., Ecker, J.R., Sun, T.-p., 2004. DELLA proteins and gibberellin-regulated seed germination and floral development in *Arabidopsis*. Plant Physiol. 135, 1008–1019.

Ueguchi-Tanaka, M., Ashikari, M., Nakajima, M., Itoh, H., Katoh, E., Kobayashi, M., Chow, T.-y., Hsing, Y.-i.C., Kitano, H., Yamaguchi, I., Matsuoka, M., 2005. *GIBBERELLIN INSENSITIVE DWARF1* encodes a soluble receptor for gibberellin. Nature 437, 693–698.

Ueguchi-Tanaka, M., Nakajima, M., Katoh, E., Ohmiya, H., Asano, K., Saji, S., Hongyu, X., Ashikari, M., Kitano, H., Yamaguchi, I., Matsuoka, M., 2007. Molecular interactions of a soluble gibberellin receptor, GID1, with a rice DELLA protein, SLR1, and gibberellin. Plant Cell 19, 2140–2155.

Vandenbussche, F., Fierro, A.C., Wiedemann, G., Reski, R., Van Der Straeten, D., 2007. Evolutionary conservation of plant gibberellin signalling pathway components. BMC Plant Biol. 7, 65.

Wang, F., Zhu, D., Huang, X., Li, S., Gong, Y., Yao, Q., Fu, X., Fan, L.M., Deng, X.W., 2009a. Biochemical insights on degradation of *Arabidopsis* DELLA proteins gained from a cell-free assay system. Plant Cell 21, 2378–2390.

Wang, J.W., Czech, B., Weigel, D., 2009b. miR156-regulated SPL transcription factors define an endogenous flowering pathway in *Arabidopsis thaliana*. Cell 138, 738–749.

Wang, Z.Y., Bai, M.Y., Oh, E., Zhu, J.Y., 2012. Brassinosteroid signaling network and regulation of photomorphogenesis. Annu. Rev. Genet. 46, 701–724.

Wasternack, C., 2007. Jasmonates: an update on biosynthesis, signal transduction and action in plant stress response, growth and development. Ann. Bot. 100, 681–697.

Wells, L., Kreppel, L.K., Comer, F.I., Wadzinski, B.E., Hart, G.W., 2004. O-GlcNAc transferase is in a functional complex with protein phosphatase 1 catalytic subunits. J. Biol. Chem. 279, 38466–38470.

Wen, C.K., Chang, C., 2002. Arabidopsis RGL1 encodes a negative regulator of gibberellin responses. Plant Cell 14, 87–100.

Weston, D.E., Elliott, R.C., Lester, D.R., Rameau, C., Reid, J.B., Murfet, I.C., Ross, J.J., 2008. The Pea DELLA proteins LA and CRY are important regulators of gibberellin synthesis and root growth. Plant Physiol. 147, 199–205.

Wild, M., Achard, P., 2013. The DELLA protein RGL3 positively contributes to jasmonate/ethylene defense responses. Plant Signal. Behav. 8, e23891.

Wild, M., Davière, J.M., Cheminant, S., Regnault, T., Baumberger, N., Heintz, D., Baltz, R., Genschik, P., Achard, P., 2012. The *Arabidopsis* DELLA *RGA-LIKE3* is a direct target of MYC2 and modulates jasmonate signaling responses. Plant Cell 24, 3307–3319.

Willige, B.C., Ghosh, S., Nill, C., Zourelidou, M., Dohmann, E.M., Maier, A., Schwechheimer, C., 2007. The DELLA domain of GA INSENSITIVE mediates the interaction with the GA INSENSITIVE DWARF1A gibberellin receptor of *Arabidopsis*. Plant Cell 19, 1209–1220.

Willige, B.C., Isono, E., Richter, R., Zourelidou, M., Schwechheimer, C., 2011. Gibberellin regulates PIN-FORMED abundance and is required for auxin transport-dependent growth and development in *Arabidopsis thaliana*. Plant Cell 23, 2184–2195.

Wilson, R.N., Somerville, C.R., 1995. Phenotypic suppression of the gibberellin-insensitive mutant (*gai*) of *Arabidopsis*. Plant Physiol. 108, 495–502.

Winkler, R.G., Freeling, M., 1994. Analysis of the autonomy of maize dwarf 1 action in genetic mosaics. J. Hered. 85, 377–380.

Yamaguchi, S., 2008. Gibberellin metabolism and its regulation. Annu. Rev. Plant Biol. 59, 225–251.

Yamamoto, Y., Hirai, T., Yamamoto, E., Kawamura, M., Sato, T., Kitano, H., Matsuoka, M., Ueguchi-Tanaka, M., 2010. A rice *gid1* suppressor mutant reveals that gibberellin is not always required for interaction between its receptor, GID1, and DELLA proteins. Plant Cell 22, 3589–3602.

Yang, D.L., Yao, J., Mei, C.S., Tong, X.H., Zeng, L.J., Li, Q., Xiao, L.T., Sun, T.P., Li, J., Deng, X.W., Lee, C.M., Thomashow, M.F., Yang, Y., He, Z., He, S.Y., 2012. Plant hormone jasmonate prioritizes defense over growth by interfering with gibberellin signaling cascade. Proc. Natl. Acad. Sci. USA 109, E1192–E1200.

Yasumura, Y., Crumpton-Taylor, M., Fuentes, S., Harberd, N.P., 2007. Step-by-step acquisition of the gibberellin–DELLA growth-regulatory mechanism during land-plant evolution. Curr. Biol. 17, 1225–1230.

Yoshida, H., Hirano, K., Sato, T., Mitsuda, N., Nomoto, M., Maeo, K., Koketsu, E., Mitani, R., Kawamura, M., Ishiguro, S., Tada, Y., Ohme-Takagi, M., Matsuoka, M., Ueguchi-Tanaka, M., 2014. DELLA protein functions as a transcriptional activator through the DNA binding of the indeterminate domain family proteins. Proc. Natl. Acad. Sci. USA 111, 7861–7866.

Yu, S., Galvão, V.C., Zhang, Y.C., Horrer, D., Zhang, T.Q., Hao, Y.H., Feng, Y.Q., Wang, S., Schmid, M., Wang, J.W., 2012. Gibberellin regulates the *Arabidopsis* floral transition through miR156-targeted SQUAMOSA promoter binding-like transcription factors. Plant Cell 24, 3320–3332.

Zentella, R., Zhang, Z.L., Park, M., Thomas, S.G., Endo, A., Murase, K., Fleet, C.M., Jikumaru, Y., Nambara, E., Kamiya, Y., Sun, T.-p., 2007. Global analysis of DELLA direct targets in early gibberellin signaling in *Arabidopsis*. Plant Cell 19, 3037–3057.

Zhang, Z.L., Ogawa, M., Fleet, C.M., Zentella, R., Hu, J., Heo, J.O., Lim, J., Kamiya, Y., Yamaguchi, S., Sun, T.P., 2011. Scarecrow-like 3 promotes gibberellin signaling by antagonizing master growth repressor DELLA in *Arabidopsis*. Proc. Natl. Acad. Sci. USA 108, 2160–2165.

Zhang, D., Iyer, L.M., Aravind, L., 2012. Bacterial GRAS domain proteins throw new light on gibberellic acid response mechanisms. Bioinformatics 28, 2407–2411.

Zhang, D., Jing, Y., Jiang, Z., Lin, R., 2014. The chromatin-remodeling factor PICKLE integrates brassinosteroid and gibberellin signaling during skotomorphogenic growth in *Arabidopsis*. Plant Cell 26, 2472–2485.

Zi, J., Mafu, S., Peters, R.J., 2014. To gibberellins and beyond! Surveying the evolution of (di)terpenoid metabolism. Annu. Rev. Plant Biol. 65, 259–286.

CHAPTER 21

bZIP and bHLH Family Members Integrate Transcriptional Responses to Light

Marçal Gallemí*, Jaime F. Martínez-García*,**

*Centre for Research in Agricultural Genomics (CRAG), Consortium CSIC-IRTA-UAB-UB, Barcelona, Spain
**Institució Catalana de Recerca i Estudis Avançats (ICREA), Barcelona, Spain

OUTLINE

21.1 The Role of Light in the Control of Plant Development: A Brief Introduction — 329
    21.1.1 Photoreceptors: The Phytochromes and Their Role in Plant Development (Photomorphogenesis and Shade Avoidance Responses) — 330
    21.1.2 Genetic Identification of Light and Phytochrome-Signaling Components Encoding Non-Transcription Factors: A Global Overview — 331
    21.1.3 Genetic Identification of Light and Phytochrome-Signaling Components Encoding Transcription Factors — 331

21.2 PIFs: Factors that Link Light Perception, Changes in Gene Expression, and Plant Development — 333
    21.2.1 Identification of PIFs — 333
    21.2.2 Analyses of PIF-Deficient Plants — 334
    21.2.3 Regulation of PIF Activity by Light: Transcriptional Controls Versus Posttranscriptional Controls — 334

21.3 HFR1 and PAR1: Atypical bHLH Factors that Act as Transcriptional Cofactors — 336
    21.3.1 Identification of the Microproteins HFR1 and PAR1 — 336
    21.3.2 Regulation of HFR1 and PAR1 Activity by Light: Transcriptional and Posttranscriptional Mechanisms of Control — 336

21.4 HY5: A Paradigm of a bZIP Member in Integrating Light Responses — 337
    21.4.1 Genetic and Molecular Analyses of HY5 — 337
    21.4.2 Regulation of HY5 Activity by Light: Transcriptional and Posttranscriptional Mechanisms of Control — 338
    21.4.3 The Antagonistic Role of PIFs and HY5 in the Integration of Light Cues — 338

21.5 Conclusions — 339

References — 339

21.1 THE ROLE OF LIGHT IN THE CONTROL OF PLANT DEVELOPMENT: A BRIEF INTRODUCTION

Light has a dual role in plants: it is an energy source for photosynthesis, and an informative signal about the environment. As an informative signal, plants perceive light characteristics and respond by modulating their development within their life cycle; for example, to initiate germination, to adapt seedling development to the absence or presence of light (photomorphogenesis), to track the day length indicative of the season of the year (photoperiodism), to direct its growth toward or away from light (phototropism), or to detect and adapt to vegetation proximity (shade avoidance syndrome).

21.1.1 Photoreceptors: The Phytochromes and Their Role in Plant Development (Photomorphogenesis and Shade Avoidance Responses)

Light characteristics are perceived by several plant photoreceptors that absorb different wavelengths. Plants have, at least, five photoreceptor families: the phototropins; the cryptochromes and the zeitlupes, which absorb ultraviolet-A (UV-A, ~320–400 nm) or blue light (B, ~400–500 nm, with a peak of absorption ~470 nm; Christie, 2007; Lin and Shalitin, 2003; Somers and Fujiwara, 2009); the phytochromes which perceive red (R, absorption peak of ~660 nm) and far-red (FR, absorption peak of ~730 nm) light (Franklin and Quail, 2010); and at least one ultraviolet-B (UV-B, ~280–315 nm) photoreceptor called "UV RESISTANCE LOCUS 8" (UVR8; Tilbrook et al., 2013).

Photoreceptors are proteins that are covalently bound to light-absorbing nonprotein factors called "chromophores." The only exception is UVR8, that has at least two tryptophan residues, Trp233 and Trp285, which constitute the chromophore for UV-B perception (Tilbrook et al., 2013). Independently of their chemical nature, the chromophore of each photoreceptor absorbs light of specific wavelengths, causing structural changes in the protein part of the photoreceptor that initiate signaling, that is, a cascade of events that implicate many components, such as several families of transcription factors, and eventually result in photoresponses (Bae and Choi, 2008; Martinez-Garcia et al., 2010; Tilbrook et al., 2013).

Among all photoreceptors, phytochromes are likely the most studied. After their initial discovery in the early 1950s by analyzing what light colors were most effective in modulating lettuce germination (Borthwick et al., 1952), subsequent work indicated that phytochromes are proteins with a bilin chromophore. Phytochromes have been found in all plants that have been analyzed. In the model plant *Arabidopsis thaliana*, these photoreceptors are encoded by a small family of five genes, designated *PHYA* to *PHYE*. The corresponding photoactive (chromophore-bound) proteins, phyA–phyE, are divided into two types depending on their light stability: type I or photolabiles (phyA), which are rapidly degraded after light perception, and type II or photostables (phyB–phyE). Phytochromes are synthesized in an inactive form, Pr, which has a maximum absorbance of R light (666 nm). In darkness, all phytochromes are on its Pr inactive form and, while phyA and phyB locate in the cytoplasm, phyC–phyE are constitutively nuclear. After light activation, the Pr form is rapidly converted into the active Pfr form, which has a maximum absorbance of FR light (730 nm). Upon activation, phyA and phyB also translocate to the nucleus. Photoconversion between Pr and Pfr upon R and FR light absorption is reversible, allowing phytochromes to act as molecular switches (Bae and Choi, 2008).

The perception of environmental light signals by phytochromes controls many aspects of plant life, from seed germination to the timing of flowering and seed set (Smith, 2000). In this chapter we will focus mostly on two processes: seedling de-etiolation and the responses to neighboring plant competition. When germinating in the dark, seedlings are not capable of photosynthesis and therefore grow heterotrophically at the expense of their seed reserves. This has a strong impact on the morphology and development of the emerging seedling: elongation of the hypocotyl is strongly promoted, cotyledons are closed protecting the apical meristem (forming the apical hook), and meristem growth is arrested. This growth mode is known as skotomorphogenic or etiolated (Arsovski et al., 2012; Leivar and Quail, 2011). When etiolated seedlings perceive light, photoreceptors become activated, resulting in a change to a deetiolated or photomorphogenic growth mode that is visualized by the opening and expansion of cotyledons, a strong inhibition of longitudinal growth, initiation of meristematic activity, and onset of photosynthesis. Subsequently, seedlings develop an autotrophic life style (Figure 21.1A). Genetic analyses have shown that phyA is exclusively responsible for controlling seedling de-etiolation under FR with phyB playing the major role in this response under R (Franklin and Quail, 2010).

Once plants are fully de-etiolated, phytochromes control responses to neighboring vegetation. Sunlight contains roughly equal proportions of R and FR light, and the resulting R to FR ratio (R:FR) is high (~1.2). Neighboring vegetation results in two different situations: (i) plant proximity (without direct vegetative shading), and (ii) direct plant canopy shade. Vegetation preferentially reflects FR light compared to other wavelengths, and hence, plant proximity reduces the R:FR around the light impinging neighbors. By contrast, under a plant canopy, the photosynthetic active radiation (PAR, 400–700 nm) is strongly absorbed by photosynthetic pigments (chlorophylls and carotenoids) whereas FR, which is poorly absorbed by the photosynthetic tissues, is transmitted through, or reflected from, vegetation. Hence, under direct plant canopy shade both the amount of PAR and R:FR are greatly reduced. In either case, the low R:FR of the incoming light becomes an informative signal for the presence of potentially competing neighboring vegetation, and activates a set of responses known as the shade avoidance syndrome (SAS; Figure 21.1B). These include promotion of hypocotyl and petiole elongation, reduced branching and, eventually, flowering induction (Martinez-Garcia et al., 2010; Smith, 2000). Under high R:FR, the Pr–Pfr photoequilibrium is displaced toward the active Pfr form, and the SAS is suppressed; under low R:FR, the photoequilibrium is displaced toward the inactive form

21.1 THE ROLE OF LIGHT IN THE CONTROL OF PLANT DEVELOPMENT: A BRIEF INTRODUCTION

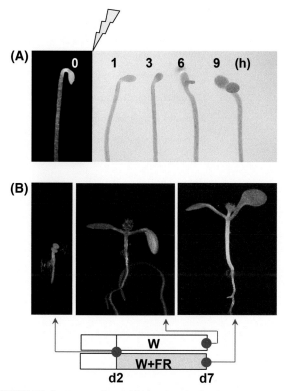

FIGURE 21.1 Perception of light signals by the photoreceptors that control plant development. (A) When seeds germinate in the dark, the resulting seedlings grow etiolated. Within hours (h) of irradiation, light perception initiates the photomorphogenic program of development. (B) Representative aspect of seedlings growing under white (W) light for 2 days (d2) after germination (left panel), and growing under W (middle panel) or W + FR (right panel) conditions for 5 additional days (d7). The most obvious effect of simulated shade treatment is the strong promotion of hypocotyl elongation.

and the SAS is induced. Plant canopy shade, with low B light content, also inactivates cryptochromes, an aspect that will not be covered here.

The absence of active phytochromes (Pfr) in the dark, or the reduced Pfr levels under low R:FR, determine the magnitude of the photoresponse. This is achieved, at least partially, because active Pfr forms are able to interact with a subgroup of proteins that belong to the basic helix–loop–helix (bHLH) family of transcription factors, called "PHYTOCHROME INTERACTING FACTORs" (PIFs; Leivar and Quail, 2011). These factors will be discussed in more detail in Section 21.2.

21.1.2 Genetic Identification of Light and Phytochrome-Signaling Components Encoding Non-Transcription Factors: A Global Overview

Several research groups have performed genetic screenings looking for regulators of the etiolated and photomorphogenic growth. These studies identified two classes of mutants. The first class refers to recessive mutants of identified *CONSTITUTIVE PHOTOMORPHOGENIC* (*COP*; Deng et al., 1991), *DE-ETIOLATED* (*DET*; Chory et al., 1991; Chory et al., 1989), and *FUSCA* (*FUS*; Misera et al., 1994) genes, whose seedlings display photomorphogenic growth in the dark (Table 21.1). *COP1/FUS1* encodes a component of a large protein complex and has E3 ubiquitin ligase activity toward some transcription factors. In dark-grown seedlings, COP1 interacts with these factors, targeting them for proteasome-mediated degradation with the involvement of the COP9 signalosome and COP10/FUS9, which is an E2 ubiquitin-conjugating enzyme variant. SUPPRESSOR OF phyA-105 1 (SPA1) is a member of a small family that includes three more proteins (SPA2, SPA3, and SPA4) shown to repress photomorphogenesis in both light-grown and dark-grown seedlings. SPA proteins, with high sequence similarity to COP1, were shown later to enhance the COP1 E3 ligase activity (Laubinger et al., 2004). DET1/FUS2 and DDB1 (the plant homolog of UV-DAMAGED DNA-BINDING PROTEIN 1), a DET1-interacting factor, have been shown to form a complex with COP10, the so-called "CDD complex," which interacts with the COP1 complex (Schroeder et al., 2002; Serino and Deng, 2003; Yanagawa et al., 2004). In summary, these mutants function as repressors of light-mediated (photomorphogenic) plant development.

21.1.3 Genetic Identification of Light and Phytochrome-Signaling Components Encoding Transcription Factors

The second class of mutants identified components involved in the processes of light perception and signaling. They were identified in genetic screenings looking for seedlings with altered hypocotyl elongation under specific light conditions (Table 21.1). For instance, *long hypocotyl 5* (*hy5*) mutant plants were isolated as seedlings with long hypocotyls under white (W) light, together with several other *hy* mutants affected in different stages of light signaling, such as *hy1*, *hy2*, and *hy6* (shown to be deficient in the biosynthesis of the chromophore for phytochromes), *phyB* (formerly designated *hy3*, deficient in a specific phytochrome), and *hy4* (deficient in the blue light photoreceptor CRY1; Ahmad and Cashmore, 1993; Koornneef and van der Veen, 1980; Parks and Quail, 1991; Reed et al., 1993). Genetic analyses indicated that HY5 acts downstream of multiple families of photoreceptors and promotes photomorphogenesis. HY5 encodes a basic domain/leucine zipper (bZIP) transcription factor (Oyama et al., 1997) which can directly interact with promoters to regulate gene expression (Ang et al., 1998; Lee et al., 2007). The *HY5 HOMOLOG* (*HYH*) encodes a close homolog of HY5 that also has a role in the inhibition of hypocotyl elongation. The phenotype of the *hyh* mutant is evident only in B light; however, increased levels of *HYH* can suppress the elongated phenotype of *hy5* in W light (Holm et al., 2002; Sibout et al., 2006).

TABLE 21.1 Principal Factors Cited in this Article

Acronym	Protein type/activity	Effect on development
PIF1	bHLH transcription factor	Promotes elongation growth
PIF3	bHLH transcription factor	Promotes elongation growth
PIF4	bHLH transcription factor	Promotes elongation growth
PIF5/PIL6	bHLH transcription factor	Promotes elongation growth
PIF6/PIL2	bHLH transcription factor	Promotes elongation growth
PIF7	bHLH transcription factor	Promotes elongation growth
PIF8	bHLH transcription factor	Unknown
PIL1	bHLH transcription factor	Promotes elongation growth
SPT	bHLH transcription factor	Promotes elongation growth
BIM1	bHLH transcription factor	Promotes elongation growth
BEE1, BEE2, BEE3	bHLH transcription factors	Promotes elongation growth
HFR1	bHLH transcription cofactor	Inhibits elongation growth
PAR1	bHLH transcription cofactor	Inhibits elongation growth
PAR2	bHLH transcription cofactor	Inhibits elongation growth
BNQ1/PRE1	bHLH transcription cofactor	Promotes elongation growth
BNQ2	bHLH transcription cofactor	Promotes elongation growth
BNQ3	bHLH transcription cofactor	Promotes elongation growth
PRE6/KIDARI	bHLH transcription cofactor	Promotes elongation growth
HY5	bZIP transcription factor	Promotes elongation growth
HYH	bZIP transcription factor	Promotes elongation growth
ATHB2	HD–Zip II transcription factor	Promotes elongation growth
ATHB4	HD–Zip II transcription factor	Promotes elongation growth
DET1/FUS2	Catalytic	Represses photomorphogenesis
DDB1	Catalytic	Represses photomorphogenesis
COP1/FUS1	Catalytic, E3 ubiquitin–protein ligase	Represses photomorphogenesis
SPA1, SPA2, SPA3, SPA4	Catalytic, enhancers of COP1	Represses photomorphogenesis

The *long hypocotyl in far-red light 1* (*hfr1*) mutant was first identified because of its long hypocotyls under FR light but not under R light (Fairchild et al., 2000; Fankhauser and Chory, 2000; Soh et al., 2000). HFR1 also has a role in cry1-mediated B light signaling, indicating this factor promotes photomorphogenesis integrating information from both phyA and cry1 pathways (Duek and Fankhauser, 2003). *HFR1* encodes an atypical member of the bHLH family of transcription factors (Section 21.3; Fairchild et al., 2000). Mutant *long after far-red light 1* (*laf1*) has an elongated hypocotyl specifically under FR light, suggesting that LAF1 is involved in phyA signaling. *LAF1* encodes a nuclear protein with homology to members of the R2R3–MYB family of DNA-binding proteins (Ballesteros et al., 2001).

The abundance of HY5, HYH, LAF1, HFR1, and phyA itself is directly controlled by COP1. COP1 abundance and subcellular localization are also light regulated: in darkness, it accumulates in the nucleus, whereas in light it translocates to the cytoplasm (von Arnim et al., 1997). Consistent with COP1 activity, these transcription factors are found at low concentrations in darkness and accumulate in the light, promoting photomorphogenesis (Ang et al., 1998; Holm et al., 2002; Jang et al., 2005; Seo et al., 2003). In the case of phyA, which is known to accumulate at high levels in the dark, this system would allow the Pfr form to degrade because the rapid nuclear translocation of phyA occurs only after irradiation, at a time when COP1 is still nuclear, acting as a mechanism to quickly desensitize the plant after initial exposure to light (termination of signaling through this pathway ensures that responses to a single stimulatory event are not perpetuated indefinitely; Seo et al., 2004).

The search for regulatory components of the SAS in *Arabidopsis* was focused on genes that (i) were rapidly upregulated after illumination with simulated shade (low R:FR light), and (ii) encoded transcription factors. The expression of *PHYTOCHROME RAPIDLY REGULATED* (*PAR*) genes is rapidly regulated by phytochrome action after exposure to simulated shade (Roig-Villanova et al., 2006). Initial studies described genes such as *ARABIDOPSIS THALIANA HOMEOBOX 2* (*ATHB2*), *ATHB4*, and *PIF3-LIKE 1* (*PIL1*; Carabelli et al., 1996; Carabelli et al., 1993; Salter et al., 2003). This list was later extended with *HFR1*, *PAR1*, *PAR2*, and a few *B-BOX CONTAINING* (BBX) genes (Table 21.1). Those *PAR* genes encoded members of three different families of transcriptional regulators: bHLHs (*HFR1*, *PAR1*, *PAR2*, *PIL1*), homeodomain–leucine zipper (HD–ZIP) subfamily II (*ATHB2*, *ATHB4*; Chapter 7), and the BBX family of proteins. Genetic analyses demonstrated positive and negative roles in SAS regulation for several of these *PAR* genes encoding transcriptional regulators, basically by analyzing hypocotyl elongation. Their role in the control of transcriptional responses to light will be detailed in Section 21.3.

21.2 PIFs: FACTORS THAT LINK LIGHT PERCEPTION, CHANGES IN GENE EXPRESSION, AND PLANT DEVELOPMENT

21.2.1 Identification of PIFs

Active phytochromes interact with PIFs, a subset of bHLH transcription factors that negatively regulate photomorphogenesis (Castillon et al., 2007; Leivar and Quail, 2011). The founder member of the PIF family was PIF3, identified in a yeast two-hybrid screening using phyB as bait. *In vitro* binding assays showed that both phyA and phyB active forms (i.e., PfrA and PfrB) could interact with this factor (Martinez-Garcia et al., 2000; Ni et al., 1998).

A search for PIL protein sequences in *Arabidopsis* resulted in the identification of 14 proteins, 7 of which were true PIFs (Table 21.1). From these all but PIF8 have a demonstrated role in light-mediated development: *PIF1* (initially named *PIL5*), *PIF4*, *PIF5*/*PIL6*, *PIF6*/*PIL2*, *PIF7*, *PIF8*, and *PIL1* (Huq et al., 2004; Khanna et al., 2004; Leivar et al., 2008a; Leivar and Quail, 2011; Luo et al., 2014; Yamashino et al., 2003). PIF4, PIF5, PIF6, PIF7, PIF8, and PIL1 interact with the active Pfr form of phyB, while PIF1 (like PIF3) also binds the Pfr form of phyA (Khanna et al., 2004; Leivar and Quail, 2011; Luo et al., 2014). All characterized PIFs are transcription factors that contain the conserved HLH domain and an adjacent basic region that provides the ability to bind specific DNA sequences such as dimers; a functional nuclear localization signal (NLS) is embedded in the basic region. To bind to the active Pfr phyB form, PIFs have a conserved sequence designated the active phyB-binding (APB) motif. PIF1 and PIF3 also have a less conserved region, the active phyA-binding (APA) region, which is necessary for phyA binding (Figure 21.2; Khanna et al., 2004; Leivar and Quail, 2011; Ni et al., 1998). *In vitro* electrophoretic mobility shift assays (EMSAs) also indicated that PIF3 bound to PfrB specifically binds DNA, supporting the idea that PIF3 can rapidly and directly transduce changes in light conditions into changes in gene expression (Martinez-Garcia et al., 2000).

The other seven PIL members of the subfamily were not demonstrated to bind to phytochromes. This is the

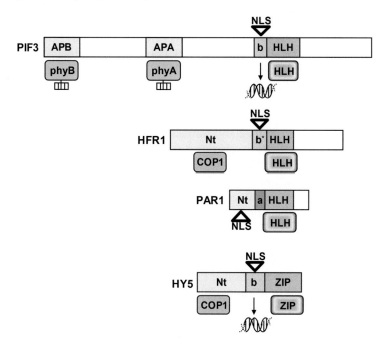

FIGURE 21.2 Schematic representation of transcriptional regulators belonging to the bHLH (PIF3, HFR1, PAR1) or bZIP (HY5) families involved in light signaling. (a) Acidic region; APA, active phyA binding domain; APB, active phyB binding domain; (b) basic region; b*, non canonical basic region; HLH, helix-loop-helix domain; Nt, N-terminal region; NLS, nuclear localization signal; ZIP, leucine-zipper. The molecular function of each region or domain is represented, such as binding to a known protein (phyA, phyB, COP1), protein interaction domain (HLH, ZIP), or DNA-binding activity.

case for *HFR1*, *SPT* (*SPATULA*), *ALC* (*ALCATRAZ*), *BHLH23*, *BHLH56*, *BHLH119*, and *BHLH127* (Leivar and Quail, 2011). Apart from HFR1, SPT is the only member of this group that regulates a range of light-mediated responses, such as germination, hypocotyl elongation, cotyledon expansion, or flower development. In addition to heterodimerizing with PIFs, SPT was shown to bind to the same direct target promoters of PIFs and regulate their transcription (Foreman et al., 2011b; Josse et al., 2011; Penfield et al., 2005; Reymond et al., 2012).

21.2.2 Analyses of PIF-Deficient Plants

Phenotypic characterization of PIF-deficient plants indicated that these proteins act as negative regulators of photomorphogenesis. Monogenic *pif1*, *pif3*, *pif4*, *pif5*, or *pif7* mutants exhibit light-hypersensitive seedling phenotypes in response to monochromatic R and/or FR light, with shorter hypocotyls and larger cotyledons than its wild type (Huq et al., 2004; Huq and Quail, 2002; Khanna et al., 2007; Kim et al., 2003; Leivar et al., 2008a; Lorrain et al., 2008; Oh et al., 2004). Single PIF mutants display different degrees of de-etiolated growth in the dark. The PIF quadruple-mutant (*pif1–pif3–pif4–pif5*, renamed as *pifq*) shows a stronger phenotype, with a mild constitutive photomorphogenic growth, having shorter hypocotyls and opened cotyledons when growing in the dark (Leivar et al., 2008b; Shin et al., 2009). The mild photomorphogenic phenotype of the dark-grown *pifq* mutant suggests that other factors, such as PIF6, PIF7, and PIF8, may participate in promoting seedling skotomorphogenesis. In addition, *pifq* seedlings growing in darkness display a gene expression pattern similar to that of light-grown wild-type seedlings (Leivar and Quail, 2011). The long-hypocotyl phenotype of *phyB* mutant seedlings (which resembles partial etiolated growth) is reversed by the *pifq* mutations (Leivar et al., 2012a), and transgenic lines overexpressing *PIF3* display longer hypocotyls and smaller cotyledons under R light and reduced cotyledon opening under both R or FR light (Kim et al., 2003). Similarly, overexpression of *PIF4* and *PIF5* also promote hypocotyl elongation and induce flowering (Fujimori et al., 2004; Khanna et al., 2004; Kumar et al., 2012; Lorrain et al., 2008). All these traits observed in dark- and light-grown plants provide support for the idea that PIFQ promotes skotomorphogenesis in the absence of active phytochromes and represses, additively and in a redundant manner, the photomorphogenic growth in the presence of active phytochromes.

PIFs also work in a partially redundant manner promoting other light responses such as those of the SAS. PIF4 and PIF5 are required for hypocotyl and petiole elongation in response to shade (Lorrain et al., 2008); PIF1 and PIF3 also contribute modestly to this shade-induced growth response (Leivar et al., 2012b; Sellaro et al., 2012). Although the PIFQ positively contribute to the SAS hypocotyl response, the single *pif7* mutant shows a strongly attenuated hypocotyl elongation in response to low R:FR light, suggesting that PIF7 plays a dominant role in this shade response (Li et al., 2012). Genetic analyses indicate that PIFQ and PIF7 promote growth under different light conditions.

Different reasons could explain why some PIFs are more relevant than others in regulating some responses, even in those cases when they display similar expression patterns. On one hand, PIF activity could be controlled by different phytochromes (e.g., phyA binds to just two of the PIF members) or by the same phytochrome but with different affinities. On the other hand, despite the similarity in the DNA-binding motif recognition between the PIFQ members (G-boxes, CACGTG, and a related DNA motif called the "PIF binding E-box"), they have shared but distinct *in vivo* target genes, which suggests that PIFs have an intrinsic differential activity (Pfeiffer et al., 2014). Since different PIFs and PILs can also dimerize (Huq and Quail, 2002; Luo et al., 2014) and they can promote (HFR1, PIL1) or repress (PIFQ) photomorphogenesis, it is expected that heterodimeric interactions between them may provide a combinatorial mechanism for the fine control of photomorphogenesis or that of different transcriptional pathways in a time- and tissue-specific manner.

21.2.3 Regulation of PIF Activity by Light: Transcriptional Controls Versus Posttranscriptional Controls

The expression patterns of *PIF* mRNAs are partly responsible for their shared and distinct roles, being affected differently by endogenous or environmental signals. For instance, *PIF4* (but not other *PIFs*) have increased expression at high temperature, inducing hypocotyl elongation and early flowering (Franklin et al., 2011; Koini et al., 2009; Kumar et al., 2012). Photoactive phytochromes regulate the expression of *PIF4*, *PIF5*, and *PIF7* (but not that of *PIF1* or *PIF3*) indirectly, through the circadian clock in such manner that *PIF4* and *PIF5* have peaks of expression during the day, and low expression during the night (Kidokoro et al., 2009; Yamashino et al., 2003).

PIF activity is also under posttranslational control (Figure 21.3). When a plant is exposed to R, FR, or W light, the interaction of the PIFQ with the active form of the phytochromes results in their rapid phosphorylation which induces their degradation via the 26S proteasome, with half-lives of 5–20 min (Al-Sady et al., 2006; Lorrain and Fankhauser, 2012; Nozue et al., 2007; Shen et al., 2007). In contrast with these photolabile PIFs, the rapid and reversible phosphorylation of PIF7 by active phytochromes does not result in its degradation (Leivar et al., 2008a). Therefore, PIF7 can be considered photostable. Phosphorylation of PIF7 does have regulatory consequences for PIF7: it

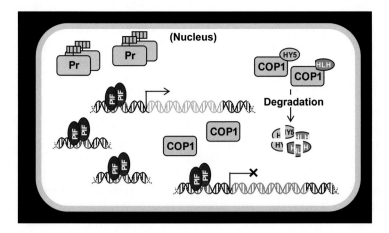
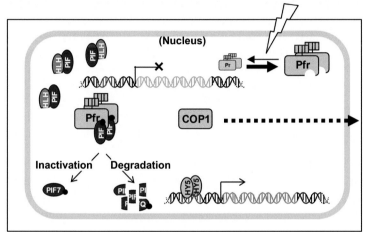
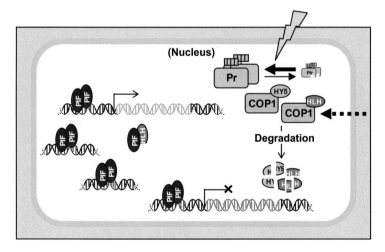

FIGURE 21.3 Simplified model showing light-signaling components in the nuclei of dark (upper panel), sunlight (high R:FR, middle panel) and shade (low R:FR, lower panel) growing plants. For simplicity, only a few transcriptional regulators are shown. In darkness, PIFs promote etiolated development (and hypocotyl elongation), while HY5 and HFR1 are degraded by nuclear-accumulated COP1. In sunlight, active phytochromes mediate PIF phosphorylation, which results in PIF degradation and/or inactivation; HY5 and HFR1 are stabilized due to COP1 nuclear export. Together, these molecular changes repress elongation. Under shade, the partial inactivation of the phytochromes triggers the stabilization and/or activation of PIFs, while HY5 and HFR1 are destabilized; at the same time, *HFR1* expression is increased (not represented in the figure) which allow its accumulation and some inhibition of PIF activity; together these changes promote hypocotyl elongation. Pr, inactive phytochromes; Pfr, active phytochromes. Black dots represent phosphorylation (only phytochrome-mediated phosphorylation of PIFs is shown, although both HY5 and HFR1 are phosphorylated).

reduces DNA-binding activity toward its target genes (Li et al., 2012). The DNA-binding activity of PIF1 and PIF3 is also reduced by active phytochromes, a process considered to be independent of their protein degradation (Park et al., 2012). These observations suggest that PIFQ DNA-binding activity can mediate the phytochrome-modulated repression of gene expression, such as that of *PIL1*. It does not explain, however, phytochrome-mediated activation of gene expression. The initial *in vitro* evidence showing formation of a ternary complex of active phyB, PIF3, and DNA (Martinez-Garcia et al., 2000), together with recent *in vivo* data showing that phyA can be directly recruited to gene promoters and coactivate transcription, suggests the escort model, which proposes that phyA and phyB directly associate with gene promoters by interacting with transcription factors and coregulators (Chen et al., 2014).

Genetic analyses indicate that PIFs also regulate a range of downstream light-regulated processes such as carpel formation or synthesis of photosynthetic pigments (Reymond et al., 2012; Toledo-Ortiz et al., 2010). PIFs also serve as nodes to integrate different pathways such as light and temperature (Koini et al., 2009; Kumar et al., 2012; Stavang et al., 2009). Indeed, environmental cues, such as warm temperatures (25°C) and low R:FR light also promote the accumulation of phosphorylated PIF4, which increases its stability, conditions that also promote growth such as the hypocotyl or petiole elongation (Foreman et al., 2011a; Lorrain et al., 2008). Together, PIFs have the ability to integrate endogenous (i.e., circadian time, hormonal) and environmental (i.e., temperature, light) information.

An additional mechanism by which active phytochromes modulate PIF activity is by increasing the activity of microproteins (miPs) that interact with PIFs and alter some of their molecular activities (Section 21.3.1).

21.3 HFR1 AND PAR1: ATYPICAL bHLH FACTORS THAT ACT AS TRANSCRIPTIONAL COFACTORS

21.3.1 Identification of the Microproteins HFR1 and PAR1

In addition to its role in promoting phyA-mediated photomorphogenesis (Section 21.1.3), genetic analyses indicate that HFR1 also has a negative role in regulating SAS responses, as deduced from the increased elongation of mutant hypocotyls under simulated or canopy shade (Roig-Villanova et al., 2007; Sessa et al., 2005). Genetic analyses indicated that, under shade, PAR1, PAR2, and HFR1 (Section 21.1.3) form a negative feedback loop to prevent an exaggerated elongation response to shade (Roig-Villanova et al., 2007; Sessa et al., 2005). *PAR1* and *PAR2* are both early phytochrome target genes that are also rapidly repressed during seedling deetiolation under FR and R (Roig-Villanova et al., 2007; Roig-Villanova et al., 2006). PAR1 and PAR2 also act as positive factors in seedling de-etiolation, mediating FR, R, and B light signaling (Hao et al., 2012; Roig-Villanova et al., 2007; Zhou et al., 2014). *HFR1*, *PAR1*, and *PAR2* encode atypical bHLHs (Fairchild et al., 2000; Galstyan et al., 2011; Roig-Villanova et al., 2007). Typical bHLH proteins contain a conserved HLH region and a basic domain involved in their binding to E-box or G-box motifs as dimers. Unlike them, HFR1 shows a nonconsensual basic domain, and PAR1 and PAR2 both have an acidic (instead of basic) domain (Figure 21.2). By overexpressing truncated derivatives of *PAR1* and *HFR1* in plants, it was shown that these proteins are actually transcriptional cofactors that do not bind DNA to directly regulate transcription (Galstyan et al., 2012; Galstyan et al., 2011). The inability of HFR1 to bind to any DNA sequence was further confirmed by protein-binding arrays (Hornitschek et al., 2012). These findings led to the conclusion that protein–protein interactions involving the HLH domain of PAR1 and HFR1 are a fundamental aspect of the mechanism by which these proteins regulate gene expression (Galstyan et al., 2011).

HFR1, which heterodimerizes with PIF1, PIF3, PIF4, and PIF5, functions by trapping these PIFs into nonfunctional complexes that cannot bind to the DNA regulatory motifs of their target genes, resulting in the loss of PIF activity (Fairchild et al., 2000; Hornitschek et al., 2009; Shi et al., 2013). Consequently, the growth-promoting action of PIFs is limited by HFR1. PAR1–PIF4 and PAR2–PIF4 heterodimers form a complex HLH network regulating cell elongation and plant development (Hao et al., 2012). PAR1 also interacts with BES1-INTERACTING MYC-LIKE 1 (BIM1), BRASSINOSTEROID-ENHANCED EXPRESSION 1 (BEE1), BEE2, and BEE3, which were previously identified as participating in different aspects of brassinosteroid signaling. *BIM1* and *BEE* genes, whose expression is also rapidly induced after simulated shade perception, have a positive role in SAS regulation. Like HFR1, PAR1, and PAR2 function by trapping BIM and BEE factors inside nonfunctional complexes that cannot bind to DNA, resulting in inhibition of their transcriptional activity (Cifuentes-Esquivel et al., 2013). PAR1, PAR2, and HFR1 were recently coined as "microproteins" (miPs), a term that refers to proteins that perturb the formation of functional protein dimers by forming nonfunctional, homotypic protein complexes with their targets, which they regulate in a dominant-negative manner (Staudt and Wenkel, 2011). The PIF, BIM, and BEE bHLH transcription factors are positive regulators of shade avoidance responses, whereas the miPs HFR1, PAR1, and PAR2, acting as posttranslational regulators, form a negative feedback loop to prevent exaggerated shade growth responses (Roig-Villanova et al., 2007; Sessa et al., 2005).

A similar mechanism of inhibition of PIF activity has been shown to work for non-bHLH members, like the DELLA repressors of gibberellin signaling (Chapter 20). Like HFR1 or PAR1, DELLAs are not able to bind DNA, but regulate gene expression of DNA-binding PIF factors. Interaction of PIF3 and PIF4 with DELLAs prevents PIFs from binding to DNA, consequently modulating PIF transcriptional activity (de Lucas et al., 2008; Feng et al., 2008; Josse et al., 2011).

21.3.2 Regulation of HFR1 and PAR1 Activity by Light: Transcriptional and Posttranscriptional Mechanisms of Control

HFR1, *PAR1*, and *PAR2* expression is upregulated after shade perception (Section 21.1.3). In other light-regulated responses, such as seedling de-etiolation, expression of these genes is also subject to strict transcriptional control. COP1 and DET1 also repress shade-induced *PAR1* transcription in seedlings (Roig-Villanova et al., 2006); in addition, seedlings with reduced levels of PAR1 and PAR2 partially suppressed the constitutive photomorphogenic *cop1-4* phenotype in darkness, suggesting that PAR1 and PAR2 act downstream of COP1 (Zhou et al., 2014). *HFR1* expression shows more than two fold induction in wild-type seedlings grown in FR light but a 14-fold decrease in R-grown seedlings relative to those grown in darkness (Fairchild et al., 2000). *PAR1* and *PAR2* transcript abundances are also rapidly repressed by phyA, phyB, and cry1 after transfer to FR, R, and B light conditions, respectively (Tepperman et al., 2001; Zhou et al., 2014). Compared with FR or B light, both R and W light conditions repress *PAR1* and *PAR2* transcription more strongly. These expression data are consistent with phenotypes shown by plants with altered levels of these components (Section 21.3.1).

HFR1 and PAR1 proteins are also subject to post-translational control. HFR1 is degraded in the dark and its accumulation in the light correlates with the fluence rate, an effect that depends on COP1 (Figure 21.3). In the dark, COP1 accumulates in the nucleus, mediating HFR1 degradation. In light, COP1 translocates to the cytosol, which allows the accumulation of HFR1 to promote photomorphogenesis (Duek et al., 2004; Jang et al., 2005; Yang et al., 2005). Degradation of HFR1 is mediated by sequences found in the N-terminus of HFR1. Overexpression of a truncated version of HFR1 (from which the N-terminus has been deleted) stabilizes the protein in the dark and shows a constitutively photomorphogenic phenotype similar to the one reported for the *pifq* mutant (Yang et al., 2005, Yang et al., 2003). PAR1 and PAR2 levels also increase when etiolated seedlings are irradiated with W light, an effect that also requires COP1, which mediates the degradation of PAR1 and PAR2 via the 26S proteasome (Hao et al., 2012; Zhou et al., 2014). COP1 has been found to rapidly accumulate in the nucleus under natural or simulated shade. Hence, nuclear COP1 might reduce HFR1, PAR1, and PAR2 abundance to promote SAS responses (Pacin et al., 2013).

HFR1 and PAR1 activities may also be antagonized by atypical bHLH proteins. This is the case for PA-CLOBUTRAZOL RESISTANCE 1 (PRE1)/BANQUO1 (BNQ1), BNQ2, and BNQ3 proteins. Plants that overexpress *BNQ* genes show elongated hypocotyls in R light. Because *BNQ* genes can suppress the short-hypocotyl phenotype of seedlings overexpressing HFR1, it is likely that they do so by blocking the inhibitory effect of HFR1 over PIF proteins (Mara et al., 2010). Overexpression of *PRE1/BNQ1* completely suppressed the growth defect of overexpressed *PAR1*, indicating that PRE1 directly suppresses PAR1, leading to derepression of PIF4 and activation of cell elongation (Hao et al., 2012). Another member of this family, PRE6/KIDARI, was shown to interact with HFR1 (Hyun and Lee, 2006). In summary, interactions between non-DNA-binding bHLH proteins, such as HFR1, PAR1, BNQ, KIDARI, or PRE1, add an additional level of regulatory possibilities, very likely acquired to fine-tune biological responses to changing environmental conditions.

21.4 HY5: A PARADIGM OF A bZIP MEMBER IN INTEGRATING LIGHT RESPONSES

21.4.1 Genetic and Molecular Analyses of HY5

In addition to elongated light-grown hypocotyls (Section 21.1.3), *hy5* mutant seedlings show reduced chlorophyll and anthocyanin accumulation, and reduced chloroplast development in greening hypocotyls. Moreover, the *HY5* gene is responsible for the regulation of fundamental developmental processes, like cell elongation and cell proliferation (Ang et al., 1998; Oyama et al., 1997). All these characters are consistent with aberrant light-mediated signaling, confirming its role as a positive regulator of photomorphogenesis. HY5 was also found to be involved in several hormonal signaling pathways, like those regulated by auxin, cytokinin, ethylene, and jasmonic acid, which led to the proposition that HY5 acts as a node that links various signaling pathways to coordinate growth and development (Alabadi and Blazquez, 2009; Cluis et al., 2004; Jiao et al., 2007; Lee et al., 2007).

The *HY5* gene encodes a relatively small-sized protein (168 amino acids) that contains a bZIP functional domain involved in DNA-binding as a dimer (Figure 21.2). Genome-wide analyses indicated that HY5 directly binds to DNA motifs preferentially located in the promoter regions. HY5-binding consensus sequences are the so-called ACGT-containing element (ACE) motifs that include the Z- (AT**ACGT**GT), C- (G**ACGT**C), and G- (C**ACGT**G), as well as the hybrid C/G- (G**ACGT**G) and C/A- (G**ACGT**A) boxes (Lee et al., 2007; Zhang et al., 2011). Although HY5 was identified as a transcriptional activator of the *CHALCONE SYNTHASE* (*CHS*) gene, which encodes the first committed step for anthocyanin biosynthesis (Ang et al., 1998), it binds and affects the expression, either positively (like *CHS*) or negatively, of over 1100 genes. For instance, HY5 represses *FAR-RED ELONGATED HYPOCOTYL1* (*FHY1*) and *FHY1-LIKE* (*FHL*) expression, two crucial components in phyA signaling. It does this by directly binding to specific ACE motifs found in the *FHY1/FHL* promoters. In the dark, the transcription factors FHY3 and FAR-RED IMPAIRED RESPONSE 1 (FAR1), act together to directly activate the transcription of *FHY1* and *FHL*. After exposure to FR light, HY5 physically interacts with the FHY3/FAR1 through their respective DNA-binding domains and negatively regulates FHY3/FAR1-activated *FHY1/FHL* expression. Therefore, HY5 provides a dual mechanism to act as an activator of light-induced genes (e.g., *CHS*) and a repressor of FR light-repressed genes (e.g., *FHY1/FHL*) for fine-tuning phyA signaling homeostasis (Li et al., 2010).

21.4.2 Regulation of HY5 Activity by Light: Transcriptional and Posttranscriptional Mechanisms of Control

HY5 is expressed in various tissues, including roots, hypocotyls, cotyledons, leaves, stems, and flowers. The activity of the *HY5* promoter is strong in light-grown seedlings and weak in dark-grown seedlings, and is mediated by the direct binding of HY5 and CAM7/ZBF3 (a Z-box binding transcription factor) in all stages and light conditions analyzed (Abbas et al., 2014; Oyama et al., 1997). *HY5* expression is rapidly increased (1 h) after exposure to UV-B, a response that requires participation of the UV-B receptor UVR8 (Brown et al., 2009), and more slowly (4–8 h) after exposure to simulated shade, a response that requires the participation of phyA (Ciolfi et al., 2013). In the morning, *HY5* expression increases in the transition between night and day. In the afternoon, *HY5* expression is low, but transfer from shade light to sunfleck (i.e., brief periods of exposure to unfiltered sunlight conditions) also promotes its expression (Sellaro et al., 2011).

The light-related changes in *HY5* expression reflect changes in chromatin structure. A systematic comparison of the global histone modification patterns between dark-grown and light-exposed (6 h) seedlings revealed that only H3K9 acetylation (H3K9ac), a histone mark associated with gene activation, showed a massive peak in the transcribed region of *HY5* following transition to light, while no peak was detected in dark-grown seedlings. More than fifty percent of the putative HY5 target genes were also targeted by H3K9ac following transition to light (Charron et al., 2009). Therefore, chromatin marks might act synergistically with HY5 to properly regulate the transcription of downstream effectors.

HY5 protein is regulated posttranscriptionally by degradation in the dark, and its abundance directly correlates with the degree of photomorphogenic development. HY5 is largely controlled by COP1, a repressor of photomorphogenesis (Section 21.1.2). HY5 and COP1 interact physically *via* a motif present in the HY5 N-terminal region (Figure 21.2), and this interaction leads to HY5 degradation (Figure 21.3; Ang et al., 1998; Bae and Choi, 2008; Hardtke et al., 2000; Osterlund et al., 2000). Consistently, overexpression of full-length *HY5* showed a wild-type phenotype under both light and dark conditions, whereas overexpression of a truncated HY5, missing the first 77 amino acids in the N-terminal region, resulted in a hypersensitive phenotype in seedlings grown in W, FR, B, and R light (Ang et al., 1998). Although *HY5* mRNA levels are also affected according to light- and dark-grown seedlings (2–3-fold), this effect is considerably less important than the 15–20-fold difference seen in HY5 protein levels. Therefore, mRNA levels could contribute to differential accumulation of the HY5 protein, but translational or posttranslational regulation may be primarily responsible for regulating its abundance.

21.4.3 The Antagonistic Role of PIFs and HY5 in the Integration of Light Cues

It is generally established that HY5 acts antagonistically with the PIFs. For instance, *hy5* suppresses both the morphological and molecular phenotypes of *pifq* in the dark, which shows that the PIFQ also mediates the repression of photomorphogenesis through destabilization of HY5 in the dark (Xu et al., 2014). As mentioned, HY5 promotes photomorphogenic development by activating genes that promote photosynthetic machinery assembly, photopigment production, chloroplast development, and seedling cotyledon expansion, while PIFs suppress these responses to maintain etiolated growth. As mentioned, the PIFQ is photolabile and HY5 is stabilized by light, which means that they exhibit opposing diurnal expression patterns. ChIP, EMSA, and transcript analysis have established that HY5 and PIFs impart antagonistic regulation to common gene targets through direct binding to the same G-box *cis*-element. One of these common targets is *PHYTOENE SYNTHASE* (*PSY*), which encodes an enzyme for a rate-limiting step in the carotenoid biosynthetic pathway (Ruiz-Sola and Rodriguez-Concepcion, 2012). PIFs restrict the accumulation of carotenoids during deetiolation, in part by negatively regulating the expression of *PSY* (Toledo-Ortiz et al., 2010). HY5 was proposed to be an activator of *PSY*, having an opposite role to PIFs in carotenoid biosynthesis and regulation of *PSY* expression during seedling de-etiolation through direct binding to a G-box of its promoter. It seems, therefore, that G-box *cis*-element convergence appears to represent a common mechanism through which HY5 and PIFs regulate some carotenoid biosynthetic genes. It was proposed that HY5 and PIFs form a dynamic activation–suppression transcriptional module that provides a simple and direct mechanism through which environmental changes can redirect the transcriptional control of genes required for photosynthesis and photoprotection (Toledo-Ortiz et al., 2014).

HY5 and PIFs do not always act antagonistically. In the case of anthocyanin biosynthesis, PIF3 and HY5 both positively regulate the pathway by activating transcription of the same biosynthetic genes by binding to distinct *cis*-promoter elements (PIF3 binding to a G-box and HY5 to an ACE motif; Shin et al., 2007). In this case, PIF3 binding is facilitated by the presence of HY5, suggesting these factors act in a cooperative manner in the regulation of this response.

21.5 CONCLUSIONS

To adapt to prevailing light conditions, plants employ photoreceptors that transform environmental signals into biochemical and cellular information by affecting the activity of an array of transcriptional regulators that either promote or repress growth. The underlying molecular mechanisms involve protein degradation and modulation of transcriptional activity by phosphorylation and/or negative interference of miPs. Their combined action results in massive changes in gene expression that eventually enable plant growth to be induced or repressed to optimally adapt to changes in light conditions.

Acknowledgments

The authors acknowledge Briardo Llorente and Irma Roig-Villanova for their helpful comments and suggestions on the manuscript. M.G. acknowledges receipt of an FPI fellowship from the Spanish Ministry of Economy and Competitivity (MINECO). The authors acknowledge the support of grants from the Generalitat de Catalunya and from MINECO's Fondo Europeo de Desarrollo Regional to J.F.M.G. (2014-SGR447, Xarba and BIO2011-23489).

References

Abbas, N., Maurya, J.P., Senapati, D., Gangappa, S.N., Chattopadhyay, S., 2014. *Arabidopsis* CAM7 and HY5 physically interact and directly bind to the HY5 promoter to regulate its expression and thereby promote photomorphogenesis. Plant Cell 26, 1036–1052.

Ahmad, M., Cashmore, A.R., 1993. HY4 gene of *A. thaliana* encodes a protein with characteristics of a blue-light photoreceptor. Nature 366, 162–166.

Alabadi, D., Blazquez, M.A., 2009. Molecular interactions between light and hormone signaling to control plant growth. Plant Mol. Biol. 69, 409–417.

Al-Sady, B., Ni, W., Kircher, S., Schafer, E., Quail, P.H., 2006. Photoactivated phytochrome induces rapid PIF3 phosphorylation prior to proteasome-mediated degradation. Mol. Cell 23, 439–446.

Ang, L.H., Chattopadhyay, S., Wei, N., Oyama, T., Okada, K., Batschauer, A., et al., 1998. Molecular interaction between COP1 and HY5 defines a regulatory switch for light control of *Arabidopsis* development. Mol. Cell 1, 213–222.

Arsovski, A.A., Galstyan, A., Guseman, J.M., Nemhauser, J.L., 2012. Photomorphogenesis. Arabidopsis Book 10, e0147.

Bae, G., Choi, G., 2008. Decoding of light signals by plant phytochromes and their interacting proteins. Annu. Rev. Plant Biol. 59, 281–311.

Ballesteros, M.L., Bolle, C., Lois, L.M., Moore, J.M., Vielle-Calzada, J.P., Grossniklaus, U., et al., 2001. LAF1, a MYB transcription activator for phytochrome A signaling. Genes Dev. 15, 2613–2625.

Borthwick, H.A., Hendricks, S.B., Parker, M.W., Toole, E.H., Toole, V.K., 1952. A reversible photoreaction controlling seed germination. Proc. Natl. Acad. Sci. USA 38, 662–666.

Brown, B.A., Headland, L.R., Jenkins, G.I., 2009. UV-B action spectrum for UVR8-mediated HY5 transcript accumulation in *Arabidopsis*. Photochem. Photobiol. 85, 1147–1155.

Carabelli, M., Sessa, G., Baima, S., Morelli, G., Ruberti, I., 1993. The *Arabidopsis* Athb-2 and -4 genes are strongly induced by far-red-rich light. Plant J. 4, 469–479.

Carabelli, M., Morelli, G., Whitelam, G., Ruberti, I., 1996. Twilight-zone and canopy shade induction of the Athb-2 homeobox gene in green plants. Proc. Natl. Acad. Sci. USA 93, 3530–3535.

Castillon, A., Shen, H., Huq, E., 2007. Phytochrome interacting factors: central players in phytochrome-mediated light signaling networks. Trends Plant Sci. 12, 514–521.

Charron, J.B., He, H., Elling, A.A., Deng, X.W., 2009. Dynamic landscapes of four histone modifications during deetiolation in *Arabidopsis*. Plant Cell 21, 3732–3748.

Chen, F., Li, B., Li, G., Charron, J.B., Dai, M., Shi, X., et al., 2014. *Arabidopsis* phytochrome A directly targets numerous promoters for individualized modulation of genes in a wide range of pathways. Plant Cell 26, 1949–1966.

Chory, J., Peto, C., Feinbaum, R., Pratt, L., Ausubel, F., 1989. *Arabidopsis thaliana* mutant that develops as a light-grown plant in the absence of light. Cell 58, 991–999.

Chory, J., Nagpal, P., Peto, C.A., 1991. Phenotypic and genetic analysis of det2, a new mutant that affects light-regulated seedling development in *Arabidopsis*. Plant Cell 3, 445–459.

Christie, J.M., 2007. Phototropin blue-light receptors. Annu. Rev. Plant Biol. 58, 21–45.

Cifuentes-Esquivel, N., Bou-Torrent, J., Galstyan, A., Gallemi, M., Sessa, G., Salla Martret, M., et al., 2013. The bHLH proteins BEE and BIM positively modulate the shade avoidance syndrome in *Arabidopsis* seedlings. Plant J. 75, 989–1002.

Ciolfi, A., Sessa, G., Sassi, M., Possenti, M., Salvucci, S., Carabelli, M., et al., 2013. Dynamics of the shade-avoidance response in *Arabidopsis*. Plant Physiol. 163, 331–353.

Cluis, C.P., Mouchel, C.F., Hardtke, C.S., 2004. The *Arabidopsis* transcription factor HY5 integrates light and hormone signaling pathways. Plant J. 38, 332–347.

de Lucas, M., Daviere, J.M., Rodriguez-Falcon, M., Pontin, M., Iglesias-Pedraz, J.M., Lorrain, S., et al., 2008. A molecular framework for light and gibberellin control of cell elongation. Nature 451, 480–484.

Deng, X.W., Caspar, T., Quail, P.H., 1991. cop1: a regulatory locus involved in light-controlled development and gene expression in *Arabidopsis*. Genes Dev. 5, 1172–1182.

Duek, P.D., Fankhauser, C., 2003. HFR1, a putative bHLH transcription factor, mediates both phytochrome A and cryptochrome signalling. Plant J. 34, 827–836.

Duek, P.D., Elmer, M.V., van Oosten, V.R., Fankhauser, C., 2004. The degradation of HFR1, a putative bHLH class transcription factor involved in light signaling, is regulated by phosphorylation and requires COP1. Curr. Biol. 14, 2296–2301.

Fairchild, C.D., Schumaker, M.A., Quail, P.H., 2000. HFR1 encodes an atypical bHLH protein that acts in phytochrome A signal transduction. Genes Dev. 14, 2377–2391.

Fankhauser, C., Chory, J., 2000. RSF1, an *Arabidopsis* locus implicated in phytochrome A signaling. Plant Physiol. 124, 39–45.

Feng, S., Martinez, C., Gusmaroli, G., Wang, Y., Zhou, J., Wang, F., et al., 2008. Coordinated regulation of *Arabidopsis thaliana* development by light and gibberellins. Nature 451, 475–479.

Foreman, J., Johansson, H., Hornitschek, P., Josse, E.M., Fankhauser, C., Halliday, K.J., 2011a. Light receptor action is critical for maintaining plant biomass at warm ambient temperatures. Plant J. 65, 441–452.

Foreman, J., White, J., Graham, I., Halliday, K., Josse, E.M., 2011b. Shedding light on flower development: phytochrome B regulates gynoecium formation in association with the transcription factor SPATULA. Plant Signal. Behav. 6, 471–476.

Franklin, K.A., Quail, P.H., 2010. Phytochrome functions in *Arabidopsis* development. J. Exp. Bot. 61, 11–24.

Franklin, K.A., Lee, S.H., Patel, D., Kumar, S.V., Spartz, A.K., Gu, C., et al., 2011. Phytochrome-interacting factor 4 (PIF4) regulates auxin biosynthesis at high temperature. Proc. Natl. Acad. Sci. USA 108, 20231–20235.

Fujimori, T., Yamashino, T., Kato, T., Mizuno, T., 2004. Circadian-controlled basic/helix–loop–helix factor, PIL6, implicated in light-signal transduction in *Arabidopsis thaliana*. Plant Cell Physiol. 45, 1078–1086.

Galstyan, A., Cifuentes-Esquivel, N., Bou-Torrent, J., Martinez-Garcia, J.F., 2011. The shade avoidance syndrome in *Arabidopsis*: a fundamental role for atypical basic helix–loop–helix proteins as transcriptional cofactors. Plant J. 66, 258–267.

Galstyan, A., Bou-Torrent, J., Roig-Villanova, I., Martinez-Garcia, J.F., 2012. A dual mechanism controls nuclear localization in the atypical basic-helix–loop–helix protein PAR1 of *Arabidopsis thaliana*. Mol. Plant 5, 669–677.

Hao, Y., Oh, E., Choi, G., Liang, Z., Wang, Z.Y., 2012. Interactions between HLH and bHLH factors modulate light-regulated plant development. Mol. Plant 5, 688–697.

Hardtke, C.S., Gohda, K., Osterlund, M.T., Oyama, T., Okada, K., Deng, X.W., 2000. HY5 stability and activity in *Arabidopsis* is regulated by phosphorylation in its COP1 binding domain. EMBO J. 19, 4997–5006.

Holm, M., Ma, L.G., Qu, L.J., Deng, X.W., 2002. Two interacting bZIP proteins are direct targets of COP1-mediated control of light-dependent gene expression in *Arabidopsis*. Genes Dev. 16, 1247–1259.

Hornitschek, P., Lorrain, S., Zoete, V., Michielin, O., Fankhauser, C., 2009. Inhibition of the shade avoidance response by formation of non-DNA binding bHLH heterodimers. EMBO J. 28, 3893–3902.

Hornitschek, P., Kohnen, M.V., Lorrain, S., Rougemont, J., Ljung, K., Lopez-Vidriero, I., et al., 2012. Phytochrome interacting factors 4 and 5 control seedling growth in changing light conditions by directly controlling auxin signaling. Plant J. 71, 699–711.

Huq, E., Quail, P.H., 2002. PIF4, a phytochrome-interacting bHLH factor, functions as a negative regulator of phytochrome B signaling in *Arabidopsis*. EMBO J. 21, 2441–2450.

Huq, E., Al-Sady, B., Hudson, M., Kim, C., Apel, K., Quail, P.H., 2004. Phytochrome-interacting factor 1 is a critical bHLH regulator of chlorophyll biosynthesis. Science 305, 1937–1941.

Hyun, Y., Lee, I., 2006. KIDARI, encoding a non-DNA binding bHLH protein, represses light signal transduction in *Arabidopsis thaliana*. Plant Mol. Biol. 61, 283–296.

Jang, I.C., Yang, J.Y., Seo, H.S., Chua, N.H., 2005. HFR1 is targeted by COP1 E3 ligase for post-translational proteolysis during phytochrome A signaling. Genes Dev. 19, 593–602.

Jiao, Y., Lau, O.S., Deng, X.W., 2007. Light-regulated transcriptional networks in higher plants. Nat. Rev. Genet. 8, 217–230.

Josse, E.M., Gan, Y., Bou-Torrent, J., Stewart, K.L., Gilday, A.D., Jeffree, C.E., et al., 2011. A DELLA in disguise: SPATULA restrains the growth of the developing *Arabidopsis* seedling. Plant Cell 23, 1337–1351.

Khanna, R., Huq, E., Kikis, E.A., Al-Sady, B., Lanzatella, C., Quail, P.H., 2004. A novel molecular recognition motif necessary for targeting photoactivated phytochrome signaling to specific basic helix–loop–helix transcription factors. Plant Cell 16, 3033–3044.

Khanna, R., Shen, Y., Marion, C.M., Tsuchisaka, A., Theologis, A., Schafer, E., et al., 2007. The basic helix–loop–helix transcription factor PIF5 acts on ethylene biosynthesis and phytochrome signaling by distinct mechanisms. Plant Cell 19, 3915–3929.

Kidokoro, S., Maruyama, K., Nakashima, K., Imura, Y., Narusaka, Y., Shinwari, Z.K., et al., 2009. The phytochrome-interacting factor PIF7 negatively regulates DREB1 expression under circadian control in *Arabidopsis*. Plant Physiol. 151, 2046–2057.

Kim, J., Yi, H., Choi, G., Shin, B., Song, P.S., 2003. Functional characterization of phytochrome interacting factor 3 in phytochrome-mediated light signal transduction. Plant Cell 15, 2399–2407.

Koini, M.A., Alvey, L., Allen, T., Tilley, C.A., Harberd, N.P., Whitelam, G.C., et al., 2009. High temperature-mediated adaptations in plant architecture require the bHLH transcription factor PIF4. Curr. Biol. 19, 408–413.

Koornneef, M., van der Veen, J.H., 1980. Induction and analysis of gibberellin sensitive mutants in *Arabidopsis thaliana* (L.) heynh. Theor. Appl. Genet. 58, 257–263.

Kumar, S.V., Lucyshyn, D., Jaeger, K.E., Alos, E., Alvey, E., Harberd, N.P., et al., 2012. Transcription factor PIF4 controls the thermosensory activation of flowering. Nature 484, 242–245.

Laubinger, S., Fittinghoff, K., Hoecker, U., 2004. The SPA quartet: a family of WD-repeat proteins with a central role in suppression of photomorphogenesis in *Arabidopsis*. Plant Cell 16, 2293–2306.

Lee, J., He, K., Stolc, V., Lee, H., Figueroa, P., Gao, Y., et al., 2007. Analysis of transcription factor HY5 genomic binding sites revealed its hierarchical role in light regulation of development. Plant Cell 19, 731–749.

Leivar, P., Quail, P.H., 2011. PIFs: pivotal components in a cellular signaling hub. Trends Plant Sci. 16, 19–28.

Leivar, P., Monte, E., Al-Sady, B., Carle, C., Storer, A., Alonso, J.M., et al., 2008a. The *Arabidopsis* phytochrome-interacting factor PIF7, together with PIF3 and PIF4, regulates responses to prolonged red light by modulating phyB levels. Plant Cell 20, 337–352.

Leivar, P., Monte, E., Oka, Y., Liu, T., Carle, C., Castillon, A., et al., 2008b. Multiple phytochrome-interacting bHLH transcription factors repress premature seedling photomorphogenesis in darkness. Curr. Biol. 18, 1815–1823.

Leivar, P., Monte, E., Cohn, M.M., Quail, P.H., 2012a. Phytochrome signaling in green *Arabidopsis* seedlings: impact assessment of a mutually negative phyB–PIF feedback loop. Mol. Plant 5, 734–749.

Leivar, P., Tepperman, J.M., Cohn, M.M., Monte, E., Al-Sady, B., Erickson, E., et al., 2012b. Dynamic antagonism between phytochromes and PIF family basic helix–loop–helix factors induces selective reciprocal responses to light and shade in a rapidly responsive transcriptional network in *Arabidopsis*. Plant Cell 24, 1398–1419.

Li, J., Li, G., Gao, S., Martinez, C., He, G., Zhou, Z., et al., 2010. *Arabidopsis* transcription factor ELONGATED HYPOCOTYL5 plays a role in the feedback regulation of phytochrome A signaling. Plant Cell 22, 3634–3649.

Li, L., Ljung, K., Breton, G., Schmitz, R.J., Pruneda-Paz, J., Cowing-Zitron, C., et al., 2012. Linking photoreceptor excitation to changes in plant architecture. Genes Dev. 26, 785–790.

Lin, C., Shalitin, D., 2003. Cryptochrome structure and signal transduction. Annu. Rev. Plant Biol. 54, 469–496.

Lorrain, S., Fankhauser, C., 2012. Plant development: should I stop or should I grow? Curr. Biol. 22, R645–R647.

Lorrain, S., Allen, T., Duek, P.D., Whitelam, G.C., Fankhauser, C., 2008. Phytochrome-mediated inhibition of shade avoidance involves degradation of growth-promoting bHLH transcription factors. Plant J. 53, 312–323.

Luo, Q., Lian, H.L., He, S.B., Li, L., Jia, K.P., Yang, H.Q., 2014. COP1 and phyB physically interact with PIL1 to regulate its stability and photomorphogenic development in *Arabidopsis*. Plant Cell 26, 2441–2456.

Mara, C.D., Huang, T., Irish, V.F., 2010. The *Arabidopsis* floral homeotic proteins APETALA3 and PISTILLATA negatively regulate the BANQUO genes implicated in light signaling. Plant Cell 22, 690–702.

Martinez-Garcia, J.F., Huq, E., Quail, P.H., 2000. Direct targeting of light signals to a promoter element-bound transcription factor. Science 288, 859–863.

Martinez-Garcia, J.F., Galstyan, A., Salla-Martret, M., Cifuentes-Esquivel, N., Gallemí, M., Bou-Torrent, J., 2010. Regulatory components of shade avoidance syndrome. Adv. Bot. Res. 53, 65–116.

Misera, S., Muller, A.J., Weiland-Heidecker, U., Jurgens, G., 1994. The FUSCA genes of *Arabidopsis*: negative regulators of light responses. Mol. Gen. Genet. 244, 242–252.

Ni, M., Tepperman, J.M., Quail, P.H., 1998. PIF3, a phytochrome-interacting factor necessary for normal photoinduced signal transduction, is a novel basic helix–loop–helix protein. Cell 95, 657–667.

Nozue, K., Covington, M.F., Duek, P.D., Lorrain, S., Fankhauser, C., Harmer, S.L., et al., 2007. Rhythmic growth explained by coincidence between internal and external cues. Nature 448, 358–361.

Oh, E., Kim, J., Park, E., Kim, J.I., Kang, C., Choi, G., 2004. PIL5, a phytochrome-interacting basic helix–loop–helix protein, is a key negative regulator of seed germination in *Arabidopsis thaliana*. Plant Cell 16, 3045–3058.

Osterlund, M.T., Hardtke, C.S., Wei, N., Deng, X.W., 2000. Targeted destabilization of HY5 during light-regulated development of *Arabidopsis*. Nature 405, 462–466.

Oyama, T., Shimura, Y., Okada, K., 1997. The *Arabidopsis* HY5 gene encodes a bZIP protein that regulates stimulus-induced development of root and hypocotyl. Genes Dev. 11, 2983–2995.

Pacin, M., Legris, M., Casal, J.J., 2013. COP1 re-accumulates in the nucleus under shade. Plant J. 75, 631–641.

Park, E., Park, J., Kim, J., Nagatani, A., Lagarias, J.C., Choi, G., 2012. Phytochrome B inhibits binding of phytochrome-interacting factors to their target promoters. Plant J. 72, 537–546.

Parks, B.M., Quail, P.H., 1991. Phytochrome-deficient hy1 and hy2 long hypocotyl mutants of *Arabidopsis* are defective in phytochrome chromophore biosynthesis. Plant Cell 3, 1177–1186.

Penfield, S., Josse, E.M., Kannangara, R., Gilday, A.D., Halliday, K.J., Graham, I.A., 2005. Cold and light control seed germination through the bHLH transcription factor SPATULA. Curr. Biol. 15, 1998–2006.

Pfeiffer, A., Shi, H., Tepperman, J.M., Zhang, Y., Quail, P.H., 2014. Combinatorial complexity in a transcriptionally centered signaling hub in *Arabidopsis*. Mol. Plant 7, 1598–1618.

Reed, J.W., Nagpal, P., Poole, D.S., Furuya, M., Chory, J., 1993. Mutations in the gene for the red/far-red light receptor phytochrome B alter cell elongation and physiological responses throughout *Arabidopsis* development. Plant Cell 5, 147–157.

Reymond, M.C., Brunoud, G., Chauvet, A., Martinez-Garcia, J.F., Martin-Magniette, M.L., Moneger, F., et al., 2012. A light-regulated genetic module was recruited to carpel development in *Arabidopsis* following a structural change to SPATULA. Plant Cell 24, 2812–2825.

Roig-Villanova, I., Bou, J., Sorin, C., Devlin, P.F., Martinez-Garcia, J.F., 2006. Identification of primary target genes of phytochrome signaling. Early transcriptional control during shade avoidance responses in *Arabidopsis*. Plant Physiol. 141, 85–96.

Roig-Villanova, I., Bou-Torrent, J., Galstyan, A., Carretero-Paulet, L., Portoles, S., Rodriguez-Concepcion, M., et al., 2007. Interaction of shade avoidance and auxin responses: a role for two novel atypical bHLH proteins. EMBO J. 26, 4756–4767.

Ruiz-Sola, M.A., Rodriguez-Concepcion, M., 2012. Carotenoid biosynthesis in *Arabidopsis*: a colorful pathway. Arabidopsis Book 10, e0158.

Salter, M.G., Franklin, K.A., Whitelam, G.C., 2003. Gating of the rapid shade-avoidance response by the circadian clock in plants. Nature 426, 680–683.

Schroeder, D.F., Gahrtz, M., Maxwell, B.B., Cook, R.K., Kan, J.M., Alonso, J.M., et al., 2002. De-etiolated 1 and damaged DNA binding protein 1 interact to regulate *Arabidopsis* photomorphogenesis. Curr. Biol. 12, 1462–1472.

Sellaro, R., Yanovsky, M.J., Casal, J.J., 2011. Repression of shade-avoidance reactions by sunfleck induction of HY5 expression in *Arabidopsis*. Plant J. 68, 919–928.

Sellaro, R., Pacin, M., Casal, J.J., 2012. Diurnal dependence of growth responses to shade in *Arabidopsis*: role of hormone, clock, and light signaling. Mol. Plant 5, 619–628.

Seo, H.S., Yang, J.Y., Ishikawa, M., Bolle, C., Ballesteros, M.L., Chua, N.H., 2003. LAF1 ubiquitination by COP1 controls photomorphogenesis and is stimulated by SPA1. Nature 423, 995–999.

Seo, H.S., Watanabe, E., Tokutomi, S., Nagatani, A., Chua, N.H., 2004. Photoreceptor ubiquitination by COP1 E3 ligase desensitizes phytochrome A signaling. Genes Dev. 18, 617–622.

Serino, G., Deng, X.W., 2003. The COP9 signalosome: regulating plant development through the control of proteolysis. Annu. Rev. Plant Biol. 54, 165–182.

Sessa, G., Carabelli, M., Sassi, M., Ciolfi, A., Possenti, M., Mittempergher, F., et al., 2005. A dynamic balance between gene activation and repression regulates the shade avoidance response in *Arabidopsis*. Genes Dev. 19, 2811–2815.

Shen, Y., Khanna, R., Carle, C.M., Quail, P.H., 2007. Phytochrome induces rapid PIF5 phosphorylation and degradation in response to red-light activation. Plant Physiol. 145, 1043–1051.

Shi, H., Zhong, S., Mo, X., Liu, N., Nezames, C.D., Deng, X.W., 2013. HFR1 sequesters PIF1 to govern the transcriptional network underlying light-initiated seed germination in *Arabidopsis*. Plant Cell 25, 3770–3784.

Shin, J., Park, E., Choi, G., 2007. PIF3 regulates anthocyanin biosynthesis in an HY5-dependent manner with both factors directly binding anthocyanin biosynthetic gene promoters in *Arabidopsis*. Plant J. 49, 981–994.

Shin, J., Kim, K., Kang, H., Zulfugarov, I.S., Bae, G., Lee, C.H., et al., 2009. Phytochromes promote seedling light responses by inhibiting four negatively-acting phytochrome-interacting factors. Proc. Natl. Acad. Sci. USA 106, 7660–7665.

Sibout, R., Sukumar, P., Hettiarachchi, C., Holm, M., Muday, G.K., Hardtke, C.S., 2006. Opposite root growth phenotypes of hy5 versus hy5 hyh mutants correlate with increased constitutive auxin signaling. PLoS Genet. 2, e202.

Smith, H., 2000. Phytochromes and light signal perception by plants – an emerging synthesis. Nature 407, 585–591.

Soh, M.S., Kim, Y.M., Han, S.J., Song, P.S., 2000. REP1, a basic helix–loop–helix protein, is required for a branch pathway of phytochrome A signaling in *Arabidopsis*. Plant Cell 12, 2061–2074.

Somers, D.E., Fujiwara, S., 2009. Thinking outside the F-box: novel ligands for novel receptors. Trends Plant Sci. 14, 206–213.

Staudt, A.C., Wenkel, S., 2011. Regulation of protein function by 'microProteins'. EMBO Rep. 12, 35–42.

Stavang, J.A., Gallego-Bartolome, J., Gomez, M.D., Yoshida, S., Asami, T., Olsen, J.E., et al., 2009. Hormonal regulation of temperature-induced growth in *Arabidopsis*. Plant J. 60, 589–601.

Tepperman, J.M., Zhu, T., Chang, H.S., Wang, X., Quail, P.H., 2001. Multiple transcription-factor genes are early targets of phytochrome A signaling. Proc. Natl. Acad. Sci. USA 98, 9437–9442.

Tilbrook, K., Arongaus, A.B., Binkert, M., Heijde, M., Yin, R., Ulm, R., 2013. The UVR8 UV-B photoreceptor: perception, signaling and response. Arabidopsis Book 11, e0164.

Toledo-Ortiz, G., Huq, E., Rodriguez-Concepcion, M., 2010. Direct regulation of phytoene synthase gene expression and carotenoid biosynthesis by phytochrome-interacting factors. Proc. Natl. Acad. Sci. USA 107, 11626–11631.

Toledo-Ortiz, G., Johansson, H., Lee, K.P., Bou-Torrent, J., Stewart, K., Steel, G., et al., 2014. The HY5–PIF regulatory module coordinates light and temperature control of photosynthetic gene transcription. PLoS Genet. 10, e1004416.

von Arnim, A.G., Osterlund, M.T., Kwok, S.F., Deng, X.W., 1997. Genetic and developmental control of nuclear accumulation of COP1, a repressor of photomorphogenesis in *Arabidopsis*. Plant Physiol. 114, 779–788.

Xu, X., Paik, I., Zhu, L., Bu, Q., Huang, X., Deng, X.W., et al., 2014. PHYTOCHROME INTERACTING FACTOR1 enhances the E3 ligase activity of CONSTITUTIVE PHOTOMORPHOGENIC1 to synergistically repress photomorphogenesis in *Arabidopsis*. Plant Cell 26, 1992–2006.

Yamashino, T., Matsushika, A., Fujimori, T., Sato, S., Kato, T., Tabata, S., et al., 2003. A link between circadian-controlled bHLH factors and the APRR1/TOC1 quintet in *Arabidopsis thaliana*. Plant Cell Physiol. 44, 619–629.

Yanagawa, Y., Sullivan, J.A., Komatsu, S., Gusmaroli, G., Suzuki, G., Yin, J., et al., 2004. *Arabidopsis* COP10 forms a complex with DDB1 and DET1 *in vivo* and enhances the activity of ubiquitin conjugating enzymes. Genes Dev. 18, 2172–2181.

Yang, K.Y., Kim, Y.M., Lee, S., Song, P.S., Soh, M.S., 2003. Overexpression of a mutant basic helix–loop–helix protein HFR1, HFR1–deltaN105, activates a branch pathway of light signaling in *Arabidopsis*. Plant Physiol. 133, 1630–1642.

Yang, J., Lin, R., Sullivan, J., Hoecker, U., Liu, B., Xu, L., et al., 2005. Light regulates COP1-mediated degradation of HFR1, a transcription factor essential for light signaling in *Arabidopsis*. Plant Cell 17, 804–821.

Zhang, H., He, H., Wang, X., Yang, X., Li, L., Deng, X.W., 2011. Genome-wide mapping of the HY5-mediated gene networks in *Arabidopsis* that involve both transcriptional and post-transcriptional regulation. Plant J. 65, 346–358.

Zhou, P., Song, M., Yang, Q., Su, L., Hou, P., Guo, L., et al., 2014. Both PHYTOCHROME RAPIDLY REGULATED1 (PAR1) and PAR2 promote seedling photomorphogenesis in multiple light signaling pathways. Plant Physiol. 164, 841–852.

CHAPTER 22

What Do We Know about Homeodomain–Leucine Zipper I Transcription Factors? Functional and Biotechnological Considerations

Pamela A. Ribone, Matías Capella, Agustín L. Arce, Raquel L. Chan

Instituto de Agrobiotecnología del Litoral, Universidad Nacional del Litoral, CONICET, Santa Fe, Argentina

OUTLINE

22.1 HD–Zip Transcription Factors are Unique to Plants	344
22.2 Brief History of the Discovery of HD-Zip Transcription Factors	344
22.3 Expression Patterns of HD-Zip I Genes	344
22.4 Environmental Factors Regulate the Expression of HD-Zip I Encoding Genes	346
22.5 The Function of HD-Zip I TFs from Model Plants	346
22.5.1 Members of Clade I	346
22.5.2 Members of Clade II	348
22.5.3 Members of Clade III	348
22.5.4 Members of Clade V	348
22.5.5 Other HD-Zip I TFs	349
22.6 HD-Zip I TFs from Nonmodel Species	349
22.6.1 PhHD-ZIP and RhHB1 from Clade I	350
22.6.2 LeHB-1 from Clade III	350
22.6.3 CpHB6 and CpHB7 from Clade IV	350
22.6.4 HaHB1, VvHB13, PgHZ1, and GhHB1 from Clade V	350
22.6.5 Vrs1 from Clade VI	351
22.6.6 Other HD-Zip TFs	351
22.7 Divergent HD-Zip I Proteins from Nonmodel Plants	351
22.8 Knowledge Acquired from Ectopic Expressors	352
22.9 HD-Zip I TFs in Biotechnology	352
22.9.1 Improved Development and Yield Increase	353
22.9.2 Tolerance to Abiotic Stress	353
22.9.3 Tolerance to Abiotic Stress and Yield Improvement	353
22.10 Concluding Remarks	353
References	354

22.1 HD–ZIP TRANSCRIPTION FACTORS ARE UNIQUE TO PLANTS

Homeodomain (HD)-containing transcription factors (TFs) were first discovered in *Drosophila melanogaster* in 1983 (Gehring, 1987) and have subsequently been identified in many other species from all eukaryotic kingdoms. They were named after the homeotic abnormalities caused by mutations in these genes; however, many of them are not homeotic (Bürglin, 2011).

Knotted1, identified and isolated from maize (*Zea mays*) two decades ago, was the first homeobox-containing gene discovered in plants (Vollbrecht et al., 1991; Chapter 14). The name *Knotted* was given after its constitutive expression, which induces areas of irregular cell division along secondary veins of the lamina, generating knotted leaves.

After this discovery, a large number of HD-containing proteins were identified in many plant species. These genes were classified into 14 families according to the sequence similarity of the HDs and their unique codomains (Mukherjee et al., 2009; Chapter 6). One of these classes, termed the "homeodomain–leucine zipper" (HD-Zip), comprises proteins, which have a HD adjacent to a leucine–zipper (LZ) domain. Although these two motifs are widely spread in different eukaryotic kingdoms, this particular structure association with both motifs in a single molecule is unique to plants (Ruberti et al., 1991). The function of both domains is well known: the LZ enables proteins to form homo- or heterodimers; while the HD, immediately upstream of the LZ, is responsible for protein–DNA-specific recognition and interaction.

HD-Zip TFs are classified in four subfamilies, named I–IV, according to conservation of the HD-Zip domain and the presence of additional conserved motifs outside this structure (Chapter 7). The HD-Zip I subfamily, in particular, is composed of genes with a similar intron/exon distribution, a conserved homeobox (HB) sequence, and a more divergent LZ-encoding sequence. *In vitro* assays showed that these proteins bind the pseudopalindromic sequence CAAT(A/T)ATTG and that the dimerization through the LZ is absolutely necessary for binding. The absence of the LZ completely abolishes DNA-binding ability (Palena et al., 1999; Johannesson et al., 2001).

Other motifs outside the conserved HD-Zip have been discovered (Arce et al., 2011), leading to a new classification of HD-Zip I proteins based on the presence of conserved motifs in their carboxy-terminal regions (CTR) and amino-terminal regions (NTR; Figure 22.1). A previous classification was based according to the intron/exon distribution (Henriksson et al., 2005); for details of these structures see Chapter 7.

In *Arabidopsis*, the HD-Zip I subfamily has 17 members (*ATHB1/HAT5, ATHB3/HAT7, ATHB5–ATHB7, ATHB12, ATHB13, ATHB16, ATHB20–ATHB23, ATHB40, ATHB51–ATHB54*) while 14 members have been identified in rice (*OsHox4–OsHox6, OsHox8, OsHox12–OsHox14, OsHox16, OsHox20–OsHox25*; Ariel et al., 2007; Agalou et al., 2008).

22.2 BRIEF HISTORY OF THE DISCOVERY OF HD-ZIP TRANSCRIPTION FACTORS

In 1991, Ruberti and coworkers employed a strategy that was successful in isolating HB genes from *Caenorhabditis elegans* (Burglin et al., 1989), and isolated two HB-containing sequences from *Arabidopsis thaliana*. Curiously, in these genes, the HD-encoding sequence was immediately adjacent to another well-known motif, a LZ (Ruberti et al., 1991).

Since then, several HD-Zip-encoding genes have been isolated from *A. thaliana* and connected to different aspects of plant development (Mattsson et al., 1992; Schena and Davis, 1992; Carabelli et al., 1993; Schena et al., 1993). By 1994, *HaHB-1* had been isolated from *Helianthus annuus*, becoming the first HD-Zip gene described in a species other than *Arabidopsis* (Chan and Gonzalez, 1994). Many HD-Zip sequences were further identified in a wide variety of species like tomato (*Lycopersicum esculentum*; Meissener and Theres, 1995), carrot (*Daucus carota*; Kawahara et al., 1995), and rice (*Oryza sativa*; Meijer et al., 1997), among others.

Only in 2005, almost 15 years after the discovery of the HB gene *Knotted1* and the HD-Zip I *AtHB1*, the first genome-wide analysis of HD-Zip TFs was carried out using the rice genome completed in 2002 (Goff et al., 2002). Since then, studies describing the complement of HD-Zip proteins from a particular species, including *Arabidopsis*, have increased (Henriksson et al., 2005; Agalou et al., 2008; Hu et al., 2012; Xu et al., 2012). Nowadays, with the increasing abundance of sequenced plant genomes, it has become easier to identify, classify, and study the whole set of HD-Zip TF sequences in a species, rapidly increasing our knowledge about these proteins, their function, and their possible biotechnological applications.

22.3 EXPRESSION PATTERNS OF HD-ZIP I GENES

Usually, the first approach to studying a protein of unknown function is the analysis of its expression pattern. Both spatial and temporal expression patterns are useful, as well as monitoring the transcript levels of a gene under different conditions and after various treatments. The expression patterns of many HD-Zip I TFs have been well characterized, allowing us to identify the similarities and differences between them. Figure 22.2 schematizes the localization of expression of several *A. thaliana* HD-Zip I TFs under standard growth conditions.

Examples of similarities and differences can be observed, for instance, with *AtHB1*, *AtHB5*, *AtHB6*, and

22.3 EXPRESSION PATTERNS OF HD-ZIP I GENES

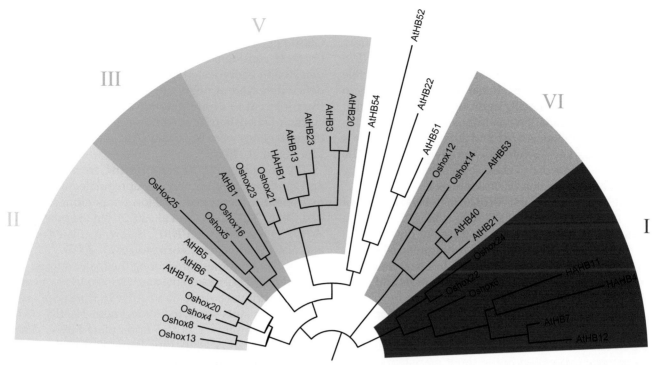

FIGURE 22.1 **Phylogenetic tree of HD-Zip I TFs from *A. thaliana*, *O. sativa*, and *H. annuus*.** The tree was constructed with full-length amino acid sequences. The most important HD-Zip I clades are described in detail in Arce et al. (2011).

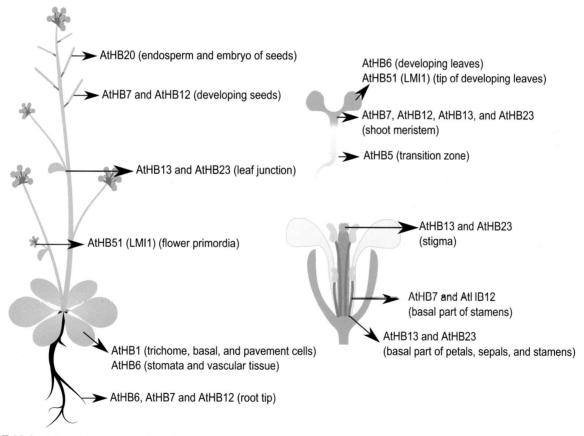

FIGURE 22.2 Schematic representation of a mature plant, a young seedling and a flower from *A. thaliana* showing the expression patterns of HD-Zip I TFs in this species according to the published literature and public databases.

C. FUNCTIONAL ASPECTS OF PLANT TRANSCRIPTION FACTOR ACTION

AtHB16 (all members of the *Arabidopsis* β-clade; Henriksson et al., 2005). These genes are ubiquitously found in all plant organs and differentially regulated by salinity stress and low temperatures: *AtHB1*, *AtHB5*, and *AtHB16* are downregulated by NaCl and cold temperatures while *AtHB6* is upregulated by these factors (Henriksson et al., 2005) and additionally by drought stress (Söderman et al., 1999). Illumination conditions also affect HD-Zip I protein expression: *AtHB1* and *AtHB16* expression is induced by darkness while blue light represses *AtHB16*, but other members of this clade seem to be unaffected by this kind of illumination (Henriksson et al., 2005).

Interestingly, *AtHB5* and *AtHB20*, members of different clades (Henriksson et al., 2005; Arce et al., 2011), share a common expression pattern with transient downregulation during seed development reverting to high induction 24–48 h after seedling imbibition, located around the transition zone, the edge between hypocotyls and roots (Johannesson et al., 2003; Barrero et al., 2009).

Paralogous genes deserve a special mention because they do not always show the same behavior. For example, the pair *AtHB13* and *AtHB23* exhibit the same expression pattern: expression in the shoot meristem region, leaf junction, basal part of petals, sepals, and stamens and within the stigma (Hanson et al., 2002; Kim et al., 2007), while a second pair of paralogous genes, *AtHB12* and *AtHB7*, exhibit a clearly different pattern: *AtHB12* is expressed during early seedling development whereas *AtHB7* is not. Higher levels of *AtHB7* were detected at later developmental stages and a complex mechanism in which the expression of both genes affects the expression of one another has been recently reported (Ré et al., 2014).

22.4 ENVIRONMENTAL FACTORS REGULATE THE EXPRESSION OF HD-ZIP I ENCODING GENES

Plants are sessile organisms and they need to adapt themselves to changing environmental conditions. Many different kinds of molecules are responsible for sensing the habitat, triggering a biochemical chain of cellular events leading to changes in the transcriptome, proteome, and metabolome. TFs are essential proteins responsible for repressing or activating signal transduction pathways, but first, their own transcription is regulated or, in other cases, they need posttranslational modifications. HD-Zip I TFs are involved in various stress responses and, therefore, are themselves regulated by stress-generating conditions.

Changes in illumination conditions, both quality and intensity, as well as water availability and extreme temperatures, alter the expression of most HD-Zip I proteins (Figure 22.3). For example, *AtHB12* and *AtHB7* are dramatically upregulated, especially in the vasculature, after abscisic acid (ABA) treatment or in plants subjected to drought (Söderman et al., 1996; Hjellström et al., 2003; Olsson et al., 2004). In rice, transcripts of *OsHox22* and *OsHox24*, considered *AtHB7* and *AtHB12* homologs, were detected in every tested tissue. However, after a water deficit they were strongly upregulated only in a rice cultivar tolerant to drought, while in a drought-sensitive cultivar (Zhenshan97) no differences in the expression of these genes were observed (Agalou et al., 2008; Nakashima et al., 2014).

The homologs *AtHB13* and *HaHB1*, from *Arabidopsis* and sunflower, respectively, are upregulated in response to drought, salinity stress, and low temperatures (Cabello and Chan, 2012; Cabello et al., 2012). On the other hand, *AtHB23*, the paralog of *AtHB13*, is downregulated by ABA and NaCl treatments (Henriksson et al., 2005).

Light and darkness are among the major stimuli able to modify the expression of some HD-Zip I TFs. As mentioned previously, *AtHB1* and *AtHB16* transcript levels increase in plants subjected to darkness. *HaHB4*, a HD-Zip I from sunflower, is also upregulated by darkness (Manavella et al., 2008b). Blue light, on the other hand, represses the expression levels of *AtHB16* while *AtHB1* is unaffected (Henriksson et al., 2005).

Some biotic factors also affect the expression of HD-Zip I genes. *HaHB4* and *NaHD20* expression levels increased after wounding and exogenous jasmonate (JA) application (Manavella et al., 2008a; Ré et al., 2011).

22.5 THE FUNCTION OF HD-ZIP I TFs FROM MODEL PLANTS

Two major organisms have been taken as model plants due to their relative small genomes, short life cycles, and ability to be transformed: *Arabidopsis* (*A. thaliana*) as dicotyledonous and rice (*O. sativa*) as monocotyledonous. It is worth mentioning that both species fix carbon via a C3 metabolic pathway. *Arabidopsis* is an annual weed with no agricultural use, belonging to the crucifer family (Brassicaceae), and closely related to agricultural plants like broccoli, cabbage, and radish. On the other hand, rice is a monocotyledonous crop with immense economic value and closely related to other crops such as wheat, maize, and barley.

For years, HD-Zip I proteins have been considered to be related to abiotic stress responses; however, current knowledge indicates that these kind of TFs are also involved in several developmental processes, and hormone related pathways, not necessarily associated with stress. Figure 22.4 summarizes the known functions of *Arabidopsis* HD-Zip I TFs.

22.5.1 Members of Clade I

HD-Zip I TFs from Clade I (Arce et al., 2011) have been the most studied, all being involved in abiotic stress

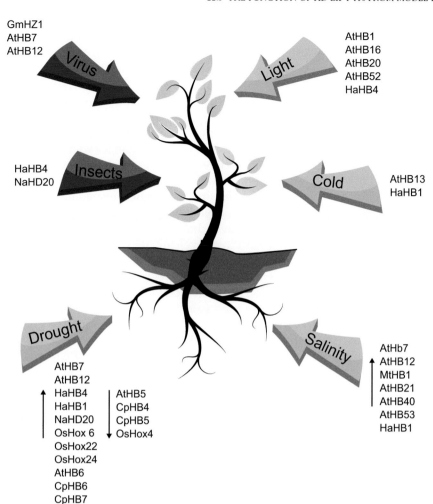

FIGURE 22.3 **Schematic representation showing the most common environmental factors affecting HD-Zip I TF expression.** Red and green arrows show biotic and abiotic stresses, respectively. ↑ represents upregulation, ↓ represents downregulation.

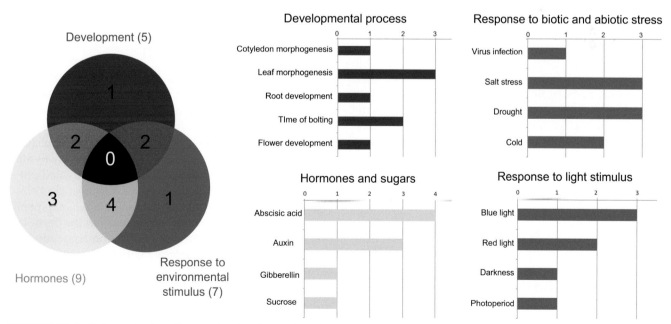

FIGURE 22.4 Left panels: Venn diagram showing the number of HD-Zip I TFs with a known function involved in plant development (red), response to environmental stimuli (blue), hormone signalization pathways (yellow) and their intersections. Right panel: Different functions and number of HD–Zip I members included in each of the previous categories.

responses, particularly those mediated by ABA (desiccation, osmotic, and salinity stresses). *AtHB7* and *AtHB12*, paralogous genes with high sequence similarity (about 80%), are strongly upregulated by ABA. When overexpressed in *Arabidopsis*, both confer a particular ABA-associated phenotype characterized by a hypersensitive response to this hormone in root elongation assays, a delay in inflorescence stem elongation, rounder rosette leaves, shorter petioles, and enhanced branching in the inflorescence stem (Olsson et al., 2004).

Two more putative functions have been proposed for *AtHB12*. The first one is as a TF involved in gibberellin-mediated responses; mutant *athb12* plants in a Wassilewskija background exhibit longer inflorescence stems and higher levels of *GA20ox2* than controls (Son et al., 2010). The second probable function assigned to *AtHB12* is in the activation of the response associated with the infection of beet severe curly top virus (BSCTV), a geminivirus affecting a number of monocotyledonous and dicotyledonous plants (Park et al., 2011).

OsHox22, *OsHox24*, and *OsHox6*, the rice members of this clade, are induced after water deficit in a rice cultivar tolerant to drought, while in a drought-sensitive one, no regulation at all was observed (Agalou et al., 2008; Nakashima et al., 2014). *OsHox22* is also upregulated by salt stress and ABA treatments (Zhang et al., 2012); and mutant plants in which it was knocked-down showed enhanced drought and salt tolerances, a lower ABA content than in controls, and hyposensitivity to exogenous ABA application. Altogether, these observations led to the conclusion that members of clade I have functions in both drought and salt responses that are mediated by ABA.

Two other HD-Zip I encoding genes from model organisms and homologs of *AtHB7* and *AtHB12* have been described: *MtHB1* from *Medicago truncatula* and *NaHD20* from *Nicotiana attenuata*. *MtHB1* is involved in root development, expressed in root meristems, and acts as a repressor of *MtLBD1* during environmental stress, inhibiting lateral root initiation in this species (Ariel et al., 2010a, 2010b). *NaHD20* is also involved in ABA-mediated signaling of water stress responses and has a crucial role in the emission of benzylacetone from flowers (Ré et al., 2011, 2012).

22.5.2 Members of Clade II

In this clade, we can find three HD-Zip I TFs from *A. thaliana* (*AtHB5*, *AtHB6*, and *AtHB16*) and four from rice (*OsHox3*, *OsHox4*, *OsHox8*, and *OsHox20*). As well as HD-Zip I TFs from clade I, *AtHB5* and *AtHB6* have been described as mediators in ABA responses. The former has been proposed as a positive regulator of the ABA response during seedling establishment in view of the increased sensitivity to ABA displayed by seedlings overexpressing *AtHB5* (Johannesson et al., 2003). On the other hand, *AtHB6* has been described as a negative regulator of the ABA signalization pathway (Himmelbach et al., 2002). It has been reported that *AtHB5* is a negative regulator of *BDL* expression, an AUXIN/INDOLE-3-ACETIC ACID (AUX/IAA) inhibitor involved in auxin responses (De Smet et al., 2013). However, the only hormone, among many different ones tested, which was able to down-regulate *AtHB5* was ABA, and no changes were observed after auxin, gibberellins (GA), brassinosteroid, or ethylene treatments (Johannesson et al., 2003).

AtHB16 has been linked to light sensing and light responses. Transgenic plants overexpressing *AtHB16* show smaller rosette leaves due to reduced leaf cell expansion and plants with low levels of *AtHB16* transcripts have the opposite phenotype. These observations indicate a role for *AtHB16* as a growth regulator. Transgenic plants of *AtHB16* also show an increased response in flowering time to photoperiod (considering that plants with high *AtHB16* transcript levels enter the reproductive phase later than WT plants in a long-day photoperiod, but earlier under short-day conditions). The opposite happens with transgenic plants under reduced levels of *AtHB16*. Altogether, this indicates a relationship between this TF and photoperiod sensing (Wang et al., 2003).

OsHox4, one of the rice members of this clade, was functionally characterized by Dai and coworkers in 2008. Based on analysis of rice plants overexpressing *OsHox4*, they discovered that this gene is involved in the negative regulation of GA-signaling pathways because transgenic plants showed a semidwarf phenotype despite accumulating high levels of bioactive GA. Moreover, overexpressing plants arrest the GA-dependent production of α-amylase (Dai et al., 2008).

22.5.3 Members of Clade III

Clade III has only one member in *A. thaliana*, *AtHB1/HAT5*, and three in rice, *OsHox5*, *OsHox16*, and *OsHox25*, but their function is largely unknown so far.

AtHB1 has been suggested to be involved in leaf development because overexpression of this gene in tobacco plants affected the development of palisade parenchyma. The same tobacco plants also showed a de-etiolated phenotype when growing in darkness (Aoyama et al., 1995). Hence, *AtHB1* has been proposed as a fundamental protein in the light-sensing mechanism that leads to the deetiolation of seedlings (Aoyama et al., 1995). Another study has reported that *AtHB1* can interact with TBP, a component of the basal transcription machinery (Capella et al., 2014).

22.5.4 Members of Clade V

Members of Clade V have been less studied than those of Clade I. Little is known about the function of *OsHox21* and *OsHox23*, the two rice members of this clade, with only a few studies of the *Arabidopsis* members published so far. *AtHB13* and *AtHB23* are encoded by paralogous genes and share 78% identity in the HD-Zip domain. Despite their high similarity, these two TFs seem to have very different functions. Studies carried out with plants overexpressing *AtHB13* showed that it is involved in the regulation of cotyledon and leaf development in response to carbon availability, quantified by sucrose concentration in early developmental stages. Overexpressing plants exhibit reduced lateral cell expansion in cotyledons only when sugar is present in the growth medium (Hanson et al., 2002). Other studies have shown that *AtHB13* is involved in abiotic stress responses; plants ectopically expressing this gene were able to tolerate freezing temperatures, severe drought, and salinity stresses by stabilizing cell membranes (Cabello and Chan, 2012; Cabello et al., 2012). On the other hand, *AtHB23* has been reported as a phytochrome B-interacting protein, involved in the PhyB-mediated red light signaling pathway. Mutant *athb23* plants displayed a particular phenotype characterized by long hypocotyls, defects in seed germination, and cotyledon expansion under red light. Thus, *AtHB23* has been suggested to be involved in red light sensing mediated by PhyB (Choi et al., 2013).

Two other genes from *A. thaliana* form this group, *AtHB3* and *AtHB20*, but they have been less studied than the others. *AtHB20* has been reported as an important protein in breaking seed dormancy and in ABA sensing, because *athb20* mutants displayed lower germination rates and increased sensitivity to ABA during seed germination (Barrero et al., 2010). *AtHB20* has also been also implicated in the organization of vascular tissues in response to auxin signals, since histochemical assays, performed with the *AtHB20* promoter region fused to *GUS*, showed staining around the emerging veins and rapid upregulation by IAA treatment (Mattsson et al., 2003). Almost nothing is known about the function of *AtHB3/HAT7*. It has been reported a rapid downregulation of *AtHB3* after seedling deetiolation under far-red (FR) light, but no further studies have been published so far (Roig-Villanova et al., 2006).

22.5.5 Other HD-Zip I TFs

AtHB51, also named *LATE MERISTEM IDENTITY 1* (*LMI1*), is a particular member of the subfamily that has not been resolved in any of the groups. It acts downstream of *LFY* as a meristem-identity gene, activating *CAL* expression (Saddic et al., 2006). It has been assigned a second role, independent of *LFY*, in leaf and bract development (Saddic et al., 2006). Further molecular analyses showed that *LMI1* is a positive regulator of *AGL24* and *SVP*, two genes controlling floral meristem identity (Grandi et al., 2012).

Another curious member of this subfamily is *AtHB52*, not resolved in any of the six clades. It was described as an important protein in the induction of callus formation and plant regeneration. *AtHB52* is upregulated in both aerial and root explants after 96 h in a callus-inducing medium, and overexpressing plants formed callus-like structures without exogenous addition of phytohormones (Xu et al., 2012).

22.6 HD-ZIP I TFs FROM NONMODEL SPECIES

It is accepted that HD-Zip proteins are unique to land plants since, so far, no HD-Zip genes have been found in species belonging to other kingdoms. Moreover, these genes were found in all divisions of vascular plants and in bryophytes (*Physcomitrella patens*; Sakakibara et al., 2001; Hu et al., 2012). In the moss *P. patens*, 33 HD-Zip genes have been identified including representatives of each of the four subfamilies, indicating that the divergence of vascular plants and moss lineages happened after the origin of these subfamilies (Sakakibara et al., 2001; Hu et al., 2012). On the other hand, the presence of HD-Zip genes has also been reported in the fern *Ceratopteris richardii*, in which 6 out of 11 HD-Zip TFs identified belong to the HD-Zip I subfamily (Aso et al., 1999).

Regarding plant species in which this kind of TF has been identified and, in some cases, functionally characterized (besides model plants), HD-Zip I proteins have been isolated from a broad variety of species including crops with great economical and/or nutritional importance such as tomato, sunflower, cotton, maize, soybean, poplar, and barley, among others (Chan and Gonzalez, 1994; Sterky et al., 1998; Wang et al., 2005; Komatsuda et al., 2007a; Lin et al., 2008; Ni et al., 2008; Zhao et al., 2011).

Taking advantage of wholly sequenced genomes, several "genome-wide scale" studies have been carried out on plants of economic interest such as maize, soybean, sorghum, grape, and poplar. Table 22.1 summarizes the number of HD-Zip I TFs found in each of these species. Four of them (maize, poplar, soybean, and grape) show segmental duplication leading to the expansion of the HD-Zip family because almost all the subfamily members found in these crops have a paralogous gene (Zhao et al., 2011; Hu et al., 2012; Chen et al., 2014; Wang et al., 2014).

As observed in model plants, many HD-Zip I genes from nonmodel species are induced or repressed by abiotic stress factors. For example, 21 of 23 HD-Zip I genes tested in soybean were found to be regulated by drought or salinity stresses (Chen et al., 2014); in maize, all the 17 HD-Zip I genes changed their expression levels in plants subjected to drought stress. Moreover, paralogous genes within a pair showed similar expression patterns

TABLE 22.1 Number of HD-Zip I Transcription Factors in Different Species

	Plant species	Number of HD-Zip I proteins	References
Bryophytes	*P. patens* (moss)	17	Hu et al. (2012)
Pteridophytes	*C. richardii* (fern)	6	Aso et al. (1999)
Monocots	*Brachypodium distachyon*	12	Hu et al. (2012)
	Sorghum bicolor (sorghum)	13	Hu et al. (2012)
	O. sativa (rice)	14	Agalou et al. (2008)
	Z. mays L. (maize)	17	Zhao et al. (2011)
Dicots	*A. thaliana*	17	Henriksson et al. (2005)
	M. truncatula	9	Hu et al. (2012)
	Populus trichocarpa (poplar)	21	Hu et al. (2012)
	Vitis vinifera (grape)	27	Hu et al. (2012)
	Solanum lycopersicum (tomato)	22	Zhang et al. (2014)
	G. max (soybean)	30	Chen et al. (2014)

(Zhao et al., 2011). All tomato HD-Zip I genes altered their expression in the face of cold stress; some were upregulated (13 of 22), while the rest were downregulated. Paralogous genes in species like maize showed similar behavior at low temperatures (Zhang et al., 2014).

Unfortunately, the present situation is that only one of the genes identified by these genome-wide scale studies was further characterized. *Zmhdz10* was isolated from maize and ectopically expressed in rice and *Arabidopsis* plants, in both cases bringing about enhanced tolerance to drought and salt stress and hypersensitivity to ABA (Zhao et al., 2014).

22.6.1 PhHD-ZIP and RhHB1 from Clade I

Several HD-Zip I TFs from nonmodel plants are involved in inflorescence architecture and flower development. For example, in maize, a homolog to *AtHB6* and a homolog to *AtHB21* are targets of *RA1* and *KN1*, two genes involved in the determination of inflorescence architecture in this species (Eveland et al., 2014).

In *Petunia hybrida* (petunia) and *Rosa hybrida* (rose), two HD-Zip I TFs involved in flower senescence have been found, one in each species (*PhHD-ZIP* and *RhHB1*, respectively). The silencing of both TFs independently produced a delay in flower senescence, a reduction in ethylene production and ABA synthesis, and lower expression of senescence-related genes like *SAG12* and *SAG29*. Both *PhHD-ZIP* and *RhHB1* are upregulated during flower senescence and, curiously, both are closely related to *AtHB12* and *AtHB7*, involved in ABA-mediated responses in *Arabidopsis* (Chang et al., 2014; Lü et al., 2014; Ré et al., 2014).

22.6.2 LeHB-1 from Clade III

LeHB-1 from tomato is an HD-Zip I TF that was associated with fruit morphogenesis. *LeHB-1* is an activator of *LeACO1*, an ACC oxidase gene that ultimately allows the ripening of tomato fruits. Ectopic expression of *LeHB-1* in tomato alters floral organ morphology while reduction of *LeHB-1* transcript levels inhibits fruit ripening indicating the crucial role played by this gene in this process (Lin et al., 2008).

22.6.3 CpHB6 and CpHB7 from Clade IV

The resurrection plant *Craterostigma plantagineum* exhibits extreme dehydration tolerance; so, study of this species is one of the strategies used to understand plant drought responses and protective mechanisms. *C. plantagineum* has four HD-Zip I TFs (*CpHB4–7*) and their expression is modulated by water deficit and exogenous ABA application. However, their behavior under dehydration is different; while *CpHB6* and *CpHB7* are upregulated, *CpHB4* and *CpHB5* are downregulated and no changes were observed after ABA treatment (Deng et al., 2002). These results suggested that *C. plantagineum* HD-Zip I TFs are involved in the response pathway that leads to desiccation tolerance. *CpHB7* was further studied and proposed as a negative regulator of the ABA response pathway, involved in early organ development (Deng et al., 2006).

22.6.4 HaHB1, VvHB13, PgHZ1, and GhHB1 from Clade V

HaHB1, from *H. annuus* (sunflower), is a well-characterized HD-Zip I that, as well as its homolog *AtHB13*, is involved in abiotic stress responses. The ectopic overexpression of any of these two genes in *Arabidopsis* plants induced the expression of proteins that stabilize membranes bringing about enhanced tolerance to drought, salinity, and low temperatures (Cabello and Chan, 2012; Cabello et al., 2012). Although no sunflower mutants are available to discard an artificial effect caused by ectopic expression, *athb13* mutants showed the opposite scenario when regulated genes and some phenotypical features observed in *HaHB1* or *AtHB13* overexpressors are considered (Cabello and Chan, 2012).

A further three genes encoding HD-Zip I TFs from nonmodel plants belonging to Clade V have been characterized: *VvHB13* from grape, *PgHZ1* from *Picea glauca*, and *GhHB1* from cotton. *PgHZ1* and *GhHB1* are both upregulated by ABA and transcripts of *GhHB1* are dramatically increased in roots when plants are subjected to salinity stress, suggesting that these genes are involved in abiotic stress responses like their homologs *AtHB13* or *HaHB1* (Ni et al., 2008; Tahir et al., 2008). *PgHZ1* may also be involved in embryonic development and growth (Tahir et al., 2008). On the other hand, *VvHB13* is highly and specifically expressed in fruit of grapes, particularly at those stages and in tissues that experience intense cell division (Fernandez et al., 2007).

In *Gossypium hirsutum* (cotton) a HD-Zip I TF, *GhHB1*, was isolated and characterized. Based on the expression pattern of this gene, it has been proposed to be involved in responses to salt stress and ABA signaling. *GhHB1* is mainly expressed in young roots and its expression decays with root development. On the other hand, treatments with NaCl or ABA dramatically increase *GhHB1* expression (Ni et al., 2008)

22.6.5 Vrs1 from Clade VI

A well-studied gene involved in the determination of inflorescence shape is *Vrs1* (*HvHOX1*) from barley. The barley progenitor *Hordeum vulgare* ssp. *spontaneum* has a two-rowed spike phenotype. During domestication of this crop, a six-rowed spike phenotype (*H. vulgare* ssp. *vulgare*) was selected because it yielded three times more grains. Molecular analysis of these varieties revealed that mutations in *Vrs1* were responsible for the six-rowed spike phenotype, suggesting that this TF is involved in the suppression of lateral spikelets in the two-rowed spike barley (Komatsuda et al., 2007a). Moreover, *Vrs1* has a paralog named *HvHox2*, both generated after duplication of an ancestral gene; however, while *HvHox2* is conserved in all cereals, *Vrs1* acquired its function during the evolution of barley (Sakuma et al., 2010). It has been found that both TFs are transcriptional activators capable of inducing expression of the same target genes in a yeast one-hybrid (Y1H) system. However, because their expression patterns have diverged, they exhibit different functions (Sakuma et al., 2013).

22.6.6 Other HD-Zip TFs

Two other HD-Zip I TFs from tomato, *VAHOX1* and *H52*, have been characterized. *VAHOX1* is expressed during secondary growth of phloem tissues (Tornero et al., 1996) while *H52* is upregulated after infection. Its function seems to limit the spread of programmed cell death, because transgenic plants with lower levels of *H52* exhibited a spontaneous initiation and propagation of cell death in leaves, accompanied by an accumulation of ethylene and salicylic acid (Mayda et al., 1999).

On the other hand, Wang and coworkers have isolated and characterized a HD-Zip I gene from *Glycine max* (soybean) and named it *GmHZ1*. When soybean plants were inoculated with the soybean mosaic virus (SMV) strain N3, *GmHZ1* expression increased in a variety susceptible to that virus, while in a resistant variety expression of this gene diminished, suggesting that *GmHZ1* could function as an activator of soybean responses to SMV infection (Wang et al., 2005).

22.7 DIVERGENT HD-ZIP I PROTEINS FROM NONMODEL PLANTS

Thanks to the increasing availability of sequence data, especially from nonmodel species, new phylogenetic trees using HD-Zip I proteins were constructed that resolved these proteins in six groups (Arce et al., 2011). The novelty of these new trees was the use of whole protein sequences that considered the regions outside conserved HD-Zip domains. Interestingly, model plants have HD-Zip I members in each group and all the groups have proteins from mono and dicotyledonous plants. However, a few TFs from this subfamily have not been resolved in any group. Among these divergent proteins, *HaHB4* and *HaHB11*, from the sunflower, were close to Group I, which is made up of divergent members. Their carboxy termini are shorter than those from model plants and did not present the same conserved motifs. This structural divergence suggests different functions for these sunflower TFs.

HaHB4, from *H. annuus* (sunflower), is probably the most studied HD-Zip I transcription factor apart from those from *Arabidopsis*. *HaHB4* is weakly expressed in several tissues/organs including roots, leaves, and flowers but it is noticeably upregulated by many treatments and hormones. Drought, osmotic stress, mechanical damage, darkness, salinity, abscisic acid, methyl-jasmonic acid, and ethylene are among external factors inducing expression of this divergent sunflower gene (Gago et al., 2002; Dezar et al., 2005; Manavella et al., 2006, 2008a, 2008b, 2008c).

Most of our knowledge of *HaHB4* was provided by the use of transgenic *A. thaliana* plants ectopically expressing this gene. These plants exhibit a differential phenotype including morphological and developmental traits as well as different stress responses. Transgenic *HaHB4* plants have shorter stems and internodes, rounder leaves, and compacter inflorescences than controls. In addition, they display strong tolerance to drought stress, senescence delay, less sensitivity to external ethylene, lower levels of chlorophyll, and higher transcript levels of defense-related genes conferring limited resistance to herbivory attack (Manavella et al., 2008a). Transcriptomic changes caused by overexpression or ectopic expression of *HaHB4* were validated by transient transformation of sunflower leaf disks with constructs able to overexpress or silence this gene. These experiments demonstrate the repressive effect of *HaHB4* over genes related to ethylene synthesis and signaling (*ACO*, *SAM*, *ERF2*, *ERF5*) and on photosynthesis-related genes, as well as an inductive effect on genes related to herbivory attack response (Manavella et al., 2006, 2008a, 2008b).

Based on this evidence, *HaHB4* has been proposed as a TF involved in a mechanism related to ethylene-mediated senescence, which functions to improve desiccation tolerance, as well as in the regulation of the photosynthetic machinery in darkness, and in the coordination of the production of phytohormones related to biotic stress.

HaHB11 is another sunflower divergent gene encoding a 175-amino-acid protein. Phylogenetic analysis shows that *AtHB7* and *AtHB12* are its closest *Arabidopsis* genes (Arce et al., 2011). *HaHB11* and *HaHB4* present 51% similarity in the HD-Zip domain between them. When considering the whole protein this similarity is only 31%.

HaHB11 was ectopically expressed in transgenic *Arabidopsis* plants conferring them multiple abiotic stress tolerance (Cabello et al., 2013). These transgenic plants performed better than controls when subjected to drought, salinity, waterlogging, and flooding. They had longer roots, larger rosettes, and improved yield than controls growing both under standard conditions or subjected to the abovementioned stresses (Cabello et al., 2013).

22.8 KNOWLEDGE ACQUIRED FROM ECTOPIC EXPRESSORS

A gene is ectopically expressed in a tissue/organ in which it is not programmed. Ectopic expression can be due to a disease or abnormality. It became an important tool in determination of the function of a certain gene. In plants the *35S Cauliflower Mosaic Virus* and ubiquitin promoters are most widely used to obtain constitutive and/or ectopic expression of a particular gene of interest.

Ectopic and/or constitutive expression of genes constituted a very important tool in the study of HD-Zip I TFs. For example, ectopic expression of *AtHB6* in *Arabidopsis* causes a reduction in stomatal closure and diminishes the inhibition of germination by ABA. These observations allowed Himmelbach et al. to propose *AtHB6* as a protein involved in ABA response pathways (Himmelbach et al., 2002). In fact, it was demonstrated that *AtHB6* is a target of *ABI1*, acting downstream of *ABI1* and *ABI2* (Himmelbach et al., 2002).

Another example is the study of transgenic plants harboring the construct *35S::AtHB52*, which allowed the authors to assign a role to this HD-Zip I TF in the regulatory network controlling the formation of pluripotent cells. This was because callus-like structures were spontaneously formed without the addition of plant hormones in these overexpressing plants (Xu et al., 2012).

Ectopic expression was also important for the investigation of proteins from nonmodel plants. Since most nonmodel plants are not easy to transform, the heterologous approach helped make a first approximation to functional characterization of a TF. An example of this is the inspection of the *CpHB7* function carried out by Deng et al. in 2006. *CpHB7* was ectopically expressed in tobacco plants revealing that this gene is involved in ABA sensing during germination and stomata closure (Deng et al., 2006). Another example is *Zmhdz10* from maize, which was ectopically expressed in both rice and *Arabidopsis* in order to determine its function (Zhao et al., 2014).

It is important to note that the conclusions drawn from this kind of study are limited. The ectopic and sometimes constitutive expression of a gene can cause some artificial effects, which can or cannot be related with the native function of the gene within the source plant.

22.9 HD-ZIP I TFs IN BIOTECHNOLOGY

The overexpression and transgenic expression of TFs trigger complex phenotypes in plants, probably as a result of the simultaneous regulation of multiple pathways. Some TFs seem to be potential biotechnological tools for improving agronomic crops, because the complex phenotypes generated sometimes included desirable agronomic traits. This is the reason in the last decades many patents involving the transgenic expression or silencing of plant TFs have been presented for uses including yield improvement, tolerance to a variety of biotic and abiotic stresses, and, of course, an assorted combination of these beneficial traits (Arce et al., 2008).

HD-Zip I TFs are not an exception to this trend and several patent applications include these genes as a core part of their inventions. The main claims in HD-Zip I patents are tolerance to abiotic stress, better development and yield, or a combination of both (Chan et al., 2004; Komatsuda et al., 2007b; Altpeter and Zhang, 2008; Sanz Molinero, 2008; Cabello et al., 2010, 2013; Lopato et al., 2010; Alves Ferreira et al., 2012; Guo et al., 2012; Chan and Gonzalez, 2013; Komatsuda et al., 2014).

22.9.1 Improved Development and Yield Increase

As an example of patents that include HD-Zip I TFs, there is a method for yield increase by silencing *Vrs1* in barley and its homolog in wheat. The invention describes a siRNA able to silence *Vrs1* and, as described earlier, this technique transformed the two-rowed spikelet into a six-rowed one, increasing grain production (Komatsuda et al., 2007b, 2014). A good example of using silencing as a strategy is the patent protecting a *ZmME293* RNAi to downregulate this maize gene. Plants ectopically expressing this silencing construct showed increased ear number per plant and faster leaf senescence and ear dry-down, leading to shorter life cycles without a compromise in yield (Guo et al., 2012).

Another example of a patent application involving a HD-Zip TF and a developmental trait is the one regarding *TaHDZipI-2* from *Triticum aestivum*, a gene with high sequence similarity to *AtHB13* from *A. thaliana*. Its overexpression is capable of modulating several aspects of cell wall development, mainly secondary cell wall deposition (Lopato et al., 2010).

Finally, a method for grass quality improvement has been developed involving the overexpression of *AtHB16* from *Arabidopsis*. It has been found that *AtHB16*, ectopically expressed in bahiagrass and related species, suppresses or reduces the formation of seedheads and increases the number of vegetative tillers per plant; moreover, it improves tolerance to abiotic stresses (Altpeter and Zhang, 2008).

22.9.2 Tolerance to Abiotic Stress

Among patents involving HD-Zip I TFs claiming tools or methods to improve crop tolerance to abiotic stress, are those protecting the ectopic expression of *Arabidopsis AtHB12* (Jong-Yoon Chun and Yong-Hun Lee, 1999), rice *OsHox5* and its homologs (Sanz Molinero, 2008), and *Coffea arabica CaHB12* (Alves Ferreira et al., 2012). In all these cases, overexpression of these genes enhanced tolerance to drought, salinity, or general abiotic stresses.

22.9.3 Tolerance to Abiotic Stress and Yield Improvement

The sunflower *HaHB4* HD-Zip gene was protected in 2003; this gene confers strong tolerance to drought both when expressed under the control of a constitutive promoter or its own promoter. A modified construct bearing a mutated *HaHB4* has been shown to confer not only tolerance to abiotic stresses but also increased yield under any condition (stress or standard) in three different crops: maize, wheat, and soybean (Chan and Gonzalez, 2013).

HaHB11, the other divergent HD-Zip gene from sunflower, described earlier, also has a dual function, improving plant tolerance to flooding, salinity, and drought and at the same time improving biomass and grain yield (Chan et al., 2013).

HaHB1, again from sunflower, has been found to be a protein capable of modifying plant response to freezing, drought, and salinity without yield loss; it too was protected along with its homolog *AtHB13* (Cabello et al., 2010).

Although the use of HD-Zip I TFs as biotechnological tools for a variety of purposes has been postulated, and many related patents have been presented, to date none of these TFs form part of a commercialized crop. It is tempting to speculate that in the near future these products will be the first choice for breeders for two main reasons: first, their efficacy conferring desirable traits to transgenic crops has been demonstrated; second, transgenic expression of a plant gene would be more accepted by the public than the use of genes from unrelated kingdoms.

22.10 CONCLUDING REMARKS

Although HD-Zip TFs were discovered more than two decades ago and extensively studied, their detailed functions and mechanisms of action in plants remain largely unknown.

So far, the evidence describes HD-Zip I TFs as key players in the interplay between environmental stimuli and modifications in growth patterns occurring in response to them. These stimuli could be both changes in the environment, sensed by plants during their normal growth, or changes in the environment that cause a stress, triggering plant responses to minimize the impact of this stress. For example, cold temperatures induce the expression of *HaHB1* and this allows the expression of proteins that stabilize membranes (Cabello et al., 2012).

The mechanistic aspects of HD-Zip I actions remain largely unknown. Little is known about the target genes of these TFs or the regulatory pathways in which they participate. It is known that almost all of these proteins, under *in vitro* conditions, can bind the pseudopalindromic sequence CAAT(A/T)ATTG, but further studies are needed in order to understand why, despite exhibiting similar expression patterns, two different HD-Zip I proteins can exert different functions. What else defines their target genes? Are posttranslational modifications involved? Is the interaction with other molecules what defines the promoters recognized by HD-Zip I TFs? Are other TFs the proteins that interact with them? Are changes in the nature of the members of the HD-Zip pair the decisive factor in the selection of target promoters?

Numerous questions still remain unanswered, and elucidation of their answers could be very useful in developing new biotechnological tools, like the ones described previously, as well as improving existing ones.

References

Agalou, A., Purwantomo, S., Övernäs, E., Johannesson, H., Zhu, X., Estiati, A., Kam, R., Engström, P., Slamet-Loedin, I., Zhu, Z., Wang, M., Xiong, L., Mijer, A., Ouwerkerk, P., 2008. A genome-wide survey of HD-Zip genes in rice and analysis of drought-responsive family members. Plant Mol. Biol. 66, 87–103.

Altpeter, F., Zhang, H., 2008. Materials and methods for improving quality and characteristics of grasses. WO2008021397A1.

Alves Ferreira, M., Pinheiro Da Cruz Waltenberg, F., Romano De Campos Pinto, E., Grossi de Sá, M., 2012. Use of the coffee homeobox gene CaHB12 to produce transgenic plants with greater tolerance to water scarcity and salt stress. WO2012061911A2.

Aoyama, T., Dong, C., Wu, Y., Carabelli, M., Sessa, G., Ruberti, I., Morelli, G., Chua, N., 1995. Ectopic expression of the Arabidopsis transcriptional activator Athb-1 alters leaf cell fate in tobacco. Plant Cell 7, 1773–1785.

Arce, A.L., Cabello, J.V., Chan, R.L., 2008. Patents on plant transcription factors. Recent Pat. Biotechnol. 2, 209–217.

Arce, A.L., Raineri, J., Capella, M., Cabello, J.V., Chan, R.L., 2011. Uncharacterized conserved motifs outside the HD-Zip domain in HD-Zip subfamily I transcription factors; a potential source of functional diversity. BMC Plant Biol. 11, 42.

Ariel, F., Manavella, P., Dezar, C., Chan, R., 2007. The true story of the HD-Zip family. Trends Plant Sci. 12, 419–426.

Ariel, F., Diet, A., Verdenaud, M., Gruber, V., Frugier, F., Chan, R., Crespi, M., 2010a. Environmental regulation of lateral root emergence in Medicago truncatula requires the HD-Zip I transcription factor HB1. Plant Cell 22, 2171–2183.

Ariel, F., Diet, A., Crespi, M., Chan, R., 2010b. The LOB-like transcription factor MtLBD1 controls Medicago truncatula root architecture under salt stress. Plant Signal. Behav. 5, 1666–1668.

Aso, K., Kato, M., Banks, J., Hasebe, M., 1999. Characterization of homeodomain–leucine zipper genes in the fern Ceratopteris richardii and the evolution of the homeodomain–leucine zipper gene family in vascular plants. Mol. Biol. Evol. 16, 544–552.

Barrero, J., Millar, A., Griffiths, J., Czechowski, T., Scheible, W., Udvardi, M., Reid, J., Ross, J., Jacobsen, J., Gubler, F., 2010. Gene expression profiling identifies two regulatory genes controlling dormancy and ABA sensitivity in Arabidopsis seeds. Plant J. 61, 611–622.

Barrero, J.M., Talbot, M.J., White, R.G., Jacobsen, J.V., Gubler, F., 2009. Anatomical and transcriptomic studies of the coleorhiza reveal the importance of this tissue in regulating dormancy in barley. Plant Physiol. 150, 1006–1021.

Bürglin, T.R., 2011. Homeodomain subtypes and functional diversity. Subcell. Biochem. 52, 95–122.

Bürglin, T.R., Finney, M., Coulson, A., Ruvkun, G., 1989. Caenorhabditis elegans has scores of homoeobox-containing genes. Nature 341, 239–243.

Cabello, J.V., Chan, R.L., 2012. The homologous homeodomain–leucine zipper transcription factors HaHB1 and AtHB13 confer tolerance to drought and salinity stresses via the induction of proteins that stabilize membranes. Plant Biotechnol. J. 10, 815–825.

Cabello, J.V., Arce, A.L., Chan, R.L., 2010. Methods and compositions for stress tolerance in plants. WO2010139993A1.

Cabello, J.V., Arce, A.L., Chan, R.L., 2012. The homologous HD-Zip I transcription factors HaHB1 and AtHB13 confer cold tolerance via the induction of pathogenesis-related and glucanase proteins. Plant J. 69, 141–153.

Cabello, J.V., Giacomelli, J.I., Chan, R.L., 2013. HaHB11 provides improved plant yield and tolerance to abiotic stress. WO2013116750A1.

Capella, M., Ré, D.A., Arce, A.L., Chan, R.L., 2014. Plant homeodomain–leucine zipper I transcription factors exhibit different functional AHA motifs that selectively interact with TBP or/and TFIIB. Plant Cell Rep. 33, 955–967.

Carabelli, M., Sessa, G., Baima, S., Morelli, G., Ruberti, I., 1993. The Arabidopsis Athb-2 and -4 genes are strongly induced by far-red-rich light. Plant J. 4, 469–479.

Chan, R.L., Gonzalez, D.H., 1994. A cDNA encoding an HD-Zip protein from sunflower. Plant Physiol. 106, 1687–1688.

Chan, R.L., Gonzalez, D.H., 2013. Modified Helianthus annuus transcription factor improves yield. WO2013126451A1.

Chan, R., Gonzalez, D., Dezar, C., Gago, G., 2004. Transcription factor gene induced by water deficit conditions and abscisic acid from Helianthus annuus, promoter and transgenic plants, WO20040993652A2.

Chan, R.L., Cabello, J.V., Giacomelli, J.I., 2013. HaHB11 provides improved plant yield and tolerance to abiotic stress. WO2013116750A1.

Chang, X., Donelly, L., Sun, D., Rao, J., Reid, M., Jiang, C., 2014. A petunia homeodomain–leucine zipper protein, PhHD-Zip, plays an important role in flower senescence. PLoS ONE 9, e88320.

Chen, X., Chen, Z., Zhao, H., Zhao, Y., Chang, B., Xiang, Y., 2014. Genome-wide analysis of soybean HD-Zip gene family and expression profiling under salinity and drought treatments. PLoS ONE 9, e87156.

Choi, H., Jeong, S., Kim, D., Na, H., Ryu, J., Lee, S., Nam, H., Lim, P., Woo, H., 2013. The homeodomain–leucine zipper ATHB23, a phytochrome B-interacting protein, is important for phytochrome B-mediated red light signaling. Physiol. Plant. 150, 308–320.

Dai, M., Hu, Y., Ma, Q., Zhao, Y., Zhou, D.X., 2008. Functional analysis of rice HOMEOBOX4 (Oshox4) gene reveals a negative function in gibberellin responses. Plant Mol. Biol. 66, 289–301.

De Smet, I., Lau, S., Ehrismann, J., Axiotis, I., Kolb, M., Kientz, M., Wiejers, D., Jürgens, G., 2013. Transcriptional repression of BODENLOS by HD-ZIP transcription factor HB5 in Arabidopsis thaliana. J. Exp. Bot. 64, 3009–3019.

Deng, X., Phillips, J., Meijer, A., Salamini, F., Bartels, D., 2002. Characterization of five novel dehydration-responsive homeodomain leucine zipper genes from the resurrection plant Craterostigma plantagineum. Plant Mol. Biol. 49, 601–610.

Deng, X., Phillips, J., Bräutigam, A., Engström, P., Johannesson, H., Ouwerkerk, P., Ruberti, I., Salinas, J., Vera, P., Iannacone, R., Meijer, A., Bartels, D., 2006. A homeodomain leucine zipper gene from Craterostigma plantagineum regulates abscisic acid responsive gene expression and physiological responses. Plant Mol. Biol. 61, 469–489.

Dezar, C.A., Gago, G., Gonzalez, D.H., Chan, R.L., 2005. Hahb-4, a sunflower homeobox–leucine zipper gene, is a developmental regulator and confers drought tolerance to Arabidopsis thaliana plants. Transgenic Res. 14, 429–440.

Eveland, A., Goldshmidt, A., Pautler, M., Morohashi, K., Liseron-Monfils, C., Lewis, M., Kumari, S., Hiraga, S., Yang, F., Unger-Wallace, E., Olson, A., Hake, S., Vollbrecht, E., Grotewold, E., Ware, D., Jackson, D., 2014. Regulatory modules controlling maize inflorescence architecture. Genome Res. 24, 431–443.

Fernandez, L., Torregrosa, L., Terrier, N., Sreekantan, L., Grimplet, J., Davies, C., Thomas, M., Romieu, C., Ageorges, A., 2007. Identification of genes associated with flesh morphogenesis during grapevine fruit development. Plant Mol. Biol. 63, 307–323.

Gago, G.M., Almoguera, C., Jordan, J., González, D.H., Chan, R.L., 2002. Hahb-4, a homeobox–leucine zipper gene potentially involved in ABA-dependent responses to water stress in sunflower. Plant Cell Environ. 25, 633–640.

Gehring, W.J., 1987. Homeoboxes in the study of development. Science 236, 1245–1252.

Goff, S.A., Ricke, D., Lan, T.H., Presting, G., Wang, R., Dunn, M., Glazebrook, J., Sessions, A., Oeller, P., Varma, H., Hadley, D., Hutchison, D., Martin, C., Katagiri, F., Lange, B.M., Moughamer, T., Xia, Y., Budworth, P., Zhong, J., Miguel, T., Paszkowski, U., Zhang, S., Colbert, M., Sun, W.L., Chen, L., Cooper, B., Park, S., Wood, T.C., Mao, L., Quail, P., Wing, R., Dean, R., Yu, Y., Zharkikh, A., Shen, R., Sahasrabudhe, S., Thomas, A., Cannings, R., Gutin, A., Pruss, D.,

Reid, J., Tavtigian, S., Mitchell, J., Eldredge, G., Scholl, T., Miller, R.M., Bhatnagar, S., Adey, N., Rubano, T., Tusneem, N., Robinson, R., Feldhaus, J., Macalma, T., Oliphant, A., Briggs, S., 2002. A draft sequence of the rice genome (Oryza sativa L. ssp. japonica). Science 296, 79–92.

Grandi, V., Gregis, V., Kater, M.M., 2012. Uncovering genetic and molecular interactions among floral meristem identity genes in Arabidopsis thaliana. Plant J. 69, 881–893.

Guo, M., Niu, X., Rupe, M., Schussler, J., 2012. Down-regulation of a homeodomain–leucine zipper I-class homeobox gene for improved plant performance. WO2012148835A1.

Hanson, J., Regan, S., Engström, P., 2002. The expression pattern of the homeobox gene ATHB13 reveals a conservation of transcriptional regulatory mechanisms between Arabidopsis and hybrid aspen. Plant Cell Rep. 21, 81–89.

Henriksson, E., Olsson, A., Johannesson, H., Johansson, H., Hanson, J., Engström, P., Söderman, E., 2005. Homeodomain–leucine zipper class I genes in Arabidopsis. Expression patterns and phylogenetic relationships. Plant Physiol. 139, 509–518.

Himmelbach, A., Hoffmann, T., Leube, M., Höhener, B., Grill, E., 2002. Homeodomain protein ATHB6 is a target of the protein phosphatase ABI1 and regulates hormone responses in Arabidopsis. EMBO J. 21, 3029–3038.

Hjellström, M., Olsson, A., Engström, P., Söderman, E., 2003. Constitutive expression of the water deficit-inducible homeobox gene ATHB7 in transgenic Arabidopsis causes a suppression of stem elongation growth. Plant Cell Environ. 26, 1127–1136.

Hu, R., Chi, X., Chai, G., Kong, Y., He, G., Wang, X., Shi, D., Zhang, D., Zhou, G., 2012. Genome-wide identification, evolutionary expansion, and expression profile of homeodomain–leucine zipper gene family in poplar (Populus trichocarpa). PLoS ONE 7, e31149.

Johannesson, H., Wang, Y., Engström, P., 2001. DNA-binding and dimerization preferences of Arabidopsis homeodomain–leucine zipper transcription factors in vitro. Plant Mol. Biol. 45, 63–73.

Johannesson, H., Wang, Y., Hanson, J., Engström, P., 2003. The Arabidopsis thaliana homeobox gene ATHB5 is a potential regulator of abscisic acid responsiveness in developing seedlings. Plant Mol. Biol. 51, 719–729.

Jong-Yoon Chun, K., Yong-Hun Lee, S., 1999. Transcription factor gene induced by water deficit and abscisic acid isolated from Arabidopsis thaliana. US patent 5,981,729.

Kawahara, R., Komamine, A., Fukuda, H., 1995. Isolation and characterization of homeobox-containing genes of carrot. Plant Mol. Biol. 27, 155–164.

Kim, Y.K., Son, O., Kim, M.R., Nam, K.H., Kim, G.T., Lee, M.S., Choi, S.Y., Cheon, C.I., 2007. ATHB23, an Arabidopsis class I homeodomain-leucine zipper gene, is expressed in the adaxial region of young leaves. Plant Cell Rep. 26, 1179–1185.

Komatsuda, T., Pourkheirandish, M., He, C., Azhaguvel, P., Kanamori, H., Perovic, D., Stein, N., Grane, A., Wicker, T., Tagiri, A., Lundqvist, U., Fujimura, T., Matsuoka, M., Matsumoto, T., Yano, M., 2007a. Six-rowed barley originated from a mutation in a homeodomain–leucine zipper I-class homeobox gene. Proc. Natl. Acad. Sci. USA 104, 1424–1429.

Komatsuda, T., Yano, M., Matsumoto, T., 2007b. Barley row type gene and use thereof. WO2007069677A1.

Komatsuda, T., Matsumoto, T., Sakuma, S., Ogawa, T., 2014. Wheat with increased grain-bearing number and method for producing same, and chemical for increasing grain-bearing number of wheat. WO2014007396A1.

Lin, Z., Hong, Y., Yin, M., Li, C., Zhang, K., Grierson, D., 2008. A tomato HD-Zip homeobox protein, LeHB-1, plays an important role in floral organogenesis and ripening. Plant J. 55, 301–310.

Lopato, S., Kovalshuk, N., Langridge, P., Shirley, N., 2010. Modulation of plant cell wall deposition via HDZIPI. WO2010034066A1.

Lü, P., Zhang, C., Liu, J., Liu, X., Jiang, G., Jiang, X., Khan, M., Wang, L., Hong, B., Gao, J., 2014. RhHB1 mediates the antagonism of gibberellins to ABA and ethylene during rose (Rosa hybrida) petal senescence. Plant J. 78, 578–590.

Manavella, P.A., Arce, A.L., Dezar, C.A., Bitton, F., Renou, J., Crespi, M., Chan, R.L., 2006. Cross-talk between ethylene and drought signalling pathways is mediated by the sunflower Hahb-4 transcription factor. Plant J. 48, 125–137.

Manavella, P.A., Dezar, C.A., Ariel, F.D., Drincovich, M.F., Chan, R.L., 2008a. The sunflower HD-Zip transcription factor HAHB4 is up-regulated in darkness, reducing the transcription of photosynthesis-related genes. J. Exp. Bot. 59, 3143–3155.

Manavella, P.A., Dezar, C.A., Ariel, F.D., Chan, R.L., 2008b. HAHB4, a sunflower HD-Zip protein, integrates signals from the jasmonic acid and ethylene pathways during wounding and biotic stress responses. Plant J. 56, 376–388.

Manavella, P.A., Dezar, C.A., Chan, R.L., 2008c. Two ABREs, two redundant root-specific and one W-box cis-acting elements are functional in the sunflower HAHB4 promoter. Plant Physiol. Biochem. 46, 860–867.

Mattsson, J., Söderman, E., Svenson, M., Borkird, C., Engström, P., 1992. A new homeobox–leucine zipper gene from Arabidopsis thaliana. Plant Mol. Biol. 18, 1019–1022.

Mattsson, J., Ckurshumova, W., Berleth, T., 2003. Auxin signaling in Arabidopsis leaf vascular development. Plant Physiol. 131, 1327–1329.

Mayda, E., Tornero, P., Conejero, V., Vera, P., 1999. A tomato homeobox gene (HD-Zip) is involved in limiting the spread of programmed cell death. Plant J. 20, 591–600.

Meijer, A., Scarpella, E., Van Dijk, E., Qin, L., Taal, A., Rued, S., Harrington, S., McCouch, S., Schilperoort, R., Hoge, J., 1997. Transcriptional repression by Oshox1, a novel homeodomain leucine zipper protein from rice. Plant J. 11, 263–276.

Meissner, R., Theres, K., 1995. Isolation and characterization of the tomato homeobox gene THOM1. Planta 195, 541–547.

Mukherjee, K., Brocchieri, L., Bürglin, T., 2009. A comprehensive classification and evolutionary analysis of plant homeobox genes. Mol. Biol. Evol. 26, 2775–2794.

Nakashima, K., Jan, A., Todaka, D., Maruyama, K., Goto, S., Shinozaki, K., Yamaguchi-Shinozaki, K., 2014. Comparative functional analysis of six drought-responsive promoters in transgenic rice. Planta 239, 47–60.

Ni, Y., Wang, X., Li, D., Wu, Y., Xu, W., Li, X., 2008. Novel cotton homeobox gene and its expression profiling in root development and in response to stresses and phytohormones. Acta Biochim. Biophys. Sin. Shanghai 40, 78–84.

Olsson, A., Engström, P., Söderman, E., 2004. The homeobox genes ATHB12 and ATHB7 encode potential regulators of growth in response to water deficit in Arabidopsis. Plant Mol. Biol. 55, 663–677.

Palena, C., Gonzalez, D.H., Chan, R.L., 1999. A monomer–dimer equilibrium modulates the interaction of the sunflower homeodomain leucine–zipper protein Hahb-4 with DNA. Biochem. J. 341, 81–87.

Park, J., Lee, H., Cheon, C., Kim, S., Hur, Y., Auh, C., Im, K., Yun, D., Lee, S., Davis, K., 2011. The Arabidopsis thaliana homeobox gene AtHB12 is involved in symptom development caused by geminivirus infection. PLoS ONE 6, e20054.

Ré, D.A., Dezar, C.A., Chan, R.L., Baldwin, I., Bonaventure, G., 2011. Nicotiana attenuata NaHD20 plays a role in leaf ABA accumulation during water stress, benzylacetone emission from flowers, and the timing of bolting and flower transitions. J. Exp. Bot. 62, 155–166.

Ré, D.A., Raud, B., Chan, R.L., Baldwin, I., Bonaventure, G., 2012. RNAi-mediated silencing of the HD-Zip gene HD20 in Nicotiana attenuata affects benzyl acetone emission from corollas via ABA levels and the expression of metabolic genes. BMC Plant Biol. 12, 60.

Ré, D.A., Capella, M., Bonaventure, G., Chan, R.L., 2014. Arabidopsis AtHB7 and AtHB12 evolved divergently to fine-tune processes associated with growth and responses to water stress. BMC Plant Biol. 14, 150.

Roig-Villanova, I., Bou, J., Sorin, C., Devlin, P.F., Martínez-García, J.F., 2006. Identification of primary target genes of phytochrome signaling. Early transcriptional control during shade avoidance responses in Arabidopsis. Plant Physiol. 141, 85–96.

Ruberti, I., Sessa, G., Luchetti, S., Morelli, G., 1991. A novel class of plant proteins containing a homeodomain with a closely linked leucine zipper motif. EMBO J. 10, 1787–1791.

Saddic, L., Huvermann, B., Bazhani, S., Su, Y., Winter, C., Kwon, C., Collum, R., Wagner, D., 2006. The LEAFY target LMI1 is a meristem identity regulator and acts together with LEAFY to regulate expression of *CAULIFLOWER*. Development 133, 1673–1682.

Sakakibara, K., Nishiyama, T., Kato, M., Hasebe, M., 2001. Isolation of homeodomain–leucine zipper genes from the moss *Physcomitrella patens* and the evolution of homeodomain–leucine zipper genes in land plants. Mol. Biol. Evol. 18, 491–502.

Sakuma, S., Pourkheirandish, M., Matsumoto, T., Koba, T., Komatsuda, T., 2010. Duplication of a well-conserved homeodomain–leucine zipper transcription factor gene in barley generates a copy with more specific functions. Funct. Integr. Genomics 10, 123–133.

Sakuma, S., Pourkheirandish, M., Hensel, G., Kumlehn, J., Stein, N., Tagiri, A., Yamaji, N., Ma, J., Sassa, H., Koba, T., Komatsuda, T., 2013. Divergence of expression pattern contributed to neofunctionalization of duplicated HD-Zip I transcription factor in barley. New Phytol. 197, 939–948.

Sanz Molinero, A.I., 2008. Plants having improved growth characteristics under reduced nutrient availability and a method for making the same. WO2008132231A1.

Schena, M., Davis, R.W., 1992. HD-Zip proteins: members of an *Arabidopsis* homeodomain protein superfamily. Proc. Natl. Acad. Sci. USA 89, 3894–3898.

Schena, M., Lloyd, A.M., Davis, R.W., 1993. The *HAT4* gene of *Arabidopsis* encodes a developmental regulator. Genes Dev. 7, 367–379.

Söderman, E., Mattsson, J., Engström, P., 1996. The *Arabidopsis* homeobox gene *AtHB-7* is induced by water deficit and by abscisic acid. Plant J. 10, 375–381.

Söderman, E., Hjellström, M., Fahleson, J., Engström, P., 1999. The HD-Zip gene *ATHB6* in *Arabidopsis* is expressed in developing leaves, roots and carpels and up-regulated by water deficit conditions. Plant Mol. Biol. 40, 1073–1083.

Son, O., Hur, Y.S., Kim, Y.K., Lee, H.J., Kim, S., Kim, M.R., Nam, K.H., Lee, M.S., Kim, B.Y., Park, J., Park, J., Lee, S.C., Hanada, A., Yamaguchi, S., Lee, I.J., Kim, S.K., Yun, D.J., Söderman, E., Cheon, C.I., 2010. ATHB12, an ABA-inducible homeodomain–leucine zipper (HD-Zip) protein of *Arabidopsis*, negatively regulates the growth of the inflorescence stem by decreasing the expression of a gibberellin 20-oxidase gene. Plant Cell Physiol. 51, 1537–1547.

Sterky, F., Regan, S., Karlsson, J., Hertzberg, M., Rohde, A., Holmberg, A., Amini, B., Bhaleraos, R., Larsson, M., Villarroel, R., Van Montagu, M., Sandberg, G., Olsson, O., Teeri, T., Boerjan, W., Gustafsson, P., Uhlen, M., Sundberg, B., Lundeberg, J., 1998. Gene discovery in the wood-forming tissues of poplar: analysis of 5,692 expressed sequence tags. Proc. Natl. Acad. Sci. USA 95, 13330–13335.

Tahir, M., Belmonte, M., Elhiti, M., Flood, H., Stasolla, C., 2008. Identification and characterization of PgHZ1, a novel homeodomain leucine–zipper gene isolated from white spruce (*Picea glauca*) tissue. Plant Physiol. Biochem. 46, 1031–1039.

Tornero, P., Conejero, V., Vera, P., 1996. Phloem-specific expression of a plant homeobox gene during secondary phases of vascular development. Plant J. 9, 639–648.

Vollbrecht, E., Veit, B., Sinha, N., Hake, S., 1991. The developmental gene *Knotted-1* is a member of a maize homeobox gene family. Nature 350, 241–243.

Wang, Y., Henriksson, E., Söderman, E., Henriksson, K., Sundberg, E., Engström, P., 2003. The *Arabidopsis* homeobox gene, *ATHB16*, regulates leaf development and the sensitivity to photoperiod in *Arabidopsis*. Dev. Biol. 264, 228–239.

Wang, Y.J., Li, Y., Luo, G., Tian, A., Wang, H., Zhang, J., Chen, S., 2005. Cloning and characterization of an HDZip I gene *GmHZ1* from soybean. Planta 221, 831–843.

Wang, H., Yin, X., Li, X., Wang, L., Zheng, Y., Xu, X., Zhang, Y., Wang, X., 2014. Genome-wide identification, evolution and expression analysis of the grape (*Vitis vinifera* L.) zinc finger–homeodomain gene family. Int. J. Mol. Sci. 15, 5730–5748.

Xu, K., Liu, J., Fan, M., Xin, W., Hu, Y., Xu, C., 2012. A genome-wide transcriptome profiling reveals the early molecular events during callus initiation in *Arabidopsis* multiple organs. Genomics 100, 116–124.

Zhang, S., Haider, I., Kohlen, W., Jiang, L., Bouwmeester, H., Meijer, A., Schluepmann, H., Liu, C., Ouwerkerk, P., 2012. Function of the HD-Zip I gene *Oshox22* in ABA-mediated drought and salt tolerances in rice. Plant Mol. Biol. 80, 571–585.

Zhang, Z., Chen, X., Guan, X., Liu, Y., Chen, H., Wang, T., Mouekouba, L., Li, J., Wang, A., 2014. A genome-wide survey of homeodomain–leucine zipper genes and analysis of cold-responsive HD-Zip I members' expression in tomato. Biosci. Biotechnol. Biochem., 1–13.

Zhao, Y., Zhou, Y., Jiang, H., Li, X., Gan, D., Peng, X., Zhu, S., Cheng, B., 2011. Systematic analysis of sequences and expression patterns of drought-responsive members of the HD-Zip gene family in maize. PLoS ONE 6, e28488.

Zhao, Y., Ma, Q., Jin, X., Peng, X., Liu, J., Deng, L., Yan, H., Sheng, L., Jiang, H., Cheng, B., 2014. A novel maize homeodomain–leucine zipper (HD-Zip) gene, *Zmhdz10*, positively regulates drought and salt tolerance in both rice and *Arabidopsis*. Plant Cell Physiol. 55, 1142–1156.

SECTION D

MODULATION OF PLANT TRANSCRIPTION FACTOR ACTION

23 *Intercellular movement of plant transcription factors, coregulators, and their mRNAs* — *359*

24 *Redox-regulated plant transcription factors* — *373*

25 *Membrane-bound transcription factors in plants: physiological roles and mechanisms of action* — *385*

26 *Ubiquitination of plant transcription factors* — *395*

CHAPTER 23

Intercellular Movement of Plant Transcription Factors, Coregulators, and Their mRNAs

David J. Hannapel

Plant Biology Major, Iowa State University, Ames, Iowa, USA

OUTLINE

23.1 Introduction to Noncell-autonomous Mobile Signals 359	23.5 Full-length Mobile mRNAs and their Roles in Development 363
23.2 Mobile Transcription Factors of the Shoot Apex in Protein Form 360	23.5.1 Long-Distance Transport of GA INSENSITIVE 363
23.2.1 Intercellular Movement of KNOTTED1 360	23.5.2 ATC mRNA is Mobile Through the Phloem 363
23.2.2 WUSCHEL in the Shoot Apical Meristem 360	23.5.3 FT mRNA Moves to the Shoot Apex 365
23.3 Mobile Root Transcription Factors 360	23.5.4 StBEL5, a Mobile RNA of Potato 367
23.4 Transcription Factors and Coregulators that Move Long Distance Through the Sieve Element System 361	23.6 Conclusions 369
23.4.1 FLOWERING LOCUS T 361	References 369
23.4.2 The Long-Distance Tuberization Signal 362	
23.4.3 StFT/StSP6A 362	

23.1 INTRODUCTION TO NONCELL-AUTONOMOUS MOBILE SIGNALS

In response to ever-changing environmental conditions, plants have evolved numerous distance signaling pathways, many of which involve transcription factors (TFs) or coregulators. As an extension of the endoplasmic reticulum, plasmodesmata (PD) function as transport channels through the cell wall. These connections are operational in the movement of photosynthate, small RNAs, viruses, and numerous signaling molecules (Lucas et al., 2009). The companion cell/sieve element system functions as a remarkably efficient conduit for transferring numerous signal molecules throughout the plant.

Proteins, metabolites, small RNAs, and messenger RNAs move readily through the sieve element system to regulate development and respond to environmental cues. Research over the past 10 years has demonstrated the movement of numerous TFs through the PD in the form of both proteins and RNAs. This transport can be categorized as either short range from cell to cell or long range throughout the plant via the sieve element system.

Some of the best examples of short-range movement include KNOTTED1 (KN1), WUSCHEL (WUS), and SHORT-ROOT (SHR). Prominent examples of long-distance movement are characterized by *Flowering Locus T* (FT) protein that mediates flowering (reviewed by Turck et al., 2008) and tuberization (Navarro et al., 2011) and *GA*

Plant Transcription Factors. http://dx.doi.org/10.1016/B978-0-12-800854-6.00023-3
Copyright © 2016 Elsevier Inc. All rights reserved.

INSENSITIVE (*GAI*) and *StBEL5* transcripts that regulate leaf architecture and tuberization, respectively (Haywood et al., 2005; Banerjee et al., 2006). This chapter will provide an overview of TFs that move short range from cell to cell and a coregulator that traffics long distance through the phloem. In most cases, the protein is mobile but reports continue to accumulate on examples of the transport of full-length mRNAs that regulate development. The first three sections address three prominent models on noncell-autonomous TFs. The remaining sections address models on the movement of the coregulator, FT, and mRNAs that encode both coregulators and TFs.

23.2 MOBILE TRANSCRIPTION FACTORS OF THE SHOOT APEX IN PROTEIN FORM

23.2.1 Intercellular Movement of KNOTTED1

KN1, a maize homeodomain protein, functions in regulating meristem organization and plant architecture and was the first plant protein found to traffic cell to cell (Lucas et al., 1995; Kim et al., 2005; Chapter 14). Subsequent studies have shown that class I KNOX proteins promote transport of the KN1 mRNA and that this trafficking occurs through PD (Lucas et al., 1995). This transport may occur in layers of the shoot meristem and from internal tissues of the leaf to the epidermis and is tightly regulated. Movement through the PD is mediated by unfolding of the protein through an interaction with the CCT8 chaperonin protein (Kragler et al., 1998; Xu et al., 2011) and by an interaction with a protein partner that binds to the trihelical homeodomain region of KN1 (Bolduc et al., 2008). Movement protein-binding protein 2C was identified as a protein that interacts with the KN1 homeodomain and negatively regulates the cell-to-cell trafficking of KN1 by targeting the protein to microtubules (Winter et al., 2007; Bolduc et al., 2008). These results suggest that KN1 has multiple potential cellular locations, each of which is controlled by its homeodomain. Chen et al. (2014) have shed light on the three-dimensional protein structure required for cell-to-cell movement and have identified conserved amino acid residues in the homeodomain, an arginine in helix α1 and a leucine in helix α3, as essential for intercellular trafficking. Other KN1-like TFs, BREVIPEDICELLUS and SHOOT MERISTEMLESS (STM), in fusion with GREEN FLUORESCENT PROTEIN (GFP), were also capable of moving from the L1 to the L2/L3 layers of the meristem (Kim et al., 2003). The developmental activity of these KN1-like TFs appears to be dependent on their cell-to-cell trafficking.

23.2.2 WUSCHEL in the Shoot Apical Meristem

WUS is a homeodomain TF synthesized in cells of the organizing center of shoot apical meristems (Laux et al., 1996; Chapter 14). WUS specifies stem cell fate and controls its own levels by inducing the transcription of a negative regulator, CLAVATA3 (CLV3), in adjacent cells of the central zone. CLV3 is a small secreted peptide that binds to CLV1 and possibly to CLV1-related receptors to activate signaling that suppresses WUS transcription. Yadav et al. (2011) have confirmed that a GFP:WUS fusion protein was mobile, moving cell to cell from its source in the L3 layer into the L1 and L2 layers of the meristem where it activates CLV3 by binding to *cis*-elements in its promoter. Mobility of WUS decreased by adding a nuclear localization signal to the protein and restriction of WUS movement inhibited WUS function (Yadav et al., 2011). The WUS/CLV3 feedback network establishes the cell-to-cell communication ongoing during stem cell maintenance, and movement of WUS explains how this communication occurs. These results demonstrate that WUS short-range mobility establishes a gradient in the central zone of the shoot apical meristem that is essential for maintaining a constant number of cells in the stem cell niche (Yadav and Reddy, 2012).

23.3 MOBILE ROOT TRANSCRIPTION FACTORS

It is now clearly established that numerous TFs traffic from cell to cell in roots (Petricka et al., 2012). To identify noncell-autonomous TFs (NCATFs) in roots, a genome-wide screen was performed using the GAL4–UAS transactivation expression system, in combination with transgenic lines expressing dual fluorescent reporter proteins (Rim et al., 2011). Their screen identified 22 TFs (out of a total of 76), belonging to 17 TF families that were classified as NCATFs. This study showed that protein size and subcellular localization are significant factors controlling intercellular movement of TFs in roots. Specific TFs that traffic cell to cell in roots include SHR, GLABRA3, UPBEAT1, and CAPRICE. The SHR model in particular has proven to be invaluable in our understanding of NCATFs and their role in controlling development. The SHR gene encodes a TF belonging to the GRAS gene family (Di Laurenzio et al., 1996; Helariutta et al., 2000; Chapter 10) and, with various other proteins, is involved in the maintenance of the root apical meristem (Benfey et al., 1993) and in the regulation of primary, lateral, and adventitious root development in *Arabidopsis* (Lucas et al., 2011). SHR is transcribed in the stele and its protein migrates outward one cell layer to the adjacent quiescent center cortex–endodermis initial cell, and the endodermis layer, where it activates the expression of SCARECROW (Helariutta et al., 2000; Nakajima et al., 2001). Movement of SHR is regulated by accumulation of callose at the PD (Vatén et al., 2011) and by a protein partner, SHORT-ROOT INTERACTING EMBRYONIC LETHAL (SIEL; Koizumi et al., 2011). SIEL is an endosome-associated protein that promotes intercellular movement of both SHR and CAPRICE. At another level of movement control, in the endodermis, SHR activates transcription of SCARECROW (Cui et al., 2007) and JACKDAW, which in turn inhibits the movement of SHR from the endodermis (Welch et al., 2007).

23.4 TRANSCRIPTION FACTORS AND COREGULATORS THAT MOVE LONG DISTANCE THROUGH THE SIEVE ELEMENT SYSTEM

Consider the proteomic profile of pumpkin sap, which is rich in proteins involved in translation and interactions with RNA (Lin et al., 2009). Despite the functional complexity of this profile (1209 proteins were identified), there are very few TFs detected and several of these are coregulators including the pumpkin ortholog of FT. Through these proteomics analyses, TFs such as BTF3b, HAP5A, HAP5B, MBF1B, and ZFP30 have been identified in pumpkin and rice (Aki et al., 2008; Lin et al., 2009). Transcripts that encode for TFs, on the other hand, are relatively abundant in phloem sap. Phloem sap from melon (*Cucumis melo*) yielded 986 unique transcripts and approximately 3% of these were TFs (Omid et al., 2007).

Of 2417 unique transcripts identified from phloem sap exudate harvested from leaf petioles of *Arabidopsis*, approximately 90 (3.7%) encoded TFs (Deeken et al., 2008). Among these mRNAs present in sieve elements were TFs representing numerous families including ethylene response factors such as APETALA2 (AP2)-like, MYB, bHLH, Zn finger, AUX/IAA, bZIP, NAC, and WRKY types. Beyond the rather obvious issue of detection limits, these observations suggest that the long-distance transport of TF proteins through the sieve element system is not a common mechanism for signaling, whereas transport of their mRNAs is widespread. The most prominent groups of known mobile RNAs encode for TFs or coregulators (Table 23.1). For example, of the seven RNAs identified in the RNA-binding protein 50 (RBP50)/RNA complex in pumpkin, six were TFs (Ham et al., 2009).

23.4.1 FLOWERING LOCUS T

Perhaps the best example of a long-distance mobile signal is FT. Under the control of CONSTANS, FT is transcribed and translated in the leaf under conditions inductive for flowering. It then moves from companion cells through the PD into the sieve element system where it is transported to the shoot apical meristem (reviewed by Turck et al., 2008). FT movement from companion

TABLE 23.1 Phloem-Mobile mRNAs of TFs that Move Across Heterografts

RNA	Annotation	Putative function	References
MpSLR/IAA14	Auxin response factor	Transcriptional repressor	Kanehira et al. (2010)
CmSCL14P	Scarecrow-like	TF	Ham et al. (2009)
CmSTM	Shoot meristemless	Meristem regulator	Ham et al. (2009)
CmERF	Ethylene response factor	Ethylene signaling	Ham et al. (2009)
CmNAC	NAM, ATAF1/2, and CUC2	Meristem development	Ruiz-Medrano et al. (1999)
CmMyb	Myb-like TF	Transcriptional activator	Ham et al. (2009)
BoFVE	Mammalian retinoblastoma-associated protein	Floral regulator	Yang and Yu (2010)
BoAGL24	Agamous-like	Floral regulator	Yang and Yu (2010)
AtAux/IAA18 and -28*	Auxin response factor	Auxin signaling ⇓	Notaguchi et al. (2012)
CmGAI*	GA insensitive	Leaf morphology ⇑	Haywood et al. (2005)
StBEL5*	Potato BEL1-like family	Tuber growth ⇓	Banerjee et al. (2006)
POTH1*	Potato Knotted1-type	Vegetative growth ⇓	Mahajan et al. (2012)
PFP–LeT6*	Tomato Knotted1-type fusion	Leaf morphology ⇑	Kim et al. (2001)
FT*	*Arabidopsis* FT	Activates flowering ⇑	Li et al. (2011)
ATC*	*Arabidopsis* CENTRORADIALIS	Represses flowering ⇑	Huang et al. (2012)

An asterisk indicates that movement of the RNA is associated with a phenotype. At, *Arabidopsis thaliana*; Cm, *Cucurbita maxima*; Cme, *Cucumis melo*; Le, *Lycopersicon esculentum*; Mp, *Malus prunifolia*; Bo, *Brassica oleracea*; St, *Solanum tuberosum*; PFP, pyrophosphate-dependent fructose 6-phosphate phosphotransferase. Arrows in the function column of the last seven RNAs indicate the prominent direction of the mobile transcript through a graft union.

cells to sieve elements is regulated by an ER membrane protein, FT-INTERACTING PROTEIN 1 (FTIP1). FTIP1 is essential for the transport of FT into the sieve elements and contains three C2 domains at its amino-terminus and a membrane-targeted phosphoribosyltransferase carboxy-terminal domain. The carboxy-terminal domain is likely anchored to the desmotubule of the PD, whereas the amino-terminal C2 domains likely interact with FT (Liu et al., 2012).

In the shoot apex, FT binds to the bZIP TF, FD, which in turn activates the floral pathway in numerous plant species (Wigge et al., 2005; Corbesier et al., 2007; Tamaki et al., 2007; Yoo et al., 2013). The functional role of FT in this heterodimer is currently unclear. In tandem with FD, this complex activates floral genes like APETALA1 (AP1) and LEAFY leading to floral development. FT is a member of the phosphatidylethanolamine-binding protein (PEBP) family whose crystal structure is similar to that of mammalian PEBPs (Ahn et al., 2006). PEBPs have an anion-binding pocket composed of highly conserved amino acids. FT and its homologs contain this anion-binding pocket, but binding to anions, phosphate groups, and phospholipids common to mammalian PEBPs has not been observed for FT (Banfield and Brady, 2000; Ahn et al., 2006). Most likely, FT serves as a coactivator of transcription, and accumulating evidence supports this idea (Abe et al., 2005; Wigge et al., 2005; Taoka et al., 2011). Despite these observations, no DNA-binding sequences have been identified in the FT protein. Perhaps the most intriguing recent development on the mechanism of FT activity is a proposed model for a floral activation complex (FAC). In this model, FT functions in a hexameric complex composed of three homodimers of FT, FD, and a 14-3-3 protein which functions as a scaffold (Taoka et al., 2011). In rice the FT homolog, Hd3a, interacts with 14-3-3 proteins in the apical meristem, producing a complex that moves into the nucleus and binds to *Oryza sativa* FD1. This complex then induces transcription of *OsMADS15* (the rice *AP1* homolog) and activates the floral pathway (Taoka et al., 2011).

23.4.2 The Long-Distance Tuberization Signal

Photoperiod is critical in plants as an environmental cue for regulating numerous developmental processes including both flowering and tuber formation. Several reports have addressed the similarities between photoperiodic signaling in both flowering and tuberization (Suárez-López, 2005; Rodríguez-Falcón et al., 2006; Abelenda et al., 2011). Both involve phloem-mobile signals that are activated and mobilized by day length cues. The physiology of the mobile signal of potato has been the focus of much research over several decades (Bernard, 1902; Garner and Allard, 1920; Gregory, 1956; Chapman, 1958). These studies showed that under conditions of low temperature and short days, a graft-transmissible signal produced in leaves moves down the phloem system into stolons to induce tuberization. Using a tobacco/potato heterograft, the floral signal from a tobacco scion can induce tuberization in a noninduced potato stock (Chailakhyan et al., 1981). Currently, the best candidates for phloem-mobile signals that regulate tuber formation and that will be addressed in this chapter are StFT/SP6A protein and StBEL5 protein derived from mobile RNA (see Section 23.5.4). Similar to FT, SP6A functions as a coregulator, whereas StBEL5 is a TF from the homeodomain three-amino-loop extension (TALE) superfamily (Sharma et al., 2014; Chapter 6).

23.4.3 StFT/StSP6A

As discussed previously, FT functions as the main component in controlling flowering in several species and has been established as the universal flowering signal (Corbesier et al., 2007; Lin et al., 2007; Tamaki et al., 2007; Turck et al., 2008). Recent reports have confirmed, however, that FT-like genes function in a wide range of developmental events beyond flowering (Pin and Nilsson, 2012; Ando et al., 2013; Karlgren et al., 2013; Niwa et al., 2013). Consistent with the premise that StFT/SP6A functions as the tuberization signal, transgenic overexpression lines (StSP6Aox) tuberized under noninductive, long-day conditions, whereas transgenic suppression lines exhibited a strong reduction in tuber yield under short-day conditions (Navarro et al., 2011). *Solanum tuberosum* CONSTANS (StCO) which has a negative effect on tuberization (González-Schain et al., 2012) also represses *StSP6A* expression under LDs. Whereas StSP6Aox scions grafted onto wild-type stocks induced the stocks to tuberize, there was no direct evidence of StSP6A protein moving through the graft unions. Hd3a–GFP fusion, however, was shown to move through a graft into stolons (Figure 23.1). Hd3a is the rice ortholog of FT. RT–PCR results confirmed that *Hd3a* transcripts did not move through the heterograft. Supporting the common theme with flowering, this Hd3a construct was able to increase tuber production in OE lines under long-day conditions as well as through heterografts with WT stocks (Navarro et al., 2011; Figure 23.1A). Local induction of *StSP6A* transcripts in stolons utilizing an alcohol-inducible promoter activated several tuber-identity genes including *StGA2ox1* (Navarro et al., 2011). Evidence was provided showing that two different FT-like paralogs, StSP3D and StSP6A, control the potato floral and tuberization transitions, respectively. Despite the proposed mechanism for trafficking the SP6A signal in protein form, transcription of *SP6A* occurs locally in stolons in a putative autorelay mechanism involving control by CONSTANS (Navarro et al., 2011). Overexpression of *StCO* represses *SP6A* transcription in both leaves and stolons from plants grown under inductive short-day conditions. As the canonical tuber signal in protein form is transported from leaf to the stolon sink, *SP6A* transcription

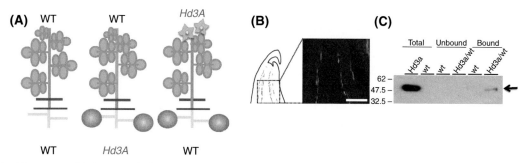

FIGURE 23.1 **Movement of Hd3A:GFP fusion protein through a heterograft and its effect on tuberization and flowering.** (A) Localization of GFP in stolons of WT stocks of Hd3a/WT heterografts was confirmed by visualization of GFP using confocal microscopy (B) and by immunoblot using GFP antibody (C, arrow). Hd3a–GFP fluorescence is detected in phloem companion cells. Scale bar in (B): 250 μm. Extracts in (C) were concentrated by affinity binding to an anti-GFP agarose matrix (bound). The unbound fraction (unbound) was loaded to test for binding efficiency. RT–qPCR confirmed that *Hd3a* mRNA was not transported across either graft union. *Reprinted from Navarro et al. (2011) with the permission of Prof Salomé Prat.*

in stolons may function to augment the mobile signal similar to *StBEL5* autoregulation (Lin et al., 2013).

The information available on StSP6A strongly suggests that it is a prominent candidate for a mobile tuberization signal. However, some questions remain. Similar to FT, StSP6A is a member of the phosphatidylethanolamine-binding protein family and exhibits no known DNA-binding activity. It is a coregulator and must form a dimer with a partner TF like FD. It is not yet clear if there is a tuber-specific StFD that complements StSP6A activity, or some other undiscovered transcriptional partner that induces tuber-specific gene expression. Another unanswered question is how is the downward direction of SP6A transport regulated? Floral FT moves acropetally toward the shoot apex. What proteins (or other partners) act as chaperones to transport SP6A to stolon tips?

23.5 FULL-LENGTH MOBILE mRNAs AND THEIR ROLES IN DEVELOPMENT

Either by phloem cell microdissection or analysis of phloem sap, the transcriptome of phloem includes thousands of full-length mRNAs with a diverse range of potential functions (Omid et al., 2007; Deeken et al., 2008; Kehr and Buhtz, 2008). Despite the observation that so many mRNAs can be detected in phloem sap, movement of only a few has been confirmed (Table 23.1) and even fewer have been associated with a phenotype. This latter group includes *StBEL5* (Banerjee et al., 2006) and *POTH1* (Mahajan et al., 2012) from potato, *CmGAI* from pumpkin (Haywood et al., 2005), *PFP–LeT6* from tomato (Kim et al., 2001), and *AUX/IAA* (Notaguchi et al., 2012), *FT*, and *CENTRORADIALIS* (Li et al., 2011; Huang et al., 2012; Lu et al., 2012) from *Arabidopsis*. Remarkably, all seven of these mobile RNAs function as TFs or coregulators (Table 23.1). For the purposes of this chapter, the mobile RNAs of GAI, ATC (*Arabidopsis thaliana* CENTRORADIALIS), FT, and StBEL5 will be discussed in detail.

23.5.1 Long-Distance Transport of GA INSENSITIVE

Long-distance transport of the transcript for *GAI*, encoding a protein that belongs to the DELLA subfamily of GRAS TFs (Pysh et al., 1999; Chapter 20), has been confirmed in several plant species including cucumber, tomato, pumpkin (Haywood et al., 2005; Ham et al., 2009), apple (Xu et al., 2010), and *Arabidopsis* (Huang and Yu, 2009). GAI functions in the negative regulation of gibberellic acid responses (Silverstone et al., 1998; Dill and Sun, 2001; Wen and Chang, 2002). *AtGAI* was the first mobile RNA associated with a phenotype (Haywood et al. 2005) and *CmGAI* of pumpkin was the first phloem-mobile mRNA identified in an RNA/protein complex (Ham et al., 2009). Using heterografts of tomato, *GAI* RNA movement was shown to mediate phenotypic changes in tomato leaf morphology (Haywood et al. 2005). Ham et al. (2009) identified specific cytosine/uracil sequence runs in the *CmGAI* transcript that interacted with RBP50, a 50 kD phloem RNA-binding protein of pumpkin, related to animal polypyrimidine tract-binding proteins (Auweter and Allain, 2008). RBP50 was the core protein in a phloem-mobile RNA/protein complex consisting of 6 mRNAs and up to 16 proteins (Ham et al. 2009). By utilizing a series of deletion mutants and movement assays with grafts of *Arabidopsis*, Huang and Yu (2009) showed that the untranslated regions (UTRs) of *GAI* were essential for movement of the full-length *GAI* transcript. Overall, these results demonstrate that phloem transport of *GAI* is mediated by sequence motifs recognized by specific RBPs and conserved among plant families.

23.5.2 ATC mRNA is Mobile Through the Phloem

Genetic analyses has demonstrated that ATC, an *Arabidopsis* FT homolog similar to TF1, antagonizes FT activity and functions as a floral inhibitor as both FT and ATC affect AP1 activity in opposition (Mimida et al., 2001). FT is induced under long-day conditions mainly in companion cells of leaves (Takada and Goto, 2003), whereas ATC is induced under short-day conditions (Yoo et al., 2010). FT is then transported to the apex where it interacts with FD, a bZIP TF, and activates floral identity genes like AP1 (Abe et al., 2005; Wigge et al., 2005). Both ATC and FT interact with FD in the apex. ATC promoter activity was detected in the vasculature of the leaf and stem, most prominently in the phloem, but was not detected in the shoot apex (Figure 23.2A,B; Huang et al., 2012). This observation suggests that ATC may be transported from the phloem to the shoot apex to regulate flowering. In opposition to FT, suppression of ATC enhances flowering, whereas overexpression delays it. Grafting experiments with *Arabidopsis* seedlings demonstrated that *ATC* RNA moved long distance toward the shoot apex and that floral inhibition by ATC is graft transmissible (Figure 23.2C–E;

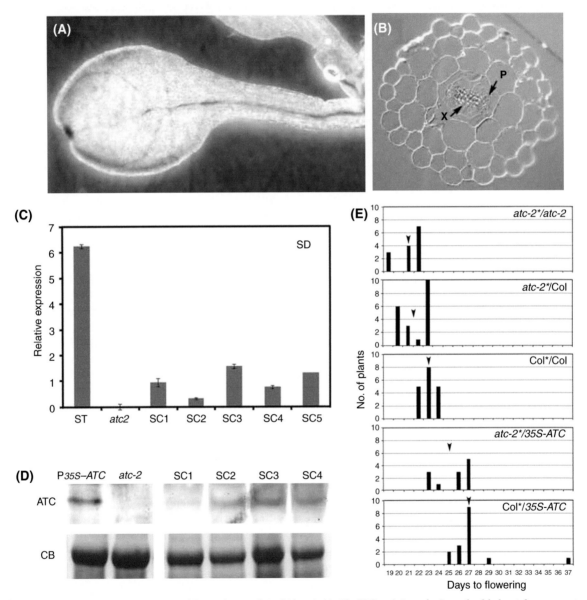

FIGURE 23.2 Promoter activity of ATC in phloem tissue of *Arabidopsis* (A–B). GUS staining of a 2-week-old short-day grown transgenic *Arabidopsis* plant carrying the ATC promoter driving GUS expression from (A) a cotyledon and (B) transverse section of hypocotyls. Long-distance movement of ATC in *Arabidopsis* seedling grafting experiments (C–E). Real-time RT–PCR analysis of individual scion samples grafted onto wild-type or atc-2 stock (C). The scion samples were harvested from atc-2 scion grafted onto atc-2 stock (atc2), or atc-2 scions grafted onto wild-type stocks (SC1–SC5), and wild-type stock of SC1 (ST). (D) Immunoblot analyses of ATC protein detected in atc-2 scions grafted onto P35S-ATC stocks. The upper panel is *Arabidopsis* P35S–ATC transformants and atc-2, or atc-2 scions grafted onto P35S–ATC stocks. The lower panel is Coomassie blue (CB) staining for a loading control. (E) Flowering inhibition by ATC is graft transmissible. Flowering time of individual scions (asterisks) grafted onto atc-2, Col, or P35S–ATC transgenic stocks under long-day conditions. The mean of the flowering time is indicated by arrowhead. The distribution of flowering time of scions was shifted after being grafted onto P35S–ATC transformants. *From Huang et al. (2012) with permission of Tien-Shin Yu.*

Huang et al., 2012). Although *ATC* trafficking from the wild-type stock to the atc-2 scion has been clearly established (Huang et al., 2012), the possibility that both the *ATC* RNA and protein move through the sieve element system to regulate flowering cannot be ruled out. For example, recent reports suggest that both the protein and RNA of FT are transported through the phloem to the shoot apex (see Section 23.5.3; Li et al., 2011). Overall, these results suggest that ATC functions antagonistically to FT in a non-cell autonomous manner to inhibit floral initiation.

23.5.3 FT mRNA Moves to the Shoot Apex

As previously discussed, the movement of the FT protein from source leaf to the shoot apex to initiate flowering has been widely established in several plant species. To add to the story, recent reports provide evidence that the full-length transcript of FT is transported from leaf to shoot apex. Whereas there is considerable controversy surrounding these observations (Huang et al., 2005; Lifschitz et al., 2006; Böhlenius et al., 2007; Jaeger and Wigge, 2007; Lin et al., 2007), the phloem mobility of *FT* RNA associated with floral induction has been confirmed by two independent research groups. By using *Arabidopsis* cleft-grafting experiments, Lu et al. (2012) showed that *FT* RNA of *Arabidopsis* undergoes long-distance movement from the stock to the scion apex in both FT transformants and nontransformants. Movement assays for FT sequences fused to GFP confirmed the first 210 nt of the coding sequence as sufficient for long-distance trafficking of this cell-autonomous mRNA. Fusion constructs with FT and RED FLUORESCENT PROTEIN (RFP) and heterografts were utilized to uncouple protein and RNA movement (Lu et al., 2012). The RFP protein is cell autonomous, whereas its RNA is not. Whereas RFP–FT protein was retained in companion cells, the detection of RFP–FT RNA was correlated with floral promotion in the scion (Figure 23.3). Further degradation of the graft translocated RFP–FT RNA by RNAi, or artificial miRNA against *FT*, suppressed floral development (Figure 23.4), indicating that the translocated *FT* RNA acts as a phloem-mobile floral signal.

More evidence for the long-distance transport of *FT* RNA to the shoot apex was provided using a model transport system that utilizes two distinct movement-defective plant viruses, *Potato virus X* (PVX) and *turnip crinkle virus*, and an agroinfiltration mobility assay (Li et al., 2009). This study demonstrated that *FT* mRNA that contained nonsense mutations, preventing FT protein synthesis, was able to move and promoted long-distance trafficking of heterologous GFP. Both *FT* and *nonsense FT* possessed comparable abilities to facilitate the spread of heterologous RNA fusions. Consistent with the Lu et al. (2012) study, the sequence essential for transcript movement was mapped to nucleotides 1–102 of the *FT* mRNA coding sequence. Viral ectopic expression of *FT* in leaves followed by movement of its RNA activated flowering in *Nicotiana tabacum* cv. Maryland Mammoth under noninductive (long-day) conditions (Li et al., 2009). This induced flowering was

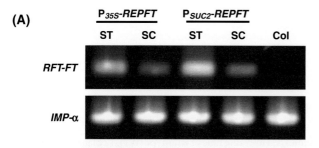

FIGURE 23.3 **Long-distance movement of RFP–FT RNA is correlated with floral promotion.** (A) RT–PCR detection of RNA of RFP–FT in wild-type scions grafted onto P35S–RFP–FT or PSUC2–RFP–FT stocks. RNA was extracted from P35S–RFP–FT or PSUC2–RFP–FT stocks (ST), wild-type scions (SC), or wild-type control (Col). PCR was conducted with gene-specific primers for RFP and FT. IMPORTIN-α (IMP-α) was used as the loading control. The gel was stained with SYBR green for visualization. (B) *Arabidopsis* transformants carrying PSUC2–RFP–FT or P35S–RFP–FT (T2 generations) displayed an early flowering phenotype when grown under long-day conditions. *From Lu et al. (2012) with permission of Tien-Shin Yu.*

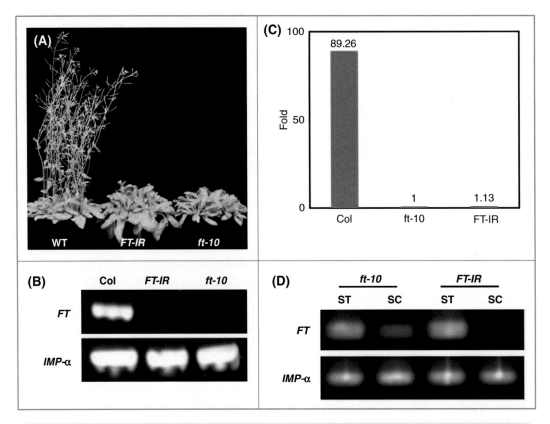

FIGURE 23.4 **The depletion of FT RNA from scion delays scion floral initiation.** (A) Flowering time of wild-type (WT), T2 plants of P35S–FT-IR transformant (FT-IR) and ft-10 mutant under long-day conditions. (B) RT–PCR analysis of FT RNA accumulation. RNA was extracted from wild-type plant (Col), P35S–FT-IR transformant (FT-IR) or ft-10 mutant plants. IMPORTIN-α (IMP-α) was used as the loading control. The gel was stained with SYBR green for visualization. (C) Quantitative RT–PCR analysis of wild-type plant (Col), P35S–FT-IR transformant (FT-IR) and ft-10 mutant. The RNA fold change is relative to the level of ft-10. (D) RT–PCR analysis of *Arabidopsis* grafting results. The ft-10 mutants (ft-10) or P35S–FT-IR transformants (FT-IR) were used as scions and grafted onto the P35S–RFP–FT stocks. PCR was conducted with gene-specific primers for FT and NOS terminator. IMPORTIN-α (IMP-α) was used as the loading control. In heterografting experiments (E), with ft-10 or P35S–FT-IR scions grafted onto ft-10 stocks, both scions flowered at approx. 49 d (E). However, with ft-10 scions grafted onto P35S–RFP–FT or PSUC2–RFP–FT stocks, the flowering time of ft-10 scions was enhanced. With P35S–FT-IR transformants grafted onto P35S–RFP-FT stocks, the flowering time of the P35S–FT-IR scions was delayed to 48 d (E). RT–PCR of these grafting lines detected RFP–FT RNA only in P35S–RFP–FT stocks, but not in P35S–FT-IR scions, which suggests that the translocated RFP–FT RNA was degraded in the P35S–FT-IR scions (D), leading to delayed flowering. *From Lu et al. (2012) with permission of Tien-Shin Yu.*

positively correlated with the accumulation of functional *FT* transcripts in emerging leaves arising from the shoot apex (Li et al., 2009). *Arabidopsis FT* mRNA, independent of the FT protein, readily moved into the shoot apical meristem (Li et al., 2011). Both sense and mutated nontranslatable *FT* mRNA can facilitate delivery of the PVX sequence into the SAM, whereas PVX fused to GFP sequence alone could not (Li et al., 2011).

These results demonstrate that *Arabidopsis FT* mRNA can move through the surveillance system that excludes viral RNAs from the SAM (Foster et al., 2002; Schwach et al., 2005). These combined experiments demonstrate that, along with FT protein, *FT* RNA also moves long distance from source leaves through phloem cells to the shoot apex to regulate the photoperiodic floral signal. Based on previous examples (Lucas et al., 1995; Xoconostle-Cázares et al., 1999; Ham et al., 2009), it is very likely that an escort protein(s) binds to *FT* mRNA to facilitate its stability and localization. The best example of a core protein involved in a mobile RNA/protein complex is the aforementioned CmRBP50 (Ham et al., 2009). RBP50 is a polypyrimidine tract binding (PTB) protein that binds to cytosine/uracil-rich motifs predominately located in the 3′ UTRs in the RNA sequences of *GAI* (Ham et al., 2009) and *StBEL5* (Mahajan et al., 2012; Hannapel, 2013a). Coincidentally, the 102 nt region identified as sufficient for transporting *FT* transcripts is also rich in cytosine/uracil sequences.

23.5.4 StBEL5, a Mobile RNA of Potato

From both a functional and mechanistic perspective, arguably the most comprehensive example of a phloem-mobile mRNA is the *StBEL5* model (Banerjee et al., 2006; Hannapel, 2010). StBEL5 is a TF from the ubiquitous BEL1-like family that regulates numerous aspects of development (Chen et al., 2003; Bhatt et al., 2004; Smith et al., 2004; Bencivenga et al., 2012). BEL1 TFs interact with KN1 types to regulate transcription in thousands of target genes (Bolduc et al., 2012; Chapter 14). Both BELs and KN1 types belong to the TALE superfamily characterized by the proline/tyrosine/proline TALE between helices 1 and 2 of the homeodomain conserved in all members (Sharma et al., 2014; Chapter 6). The tandem BEL/KNOX complex recognizes a double TTGAC core element present in the upstream sequence of target genes with each TF binding to a separate TTGAC element (Chen et al., 2004). Transcription assays and point mutation analyses have demonstrated that the binding of both TFs is required to regulate activity of the target gene (Figure 23.5).

Overexpression and accumulation of *StBEL5* RNA in whole plants and stocks of heterografts has consistently been associated with enhanced root and tuber growth. Tuber yield increases have ranged from 1.5-fold in heterografts (Banerjee et al., 2006) to almost 7-fold in transgenic soil-grown plants (Banerjee et al., 2009). Overexpression of BEL5 even resulted in tuber production under long days in potato plants with a short-day requirement for tuberization (Chen et al., 2003). In roots, StBEL5 appears to regulate secondary root growth and stele development by modulating hormone levels (Lin et al., 2013). Using both a phloem transport induction system for StBEL5, gel shift assays, and the potato genome as a resource, numerous target genes of the StBEL5/POTH1 complex that contain the tandem TTGAC motif in their upstream sequence have been identified (Hannapel, 2013; Lin et al., 2013; Sharma et al., 2014). Included in this short list are several genes involved in hormone metabolism

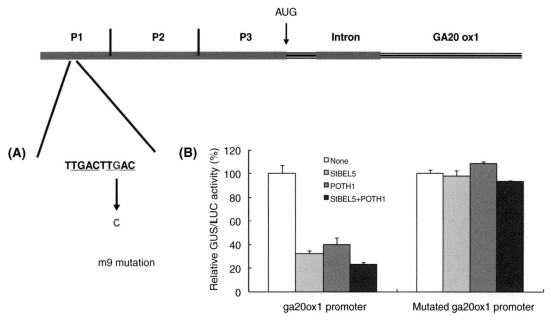

FIGURE 23.5 **A single point mutation (m9) abolishes repression activity.** A single point mutation in the StBEL5–POTH1 heterodimer binding site blocked repression activity on the *ga20ox1* promoter mediated by StBEL5 or POTH1. POTH1 is a KNOX-type TF of potato. (A) The single-point mutation in the 10 bp binding motif. (B) Relative GUS/LUC activity in the transcription assay. The construct with the luciferase gene under the CaMV 35S promoter was used as a control. Each transfection was performed three times. Relative GUS–LUC activity was calculated with the reporter gene alone set as the 100% standard. Data are means ± SE. *Reprinted from Chen et al. (2004).*

such as GA20 oxidase1, GA2 oxidase1, YUCCA1, StPINs, and LONELY GUY. In concordance with hormone measurements and previous studies on TALE TFs (Tanaka-Ueguchi et al., 1998; Chen et al., 2003; Rosin et al., 2003; Bolduc and Hake, 2009), these results strongly suggest that StBEL5, and its KNOX partners, regulate growth by directly controlling the transcriptional activity of genes involved in hormone synthesis.

Using *in situ* hybridization and laser capture microdissection coupled with RT–PCR, *StBEL5* transcripts were detected in phloem cells of both the stem and the stolon (Banerjee et al., 2006; Yu et al., 2007). To identify the source of *StBEL5* transcripts, promoter activity was analyzed with a GUS marker and was observed in leaf veins, phloem cells of the petiole, roots, stolons, and new tubers. Despite an abundance of RNA in stems, StBEL5 transcriptional activity was not detected in stems (Banerjee et al., 2006). Whereas in leaves, *StBEL5* promoter activity is induced by light, in roots, stolons, and new tubers, StBEL5 autoregulates its own activity to augment this long-distance signal. The *StBEL5* promoter contains the tandem TTGAC motif 820 nucleotides upstream from the transcription start site that controls this autoregulation (Lin et al., 2013). Crossregulation of transcription is also prevalent among members of the StBEL family (Sharma et al., 2014).

RNA mobility assays and heterografting experiments demonstrated that *BEL5* transcripts are present in the phloem cells of leaves and move across a graft union to localize in stolon tips, the site of tuber induction (Figure 23.6). This movement is enhanced under a short-day photoperiod (inductive for tuber formation) and mediated by its UTRs. Fusing both the 5' and 3' UTRs of *StBEL5* to another nonmobile *StBEL* RNA enhanced its mobility and targeted localization to stolons (Banerjee et al., 2009). Noncell-autonomous mRNAs interact with proteins that stabilize the RNA, regulate translation, and direct its movement (Lucas et al., 1995; Xoconostle-Cázares et al., 1999; Ham et al., 2009). Established as a chaperone for mobile RNAs in pumpkin, two similar PTB proteins of potato bind to specific cytosine/uracil-rich sequences in the 3' UTRs of *StBEL5* and are expressed in

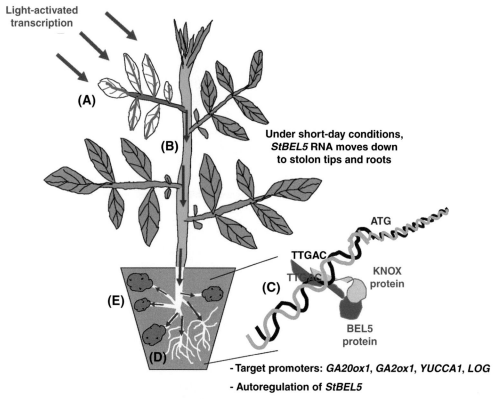

FIGURE 23.6 **Model showing the impact of mobile *StBEL5* RNA on root and tuber development.** The long-distance transport of *StBEL5* RNA is strongly correlated with the induction of tuber formation in potato (Banerjee et al., 2006). This signaling pathway is based on the initial activation of transcription by light (A, red arrows) of the *StBEL5* gene in the veins of leaves and petioles (A, blue). A short-day photoperiod facilitates movement of the *StBEL5* RNA to stolon tips, whereas movement to roots occurs regardless of day length (B, red arrows). Under these conditions, RNA is escorted to site-specific targets, like stolon tips and roots, via protein chaperones. Enhanced translation then occurs in the stolon tip or root followed by binding to a KNOX protein partner (C) and subsequent activation of transcription and regulation of select genes (e.g., *GA20ox1*, *GA2ox1*, *YUCCA1*, *IPT*, *LOG*, and *StBEL5*) by binding to the tandem TTGAC core motif of the target promoter. In this model, transcriptional regulation then leads to enhanced growth of roots (D) and tubers (E) modulated by hormone levels. *Modified from Figure 10.4 of Hannapel (2013b). This material is reproduced with permission of John Wiley & Sons, Inc.*

companion cells of the phloem (Butler and Hannapel, 2012; Mahajan et al., 2012). Overexpression of both PTB types in transgenic lines resulted in enhanced tuber production correlated with increased stability of *StBEL5*. Recent preliminary results confirmed that two other *StBEL* RNAs related to *BEL5* also move through the sieve element system. These results suggest that StBEL5-like TFs are involved in a complex developmental network that regulates expression in target organs through the long-distance transport of select full-length mRNAs.

23.6 CONCLUSIONS

Due to their sessile life style, and in response to environmental cues, plants have evolved a complex network of noncell-autonomous signals in the form of proteins and RNAs as a unique feature of long-distance communication. As this emerging field continues to grow, we now understand that both cell-to-cell and long-distance phloem-mobile signals are ubiquitous in the plant kingdom. What is remarkable is that a large proportion of these signals are TFs or coregulators and play significant roles in the control of flowering, tuberization, and leaf and root morphology. Despite the steady increase in the identification of putative signals, clearly the key to our understanding of this process is inherent in the mechanisms and chaperones that facilitate and regulate these transport networks. As discussed here and in other reviews, important examples include FTIP1, chaperonin complexes (Xu et al., 2011), SIEL, and PTB proteins. These protein partners function to stabilize and traffic signals, and to recognize surveillance systems. The PTB protein of pumpkin, CmRBP50, effectively illustrates this importance as it functions to escort mRNAs by binding to motifs in RNA sequences rich in uracils and cytosines. As the core member of a phloem-mobile RNA/protein complex, RBP50 regulates the stability and transport of full-length mRNAs, many of which encode for TFs. Much more work is still needed, however, to better understand the processes of recognition, delivery, and release that are operational in the transport of mobile protein and RNA signals. Except for a few examples (Kragler et al., 1998; Bolduc et al., 2008; Xu et al., 2011), the mechanisms that facilitate signal access to the PD and transport into the sieve elements through the cytoplasmic sleeve of the PD or the lumen of the desmotubule are still largely unknown.

Acknowledgments

The author acknowledges Salomé Prat and Tien-Shin Yu for allowing use of their published work for this book chapter. He further acknowledges Anjan Banerjee and Hao Chen for their dedicated contributions to the StBEL5 story and Pooja Sharma for her critical reading of the chapter.

References

Abe, M., Kobayashi, Y., Yamamoto, S., Daimon, Y., Yamaguchi, A., Ikeda, Y., Ichinoki, H., Notaguchi, M., Goto, K., Araki, T., 2005. FD, a bZIP protein mediating signals from the floral pathway integrator FT at the shoot apex. Science 309, 1052–1056.

Abelenda, J.A., Navarro, C., Prat, S., 2011. From the model to the crop: genes controlling tuber formation in potato. Curr. Opin. Biotech. 22, 287–292.

Ahn, J.H., Miller, D., Winter, V.J., Banfield, M.J., Lee, J.H., Yoo, S.Y., Henz, S.R., Brady, R.L., Weigel, D., 2006. A divergent external loop confers antagonistic activity on floral regulators FT and TFL1. EMBO J. 25, 605–614.

Aki, T., Shigyo, M., Nakano, R., Yoneyama, T., Yanagisawa, S., 2008. Nano scale proteomics revealed the presence of regulatory proteins including three FT-like proteins in phloem and xylem saps from rice. Plant Cell Physiol. 49, 767–790.

Ando, E., Ohnishi, M., Wang, Y., Matsushita, T., Watanabe, A., Hayashi, Y., Fujii, M., Ma, J.F., Inoue, S., Kinoshita, T., 2013. TWIN SISTER OF FT, GIGANTEA, and CONSTANS have a positive but indirect effect on blue light-induced stomatal opening in *Arabidopsis*. Plant Physiol. 162, 1529–1538.

Auweter, S.D., Allain, F.H.T., 2008. Structure–function relationships of the polypyrimidine tract binding protein. Cell. Mol. Life Sci. 65, 516–527.

Banerjee, A.K., Chatterjee, M., Yu, Y., Suh, S.G., Miller, W.A., Hannapel, D.J., 2006. Dynamics of a mobile RNA of potato involved in a long-distance signaling pathway. Plant Cell 18, 3443–3457.

Banerjee, A.K., Lin, T., Hannapel, D.J., 2009. Untranslated regions of a mobile transcript mediate RNA metabolism. Plant Physiol. 151, 1831–1843.

Banfield, M.J., Brady, R.L., 2000. The structure of *Antirrhinum centroradialis* protein (CEN) suggests a role as a kinase regulator. J. Mol. Biol. 297, 1159–1170.

Bencivenga, S., Simonini, S., Benková, E., Colombo, L., 2012. The transcription factors BEL1 and SPL are required for cytokinin and auxin signaling during ovule development in *Arabidopsis*. Plant Cell 24, 2886–2897.

Benfey, P.N., Linstead, P.J., Roberts, K., Schiefelbein, J.W., Hauser, M.T., Aeschbacher, R.A., 1993. Root development in *Arabidopsis*: four mutants with dramatically altered root morphogenesis. Development 119, 57–70.

Bernard, N., 1902. Studies of tuberization. Rev. Gen. Bot. 14, 5–45.

Bhatt, A.M., Etchells, J.P., Canales, C., Lagodienko, A., Dickinson, H., 2004. VAAMANA, A BEL1-like homeodomain protein, interacts with KNOX proteins BP and STM and regulates inflorescence stem growth in *Arabidopsis*. Gene 328, 103–111.

Böhlenius, H., Eriksson, S., Parcy, F., Nilsson, O., 2007. Retraction. Science 316, 367.

Bolduc, N., Hake, S., 2009. The maize transcription factor KNOTTED1 directly regulates the gibberellin catabolism gene ga2ox1. Plant Cell 21, 1647–1658.

Bolduc, N., Hake, S., Jackson, D., 2008. Dual functions of the KNOTTED1 homeodomain: sequence-specific DNA binding and regulation of cell-to-cell transport. Sci. Signal. 1 (23), pe28.

Bolduc, N., Yilmaz, A., Mejia-Guerra, M.K., Morohashi, K., O'Connor, D., Grotewold, E., Hake, S., 2012. Unraveling the KNOTTED1 regulatory network in maize meristems. Genes Dev. 26, 1685–1690.

Butler, N.M., Hannapel, D.J., 2012. Promoter activity of polypyrimidine tract-binding protein genes of potato responds to environmental cues. Planta 236, 1747–1755.

Chailakhyan, M.K., Yanina, L.I., Devedzhyan, A.G., Lotova, G.N., 1981. Photoperiodism and tuber formation in grafting of tobacco onto potato. Dokl. Akad. Nauk 257, 1276–1280.

Chapman, H.W., 1958. Tuberization in the potato plant. Physiol. Plant. 11, 215–224.

Chen, H., Banerjee, A.K., Hannapel, D.J., 2004. The tandem complex of BEL and KNOX partners is required for transcriptional repression of *ga20ox1*. Plant J. 38, 276–284.

Chen, H., Jackson, D., Kim, J.Y., 2014. Identification of evolutionarily conserved amino acid residues in homeodomain of KNOX proteins for intercellular trafficking. Plant Signal. Behav. 9 (2), e28355.

Chen, H., Rosin, F.M., Prat, S., Hannapel, D.J., 2003. Interacting transcription factors from the three-amino acid loop superclass regulate tuber formation. Plant Physiol. 132, 1391–1404.

Corbesier, L., Vincent, C., Jang, S., Fornara, F., Fan, Q., Searle, I., Giakountis, A., Farrona, S., Gissot, L., Turnbull, C., Coupland, G., 2007. FT protein movement contributes to long-distance signaling in floral induction of *Arabidopsis*. Science 316, 1030–1033.

Cui, H., Levesque, M.P., Vernoux, T., Jung, J.W., Paquette, A.J., Gallagher, K.L., Wang, J.Y., Blilou, I., Scheres, B., Benfey, P.N., 2007. An evolutionarily conserved mechanism delimiting SHR movement defines a single layer of endodermis in plants. Science 316, 421–425.

Deeken, R., Ache, P., Kajahn, I., Klinkenberg, J., Bringmann, G., Hedrich, R., 2008. Identification of *Arabidopsis thaliana* phloem RNAs provides a search criterion for phloem-based transcripts hidden in complex datasets of microarray experiments. Plant J. 55, 746–759.

Di Laurenzio, L., Wysocka-Diller, J., Malamy, J.E., Pysh, L., Helariutta, Y., Freshour, G., Hahn, M.G., Feldmann, K.A., Benfey, P.N., 1996. The SCARECROW gene regulates an asymmetric cell division that is essential for generating the radial organization of the *Arabidopsis* root. Cell 86, 423–433.

Dill, A., Sun, T.P., 2001. Synergistic derepression of gibberellin signaling by removing RGA and GAI function in *Arabidopsis thaliana*. Genetics 159, 777–785.

Foster, T.M., Lough, T.J., Emerson, S.J., Lee, R.H., Bowman, J.L., Forster, R.L.S., Lucas, W.J., 2002. A surveillance system regulates selective entry of RNA into the shoot apex. Plant Cell 14, 1497–1508.

Garner, W.W., Allard, H.A., 1920. Effect of length of day and other factors of the environment on growth and reproduction in plants. J. Agric. Res. 18, 553–578.

González-Schain, N.D., Díaz-Mendoza, M., Zurczak, M., Suárez-López, P., 2012. Potato CONSTANS is involved in photoperiodic tuberization in a graft-transmissible manner. Plant J. 70, 678–690.

Gregory, L.E., 1956. Some factors for tuberization in the potato plant. Am. J. Bot. 43, 281–288.

Ham, B.K., Brandom, J.L., Xoconostle-Cazares, B., Ringgold, V., Lough, T.L., Lucas, W.J., 2009. A polypyrimidine tract binding protein, pumpkin RBP50, forms the basis of a phloem-mobile ribonucleoprotein complex. Plant Cell 21, 197–215.

Hannapel, D.J., 2010. A model system of development regulated by the long-distance transport of mRNA. J. Integr. Plant Biol. 52, 40–52.

Hannapel, D.J., 2013a. Long-distance signaling via mobile RNAs. In: Baluška, František (Ed.), Long-Distance Systemic Signaling and Communication, vol. 19, Springer-Verlag, Berlin, pp. 53–70.

Hannapel, D.J., 2013b. The effect of long-distance signaling on development. In: van Bel, A.J.E., Thompson, G.A. (Eds.), Phloem: Molecular Cell Biology, Systemic Communication, Biotic Interactions. John Wiley & Sons, New York, pp. 209–226.

Hannapel, D.J., Sharma, P., Lin, T., 2013. Phloem-mobile messenger RNAs and root development. Front. Plant Sci. 4, 257.

Haywood, V., Yu, T.S., Huang, N.C., Lucas, W.J., 2005. Phloem long-distance trafficking of GIBBERELLIC ACID-INSENSITIVE RNA regulates leaf development. Plant J. 42, 49–68.

Helariutta, Y., Fukaki, H., Wysocka-Diller, J., Nakajima, K., Jung, J., Sena, G., Hauser, M.T., Benfey, P.N., 2000. The SHORT-ROOT gene controls radial patterning of the *Arabidopsis* root through radial signaling. Cell 101, 555–567.

Huang, N.C., Yu, T.S., 2009. The sequences of *Arabidopsis* GA-INSENSITIVE RNA constitute the motifs that are necessary and sufficient for RNA long-distance trafficking. Plant J. 59, 921–929.

Huang, N.C., Jane, W.N., Chen, J., Yu, T.S., 2012. *Arabidopsis* CENTRORADIALIS homologue acts systemically to inhibit floral initiation in *Arabidopsis*. Plant J. 72, 175–184.

Huang, T., Böhlenius, H., Eriksson, S., Parcy, F., Nilsson, O., 2005. The mRNA of the *Arabidopsis* gene *FT* moves from leaf to shoot apex and induces flowering. Science 309, 1694–1696.

Jaeger, K.E., Wigge, P.A., 2007. FT protein acts as a long-range signal in *Arabidopsis*. Curr. Biol. 17, 1050–1054.

Karlgren, A., Gyllenstrand, N., Clapham, D., Lagercrantz, U., 2013. FLOWERING LOCUS T/TERMINAL FLOWER1-like genes affect growth rhythm and bud set in Norway spruce. Plant Physiol. 163, 792–803.

Kanehira, A., Yamada, K., Iwaya, T., Tsuwamoto, R., Kasai, A., Nakazono, M., Harada, T., 2010. Apple phloem cells contain some mRNAs transported over long distances. Tree Genet. Genomes 5, 635–642.

Kehr, J., Buhtz, A., 2008. Long distance transport and movement of RNA through the phloem. J. Exp. Bot. 59, 85–92.

Kim, M., Canio, W., Kessler, S., Sinha, N., 2001. Developmental changes due to long-distance movement of a homeobox fusion transcript in tomato. Science 293, 287–289.

Kim, J.Y., Yuan, Z., Jackson, D., 2003. Developmental regulation and significance of KNOX protein trafficking in *Arabidopsis*. Development 130, 4351–4362.

Kim, J.Y., Rim, Y., Wang, J., Jackson, D., 2005. A novel cell-to-cell trafficking assay indicates that the KNOX homeodomain is necessary and sufficient for intercellular protein and mRNA trafficking. Genes Dev. 19, 788–793.

Koizumi, K., Wu, S., MacRae-Crerar, A., Gallagher, K.L., 2011. An essential protein that interacts with endosomes and promotes movement of the SHORT-ROOT transcription factor. Curr. Biol. 21, 1559–1564.

Kragler, F., Monzer, J., Shash, K., Xoconostle-Cazares, B., Lucas, W.J., 1998. Cell-to-cell transport of proteins: requirement for unfolding and characterization of binding to a putative plasmodesmal receptor. Plant J. 15, 367–381.

Laux, T., Mayer, K.F., Berger, J., Jurgens, G., 1996. The WUSCHEL gene is required for shoot and floral meristem integrity in *Arabidopsis*. Development 122, 87–96.

Li, C., Zhang, K., Zeng, X., Jackson, S., Zhou, Y., Hong, Y., 2009. A cis element within *flowering locus T* mRNA determines its mobility and facilitates trafficking of heterologous viral RNA. J. Virol. 83, 3540–3548.

Li, C., Gu, M., Shi, N., Zhang, H., Yang, X., Osman, T., Liu, Y., Wang, H., Vatish, M., Jackson, S., Hong, Y., 2011. Mobile FT mRNA contributes to the systemic florigen signalling in floral induction. Sci. Rep. 1, 73.

Lifschitz, E., Eviatar, T., Rozman, A., Shalit, A., Goldshmidt, A., Amsellem, Z., Alvarez, J.P., Eshed, Y., 2006. The tomato FT ortholog triggers systemic signals that regulate growth and flowering and substitute for diverse environmental stimuli. Proc. Natl. Acad. Sci. USA 103, 6398–6403.

Lin, M.K., Belanger, H., Lee, Y.J., Varkonyi-Gasic, E., Taoka, K., Miura, E., Xoconostle-Cázares, B., Gendler, K., Jorgensen, R.A., Phinney, B., Lough, T.J., Lucas, W.J., 2007. FLOWERING LOCUS T protein may act as the long-distance florigenic signal in the cucurbits. Plant Cell 19, 1488–1506.

Lin, M.K., Lee, Y.J., Lough, T.J., Phinney, B.S., Lucas, W.J., 2009. Analysis of the pumpkin phloem proteome provides insights into angiosperm sieve tube function. Mol. Cell Proteomics 8, 343–356.

Lin, T., Sharma, P., Gonzalez, D.H., Viola, I.L., Hannapel, D.J., 2013. The impact of the long-distance transport of a *BEL1*-like mRNA on development. Plant Physiol. 161, 760–772.

Liu, L., Liu, C., Hou, X., Xi, W., Shen, L., Tao, Z., Wang, Y., Yu, H., 2012. FTIP1 is an essential regulator required for florigen transport. PLoS Biol. 10, e1001313.

Lu, K.J., Huang, N.C., Liu, Y.S., Lu, C.A., Yu, T.S., 2012. Long-distance movement of *Arabidopsis* FLOWERING LOCUS T RNA participates in systemic floral regulation. RNA Biol. 9 (5), 653–662.

Lucas, W.J., Bouché-Pillon, S., Jackson, D.P., Nguyen, L., Baker, L., Ding, B., Hake, S., 1995. Selective trafficking of KNOTTED1 homeodomain protein and its mRNA through plasmodesmata. Science 270, 1980–1983.

Lucas, W.J., Ham, B.K., Kim, J.Y., 2009. Plasmodesmata – bridging the gap between neighboring plant cells. Trends Cell Biol. 19, 495–503.

Lucas, M., Swarup, R., Paponov, I.A., Swarup, K., Casimiro, I., Lake, D., Peret, B., Zappala, S., Mairhofer, S., Whitworth, M., Wang, J., Ljung, K., Marchant, A., Sandberg, G., Holdsworth, M.J., Palme, K., Pridmore, T., Mooney, S., Bennett, M.J., 2011. Short-root regulates primary, lateral, and adventitious root development in *Arabidopsis*. Plant Physiol. 155, 384–398.

Mahajan, A., Bhogale, S., Kang, I.H., Hannapel, D.J., Banerjee, A.K., 2012. The mRNA of a Knotted1-like transcription factor of potato is phloem mobile. Plant Mol. Biol. 79, 595–608.

Mimida, N., Goto, K., Kobayashi, Y., Araki, T., Ahn, J.H., Weigel, D., Murata, M., Motoyoshi, F., Sakamoto, W., 2001. Functional divergence of the TFL1-like gene family in *Arabidopsis* revealed by characterization of a novel homologue. Genes Cells 6, 327–336.

Nakajima, K., Sena, G., Nawy, T., Benfey, P.N., 2001. Intercellular movement of the putative transcription factor SHR in root patterning. Nature 413, 307–311.

Navarro, C., Abelenda, J.A., Cruz-Oró, E., Cuéllar, C.A., Tamaki, S., Silva, J., Shimamoto, K., Prat, S., 2011. Control of flowering and storage organ formation in potato by *FLOWERING LOCUS T*. Nature 478, 119–122.

Niwa, M., Endo, M., Araki, T., 2013. Florigen is involved in axillary bud development at multiple stages in *Arabidopsis*. Plant Signal. Behav. 8 (11), e27167.

Notaguchi, M., Wolf, S., Lucas, W.J., 2012. Phloem-mobile *Aux/IAA* transcripts target to the root tip and modify root architecture. J. Integr. Plant Biol. 54, 760–772.

Omid, A., Keilin, T., Glass, A., Leshkowitz, D., Wolf, S., 2007. Characterization of phloem-sap transcription profile in melon plants. J. Exp. Bot. 58, 3645–3656.

Petricka, J.J., Winter, C.M., Benfey, P.N., 2012. Control of *Arabidopsis* root development. Annu. Rev. Plant Biol. 63, 563–590.

Pin, P.A., Nilsson, O., 2012. The multifaceted roles of *FLOWERING LOCUS T* in plant development. Plant Cell Environ. 35, 1742–1755.

Pysh, L.D., Wysocka-Diller, J.W., Camilleri, C., Bouchez, D., Benfey, P.N., 1999. The GRAS gene family in *Arabidopsis*: sequence characterization and basic expression analysis of the SCARECROW-LIKE genes. Plant J. 18, 111–119.

Rim, Y., Huang, L., Chu, H., Han, X., Cho, W.K., Jeon, C.O., Kim, H.J., Hong, J.C., Lucas, W.J., Kim, J.Y., 2011. Analysis of *Arabidopsis* transcription factor families revealed extensive capacity for cell-to-cell movement as well as discrete trafficking patterns. Mol. Cells 32, 519–526.

Rodríguez-Falcón, M., Bou, J., Prat, S., 2006. Seasonal control of tuberization in potato: conserved elements with the flowering response. Annu. Rev. Plant Biol. 57, 151–180.

Rosin, F.M., Hart, J.K., Horner, H.T., Davies, P.J., Hannapel, D.J., 2003. Overexpression of a knotted-like homeobox gene of potato alters vegetative development by decreasing gibberellin accumulation. Plant Physiol. 132, 106–117.

Ruiz-Medrano, R., Xoconostle-Cazares, B., Lucas, W.J., 1999. Phloem long-distance transport of CmNACP mRNA: implications for supracellular regulation in plants. Development 126, 4405–4419.

Schwach, F., Vaistij, F.E., Jones, L., Baulcombe, D.C., 2005. An RNA-dependent RNA polymerase prevents meristem invasion by potato virus X and is required for the activity but not the production of a systemic silencing signal. Plant Physiol. 138, 1842–1852.

Sharma, P., Lin, T., Grandellis, C., Yu, M., Hannapel, D.J., 2014. The BEL1-like family of transcription factors in potato. J. Exp. Bot. 65, 709–723.

Silverstone, A.L., Ciampaglio, C.N., Sun, T.-P., 1998. The *Arabidopsis* RGA gene encodes a transcriptional regulator repressing the gibberellin signal transduction pathway. Plant Cell 10, 155–169.

Smith, H.M., Campbell, B.C., Hake, S., 2004. Competence to respond to floral inductive signals requires the homeobox genes PENNYWISE and POUND-FOOLISH. Curr. Biol. 14, 812–817.

Suárez-López, P., 2005. Long-range signalling in plant reproductive development. Int. J. Dev. Biol. 49, 761–771.

Takada, S., Goto, K., 2003. TERMINAL FLOWER2, an *Arabidopsis* homolog of HETEROCHROMATIN PROTEIN1, counteracts the activation of *FLOWERING LOCUS T* by CONSTANS in the vascular tissues of leaves to regulate flowering time. Plant Cell 15, 2856–2865.

Tamaki, S., Matsuo, S., Wong, H.L., Yokoi, S., Shimamoto, K., 2007. Hd3a protein is a mobile flowering signal in rice. Science 316, 1033–1036.

Tanaka-Ueguchi, M., Itoh, H., Oyama, N., Koshioka, M., Matsuoka, M., 1998. Over-expression of a tobacco homeobox gene, NTH15, decreases the expression of a gibberellin biosynthetic gene encoding GA 20-oxidase. Plant J. 15, 391–400.

Taoka, K., Ohki, I., Tsuji, H., Furuita, K., Hayashi, K., Yanase, T., Yamaguchi, M., Nakashima, C., Purwestri, Y.A., Tamaki, S., Ogaki, Y., Shimada, C., Nakagawa, A., Kojima, C., Shimamoto, K., 2011. 14-3-3 proteins act as intracellular receptors for rice Hd3a florigen. Nature 476, 332–335.

Turck, F., Fornara, F., Coupland, G., 2008. Regulation and identity of florigen: *FLOWERING LOCUS T* moves center stage. Annu. Rev. Plant Biol. 59, 573–594.

Vatén, A., Dettmer, J., Wu, S., Stierhof, Y.D., Miyashima, S., Yadav, S.R., Roberts, C.J., Campilho, A., Bulone, V., Lichtenberger, R., Lehesranta, S., Mähönen, A.P., Kim, J.Y., Jokitalo, E., Sauer, N., Scheres, B., Nakajima, K., Carlsbecker, A., Gallagher, K.L., Helariutta, Y., 2011. Callose biosynthesis regulates symplastic trafficking during root development. Dev. Cell 21, 1144–1155.

Welch, D., Hassan, H., Blilou, I., Immink, R., Heidstra, R., Scheres, B., 2007. *Arabidopsis* JACKDAW and MAGPIE zinc finger proteins delimit asymmetric cell division and stabilize tissue boundaries by restricting SHORT-ROOT action. Genes Dev. 21, 2196–2204.

Wen, C.-K., Chang, C., 2002. *Arabidopsis* RGL1 encodes a negative regulator of gibberellin responses. Plant Cell 14, 87–100.

Wigge, P.A., Kim, M.C., Jaeger, K.E., Busch, W., Schmid, M., Lohmann, J.U., Weigel, D., 2005. Integration of spatial and temporal information during floral induction in *Arabidopsis*. Science 309, 1056–1059.

Winter, N., Kollwig, G., Zhang, S., Kragler, F., 2007. MPB2C, a microtubule-associated protein, regulates non-cell-autonomy of the homeodomain protein KNOTTED1. Plant Cell 19, 3001–3018.

Xoconostle-Cázares, B., Xiang, Y., Ruiz-Medrano, R., Wang, H.L., Monzer, J., Yoo, B.C., McFarland, K.C., Franceschi, V.R., Lucas, W.J., 1999. Plant paralog to viral movement protein that potentiates transport of mRNA into the phloem. Science 283, 94–98.

Xu, H., Zhang, W., Li, M., Harada, T., Han, Z., Li, T., 2010. *GAI* mRNA transport in both directions between stock and scion in *Malus*. Tree Genet. Genomes 6, 1013–1019.

Xu, X.M., Wang, J., Xuan, Z., Goldshmidt, A., Borrill, P.G., Hariharan, N., Kim, J.Y., Jackson, D., 2011. Chaperonins facilitate KNOTTED1 cell-to-cell trafficking and stem cell function. Science 333, 1141–1144.

Yadav, R.K., Reddy, G.V., 2012. WUSCHEL protein movement and stem cell homeostasis. Plant Signal. Behav. 7, 592–594.

Yadav, R.K., Perales, M., Gruel, J., Girke, T., Jonsson, H., Reddy, G.V., 2011. WUSCHEL protein movement mediates stem cell homeostasis in the *Arabidopsis* shoot apex. Genes Dev. 25, 2025–2030.

Yang, H.W., Yu, T.S., 2010. *Arabidopsis* floral regulators *FVE* and *AGL24* are phloem-mobile RNAs. Bot. Stud. 51, 17–26.

Yoo, S.J., Chung, K.S., Jung, S.H., Yoo, S.Y., Lee, J.S., Ahn, J.H., 2010. BROTHER OF AT AND TFL1 (BFT) has TFL1-like activity and functions redundantly with TFL1 in inflorescence meristem development in *Arabidopsis*. Plant J. 63, 241–253.

Yoo, S.C., Chen, C., Rojas, M., Daimon, Y., Ham, B.K., Araki, T., Lucas, W.J., 2013. Phloem long-distance delivery of *FLOWERING LOCUS T* (FT) to the apex. Plant J. 75, 456–468.

Yu, Y., Lashbrook, C.C., Hannapel, D.J., 2007. Tissue integrity and RNA quality of laser microdissected phloem of potato. Planta 226, 797–803.

CHAPTER 24

Redox-Regulated Plant Transcription Factors

Yuan Li, Gary J. Loake

Institute of Molecular Plant Sciences, School of Biological Sciences, University of Edinburgh, King's Buildings, Edinburgh, UK

OUTLINE

24.1 Introduction	373
24.2 Concept of Redox Regulation	374
24.2.1 Production of ROS and NO in Plants	374
24.2.2 Mechanisms of Redox Signaling	376
24.3 Redox Regulation of NPR1 During Plant Immunity	377
24.4 Redox Regulation of Basic Leucine Zipper Transcription Factors	378
24.5 Redox Regulation of MYB Transcription Factors	379
24.6 Redox Regulation of Homeodomain-leucine Zipper Transcription Factors	380
24.7 Rap2.4a is Under Redox Regulation	381
24.8 Redox Regulation of Class I TCP Transcription Factors	381
24.9 Conclusion	382
References	382

24.1 INTRODUCTION

Plants face a wide variety of biotic and abiotic stresses during their growth and development. Transcription factor (TF) activity is instrumental for the deployment of specific gene expression patterns in response to a given stress (Liu et al., 1999). TFs induce or suppress gene expression by binding to specific motifs present on target gene promoters. The function of TFs can be regulated at the transcriptional level, such as maize *C1* (Kao et al., 1996) and/or at the posttranscriptional level, for instance maize *P* (Grotewold et al., 1991). While prototypic posttranslational modifications such as phosphorylation and dephosphorylation are well-established regulators of plant TF function (Klimczak et al., 1992), oxidation reduction (redox) regulation is also now emerging as an important mechanism for controlling TF activity (Table 24.1).

Redox regulation is involved in fine-tuning many metabolic reactions, developmental processes, and defense mechanisms in plants. Triggered as a stress response, redox regulation means reactive oxygen species (ROS) and reactive nitrogen species (RNS) are produced, which can function as redox-active signaling molecules (Apel and Hirt, 2004). However, a high concentration of these molecules can be potentially harmful for plant cells, due to their capacity to mediate cellular damage (Lorrain et al., 2003). To ameliorate this potential outcome, a plethora of antioxidant mechanisms have evolved to protect plant cells from such damage (Trachootham et al., 2008). In the last decade, it has become apparent that redox regulation is integral to control the function of some key TFs *in vivo*.

TABLE 24.1 Redox-Regulated Plant TFs

TFs	Regulatory events	Physiological consequences	References
NPR1	S-nitrosylation on Cys156	Facilitates oligomerization of NPR1	Tada et al. (2008)
	Disulfide bond formation	Facilitates NPR1 oligomer formation, resulting in cytoplasmic sequestration.	Mou et al. (2003)
TGA1	Disulfide bond reduction	Critical for TGA1 target site binding	Després et al. (2003)
	S-nitrosylation and S-gultathionation on Cys260 and Cys266	Enhances binding to *as-1* motif and prevents disulfide bond formation	Lindermayr et al. (2010)
AtbZIP16, AtbZIP68, GBF1	Disulfide bond formation on Cys330	Supports dimerization, which is critical for function	Shaikhali et al. (2012)
P1	Disulfide bond formation between Cys49 and Cys53	DNA binding blunted	Heine et al. (2004)
AtMYB2	S-nitrosylation of Cys53	DNA binding blunted	Serpa et al. (2007)
HAHR1	Disulfide bond formation between Cys156, Cys162, and Cys163	Intermolecular disulfide bond formation between Cys156 results in active dimer. Intermolecular disulfide bond formation between Cys162 and Cys163 results in inactive dimer	Tron et al. (2002)
Athb-9	Disulfide bond formation	DNA binding blunted	Comelli and Gonzalez (2007)
Rap2.4a	Oxidation/reduction in response to cellular redox change	Rap2.4a exist as monomer, dimer, and oligomer depending on cellular redox state. Rap2.4a dimer is active	Shaikhali et al. (2008)
Class I TCP TFs	Intermolecular disulfide bond formation at Cys20 in TCP domain	Reducing environment is required for DNA binding	Viola et al. (2013)

24.2 CONCEPT OF REDOX REGULATION

A redox reaction involves the movement of electrons. A complete redox reaction is comprised of two half-reactions, $A + 2H^+ + 2e^- \rightarrow AH_2$. Thus, the term redox potential is defined by the tendency of a substrate to gain an electron. The bigger the redox potential, the greater the tendency of a substrate to be reduced. The ability of an electron donor to function in a redox reaction depends solely on its redox potential.

The redox-active small molecule, reduced glutathione (GSH), is subsequently oxidized following redox reactions to oxidized glutathione (GSSG), and GSSG is then recycled by NADPH-dependent glutathione reductase (GR). GSSG gains an electron from NADPH in this process while NADPH loses an electron to form nicotinamide adenine dinucleotide phosphate (NADP$^+$; May et al., 1998). NADPH not only acts as an electron donor, but also as an electron acceptor. In photosynthesis, ferredoxin is the acceptor in the photosynthetic electron transport chain with a redox potential of –0.432 V (Segel, 1976), which is more negative than the –0.3 V of NADP$^+$ (Segel, 1976).

Thus, ferredoxin can donate an electron to NADP$^+$ and is catalyzed by ferredoxin–NADP$^+$ reductase.

ROS signaling is central to cellular communication (Apel and Hirt, 2004) and is a key factor in plant responses to both biotic and abiotic stress (Jacquot et al., 2013). It is well established that ROS can interact with protein cysteine residues (D'Autréaux and Toledano, 2007). In addition, production of another plant signaling molecule, nitric oxide (NO), is usually linked to the production of ROS (Neill et al., 2002). In order to protect cells from excessive amounts of ROS antioxidant enzymes such as catalase and ascorbate peroxidase (APX), and low-molecular-weight antioxidants like ascorbate and glutathione, are involved in ROS detoxification (Foyer and Shigeoka, 2011).

24.2.1 Production of ROS and NO in Plants

Photosynthesis in chloroplasts is a major source of ROS in green plant tissues (Dietz et al., 2010). Electrons are transferred from water to oxidized NADP$^+$ to form NADPH via linear electron transfer (LET) via photosystem II (PSII), plastoquinone (Pq), cytochrome b$_6$f complex

(Cytb₆f), plastocyanin, and photosystem I (PSI). Electron transfer through Cytb$_6$f is coupled with photon pumping into the thylakoid to produce a transthylakoid proton motive force to drive the synthesis of ATP (Sierla et al., 2013; Figure 24.1). Unlike LET, cyclic electron transport (CET) redirects electrons from PSI back to Cytb$_6$f, while a proton gradient is built across the thylakoid membrane (Johnson, 2011). In the stroma, electrons are transferred through PSI out of the thylakoid and form H_2O_2 with molecular oxygen (Asada, 1999), which then interacts with the reaction center of chlorophyll P680 in PSII to form singlet oxygen ($^1O_2$). Previous research has linked singlet oxygen to light-induced oxidative damage and photoinhibition of PSII (Hideg et al., 1998). Increasing evidence has suggested that singlet oxygen also plays a major signaling role in response to both excessive and mild light (Alboresi et al., 2011; Kim et al., 2012; Ramel et al., 2012) and during plant immunity (Nomura et al., 2012).

ROS production is also an important early event during plant immune responses (Grant and Loake, 2000; Skelly and Loake, 2013). The major source of ROS under such circumstances is catalyzed by respiratory burst oxidase homologs (RBOHs; Figure 24.2), analogous to the enzymes found in mammalian phagocytes (Grant and Loake, 2000; Lambeth, 2004). The plant genome contains 10 RBOH genes (*RBOHA–RBOHJ*; Torres and Dangl, 2005). By analyzing the single- and double-mutant of RBOH in *Arabidopsis*, RBOHD and RBOHF were found to be the main sources generating ROS after recognition of pathogen-associated molecular patterns (PAMPS) and avirulent pathogens (Torres et al., 2002). In addition to this, RBOHA and RBOHB in tobacco are also required to drive H_2O_2 accumulation

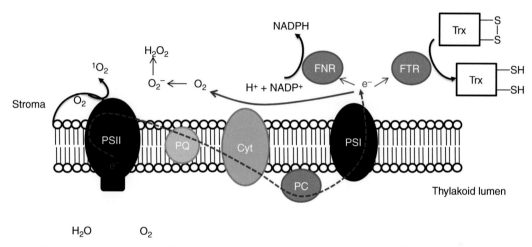

FIGURE 24.1 **Overview of electron flow in the chloroplast during photosynthesis and formation of ROS.** In linear electron transfer, an electron extracted from water in the thylakoid lumen is transferred through photosystem II (PSII), plastoquinone (PQ), cytochrome b$_6$f (Cyt), plastocyanin (PC), and photosystem I (PSI) into the chloroplast stroma. Afterward, the electron acceptor of PSI may redirect electrons to ferredoxin–thioredoxin reductase (FTR) to form reduced thioredoxin, to ferredoxin–NADPH reductase (FNR) to generate NADPH, or to molecular oxygen to form hydrogen peroxide (H_2O_2). In PSII, electron transfer may lead to formation of a singlet oxygen ($^1O_2$) at the stroma.

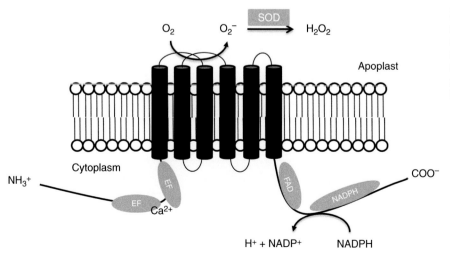

FIGURE 24.2 **Schematic representation of a plant respiratory burst oxidase homolog (RBOH) protein.** Plant RBOHs are located on the plasma membrane and synthesize ROS into the apoplast. NADPH and FAD binding occurs at the C-terminal domain. Two calcium binding EF-hand domains are located at the N-terminus.

in response to an attempted *Phytophthora infestans* infection (Yoshioka et al., 2003). These transmembrane proteins reduce apoplastic oxygen to the superoxide radical, O_2^-. Furthermore, these proteins have an N-terminal EF hand domain, which presumably functions as a calcium-binding domain (Keller et al., 1998). Interestingly, the activity of RBOHD might be under redox regulation during the plant immune response (Yun et al., 2011) and similar findings have also been reported for these cognate NADPH oxidases in mammals (Qian et al., 2012).

Although the emerging evidence suggests that NO has an important role in plant growth, development, and stress response (Yu et al., 2014), the source of NO remains controversial (Yu et al., 2014). In mammals, three isoforms of NO synthase (NOS) have been characterized (nNOS, eNOS, and iNOS; Alderton et al., 2001). These NADP(H)-dependent enzymes catalyze the oxidation of L-arginine to NO and citrulline. However, a NOS similar to those found in mammals has not been found in higher plants. However, a NOS with significant sequence similarity to human eNOS has been found in the unicellular marine green algae, *Ostreococcus tauri* (Foresi et al., 2010). In the model plant *Arabidopsis*, *NIA1* and *NIA2* are genes that encode nitrate reductase (NR), with NIA2 being responsible for most of the NR activity (Wilkinson and Crawford, 1991). It has been reported that NR can convert NO_2^- to NO under low oxygen tensions and high nitrate concentrations with very low efficiency *in vitro* (Rockel et al., 2002). Thus, NR might function as a source of NO in plants.

24.2.2 Mechanisms of Redox Signaling

ROS and RNS target proteins at specific amino acids mediating covalent modifications (Nathan, 2003). It has been reported that rare reactive cysteines with a low pKa sulfhydryl (SH) group are susceptible to a wide variety of oxidizing modifications (Figure 24.3). These modifications include S-nitrosylation, which is the addition of a NO moiety to a Cys thiol to form an S-nitrosothiol (SNO; Yu et al., 2012; Yu et al., 2014). Furthermore, S-sulfenation (SOH) is the oxidation of a Cys thiol to form sulfenic acid. Protein sulfenic acid is relatively unstable and a disulfide bond (S–S) can be formed between two sulfenic acid residues by further oxidization. Disulfide bonds can be either intra or intermolecular and this level of oxidation is important in controlling protein folding and multimerization (D'Autréaux and Toledano, 2007). S-glutathionylation (SSG) is the formation of a disulfide between GSH and a reactive protein Cys thiol residue. S-sulfination (SO_2H) and irreversible S-sulfonation (SO_3H) are more extreme oxidations of protein cysteine thiols (Spadaroa et al., 2010). Many of these modifications can be reversed either in response to changes in the cellular redox environment or enzymatically (Benhar et al., 2008; Tada et al., 2008), thus providing different strategies to plants to regulate protein function in order to adapt to environmental changes.

GSH is the major low-molecular-weight thiol of plant cells. Two reduced GSH molecules can be oxidized to the disulfide GSSG, and glutathione reductase (GR) catalyzes the reduction of GSSG to reform GSH (May et al., 1998). As an important antioxidant, GSH effectively scavenges free radicals, ROS and RNS, directly or indirectly through enzymatic reactions. In particular, scavenging of free radicals by GSH turns GSH into unstable GS^- and subsequently generates $GSSG^-$; together with O_2, $GSSG^-$ can form GSSG and O_2^-. Additionally, O_2^- can be turned over by superoxide dismutase (SOD) or glutathione peroxidase (GPX) to complete the scavenging (Winterbourn, 1993; Aquilano et al., 2014). The cellular GSH/GSSH pool is in highly reduced state in the absence of stress (Noctor et al., 2012).

Glutathione also serves as a carrier of NO, in the form of S-nitrosoglutathione (GSNO). This redox-active small molecule can S-nitrosylate appropriate targets, either by directly transferring NO to a target cysteine or indirectly releasing NO into the cellular environment (Pawloski et al., 2001).

GSNO reductase (GSNOR), the enzyme that turns over GSNO, has been linked to plant immunity (Feechan et al., 2005). *Arabidopsis* plants with increased GSNOR activity have a lower global S-nitrosylation level and exhibit enhanced disease resistance. In contrast, *gsnor* mutants displayed increased total cellular S-nitrosylation and were compromised in multiple modes of disease resistance (Feechan et al., 2005). Additionally, studies on GSNOR also revealed its role in cell death (Chen et al., 2009) and thermotolerance (Lee et al., 2008).

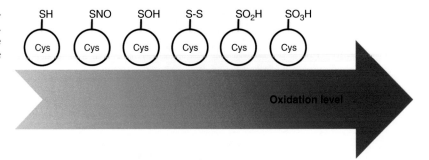

FIGURE 24.3 **Schematic sketch of cysteine modifications according to an increasing level of oxidation.** S-nitrosothiol (SNO), sulfenic acid (SOH), disulfide bond (S–S), sulfinic acid (SO_2H), and the irreversible modification, sulfonic acid (SO_3H).

Thioredoxins (TRX) are important enzymatic antioxidants found in plants. TRX has a conserved active site motif (WCG/PPC) with one or two cysteines. It is known that TRXs are low-redox-potential proteins, and their reactive cysteines at the active site switch form from a reduced state (–SH) to an oxidized form (-S–S-) as a result of interacting with, and reducing, a target protein (Serrato et al., 2013). With more than 40 existing members found in plants, TRXs have been divided into 15 subgroups based on their subcellular location and sequence similarity (Meyer et al., 2012). Two well-defined compartment-specific systems have been found as electron donors for TRX, one is the ferredoxin–thioredoxin system (FTS), which is catalyzed by the enzyme ferredoxin–thioredoxin reducatase (FTR) resulting in reduction of TRX and oxidation of ferredoxin; the other is the NADP–thioredoxin system (NTS), in which electrons are transferred from NADPH to TRX h and o in the cytosol or nucleus, catalyzed by NADPH–thioredoxin reductase (NTR; Serrato et al., 2013).

TRX h3 and h5 have been reported to reduce the disulfide bonds integral to the control of plant immunity (Tada et al., 2008). Furthermore, NTR, which functions to recycle cytosolic TRXs is also required for plant immune function. Apart from denitrosylation, TRX has also been reported to reduce the disulfide bond in GSSG (Marty et al., 2009). It has been suggested that the cytosolic TRX h3/NTRA system operates as a backup in the reduction of GSSG to GSH in the absence of GLUTAHIONE REDUCTASE activity, albeit with lower efficiency. Collectively, the emerging data suggest that redox signaling underpins many features associated with plant growth, development, immunity, and environmental interactions (Homem and Loake, 2013; Yu et al., 2014).

24.3 REDOX REGULATION OF NPR1 DURING PLANT IMMUNITY

Non-Expresser of Pathogenesis Related Gene 1 (NPR1) is a key transcriptional coactivator that regulates the salicylic acid (SA)-dependent plant immune response (Grant and Loake, 2007). A genetic screen found that *npr1* mutants showed an inability to induce *pathogenesis related* (*PR*) genes and cognate systemic acquired resistance (SAR) upon either pathogen challenge or treatment with the plant immune activator, SA (Cao et al., 1994). A key feature of NPR1 is its apparent regulation by redox cues. In this context, conserved cysteine (Cys) residues in NPR1 have been proposed to form intermolecular disulfide bonds, resulting in the formation of a cytosolic NPR1 oligomer (Mou et al., 2003). Furthermore, the formation of this structure may suppress the movement of NPR1 from the cytoplasm into the nucleus, thereby inhibiting its potential interaction with transcriptional regulators integral to the expression of *PR* genes and the subsequent establishment of SAR.

After pathogen or elicitor induction, SA-induced redox changes reduce NPR1 from an oligomer to monomers, facilitating the translocation of NPR1 from the cytosol into the nucleus, which is followed by target gene activation and NPR1-dependent immune responses (Figure 24.4). Further studies have provided an insight into the complexity of

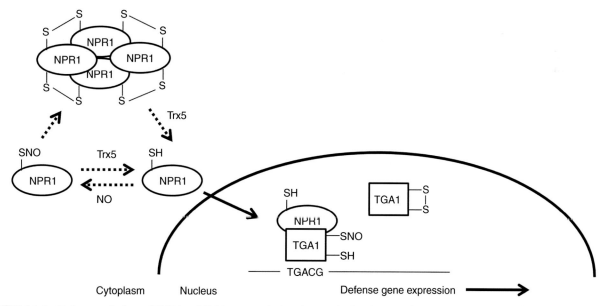

FIGURE 24.4 Redox regulation of NPR1 and TGA1 transcription factors. In the absence of pathogen challenge, NPR1 exists as an oligomer in the cytoplasm with intermolecular disulfide bonds between monomers. TGA1 is located in the nucleus, with intramolecular disulfide bonds preventing binding to its promoter target site. Upon pathogen challenge, a cellular redox change reduces NPR1 oligomer to monomer, facilitating its translocation into the nucleus. NPR1 monomers can then interact with TGA1, which subsequently can bind to its target site, driving defense gene expression. S-nitrosylation of NPR1 monomer promotes its oligomerization in the cytoplasm, while Trx5 can denitrosylate SNO–NPR1 promoting its nuclear translocation. Further, S-nitrosylation of TGA1 may protect against disulfide bond formation, thus increasing its DNA-binding activity.

how NPR1 translocation is regulated by redox change upon pathogen challenge (Mou et al., 2003; Tada et al., 2008). Cys82 and Cys216 are thought to be critical for intermolecular disulfide bond formation facilitating NPR1 oligomer generation. The mutation of both cysteines resulted in the constitutive nuclear localization of the NPR1 monomer and cognate *PR-1* expression (Mou et al., 2003).

It is worth noting that reduction of NPR1 disulfide bonds is synchronized with a transient increase of cellular redox potential as an outcome of SA accumulation upon pathogen attack. By monitoring the levels and ratio of GSH to GSSG after treatment with the SA analog, 2,6-dichloroisonicotinic acid (INA), it was shown that, after an initial oxidative phase, cellular redox potential is increased (Mou et al., 2003). A later study also revealed that, in addition to Cys82 and Cys216, another conserved cysteine, Cys156, is also important in driving the formation of the NPR1 oligomer (Tada et al., 2008; Figure 24.4). Thus, redox control appears to be an important feature in the regulation of plant immune function (Frederickson Matika and Loake, 2014; Hong et al., 2008).

Cys156 in NPR1 can be S-nitrosylated by GSNO. S-nitrosylation of NPR1 has been proposed to facilitate NPR1 oligomerization by promoting disulfide bond formation. Informatively, mutating cysteine 156 in NPR1 leads to constitutive nuclear localization of biological active monomers and enhanced immunity. Unlike the corresponding wild-type plant, a transgenic line expressing a NPR1 C156A mutant does not show boosted immunity in response to SA treatment. Rather, pathogen resistance is compromised as NPR1 C156A was depleted 48 h after SA treatment (Tada et al., 2008). These findings suggest S-nitrosylation of NPR1 Cys156 is integral to the maintenance of NPR1 homeostasis, which promotes sustained immunity. In *gsnor* mutant plants, elevated global SNO levels were found in *Arabidopsis*, and SA signaling was blunted; it was proposed that increased S-nitrosylation of NPR1 promotes its oligomerization, resulting in compromised SA signaling and reduced plant immunity (Feechan et al., 2005). However, in contrast, GSNO has also been reported to promote the translocation of NPR1 from the cytosol to the nucleus (Lindermayr et al., 2010), rather than facilitating NPR1 oligomer formation. These discrepancies remain to be fully resolved; the relatively long timeline between GSNO treatment and subsequent analysis, reported by Lindermayr et al. (2010), suggests these observations may result from the indirect effects of GSNO, such as the induction of SA synthesis.

Interestingly, cytosolic thioredoxins, TRX-h3 and TRX-h5, appear to counter GSNO, by virtue of their ability to reduce disulfide bonds, facilitating NPR1 monomerization (Tada et al., 2008). *TRX-h3* is constitutively expressed while *TRX-h5* expression is upregulated upon pathogen infection (Laloi et al., 2004); both of these genes are required for full induction of *PR* expression. Pull-down experiments have shown that both TRX-h3 and TRX-h5 interact with NPR1 to reduce its oligomerization, and this interaction is enhanced by pathogen infection or SA treatment. Furthermore, the binding affinity between TRX and NPR1 is inversely correlated to its enzymatic activity (Tada et al., 2008). In *trx* mutants, NPR1-dependent SAR is partially impaired; however, in a TRX reductase mutant, *ntra*, which is unable to regenerate active cytosolic TRXs, SAR is strikingly reduced (Tada et al. 2008; Reichheld et al., 2007). Important recent data also suggest that TRXh5 can function as a denitrosylase enzyme by directly converting SNOs in NPR1 to thiols, thus reversing the effect of S-nitrosylation. Hence, TRXh5, and possibly other TRX enzymes, can provide previously unrecognized specificity and reversibility to protein–SNO signaling in plant immunity (Kneeshaw et al., 2014).

Nuclear localization of the NPR1 monomer facilitates its interaction with TGA TFs, which can bind to *PR* gene promoters (Després et al., 2000; Zhou et al., 2000; Fobert and Després, 2005). However, it has been reported that TGA1 does not interact with NPR1 *in vitro* (Després et al., 2003). It has been proposed that cysteines in TGA1 form intramolecular disulfide bonds in an oxidizing environment that preclude the interaction between NPR1 and TGA1 (Després et al., 2003). This finding suggests that, in order to interact with NPR1, the disulfide bonds in TGA1 need to be reduced. In this context, it has been proposed that the in *vivo* accumulation of SA changes the cellular redox potential and reduces the Cys residues in TGA1, stimulating NPR1–TGA1 interaction (Després et al., 2003). It has been revealed that TGA1 is also regulated by GSNO. Cys260 and Cys266 in TGA1 have been reported to be the site of both S-nitrosylation and S-glutathionation, and treatment of TGA1 with GSNO has been reported to enhance its binding activity to the *as-1* motif, a key *cis*-element within the *PR1* promoter and the promoters of other SA-regulated genes (Lindermayr et al., 2010). These data imply that GSNO, by driving the S-nitrosylation/S-glutathionation of TGA1 Cys260 and Cys266, might protect TGA1 from forming intramolecular disulfide bonds in an oxidizing environment. Consequently, this would facilitate the interaction of TGA1 with NPR1 promoting the expression of *PR* genes and the establishment of SAR. Collectively, the emerging data suggest that redox regulation operates at a number of different nodes during the development of plant immunity.

24.4 REDOX REGULATION OF BASIC LEUCINE ZIPPER TRANSCRIPTION FACTORS

AtbZIP16, AtbZIP68, and GBF1 are G-group basic leucine zipper (bZIP) TFs that have been found to respond to redox changes in the chloroplast following exposure to high light (Shaikhali et al., 2012). All three TFs are able to bind to G-box elements within the promoter of *light-harvesting chlorophyll a/b-binding protein 2.4* (*LHCB2.4*; Shaikhali et al., 2012). LHCB2.4 functions in photosystem II, together with another LHCB1 subunit, LHCB1.1. It has been shown that misregulation of both proteins leads to a high-irradiance-sensitive phenotype and photoinactivation of photosystem II (Kindgren et al., 2012). Previous data have suggested that the expression of *LHCB2.4* is reduced in excessive light. Furthermore, the same study also indicated that expression of *LHCB2.4* is either up- or downregulated in response to different redox states (Kindgren et al., 2012).

Given that bZIP16 can bind to the promoter region of *LHCB2.4* and bZIP TFs function as dimers, Shaikhali et al. (2012) studied the mechanism that regulates bZIP16, bZIP68, and GBF1 binding to the *LHCB2.4* promoter. Gel shift assays revealed that bZIP16, bZIP68, and GBF1 are capable of binding to the G-box sequence on *LHCB2.4*, and the affinity of binding increased after the addition of the reducing agent, dithiothreitol (DTT), as this agent stimulated the reduction of bZIP16 oligomers. Conversely, adding H_2O_2 decreased the affinity of this bZIP for the G-box sequence. Similar results were also obtained from analyses of bZIP68 and GBF1. These observations suggest the binding activities of bZIP16, bZIP68, and GBF1 are controlled by redox regulation. These bZIPs have high sequence homology and share the same binding motif. In addition, yeast two-hybrid (Y2H) analysis demonstrated that bZIP16 interacts with bZIP68 and GBF1, which is consistent with previous research (Shen et al., 2008). As bZIP TFs form dimers to function, a gel shift assay using a truncated bZIP16, which only contained the DNA-binding domain, precluding dimerization via the leucine zipper domain, demonstrated DNA binding can be performed by a bZIP16 monomer. This implies that dimerization might occur subsequent to DNA binding. By using modeling tools, a conserved Cys330 has been proposed to be the target of disulfide bridge formation, which is thought to be critical for dimerization (Shaikhali et al., 2012).

Transgenic *Arabidopsis* plants that overexpress *bZIP16* exhibit suppression of *LHCB2.4* expression. However, overexpressing the bZIP16 Cys330 mutant failed to establish a similar phenotype, suggesting bZIP16 homodimers function as repressors of *LHCB2.4*. Interestingly, *bZIP68* and *GBF1* mutants also reduce *LHCB2.4* expression levels in response to light, indicating that bZIP68 and GBF1 are activators. Therefore, bZIP16 heterodimers function differently from homodimers. The combinatorial interactions between bZIP16, bZIP68, and GBF1 may therefore depend on the redox environment and the associated regulation of conserved Cys residues (Shaikhali et al., 2012).

24.5 REDOX REGULATION OF MYB TRANSCRIPTION FACTORS

MYB TFs possess a highly conserved MYB domain required for DNA binding. MYB proteins are classified according to the numbers of imperfect amino acid repeats (R) found within the MYB domain. MYB protein classes include R1, R2, and R3. Each repeat contains three α-helices, with the third helix of each repeat responsible for direct contact with DNA during binding (Dubos et al., 2010).

The most abundant MYB class found in plants is R2R3 MYB. These proteins participate in a large variety of biological processes, including regulating the primary and secondary metabolism of *Arabidopsis*, control of its cell fate, and plant development (Dubos et al., 2010). In animals, Cys130 c-MYB is a highly conserved Cys residue that acts as a redox sensor. Studies have shown that by mutating this Cys, c-MYB DNA-binding activity was abolished. In addition, reduction of this Cys is essential for c-MYB DNA-binding (Guehmann et al., 1992).

In plants, the maize R2R3 MYB protein P1 requires a strong reducing environment in order for DNA binding to occur (Williams and Grotewold, 1997).

Plant R2R3 MYB proteins contain one highly conserved Cys, Cys53, which is believed to correspond to Cys130 in the human c-MYB, which functions as an NO sensor. Another Cys, Cys49, is also conserved in typical R2R3 MYB proteins. Maize P1 contains both conserved Cys49 and Cys53, acting as a R2R3 MYB protein, showing P1 binding to the *a1* promoter (Grotewold et al., 1994). Mutation studies of these two Cys residues have shown that the conserved Cys53 is not essential for the DNA-binding activity of P1, and Cys49 is more important for redox sensing *in vivo* (Heine et al., 2004). In the presence of DTT, the C53A P1 MYB mutant has similar binding affinity to wild-type P1, and C53A is also capable of DNA binding, albeit with less affinity. Furthermore, in a nonreducing environment, C53A still exhibits DNA binding while C53S and wild-type P1 does not. Subsequent tryptophan fluorescence revealed that compared with the wild type, C53A has a more closed conformation in a nonreducing environment, which may result in the formation of a hydrophobic core facilitating DNA binding. A similar conformational change is thought to occur in wild-type P1 and C53S under a reducing environment, while C53A only slightly responds to this redox change (Heine et al., 2004). Additionally, both C53A and C53S are capable of binding the *a1* promoter *in vivo*.

In contrast with Cys53 mutants, C49A and C49S both show binding affinities to DNA and do not respond to redox changes. However, the C49I mutant, even though it exhibits weaker binding compared with wild-type P1, does respond to changes in the redox environment, behaving in a similar fashion to wild-type protein. It is also

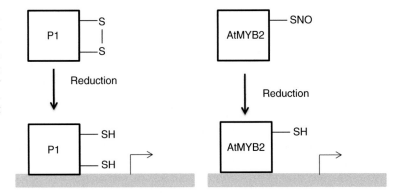

FIGURE 24.5 **Overview of MYB R2R3 TF activity in response to the cellular redox level.** The R2R3 MYB TF requires a relatively reduced environment to function. P1 possesses an intermolecular disulfide bond, while AtMYB2 is S-nitrosylated at Cys53, with both modifications inhibiting the DNA binding of these TFs. In a more reducing cellular environment, the disulfide bond of P1 and the S-nitrosylated thiol group of AtMYB2 are reduced, resulting in the binding of these TFs to their target sites.

worth noting that in the Cys53 mutant an intermolecular disulfide bond forms at two adjacent Cys49 residues in P1 proteins under oxidizing conditions and blocks P1 DNA binding. In wild-type P1, the intramolecular disulfide bond forms between Cys49 and Cys53, which also blocks the binding of DNA (Heine et al., 2004; Figure 24.5).

For atypical R2R3 MYB proteins, the data suggest an inability to form intermolecular disulfide bonds. In this case, S-nitrosylation has been proposed to occur in order to inhibit MYB-binding activity (Serpa et al., 2007). AtMYB2 is a TF involved in controlling salt and dehydration responses in *Arabidopsis* (Abe et al., 2003). Unlike P1, AtMYB2 only contains the conserved Cys53 and not Cys49, which abolishes its ability to form a disulfide bond which would inhibit its binding activity. It has been shown that S-nitrosylation by endogenous NO of AtMYB2 also inhibits its binding activity *in vitro*. Using gel shift assays, it was shown that, unlike typical R2R3 MYB proteins, DNA binding was not facilitated by DTT, which may be due to the fact that its DNA-binding domain is originally reduced. Furthermore, GSNO was found to inhibit its binding activity. Subsequent analysis of AtMYB2 revealed that inhibition of DNA binding correlated with S-nitrosylation of the Cys53 residue (Serpa et al., 2007; Figure 24.5). Collectively, these findings have shown that both oxidation and S-nitrosylation can negatively control the DNA-binding activity of R2R3 MYB TFs.

24.6 REDOX REGULATION OF HOMEODOMAIN-LEUCINE ZIPPER TRANSCRIPTION FACTORS

The homeodomain (HD) is a conserved 60-amino-acid motif that can be found in many TFs throughout eukaryotic organisms. In this domain a three α-helix structure is established by 60 amino acids, which lets these proteins interact with DNA; in this context the third helix functions as a DNA recognition helix (Ariel et al., 2007; Chapter 6).

Among all TFs with an HD, transcriptional regulators with a HD in association with a leucine zipper (HD–ZIP) belong to a family that is unique to plants (Schena and Davis, 1992; Chapter 7). Like other bZIP TFs, the leucine zipper domain in HD–ZIP proteins also functions as a dimerization domain, and HD–ZIP proteins bind DNA as a dimer (Sessa et al., 1993). A different group of HD TFs, Glabra2, contain an N-terminal HD. These proteins also bind DNA as a dimer. In this case, instead of having a complete leucine zipper, the dimerization domain of these HD proteins is comprised of an amino sequence adjacent to the C-terminus of the HD (Di Cristina et al., 1996; Palena et al., 1997). Deleting the ZIP domain in HD–ZIP proteins significantly reduces their ability to bind DNA, this indicates that dimerization is likely to be essential for DNA binding (Palena et al., 1999).

Using sequence alignments, two conserved Cys residues (Cys185 and Cys188) have been identified in the dimerization domain among seven Glabra2 proteins, including HAHR1 (Tron et al., 2002). This study has also shown that HAHR1 requires a reducing environment in order to bind DNA. As shown in gel shift assays, both GSH and DTT were able to facilitate HAHR1 DNA binding. In contrast, no binding was detected by adding water or diamide, which establishes either a neutral or an oxidizing condition, respectively. Mutating both Cys residues into serines results in HAHR1 becoming insensitive to redox changes, exhibiting persistent binding to DNA under changing redox tones. Additionally, an intermolecular disulfide bond was identified in HAHR1 proteins in an oxidized environment. Moreover, HAHR1 monomers rather than dimers were observed in nuclear extracts under reducing environments, but were not observed following exposure to oxidizing conditions (Tron et al., 2002).

Five conserved Cys residues (Cys156, Cys162, Cys163, Cys201, and Cys204) in HD–ZIP II proteins have been identified in the same alignment. Cys156 and Cys163 are conserved in all HD–ZIP II proteins, while Cys162 is present in most. In gel shift assays, Hahb-10 showed similar properties to HAHR1; the binding activity of Hahb-10 was weakened under oxidizing conditions, while GSH and DTT increased their DNA-binding activity. Furthermore, unless all Cys were mutated, Hahb-10 formed intermolecular disulfide bonds under oxidizing conditions. Studies of different combinations of Hahb-10 Cys mutants demonstrated that Cys162 and Cys163 mainly determine redox sensitivity with the help of Cys201 and Cys204, while oxidation of Cys156 does not have an impact on DNA binding. As regards dimerization, Cys156, Cys162, and Cys163 are responsible for dimer and oligomer formation; intramolecular disulfide bonds formed by two adjacent Cys156 residues result in an active dimer, while a dimer formed between Cys162 and Cys163 may result in an inactive protein. Furthermore, it has also been suggested that reduced thioredoxin can activate HAHR1 and Hahb-10 (Tron et al., 2002).

HD–ZIP class III protein, Athb-9, contains four cysteines. Consistent with other HD–ZIP proteins, Athb-9 was only found to be active in the presence of the reducing agents DTT and GSH. Furthermore, reduced thioredoxin could also catalyze the activation of Athb-9 (Comelli and Gonzalez, 2007). Additionally, as reported for other HD–ZIP proteins, Athb-9 forms intermolecular disulfide bonds in oxidized environments, which can be reduced by GSH and DTT. Treating cauliflower nuclear protein extracts with DTT and TRX facilitates protein binding to Athb-9 DNA target sequences, while oxidizing reagents do not promote this reaction (Comelli and Gonzalez, 2007). All single Cys mutants show the same response to redox changes as the wild type and are capable of forming intramolecular disulfide bonds. Therefore, more than one Cys may be required to establish the redox sensitivity of Athb-9. By testing different combinations of Cys mutations, it was established that Cys23 and Cys38 are responsible for the redox sensitivity of Athb-9; Cys42 is not conserved among all HD–ZIP III proteins. In addition, Cys58 is located in the DNA recognition helix of the protein, and gel shift assays suggest that mutation of Cys58 may also alter the DNA-binding specificity of this protein (Comelli and Gonzalez, 2007). Thus, redox control appears to be integral to HD–ZIP protein activity.

24.7 RAP2.4A IS UNDER REDOX REGULATION

Rap2.4a is a TF, isolated recently following a yeast one-hybrid (Y1H) screen, which binds to the 2-Cys peroxiredoxin-A (*2CPA*) gene promoter. 2CPA is a well-characterized chloroplast-located antioxidant enzyme (Shaikhali et al., 2008). Rap2.4a was shown to bind to the CAGGCGATTC sequence within the *2CPA* promoter region between −438 bp and −451 bp. Rap2.4a functions as an activator, and this activity is elevated in oxidizing conditions. In contrast, its transcriptional activator activity was decreased in response to reducing agents such as DTT and ascorbate. Therefore, the activity of Rap2.4 appears to be redox dependent (Shaikhali et al., 2008). SDS–PAGE suggested Rap2.4a exists as a monomer, dimer, and oligomer. Oxidizing reagents such as H_2O_2 can facilitate its oligomerization and, in contrast, reducing reagents like DTT can reduce Rap2.4a into individual monomers. *In vitro* binding experiments have indicated that under high oxidizing or reducing environments the activity of Rap2.4a is inhibited. These results suggest that Rap2.4a functions as a dimer, is required to drive the expression of *2CPA*, and is under redox control (Shaikhali et al., 2008; Figure 24.6).

24.8 REDOX REGULATION OF CLASS I TCP TRANSCRIPTION FACTORS

Recent data imply that redox regulation is an important feature of plant developmental mechanisms (Homem and Loake, 2013; Kwon et al., 2012). TEOSINTE BRANCHED1, CYCLOIDEA, and PROLIFERATING CELL FACTOR1 (TCP) proteins are a family of plant TFs that function in plant developmental processes associated with cell proliferation and growth (Martín-Trillo and Cubas, 2010; Uberti Manassero et al., 2013; Chapter 16). TCP proteins contain a highly conserved domain (TCP domain) which is involved in DNA binding and dimerization. The TCP domain contains an N-terminal feature enriched with basic amino acids and a helix–loop–helix motif similar to basic helix–loop–helix (bHLH) TFs (Martín-Trillo and Cubas, 2010; Aggarwal et al., 2010; Chapter 9). However, compared with bHLH proteins, their target recognition sites are different (Kosugi and Ohashi, 2002). Based on sequence homology, TCP proteins have been divided into two classes. Despite differences in the size of TCP proteins, a conserved Cys has been found at position 20 within the TCP domain of class I TCP TFs (Viola et al., 2013). The binding activity of class I TCP proteins has been shown to be altered as redox conditions change. Furthermore, Cys20 in the class I TCP domain has been shown to be a redox sensor, as a mutation of this Cys does not affect DNA-binding ability but rather eliminates its sensitivity to redox change (Viola et al., 2013). Furthermore, studies have shown that a class I TCP protein, with Cys20 mutation, exists as a monomer under all conditions, which suggests this residue is responsible for the formation of an intermolecular disulfide bond under an oxidized environment, affecting the binding activity of class I TCP proteins to DNA. In addition, subsequent *in vivo* experiments implied that the DNA-binding activity of TCP proteins is modulated by redox treatment (Viola et al., 2013). Therefore, redox regulation is emerging as an important feature of TCP protein function.

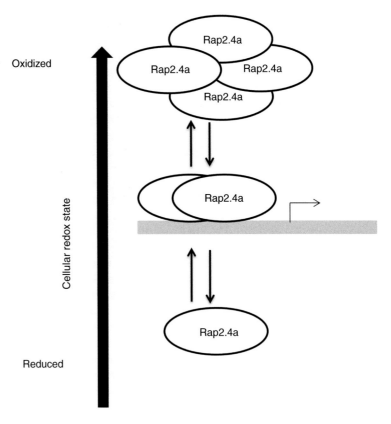

FIGURE 24.6 **Activity of Rap2.4a in response to cellular redox change.** Rap2.4a does not bind to target genes as a monomer. As the cellular redox state changes to a more oxidized environment, Rap2.4a forms dimers and subsequently DNA binding occurs, promoting target gene expression. A further increase in the cellular oxidation state results in Rap2.4a oligomer formation, which inhibits the binding of this regulator to its DNA target site. Figure modified from Shaikhali et al. (2008).

24.9 CONCLUSION

TFs are important regulators that control gene expression in plant growth, development, and response to environmental stress. When considering their importance, it is vital to stringently control their activity. Emerging evidence suggests that an increasing number of TFs are controlled by redox regulation, including bZIP, MYB, and HD–ZIP proteins. These TFs function in the control of photosynthesis, immunity, and abiotic stress responses. Further insights into the redox regulation of plant TFs may help guide future breeding and crop design strategies.

Acknowledgments

G.J.L. acknowledges support of a Biotechnology and Biological Sciences Research Council research grant BB/DO11809/1 to help work on redox regulation in the Loake Laboratory.

References

Abe, H., Urao, T., Ito, T., Seki, M., Shinozaki, K., Yamaguchi-Shinozaki, K., 2003. Arabidopsis AtMYC2 (bHLH) and AtMYB2 (MYB) function as transcriptional activators in abscisic acid signaling. Plant Cell 15, 63–78.

Aggarwal, P., Das Gupta, M., Joseph, A.P., Chatterjee, N., Srinivasan, N., Nath, U., 2010. Identification of specific DNA binding residues in the TCP family of transcription factors in Arabidopsis. Plant Cell 22, 1174–1189.

Alboresi, A., Dall'osto, L., Aprile, A., Carillo, P., Roncaglia, E., Cattivelli, L., Bassi, R., 2011. Reactive oxygen species and transcript analysis upon excess light treatment in wild-type Arabidopsis thaliana vs. a photosensitive mutant lacking zeaxanthin and lutein. BMC Plant Biol. 11, 62.

Alderton, W.K., Cooper, C.E., Knowles, R.G., 2001. Nitric oxide synthases: structure, function and inhibition. Biochem. J. 357, 593–615.

Apel, K., Hirt, H., 2004. Reactive oxygen species: metabolism, oxidative stress, and signal transduction. Annu. Rev. Plant. Biol. 55, 373–399.

Aquilano, K., Baldelli, S., Ciriolo, M.R., 2014. Glutathione: new roles in redox signaling for an old antioxidant. Front. Pharmacol. 5, 196.

Ariel, A.D., Manavella, P.A., Dezar, C.A., Chan, R.C., 2007. The true story of the HD-Zip family. Trends Plant Sci. 12, 419–426.

Asada, K., 1999. The water–water cycle in chloroplasts: scavenging of active oxygens and dissipation of excess photons. Annu. Rev. Plant Physiol. Plant Mol. Biol. 50, 601–639.

Benhar, M., Forrester, M.T., Hess, D.T., Stamler, J.S., 2008. Regulated protein denitrosylation by cytosolic and mitochondrial thioredoxins. Science 320, 1050–1054.

Cao, H., Bowling, S.A., Gordon, S., Dong, X., 1994. Characterization of an Arabidopsis mutant that is nonresponsive to inducers of systemic acquired resistance. Plant Cell 6, 1583–1592.

Chen, R., Sun, S., Wang, C., Li, Y., Liang, Y., An, F., Li, C., Dong, H., Yang, X., Zhang, J., Zuo, J., 2009. The Arabidopsis PARAQUAT RESISTANT2 gene encodes an S-nitrosoglutathione reductase that is a key regulator of cell death. Cell Res. 19, 1377–1387.

Comelli, R.N., Gonzalez, D.H., 2007. Conserved homeodomain cysteines confer redox sensitivity and influence the DNA binding properties of plant class III HD-Zip proteins. Arch. Biochem. Biophys. 467, 41–47.

D'Autréaux, B., Toledano, M.B., 2007. ROS as signalling molecules: mechanisms that generate specificity in ROS homeostasis. Nat. Rev. Mol. Cell Biol. 8, 813–824.

Després, C., DeLong, C., Glaze, S., Liu, E., Fobert, P.R., 2000. The *Arabidopsis* NPR1/NIM1 protein enhances the DNA binding activity of a subgroup of the TGA family of bZIP transcription factors. Plant Cell 12, 279–290.

Després, C., Chubak, C., Rochon, A., Clark, R., Bethune, T., Desveaux, D., Fobert, P.R., 2003. The *Arabidopsis* NPR1 disease resistance protein is a novel cofactor that confers redox regulation of DNA binding activity to the basic domain/leucine zipper transcription factor TGA1. Plant Cell 15, 2181–2191.

Di Cristina, M., Sessa, G., Dolan, L., Linstead, P., Baima, S., Ruberti, I., Morelli, G., 1996. The *Arabidopsis* Athb-10 (GLABRA2) is an HD-Zip protein required for regulation of root hair development. Plant J. 10, 393–402.

Dietz, K.J., Jacquot, J.P., Harris, G., 2010. Hubs and bottlenecks in plant molecular signalling networks. New Phytol. 188, 919–938.

Dubos, C., Stracke, R., Grotewold, E., Weisshaar, B., Martin, C., Lepiniec, L., 2010. MYB transcription factors in *Arabidopsis*. Trends Plant Sci. 15, 573–581.

Feechan, A., Kwon, E., Yun, B.W., Wang, Y.Q., Pallas, J.A., Loake, G.J., 2005. A central role for S-nitrosothiols in plant disease resistance. Proc. Natl. Acad. Sci. USA 102, 8054–8059.

Fobert, P.R., Després, C., 2005. Redox control of systemic acquired resistance. Curr. Opin. Plant Biol. 8, 378–382.

Foresi, N., Correa-Aragunde, N., Parisi, G., Calo, G., Salerno, G., Lamattina, L., 2010. Characterization of a nitric oxide synthase from the plant kingdom: NO generation from the green alga *Ostreococcus tauri* is light irradiance and growth phase dependent. Plant Cell 22, 3816–3830.

Foyer, C.H., Shigeoka, S., 2011. Understanding oxidative stress and antioxidant functions to enhance photosynthesis. Plant Physiol. 155, 93–100.

Frederickson Matika, D.E., Loake, G.J., 2014. Redox regulation in plant immune function. Antioxid. Redox Signal. 21, 1373–1388.

Grant, J., Loake, G.J., 2000. Role of reactive oxygen intermediates and cognate redox signalling in disease resistance. Plant Physiol. 124, 21–29.

Grant, M., Loake, G.J., 2007. Recent advances in salicylic acid signalling. Curr. Opin. Plant Biol. 10, 466–472.

Grotewold, E., Athma, P., Peterson, T., 1991. Alternatively spliced products of the maize P gene encode proteins with homology to the DNA-binding domains of myb-like transcription factors. Proc. Natl. Acad. Sci. USA 88, 4587–4591.

Grotewold, E., Drummond, B.J., Bowen, B., Peterson, T., 1994. The myb-homologous P gene controls phlobaphene pigmentation in maize floral organs by directly activating a flavonoid biosynthetic gene subset. Cell 76, 543–553.

Guehmann, S., Vorbrueggen, G., Kalkbrenner, F., Moelling, K., 1992. Reduction of a conserved Cys is essential for Myb DNA-binding. Nucleic Acids Res. 20, 2279–2286.

Heine, G.F., Hernandez, J.M., Grotewold, E., 2004. Two cysteines in plant R2R3 MYB domains participate in REDOX-dependent DNA binding. J. Biol. Chem. 279, 37878–37885.

Hideg, E., Kálai, T., Hideg, K., Vass, I., 1998. Photoinhibition of photosynthesis *in vivo* results in singlet oxygen production detection via nitroxide-induced fluorescence quenching in broad bean leaves. Biochemistry 37, 11405–11411.

Homem, R.A., Loake, G.J., 2013. Orchestrating plant development, metabolism and plant –microbe interactions NO problem! New Phytol. 197, 1035–1038.

Hong, J.-K., Yun, B.W., Kang, J.G., Raja, M.U., Kwon, E.-J., Sorhagen, S., Chu, C., Wang, Y.-Q., Loake, G.J., 2008. Nitric oxide function and signalling in plant disease resistance. J. Exp. Bot. 59, 147–154.

Jacquot, J.P., Dietz, K.J., Rouhier, N., Meux, E., Lallement, P.A., Selles, B., Hecker, A., 2013. Redox regulation in plants: glutathione and "redoxin"-related families. In: Jakob, U., Reichmann, D. (Eds.), Oxidative Stress and Redox Regulation. Springer Science Business Media, Dordrecht, pp. 213–231.

Johnson, G.N., 2011. Reprint of: Physiology of PSI cyclic electron transport in higher plants. Biochim. Biophys. Acta 1807, 906–911.

Kao, C.Y., Cocciolone, S.M., Vasil, I.K., McCarthy, D.R., 1996. Localization and interaction of the *cis*-acting elements for abscisic acid, Viviparous1 and light activation of the C1 gene of maize. Plant Cell 8, 1171–1179.

Keller, T., Damude, H.G., Werner, D., Doerner, P., Dixon, R.A., Lamb, C., 1998. A plant homolog of the neutrophil NADPH oxidase gp91[phox] subunit gene encodes a plasma membrane protein with Ca^{2+} binding motifs. Plant Cell 10, 255–266.

Kim, C., Meskauskiene, R., Zhang, S., Lee, K.P., Lakshmanan Ashok, M., Blajecka, K., Herrfurth, C., Feussner, I., Apel, K., 2012. Chloroplasts of *Arabidopsis* are the source and a primary target of a plant-specific programmed cell death signaling pathway. Plant Cell 24, 3026–3039.

Kindgren, P., Kremnev, D., Blanco, N.E., de Dios Barajas Lopez, J., Fernandez, A.P., Tellgren-Roth, C., Small, I., Strand, A., 2012. The plastid redox insensitive 2 mutant of *Arabidopsis* is impaired in PEP activity and high light-dependent plastid redox signalling to the nucleus. Plant J. 70, 279–291.

Klimczak, L.J., Schindler, U., Cashmore, A.R., 1992. DNA-binding activity of the *Arabidopsis* G-box binding factor GBF1 is stimulated by phosphorylation by casein kinase II from broccoli. Plant Cell 4, 87–98.

Kneeshaw, S., Gelineau, S., Tada, Y., Loake, G.J., Spoel, S.H., 2014. Selective protein denitrosylation activity of thioredoxin-*h5* modulates plant immunity. Mol. Cell 56, 153–162.

Kosugi, S., Ohashi, Y., 2002. DNA binding and dimerization specificity and potential targets for the TCP protein family. Plant J. 30, 337–348.

Kwon, E., Feechan, A., Yun, B.W., Hwang, B.H., Pallas, J.A., Kang, J.-G., Loake, G.J., 2012. AtGSNOR1 function is required for multiple developmental programs in *Arabidopsis*. Planta 236, 887–900.

Laloi, C., Mestres-Ortega, D., Marco, Y., Meyer, Y., Reichheld, J.P., 2004. The *Arabidopsis* cytosolic thioredoxin *h5* gene induction by oxidative stress and its W-box-mediated response to pathogen elicitor. Plant Physiol. 134, 1006–1016.

Lambeth, J.D., 2004. NOX enzymes and the biology of reactive oxygen. Nat. Rev. Immunol. 4, 181–189.

Lee, U., Wie, C., Fernandez, B.O., Feelisch, M., Vierling, E., 2008. Modulation of nitrosative stress by S-nitrosoglutathione reductase is critical for thermotolerance and plant growth in *Arabidopsis*. Plant Cell 20, 786–802.

Lindermayr, C., Sell, S., Muller, B., Leister, D., Durner, J., 2010. Redox regulation of the NPR1–TGA1 system of *Arabidopsis thaliana* by nitric oxide. Plant Cell 22, 2894–2907.

Liu, L., White, M.J., MacRae, T.H., 1999. Transcription factors and their genes in higher plants: functional domains, evolution and regulation. Eur. J. Biochem. 262, 247–257.

Lorrain, S., Vailleau, F., Balague, C., Roby, D., 2003. Lesion mimic mutants: keys for deciphering cell death and defense pathways in plants? Trends Plant Sci. 8, 263–271.

Martín-Trillo, M., Cubas, P., 2010. TCP genes: a family snapshot ten years later. Trends Plant Sci. 15, 31–39.

Marty, L., Siala, W., Schwarzlander, M., Fricker, M.D., Wirtz, M., Sweetlove, L.J., Meyer, Y., Meyer, A.J., Reichheld, J.P., Hell, R., 2009. The NADPH-dependent thioredoxin system constitutes a functional backup for cytosolic glutathione reductase in *Arabidopsis*. Proc. Natl. Acad. Sci. USA 106, 9109–9114.

May, M.J., Vernoux, T., Leaver, C., Van Montagu, M., Inze, D., 1998. Glutathione homeostasis in plants: implications for environmental sensing and plant development. J. Exp. Bot. 49, 649–667.

Meyer, Y., Belin, C., Delorme-Hinoux, V., Reichheld, J.P., Riondet, C., 2012. Thioredoxin and glutaredoxin systems in plants: molecular mechanisms, crosstalks, and functional significance. Antioxid. Redox Signal. 17, 1124–1160.

Mou, Z., Fan, W., Dong, X., 2003. Inducers of plant systemic acquired resistance regulate NPR1 function through redox changes. Cell 113, 935–944.

Nathan, C., 2003. Specificity of a third kind: reactive oxygen and nitrogen intermediates in cell signaling. J. Clin. Invest. 111, 769–778.

Neill, S., Desikan, R., Hancock, J., 2002. Hydrogen peroxide signalling. Curr. Opin. Plant Biol. 5, 388–395.

Noctor, G., Mhamdi, A., Chaouch, S., Han, Y., Neukermans, J., Queval, G., Foyer, C.H., 2012. Glutathione in plants: an integrated overview. Plant Cell Environ. 35, 454–458.

Nomura, H., Komori, T., Uemura, S., Kanda, Y., Shimotani, K., Nakai, K., Furuichi, T., Takebayashi, K., Sugimoto, T., Sano, S., Suwastika, I.N., Fukusaki, E., Yoshioka, H., Nakahira, Y., Shiina, T., 2012. Chloroplast-mediated activation of plant immune signaling in *Arabidopsis*. Nat. Commun. 3, 926.

Palena, C.M., Chan, R.L., Gonzalez, D.H., 1997. A novel type of dimerization motif, related to leucine zippers, is present in plant homeodomain proteins. Biochim. Biophys. Acta 1352, 203–212.

Palena, C.M., Gonzalez, D.H., Chan, R.L., 1999. A monomer–dimer equilibrium modulates the interaction of the sunflower homeodomain leucine–zipper protein Hahb-4 with DNA. Biochem J. 341, 81–87.

Pawloski, J.R., Hess, D.T., Stamler, J.S., 2001. Export by red blood cells of nitric oxide bioactivity. Nature 409, 622–626.

Qian, J., Chen, F., Kovalenkov, Y., Pandey, D., Moseley, M.A., Foster, M.W., Black, S.M., Venema, R.C., Stepp, D.W., Fulton, D.J., 2012. Nitric oxide reduces NADPH oxidase 5 (Nox5) activity by reversible S-nitrosylation. Free Radic. Biol. Med. 52, 1806–1819.

Ramel, F., Birtic, S., Ginies, C., Soubigou-Taconnat, L., Triantaphylides, C., Havaux, M., 2012. Carotenoid oxidation products are stress signals that mediate gene responses to singlet oxygen in plants. Proc. Natl. Acad. Sci. USA 109, 5535–5540.

Reichheld, J.P., Khafif, M., Riondet, C., Droux, M., Bonnard, G., Meyer, Y., 2007. Inactivation of thioredoxin reductases reveals a complex interplay between thioredoxin and glutathione pathway in *Arabidopsis* development. Plant Cell 19, 1851–1865.

Rockel, P., Strube, F., Rockel, A., Wildt, J., Kaiser, W.M., 2002. Regulation of nitric oxide (NO) production by plant nitrate reductase *in vivo* and *in vitro*. J. Exp. Bot. 53, 103–110.

Schena, M., Davis, R.W., 1992. HD-Zip protein members of *Arabidopsis* homeodomain protein superfamily. Proc. Natl. Acad. Sci. USA 89, 3894–3898.

Segel, I.H., 1976. Biochemical Calculations. John Wiley & Sons, New York.

Serpa, V., Vernal, J., Lamattina, L., Grotewold, E., Cassia, R., Terenzi, H., 2007. Inhibition of AtMYB2 DNA-binding by nitric oxide involves cysteine S-nitrosylation. Biochem. Biophys. Res. Commun. 361, 1048–1053.

Serrato, A.J., Fernández-Trijueque, J., Barajas-López, J.-D., Chueca, A., Sahrawy, M., 2013. Plastid thioredoxins: a "one-for-all" redox-signaling system in plants. Front. Plant Sci. 4, 463.

Sessa, G., Morelli, G., Ruberti, I., 1993. The Athb-1 and -2 HD-Zip domains homodimerize forming complexes of different DNA binding specificities. EMBO J. 12, 3507–3517.

Shaikhali, J., Heiber, I., Seidel, T., Stroher, E., Hiltscher, H., Birkmann, S., Dietz, K.J., Baier, M., 2008. The redox-sensitive transcription factor Rap2.4a controls nuclear expression of 2-Cys peroxiredoxin A and other chloroplast antioxidant enzymes. BMC Plant Biol. 8, 48.

Shaikhali, J., Norén, L., de Dios Barajas-López, J., Srivastava, V., König, J., Sauer, U.H., Wingsle, G., Dietz, K.J., Strand, A., 2012. Redox-mediated mechanisms regulate DNA binding activity of the G-group of basic region leucine zipper (bZIP) transcription factors in *Arabidopsis*. J. Biol. Chem. 287, 27510–27525.

Shen, H., Cao, K., Wang, X., 2008. AtbZIP16 and AtbZIP68, two new members of GBFs, can interact with other G group bZIPs in *Arabidopsis thaliana*. BMB Rep. 41, 132–138.

Sierla, M., Rahikainen, M., Salojärvi, J., Kangasjärvi, J., Kangasjärvi, S., 2013. Apoplastic and chloroplastic redox signaling networks in plant stress responses. Antioxid. Redox Signal. 18, 2220–2239.

Skelly, M., Loake, G.J., 2013. Synthesis of redox-active molecules and their signaling functions during the expression of plant disease resistance. Antioxid. Redox Signal. 19, 990–997.

Spadaroa, D., Yun, B.Y., Spoel, S.H., Chuc, C., Wang, Y.Q., Loake, G.J., 2010. The redox switch: dynamic regulation of protein function by cysteine modifications. Physiol. Plant. 138, 360–371.

Tada, Y., Spoel, S.H., Pajerowska-Mukhtar, K., Mou, Z., Song, J., Wang, C., Zuo, J., Dong, X., 2008. Plant immunity requires conformational charges of NPR1 via S-nitrosylation and thioredoxins. Science 321, 952–956.

Torres, M.A., Dangl, J.L., 2005. Functions of the respiratory burst oxidase in biotic interactions, abiotic stress and development. Curr. Opin. Plant. Biol. 8, 397–403.

Torres, M.A., Dangl, J.L., Jones, J.D., 2002. *Arabidopsis* gp91^{phox} homologues AtrbohD and AtrbohF are required for accumulation of reactive oxygen intermediates in the plant defense response. Proc. Natl. Acad. Sci. USA 99, 517–522.

Trachootham, D., Lu, W., Ogasawara, M.A., Nilsa, R.D., Huang, P., 2008. Redox regulation of cell survival. Antioxid. Redox Signal. 10 (8), 1343–1374.

Tron, A.E., Bertoncini, C.W., Chan, R.L., Gonzalez, D.H., 2002. Redox regulation of plant homeodomain transcription factors. J. Biol. Chem. 277, 34800–34807.

Uberti Manassero, N.G., Viola, I.L., Welchen, E., Gonzalez, D.H., 2013. TCP transcription factors: architectures of plant form. Biomol. Concepts 4, 111–127.

Viola, I.L., Güttlein, L.N., Gonzalez, D.H., 2013. Redox modulation of plant developmental regulators from the class I TCP transcription factor family. Plant Physiol. 162, 1434–1447.

Wilkinson, J.Q., Crawford, N.M., 1991. Identification of the *Arabidopsis* CHL3 gene as the nitrate reductase structural gene NIA2. Plant Cell 3, 461–471.

Williams, C.E., Grotewold, E., 1997. Differences between plant and animal Myb domains are fundamental for DNA binding activity and chimeric Myb domains have novel DNA-binding specificities. J. Biol. Chem. 272, 563–571.

Winterbourn, C.C., 1993. Superoxide as an intracellular radical sink. Free Radic. Biol. Med. 14, 85–90.

Yoshioka, H., Numata, N., Nakajima, K., Katou, S., Kawakita, K., Rowland, O., Jones, J.D., Doke, N., 2003. *Nicotiana benthamiana* gp91 homologs NbrbohA and NbrbohB participate in H$_2$O$_2$ accumulation and resistance to *Phytophthora infestans*. Plant Cell 15, 706–718.

Yu, M., Yun, B.W., Spoel, S.H., Loake, G.J., 2012. A sleigh ride through the SNO: regulation of plant immune function by S-nitrosylation. Curr. Opin. Plant Biol. 15, 424–431.

Yu, M., Lamattina, L., Spoel, S.H., Loake, G.J., 2014. Nitric oxide function in plant biology: a redox cue in deconvolution. New Phytol. 202, 1142–1156.

Yun, B.-W., Feechan, A., Yin, M., Saidi, N.B., Le Bihan, T., Yu, M., Moore, J.W., Kang, J.G., Kwon, E., Spoel, S.H., Pallas, J.A., Loake, G.J., 2011. S-nitrosylation of NADPH oxidase regulates cell death in plant immunity. Nature 478, 264–268.

Zhou, J.M., Trifa, Y., Silva, H., Pontier, D., Lam, E., Shah, J., Klessig, D.F., 2000. NPR1 differentially interacts with members of the TGA/OBF family of transcription factors that bind an element of the *PR-1* gene required for induction by salicylic acid. Mol. Plant Microbe Interact. 13, 191–202.

CHAPTER 25

Membrane-Bound Transcription Factors in Plants: Physiological Roles and Mechanisms of Action

Yuji Iwata, Nozomu Koizumi

Graduate School of Life and Environmental Sciences, Osaka Prefecture University, Nakaku, Sakai, Osaka, Japan

OUTLINE

25.1 Introduction	385	25.3.2 NTL6	389
25.2 bZIP Transcription Factors	386	25.3.3 ANAC089	390
25.2.1 bZIP28	386	25.3.4 ANAC060	390
25.2.2 bZIP17	387	25.3.5 ANAC013 and ANAC017	390
25.2.3 bZIP60	387	25.4 Conclusions and Future Perspectives	391
25.3 NAC Transcription Factors	389	References	393
25.3.1 NTM1	389		

25.1 INTRODUCTION

The activity of transcription factors needs to be tightly controlled to regulate gene expression at the correct developmental timing or in a response to environmental stress. Posttranslational modifications modulate the activity of transcription factors through different mechanisms. For example, the phosphorylation status of transcription factors affects subcellular localization so that they only translocate into the nucleus, where gene expression occurs, when needed. In some other instances, the ability of transcription factors to recognize and bind to a DNA sequence on the promoters of target genes is affected, so that they activate genes only when required. Additionally, this also regulates other aspects of transcription factors, including stability and the interaction with cofactors. These mechanisms provide a rapid means of regulating a transcriptional response via transcription factors.

A recently emerging mechanism of transcription factor activation is the liberation of a membrane-bound transcription factor. This is where a transcription factor resides on an intracellular membrane by virtue of a transmembrane domain (TMD) as an inactive precursor form being excluded from the nucleus. Upon internal or environmental stimuli, the transcription factor is freed from the membrane and translocates into the nucleus where gene expression occurs. In most cases, this is mediated *via* proteolytic cleavage by site-specific membrane-bound proteases, which allow the cytoplasmic side of the processed peptide fragments containing a DNA-binding domain to be liberated from the membrane. Of interest is a mechanism by which proteolytic processing occurs within a TMD of membrane-bound transcription factors called regulated intramembrane proteolysis (RIP). This is widely conserved in diverse organisms ranging from bacteria, archaea, yeast, animals, and plants (Adam, 2013; Weihofen and Martoglio, 2003; Seo et al., 2008; Gao et al., 2008).

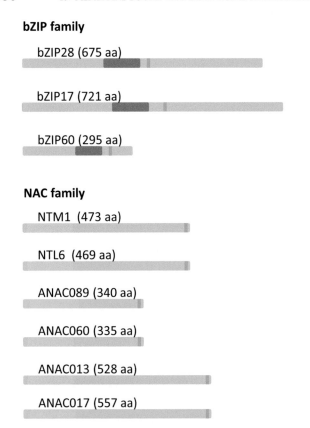

FIGURE 25.1 Schematic representation of the domain structure of membrane-bound transcription factors described in this review. Locations of bZIP domain (orange), NAC domain (yellow), and TMD (light blue) are indicated. The numbers of amino acid residues for each transcription factor are also indicated. Shown are those from the Col accession except for ANAC060 from the C24 accession.

Bioinformatics analysis has shown that plant genomes encode a significant number of genes for transcription factors with putative TMD (Kim et al., 2010b). Indeed, in the past several years, plant biologists have witnessed the important roles played by membrane-bound transcription factors in diverse physiological phenomena, such as cell division, brassinosteroid signaling, and salt and drought stresses, to name but a few. To date, several membrane-bound transcription factors with a basic leucine zipper (bZIP) domain or NAM/ATAF1/2/CUC2 (NAC) domain from *Arabidopsis thaliana* have been characterized. These will therefore be the main focus in the following sections of this chapter (Figure 25.1). We will summarize their physiological roles in development and stress responses as well as their molecular mechanisms of activation.

25.2 bZIP TRANSCRIPTION FACTORS

25.2.1 bZIP28

Arabidopsis bZIP28 is probably the best characterized plant membrane-bound transcription factor whose activity is regulated by RIP. bZIP28 is activated by the accumulation of unfolded proteins in the endoplasmic reticulum (ER), a condition called ER stress. ER stress arises due to various circumstances, including developmental processes and stress responses that require a large amount of secretory and membrane protein synthesis. This is overcome by activating genes encoding components for ER protein quality control such as the ER-resident molecular chaperones that assist protein folding and maturation in the ER. This cellular response is referred to as the unfolded protein response or the ER stress response (Iwata and Koizumi, 2012; Howell, 2013; Korennykh and Walter, 2012; Walter and Ron, 2011; Vitale and Boston, 2008). The signal transduction pathways of the plant ER stress response have been extensively characterized in the last decade using pharmacological agents that induce ER stress, such as the asparagine-linked glycosylation inhibitor tunicamycin and the reducing agent dithiothreitol.

bZIP28 resides in the ER membrane through its TMD with a bZIP domain in the cytoplasm and a C-terminal stress-sensing domain in the ER lumen (Tajima et al., 2008; Liu et al., 2007a; Liu and Howell, 2010a). It has been proposed that, upon receiving ER stress, bZIP28 is translocated *via* vesicle transport to the Golgi where sequential cleavage occurs by two membrane-bound proteases, site-1 protease (S1P) and site-2 protease (S2P), both of which are found in many eukaryotic cells including those of animals and plants (Iwata and Koizumi, 2012; Liu and Howell, 2010b; Kroos and Akiyama, 2013; Brown and Goldstein, 1999). The first cleavage is believed to be carried out by S1P, a serine protease with a single TMD, presumably at the canonical S1P recognition site, RxxL or RxL, in the C-terminal domain of bZIP28, which shortens the length of the C-terminal domain and is believed to allow S2P to access bZIP28. The second cleavage is by the multispanning membrane metalloprotease S2P (Brown and Goldstein, 1999; Kroos and Akiyama, 2013). Interestingly, the crystal structure of the TMD core of S2P in *Methanocaldococcus jannaschii* demonstrated that S2P contains catalytic residues in its TMD and cleaves the peptide bond within the TMD (Feng et al., 2007). This two-step proteolytic cleavage allows the N-terminal domain of bZIP28, including its bZIP domain, to translocate into the nucleus where it directly activates transcription of ER stress-responsive genes (Figure 25.2; Liu and Howell, 2010a; Liu et al., 2007a; Tajima et al., 2008).

The relocation of bZIP28 from the ER to the Golgi is a critical initial step of bZIP28 activation (Srivastava et al., 2012, 2013). Under unstressed conditions, bZIP28 is retained in the ER by the interaction of the C-terminal domain of bZIP28 with BiP, an ER-resident HSP70 cognate that assists protein folding by binding to the hydrophobic peptides of unfolded proteins (Srivastava et al., 2013). Under ER stress, BiP dissociates from bZIP28 and binds to unfolded proteins triggering the vesicle transport of

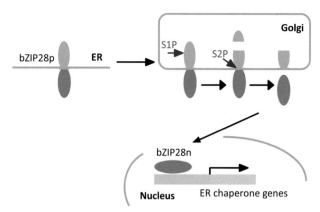

FIGURE 25.2 **Model for the activation mechanism of bZIP28 by RIP.** bZIP28 resides in the ER membrane as an inactive form, termed bZIP28p, under unstressed conditions. In response to ER stress, bZIP28 is translocated *via* vesicle transport to the Golgi where two sequential cleavages by S1P and S2P proteases occur. The first cleavage by S1P shortens the C-terminal domain of bZIP28, which is believed to allow S2P to access the TMD of bZIP28. The second cleavage by S2P liberates the cytoplasmic domain containing the bZIP domain, termed bZIP28n, from the membrane. bZIP28n then moves to the nucleus and activates ER stress-responsive genes such as those encoding ER chaperones.

bZIP28 to the Golgi, where two-step processing occurs. Indeed, an *Arabidopsis* plant expressing a mutant form of bZIP28, in which the interaction domain with BiP is deleted, does not bind to BiP and is not retained in the ER even under unstressed conditions (Srivastava et al., 2013). It has also been reported that the dibasic amino acid residues on the cytoplasmic side of bZIP28, in a region close to its TMD, are important for relocation from the ER to the Golgi. This is because the mutation of the dibasic residues abolished the interaction between Sar1 and Sec12, components of the COPII machinery, and their migration to the Golgi (Srivastava et al., 2012).

In addition to ER stress induced by pharmacological agents, bZIP28 has also been reported to be activated in response to heat shock (Gao et al., 2008). Proteolytic cleavage and nuclear relocation of GFP–bZIP28 have been demonstrated in response to heat shock. Transcriptional induction of some heat shock-inducible genes, *BiP2* and *HSP26.5-P*, was abolished in a *bzip28* mutant. Furthermore, the *bzip28* mutant shows higher sensitivity to heat shock (Gao et al., 2008). These results indicate that bZIP28 plays an important role in heat shock response. bZIP28 is also implicated in brassinosteroid signaling (Che et al., 2010). Molecular and genetic evidence has suggested that S2P-activated bZIP28 induces brassinosteroid signaling through increasing ER chaperones that facilitate the folding and maturation of the brassinosteroid receptor BRI1 and its subsequent translocation to the plasma membrane (Che et al., 2010).

Application of exogenous salicylic acid has also been reported to trigger bZIP28 proteolysis as well as activation of bZIP60 (Nagashima et al., 2014; see Section 25.2.3). The transcriptional induction of ER chaperone genes by salicylic acid treatment has been reported in *Arabidopsis* (Krinke et al., 2007; Wang et al., 2005), and it has been proposed that salicylic-acid-mediated activation of ER chaperone genes, during pathogen response, is required to increase the capacity plants have to secrete a large amount of pathogenesis-related (PR) proteins (Wang et al., 2005). Indeed, mutants defective in an ER chaperone gene secrete less PR1 protein than the wild-type plants, and exhibit susceptibility to the bacterial pathogen *Pseudomonas syringae* pv. *maculicola* ES4326 (Wang et al., 2005).

25.2.2 bZIP17

bZIP17 is an *Arabidopsis* membrane-bound transcription factor with the same domain structure as bZIP28. Its cytoplasmic domain, containing a bZIP domain, is released by RIP. Distinctive from bZIP28, RIP-mediated bZIP17 activation is triggered by salt stress (Liu et al., 2007b). It has been shown that transgenic *Arabidopsis* plants, expressing a GFP–bZIP17 fusion protein, exhibit fluorescence in the ER under unstressed conditions and in the nucleus under salt stress conditions. It has also been shown that the *bzip17* mutant is defective in the transcriptional responses of several salt stress response genes, including those encoding the homeodomain transcription factor ATHB-7 and the calcium-binding ER hand protein RD20 (Takahashi et al., 2000; Soderman et al., 1996).

Salt-stress-induced bZIP17 activation has been shown to be dependent on the protease S1P (Liu et al., 2007b). It has been shown that the proteolytic processing of myc-tagged bZIP17, expressed under the CaMV35S promoter, occurs in the wild-type plant but does not occur in the *s1p* mutant (Liu et al., 2007b). S1P was also shown to be able to cleave a C-terminal fragment of bZIP17 *in vitro*, presumably at a canonical S1P cleavage site. To support the notion that S1P is necessary for bZIP17 activation, both *bzip17* and *s1p* mutants show higher sensitivity to salt stress (Liu et al., 2007b). Furthermore, stress-inducible expression of the active form of bZIP17, in which the TMD and the C-terminal domain are deleted, conferred tolerance to salt stress, with less chlorophyll bleaching and an enhanced survival rate (Liu et al., 2008).

25.2.3 bZIP60

Although the activation of membrane-bound transcription factors is carried out via proteolytic processing in most cases, there is a notable exception. bZIP60 is an *Arabidopsis* transcription factor with a TMD that activates the expression of ER chaperone genes in response to ER stress, as does bZIP28. Due to the presence of a TMD, it was first postulated that bZIP60 is activated by RIP (Iwata and Koizumi, 2005; Iwata et al., 2008, 2009a, 2009b), but later studies demonstrated that its activation is mediated through cytoplasmic splicing, atypical splicing that occurs in the cytoplasm, which is different from conventional spliceosome-dependent splicing that occurs in the nucleus (Deng et al., 2011; Nagashima et al., 2011).

bZIP60u protein (u for unspliced) resides in the ER membrane as an inactive form and is activated via the cytoplasmic splicing of *bZIP60u* mRNA mediated by IRE1, an ER membrane-localized protein kinase/ribonuclease that is conserved among eukaryotic cells including plants (Koizumi et al., 2001; Okushima et al., 2002). Upon sensing ER stress, IRE1 activates its ribonuclease activity and cleaves two stem-loop structures of *bZIP60u* mRNA and removes an intron 23 nucleotides in length (Deng et al., 2011; Nagashima et al., 2011). Two halves of *bZIP60* mRNA are joined, by a yet unidentified RNA ligase, to generate *bZIP60s* mRNA (s for spliced; Figure 25.3A). This splicing causes a frameshift, resulting in translation of a new open reading frame (ORF) that does not contain a TMD (Figure 25.3B). The resulting bZIP60s protein, without a TMD, moves to the nucleus (Deng et al., 2011; Nagashima et al., 2011). Importantly, it is likely that the TMD of bZIP60u is required not only for retaining bZIP60u in the ER membrane, but also for the localization of *bZIP60u* mRNA onto the ER membrane where IRE1-mediated cytoplasmic splicing takes place, as in the case of XBP1, an animal ortholog of bZIP60 (Yanagitani et al., 2009).

Whether or not the new ORF of *Arabidopsis* bZIP60, translated after cytoplasmic splicing, has a function remains unclear. In this respect, it is worth mentioning OsbZIP50, a rice ortholog of *Arabidopsis* bZIP60. OsbZIP50 is a membrane-bound transcription factor whose activity is regulated by cytoplasmic splicing as *Arabidopsis* bZIP60 (Hayashi et al., 2012). Interestingly, a new ORF, translated after cytoplasmic splicing, contains a nuclear localization signal and facilitates the translocation of ObZIP50s into the nucleus (Hayashi et al., 2012).

In addition to ER stress, induced by pharmacological agents, bZIP60 has been implicated in diverse physiological phenomena. bZIP60s protein has been observed in anthers, structures that contain pollen grains and tapetal cells (Iwata et al., 2008). The tapetum is a highly secretory tissue that produces pollen surface proteins and lipids, with the pollen grains being poised for the high secretory activity that supports rapid pollen tube growth. Furthermore, a promoter–reporter assay suggests that *bZIP60* transcripts are highly abundant in these tissues (Iwata et al., 2008). These observations suggest that aspects of the ER stress response, regulated by bZIP60, are important in the development and/or function of such secretory cells.

bZIP60 has also been implicated in the response to abiotic and biotic stresses. It has been reported that cytoplasmic splicing of *bZIP60* mRNA is triggered by heat shock (Deng et al., 2011), as is proteolytic activation of bZIP28 (Gao et al., 2008). Exogenous salicylic acid also triggers cytoplasmic splicing of bZIP60 mRNA as well as

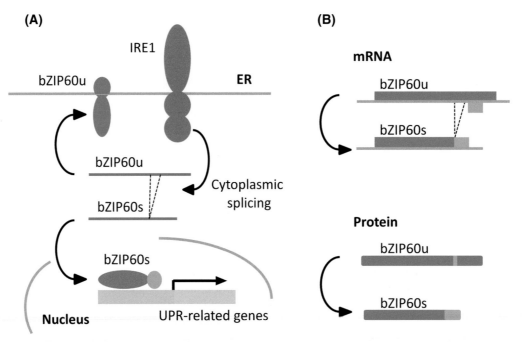

FIGURE 25.3 **Model for the activation mechanism of bZIP60 by cytoplasmic splicing.** (A) bZIP60 resides in the ER membrane as an inactive form, termed bZIP60u, as does bZIP28. *bZIP60u* mRNA is spliced by cytoplasmic splicing mediated by the ER membrane protein IRE1. It removes a 23-nucleotide-long intron and causes a frameshift, resulting in the translation of a new ORF that does not contain a TMD. bZIP60s protein, translated from *bZIP60s* mRNA, can translocate into the nucleus and activate ER stress-responsive genes such as those encoding ER chaperones. (B) Schematic representation of unspliced and spliced *bZIP60* mRNA and their protein products. The TMD and the new ORF translated after splicing are indicated by light blue and green, respectively.

proteolytic activation of bZIP28 (Nagashima et al., 2014). Indeed, the *ire1* mutants, which exhibit defective *bZIP60* mRNA splicing, have been reported to be more susceptible to bacterial pathogens (Moreno et al., 2012). A bZIP60 homolog in *Nicotiana benthamiana* has also been implicated in the defense response against bacterial pathogens (Tateda et al., 2008). Expression of *bZIP60* in *N. benthamiana* is significantly upregulated upon infection by the nonhost pathogen *Pseudomonas cichorii*. Furthermore, silencing of *bZIP60* in *N. benthamiana* plants allows higher multiplication of *P. cichorii* than of control plants (Tateda et al., 2008).

25.3 NAC TRANSCRIPTION FACTORS

NAC transcription factors are specific to plants and comprise a large gene family in flowering plants (Olsen et al., 2005; Chapter 13). The *Arabidopsis* genome encodes over 100 genes including NAC transcription factors, which have been shown to be involved in various growth and developmental processes such as shoot apical meristem formation, floral development, and stress- and hormone-signaling pathways. A bioinformatics survey demonstrated that many NAC transcription factors in *Arabidopsis* and rice contain a putative TMD and are predicted to be membrane bound (Kim et al., 2007b, 2010b). In this section we review some of the membrane-bound NAC transcription factors that have been characterized in the literature.

25.3.1 NTM1

NAC with Transmembrane Motif1 (NTM1), from *Arabidopsis*, is the first characterized NAC transcription factor found to have a TMD (Kim et al., 2006). The *ntm1-D* mutant contains a T-DNA insertion in the fourth exon of the *NTM1* gene and produces a truncated form of NTM1 that lacks the TMD at the C-terminus. The *ntm1-D* mutant therefore expresses a constitutively active NTM1 protein. The *ntm1-D* mutant exhibits pleiotropic phenotypes, including serrated leaves, abnormally developed petals, short stems, and reduced primary root growth and lateral root formation. Further analysis demonstrated that the *ntm1-D* mutant shows reduced cell division, accumulation of transcripts for CDK inhibitor genes, *KRP*s (De Veylder et al., 2003), and drastically reduced expression of the histone H4 gene, which is highly expressed in actively dividing cells (Riou-Khamlichi et al., 1999). In contrast, *ntm1-1*, a knockout mutant, exhibits elongated hypocotyls and reduced expression of G1/S cell cycle regulators, *KRP2* and *KRP7*, as well as *CYCD3;1* (Riou-Khamlichi et al., 1999). These results indicate that NTM1 is involved in the regulation of cell division.

NTM1 is associated with nuclear and ER membranes, with the TMD at the C-terminus and the N-terminal fragment containing the NAC domain produced by proteolytic processing (Kim et al., 2006). Although direct evidence has not been shown, this proteolytic processing is likely mediated by a protease called calpain (Croall and Ersfeld, 2007). Evidence for this being that processing is abolished when treated with a calpain inhibitor, *N*-acetyl-leucinyl-leucinyl-norleucinal (Kim et al., 2006). The *Arabidopsis* genome possesses a single calpain gene, *Defective Kernel1* (*DEK1*) and, notably, the DEK1 protein harbors a TMD that is likely membrane anchored. The notion that DEK1 cleaves and activates NTM1 is also consistent with the observation that transgenic *Arabidopsis* plants overexpressing *DEK1* showed serrated leaves and floral defects (Lid et al., 2005), reminiscent of the *ntm1-D* mutant phenotype.

25.3.2 NTL6

Several other membrane-bound NAC transcription factors, designated NTM-likes (NTLs), have been identified in *Arabidopsis* (Kim et al., 2007b). One such NTL, NTL6, has been shown to be proteolytically processed in response to cold (Seo et al., 2010a, 2010b). Unlike NTM1 and membrane-bound bZIP transcription factors, which reside at the ER membrane, NTL6 is localized at the plasma membrane. Proteolytic processing of NTL6 is initiated within 30 min of cold treatment and steadily increases for 15 h after cold exposure. Upon cold treatments, NTL6 translocates into the nucleus and activates the expression of a subset of *PR* genes such as *PR1*, *PR2*, and *PR5* through direct binding to their promoters. Notably, this *PR* gene induction by NTL6 is independent of salicylic acid. Overexpression of the truncated NTL6 in which the TMD is deleted activates expression of these genes and brings about enhanced resistance to *Pseudomonas syringae* infection (Seo et al., 2010a). Conversely, RNAi plants with less NTL6 activity are defective in cold-induced *PR1* gene expression and are more susceptible to the same pathogen at low temperatures (Seo et al., 2010a).

NTL6 is unlikely to be processed by a calpain protease, which is likely to structurally process the similar NTM1, because NTL6 processing is unaffected in the presence of the calpain inhibitor *N*-acetyl-leucinyl-leucinyl-norleucinal (Seo et al., 2010b). NTL6 is also unlikely to be processed by the S1P protease because NTL6 processing is unaffected in *s1p* mutants. Instead of this, NTL6 processing is inhibited by treatment with a metalloprotease inhibitor, 1,10-phenanthroline (Seo et al., 2010b), indicating that metalloprotease activity is responsible for NTL6 processing.

NTL6 processing is proposed to be regulated by cold-induced remodeling of membrane fluidity (Seo et al., 2010b). The effect of cold on NTL6 processing is significantly reduced in cold-acclimatized plants. In addition, NTL6 processing was stimulated by treatment with 6–12% dimethyl sulfoxide

(DMSO), which is known to reduce membrane fluidity and therefore mimic the effects of cold (Orvar et al., 2000). Furthermore, the pattern of NTL6 processing is affected in mutants with altered membrane lipid compositions. In the *ssi2* mutant, which is defective in a gene encoding a stearoyl–acyl carrier protein–desaturase that regulates the levels of unsaturated fatty acids in the cells (Kachroo et al., 2005; Shah et al., 2001), NTL6 processing was increased even under ambient temperatures (Seo et al., 2010b). The *ssi2* mutant also accumulated more processed NTL6 fragments upon cold treatment than did the wild-type plant (Seo et al., 2010b). Consistent with this observation, the *ssi2* mutant has been reported to constitutively express NTL6-target genes, *PR1*, *PR2*, and *PR5* (Shah et al., 2001).

It is interesting to note that the translocation of active NTL6 also requires phosphorylation to induce a drought-resistant response, in addition to proteolytic processing (Kim et al., 2012). The cytosolic kinase SnRK2.8 phosphorylates NTL6 at a threonine residue at position 142 and facilitates translocation of NTL6 into the nucleus to induce a drought-resistant response. A mutant form of NTL6, with a mutation of threonine at position 142 to alanine, was poorly phosphorylated and failed to move into the nucleus (Kim et al., 2012). Conversely, the dehydration-resistant phenotype of NTL6-overexpressing plants was suppressed by the introduction of mutations in *SnRK2.8* (Kim et al., 2012). This is an interesting example in which two-step activation ensures tight regulation of membrane-bound transcription factors.

25.3.3 ANAC089

ANAC089 was identified by analyzing an *Arabidopsis* Landsberg *erecta* (L*er*)/Cape Verde Islands (Cvi) recombinant inbred line (RIL) population to investigate fructose-signaling pathways (Li et al., 2011). The Cvi accession is more sensitive than the L*er* accession to high fructose levels when grown on an agar-solidified medium. Sensitivity of seedling development to fructose was quantified in the L*er*/Cvi recombinant inbred line population, and a total of eight fructose-sensing quantitative trait loci (QTLs) were identified. A map-based approach showed that one of the QTLs encodes a gene encoding a NAC transcription factor with TMD, ANAC089. Further analysis demonstrated that whereas the Columbia (Col) and L*er* alleles encode an ANAC089 protein of 340 amino acid residues, the Cvi allele encodes a truncated protein of 224 amino acid residues due to a 1 bp (base pair) deletion, resulting in a premature stop codon. The Cvi accession therefore expresses the truncated ANAC089 protein without a TMD, which functions as a constitutively active transcription factor and confers altered fructose sensitivity. Such a 1 bp deletion is rare in natural *Arabidopsis* accessions and therefore may be specific to the Cvi accession because it was not found in the *Arabidopsis* genome 1001 sequence collection (http://signal-genet.salk.edu/atg1001/3.0/gebrowser.php; Li et al., 2011).

Independent research groups identified ANAC089 as a redox-dependent suppressor of stromal ascorbate peroxidase (sAPX) gene expression (Klein et al., 2012). sAPX functions as a central element of the chloroplast antioxidant defense system, and its expression in the nucleus must be under retrograde control to cope with the demands for antioxidant capacity. It was shown that ANAC089 binds to the promoter of the *sAPX* gene both in a yeast one-hybrid system and in electrophoretic mobility shift assays. *Arabidopsis* mesophyll protoplasts expressing a GFP–ANAC089 fusion protein showed fluorescence in the ER and trans-Golgi network under unstressed conditions, and in the nucleus within 30 min of treatment with reducing agents. The reporter assay showed that ANAC089 functions as a transcriptional repressor. After consideration of the evidence, it was proposed that ANAC089 functions as a repressor of *sAPX* gene expression under highly reducing conditions, when high expression of the *sAPX* gene is unnecessary (Klein et al., 2012).

25.3.4 ANAC060

ANAC060 was identified by QTL analysis for seedling sensitivity to high sugar levels in a Col/C24 F2 population of *Arabidopsis* (Li et al., 2014). A glucose- and fructose-sensing QTL was found to lie in the *ANAC060* gene. Further analyses showed that a single nucleotide polymorphism in the *ANAC060* gene of the Col accession affects the splicing pattern, resulting in an extra 20 nucleotide sequence, which contains an in-frame stop codon at the 5′ end of the fourth exon. *ANAC060* mRNA of the Col accession therefore produces a truncated version of the ANAC060 protein without the TMD. Indeed, GFP–ANAC060(Col) was localized in the nucleus, whereas GFP–ANAC060(C24) was retained in the endomembrane system in *Arabidopsis* mesophyll protoplasts. The study also indicated that ANAC060(Col) is involved in a novel negative feedback loop in the sugar-abscisic acid (ABA)-signaling pathway (Leon et al., 2012; Wind et al., 2013). In this feedback loop, the ABA-induced ERF/AP2 transcription factor ABI4 activates ANAC060 expression, but the nuclear form of ANAC060(Col) suppresses glucose-induced ABA accumulation and *ABI4* expression, thereby reducing responsiveness to sugar signals (Li et al., 2014). Unlike the 1-bp deletion of the *ANAC089* gene found only in the Cvi accession, this single nucleotide substitution resulting in a truncated version of ANAC060 was found in ~12% of natural *Arabidopsis* accessions (Li et al., 2014).

25.3.5 ANAC013 and ANAC017

Two *Arabidopsis* membrane-bound NAC transcription factors with TMD, ANAC013, and ANAC017, play an essential role in mitochondrial retrograde regulation of the oxidative stress response (De Clercq et al., 2013; Ng et al., 2013). Retrograde signaling from mitochondria regulates nuclear gene expression to coordinate the production of organelle constituents with developmental and physiological status (Schertl and Braun, 2014; Schwarzlander and Finkemeier, 2013). Upon disruption of mitochondrial function, it activates expression of genes encoding components of the alternative oxidative respiratory chain, including alternative oxidase (AOX).

A yeast one-hybrid screening identified ANAC013 as a protein that binds to the cis-element mitochondrial dysfunction motif (MDM) present in a set of promoters whose genes are upregulated by mitochondrial retrograde signaling, including the AOX1a gene (De Clercq et al., 2013). ANAC013 activates the expression of several mitochondrial stress response genes, including AOX1a and UGT74E2, through direct binding to the MDM element on their promoters. Consistent with its role in mitochondrial retrograde signaling, artificial microRNA-mediated repression of ANAC013 expression reduced the tolerance of Arabidopsis to oxidative stress, whereas transgenic Arabidopsis plants overexpressing ANAC013 were less sensitive to oxidative stress (De Clercq et al., 2013).

Another research group identified a closely related membrane-bound transcription factor, ANAC017, as a positive regulator for transcriptional activation of the AOX1a gene, through forward genetic screening that characterized mutants as showing a defective response to mitochondrial dysfunction (Ng et al., 2013). The anac017 null mutants exhibit less accumulation of AOX1a transcripts after treatment with antimycin A or hydrogen peroxide than do wild-type plants, whereas a mutant that contains a T-DNA insertion before the TMD, and therefore expresses a truncated nuclear ANAC017 protein, accumulates AOX1a transcripts even under unstressed conditions. Consistent with these observations, anac017 null mutants show more sensitivity to moderate light and drought stress than does the aox1a mutant (Ng et al., 2013).

As with other membrane-bound transcription factors, ANAC013 and ANAC017 are localized at the membranes and activated by proteolytic cleavage, which allows the N-terminal fragment containing the NAC domain to move to the nucleus (De Clercq et al., 2013; Ng et al., 2013). It is interesting to note that both ANAC013 and ANAC017 are localized at the ER membrane, rather than the mitochondria, under unstressed conditions. This indicates that the mitochondrial dysfunction signal is transmitted to the nucleus via the ER, where ANAC013 and ANAC017 activation occurs. This is an interesting example in which the ER can serve as an intermediary organelle in the signaling pathway from the mitochondria to the nucleus.

It is suggested that ANAC017 is cleaved within its TMD by a protease called rhomboid (Ng et al., 2013). Rhomboid is a serine protease with multiple TMDs found in bacteria, archaea, animals, and plants (Weihofen and Martoglio, 2003; Adam, 2013). The first rhomboid substrate identified is Spitz, an epidermal growth factor ligand in Drosophila melanogaster (Urban et al., 2001). Spitz contains a 7-amino acid sequence motif in its TMD which is essential for cleavage by the rhomboid protease, and this sequence motif is highly conserved in other rhomboid substrates (Strisovsky et al., 2009; Strisovsky, 2013). Interestingly, the TMD of ANAC017 contains an amino acid sequence similar to this consensus motif, indicative of its cleavage by a rhomboid protease (Ng et al., 2013). Indeed, antimycin-A-induced transcriptional response of AOX1a was reduced by pretreatment of Arabidopsis suspension cells with N-p-tosyl-L-phenylalanyl chloromethyl (TPCK), a rhomboid protease inhibitor (Urban et al., 2001). Conservation of the amino acid sequence is also found in the TMD of ANAC013, implicating the involvement of the rhomboid protease in the proteolytic activation of ANAC013, as well.

25.4 CONCLUSIONS AND FUTURE PERSPECTIVES

In this chapter we have described some examples of plant membrane-bound transcription factors. There are other NAC transcription factors with TMD whose involvement in biological processes has been implicated, including reactive oxygen species (ROS) production during drought-induced leaf senescence, flavonoid biosynthesis under high light, salt stress signaling, and anther dehiscence (Table 25.1; Lee et al., 2012; Morishita et al., 2009; Yabuta et al., 2010; Park et al., 2011; Kim et al., 2007a, 2008, 2010a; Shih et al., 2014). Although the majority of currently characterized membrane-bound transcription factors in plants harbor bZIP or NAC domains, ones with plant homeodomain (PHD) and plant-specific R2R3–MYB domain have also been characterized (Table 25.1; Sun et al., 2011; Slabaugh et al., 2011). We foresee the elucidation of functions of many more membrane-bound transcription factors in diverse physiological phenomena.

It should be mentioned that our current knowledge of the activation mechanism of plant membrane-bound transcription factors is limited. Four families of multispanning membrane proteases, S2P, rhomboid, presenilin, and signal peptide peptidase (SPP), have been well characterized in animals. Although homologous genes for each protease family have been found in sequenced plant genomes, most of their substrates remain to be discovered, except for S2P, which cleaves the TMD of bZIP28 and bZIP17 (Iwata and Koizumi, 2012; Liu and Howell, 2010b). Rhomboid is encoded by multiple genes in the Arabidopsis genome (Kanaoka et al., 2005) and is proposed to be required

TABLE 25.1 *Arabidopsis* Membrane-Bound Transcription Factors that have been Characterized in the Literature

Transcription factor	DNA-binding domain	Subcellular location	Activation mechanism	Related biological phenomenon	References
bZIP28	bZIP	ER	Proteolysis by S1P and S2P	ER stress, heat stress, brassinosteroid signaling, salicylic acid signaling	Liu et al. (2007a); Tajima et al. (2008); Gao et al. (2008); Che et al. (2010); Srivastava et al. (2012); Srivastava et al. (2013)
bZIP17	bZIP	ER	Proteolysis by S1P and S2P	Salt stress	Liu et al. (2007b); Liu et al. (2008)
bZIP60	bZIP	ER	Cytoplasmic splicing by IRE1	ER stress, heat stress, salicylic acid signaling	Iwata and Koizumi (2005); Iwata et al. (2008); Iwata et al. (2009b); Nagashima et al. (2011); Deng et al. (2011); Moreno et al. (2012); Nagashima et al. (2014)
NTM1	NAC	Nuclear/ER	Proteolysis by calpain	Cell division	Kim et al. (2006)
NTL6	NAC	PM	Proteolysis, protease unknown	Cold stress	Kim et al. (2007b); Seo et al. (2010a); Seo et al. (2010b)
ANAC089	NAC	ER/trans-Golgi	Proteolysis, protease unknown	Fructose signaling	Li et al. (2011); Klein et al. (2012)
ANAC060	NAC	Endomembrane	Proteolysis, protease unknown	Sugar-ABA signaling	Li et al. (2014)
ANAC013	NAC	ER	Proteolysis by rhomboid	Mitochondrial retrograde regulation	De Clercq et al. (2013)
ANAC017	NAC	ER	Proteolysis by rhomboid	Mitochondrial retrograde regulation	Ng et al. (2013)
NTL8	NAC	PM	Proteolysis, protease unknown	Salt stress	Kim et al. (2007a); Kim et al. (2008)
NTM2/ANAC060	NAC	PM	Proteolysis, protease unknown	Salt stress	Park et al. (2011)
NTL4	NAC	PM	Proteolysis, protease unknown	Drought-induced leaf senescence	Kim et al. (2010a); Lee et al. (2012)
AIF	NAC	Endomembrane	Proteolysis, protease unknown	Anther dehiscence	Shih et al. (2014)
ANAC078	NAC	Unclear	Proteolysis, protease unknown	Flavonoid biosynthesis	Morishita et al. (2009); Yabuta et al. (2010)
PTM	PHD	Chloroplast envelope	Proteolysis, protease unknown	Retrograde chloroplast signaling	Sun et al. (2011)
maMYB	R2R3–MYB	ER	Proteolysis, protease unknown	Root hair elongation	Slabaugh et al. (2011)

for proteolytic cleavage of ANAC017 (Ng et al., 2013) as described in Section 25.3.5. However, it is unclear whether multiple rhomboid proteases exhibit redundancy in substrate recognition or whether they possess different substrate specificity. Substrates of two other multispanning membrane proteases, presenilin, and SPP (Smolarkiewicz et al., 2014; Han et al., 2009; Hoshi et al., 2013; Tamura et al., 2009), remain to be identified in plants. Identification of substrates of these RIP proteases would facilitate our understanding of the proteolytic activation of membrane-bound transcription factors in plants.

References

Adam, Z., 2013. Emerging roles for diverse intramembrane proteases in plant biology. Biochim. Biophys. Acta 1828, 2933–2936.

Brown, M.S., Goldstein, J.L., 1999. A proteolytic pathway that controls the cholesterol content of membranes, cells, and blood. Proc. Natl. Acad. Sci. USA 96, 11041–11048.

Che, P., Bussell, J.D., Zhou, W., Estavillo, G.M., Pogson, B.J., Smith, S.M., 2010. Signaling from the endoplasmic reticulum activates brassinosteroid signaling and promotes acclimation to stress in Arabidopsis. Sci. Signal. 3, ra69.

Croall, D.E., Ersfeld, K., 2007. The calpains: modular designs and functional diversity. Genome Biol. 8, 218.

De Clercq, I., Vermeirssen, V., Van Aken, O., Vandepoele, K., Murcha, M.W., Law, S.R., Inze, A., Ng, S., Ivanova, A., Rombaut, D., Van De Cotte, B., Jaspers, P., Van De Peer, Y., Kangasjarvi, J., Whelan, J., Van Breusegem, F., 2013. The membrane-bound NAC transcription factor ANAC013 functions in mitochondrial retrograde regulation of the oxidative stress response in Arabidopsis. Plant Cell 25, 3472–3490.

De Veylder, L., Joubes, J., Inze, D., 2003. Plant cell cycle transitions. Curr. Opin. Plant Biol. 6, 536–543.

Deng, Y., Humbert, S., Liu, J.X., Srivastava, R., Rothstein, S.J., Howell, S.H., 2011. Heat induces the splicing by IRE1 of a mRNA encoding a transcription factor involved in the unfolded protein response in Arabidopsis. Proc. Natl. Acad. Sci. USA 108, 7247–7252.

Feng, L., Yan, H., Wu, Z., Yan, N., Wang, Z., Jeffrey, P.D., Shi, Y., 2007. Structure of a site-2 protease family intramembrane metalloprotease. Science 318, 1608–1612.

Gao, H., Brandizzi, F., Benning, C., Larkin, R.M., 2008. A membrane-tethered transcription factor defines a branch of the heat stress response in Arabidopsis thaliana. Proc. Natl. Acad. Sci. USA 105, 16398–16403.

Han, S., Green, L., Schnell, D.J., 2009. The signal peptide peptidase is required for pollen function in Arabidopsis. Plant Physiol. 149, 1289–1301.

Hayashi, S., Wakasa, Y., Takahashi, H., Kawakatsu, T., Takaiwa, F., 2012. Signal transduction by IRE1-mediated splicing of bZIP50 and other stress sensors in the endoplasmic reticulum stress response of rice. Plant J. 69, 946–956.

Hoshi, M., Ohki, Y., Ito, K., Tomita, T., Iwatsubo, T., Ishimaru, Y., Abe, K., Asakura, T., 2013. Experimental detection of proteolytic activity in a signal peptide peptidase of Arabidopsis thaliana. BMC Biochem. 14, 16.

Howell, S.H., 2013. Endoplasmic reticulum stress responses in plants. Annu. Rev. Plant Biol. 64, 477–499.

Iwata, Y., Koizumi, N., 2005. An Arabidopsis transcription factor, AtbZIP60, regulates the endoplasmic reticulum stress response in a manner unique to plants. Proc. Natl. Acad. Sci. USA 102, 5280–5285.

Iwata, Y., Koizumi, N., 2012. Plant transducers of the endoplasmic reticulum unfolded protein response. Trends Plant Sci. 17, 720–727.

Iwata, Y., Fedoroff, N.V., Koizumi, N., 2008. Arabidopsis bZIP60 is a proteolysis-activated transcription factor involved in the endoplasmic reticulum stress response. Plant Cell 20, 3107–3121.

Iwata, Y., Fedoroff, N.V., Koizumi, N., 2009a. The Arabidopsis membrane-bound transcription factor AtbZIP60 is a novel plant-specific endoplasmic reticulum stress transducer. Plant Signal. Behav. 4, 514–516.

Iwata, Y., Yoneda, M., Yanagawa, Y., Koizumi, N., 2009b. Characteristics of the nuclear form of the Arabidopsis transcription factor AtbZIP60 during the endoplasmic reticulum stress response. Biosci. Biotechnol. Biochem. 73, 865–869.

Kachroo, P., Venugopal, S.C., Navarre, D.A., Lapchyk, L., Kachroo, A., 2005. Role of salicylic acid and fatty acid desaturation pathways in ssi2-mediated signaling. Plant Physiol. 139, 1717–1735.

Kanaoka, M.M., Urban, S., Freeman, M., Okada, K., 2005. An Arabidopsis rhomboid homolog is an intramembrane protease in plants. FEBS Lett. 579, 5723–5728.

Kim, Y.S., Kim, S.G., Park, J.E., Park, H.Y., Lim, M.H., Chua, N.H., Park, C.M., 2006. A membrane-bound NAC transcription factor regulates cell division in Arabidopsis. Plant Cell 18, 3132–3144.

Kim, S.G., Kim, S.Y., Park, C.M., 2007a. A membrane-associated NAC transcription factor regulates salt-responsive flowering via FLOWERING LOCUS T in Arabidopsis. Planta 226, 647–654.

Kim, S.Y., Kim, S.G., Kim, Y.S., Seo, P.J., Bae, M., Yoon, H.K., Park, C.M., 2007b. Exploring membrane-associated NAC transcription factors in Arabidopsis: implications for membrane biology in genome regulation. Nucleic Acids Res. 35, 203–213.

Kim, S.G., Lee, A.K., Yoon, H.K., Park, C.M., 2008. A membrane-bound NAC transcription factor NTL8 regulates gibberellic acid-mediated salt signaling in Arabidopsis seed germination. Plant J. 55, 77–88.

Kim, S.G., Lee, S., Ryu, J., Park, C.M., 2010a. Probing protein structural requirements for activation of membrane-bound NAC transcription factors in Arabidopsis and rice. Plant Sci. 178, 239–244.

Kim, S.G., Lee, S., Seo, P.J., Kim, S.K., Kim, J.K., Park, C.M., 2010b. Genome-scale screening and molecular characterization of membrane-bound transcription factors in Arabidopsis and rice. Genomics 95, 56–65.

Kim, M.J., Park, M.J., Seo, P.J., Song, J.S., Kim, H.J., Park, C.M., 2012. Controlled nuclear import of the transcription factor NTL6 reveals a cytoplasmic role of SnRK2.8 in the drought-stress response. Biochem J. 448, 353–363.

Klein, P., Seidel, T., Stocker, B., Dietz, K.J., 2012. The membrane-tethered transcription factor ANAC089 serves as redox-dependent suppressor of stromal ascorbate peroxidase gene expression. Front. Plant Sci. 3, 247.

Koizumi, N., Martinez, I.M., Kimata, Y., Kohno, K., Sano, H., Chrispeels, M.J., 2001. Molecular characterization of two Arabidopsis Ire1 homologs, endoplasmic reticulum-located transmembrane protein kinases. Plant Physiol. 127, 949–962.

Korennykh, A., Walter, P., 2012. Structural basis of the unfolded protein response. Annu. Rev. Cell Dev. Biol. 28, 251–277.

Krinke, O., Ruelland, E., Valentova, O., Vergnolle, C., Renou, J.P., Taconnat, L., Flemr, M., Burketova, L., Zachowski, A., 2007. Phosphatidylinositol 4-kinase activation is an early response to salicylic acid in Arabidopsis suspension cells. Plant Physiol. 144, 1347–1359.

Kroos, L., Akiyama, Y., 2013. Biochemical and structural insights into intramembrane metalloprotease mechanisms. Biochim. Biophys. Acta 1828, 2873–2885.

Lee, S., Seo, P.J., Lee, H.J., Park, C.M., 2012. A NAC transcription factor NTL4 promotes reactive oxygen species production during drought-induced leaf senescence in Arabidopsis. Plant J. 70, 831–844.

Leon, P., Gregorio, J., Cordoba, E., 2012. ABI4 and its role in chloroplast retrograde communication. Front. Plant Sci. 3, 304.

Li, P., Wind, J.J., Shi, X., Zhang, H., Hanson, J., Smeekens, S.C., Teng, S., 2011. Fructose sensitivity is suppressed in Arabidopsis by the transcription factor ANAC089 lacking the membrane-bound domain. Proc. Natl. Acad. Sci. USA 108, 3436–3441.

Li, P., Zhou, H., Shi, X., Yu, B., Zhou, Y., Chen, S., Wang, Y., Peng, Y., Meyer, R.C., Smeekens, S.C., Teng, S., 2014. The ABI4-induced Arabidopsis ANAC060 transcription factor attenuates ABA signaling and renders seedlings sugar insensitive when present in the nucleus. PLoS Genet. 10, e1004213.

Lid, S.E., Olsen, L., Nestestog, R., Aukerman, M., Brown, R.C., Lemmon, B., Mucha, M., Opsahl-Sorteberg, H.G., Olsen, O.A., 2005. Mutation in the Arabidopsis thaliana DEK1 calpain gene perturbs endosperm and embryo development while over-expression affects organ development globally. Planta 221, 339–351.

Liu, J.X., Howell, S.H., 2010a. bZIP28 and NF-Y transcription factors are activated by ER stress and assemble into a transcriptional complex to regulate stress response genes in Arabidopsis. Plant Cell 22, 782–796.

Liu, J.X., Howell, S.H., 2010b. Endoplasmic reticulum protein quality control and its relationship to environmental stress responses in plants. Plant Cell 22, 2930–2942.

Liu, J.X., Srivastava, R., Che, P., Howell, S.H., 2007a. An endoplasmic reticulum stress response in Arabidopsis is mediated by proteolytic processing and nuclear relocation of a membrane-associated transcription factor, bZIP28. Plant Cell 19, 4111–4119.

Liu, J.X., Srivastava, R., Che, P., Howell, S.H., 2007b. Salt stress responses in *Arabidopsis* utilize a signal transduction pathway related to endoplasmic reticulum stress signaling. Plant J. 51, 897–909.

Liu, J.X., Srivastava, R., Howell, S.H., 2008. Stress-induced expression of an activated form of AtbZIP17 provides protection from salt stress in *Arabidopsis*. Plant Cell Environ. 31, 1735–1743.

Moreno, A.A., Mukhtar, M.S., Blanco, F., Boatwright, J.L., Moreno, I., Jordan, M.R., Chen, Y., Brandizzi, F., Dong, X., Orellana, A., Pajerowska-Mukhtar, K.M., 2012. IRE1/bZIP60-mediated unfolded protein response plays distinct roles in plant immunity and abiotic stress responses. PLoS ONE 7, e31944.

Morishita, T., Kojima, Y., Maruta, T., Nishizawa-Yokoi, A., Yabuta, Y., Shigeoka, S., 2009. *Arabidopsis* NAC transcription factor, ANAC078, regulates flavonoid biosynthesis under high-light. Plant Cell Physiol. 50, 2210–2222.

Nagashima, Y., Mishiba, K., Suzuki, E., Shimada, Y., Iwata, Y., Koizumi, N., 2011. *Arabidopsis* IRE1 catalyses unconventional splicing of bZIP60 mRNA to produce the active transcription factor. Sci. Rep. 1, 29.

Nagashima, Y., Iwata, Y., Ashida, M., Mishiba, K.I., Koizumi, N., 2014. Exogenous salicylic acid activates two signaling arms of the unfolded protein response in *Arabidopsis*. Plant Cell Physiol. 55, 1772–1779.

Ng, S., Ivanova, A., Duncan, O., Law, S.R., Van Aken, O., De Clercq, I., Wang, Y., Carrie, C., Xu, L., Kmiec, B., Walker, H., Van Breusegem, F., Whelan, J., Giraud, E., 2013. A membrane-bound NAC transcription factor, ANAC017, mediates mitochondrial retrograde signaling in *Arabidopsis*. Plant Cell 25, 3450–3471.

Okushima, Y., Koizumi, N., Yamaguchi, Y., Kimata, Y., Kohno, K., Sano, H., 2002. Isolation and characterization of a putative transducer of endoplasmic reticulum stress in *Oryza sativa*. Plant Cell Physiol. 43, 532–539.

Olsen, A.N., Ernst, H.A., Leggio, L.L., Skriver, K., 2005. NAC transcription factors: structurally distinct, functionally diverse. Trends Plant Sci. 10, 79–87.

Orvar, B.L., Sangwan, V., Omann, F., Dhindsa, R.S., 2000. Early steps in cold sensing by plant cells: the role of actin cytoskeleton and membrane fluidity. Plant J. 23, 785–794.

Park, J., Kim, Y.S., Kim, S.G., Jung, J.H., Woo, J.C., Park, C.M., 2011. Integration of auxin and salt signals by the NAC transcription factor NTM2 during seed germination in *Arabidopsis*. Plant Physiol. 156, 537–549.

Riou-Khamlichi, C., Huntley, R., Jacqmard, A., Murray, J.A., 1999. Cytokinin activation of *Arabidopsis* cell division through a D-type cyclin. Science 283, 1541–1544.

Schertl, P., Braun, H.P., 2014. Respiratory electron transfer pathways in plant mitochondria. Front. Plant Sci. 5, 163.

Schwarzlander, M., Finkemeier, I., 2013. Mitochondrial energy and redox signaling in plants. Antioxid. Redox. Signal. 18, 2122–2144.

Seo, P.J., Kim, S.G., Park, C.M., 2008. Membrane-bound transcription factors in plants. Trends Plant Sci. 13, 550–556.

Seo, P.J., Kim, M.J., Park, J.Y., Kim, S.Y., Jeon, J., Lee, Y.H., Kim, J., Park, C.M., 2010a. Cold activation of a plasma membrane-tethered NAC transcription factor induces a pathogen resistance response in *Arabidopsis*. Plant J. 61, 661–671.

Seo, P.J., Kim, M.J., Song, J.S., Kim, Y.S., Kim, H.J., Park, C.M., 2010b. Proteolytic processing of an *Arabidopsis* membrane-bound NAC transcription factor is triggered by cold-induced changes in membrane fluidity. Biochem. J. 427, 359–367.

Shah, J., Kachroo, P., Nandi, A., Klessig, D.F., 2001. A recessive mutation in the *Arabidopsis* SSI2 gene confers SA- and NPR1-independent expression of PR genes and resistance against bacterial and oomycete pathogens. Plant J. 25, 563–574.

Shih, C.F., Hsu, W.H., Peng, Y.J., Yang, C.H., 2014. The NAC-like gene ANTHER INDEHISCENCE FACTOR acts as a repressor that controls anther dehiscence by regulating genes in the jasmonate biosynthesis pathway in *Arabidopsis*. J. Exp. Bot. 65, 621–639.

Slabaugh, E., Held, M., Brandizzi, F., 2011. Control of root hair development in *Arabidopsis thaliana* by an endoplasmic reticulum anchored member of the R2R3–MYB transcription factor family. Plant J. 67, 395–405.

Smolarkiewicz, M., Skrzypczak, T., Michalak, M., Lesniewicz, K., Walker, J.R., Ingram, G., Wojtaszek, P., 2014. Gamma-secretase subunits associate in intracellular membrane compartments in *Arabidopsis thaliana*. J. Exp. Bot. 65, 3015–3027.

Soderman, E., Mattsson, J., Engstrom, P., 1996. The *Arabidopsis* homeobox gene ATHB-7 is induced by water deficit and by abscisic acid. Plant J. 10, 375–381.

Srivastava, R., Chen, Y., Deng, Y., Brandizzi, F., Howell, S.H., 2012. Elements proximal to and within the transmembrane domain mediate the organelle-to-organelle movement of bZIP28 under ER stress conditions. Plant J. 70, 1033–1042.

Srivastava, R., Deng, Y., Shah, S., Rao, A.G., Howell, S.H., 2013. BINDING PROTEIN is a master regulator of the endoplasmic reticulum stress sensor/transducer bZIP28 in *Arabidopsis*. Plant Cell 25, 1416–1429.

Strisovsky, K., 2013. Structural and mechanistic principles of intramembrane proteolysis – lessons from rhomboids. FEBS J. 280, 1579–1603.

Strisovsky, K., Sharpe, H.J., Freeman, M., 2009. Sequence-specific intramembrane proteolysis: identification of a recognition motif in rhomboid substrates. Mol. Cell 36, 1048–1059.

Sun, X., Feng, P., Xu, X., Guo, H., Ma, J., Chi, W., Lin, R., Lu, C., Zhang, L., 2011. A chloroplast envelope-bound PHD transcription factor mediates chloroplast signals to the nucleus. Nat. Commun. 2, 477.

Tajima, H., Iwata, Y., Iwano, M., Takayama, S., Koizumi, N., 2008. Identification of an *Arabidopsis* transmembrane bZIP transcription factor involved in the endoplasmic reticulum stress response. Biochem. Biophys. Res. Commun. 374, 242–247.

Takahashi, S., Katagiri, T., Yamaguchi-Shinozaki, K., Shinozaki, K., 2000. An *Arabidopsis* gene encoding a Ca2+-binding protein is induced by abscisic acid during dehydration. Plant Cell Physiol. 41, 898–903.

Tamura, T., Kuroda, M., Oikawa, T., Kyozuka, J., Terauchi, K., Ishimaru, Y., Abe, K., Asakura, T., 2009. Signal peptide peptidases are expressed in the shoot apex of rice, localized to the endoplasmic reticulum. Plant Cell Rep. 28, 1615–1621.

Tateda, C., Ozaki, R., Onodera, Y., Takahashi, Y., Yamaguchi, K., Berberich, T., Koizumi, N., Kusano, T., 2008. NtbZIP60, an endoplasmic reticulum-localized transcription factor, plays a role in the defense response against bacterial pathogens in *Nicotiana tabacum*. J. Plant Res. 121, 603–611.

Urban, S., Lee, J.R., Freeman, M., 2001. *Drosophila* rhomboid-1 defines a family of putative intramembrane serine proteases. Cell 107, 173–182.

Vitale, A., Boston, R.S., 2008. Endoplasmic reticulum quality control and the unfolded protein response: insights from plants. Traffic 9, 1581–1588.

Walter, P., Ron, D., 2011. The unfolded protein response: from stress pathway to homeostatic regulation. Science 334, 1081–1086.

Wang, D., Weaver, N.D., Kesarwani, M., Dong, X., 2005. Induction of protein secretory pathway is required for systemic acquired resistance. Science 308, 1036–1040.

Weihofen, A., Martoglio, B., 2003. Intramembrane-cleaving proteases: controlled liberation of proteins and bioactive peptides. Trends Cell Biol. 13, 71–78.

Wind, J.J., Peviani, A., Snel, B., Hanson, J., Smeekens, S.C., 2013. ABI4: versatile activator and repressor. Trends Plant Sci. 18, 125–132.

Yabuta, Y., Morishita, T., Kojima, Y., Maruta, T., Nishizawa-Yokoi, A., Shigeoka, S., 2010. Identification of recognition sequence of ANAC078 protein by the cyclic amplification and selection of targets technique. Plant Signal. Behav. 5, 695–697.

Yanagitani, K., Imagawa, Y., Iwawaki, T., Hosoda, A., Saito, M., Kimata, Y., Kohno, K., 2009. Cotranslational targeting of XBP1 protein to the membrane promotes cytoplasmic splicing of its own mRNA. Mol. Cell 34, 191–200.

CHAPTER 26

Ubiquitination of Plant Transcription Factors

Sophia L. Stone

Department of Biology, Dalhousie University, Halifax NS, Canada

OUTLINE

26.1 The Ubiquitin Proteasome System 396
   26.1.1 Ubiquitin and Ubiquitin Conjugation 396
   26.1.2 The Ubiquitin Enzymes 397
   26.1.3 Ubiquitin Modifications and Outcomes 399
   26.1.4 The 26S Proteasome 399

26.2 The Ubiquitin Proteasome System and Regulation of Transcription Factor Function 400
   26.2.1 Control of Transcription Factor Function by Nonproteolytic Ubiquitination 400
   26.2.2 Control of Transcription Factor Function by Proteolytic Ubiquitination 401
      26.2.2.1 Regulating Transcription Factor Abundance 401
      26.2.2.2 Regulating Repressor or Inhibitor Abundance 403
      26.2.2.3 Transcription-Coupled Proteolysis 405

References 405

As sessile organisms, plants are continually exposed to adverse environmental conditions of varying intensity, including attack by pathogens, wounding by insects, exposure to ultraviolet radiation, and decrease in water and nutrient availability. It is quite remarkable that in spite of their ever-changing environment, plants are able to adapt, continue to grow, develop, and importantly remain productive. To accomplish this feat, plants rely heavily on their ability to coordinate the perception of environmental stimuli with alterations in developmental and physiological programs required for adaptation and survival. Plant response to external signals is a very complex and highly coordinated process that involves the activation of many signaling networks and changes in the expression of hundreds of genes. In addition, it is imperative that these genes are only transcribed in response to the right signal, at the right time, in the right place, and for the appropriate amount of time. Cells employ an assortment of mechanisms to regulate transcription including utilizing the various functions of ubiquitin and the 26S proteasome.

The most documented function of ubiquitin (Ub) is its role in protein turnover. Ubiquitin-dependent proteolysis involves the attachment of ubiquitin molecules to a selected protein followed by degradation of the modified protein by the 26S proteasome, a large multicatalytic protease complex. The ubiquitin proteasome system (UPS) is used extensively to control transcription at multiple levels. A major point of integration of the UPS in transcriptional control is the regulated proteolysis of transcriptional regulators such as transcription factors (TFs). The cell utilizes the UPS to control the abundance of TFs to ensure the appropriate level of gene expression and to destroy the protein when it is no longer needed. The regulatory function of ubiquitin goes beyond controlling protein stability to influencing protein activity and localization. The nonproteolytic functions of ubiquitin can regulate gene expression by modulating the subcellular localization and activity of TFs. The connections between ubiquitin, the 26S proteasome, and transcription extends to the regulation of chromatin configuration, recruitment of regulatory proteins onto chromatin, removal of stalled RNA polymerase, transcriptional elongation, and termination, as well as pre-mRNA processing and mRNA export (see reviews by Geng et al., 2012; Hammond-Martel et al., 2012). Proteolytic function of ubiquitin in controlling transcription, particularly at the level of modulating TF abundance, is well described in plants and will be the focus of this chapter.

26.1 THE UBIQUITIN PROTEASOME SYSTEM

The UPS is a major proteolytic mechanism within the cell, responsible for the degradation of an enormous variety of proteins. The system consists of two distinct and successive steps, the attachment of a chain of ubiquitin molecules to a selected protein followed by degradation of the ubiquitinated protein by the 26S proteasome (Figure 26.1). The discovery of the UPS occurred in the late 1970s, where a small heat stable protein, later identified as ubiquitin, was found to stimulate ATP-dependent proteolysis (Ciechanover et al., 1978; Wilkinson et al., 1980). The covalent conjugation of multiple ubiquitin molecules to a protein was later shown to be a prerequisite for degradation by an ATP-dependent protease (Ciechanover et al., 1980; Hershko et al., 1980, 1983, 1984). A suite of enzymes, named E1, E2, and E3, was found to conjugate ubiquitin to proteins and the 26S proteasome was identified as the ATP-dependent protease that mediates the turnover of conjugated proteins (Ciechanover et al., 1982; Hershko et al., 1983; Hough et al., 1986; Waxman et al., 1987; Pickart and Cohen, 2004). The attached ubiquitin molecules serve as a "tag" that allows for targeting of the modified protein to the 26S proteasome for degradation. Recognition of most substrate proteins by ubiquitinating enzymes is facilitated by the presence of a degradation signal or "degron". These degrons are defined as the minimal transferable element (~10 amino acids) within a protein that is sufficient for targeting to the UPS for degradation. A well-characterized degron is the conserved domain II found in most plant auxin/indole-3-acetic (AUX/IAA) repressor proteins. The conserved domain II degron is required for auxin-induced ubiquitin-dependent proteasomal degradation of the transcriptional repressors (Dreher et al., 2006; Nishimura et al., 2009). The domain II degron is transferable, and capable of triggering the degradation of proteins that it is fused to in response to auxin (Nishimura et al., 2009).

26.1.1 Ubiquitin and Ubiquitin Conjugation

Ubiquitin is an evolutionarily conserved, ubiquitously expressed eukaryotic protein. The extraordinary level of conservation is evident when comparing the amino acid sequence of ubiquitin from plants, humans, and yeast. Plant and yeast ubiquitin differ by only 2 amino acids and only 3-amino-acid differences are observed when plant ubiquitin is compared to human ubiquitin (Sharp and Li, 1987; Callis and Vierstra, 1989). Among plant species

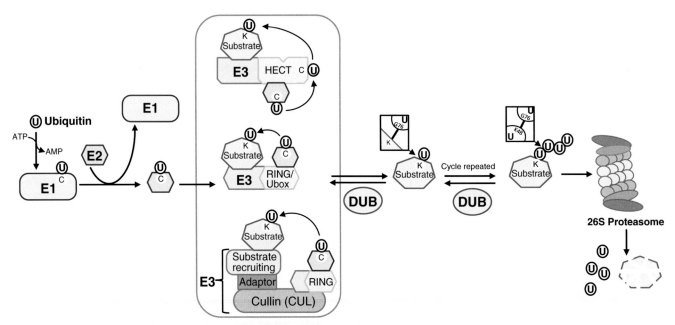

FIGURE 26.1 **The ubiquitin proteasome system.** Activation of ubiquitin (U) by the E1 is followed by transfer to the E2, forming a E2-ubiquitin intermediate, which interacts with the substrate-recruiting E3. The most common types of plant E3s are depicted. U-box and RING-domain-containing E3s, including cullin-based RING E3 complexes, mediate transfer of ubiquitin directly from the E2-ubiquitin to the substrate. HECT-domain-containing E3s form an E3-ubiquitin intermediate prior to the transfer of ubiquitin to the substrate. For polyubiquitination, the cycle is repeated to generate a ubiquitin chain. Substrates modified by the attachment of a lysine 48 (K48)-linked polyubiquitin chain are recognized and degraded by the 26S proteasome. Ubiquitin chains can be parsed or removed from modified substrates by DUBs.

the amino acid sequence of ubiquitin is identical (Callis and Vierstra, 1989; Callis et al., 1995). The *ubiquitin* gene translation product is a polyubiquitin precursor protein containing ubiquitin molecules fused in tandem. Ubiquitin is also produced as a fusion with unrelated ribosomal proteins. The fusion proteins are processed by deubiquitinating enzymes (DUBs) to produce the mature free ubiquitin monomer. DUBs are proteases that cleave isopeptide bonds between linked ubiquitin molecules to generate ubiquitin monomers (Amerik and Hochstrasser, 2004; Reyes-Turcu et al., 2009). The ubiquitin monomer is small, consisting of 76 amino acids that fold into a compact, globular, tightly hydrogen-bonded structure with a terminal diglycine sequence (Vijay-Kumar et al., 1987). Most eukaryotic genomes contain multiple *ubiquitin* genes. The genome of *Arabidopsis thaliana* (Arabidopsis), for example, contains five polyubiquitin genes (*UBQ3, UBQ4, UBQ10, UBQ11,* and *UBQ14*), each containing up to six ubiquitin-coding regions in tandem (Callis and Vierstra, 1989; Callis et al., 1995).

Ubiquitination, the process of conjugating ubiquitin to a selected protein, is highly conserved among eukaryotes. The ubiquitination pathway involves the successive action of three classes of enzymes: E1, ubiquitin activating enzyme (UBA); E2, ubiquitin conjugating enzyme (UBC); and E3 ubiquitin ligase (UBL; Figure 26.1). The three-step conjugation process culminates in the formation of an isopeptide bond between the C-terminal glycine (Gly76) of ubiquitin and the amino group of a lysine residue on the substrate protein (Figure 26.1). Instances of ubiquitin attachment occurring on a cysteine, serine, or threonine residue within a substrate protein have been reported (Wang et al., 2007; Vosper et al., 2009). Protein ubiquitination begins with the ATP-dependent attachment of ubiquitin to the E1 enzyme forming a thioester-linked E1–Ub intermediate. A conserved cysteine residue within the active site of the E1 and the Gly76 of ubiquitin is used to form the thioester bond. The activated ubiquitin is then transferred to the E2 enzyme, also forming a thioester-linked E2-Ub intermediate. Similar to the E1, the E2–Ub intermediate is formed using Gly76 of ubiquitin and a cysteine residue within a conserved region of the E2 enzyme called the UBC domain. The UBC domain is also used for interacting with the E3 enzyme. Ubiquitin transfer from the E2–Ub to the substrate protein is facilitated by the substrate-recruiting E3 enzyme. Ubiquitin is either transferred directly from the E2 to the E3-bound substrate or the E3 forms a thioester-linked E3–Ub intermediate prior to transfer of ubiquitin to the substrate protein (Figure 26.1). The attachment of a single ubiquitin molecule to the substrate protein is referred to as monoubiquitination. Following the attachment of the initial ubiquitin molecule to the substrate protein, the conjugation process can be repeated to assemble a polyubiquitin chain. Ubiquitin chain formation or polyubiquitination involves the formation of isopeptide bonds between the C-terminal Gly76 of the incoming ubiquitin molecules and one of seven lysine residues on the previously attached ubiquitin (Figure 26.1). Ubiquitin molecules can also be linked through a peptide bond between the C-terminal glycine and the N-terminal methionine of the incoming ubiquitin molecule, forming a linear ubiquitin chain. The mechanism underlying ubiquitin chain elongation is not very well understood. Proposed models suggest that ubiquitin molecules are added sequentially to the growing chain or a preassembled chain is added to the previously attached ubiquitin molecule on the substrate protein (Wang and Pickart, 2005; Hochstrasser, 2006; Li et al., 2007; Deshaies and Joazeiro, 2009; Maspero et al., 2011).

The conjugation process is reversible. In addition to generating mature free ubiquitin molecules from precursor fusion proteins, DUBs also catalyze the release of ubiquitin monomers from chains attached to polyubiquitinated substrates, or remove all ubiquitin molecules from the modified protein (Figure 26.1; Reyes-Turcu et al., 2009). The removal or remodeling of the polyubiquitin chain affects the fate of the target protein. DUBs can regulate protein stability through the release of ubiquitin molecules from the modified substrates prior to recognition by 26S proteasome, prohibiting protein degradation.

26.1.2 The Ubiquitin Enzymes

The three classes of enzymes, E1, E2, and E3, involved in the ubiquitination pathway are organized in a hierarchical order. The pathway consists of one or two E1 enzymes, tens of E2s, and hundreds of E3s. The E1 activates all ubiquitin molecules required by the cell and functions with all E2 enzymes. An E2 enzyme may function with multiple E3s, and a single E3 enzyme may be tasked with recruiting many different substrate proteins. The *Arabidopsis* genome, for example, contains two E1 coding genes that are similarly expressed in most tissues and encode proteins that share high amino acid sequence similarity (81%) and affinity for E2 enzymes (Hatfield et al., 1997). The *Arabidopsis* genome encodes for 37 E2 proteins that cluster into 16 subgroups (Kraft et al., 2005). E2s are defined mainly by the presence of the 140–150-amino-acid UBC domain, which contains a ubiquitin-accepting cysteine residue. The diversity of the E2 enzymes is reflected in their expression pattern, and their activity with ubiquitin ligases (Kraft et al., 2005). The *Arabidopsis* genome is predicted to contain over 1400 genes that encode for E3s or components of ubiquitin ligases (Craig et al., 2009). The abundance and diversity of ubiquitin ligases lends to the specificity, versatility, and pervasiveness of the ubiquitination pathway. The major types of plant ubiquitin ligases are the Homology to E6-Associated Carboxy-Terminus (HECT), U-box, and Really Interesting New Gene (RING). The HECT, U-box, and RING domains are used by the E3s to interact with the

E2–Ub intermediate (Figure 26.1). The HECT-type and U-box-type E3s are single subunit enzymes, where both the E2 and substrate-binding functions are found within the same polypeptide. The ubiquitin ligases that utilize a RING domain for E2 binding are either single or multiple subunit enzymes that have substrate recognition within the same polypeptide or on a separate protein, respectively (Figure 26.1). The release of ubiquitin molecules from modified proteins is carried out by a large family of DUBs, which are categorized into five subfamilies based on sequence similarities (Amerik and Hochstrasser, 2004). The Ubiquitin-specific Processing Protease (UBP), Ovarian Tumor-related (OUT), Ubiquitin Carboxy-terminal Hydrolases (UCH), and Ataxin-3 subfamilies are cysteine proteases and the JAMM/MPN+ proteases subfamily are metalloproteases (Amerik and Hochstrasser, 2004; Reyes-Turcu et al., 2009). The *Arabidopsis* genome, for example, is predicted to encode for at least 64 DUBs (Yang et al., 2007). The 26-member UBP and 3-member UCH subfamilies have been the most studied and have been found to regulate numerous processes in plant growth and development (Yan et al., 2000; Yang et al., 2007; Liu et al., 2008).

HECT-type E3: HECT-type E3s are unique in that they actively participate in the transfer of ubiquitin to the substrate protein by forming a thioester bonded intermediate using a cysteine residue with the conserved HECT domain (Figure 26.1). The 350 amino acid HECT domain forms a bilobed structure where the N-terminal lobe contains the E2–Ub binding site and the C-terminal lobe contains the ubiquitin accepting cysteine residue (Downes et al. 2003). The *Arabidopsis* genome is predicted to encode for seven HECT domain containing ubiquitin ligases (UPL1–UPL7) (Downes et al., 2003). The UPLs contain a C-terminal HECT domain downstream of a variety of domains such as armadillo repeats, IQ calmodulin-binding motifs, and various ubiquitin-binding domains (Downes et al., 2003).

RING-type E3: RING-type E3s do not form E3–Ub intermediates, but instead facilitate transfer of ubiquitin directly from the E2–Ub to the substrate protein (Figure 26.1). The RING domain uses an octet of cysteine (C) and histidine (H) amino acids as metal ligand residues to coordinate two zinc ions in a crossbrace structure, which is essential for ubiquitin ligase activity (Freemont et al. 1991; Freemont, 1993; Lorick et al., 1999; Lovering et al., 1993). The majority of RING-type E3s contain one of two canonical RING domains that differ in the presence of either a cysteine (C3HC4, RING-HC) or histidine (C3H2C3, RING-H2) residue at metal ligand position 5 (Deshaies and Joazeiro, 2009). RING domains can be further modified with variations in the number of amino acids between the key metal ligands or have substitutions at one or more of the metal ligand positions (Stone et al., 2005; Deshaies and Joazeiro, 2009). The RING-G domain (C3HGC3), for example, has a glycine instead of a cysteine or histidine residue at metal ligand position 5 (Dasgupta et al., 2004; Stone et al., 2005). The variability does not seem to affect function as ubiquitin ligase activity has been demonstrated for a number of the modified RING domains (Dasgupta et al., 2004; Stone et al., 2005; Kraft et al., 2005). For most eukaryotes, the family of RING-type E3s comprise the largest group of ubiquitin ligases.

The *Arabidopsis* proteome is predicted to contain a very large (approximately 476 proteins) and diverse family of RING domain-containing proteins (Stone et al., 2005). Members contain eight different types of RING domains, two canonical and six modified RING domains, and an array of additional motifs and domains, some of which may facilitate substrate protein recruitment (Stone et al., 2005). Protein–protein interaction domains found within the E3 family include Ankyrin repeats, von Willebrand factor type A (vWA), and WD40 or β-transducin repeats. A single *Arabidopsis* RING domain-containing protein, RING Box 1 (RBX1), participates as the E2–Ub binding component of the multiple subunit Cullin based RING ligases (CRLs; Smalle and Vierstra, 2004; Hotton and Callis, 2008). Three types of CRLs have been described in plants, each utilizing a different cullin (CUL) subunit (CUL1, CUL3a/3b, or CUL4). CUL proteins function as a scaffold upon which the RING domain-containing protein and the substrate-recruiting protein assemble to form the E3 complex (Figure 26.1; Hotton and Callis, 2008). Families of substrate-recruiting proteins utilized by plant CRL complexes include the F-box proteins, Broad complex Tramtrack Bric-a-Brac (BTB) proteins, and DNA-Binding Domain 1 (DDB1)-binding WD40 (DWD) proteins (Smalle and Vierstra, 2004; Hotton and Callis, 2008). CUL1-containing CRLs (also referred to as SCF for SKP1 [ASK1 in plants], Cullin, and F-box) use the adapter protein ASK1 to anchor the substrate-recruiting F-box proteins to the complex. Substrate-recruiting BTB proteins interact directly with CUL3 to assemble the E3 complex. CUL4-based RING E3s are assembled using DDB1 as an adaptor protein to bind to the substrate-recruiting DWD proteins. The diversity of substrate-recruiting subunits, and the ability to utilize one of three CUL proteins to assemble an E3, makes the CRL group the largest class of ubiquitin ligases. To illustrate this, the assembly of a CUL1-based RING E3 complex can utilize one of 700 *Arabidopsis* F-box proteins (Gagne et al., 2002; Kuroda et al., 2002).

U-box-type E3: The U-box-type E3s facilitate protein ubiquitination in a manner comparable to the RING-type ubiquitin ligases. In fact, the structure of the ~70 amino acid U-box domain is similar to that of the RING domain. The structure of the U-box domain is stabilized by salt bridges and hydrogen bonds between conserved, charged and polar, amino acids instead of zinc ion-chelating cysteine and histidine residues (Aravind and Koonin, 2000; Vander Kooi et al., 2006). Sixty-one *Arabidopsis* U-box proteins have been identified (Mudgil et al., 2004; Zeng et al., 2008; Yee and Goring, 2009). The U-box proteins are classified into eight groups based on the presence of other domains or motifs (Azevedo et al., 2001; Mudgil et al., 2004; Zeng et al., 2008). The majority of U-box-type E3s contain substrate-interacting armadillo repeats (Mudgil et al., 2004; Zeng et al., 2008). Unique to the U-box-type E3s is the presence of kinase domains in a significant number of family members (Mudgil et al., 2004; Zeng et al., 2008).

26.1.3 Ubiquitin Modifications and Outcomes

The major function of ubiquitination is to selectively target proteins for degradation. However, recent studies have greatly expanded the role of ubiquitin. In addition to regulating protein abundance, ubiquitin conjugation can influence protein activity, interaction, trafficking, and localization. The versatility of ubiquitin function makes the pathway central to virtually all cellular processes. The outcome of ubiquitin conjugation depends on the nature of the modification. A single ubiquitin molecule can be attached to one (monoubiquitination) or multiple (multimonoubiquitination) lysine (Lys) residues within the substrate protein. Ubiquitin can be linked to itself via one of seven Lys residues (Lys6, Lys11, Lys27, Lys29, Lys33, Lys48, and Lys63) producing structurally diverse polyubiquitin chains (Komander and Rape, 2012). Polyubiquitin chains can be homogenous, where the same lysine residue is used to assemble the chain, or heterogeneous, where different Lys residues are used to create ubiquitin–ubiquitin linkages within a single chain (Kirkpatrick et al., 2006; Kim et al., 2007; Ben-Saadon et al., 2006). Heterogeneous chains can also be branched or forked with two ubiquitin molecules linked to a single ubiquitin moiety (Kim et al., 2007). Lysine residue selection during polyubiquitin chain assembly is governed by the E2 and/or E3 ubiquitin enzymes. In the case of RING-type and U-box-type E3s the topology of the polyubiquitin chain is determined mainly by the E2 enzymes (Kim et al., 2007; Deshaies and Joazeiro, 2009; Xu et al., 2009; Rodrigo-Brenni et al., 2010). In contrast, the HECT-type E3 determines which lysine residue is used to create ubiquitin–ubiquitin linkages within the polyubiquitin chain (Wang and Pickart, 2005; Maspero et al., 2011). Ubiquitin molecules can also be linked "head to tail" using the C-terminal glycine of one ubiquitin molecule and the N-terminal residue of the next ubiquitin moiety (Kirisako et al., 2006). In mammalian cells, the formation of linear ubiquitin chains is carried out by a large complex, which contains two RING-type E3 ligases (Kirisako et al., 2006). Homogenous polyubiquitin chains with Lys29, Lys48, and Lys63 linkages have been detected in plants (Walsh and Sadanandom, 2014; Tomanov et al., 2014). There is very little evidence for the remaining lysine-linked polyubiquitin chains in plant systems and no evidence for the formation of linear ubiquitin chains.

The attachment of a single ubiquitin molecule to a substrate protein has been shown to be sufficient to act as a signal for internalizing membrane-bound proteins, shuttling of proteins between the cytoplasm and the nucleus, and facilitating repair of damaged DNA (Mukhopadhyay and Riezman, 2007; Komander and Rape, 2012). The functions of two types of polyubiquitination, Lys48- and Lys63-linked chains, have been extensively studied. The functions of the remaining types of polyubiquitin chains are not well characterized. In fact, very little is known about the role of Lys6-, Lys27-, and Lys33-linked chains. Proteins modified by the attachment of a Lys48-linked polyubiquitin chain are targeted for degradation by the 26S proteasome. Proteasome-dependent degradation requires the attachment of a polyubiquitin chain consisting of at least four Lys48-linked ubiquitin molecules to the substrate protein (Piotrowski et al., 1997; Thrower et al., 2000). Lys63-linked polyubiquitination has been implicated in nonproteolytic functions such as endocytosis and protein kinase activation (Mukhopadhyay and Riezman, 2007; Tomanov et al., 2014). However, Lys63-linked polyubiquitin chains have been reported to target substrates to the 26S proteasome for degradation (Saeki et al., 2009). Other types of polyubiquitin chains, such as linear, Lys11- and Lys29-linked chains, may participate in targeting substrates to the 26S proteasome for degradation (Zhao and Ulrich, 2010; Xu et al., 2009; Kravtsova-Ivantsiv and Ciechanover, 2012). Linear chains are also implicated in regulating signaling pathways and protein activation (Haas et al., 2009; Tokunaga et al. 2009). Branched polyubiquitin chains do not target substrates for degradation by the proteasome but may instead regulate protein activity (Ben-Saadon et al., 2006; Kim et al., 2007).

26.1.4 The 26S Proteasome

The proteasome is a large, ATP-dependent, multicomplex, and multifunctional protease. The 26S proteasome, named for its Svedberg sedimentation coefficient, consists of the proteolytic 20S core particle (CP) capped on one or both ends by the 19S regulatory particle (RP). The 20S CP consists of 28 subunits arranged in four stacked rings forming a large, hollow, three-chambered cylinder (Pickart and Cohen, 2004). The two internal rings, consisting of seven β subunits each, contain the protease catalytic sites, which are sequestered within the chamber. The two distal rings, each containing seven α subunits, function to gate the channel leading to the proteolytic chamber. The distal rings are also thought to function as holding chambers, maintaining substrates in an unfolded state for continued proteolysis (Ruschak et al., 2010). The 19S RP separates into two subcomplexes, the base, which abuts the 20S CP, and the lid that sits on top of the base. The 19S RP is tasked with recognizing polyubiquitin chains, unfolding substrates, and translocating them into the 20S CP for degradation, as well as recycling ubiquitin molecules. The 19S RP contains at least 17 subunits, six regulatory particle ATPase (RPTs), and 11 regulatory particle non-ATPase (RPNs) subunits. Polyubiquitin chains are recognized by two ubiquitin receptors, RPN10 and RPN13 (Walters et al., 2002; Husnjak et al., 2008). Poylubiquitinated substrates are also delivered to the 26S proteasome by shuttling ubiquitin receptors, which interact with components (e.g., RPN1 and RPN2) of the 19S RP (Rosenzweig et al., 2012). RPN11 is a DUB involved in removing ubiquitin molecules from substrates prior to unfolding and translocation into the 20S CP for degradation (Yao and Cohen, 2002). DUBs are also recruited to the 26S proteasome via interactions with components of the 19S RP (Husnjak et al., 2008). The AAA-ATPases (RTP1–RPT6) are involved in substrate unfolding, gating of the 20S CP for substrate entry, releasing peptides after proteolysis, and translocation of the substrate into the chamber (Köhler et al. 2001; Pickart and Cohen, 2004; Finley, 2009).

26.2 THE UBIQUITIN PROTEASOME SYSTEM AND REGULATION OF TRANSCRIPTION FACTOR FUNCTION

The integration of ubiquitin and the 26S proteasome into transcription is quite extensive. Cells take advantage of both the proteolytic and nonproteolytic functions of ubiquitin to control transcription. Regulating the proteins that control gene expression is a key point of participation of the UPS in the transcription system. The most apparent way in which the UPS controls transcription is through modulating the abundance, localization, and activity of TFs. (Figure 26.2).

26.2.1 Control of Transcription Factor Function by Nonproteolytic Ubiquitination

Transcription factors can be regulated by nonproteolytic ubiquitination, which mainly influences protein activity and localization (Figure 26.2). Ubiquitin conjugation can influence activity by regulating binding of TFs to target sites within the genome, processing of precursor proteins into active TFs, and regulating interaction with other regulatory proteins (Geng et al., 2012). A simple way to regulate gene expression is to sequester the TF at various locations outside of the nucleus until its function is required by the cell. Conversely, the export of TFs from the nucleus is used to inhibit function. Ubiquitination in response to external or internal signals can promote translocation of TFs into and out of the nucleus. Nonproteolytic roles for ubiquitin in controlling TF function have not yet been reported in plant systems. Mammalian Forkhead box O (FOXO) 4 and yeast Spt23 are two examples which illustrate the nonproteolytic function of ubiquitin in regulating

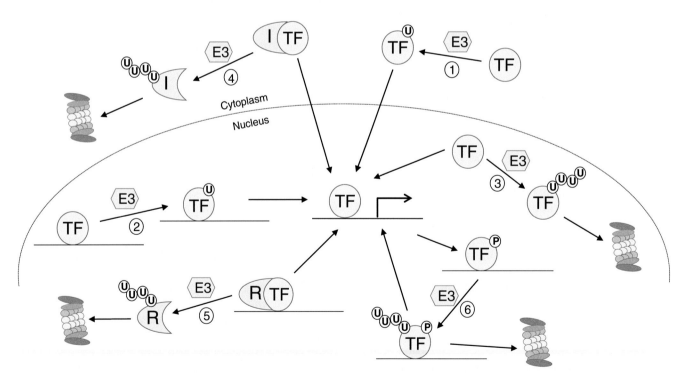

FIGURE 26.2 **Nonproteolytic (1–2) and proteolytic (3–6) functions of ubiquitin (U) in controlling TF function.** Monoubiquitination can promote translocation of TF into the nucleus (1) or activate DNA-bound TF. Continual ubiquitin-dependent proteasomal degradation (3) can maintain low levels of the TF until a stabilizing signal is received by the cell, which prohibits turnover. Ubiquitin-dependent turnover of interacting inhibitory (I) and repressor (R) proteins can promote TF activation (4 and 5). Proteolysis-coupled destruction (6) promotes gene expression. Transcriptional activation promotes phosphorylation (P) and subsequent degradation of the TF which clears the promoter allowing for binding of another TF. These modes of regulation are not mutually exclusive and may function together to control transcriptional activity.

TF function. Multimonoubiquitination of FOXO4 by the RING-type E3 Mouse double-minute 2 (Mdm2) is proposed to promote nuclear localization of the TF in response to oxidative stress (van der Horst et al., 2006; Brenkman et al., 2008). Furthermore, ubiquitinated FOXO4 is more transcriptionally active. Oxidative stress-induced deubiquitination negatively regulates FOXO4 activity and promotes translocation to the cytoplasm (van der Horst et al., 2006). The yeast TF Spt23 is produced as an inactive precursor that is tethered to the outer endoplasmic reticulum (ER) membrane (Zhang et al., 1999). Decrease in fatty acid levels triggers the release of Spt23 from the ER membrane and translocation of the TF to the nucleus where it activates the expression of genes involved in fatty acid metabolism (Hoppe et al., 2000; Rape et al., 2001; Piwko and Jentsch, 2006). The HECT type E3 RSP5 ubiquitinates the membrane-bound precursor, which leads to cleavage by the 26S proteasome to produce the active TF that is escorted to the nucleus (Hoppe et al., 2000; Shcherbik et al., 2004; Piwko and Jentsch, 2006). Plant ER and plasma membrane-bound TFs, which are processed by proteolytic cleavage to produce the active nuclear localized protein, have been described (Seo et al., 2008; Chapter 25). Many of these membrane-bound TFs are mobilized in response to ER stress (Seo et al., 2008). However, processing of membrane-bound TFs by the UPS has not been reported.

26.2.2 Control of Transcription Factor Function by Proteolytic Ubiquitination

Ubiquitin-dependent proteolysis of transcriptional regulators is well established as a way to control transcription. The UPS can, for example, inhibit transcription by limiting TF abundance or promote gene expression by targeting repressor or inhibitor proteins for degradation (Figure 26.2). Plants utilize the proteolytic function of ubiquitin in a variety of ways to control TF function. As detailed later in the chapter, regulated proteolysis features prominently in the control of gene expression in plants.

26.2.2.1 Regulating Transcription Factor Abundance

A widely used strategy employed by plants is to suppress gene expression by constitutively targeting TFs for ubiquitin-dependent degradation until the regulatory protein is required (Figure 26.2). Signal-induced prohibition of degradation allows for the accumulation of the TF and the activation of gene expression. This mode of regulation allows for rapid responses to external and internal signals by simply blocking entry of the TF into the ubiquitin proteasome system. Table 26.1 gives examples of plant TFs that are regulated in this manner. Mechanisms used to prohibit degradation include signal-induced changes in posttranslational modification (PTM) that may interfere with recognition of the TF by the E3, and relocalization or degradation of the E3 which prevents ubiquitination of the TF. The Brassinazole-Resistant 1 (BRZ1) and Brassinosteroid Insensitive EMS Suppressor 1 (BES1) are examples of TFs that are stabilized by PTM. BRZ1 and BES1 are targeted to the 26S proteasome for degradation when phosphorylated by Brassinosteroid Insensitive 2 (BIN2), a glycogen synthase kinase-3 (GSK3)-like kinase (He et al., 2002; Wang et al., 2002; Ryu et al., 2010). BRZ1 and BES1 accumulate in the nucleus as a result of brassinosteroid (BR)-induced dephosphorylation involving protein phosphatase 2A (PP2A; Ryu et al., 2010; Tang et al., 2011). The RING-type E3 Constitutive Photomorphogenic 1 (COP1) is an example of a ubiquitin ligase that is relocated to assist in increasing TF abundance. In the dark, nuclear-localized COP1 targets a number of TFs for degradation including Long Hypocotyl 5 (HY5) and Long After Far-Red Light 1 (LAF1; Jiao et al., 2007). COP1 translocation to the cytoplasm in response to light contributes to accumulation of HY5 and LAF1 and the activation of gene expression required for seedling photomorphogenesis (Osterlund et al., 2000; Seo et al., 2003; von Arnim and Deng, 1994; Chapter 21). Proteasomal degradation of the RING-type E3 Keep on Going (KEG) promotes accumulation of the TF Abscisic acid Insensitive 5 (ABI5). During seed and early seedling development, the abscisic acid (ABA)/stress-responsive ABI5 is maintained at low levels via continued proteasomal degradation mediated by KEG (Lopez-Molina et al., 2001; Stone et al., 2006; Liu and Stone, 2010). ABI5 is predicted to contain nuclear localization (NLS) and nuclear export (NES) signals, which suggests nucleocytoplasmic shuttling of the TF (Figure 26.3; Liu and Stone, 2013). KEG-mediated degradation occurs within the cytoplasm and requires interaction with the conserved C3 domain of ABI5 and Lys344 of the TF (Liu and Stone, 2013). In response to abiotic stress, ABA promotes ABI5 accumulation and activation, which allows for the transcription of genes required for stress tolerance (Lopez-Molina et al., 2001, 2002). ABA-activated kinases that phosphorylate and activate ABI5 to promote ABA responses have not yet been identified. Recent studies suggest that members of the Suc nonfermenting1-related protein kinase subfamily 2 (SnRK2), SnRK3, and Calcineurin B-like Interacting Protein Kinase (CIPK) family may be involved in ABA-dependent phosphorylation of ABI5 (Nakashima et al., 2009; Sirichandra et al., 2010; Lyzenga et al., 2013). Accumulation of ABI5 is achieved, in part, through ABA-induced self-ubiquitination and subsequent proteasomal degradation of KEG (Liu and Stone, 2010) (Figure 26.3). The mechanism linking ABA perception to KEG self-destruction is not known but may involve phosphorylation of KEG by ABA-activated kinases (Liu and Stone 2010; Lyzenga et al., 2013). KEG mediates the proteasomal degradation of other ABA-responsive TFs including ABA Responsive Element (ABRE) Binding Factors (ABF) 1 and ABF3 (Chen et al., 2013). A mechanism

TABLE 26.1 Proteolytic Function of Ubiquitin in Controlling TF Function

TF (biological function)	E3 (type)	Known or predicted mode of UPS-dependent regulation		References
		Signal/signaling pathway	Outcomes	
ABF1/3 (abiotic stress response)	KEG (RING)	ABA	UPS dependent degradation limit TF abundance, supressing gene expression Signal/signaling prohibit UPS mediated degradation allowing for accumulation of the TF and activation of gene expression	Chen et al. (2013)
ABI5 (abiotic stress response)	KEG (RING)	ABA		Liu and Stone (2010, 2013)
CO (flowering time)	COP1 (RING)	Light		Jang et al. (2008)
BRZ1, BES1 (development)	Unknown	Brassinosteroids		Ryu et al. (2010); Tang et al. (2011)
DREB2A (abiotic stress response)	DRIP1/2 (RING)	Drought and salt stresses		Qin et al. (2008); Morimoto et al. (2013)
EIN3, EIL (ethylene signalling)	EBF1/2 (CUL1-based RING)	Ethylene		Gagne et al. (2004)
HY5, LAF1 (photomorphogenesis)	COP1 (RING)	Light		Osterlund et al. (2000); Seo et al. (2003)
MYB30 (cell death)	MEIL1 (RING)	Pathogen attack		Marino et al. (2013)
PAP1/2 (anthocyanin biosynthesis)	COP1 (RING)	Light		Maier et al. (2013)
ABI3 (abiotic stress response)	AIP2 (RING)	ABA	Signal/signalling promote UPS dependent degradation of the TF, attenuating gene expression	Zhang et al. (2005)
ABI5 (abiotic stress response)	DWA1/2, ABD1 (CUL4-based RING)	ABA		Lee et al. (2010); Seo et al. (2014)
CO (flowering time)	HOS1 (RING)	Low temperature		Jung et al., (2012); Lazaro et al., (2012)
BES1 (development/shoot branching)	MAX2 (CUL1-based RING)	Strigolactones		Wang et al., (2013)
BOS1 (biotic and abiotic stress responses)	BOI (RING)	Salt stress Pathogen attack		Luo et al., (2010)
ERF53 (abiotic stress response)	RGLG1/2 (RING)	Drought stress		Cheng et al. (2012)
ICE (abiotic stress response)	HOS1 (RING)	Cold stress		Dong et al. (2006); Miura et al. (2011)
LHY (floral organ development)	UFO (CUL1-based RING)	Unknown, probably various environmental cues	'Activation by destruction' Signal/signaling promote UPS dependent degradation which promotes TF activity and gene expression	Chae et al. (2008)
MYC2 (wound/pathogen defense response)	Unknown	JA-Ile		Zhai et al. (2013)

similar to what is observed for ABI5 may participate in the ABA-dependent stabilization of ABF1 and ABF3. Unlike ABI5, the conserved C4 domain of ABF1 and ABF3 functions as a stabilizing element, which suggests differences in regulation (Sirichandra et al., 2010; Chen et al., 2013).

The UPS also targets TFs following transcriptional activation to maintain steady state levels or destroy the regulatory protein to attenuate gene expression (Figure 26.2). A significant number of plant TFs are controlled by signal-induced ubiquitin-dependent proteolysis (Table 26.1). The RING-type E3 high expression of osmotically responsive gene 1 (HOS1) is translocated into the nucleus in response to cold where it promotes the degradation of the TF Inducer of CBF Expression 1 (ICE1) which regulates the expression of cold-responsive genes (Lee et al., 2001; Dong et al., 2006; Miura et al., 2011). The TF Constans (CO), involved in regulating flowering time, is also a target for HOS1-mediated proteasomal degradation in response to low temperatures (Jung et al., 2012; Lazaro et al., 2012). Although ABA promotes the stabilization and accumulation of ABI5, the TF is turned over during ABA signaling. Two nuclear-localized CRL complexes mediate the

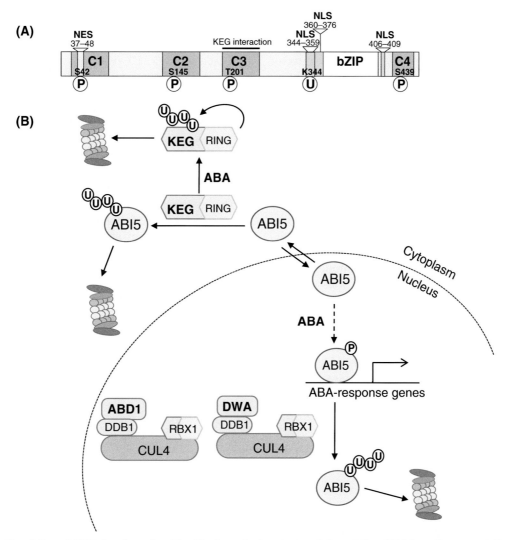

FIGURE 26.3 **Regulation of ABI5 abundance by ubiquitin-dependent proteasomal degradation.** (A) Schematic representation of ABI5 protein showing the conserved domains (C1–C4), DNA-binding bZIP domain, putative nuclear localization (NLS) and export (NES) signals as well as predicted phosphorylation (P) and ubiquitination (U) sites. (B) A simplified representation of the ABA-signaling pathway highlighting UPS control of ABI5 abundance. In the absence of the stress hormone ABA, ABI5 is continually targeted for proteasomal degradation by KEG. ABA prohibits ABI5 degradation by promoting KEG self-ubiquitination and subsequent degradation by the 26S proteasome. As a consequence, ABI5 is able to accumulate in the nucleus and becomes activated by ABA signaling. Two CUL4-based RING E3s target ABI5 for degradation in the nucleus, which is suggested to attenuate ABA signaling.

proteasomal degradation of ABI5 in the presence of ABA (Lee et al., 2010; Seo et al., 2014; Figure 26.3). The CUL4-based RING E3s recruit ABI5 using DWD hypersensitive to ABA (DWA) 1/DWA2 or ABA-hypersensitive DCAF1 (ABD1) proteins (Lee et al., 2010; Seo et al., 2014). How ABA signaling regulates substrate engagement by the CRLs to modulate or terminate ABI5 transcription activity is not known. A possible strategy may be to couple the expression of the substrate recruiting DWA1/2 and ABD1 to ABI5 transcriptional activity, ensuring that targeting of ABI5 for degradation occurs at the right time postinitiation of ABA signaling. ABA promotes the expression of ABD1, however it is not known if expression is dependent on ABI5 transcriptional activity (Seo et al., 2014). Another example is strigolactone (SL)-induced proteasomal degradation of BES1 to suppress shoot branching (Wang et al., 2013). Targeting of BES1 for degradation involves MAX2, a F-box protein that functions as the substrate recruiting component of a CUL1-based RING E3 and is also involved in branching (Stirnberg et al., 2007; Wang et al., 2013). The SL-dependent degradation of BES1 abundance is in contrast to BR regulation of BES1 stability, discussed earlier, where BR promotes accumulation of the TF.

26.2.2.2 Regulating Repressor or Inhibitor Abundance

The UPS can also control TF function by modulating the abundance of interacting regulatory proteins. One strategy that is not reported in plant systems is the regulation of TF location by controlling the abundance of an interacting inhibitory protein (Figure 26.2). This strategy is illustrated in mammalian systems by the TF nuclear factor (NF)-κB, which is held in the cytoplasm by the inhibitory protein IκB. Inflammation-induced phosphorylation of IκB promotes ubiquitination and subsequent proteasomal degradation, which allows NFF-κB to enter the nucleus and activate gene expression (Chen et al., 1995; Yaron et al., 1998).

Ubiquitin-dependent turnover of transcriptional repressors is frequently used to activate plant TFs, particularly in hormone-signaling pathways including auxin, jasmonate (JA), and gibberellin (GA; Figure 26.2). A general mechanism for gene activation is the hormone-dependent degradation of a repressor protein by a CUL1-based RING E3. In the absence of the hormone, repressors interact with, and inhibit, the activity of TFs. Hormone binding to its receptor promotes repressor interactions with the E3, proteasomal degradation of the repressor, and release of the TF which then activates gene expression. In the absence of JA, the TF Myelocytomatosis related protein 2 (MYC2) is inhibited through interaction with Jasmonate Zim-Domain (JAZ) repressor proteins which recruit the corepressor Topless (TPL) using the adaptor protein NINJA (Figure 26.4; Chini et al., 2007; Thines et al., 2007; Pauwels et al., 2010). The F-box protein Coronatine Insensitive 1 (COI1), the substrate-recruiting component of a CUL1-based RING E3, functions as a receptor for the bioactive JA, jasmonoyl-isoleucine (JA-Ile; Katsir et al., 2008; Yan et al., 2009). Hormone binding, along with the cofactor inositol pentakisphosphate (InsP$_5$), increases the affinity of COI1 for the JAZ proteins promoting degradation of the repressors (Katsir et al., 2008; Yan et al., 2009; Sheard et al., 2010). COI1-mediated degradation of JAZ results in the release of the MYC2, which promotes the expression of JA-responsive genes (Figure 26.4).

The auxin repressor proteins, AUX/IAA, are recruited to the UPS for degradation by the F-box protein Transport Inhibitor 1 (TIR1; Gray et al., 2001; Dharmasiri et al., 2005; Kepinski and Leyser, 2005; Maraschin et al. 2009). TIR1 and AUX/IAA serve as coreceptors for binding auxin and the cofactor inositol hexakisphosphate (InsP$_6$; Tan et al., 2007; Calderón Villalobos et al., 2012). At low levels of auxin, oligomers of AUX/IAA proteins inhibit auxin responsive gene expression by interacting with Auxin Response Factor (ARF) TFs (Tiwari et al., 2001; Korasick et al. 2014). An increase in auxin levels causes the proteasomal degradation of AUX/IAAs and as a consequence

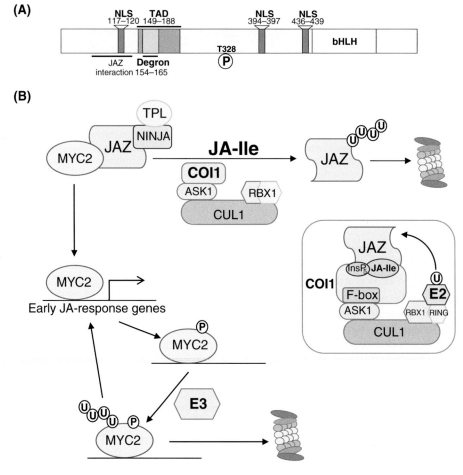

FIGURE 26.4 Control of MYC2 activity by two distinct modes of UPS-dependent regulation. (A) Schematic representation of MYC2 protein showing the DNA-binding bHLH domain, predicted TAD with overlapping degron and putative nuclear localization signals (NLS). The phosphorylation (P) site required for MYC2 turnover and transcriptional activity is shown. (B) A simplified schematic of the JA-signaling pathway highlighting regulation of MYC2 activity by the UPS. In the absence of the hormone the repressor complex, including JAZ, NINJA, and TPL proteins, interacts with and inhibits MYC2 activity. The bioactive JA, JA-Ile, promotes degradation of the JAZ repressor protein, releasing MYC2 to activate gene expression. JA-Ile and the cofactor InsP5 promote interaction between JAZ proteins and COI1, the substrate-binding subunit of a CRL. Once transcriptionally active, MYC2 becomes phosphorylated and targeted for degradation by the UPS. This is suggested to clear the binding site allowing other MYC2 proteins to bind and reinitiate transcription.

ARFs are able to activate gene expression. Activation of GA-dependent transcription requires degradation of the DELLA repressor proteins. GA bound to its receptor Gibberellin Insensitive Dwarf 1 (GID1) facilitates interactions between DELLA proteins and the F-box protein Sleepy (SLY1), allowing for degradation of the repressors and activation of gene expression (Ueguchi-Tanaka et al., 2005, 2007; Chapter 20). It is generally assumed that proteins targeted for proteasomal degradation are modified by the attachment of a Lys48-linked polyubiquitin chain. However, the DELLA protein Repressor of *ga1-3* (RGA) is modified by the attachment of a Lys29-linked chain, which is suggested to target the repressor for proteasomal degradation (Wang et al., 2009).

26.2.2.3 Transcription-Coupled Proteolysis

Another strategy used to regulate gene expression involves promotion of transcription *via* ubiquitin-dependent degradation of transcriptional regulators (Figure 26.2). TF proteolysis is generally linked to turning off gene expression. However, for this mode of regulation, TF degradation stimulates gene expression (Lipford et al., 2005; Geng et al., 2012). It is proposed that during transcriptional activation the TF becomes phosphorylated, which marks it as "spent." Phosphorylation also serves to recruit ubiquitin ligases to the TF targeting it for proteasomal degradation. Degradation of the spent TF clears the promoter, allowing "new" TFs to access the promoter and reinitiate transcription. Failure to remove the spent TF inhibits gene expression. The link between transcriptional activation and proteolysis is further suggested by the overlap in degron and transcriptional activation domain (TAD) sequences (Salghetti et al., 2000).

The "activation by destruction" mechanism is reported or proposed for only a few plant TFs, which suggests that proteolysis-coupled transcriptional activation may not be extensively used in plants (Table 26.1). The best described example for this mode of regulation is the plant TF MYC2, which regulates the expression of JA-responsive genes (Figure 26.4). MYC2 is activated as a consequence of JA-induced proteasomal degradation of JAZ repressor proteins (see previous discussion). However, degradation of MYC2 also occurs during JA signaling (Zhai et al., 2013). The proposed TAD of MYC2 contains a 12-amino-acid sequence that is required for proteolysis suggesting an overlap between the TAD and degron (Figure 26.4; Zhai et al., 2013). Deletion of the MYC2 degron stabilized the TF and prohibited the activation of MYC2-dependent genes. In addition, threonine 328 (Thr328), a potential phosphorylation site, is required for MYC2 degradation and transcription activity (Zhai et al., 2013). It is proposed that phosphorylation and proteasomal degradation are coupled to MYC2 transcriptional activation and turnover of the TF is required to promote gene expression. Another intriguing possibility for proteolysis-coupled transcription is the TF Leafy (LFY). Activation of floral homeotic gene expression by LFY requires the TF recruiting and interacting with the F-box protein Unusual Floral Organs (UFO), which is suggested to function as a transcriptional cofactor (Chae et al., 2008). LFY is ubiquitinated and as a potential component of a CUL1-based E3, UFO is proposed to also target LFY for proteasomal degradation (Chae et al., 2008). Inhibition of the proteasome prohibits LFY-dependent gene expression, suggesting that proteasomal degradation promotes transcriptional activity. The "activation by destruction" regulatory mechanism may also extend to transcriptional coactivators (Spoel et al., 2010). In response to pathogen infection, salicylic acid (SA) promotes the nuclear localization and interaction between cofactor Nonexpressor of Pathogenesis-Related genes 1 (NPR1) and TF TGA, activating defense response genes (Johnson et al., 2003). NPR1 stability is regulated by SA-induced proteasomal degradation involving a CUL3-based RING E3 (Spoel et al., 2009). It is suggested that phosphorylation-induced degradation of NPR1 clears the coactivator and interacting TF from the promoter allowing for reinitiation of transcription.

References

Amerik, A.Y., Hochstrasser, M., 2004. Mechanism and function of deubiquitinating enzymes. Biochim. Biophys. Acta 1695, 189–207.

Aravind, L., Koonin, E.V., 2000. The U box is a modified RING finger – a common domain in ubiquitination. Curr. Biol. 10, R132–R134.

Azevedo, C., Santos-Rosa, M.J., Shirasu, K., 2001. The U-box protein family in plants. Trends Plant Sci. 6, 354–358.

Ben-Saadon, R., Zaaroor, D., Ziv, T., Ciechanover, A., 2006. The polycomb protein Ring1B generates self atypical mixed ubiquitin chains required for its *in vitro* histone H2A ligase activity. Mol. Cell 24, 701–711.

Brenkman, A.B., de Keizer, P.L., van den Broek, N.J., Jochemsen, A.G., Burgering, B.M., 2008. Mdm2 induces mono-ubiquitination of FOXO4. PLoS ONE 3, e2819.

Calderón Villalobos, L.I., Lee, S., De Oliveira, C., Ivetac, A., Brandt, W., Armitage, L., Sheard, L.B., Tan, X., Parry, G., Mao, H., Zheng, N., Napier, R., Kepinski, S., Estelle, M., 2012. A combinatorial TIR1/AFB-Aux/IAA co-receptor system for differential sensing of auxin. Nat. Chem. Biol. 8, 477–485.

Callis, J., Vierstra, R.D., 1989. Ubiquitin and ubiquitin genes in higher plants. Oxf. Sum. Plant Mol. Cell Biol. 6, 1–30.

Callis, J., Carpenter, T., Sun, C.W., Vierstra, R.D., 1995. Structure and evolution of genes encoding polyubiquitin and ubiquitin-like proteins in *Arabidopsis thaliana* ecotype Columbia. Genetics 139, 921–939.

Chae, E., Tan, Q.K., Hill, T.A., Irish, V.F., 2008. An *Arabidopsis* F-box protein acts as a transcriptional co-factor to regulate floral development. Development 135, 1235–1245.

Chen, Z., Hagler, J., Palombella, V.J., Melandri, F., Scherer, D., Ballard, D., Maniatis, T., 1995. Signal-induced site-specific phosphorylation targets IκB to the ubiquitin proteasome pathway. Genes Dev. 9, 1586–1597.

Chen, Y.T., Liu, H., Stone, S., Callis, J., 2013. ABA and the ubiquitin E3 ligase KEEP ON GOING affect proteolysis of the *Arabidopsis thaliana* transcription factors ABF1 and ABF3. Plant J. 75, 965–976.

Cheng, M.C., Hsieh, E.J., Chen, J.H., Chen, H.Y., Lin, T.P., 2012. *Arabidopsis* RGLG2, functioning as a RING E3 ligase, interacts with AtERF53 and negatively regulates the plant drought stress response. Plant Physiol. 158, 363–375.

Chini, A., Fonseca, S., Fernández, G., Adie, B., Chico, J.M., Lorenzo, O., García-Casado, G., López-Vidriero, I., Lozano, F.M., Ponce, M.R., Micol, J.L., Solano, R., 2007. The JAZ family of repressors is the missing link in jasmonate signalling. Nature 448, 666–671.

Ciechanover, A., Hod, Y., Hershko, A., 1978. A heat-stable polypeptide component of an ATP-dependent proteolytic system from reticulocytes. Biochem. Biophys. Res. Commun. 81, 1100–1105.

Ciechanover, A., Heller, H., Elias, S., Haas, A.L., Hershko, A., 1980. ATP-dependent conjugation of reticulocyte proteins with the polypeptide required for protein degradation. Proc. Natl. Acad. Sci. USA 77, 1365–1368.

Ciechanover, A., Elias, S., Heller, H., Hershko, A., 1982. "Covalent affinity" purification of ubiquitin-activating enzyme. J. Biol. Chem. 257, 2537–2542.

Craig, A., Ewan, R., Mesmar, J., Gudipati, V., Sadanandom, A., 2009. E3 ubiquitin ligases and plant innate immunity. J. Exp. Bot. 60, 1123–1132.

Dasgupta, A., Ramsey, K.L., Smith, J.S., Auble, D.T., 2004. Sir Antagonist 1 (San1) is a ubiquitin ligase. J. Biol. Chem. 279, 26830–26838.

Deshaies, R.J., Joazeiro, C.A., 2009. RING domain E3 ubiquitin ligases. Annu. Rev. Biochem. 78, 399–434.

Dharmasiri, N., Dharmasiri, S., Estelle, M., 2005. The F-box protein TIR1 is an auxin receptor. Nature 435, 441–445.

Dong, C., Agarwal, M., Zhang, Y., Xie, Q., Zhu, J., 2006. The negative regulator of plant cold responses, HOS1, is a RING E3 ligase that mediates the ubiquitination and degradation of ICE1. Proc. Natl. Acad. Sci. USA 103, 8281–8286.

Downes, B.P., Stupar, R.M., Gingerich, D.J., Vierstra, R.D., 2003. The HECT ubiquitin–protein ligase (UPL) family in *Arabidopsis*: UPL3 has a specific role in trichome development. Plant J. 35, 729–742.

Dreher, K.A., Brown, J., Saw, R.E., Callis, J., 2006. The *Arabidopsis* Aux/IAA protein family has diversified in degradation and auxin responsiveness. Plant Cell 18, 699–714.

Finley, D., 2009. Recognition and processing of ubiquitin–protein conjugates by the proteasome. Annu. Rev. Biochem. 78, 477–513.

Freemont, P.S., 1993. The RING finger. A novel protein sequence motif related to the zinc finger. Ann. N. Y. Acad. Sci. 684, 174–192.

Freemont, P.S., Hanson, I.M., Trowsdale, J., 1991. A novel cysteine-rich sequence motif. Cell 64, 483–484.

Gagne, J.M., Downes, B.P., Shiu, S.H., Durski, A.M., Vierstra, R.D., 2002. The F-box subunit of the SCF E3 complex is encoded by a diverse superfamily of genes in *Arabidopsis*. Proc. Natl. Acad. Sci. USA 99, 11519–11524.

Gagne, J.M., Smalle, J., Gingerich, D.J., Walker, J.M., Yoo, S.D., Yanagisawa, S., Vierstra, R.D., 2004. *Arabidopsis* EIN3-binding F-box 1 and 2 form ubiquitin–protein ligases that repress ethylene action and promote growth by directing EIN3 degradation. Proc. Natl. Acad. Sci. USA 101, 6803–6808.

Geng, F., Wenzel, S., Tansey, W.P., 2012. Ubiquitin and proteasomes in transcription. Annu. Rev. Biochem. 81, 177–201.

Gray, W.M., Kepinski, S., Rouse, D., Leyser, O., Estelle, M., 2001. Auxin regulates SCF(TIR1)-dependent degradation of AUX/IAA proteins. Nature 414, 271–276.

Haas, T.L., Emmerich, C.H., Gerlach, B., Schmukle, A.C., Cordier, S.M., Rieser, E., Feltham, R., Vince, J., Warnken, U., Wenger, T., Koschny, R., Komander, D., Silke, J., Walczak, H., 2009. Recruitment of the linear ubiquitin chain assembly complex stabilizes the TNF–R1 signaling complex and is required for TNF-mediated gene induction. Mol. Cell 36, 831–844.

Hammond-Martel, I., Yu, H., Affar, E.B., 2012. Roles of ubiquitin signaling in transcription regulation. Cell. Signal. 24, 410–421.

Hatfield, P.M., Gosink, M.M., Carpenter, T.B., Vierstra, R.D., 1997. The ubiquitin-activating enzyme (E1) gene family in *Arabidopsis thaliana*. Plant J. 11, 213–226.

He, J.X., Gendron, J.M., Yang, Y., Li, J., Wang, Z.Y., 2002. The GSK3-like kinase BIN2 phosphorylates and destabilizes BZR1, a positive regulator of the brassinosteroid signaling pathway in *Arabidopsis*. Proc. Natl. Acad. Sci. USA 99, 10185–10190.

Hershko, A., Ciechanover, A., Heller, H., Haas, A.L., Rose, I.A., 1980. Proposed role of ATP in protein breakdown: conjugation of proteins with multiple chains of the polypeptide of ATP-dependent proteolysis. Proc. Natl. Acad. Sci. USA 77, 1783–1786.

Hershko, A., Heller, H., Elias, S., Ciechanover, A., 1983. Components of ubiquitin–protein ligase system. Resolution, affinity purification, and role in protein breakdown. J. Biol. Chem. 258, 8206–8214.

Hershko, A., Leshinsky, E., Ganoth, D., Heller, H., 1984. ATP-dependent degradation of ubiquitin–protein conjugates. Proc. Natl. Acad. Sci. USA 81, 1619–1623.

Hochstrasser, M., 2006. Lingering mysteries of ubiquitin-chain assembly. Cell 124, 27–34.

Hoppe, T., Matuschewski, K., Rape, M., Schlenker, S., Ulrich, H.D., Jentsch, S., 2000. Activation of a membrane-bound transcription factor by regulated ubiquitin/proteasome-dependent processing. Cell 102, 577–586.

Hotton, S.K., Callis, J., 2008. Regulation of cullin RING ligases. Annu. Rev. Plant Biol. 59, 467–489.

Hough, R., Pratt, G., Rechsteiner, M., 1986. Ubiquitin–lysozyme conjugates: identification and characterization of an ATP-dependent protease from rabbit reticulocyte lysates. J. Biol. Chem. 261, 2400–2408.

Husnjak, K., Elsasser, S., Zhang, N., Chen, X., Randles, L., Shi, Y., Hofmann, K., Walters, K.J., Finley, D., Dikic, I., 2008. Proteasome subunit Rpn13 is a novel ubiquitin receptor. Nature 453, 481–488.

Jang, S., Marchal, V., Panigrahi, K.C., Wenkel, S., Soppe, W., Deng, X.W., Valverde, F., Coupland, G., 2008. *Arabidopsis* COP1 shapes the temporal pattern of CO accumulation conferring a photoperiodic flowering response. EMBO J. 27, 1277–1288.

Jiao, Y., Lau, O.S., Deng, X.W., 2007. Light-regulated transcriptional networks in higher plants. Nat. Rev. Genet. 8, 217–230.

Johnson, C., Boden, E., Arias, J., 2003. Salicylic acid and NPR1 induce the recruitment of trans-activating TGA factors to a defense gene promoter in *Arabidopsis*. Plant Cell 15, 1846–1858.

Jung, J.H., Seo, P.J., Park, C.M., 2012. The E3 ubiquitin ligase HOS1 regulates *Arabidopsis* flowering by mediating CONSTANS degradation under cold stress. J. Biol. Chem. 287, 43277–43287.

Katsir, L., Schilmiller, A.L., Staswick, P.E., He, S.Y., Howe, G.A., 2008. COI1 is a critical component of a receptor for jasmonate and the bacterial virulence factor coronatine. Proc. Natl. Acad. Sci. USA 105, 7100–7105.

Kepinski, S., Leyser, O., 2005. The *Arabidopsis* F-box protein TIR1 is an auxin receptor. Nature 435, 446–451.

Kim, H.T., Kim, K.P., Lledias, F., Kisselev, A.F., Scaglione, K.M., Skowyra, D., Gygi, S.P., Goldberg, A.L., 2007. Certain pairs of ubiquitin-conjugating enzymes (E2s) and ubiquitin–protein ligases (E3s) synthesize nondegradable forked ubiquitin chains containing all possible isopeptide linkages. J. Biol. Chem. 282, 17375–17386.

Kirisako, T., Kamei, K., Murata, S., Kato, M., Fukumoto, H., Kanie, M., Sano, S., Tokunaga, F., Tanaka, K., Iwai, K., 2006. A ubiquitin ligase complex assembles linear polyubiquitin chains. EMBO J. 25, 4877–4887.

Kirkpatrick, D.S., Hathaway, N.A., Hanna, J., Elsasser, S., Rush, J., Finley, D., King, R.W., Gygi, S.P., 2006. Quantitative analysis of *in vitro* ubiquitinated cyclin B1 reveals complex chain topology. Nat. Cell Biol. 8, 700–710.

Köhler, A., Cascio, P., Leggett, D.S., Woo, K.M., Goldberg, A.L., Finley, D., 2001. The axial channel of the proteasome core particle is gated by the Rpt2 ATPase and controls both substrate entry and product release. Mol. Cell 7, 1143–1152.

Komander, D., Rape, M., 2012. The ubiquitin code. Annu. Rev. Biochem. 81, 203–229.

Korasick, D.A., Westfall, C.S., Lee, S.G., Nanao, M.H., Dumas, R., Hagen, G., Guilfoyle, T.J., Jez, J.M., Strader, L.C., 2014. Molecular basis for AUXIN RESPONSE FACTOR protein interaction and the control of auxin response repression. Proc. Natl. Acad. Sci. USA 111(14), 5427–5432.

Kraft, E., Stone, S.L., Ma, L., Su, N., Gao, Y., Lau, O., Deng, X., Callis, J., 2005. Genome analysis and functional characterization of the E2 and RING-type E3 ligase ubiquitination enzymes of *Arabidopsis*. Plant Phys. 139, 1597–1611.

Kravtsova-Ivantsiv, Y., Ciechanover, A., 2012. Non-canonical ubiquitin-based signals for proteasomal degradation. J. Cell Sci. 125, 539–548.

Kuroda, H., Takahashi, N., Shimada, H., Seki, M., Shinozaki, K., Matsui, M., 2002. Classification and expression analysis of *Arabidopsis* F-box-containing protein genes. Plant Cell Physiol. 43, 1073–1085.

Lazaro, A., Valverde, F., Piñeiro, M., Jarillo, J.A., 2012. The *Arabidopsis* E3 ubiquitin ligase HOS1 negatively regulates CONSTANS abundance in the photoperiodic control of flowering. Plant Cell 24, 982–999.

Lee, H., Xiong, L., Gong, Z., Ishitani, M., Stevenson, B., Zhu, J.K., 2001. The *Arabidopsis* HOS1 gene negatively regulates cold signal transduction and encodes a RING finger protein that displays cold-regulated nucleo-cytoplasmic partitioning. Genes Dev. 15, 912–924.

Lee, J.H., Yoon, H.J., Terzaghi, W., Martinez, C., Dai, M., Li, J., Byun, M.O., Deng, X.W., 2010. DWA1 and DWA2, two *Arabidopsis* DWD protein components of CUL4-based E3 ligases, act together as negative regulators in ABA signal transduction. Plant Cell 22, 1716–1732.

Li, W., Tu, D., Brunger, A.T., Ye, Y., 2007. A ubiquitin ligase transfers preformed polyubiquitin chains from a conjugating enzyme to a substrate. Nature 446, 333–337.

Lipford, J.R., Smith, G.T., Chi, Y., Deshaies, R.J., 2005. A putative stimulatory role for activator turnover in gene expression. Nature 438, 113–116.

Liu, H., Stone, S.L., 2010. Abscisic acid increases *Arabidopsis* ABI5 transcription factor levels by promoting KEG E3 ligase self-ubiquitination and proteasomal degradation. Plant Cell 22, 2630–2641.

Liu, H., Stone, S.L., 2013. Cytoplasmic degradation of the *Arabidopsis* transcription factor abscisic acid insensitive 5 is mediated by the RING-type E3 ligase KEEP ON GOING. J. Biol. Chem. 288, 20267–20279.

Liu, Y., Wang, F., Zhang, H., He, H., Ma, L., Deng, X.W., 2008. Functional characterization of the *Arabidopsis* ubiquitin-specific protease gene family reveals specific role and redundancy of individual members in development. Plant J. 55, 844–856.

Lopez-Molina, L., Mongrand, S., Chua, N., 2001. A postgermination developmental arrest checkpoint is mediated by abscisic acid and requires the ABI5 transcription factor in *Arabidopsis*. Proc. Natl. Acad. Sci. USA 98, 4782–4787.

Lopez-Molina, L., Mongrand, S., McLachlin, D.T., Chait, B.T., Chua, N.H., 2002. ABI5 acts downstream of ABI3 to execute an ABA-dependent growth arrest during germination. Plant J. 32, 317–328.

Lorick, K.L., Jensen, J.P., Fang, S., Ong, A.M., Hatakeyama, S., Weissman, A.M., 1999. RING fingers mediate ubiquitin-conjugating enzyme (E2)-dependent ubiquitination. Proc. Natl. Acad. Sci. USA 96, 11364–11369.

Lovering, R., Hanson, I.M., Borden, K.L., Martin, S., O'Reilly, N.J., Evan, G.I., Rahman, D., Pappin, D.J., Trowsdale, J., Freemont, P.S., 1993. Identification and preliminary characterization of a protein motif related to the zinc finger. Proc. Natl. Acad. Sci. USA 90, 2112–2116.

Luo, H., Laluk, K., Lai, Z., Veronese, P., Song, F., Mengiste, T., 2010. The Arabidopsis Botrytis Susceptible1 interactor defines a subclass of RING E3 ligases that regulate pathogen and stress responses. Plant Physiol. 154, 41766–41782.

Lyzenga, W.J., Liu, H., Schofield, A., Muise-Hennessey, A., Stone, S.L., 2013. *Arabidopsis* CIPK26 interacts with KEG, components of the ABA signalling network and is degraded by the ubiquitin–proteasome system. J. Exp. Bot. 64, 2779–2791.

Maier, A., Schrader, A., Kokkelink, L., Falke, C., Welter, B., Iniesto, E., Rubio, V., Uhrig, J.F., Hülskamp, M., Hoecker, U., 2013. Light and the E3 ubiquitin ligase COP1/SPA control the protein stability of the MYB transcription factors PAP1 and PAP2 involved in anthocyanin accumulation in *Arabidopsis*. Plant J. 74, 638–651.

Maraschin, S., Memelink, J., Offringa, R., 2009. Auxin-induced, SCF(TIR1)-mediated poly-ubiquitination marks AUX/IAA proteins for degradation. Plant J. 59, 100–109.

Marino, D., Froidure, S., Canonne, J., Ben Khaled, S., Khafif, M., Pouzet, C., Jauneau, A., Roby, D., Rivas, S., 2013. *Arabidopsis* ubiquitin ligase MIEL1 mediates degradation of the transcription factor MYB30 weakening plant defence. Nat. Commun. 4, 1476.

Maspero, E., Mari, S., Valentini, E., Musacchio, A., Fish, A., Pasqualato, S., Polo, S., 2011. Structure of the HECT: ubiquitin complex and its role in ubiquitin chain elongation. EMBO Rep. 12, 342–349.

Miura, K., Ohta, M., Nakazawa, M., Ono, M., Hasegawa, P.M., 2011. ICE1 Ser403 is necessary for protein stabilization and regulation of cold signaling and tolerance. Plant J. 67, 269–279.

Morimoto, K., Mizoi, J., Qin, F., Kim, J.S., Sato, H., et al., 2013. Stabilization of *Arabidopsis* DREB2A is required but not sufficient for the induction of target genes under conditions of stress. PLoS ONE 8, e80457.

Mudgil, Y., Shiu, S., Stone, S.L., Salt, J.N., Goring, D.R., 2004. A large complement of the predicted *Arabidopsis* ARM repeat proteins are members of the U-Box E3 ubiquitin ligase family. Plant Physiol. 134, 59–66.

Mukhopadhyay, D., Riezman, H., 2007. Proteasome-independent functions of ubiquitin in endocytosis and signaling. Science 315, 201–205.

Nakashima, K., Fujita, Y., Kanamori, N., Katagiri, T., Umezawa, T., Kidokoro, S., Maruyama, K., Yoshida, T., Ishiyama, K., Kobayashi, M., Shinozaki, K., Yamaguchi-Shinozaki, K., 2009. Three *Arabidopsis* SnRK2 protein kinases, SRK2D/SnRK2.2, SRK2E/SnRK2.6/OST1 and SRK2I/SnRK2.3, involved in ABA signaling are essential for the control of seed development and dormancy. Plant Cell Physiol. 50, 1345–1363.

Nishimura, K., Fukagawa, T., Takisawa, H., Kakimoto, T., Kanemaki, M., 2009. An auxin-based degron system for the rapid depletion of proteins in nonplant cells. Nat. Methods 6, 917–922.

Osterlund, M.T., Hardtke, C.S., Wei, N., Deng, X.W., 2000. Targeted destabilization of HY5 during light-regulated development of *Arabidopsis*. Nature 405, 462–466.

Pauwels, L., Barbero, G.F., Geerinck, J., Tilleman, S., Grunewald, W., Pérez, A.C., Chico, J.M., Bossche, R.V., Sewell, J., Gil, E., García-Casado, G., Witters, E., Inzé, D., Long, J.A., De Jaeger, G., Solano, R., Goossens, A., 2010. NINJA connects the co-repressor TOPLESS to jasmonate signalling. Nature 464, 788–791.

Pickart, C.M., Cohen, R.E., 2004. Proteasomes and their kin: proteases in the machine age. Nat. Rev. Mol. Cell Biol. 5, 177–187.

Piotrowski, J., Beal, R., Hoffman, L., Wilkinson, K.D., Cohen, R.E., Pickart, C.M., 1997. Inhibition of the 26 S proteasome by polyubiquitin chains synthesized to have defined lengths. J. Biol. Chem. 272, 23712–23721.

Piwko, W., Jentsch, S., 2006. Proteasome-mediated protein processing by bidirectional degradation initiated from an internal site. Nat. Struct. Mol. Biol. 13, 691–697.

Qin, F., Sakuma, Y., Tran, L.S., Maruyama, K., Kidokoro, S., Fujita, Y., Umezawa, T., Sawano, Y., Miyazono, K., Tanokura, M., Shinozaki, K., Yamaguchi-Shinozaki, K., 2008. *Arabidopsis* DREB2A-interacting proteins function as RING E3 ligases and negatively regulate plant drought stress-responsive gene expression. Plant Cell 20, 1693–1707.

Rape, M., Hoppe, T., Gorr, I., Kalocay, M., Richly, H., Jentsch, S., 2001. Mobilization of processed, membrane-tethered SPT23 transcription factor by CDC48(UFD1/NPL4), a ubiquitin-selective chaperone. Cell 107, 667–677

Reyes-Turcu, F.E., Ventii, K.H., Wilkinson, K.D., 2009. Regulation and cellular roles of ubiquitin-specific deubiquitinating enzymes. Annu. Rev. Biochem. 78, 363–397.

Rodrigo-Brenni, M.C., Foster, S.A., Morgan, D.O., 2010. Catalysis of lysine 48-specific ubiquitin chain assembly by residues in E2 and ubiquitin. Mol. Cell 39, 548–559.

Rosenzweig, R., Bronner, V., Zhang, D., Fushman, D., Glickman, M.H., 2012. Rpn1 and Rpn2 coordinate ubiquitin processing factors at proteasome. J. Biol. Chem. 287, 14659–14671.

Ruschak, A.M., Religa, T.L., Breuer, S., Witt, S., Kay, L.E., 2010. The proteasome antechamber maintains substrates in an unfolded state. Nature 467, 868–871.

Ryu, H., Cho, H., Kim, K., Hwang, I., 2010. Phosphorylation dependent nucleocytoplasmic shuttling of BES1 is a key regulatory event in brassinosteroid signaling. Mol. Cells 29, 283–290.

Saeki, Y., Kudo, T., Sone, T., Kikuchi, Y., Yokosawa, H., Toh-e, A., Tanaka, K., 2009. Lysine 63-linked polyubiquitin chain may serve as a targeting signal for the 26S proteasome. EMBO J. 28, 359–371.

Salghetti, S.E., Muratani, M., Wijnen, H., Futcher, B., Tansey, W.P., 2000. Functional overlap of sequences that activate transcription and signal ubiquitin-mediated proteolysis. Proc. Natl. Acad. Sci. USA 9, 3118–3123.

Seo, H.S., Yang, J.Y., Ishikawa, M., Bolle, C., Ballesteros, M.L., Chua, N.H., 2003. LAF1 ubiquitination by COP1 controls photomorphogenesis and is stimulated by SPA1. Nature 423, 995–999.

Seo, P.J., Kim, S.G., Park, C.M., 2008. Membrane-bound transcription factors in plants. Trends Plant Sci. 13, 550–556.

Seo, K.I., Lee, J.H., Nezames, C.D., Zhong, S., Song, E., Byun, M.O., Deng, X.W., 2014. ABD1 is an *Arabidopsis* DCAF substrate receptor for CUL4–DDB1-based E3 ligases that acts as a negative regulator of abscisic acid signaling. Plant Cell 26, 695–711.

Sharp, P.M., Li, W.H., 1987. Ubiquitin genes as a paradigm of concerted evolution of tandem repeats. J. Mol. Evol. 25, 58–64.

Shcherbik, N., Kee, Y., Lyon, N., Huibregtse, J.M., Haines, D.S., 2004. A single PXY motif located within the carboxyl terminus of Spt23p and Mga2p mediates a physical and functional interaction with ubiquitin ligase Rsp5p. J. Biol. Chem. 279, 53892–53898.

Sheard, L.B., Tan, X., Mao, H., Withers, J., Ben-Nissan, G., Hinds, T.R., Kobayashi, Y., Hsu, F.F., Sharon, M., Browse, J., He, S.Y., Rizo, J., Howe, G.A., Zheng, N., 2010. Jasmonate perception by inositol-phosphate-potentiated COI1–JAZ co-receptor. Nature 468, 400–405.

Sirichandra, C., Davanture, M., Turk, B.E., Zivy, M., Valot, B., Leung, J., Merlot, S., 2010. The *Arabidopsis* ABA-activated kinase OST1 phosphorylates the bZIP transcription factor ABF3 and creates a 14-3-3 binding site involved in its turnover. PLoS ONE 5, e13935.

Smalle, J., Vierstra, R.D., 2004. The ubiquitin 26S proteasome proteolytic pathway. Annu. Rev. Plant Biol. 55, 555–590.

Spoel, S.H., Mou, Z., Tada, Y., Spivey, N.W., Genschik, P., Dong, X., 2009. Proteasome-mediated turnover of the transcription coactivator NPR1 plays dual roles in regulating plant immunity. Cell 137, 860–872.

Spoel, S.H., Tada, Y., Loake, G.J., 2010. Post-translational protein modification as a tool for transcription reprogramming. New Phytol. 186, 333–339.

Stirnberg, P., Furner, I.J., Ottoline Leyser, H.M., 2007. MAX2 participates in an SCF complex which acts locally at the node to suppress shoot branching. Plant J. 50, 80–94.

Stone, S.L., Hauksdottir, H., Troy, A., Herschleb, J., Kraft, E., Callis, J., 2005. Functional analysis of the RING-type ubiquitin ligase family of *Arabidopsis*. Plant Physiol. 137, 13–30.

Stone, S.L., Williams, L.A., Farmer, L.M., Vierstra, R.D., Callis, J., 2006. KEEP ON GOING, a RING E3 ligase essential for *Arabidopsis* growth and development, is involved in abscisic acid signaling. Plant Cell 18, 3415–3428.

Tan, X., Calderon-Villalobos, L., Sharon, M., Zheng, C., Robinson, C.V., Estelle, M., Zheng, N., 2007. Mechanism of auxin perception by the TIR1 ubiquitin ligase. Nature 446, 640–645.

Tang, W., Yuan, M., Wang, R., Yang, Y., Wang, C., Oses-Prieto, J.A., Kim, T.W., Zhou, H.W., Deng, Z., Gampala, S.S., Gendron, J.M., Jonassen, E.M., Lillo, C., DeLong, A., Burlingame, A.L., Sun, Y., Wang, Z.Y., 2011. PP2A activates brassinosteroid-responsive gene expression and plant growth by dephosphorylating BZR1. Nat. Cell Biol. 13, 124–131.

Thines, B., Katsir, L., Melotto, M., Niu, Y., Mandaokar, A., Liu, G., Nomura, K., He, S.Y., Howe, G.A., Browse, J., 2007. JAZ repressor proteins are targets of the SCF(COI1) complex during jasmonate signalling. Nature 448, 661–665.

Thrower, J.S., Hoffman, L., Rechsteiner, M., Pickart, C.M., 2000. Recognition of the polyubiquitin proteolytic signal. EMBO J. 19, 94–102.

Tiwari, S.B., Wang, X.J., Hagen, G., Guilfoyle, T.J., 2001. AUX/IAA proteins are active repressors, and their stability and activity are modulated by auxin. Plant Cell 13, 2809–2822.

Tokunaga, F., Sakata, S., Saeki, Y., Satomi, Y., Kirisako, T., Kamei, K., Nakagawa, T., Kato, M., Murata, S., Yamaoka, S., Yamamoto, M., Akira, S., Takao, T., Tanaka, K., Iwai, K., 2009. Involvement of linear polyubiquitylation of NEMO in NF-kappaB activation. Nat Cell Biol. 11, 123–132.

Tomanov, K., Luschnig, C., Bachmair, A., 2014. Ubiquitin Lys 63 chains – second-most abundant, but poorly understood in plants. Front. Plant Sci. 5, 15.

Ueguchi-Tanaka, M., Ashikari, M., Nakajima, M., Itoh, H., Katoh, E., Kobayashi, M., Chow, T.Y., Hsing, Y.I., Kitano, H., Yamaguchi, I., Matsuoka, M., 2005. GIBBERELLIN INSENSITIVE DWARF1 encodes a soluble receptor for gibberellin. Nature 437, 693–698.

Ueguchi-Tanaka, M., Nakajima, M., Katoh, E., Ohmiya, H., Asano, K., Saji, S., Hongyu, X., Ashikari, M., Kitano, H., Yamaguchi, I., Matsuoka, M., 2007. Molecular interactions of a soluble gibberellin receptor, GID1, with a rice DELLA protein, SLR1, and gibberellin. Plant Cell 19, 2140–2155.

van der Horst, A., de Vries-Smits, A.M., Brenkman, A.B., van Triest, M.H., van den Broek, N., Colland, F., Maurice, M.M., Burgering, B.M., 2006. FOXO4 transcriptional activity is regulated by monoubiquitination and USP7/HAUSP. Nat. Cell Biol. 8, 1064–1073.

Vander Kooi, C.W., Ohi, M.D., Rosenberg, J.A., Oldham, M.L., Newcomer, M.E., Gould, K.L., Chazin, W.J., 2006. The Prp19 U-box crystal structure suggests a common dimeric architecture for a class of oligomeric E3 ubiquitin ligases. Biochemistry 45, 121–130.

Vijay-Kumar, S., Bugg, C.E., Cook, W.J., 1987. Structure of ubiquitin refined at 1.8 A resolution. J. Mol. Biol. 194, 531–544.

von Arnim, A.G., Deng, X.W., 1994. Light inactivation of *Arabidopsis* photomorphogenic repressor COP1 involves a cell-specific regulation of its nucleocytoplasmic partitioning. Cell 79, 1035–1045.

Vosper, J.M., McDowell, G.S., Hindley, C.J., Fiore-Heriche, C.S., Kucerova, R., Horan, I., Philpott, A., 2009. Ubiquitylation on canonical and non-canonical sites targets the transcription factor neurogenin for ubiquitin-mediated proteolysis. J. Biol. Chem. 284, 15458–15468.

Walsh, C.K., Sadanandom, A., 2014. Ubiquitin chain topology in plant cell signaling: a new facet to an evergreen story. Front. Plant Sci. 5, 122.

Walters, K.J., Kleijnen, M.F., Goh, A.M., Wagner, G., Howley, P.M., 2002. Structural studies of the interaction between ubiquitin family proteins and proteasome subunit S5a. Biochemistry 41, 1767–1777.

Wang, M., Pickart, C.M., 2005. Different HECT domain ubiquitin ligases employ distinct mechanisms of polyubiquitin chain synthesis. EMBO J. 24, 4324–4333.

Wang, Z.Y., Nakano, T., Gendron, J., He, J., Chen, M., Vafeados, D., Yang, Y., Fujioka, S., Yoshida, S., Asami, T., Chory, J., 2002. Nuclear-localized BZR1 mediates brassinosteroid-induced growth and feedback suppression of brassinosteroid biosynthesis. Dev. Cell 2, 505–513.

Wang, X., Herr, R.A., Chua, W.J., Lybarger, L., Wiertz, E.J., Hansen, T.H., 2007. Ubiquitination of serine, threonine, or lysine residues on the cytoplasmic tail can induce ERAD of MHC-I by viral E3 ligase mK3. J. Cell Biol. 177, 613–624.

Wang, F., Zhu, D., Huang, X., Li, S., Gong, Y., Yao, Q., Fu, X., Fan, L.M., Deng, X.W., 2009. Biochemical insights on degradation of *Arabidopsis* DELLA proteins gained from a cell-free assay system. Plant Cell 21, 2378–2390.

Wang, Y., Sun, S., Zhu, W., Jia, K., Yang, H., Wang, X., 2013. Strigolactone/MAX2-induced degradation of brassinosteroid transcriptional effector BES1 regulates shoot branching. Dev. Cell 27, 681–688.

Waxman, L., Fagan, J., Goldberg, A.L., 1987. Demonstration of two distinct high molecular weight proteases in rabbit reticulocytes, one of which degrades ubiquitin conjugates. J. Biol. Chem. 262, 2451–2457.

Wilkinson, K.D., Urban, M.K., Haas, A.L., 1980. Ubiquitin is the ATP-dependent proteolysis factor I of rabbit reticulocytes. J. Biol. Chem. 255, 7529–7532.

Xu, P., Duong, D.M., Seyfried, N.T., Cheng, D., Xie, Y., Robert, J., Rush, J., Hochstrasser, M., Finley, D., Peng, J., 2009. Quantitative proteomics reveals the function of unconventional ubiquitin chains in proteasomal degradation. Cell 137, 133–145.

Yan, N., Doelling, J.H., Falbel, T.G., Durski, A.M., Vierstra, R.D., 2000. The ubiquitin-specific protease family from *Arabidopsis*. AtUBP1 and 2 are required for the resistance to the amino acid analog canavanine. Plant Physiol. 124, 1828–1843.

Yan, J., Zhang, C., Gu, M., Bai, Z., Zhang, W., Qi, T., Cheng, Z., Peng, W., Luo, H., Nan, F., Wang, Z., Xie, D., 2009. The *Arabidopsis* CORONATINE INSENSITIVE1 protein is a jasmonate receptor. Plant Cell 21, 2220–2236.

Yang, P., Smalle, J., Lee, S., Yan, N., Emborg, T.J., Vierstra, R.D., 2007. Ubiquitin C-terminal hydrolases 1 and 2 affect shoot architecture in *Arabidopsis*. Plant J. 51, 441–457.

Yao, T., Cohen, R.E., 2002. A cryptic protease couples deubiquitination and degradation by the 26S proteasome. Nature 419, 403–407.

Yaron, A., Hatzubai, A., Davis, M., Lavon, I., Amit, S., Manning, A.M., Andersen, J.S., Mann, M., Mercurio, F., Ben-Neriah, Y., 1998. Identification of the receptor component of the IκBα-ubiquitin ligase. Nature 396, 590–594.

Yee, D., Goring, D.R., 2009. The diversity of plant U-box E3 ubiquitin ligases: from upstream activators to downstream target substrates. J. Exp. Bot. 60, 1109–1121.

Zeng, L.R., Park, C.H., Venu, R.C., Gough, J., Wang, G.L., 2008. Classification, expression pattern, and E3 ligase activity assay of rice U-box-containing proteins. Mol. Plant 1, 800–815.

Zhai, Q., Yan, L., Tan, D., Chen, R., Sun, J., Gao, L., Dong, M.Q., Wang, Y., Li, C., 2013. Phosphorylation-coupled proteolysis of the transcription factor MYC2 is important for jasmonate-signaled plant immunity. PLoS Genet. 9, e1003422.

Zhang, S., Skalsky, Y., Garfinkel, D.J., 1999. MGA2 or SPT23 is required for transcription of the delta9 fatty acid desaturase gene, OLE1, and nuclear membrane integrity in *Saccharomyces cerevisiae*. Genetics 151, 473–483.

Zhang, X., Garreton, V., Chua, N.H., 2005. The AIP2 E3 ligase acts as a novel negative regulator of ABA signaling by promoting ABI3 degradation. Genes Dev. 19, 1532–1543.

Zhao, S., Ulrich, H.D., 2010. Distinct consequences of posttranslational modification by linear versus K63-linked polyubiquitin chains. Proc. Natl. Acad. Sci. USA 107, 7704–7709.

Index

A

AAAG motif, 185
ABA response element (ABRE), 44
ABA-responsive element binding proteins (AREBs), 44
ABA-signaling, 208
ABI5 abundance regulation, 403
Abiotic stress
   HD-ZIP I TFS, from model plants, 353
   NAC networks, 207
ABIs (ABA insensitives), 321
Abscisic acid (ABA)
   stress hormones, 207
   treatment, 346
Abscisic acid Insensitive 5 (ABI5), 401
Abscisic acid responsive elements-binding factor3 (ABF3), 259
Acacia confusa, 282
Acetyltransferases, 36
ACGT-containing element (ACE) motifs, 337
Activation by destruction, 405
   regulatory mechanism, 405
Activation domains, 7
Activation interacting domain (ACID), 43
Active phyA-binding (APA), 333
Active phyB-binding (APB) motif, 333
Active phytochrome binding (AFB) domain, 48
Adenosine triphosphatases (ATPases), 270
Adenosine triphosphate (ATP)
   dependent chromatin remodelers, 36
   dependent protease, 396
Affinity tags, 27
Age-related resistance (ARR), 287
Amborella trichopoda, 230
Amino acids, 102, 157
   sequences, 104, 222
ANAC019
   NAC domain, 200
   structural information, 204
ANAC083-derived peptide, 205
ANAC019 dimers, 231
   DNA-bound, 202
   DNA complex structure, 200, 202
ANAC019/055/072 genes, 209
Angiosperms, NAM/CUC3 proteins, 230
Antennapedia (Ant) homeodomain
   NMR spectroscopy of, 102
Antibody-based flow cytometry, 18
Anti-RFP antibody, 27
Antirrhinum CIN genes, 261
Antirrhinum cup mutants, 235, 236
Antirrhinum majus, 235, 250
AP2/ERF family, 49
APETALA1 (AP1), 272, 285

APETALA2 (AP2), 272
Apical-to-basal pattern, 250
Arabidopsis AtHB12, 353
Arabidopsis brc1 mutants, 258
Arabidopsis brevipedicellus, 219
Arabidopsis cuc1–cuc2 double-mutants, 235
Arabidopsis cuc1–cuc2 seedlings, 234
Arabidopsis CYC/TB1 gene *TCP1*, 252
Arabidopsis genome, 48, 59, 397
   1001 sequence, 390
Arabidopsis gif1–gif2–gif3 triple-mutant, 276
Arabidopsis histidine kinase 4 (AHK4), 224
Arabidopsis membrane-bound transcription factors, 392
   NAC transcription factors, 390
Arabidopsis MIR164B gene, 239
Arabidopsis response regulator16 (ARR16), 255
Arabidopsis TCP1, 256
Arabidopsis thaliana (Arabidopsis), 127, 199, 250, 282, 313, 330, 344, 386, 396
   age and vernalization pathways, 286
   ANAC019 protein, 200
   BS cells, 300
   bZIP28, 386
   bZIP60, 388
   cell cultures
      mass spectrometry analysis, 273
      suspension, 27
   CIN genes, 261
   class M gene, 217
   core components, 39
   CUC proteins, 230, 236
   defoliation studies in, 284
   GA treatments on the gai-1 mutant, 314
   genetic and molecular analysis in, 44
   genome, 173
      sequence of, 57
   HD-containing TFs, phylogenetic tree, 116
   homologs, 235
   inflorescence meristem, 242
   leaf development, 236
   leaves, 236
   loss-of-function *brc1* mutants, 256
   mutants
      in *AtGRFs*, 275
      phenotypes and expression patterns, 234
   NAC proteins, 204
      C-termini of, 200
      schematic phylogeny, 232
   octopine synthase *(ocs)* element, 186
   PRC2 proteins CURLY LEAF (CLF), 221
   proteome, 398
   RNAPII, 42
   roots, radial patterning of, 307

   seedlings, 363
   TF families, 47
Arabidopsis thaliana GRFs (AtGRFs), 270
ARA-bidopsis thaliana homeobox 2 (ATHB2), 333
Arabidopsis WOX1, 223
Arabidopsis WUS (AtWUS) protein, 115
Arbuscular mycorrhizal fungi (AMF), 299
ARF6/ARF8 promote auxin transport, 261
ARF8 promote petal expansion, 261
Argonaute7 (AGO7)
   TCP3 promotes expression of, 255
AS1–AS2 complex, 221
Ascorbate oxidase gene binding protein (AOBP), 184
Ascorbate peroxidase (APX), 374
Aspergillus nidulans ALCR transcription factor (ALCR), 17
Ataxin-3 subfamilies, 397
AtbZIP proteins, 44
ATC, promoter activity
   phloem tissue of *Arabidopsis*, 364
AtGRF genes, 270
Autofluorescence, 29
AUX/IAA family, 51
Auxin/indole-3-acetic (AUX/IAA) repressor proteins, 396
Auxin regulates *CUC* expression, 240
Auxin repressor proteins, 404
Auxin response factors (ARFs), 51
Auxin sensitivity, 290
Auxin signaling, 221
Axillary meristems (AMs), 256
   in flower, embryo, and seed development, 301
   GRAS family proteins, functional aspects, 301
   GRAS function, general principles, 303
      DNA interaction, 303
      homo/heterodimers, 306
      interaction of, 306
      protein movement, 307
      protein–protein interactions, 306
      regulation of gene expression by mirna, 307
   in signaling, 301
      gibberellin, 301
      light, 302
      stress, 302

B

Bacillus firmus, 61
Basic domain/leucine zipper (bZIP) transcription factor, 331

Basic helix-loop-helix (bHLH)
  genes, 89
    HFR1, PAR1, PAR2, PIL1, 333
    TF family, 48
  proteins, 306
  structure, 139
  type transcription factor, 186
Basic leucine zipper (bZIP), 144
  domain, 386
  transcription factor, 206
    G-group, 378
Basic local alignment search tool
      (BLAST), 173
Basic region/leucine zipper (bZIP) family,
      44–48
  member, in integrating light responses, 337
  transcription factors, 386
    Arabidopsis bZIP28, 386
    bZIP17, 387
    bZIP60, 387
B-box (BBX)
  proteins, 50, 51
  zinc finger transcription factors, 50–51
BELL family genes
  classification, 216
  plant lineages, 216
  protein–protein interaction, 217
BEL1-like family, ubiquitous, 367
BELL1-like homeobox *(BELL/BLH)* genes, 216
BELL proteins, 216
  Gamete-specific *plus1* (Gsp1), 217
BEL proteins, 106
BES (BRI1-EMS-SUPPRESSOR), 320
Bimolecular fluorescence complementation
      (BiFC)
  assays, 262
  methodologies, 28
    multicolor BiFC (mcBiFC) vectors, 28
Biochemical assays, 21
Bioluminescence resonance energy transfer
      (BRET) method, 28
Bioluminescent luciferase (RLUC), 28
Biotin peptide (Bio), 27
BNQ genes, 337
Botrytis susceptible1 interactor
      (BOI), 321, 322
  gene1 (BRG1), 322
BRAHMA (BRM) gene, 242, 255
BRANCHED1 (BRC1), 252
Brassica napus, 276
  SCARECROW-like protein, 306
Brassinosteroid (BR), 220
Brassinosteroid Insensitive 2 (BIN2), 401
Brassinosteroid (BR) systems, 91, 238, 261
  biosynthetic pathway, 261
  signaling, AUXIN RESPONSE FACTOR2
      (ARF2) activity, 243
BRC1/TB1 genes, shoot-branching
      phenotypes, 257
BRI1-EMSSUPPRESSOR1 (BES1), 320
  strigolactone (SL)-induced proteasomal
      degradation, 402
Broad complex Tramtrack Bric-a-Brac (BTB)
      proteins, 398
Bryophytes *(Physcomitrella patens)*, 286, 349
Bulging, 235

Bundle sheath (BS) cells, 300
bZIP60u protein, 388

C

Caenorhabditis elegans, 90, 344
CAGGCG ATTC sequence, 381
Calcineurin B-like Interacting Protein Kinase
      (CIPK) family, 401
Calmodulin (CaM)-binding domain, 50
Calmodulin-binding NAC (CBNAC), 204
Calmodulin-binding peptide, 27
Calmodulin-dependent protein kinase, 299
Cape Verde Islands (Cvi), 390
Capnoides sempervirens, 259
Carboxy-terminal dimerization domain
      (CTD), 51
Carboxy-terminal regions (CTR), 344
Cardamine flexuosa, 285
Cardamine hirsuta, 238, 241, 282
CDD complex, 331
CDK8/CycC module, 42
CDK inhibitor genes, 389
cDNA library, 24, 25
Cell Division Cycle2, 299
Cell membranes, PIN1 asymmetrically
      localizes, 221
Ceratopteris richardii, 135
Charophytes, 230
Chelidomium majus (Papaveraceae), 253
Chinese cabbage *(Brassica rapa)*, 284
Chitin-inducible gibberellin-responsive
      proteins (CIGR), 157, 303
Chlamydomonas reinhardtii, 169, 188
  Cu response regulator 1 (CRR1), 64
Chloroplast, electron flow, 375
Chromatin, 13
  associated proteins, 39
  modifying enzymes, 4
  remodelers
    ATP-dependent, 36
    role of, 36–37
  remodeling, 272, 273
  structure of, 3
Chromatin immunoprecipitation (ChIP)
      assays, 21–22, 238, 303
  ChIP-chip, 22
    assay, 22
    experiments, 224
  ChIP-seq studies, 219
    ChIP-chip, 22
  protocol, 21
Chrysanthemum morifolium, 252, 261
C2H2 Zn finger genes, 89, 94
CINCIN-NATA (CIN), 250
  clade, 141
  factors, with TPL transcriptional
      corepressors, 256
  gene regulatory networks, 254
  simple leaf development, 250
Cis-element mitochondrial dysfunction motif
      (MDM), 391
CLAVATA (CLV)
  CLV1 expression, 224
  signaling pathway, 223
  in *WUS* domain, 224
CLAVATA3 (CLV3), 360

Cleavage and polyadenylation specificity
      factor (CPSF), 39
Cleavage stimulatory factor (CstF), 39
Coimmunoprecipitation, 28–29
  resonance energy transfer methods, 28–29
  in vivo split methods, 28
Co–IP experiments, 273
Co-localization experiments, 27
Commelina communis, 260
Commercial antibodies, 22
Competition assays, 20
Complex regulatory systems, 73
Conserved amino acids, 157
CONSTITUTIVE PHOTOMORPHOGENIC
      (COP), 331
Control gene expression programs, 42
Corepressors, 36, 43
Core promoter elements (CPEs), 35, 36
Cortex/endodermis initial cell (CEI), 297
Cortex-endodermis initial daughters
      (CEID), 297
Cotyledon separation, *NAM* genes, 234
CPSCE domain, 120
Craterostigma plantagineum, 350
Crosstalk
  cytokinin–auxin, 323
  GA and JA signaling, 321
  GA–ethylene, 324
C-terminal domain (CTD), 37
  of NAM/CUC3 proteins, 232
  with transcriptional regulatory activity, 200
C-terminal GRAS functional domain, 315
cuc1–cuc2 double-mutants, 234
CUC direct targets, 243
CUC genes, 219, 234
CUC1/2 genes, *miR164*-resistant variants, 235
CUC genes role, *NAM/CUC3* genes define
      boundaries, 235
  axillary meristems, 235
  floral organ boundaries, 235
  gynecium, 236
  leaf development, 236
  organ abscission, 236
CUC/miR164 genetic module, 239
CUC2 miR164 resistant cell lines, 239
cuc mutant, genetic analysis, 242
CUC regulatory network, 237
CUC transcription factors, 230
Cucumber, *GAI*, encoding protein, 363
*CUP-SHAPED COTYLEDON/Class I
      KNOTTED*-like homeobox *(CUC/
      KNOX1)* module, 253
CUPULIFORMIS (CUP) gene, 235
CYC1 genes, 141
CYC2 genes, 259
  control floral zygomorphy, in dicots, 259
    CIN genes in flower development, 261
    class I genes, 261
    floral identity of capitulum
        inflorescence, 261
    flower asymmetry in monocots, 260
    hormone regulation of flower
        development, 261
  expression, 259
Cyclamen persicum, 261
Cyclic DOF factor 1 (CDF1), 190

Cyclic electron transport (CET), 374
CYCLOIDEA/TEOSINTE BRANCHED 1 (CYC/TB1) genes, 250
CYC/TB1-like genes, 252
CYP734A genes, 220
Cys$_4$ HisCys$_3$ Zn finger motif, 107
Cys156, in NPR1 oligomer, 378
Cys residues, 379, 380
Cys thiol, 376
Cytokinin(CK)
  biosynthetic gene
    isopentenyl transferase (ipt), 220
  distribution, regulation, 224
  signaling, via *ARR* genes, 224
Cytokinin receptor *ahk2–ahk3*
    double-mutant, 238
Cytoplasmic splicing
  bZIP60, activation mechanism, 388
  of *bZIP60u* mRNA, 388

D

Dam methylase identification (DamID), 21
dcl1, 8-cell embryos
  microarray analyses, 283
DEETIOLATED (DET), 331
DEFECTIVE IN ANTHER
    DEHISCENCE1, 262
Defective Kernel1 (DEK1), 389
Degradation signal (degron), 396
Dehydration-responsive element (DRE), 59
Dehydration-responsive element binding
    protein (DREB), 59
*DEHYDRATION-RESPONSIVE ELEMENT-
    BINDING PROTEIN2A
    (DREB2A)*, 277
DELLAs (aspartic acid-glutamic acid-leucine-
    leucine-alanine) protein, 153, 313
  angiosperm species, 316
  in *Arabidopsis*, 319
  degradation of, 157
  discovery of, 314
  distribution and evolution, 316
  feedback loops, 323
  future perspectives, 324
  GA–GID1–DELLA module, in GA
      signaling, 317
  GA-independent, proteolytic signaling, 318
  GA insensitive (GAI), 298
  GA receptor GID1, 317
  GA-signaling pathways, 302, 317
  gene, 156
  green revolution, 316
  independent GA response, 318
  independent GA-signaling pathway, 318
  interfering mechanism, 320
  molecular mechanism
    interactions and target genes, 319
  motif, 159
  nonproteolytic signaling, 318
  nontranscriptional interaction, 323
  posttranslational modifications, 318
  proteolytic-dependent degradation of, 318
  structure, 315
  target genes, regulate expression, 323
  transactivation mechanism, 321
  TVHYNP motifs, 317

DENSE AND ERECT PANICLE1, 289
Deubiquitinating enzymes (DUBs), 396
*DEVELOPMENT-RELATED PCG TARGET IN
    THE APEX 4 (DPA4)* gene, 242
DICER-LIKE1 (DCL1), 283
Dimerization domain, 6, 144
Dimers formation, 6
Dimethyl sulfate, for footprinting method, 20
Dimethyl sulfoxide (DMSO), 389
Distant regulatory elements (DRE), 36
Divergent HD-ZIP I proteins
  from nonmodel plants, 351
DNA
  binding, 135
    activity, 184
    domain, 25, 96
    genes, 93
    motifs, 21, 105
    preferences, 222
    sites, 222
    structure-derived knowledge, 200
  binding proteins, 13, 21, 24, 25, 183
    R2R3–MYB family, 332
  contacting residues, 61
  fragments, 21
  HD interaction, 122
  microarrays, 21
  technology, 21
  "nonspecific" unlabeled, 18
  phosphates, 4
  recognition, by transcription factor, 4–5
  sequences, recognition of, 4
  stoichiometry, 18
DNA adenine methyltransferase (Dam), 22
DNA adenine methyltransferase
    identification (DamID), 22–24
  advantage of, 24
DNA-binding domain (DBD), 5–6, 9, 25, 44,
    144, 200
  AP2/ERF domain, 59–60
    structures of, 59
  B3 domain, 60–62
    structures of, 61
  DDB1, 398
  description of, 59–65
  endonucleases as origins of, 65–68
  lineage-specific expansion, 68–69
  NAC domain, 63–64
  of plant-specific transcription factors, 65
    evolutionary history of, 65–69
    introduction of, 57–59
  SQUAMOSA promoter-binding proteins
      (SBPs), 64–65
  structures of, 58
  WRKY domain, 62–63
    structures of, 63
DNA–protein binding reaction, 18
DNA–protein complex, 18
DNase I footprinting method, 20
Dof–bZIP interactions, 186
Dof (DNA-binding one finger) domain, 183
Dof transcription factor family
  amino acid sequences of, 184
  discovery and definition of, 183–184
  Dof genes in variety of plant species, 188
  domain as multifunctional domain, 187

domain structure of, 185
hormonal regulation, 193
model of transcriptional control, 186
molecular evolution of, 188–190
perspective, 194
phylogenetic tree, 189
physiological functions of, 190–194
  CDF-containing clade, *Arabidopsis* CDF
      and members, 190–191
  in arabidopsis, 190
  homologs in angiosperms, 191
  homologs in nonvascular plants, 191
regulation of metabolism, 192–193
response to light, 193–194
role for, 193
  regulation of development and
      differentiation, 191–192
structure and molecular characteristics,
    184–188
  domain structure of, 184
  interactions between and nuclear
      proteins, 185–187
  interactions with bHLH and zinc-finger-
      type transcription factors, 186–187
  interactions with bZIP-type transcription
      factors, 186
  interactions with chromatin-associated
      high mobility group (HMG)
      proteins, 187
  interactions with MYB-type transcription
      factors, 186
  zinc finger motif in, 185
structure, function, and evolution of, 183
transcriptional activation and repression
    domains of, 187
Donor fluorophore, 28
DORMANCY1 (DRM1), 257
DRB sensitivity-inducing factor (DSIF), 39
Drosophila homeotic mutants, 215
Drosophila melanogaster, 44, 90, 101, 114, 391
Drought-sensitive cultivar (Zhenshan97), 346
DWD hypersensitive to ABA (DWA), 402

E

Eco RII DBD/DNA complex
  structure of, 62
Ectopic expressors, 352
Ectopic ligule formation, 218
Effector-triggered immunity (ETI)
    pathway, 50
EGL3 (ENHANCER of GLABRA3), 320
EIN3-LIKE1 (EIL1), 321
Electrophoretic mobility shift assay
    (EMSA), 18
  advantage of, 18
  assay, 18
Embryophytes, 286
Endonucleases, 65
Endoplasmic reticulum (ER), 386
  membrane, 400
  stress responses, 44
  unfolded proteins, 386
Energy transfer techniques, 28
Engrailed (en) homeodomain
  x-ray crystallography of, 102
Enhance auxin sensitivity, 261

Environmental stress, 385
Epigenetic histone modification, 221
Epigenetic marks, 39
Escherichia-coli-expressed WRKY proteins, 164
ETHYLENE INSENSITIVE3 (EIN3), 321
Ethylene-responsive element binding proteins (EREBPs), 49
Ethylene-responsive factor (ERF), 256
Ethylene-responsive factor-associated amphiphilic repression (EAR) domain, 17
  like motif, 68, 222
Eucalyptus globules, 282
Eukaryotes, 36
  promoters, structure of, 4
  RNA polymerases in, 3
  transcription cycle, overview of, 36–39
    core promoter elements and general transcription factors, 36
    nucleosome-modifying enzymes and chromatin remodelers, role of, 36–37
    productive transcription initiation and elongation, 37–39
    transcription termination and reinitiation, 39
  transcription factor in, 3–4
Eukaryotic DNA, organization of, 36
Eukaryotic genes, transcription of, 52
Eukaryotic kingdoms, homeoboxes and homeodomains in, 114
Excited-state energy, 28
Exogenous salicylic acid, 387
Expressed sequence tag (EST), 154
  databases, 169

F

FAR-RED IMPAIRED RESPONSE 1 (FAR1), 337
F-box protein Unusual Floral Organs (UFO), 405
Floral activation complex (FAC), 362
Floral promotion
  RFP–FT RNA, long-distance movement of, 365
Floral stem cells, WUS expression termination, 225
FLOWERING LOCUS C (FLC), 286
FLOWERING LOCUS T (FT), 262, 285
Flowering Locus T (FT) protein, 359
Flowering time, TCP genes, 262
Flower phenotype, of class II TCP gene mutants, 260
Flower shape, TCP genes, 259
  CYC2 genes control floral zygomorphy, in dicots, 259
    CIN genes in flower development, 261
    class I genes, 261
    floral identity of capitulum inflorescence, 260
    flower asymmetry in monocots, 260
    hormone regulation of flower development, 261
  flowering time, 262
Fluorescence resonance energy transfer (FRET)
  analyses, 28, 144
  method, 28

Fluorescent proteins mutant, 28
Fluorophores
  acceptor, 28
  based *in planta* protein–protein interaction assay, 29
  donor, 28
FLYWCH proteins, 164
Footprinting assays, transcription factor–DNA complexes analysis, 20–21
Footprinting method
  dimethyl sulfate for, 20
  DNase I, 20
Formaldehyde
  advantages, 22
  use to, 22
FR light, 330
FRUITFULL (FUL), 285
FT-INTERACTING PROTEIN 1 (FTIP1), 361
FT RNA depletion, from scion delays scion floral initiation, 366
Full-length mobile mRNAs
  ATC mRNA, 363
  FT mRNA to shoot apex, 365
  GA INSENSITIVE, long-distance transport, 363
  roles in development, 363
  StBEL5 model, 367
Functional protein domains, 270
  C-terminal region, with transcriptional activation activity, 271
  DNA/protein–protein interaction domains, in N-terminal region, 270

G

GA–GID1–DELLA complex, 317
GAI-ASSOCIATED FACTOR1 (GAF1), 322
GA INSENSITIVE (GAI), 359
GA-insensitive-1 (gai-1) mutation, 313
GAL4/UASG systems, 16
Gametophyte dominant, 286
GA2ox1 expression, 220
GATC DNA sequence, 22
G-box sequence, on *LHCB2.4*, 379
Gemini-virus RepA-binding (GRAB) proteins, 205
Gene balance hypothesis, 134
Gene expression, 21, 43
  control in eukaryotes, 76
  of dominant negative forms, 16
  of dysregulated forms, 16
  of fusions to activating/repressor domains, 17
  gain-of-function mutants, 16
  overexpression/ectopic expression of, 14
  regulation of, 13, 241
  transactivation systems for, 14–16
Gene promoters, 37
General transcription factors (GTFs), 35
Gene regulatory networks (GRN), 84, 136, 207
  ANAC019/055/072, 209
  ANAC019/ANAC055/ANAC072/RD29, 207
  NAC019/055/072, 208
Gene-silencing techniques, 14
Genetic analyses, 330

Genetic identification
  light/phytochrome-signaling components encoding
    nontranscription factors, 331
    transcription factors, 331
Gene transcription, regulation of, 114
Genome, 4
  wide comparative analyses, 68
  wide scale, 349
Geranylgeranyl diphosphate (GGPP), 314
G-group basic leucine zipper (bZIP) TFs, 378
Gibberelic acid insensitive repressor of *ga1-3* scarecrow (GRAS), 313
  general principles, 303
    DNA interaction, 303
    homo/heterodimers, 306
    interaction of, 306
    protein movement, 307
    protein–protein interactions, 306
    regulation of gene expression by mirna, 307
  protein interactions, 304
Gibberella fujikuroi, 314
Gibberellic acid (GA), 298
  biosynthesis gene *SlGA20-OXIDASE1*, 255
  deactivation gene *GIBBERELLIN 2-OXIDASE 4*, 255
  deficiency defects, 315
  mediated growth, 314
  promotes flowering, 286
Gibberellic acid insensitive (GAI), 295
Gibberellin Insensitive Dwarf 1 (GID1), 315, 404
Gibberellin (GA)-regulated gene expression (GAMYB), 186
Gibberellins (GA), 220, 282, 313
  catabolism enzyme gene, 220
  overview of, 314
  signaling, 301
GIF proteins, N-terminal region of, 273
GLABRA1 (GL1), 321
Glial cells missing (GCM) domains
  structural comparison, 203
Glomus intraradices, 166
Glucocorticoid hormone receptors, 107
Glucose sensor HEXOKINASE1 (HXK1), 285
Glycogen synthase kinase-3 (GSK3)-like kinase, 401
Gossypium hirsutum (cotton), 351
Grain Size 6 (GS6), 301
GRAS family proteins, functional aspects, 295
  in axillary meristem development, 301
  in leaf development, 300
  role in development, 296
  in root development, 297
    elongation zone/lateral root formation, 299
    ground tissue, 297
    middle-cortex formation, 298
    root vascular tissue, 298
    stem cell niche maintenance, 298
  root modification, 299
  in shoot apical meristem development, 300
  in shoot development, 300

GRAS proteins, 154
    phylogenetic tree for, 296
    SCARECROW (SCR), 297
Green fluorescent protein (GFP), 27, 360
    chromophore-mutated, 28
Green Revolution, 316
Group-H bZIP proteins, 44
Growth-promoting phytohormone, 220
Growth-regulating factor (GRF) family
    GRF gene expression, analysis of, 272
    GRF–GIF complexes, 273
    GRF-interacting factors (GIFs)
        GIF1 gene, 254
    GRF–interacting factors (GIFs), 204, 273
    GRF-specific RNAi, 276
    miR396-mediated repression, 273
    in plant development, 274
    of transcription factors (TFs), 270
Growth-regulating factors, plant-specific family
    beyond organ growth
        developmental plasticity and reprogramming, 277
    development and growth of plant structures, 275
    embryo apical meristem, 275
    functional protein domains, 270
        C-terminal region, with transcriptional activation activity, 271
        DNA/protein–protein interaction domains, in N-terminal region, 270
    interacting factors, 273
    leaf development, 274
    leaf senescence, 277
    organ growth, 274
    plant development, 274
    repression by MicroRNAs, 273
    reproductive development, 276
    root development, 276
    seminal identification, 270
    shoot apical meristem, 275
    transcriptional regulation, 272
    transcription factors, 269
Gynecium, NAM/CUC3 roles, 236

H

Hahb-10 Cys mutants, 380
HAHR1 DNA binding, 380
Hairy meristem (ham), 300
Hairy meristem (HAM) proteins, 153
HB-containing genes, 114
    classification and physiological events of, 115
H2B monoubiquitination, 39
Hd3A:GFP fusion protein, 363
HD-containing proteins, 122
HD-containing transcription factors
    target sequences recognized by, 122–123
Heat shock proteins (HSPs), 49
Heat stress transcription factors (HSFs), 49, 207
HECT-type E3, 398
Helianthus annuus (sunflower), 123, 350
Heliconia stricta, 260
Helix turn-helix (HTH)
    gene, 93
    motif, 102

Herbaceous perennials, miR156 in leaf morphology, 285
Herbaceous plant Ipomoea caerulea, juvenile leaves, 284
Herpes simplex virus protein VP16, 17
HFR1 regulation
    by light, 336
    microproteins identification, 336
    transcriptional and posttranscriptional mechanisms, 336
Hidden Markov Model (HMM) databases, 165
Highthroughput in vitro selection methodologies, 21
Histone acetylases (HAT), 7
Histone chaperones, 36
    histone regulatory protein A (HIRA), 221
Histone deacetylases (HDACs), 39
Histone-modifying enzymes, 36
H3K4 trimethylated promoter-proximal nucleosomes, 36
HMG-box (HMGB) family, 187
Homeobox proteins, 101
Homeobox (HB) sequence, 344
Homeobox transcription factors, 215
Homeodomain (HD), 215
    multicellular lineages, 215
    structure of, 103
    three amino acid loop extension (TALE), 102
    transcription factors (TFs), 344
Homeodomain-leucine zipper (HD-ZIP), 333
    transcription factors, 113
    redox regulation, 380
Homeodomain-leucine zipper (HD-ZIP)-containing TFs, 136, 344
    class III protein, 381
    discovery history, 344
    evolutionary tree, circular representation of, 118
    HD-ZIP I genes, expression patterns, 344
        environmental factors, 346
    HD-ZIP I TF, 350
        expression
            A. thaliana (AtHB5 , AtHB6 , and AtHB16), 348
            schematic representation, 347
            VAHOX1 and H52, 351
            Venn diagram, 347
            illumination conditions, 344
            mature plant, schematic representation, 345
            from model plants, 346, 348
            in biotechnology, 352
            clade III members, 348
            clade II members, 348
            clade I members, 346
            clade V members, 348
            development and yield, 353
            from nonmodel species, 349
            tolerance to abiotic stress, 353
            VAHOX1 and H52, 351
            yield improvement, 353
            from model plants, clade I members
                LeHB-1 from tomato, 350
                PhHD-ZIP and RhHB1, 350

from model plants, clade IV members
    CpHB6 and CpHB7, 350
from model plants, clade VI members
    Vrs1, 351
from model plants, clade V members
    HaHB1 , VvHB13 , PgHZ1 , and GhHB1, 350
phylogenetic tree of, 345
Homeodomain-leucine zipper (HD-ZIP)-DNA complex, 105
Homeodomain-leucine zipper (HD-ZIP) family, 9, 117–122
    subfamily I, 118–120
    subfamily II, 120–121
    subfamily III, 121
    subfamily IV, 121–122
Homeodomain-leucine zipper (HD-ZIP) I genes, expression patterns, 344
    environmental factors, 346
Homeodomain-leucine zipper (HD-ZIP) proteins
    classification of, 117
    HD-Zip I proteins, 119, 351
    putative action mechanisms, schematic representation of, 119
    target sequences of, 123–124
Homeodomain proteins, 101, 104, 107
Homeodomain transcription factors
    BELL (named for the distinctive Bell domain), 117
    conserved zinc finger-like motifs associated with homeodomain (ZF-HD), 115
    DDT (named for the presence of DDT domain downstream of the homeodomain), 117
    different domains present in, 115–117
    knotted-related homeobox (KNOX), 115
    luminidependens homeobox genes (LD), 117
    nodulin homeobox genes (NDX), 117
    PINTOX, 117
    plant homeodomain (PHD) associated to a finger domain, 115
    SAWADEE (Homeodomain Associated to Sawadee domain), 117
    Wuschel-related homeobox (WOX), 115
Homology to E6-Associated Carboxy-Terminus (HECT), 397
Hordeum vulgare ssp. spontaneum, 301, 351
Hormone-binding domain, 17
Hormone-signaling pathways, 404
Hormones, KNOX downstream pathways, 220
    auxin, 220
    brassinosteroid, 220
    cytokinin, 220
    gibberellin, 220
    lignin biosynthesis, 220
Host genome, 65
HSF genes, 209
Hsp genes, 49
HTH genes, 94
HY5 gene, 337
HY5 HOMOLOG (HYH), 331
hy5 mutant seedlings, 337

Hypersensitive response (HR)
 cell death, 208
Hypocotyl 5 (HY5), 401
 genetic/molecular analyses
  antagonistic role of, 338
  bZIP member, in integrating light
   responses, 337
  integration of light cues, 338
  regulation by light, 338
 protein, 338

I

Ideal plant architecture 1 (IPA1), 258
Immune-related endonuclease (IREN), 208
Incertae sedis, 166
Indeterminate domains (IDDs), 321, 322
Indole-3-acetic acid inducible3 (IAA3), 255
Inducer of CBF Expression 1 (ICE1), 402
Inducible systems, 17–18
Intrinsically disordered proteins (IDPs), 159
Intrinsically disordered regions (IDR), 140
In vitro protein–DNA interaction assays, 21
In vitro protein–DNA interactions methods,
   18–21
 EMSA assay, 18
 microarray-based identification
  of transcription factor target genes, 21
 Selex, 18–20
 transcription factor–DNA complexes
  analysis by footprinting assays, 20–21
In vivo split methods, 28

J

Jackdaw (JKD), 297
 SCR expression, 297
Jagged and wavy-D (jaw-D), 252
Jagged lateral organ (JLO), 243
Jasmonate Zim-domain (JAZ) proteins, 51–52
Jasmonic acid (JA), 91
 biosynthetic enzyme, 262
 GA signaling crosstalk, 321
 signaling basic helix–loop–helix TF, 208
JA ZIM domain (JAZ) proteins, 51, 321
 repressor, 405

K

Klebsormidium flaccidum, 163
KN1 ChIP-seq studies, 220
kn1 loss-of-function, 220
KN1, maize homeodomain protein, 360
knotted1 (kn1) gene, 216, 344
Knotted interacting protein (KIP), 122
Knotted1-like homeobox *(KNOX)* gene, 216,
   218–220
 class I, 218, 240, 275
 classification, 216
 DNA-binding preferences, 221
 domain, 105
 downstream pathways, 219
  hormones, 220
   auxin, 220
   brassinosteroid, 220
   cytokinin, 220
   gibberellin, 220
   lignin biosynthesis, 220
  TF families, 219
 genetic analyses, 219
 KNOX – BELL pathway, 222
 KNOX-binding regions, 221
 KNOX1 genes, 255
 overexpressing tobacco plants, 220
 overexpressing transgenic plants, 220
 plant lineages, 216
 in plant shoot meristems, 218
 protein–protein interaction, 217
 regulatory network, 219
 SAM establishment, 218
 upstream regulators, 221
Knotted1 (KN1) protein, 218, 359
 expression pattern of, 218
Krüppel-associated box (KRAB) domain, 73

L

lac operator, 16
Lac repressor (LacI), 16, 76
LA gain-of-function point mutations *(La-2)*,
   253
Landsberg *erecta* (L*er*), 390
Late meristem identity 1 (LMI1), 349
Lateral organ boundaries (LOB) gene, 242
Lateral organ boundary domain (LBD)
   transcription factor, 243
Lateral organ boundary (LOB) TF, 221
Lateral suppressor (LAS) gene, 243, 301
LeafAnalyser quantitative imaging, 252
Leaf development, 242
 auxin, 255
 GRAS family proteins, functional
  aspects, 300
 TCP genes, 250
 teosinte branched1/cycloidea/proliferating
  cell factor (TCP) family, 250
  class I, 252
  compound leaf development, 253
  CUC/KNOX1 *gene pathway*, 253
  CYC/TB1 genes, 252
  gibberellin signaling, 255
  growth-regulating factors, 254
  jasmonate signaling, 256
  plants, 250
  posttranslational regulation, 256
  regulated networks, 253
  signaling, 255
   CK, 255
  simple leaves, 250
  CIN genes, 250
Leaf initiation rate (Plastochron), 287
Leaf morphology, of *A. thaliana* and
   C. flexuosa, 284
LeHB-1, from tomato, 350
Leucine heptad repeats (LHR), 315
Leucine–zipper (LZ) domain, 344
Leucine zipper transcription factors
 redox regulation, 378
Light, characteristics, 330
Light-grown hypocotyls, 337
Light-hypersensitive seedling phenoty
   pes, 334
Light-signaling, 302, 331
 components
  in nuclei of dark, 335
 hy1, *hy2*, and *hy6*, 331

O-Linked *N*-acetylglucosamine
   (O-GlcNAc), 149
Lipoxygenase2 (LOX2), 256
Long after far-red light 1 (laf1), 332
Long hypocotyl in far-red light 1 (hfr1)
   mutant, 332
Long hypocotyl 5 (hy5) mutant plants, 331
Lotus japonicus, 108, 117, 259
 orthologs, 299
LSH4 expression, 243
Lysine residue, 399
Lys33-linked chains, 399
Lys residues, 399

M

Macronucleus (MAC), 89
MADS box genes, 130, 286
 AGAMOUS (AG), 225
 evolution of
  functional aspects of, 134–136
  MIKC-type, exon–intron structure of, 131
  phylogenetic tree of, 133
  sequence and structure of, 129
 type II proteins in plants, 134–136
  MIKC^C-group proteins, 135–136
   phylogeny of, 136
  MIKC*-group proteins, 134
 type I proteins in plants, 134
MADS–box protein20 (MBP20) gene, 254
MADS domains
 diversification of, 127
 feature of, 128
 importance of, 127
 proteins, 130, 132
 remarkable feature of, 129
MAGPIE (MGP), 297
Maize
 Corngrass1 (Cg1) mutant, 285
 homeodomain protein, KN1, 360
 Kn1 dominant mutants, 218
 knotted1 (kn1) gene, 216
 liguleless1 (lg1) mutant, 287
 SWITCH complex protein 3C1, 122
 Z. mays ssp. *mays*, 290
Mammalian Forkhead box O (FOXO), 400
Mammalian glial cells missing (GCM) TF, 203
Marsilea drummondii, 285
Mass spectrometry (MS), 27
Mature plant, schematic representation, 345
Mediator (Med) complex, 36
Medicago truncatula, 235, 236, 276, 348
MEINOX protein, 108
MEKHLA domain, 105, 121
Membrane-bound transcription factors
 bZIP transcription factors, 386
  Arabidopsis bZIP28, 386
  bZIP17, 387
  bZIP60, 387
  domain structure, schematic
   representation, 386
  future perspectives, 391
 NAC transcription factors, 389
  ANAC013, 390
  ANAC017, 390
  ANAC060, 390
  ANAC089, 390

INDEX

NAC with transmembrane Motif1 (NTM1), 389
NTL6, 389
overview of, 385
Meristem formation, in *Arabidopsis* and rice, 297
Messenger RNA (mRNA) transcription, 35
Methanocaldococcus jannaschii, 386
Microarray-based identification, of transcription factor target genes, 21
Microarrays. See DNA microarrays
Micronucleus (MIC), 89
Microproteins (miPs), 336
MicroRNAs (miRNAs), 16, 273, 282
  mediated regulation, 199
  MicroRNA164 (miR164), 230
Middle cortex (MC), *scarecrow-like (scl)3* mutants, 298
miR156
  regulates vegetative phase transition flowering in *A. thaliana*, 284
  represses *SPL9* expression, 282
  SPL module, plant developmental transitions
    age-related resistance (ARR), 287
    anthocyanin biosynthesis, temporal regulation of, 289
    developmental transitions, in moss, 286
    in floral organ development, 289
    in juvenileto-adult phase transition, 283
    known functions, 288
    in leaf initiation rate (plastochron), 287
    in ligule differentiation, 287
    in plant architecture, 289
    regulates flowering time, 285
    in stress response, 287
    in timing embryonic development, 283
    trichome development on stem/floral organs, temporal regulation of, 289
mir164abc mutants
  CUC mRNA accumulation, 243
miR164 finetunes NAM gene expression, 238
  Arabidopsis, 238
  evolution and specialization, 239
  shoot development, 238
  transcriptional control, 240
miR396-GRF-GIF network, 272, 274
miR171 isoform, 307
miR396, overexpression, 275
miR171-targeted LOST MERISTEMS 1 (LOM1), 289
miR156-targeted *SPL* genes, 288
Mobile *StBEL5* RNA
  on root and tuber development, 368
Mobile transcription factors
  KNOTTED1, intercellular movement, 360
  root transcription factors, 360
  of shoot apex, in protein form, 360
  WUSCHEL, in shoot apical meristem, 360
Monocots
  TB1-like genes, SL-dependent regulation, 258
Monoculm1 (MOC1), 301
Monopteros (MP), 238
Morphogenesis, 242
Most recent common ancestor (MRCA), 132
Mouse double-minute 2 (Mdm2), 400

MS-based protein identification, 28
Multiple *WUS* regulatory loops
  for shoot stem cell maintenance, 223
  WUSCHEL–CLAVATA negative feedback loop, 223
  WUSCHEL–cytokinin positive feedback loop, 224
  WUSCHEL–HECATE negative feedback loop, 224
Multisubunit protein complexes, 39
MYB R2R3 TF activity, overview, 380
MYC2 activity, 404
  transcriptional activation, 405
  ubiquitin-proteasome system, 404
Mycorrhiza, GRAS family proteins, 299
Myeloblastosis (MYB), 289
  transcription factors, 289
    family, 48–49, 208, 379
    redox regulation, 379

N

NADPH-dependent glutathione reductase (GR), 374
nam (no apical meristem), 230
NAM/ATAF1/CUC2 (NAC) proteins, 50, 200, 202
  DNA-binding mechanisms, 231
  domains
    alignment, 231
    dimeric structures of, 64
    structural comparison, 203
  evolution of, 202
    fold, 202
    unraveling, 202
  genes, 199
    diversification and amplification, 199
  heterodimerization of, 206
  NAC–DNA complex, 201
  NAC–DNA interactions, 204
  networks, in abiotic stress responses, 207
    ANAC019, ANAC055, and ANAC072 (RD26), 207
    ANAC019/055/072 network, transcription factors, 209
    stress-associated rice, 208
  phylogeny and evolution of, 199
  structure, 200
    complex with DNA, 200, 201
    C-terminal domain, 200
    dimer, 202
    direct interactions, 205
    DNA recognition, 204
    DNA sequence and target sequences, 204
    NAC domain fold, 200
  structure of
    direct interactions, 205
    DNA recognition, 204
    DNA sequence and target sequences, 204
    features of, 200
  transcription factor (TF) family, 199
  with transmembrane motif 1 (NTM1)-like (NTL), 200, 389
NAM/ATAF1/CUC2 (NAC) transcription factors, 50, 389
  ANAC013, 390
  ANAC017, 390

ANAC060, 390
ANAC089, 390
membrane-bound transcription factors, 389
  ANAC013, 390
  ANAC017, 390
  ANAC060, 390
  ANAC089, 390
  NAC with Transmembrane Motif1 (NTM1), 389
  NTL6, 389
  NAC with Transmembrane Motif1 (NTM1), 389
  NTL6, 389
NAM/CUC3 genes, 230
  control plant development
    cellular behavior, modifications of, 242
    CUC-dependent cellular effects, 242
    CUC direct targets, 243
    CUC impact cell proliferation, 242
    KNOXI/LFY-like genes, 243
    LATERAL SUPPRESSOR (LAS) gene, 243
  define boundaries
    Arabidopsis cuc1–cuc2 seedlings, 234
    CUC genes role, in meristematic territories, 235
    axillary meristems, 235
    floral organ boundaries, 235
    gynecium, 236
    leaf development, 236
    organ abscission, 236
  dicots and monocots, 235
  identification of, 234
  in meristems, 234
  expression, transcriptional regulation
    during axillary meristem formation, 241
    chromatin modifications, 241
    during embryogenesis, 240
    GOB expression during abscission, 241
    during leaf development, 241
  multiple regulatory pathways, 236
    hormonal regulation, 237
      brassinosteroids (BRs), 238
      PIN1-generated auxin maxima, 237
NAM/CUC3 proteins, 230, 231
  alignment of, 233
  evolution/structure, 230
    NAC transcription factors, plant-specific family, 230
    Arabidopsis CUC proteins, 230
    origin/early evolution of, 230
    recent evolution, 230
  organization/domains, 231
    alignment, 233
    amino-terminal NAC domain, 231
    carboxy-terminal domain, 232
Natural/induced mutations, 16
NBS–LRR–WRKY Genes, 173
NDX homeodomain proteins, 108
Negative elongation factor (NELF), 39
Next-generation sequencing (ChIP-Seq), 272
Nicotiana attenuata, NaHD20, 348
Nicotiana benthamiana, defoliation studies in, 284
Nicotiana tabacum, 250
Nitella mirabilis, 230

NMR titration experiments, 65
Nodulation, GRAS family proteins, 299
Nodulation signaling pathway2 (NSP2), 299
Noncell-autonomous mobile signals, 359
   plasmodesmata (PD) function, 359
Noncell-autonomous TFs (NCATFs), 360
   in roots, 360
Non-expresser of pathogenesis related gene 1 (NPR1), 377
   monomer, nuclear localization, 378
   during plant immunity, 377
   redox regulation of, 377
Nonexpressor of PR1 (NPR1), 44
Nonspecific protein binding, 20
NO production, in plants, 374
Nozzle (*nzz*) mutant anthers, 289
N-*p*-tosyl-l-phenylalanyl chloromethyl (TPCK), 391
Nuclear encoded genes, 35
Nuclear export signal (NES), 49, 401
Nuclear localization signal (NLS), 49, 128, 184, 315, 333
Nuclear magnetic resonance (NMR) solution, 164
Nucleosome-modifying enzymes
   role of, 36–37
Nucleotide bases, 4
Nutcracker (NUC), 297

O

Oligomerization domain (OD), 49
Oligonucleotides, 18
One-hybrid screening of yeast (Y1H), 24
Open reading frame (ORF), 38
Oryza sativa growth-regulating factor1 (OsGRF1), 270
Oryza sativa homeobox1 (OSH1), 218
   induction, 220
OsIPT2 expression, 220
Oslate embryogenesis abundant3 (OsLEA3), 208
Osmotic stress, leaf cells in *Arabidopsis*, 303
OsNAC6/SNAC2-overexpressing transgenic plants, 208
OsSHR1 expression, 297
Ostreococcus tauri, 376
Ovarian tumor-related (OUT), 397
Oxidative stress-induced deubiquitination, 400
2-Oxoglutarate-dependent dioxygenases of GA20 (GA20ox), 314

P

Paclobutrazol (PAC), 315
Paclobutrazol resistance 1 (PRE1), 322, 337
Paralogous genes, 346
PAR1 regulation
   by light, 336
   microproteins
     identification, 336
   transcriptional and posttranscriptional mechanisms, 336
Pathogen-associated molecular patterns (PAMPS), 375
Pathogenesis related (PR) genes, 377
   HB gene A (PRHA), 115
Petroselinum (PTS), 217

Petunia, 300
Petunia hybrida (petunia), 350
Petunia *nam*, 234
Pfam (protein family), 199
PFYRE motif, 159
Phantom, DBD motif of, 63
Phaseolus vulgaris GRAS protein, 299
PHD finger proteins, 107
Phloem companion cells, Hd3a-GFP fluorescence, 363
Phloem sap, from melon (*Cucumis melo*) yield, 361
Phloem tissue, of *Arabidopsis*, 364
Phosphatidylethanolamine-binding protein (PEBP), 362
Phosphorylated Ser-2 (Ser-2P), 38
Phosphorylation, 199
Photoconversion, Pr and Pfr, 330
Photomorphogenesis, 51, 330
Photoperiod, in plants, 362
Photoperiodism, 329
Photoreceptors, 330
   light signals, 331
Photostables (phyB–phyE), 330
Photosynthesis, in chloroplasts, 374
Photosynthetic bacteria, 76
Photosynthetic pigments, 335
Photosystem I (PSI), 374
Phototropism, light, 329
PhSHOOTMERISTEMLESS expression, 300
Phylogenetic analyses, 106
Phylogeny analysis, 50
Physcomitrella patens, 74, 104, 117, 132, 169, 217, 230, 316
Phytochrome A signal transduction1 (PAT1), 299
Phytochrome B (phyB), 302
   missense allele, suppressor of, 193
   photoreceptor, 258
Phytochrome interacting factors (PIFs), 331
   analyses of deficient plants, 334
   antagonistic role of, 338
   deficient plants
     analyses of, 334
   light perception/gene expression/plant development, 333
   microproteins (miPs), 336
   photosynthetic gene expression, 320
   quadruple-mutant, 334
   regulation of activity by light, 334
   transcriptional controls *vs.* posttranscriptional controls, 334
Phytochrome rapidly regulated (PAR) genes, 333
Phytochromes, 302, 330
Phytochromes (Pfr), 331
Phytoene synthase (PSY), 338
Phytohormones, 48
Phytophthora infestans infection, 375
Picea glauca, 351
PICKLE (PKL), 322
pin1 Arabidopsis mutants, 238

Pin-formed (PIN) genes, 255
   PIN1 gene
     dependent auxin maxima, 237
     encodes, 237
     expression, 223
     PIN1-generated auxin maxima, 237
     PIN1-mediated polar auxin transport, 237
Pisum sativum, 241
Plant developmental transitions, 281
   A. thaliana, SPL gene family, 283
   future perspectives, 290
   MicroRNAs (miRNAs), 282
   miR156-SPL module
     age-related resistance (ARR), 287
     anthocyanin biosynthesis, temporal regulation of, 289
     developmental transitions, in moss, 286
     in floral organ development, 289
     in juvenileto-adult phase transition, 283
     known functions, 288
     in leaf initiation rate (plastochron), 287
     in ligule differentiation, 287
     in plant architecture, 289
     regulates flowering time, 285
     in stress response, 287
     in timing embryonic development, 283
     trichome development on stem/floral organs, temporal regulation of, 289
   photoreceptors, roles, 330
   role of light, 329
Plant genomes, 36
Plant GRAS family proteins
   domains, schematic representation of, 154
   evolution of, 153
   genomic organization (intron/exon), 156–139
   motifs identified in domains, 158
   presence in plants and other organisms, 153–139
   structure of, 153, 157–160
     additional domains, 160
     GRAS domain, 157–159
     N-terminal domain, 157–159
   unrooted phylogenetic tree with bacterial proteins, 155
Plant Gro/Tup1 family, corepressors of, 43–44
Plant growth, 242
Plant homeobox genes, 114
   evolution of, 108–110
   homeodomain, structure of, 102
   introduction, 101
   in plant genomes, distribution of, 104
   plant homeodomain families, 104–108
   specific contacts with DNA, 102–104
Plant homeodomain families, 104–108, 114–115
   and codomains, evolution of, 109
   DDT class, 107
   HD-ZIP superclass, 105
   LD class, 108
   NDX class, 108
   PHD finger homeodomain family, 107–108
   PINTOX class, 106
   PLINC Zn finger class, 107
   SAWADEE class, 108

TALE superclass, 105–106
WUSCHEL-related homeobox (WOX)
 class, 106
Plant immunity
 NPR1, redox regulation, 377
Plant interactor homeobox rice gene, 117
Plant MADS domain transcription factors
 evolution of, 132–136
  MADS box gene evolution, functional
   aspects of, 134–136
  MADS World in land plants, great
   expansion of, 132–134
  origin and early evolution of, 132
 introduction, 127–128
 proteins, structure of, 128–132
  definition of, 128–130
  features and diversity, 130–132
Plant Meds, 43
Plant respiratory burst oxidase homolog
 (RBOH) protein
 schematic representation, 375
Plant-specific transcription factors, 44, 373
 DNA-binding domains of, 57–59
 LEAFY (LFY), 285
Plant TIR-NB-LRR (toll-like/interleukin-1
 receptornucleotide-binding-leucine-
 rich repeat) immune receptors, 287
Plant transcription factors, 9, 44
 evolution and diversification of, 88
 evolution, hypothetical scheme for, 93
 gene evolution and biological function,
  94–96
 general aspects of, 35–36
 genes between plants and animals,
  comparative analysis of, 76
 genes, distinctive features of, 73–76
 genes in animals and plants
  different evolutionary methods of, 93–94
  evolution of, 74
 genes in 32 diverse organisms, comparative
  analysis of, 76–92
  evolution of prokaryotes to unicellular
   eukaryotes, 76–89
  multicellular organisms and the
   evolution of invertebrates,
   emergence of, 89–90
  seed plants, evolution of, 91–92
  slime mold, yeast, and fungi TF genes,
   evolutionary characteristics of, 89
  terrestrial plants, evolution of, 91
  vertebrates, evolution of, 90–91
 major families conserved across
  eukaryotes, 44–49
  basic region/leucine zipper (bZIP)
   family, 44–48
  bHLH TF family, 48
  heat stress transcription factors (HSFs)
   family, 49
  MYB TF family, 48–49
 new gene members during evolution,
  appearance of, 92–93
 plant-specific families, 49–50
  AP2/ERF family, 49
  NAC TF family, 50
  TCP family, 50
  WRKY TF family, 49–50

relationship map of, 95
RNAPII preinitiation complex, components
 involved in formation, 39–44
sequence specific transcription factor in
 organisms, 78
TFs without DBD but interacting with
 DBD-containing TFs, 50–52
 AUX/IAA family, 51
 B-box zinc finger transcription factors,
  50–51
 Jasmonate Zim-domain (JAZ) proteins,
  51–52
transcription cycle in eukaryotes, overview
 of, 36–39
Plasmodesmata (PD) function, 359
Polyacrylamide gel electrophoresis, 20
Polycomb repressive complex 2 (PRC2),
 242
Polyethylene glycol (PEG), 24
Polymerase chain reaction (PCR), 18
Polypyrimidine tract binding (PTB)
 protein, 366
 animal, 363
Polyubiquitin chains, 399
pOp–LhG4 system, 16
Positive transcription elongation factor b
 (P-TEFb), 38
Posttranslational modification (PTM)
 signal-induced changes, 401
Potato *(Solanum tuberosum)*, 289
Potato heterograft, 362
Potato, mobile RNA
 StBEL5 model, 367
Potato *Solanum tuberosum* CONSTANS
 (StCO), 362
PREFOLDIN3 (PFD3), 323
Preinitiation complex (PIC), 35, 39
Progenitor teosinte (Z. *mays* ssp.
 parviglumis), 290
Prokaryotic helix–turn–helix DNA-binding
 domains, 215
Prolamin-box (P-box) binding factor
 (PBF), 186
Proliferation, 235
Promoter-proximal nucleosomes, 36
Proteasome-dependent degradation, 399
Protein-binding microarrays (PBMs), 21
 advantages of, 21
Protein-coding genes, 3
 transcription of, 39
Protein–DNA interactions, 13, 22
 in vitro, methods to study, 19
 in vivo, methods to study, 21–25
  ChIP-chip and ChIP-Seq, 22
  chromatin immunoprecipitation (ChIP)
   assays, 21–22
  DNA adenine methyltransferase
   identification (DamID), 22–24
  transient assays to analyze protein–DNA
   interactions *in vivo*, 24–25
  yeast one-hybrid assay, identification of
   transcription factors, 24
Protein fragment complementation assays
 (PCA), 28
Protein fragments, 25
Protein interaction domains, 7

Protein–protein complex, identification
 methods, 25
Protein–protein detection method, 28
Protein–protein interactions, 6–8, 25, 29, 131,
 144, 148
 analysis methods, 27
 coimmunoprecipitation, 28–29
 methods for verification of, 28
 protein–protein complex, identification
  methods, 25
 tandem and one-step tag-based affinity
  purification, 27–28
 yeast two-hybrid assay (Y2H), 25–27
Protein type/activity, acronym, 332
Proteolysis-dependent GA-signaling
 model, 317
Protochlorophyllide oxidoreductase
 (POR), 300
Protozoan animals, TF genes, 88
Pumpkin, *GAI*, encoding protein, 363
Putative signal sequences, 201

Q

Quantitative trait loci (QTLs), 239, 282
Quartet formation, principle, 136
Quiescent center (QC)
 root apex, 297

R

Rabbit ears (RBE), 240
Radical induced cell death 1 (RCD1), 206
Ralstonia solanacearum, 50
Rap2.4a activity, cellular redox change, 382
Rat glucocorticoid-receptor-inducible cell
 lines, 243
RAX1–3 genes, 241
RCF2/CPL1, overexpression, 207
Reactive nitrogen species (RNS), 373
Reactive oxygen species (ROS), 373
 production in plants, 374
Receptor-like protein kinase 2 (RPK2), 223
Recognition helix, 114
Recognition motif (RM)
 Gly^{99} in ANAC019, 201
 NAC proteins, 205
 95-WKATGTDK, 200
Recombinant inbred line (RIL)
 population, 390
Red fluorescent protein (RFP), 365
Redox-regulated plant transcription factors,
 373, 374
Redox regulation, 373
 of basic leucine zipper transcription
  factors, 378
 of Class I TCP transcription factors, 381
 concept of, 374
 in fine-tuning, 373
 homeodomain-leucine zipper transcription
  factors, 380
 of MYB transcription factors, 379
 NO production, in plants, 374
 NPR1 during plant immunity, 377
 RAP2.4A, 381, 382
 redox signaling, mechanisms, 376
 ROS production, in plants, 374
Reduced Height1 and *2 (RHT1/RHT2)*, 316

Reduce membrane fluidity, 389
Regulated intramembrane proteolysis (RIP), 385
Cis-Regulatory elements, 20
RepA-binding activity, 205
Reporter genes, 24, 25
   E. coli lacZ, 25
Reporter vector, 24
Repressor motif, 17
Repressor of ga1-3 (RGA), 404
Resonance energy transfer methods, 28–29
Retinoblastoma-related protein (RBR), 298
Rhizobium etli, 299
Rhomboid protease inhibitor, 391
Rice (*Oryza sativa*), 284
   bakanae, 314
   BELL gene *qSH1*, 219
   DELLA-like *SLR1* (slender rice1), 316
   high-yielding semidwarf crop, 316
   jumonji domain 2 (JMJD2) family *jmjC* gene 706, 276
   kn1 ortholog, 218
   KNOXI genes, 275
   mutant *rdh1*, 276
   OsHox4, 348
   SNAC genes
      emerging regulatory networks, 209
      stress-responsive NAC (SNAC), 207
Rice PBF homolog (RPBF), 186
Ricinus communis, 132
RING-H2 E3 ubiquitin ligase protein, 323
RING-type E3, 398
RIP. bZIP28 resides
   bZIP28 , activation mechanism, 387
RNA
   mobility assays, 368
   silencing signals, 42
RNA-induced silencing complex (RISC) complex, 273
RNAPII enzyme, 42
RNAPII preinitiation complex
   components involved in formation, 39–44
   GTFs and RNAPII, 39–42
   mediator, 42–43
   plant Gro/Tup1 family, corepressors of, 43–44
RNAPII transcription cycle, 38
RNA polymerases, 52
   in eukaryotes, 3
   polymerase II, 3
   RNA polymerase II (RNAPII), 35
Root apical meristem (RAM), 276
Root development, 297
   elongation zone/lateral root formation, 299
   ground tissue, 297
   middle-cortex formation, 298
   root vascular tissue, 298
   stem cell niche maintenance, 298
Rosa hybrida (rose), 350
Rough sheath1 (*rs1*), 219
R protein–WRKY (RW) proteins, 179
R2R3-MYB proteins, 48, 380
R2R3 MYB TFs, DNA-binding activity, 380
RVER motif, 159

S

Saccharomyces cerevisiae, 128, 184
Salicylic acid (SA)
   immune signal, 208
   ISOCHORISMATE SYNTHASE1 (ICS1), 208
Salt-stress-induced bZIP17 activation, 387
SAWTOOTH1 (SAW1), 217
35SCaMV promoter, 14, 17
SCARECROW (SCR), 314
Scarecrow (SCR), 295
Scarecrow-like3 (SCL3), 298, 322
35S Cauliflower Mosaic Virus, 352
SCR gene (*PsySCR*) expression, 297
Seed plants
   evolution of, 91–92
      dicotyledonous plants, 92
      monocotyledonous plants, 92
      woody plants, 91
Selaginella kraussiana, 316
Selaginella moellendorffii, 104, 132, 153, 230, 270, 316
Selex, 18–20
SELEX experiments, 145
Senecio vulgaris, 260
Sequence-specific DNA, 185
Sequence-specific interactions, 18
Serine/threonine protein kinase, 318
Shade avoidance responses, 330
Shade avoidance syndrome (SAS), 258, 330
Shade-induced *PAR1* transcription in seedlings, 336
Shoot apical meristem (SAM), 230, 269
   development
      GRAS family proteins, functional aspects, 300
   formation, 219
   of vascular plants, 216
Shoot branching, TCP genes, 256
   TB1/BRC1 genes
      expression, 257
      genetic networks, 258
      up-and downstream genes, 258
      hormonal regulation of, 258
      prevent lateral branch outgrowth, 256
      sucrose regulation of, 258
Shoot development
   GRAS family proteins, functional aspects, 300
SHOOT MERISTEMLESS (STM), 234, 360
SHORT-ROOT (SHR), 297, 359
SHORTROOT INTERACTING EMBRYONIC LETHAL (SIEL), 297, 360
Sieve element system
   flowering locus T, 361
   long-distance tuberization signal, 362
   StFT/StSP6A, 362
   transcription factors and coregulators, 361
Signaling
   axillary meristem development, 301
   gibberellin signaling, 301
   light signaling, 302
   stress signaling, 302
Signal transducers and activators of transcription (STAT), 303
Single point mutation (m9) abolishes repression activity, 367
SlmiR164-resistant *GOB* variant, 238
SL-synthesis gene *CAROTENOID CLEAVAGE DIOXYGENASE8*, 258
sly1 F-box mutant, 318
Small interfering RNA (siRNA), 42
S-nitrosylation, 376
Soil-dwelling social amoeba, 89
Solanum pennellii BRC1a (*SpBRC1a*) gene, 256
Solar UV-B radiation, 277
SP6A protein, 362
Sphagnum palustre, 316
SPINDLY (SPY)
   O-GlcNAc transferase, 319
Spirodela polyrhiza, 134
Spironucleus salmonicida, 166
SPL9 (SQUAMOSA PROMOTER BINDINGLIKE 9), 320
SPL gene family, in *A. thaliana*, 283
SPL9-group genes, 286
Split reporter, 28
26S proteasome, ubiquitinated protein, 396
SQUAMOSA-promoter binding protein (SBP), 57
SQUAMOSA promoter-binding proteins (SBPs), 64
Squamosa promotor binding protein-like transcription factors (SPLs), 282, 301
START adjacent domain (SAD), 121
StBEL5 promoter, 368
StBEL5 transcripts, 359
Stem cell regulation
   WUSCHEL function, 222
Steroid hormone receptors, 17
   type I nuclear receptor, 17
Stomatal carpenter 1 (SCAP1), 192
Stress-responsive NAC (SNAC)
   in rice, 207
Stress-responsive NAC1 (SNAC1), 200
Stress signaling, 302
Strigolactones (SLs), 258
Stromal ascorbate peroxidase (sAPX) gene expression, 390
   redox-dependent suppressor, 390
StSP6A transcripts, in stolons, 362
Sucrose nonfermenting (SNF)-type chromatin-remodeling complex, 306
Sulfenic acid residues, 376
Sunflower *HaHB4* HD-Zip gene, 353
Suppressor of gamma response 1 (SOG1), 201
swi3c mutant, 322
Switch/sucrose nonfermentable (SWI/SNF) complexes, 241
Systematic evolution of ligands by exponential enrichment (SELEX), 18, 144

T

TAFs, include initiator (INR), 36
Tag-based affinity purification
   tandem and one-step, 27–28
Tagged fusion protein, 28
Tandem affinity purification (TAP)
   method, 25
   tag, 27
Target genes, 4, 13, 35, 39
   promoter, 36

TATA box, 3
TATA-box binding protein (TBP), 3, 36
Tb1 alleles, 258
TB1/BRC1 genes
  downstream genes, 259
  expression, 257, 258
  shoot branching
    expression, 257
    genetic networks, 258
      up-and downstream genes, 258
    hormonal regulation of, 258
    prevent lateral branch outgrowth, 256
    sucrose regulation of, 258
TCP1 gain-of-function allele *(tcp1-1D)*, 261
TCP5 proteins, 255
tcp14 – tcp15 double-mutants, 255
tcp2–tcp3–tcp4–tcp5–tcp10–tcp13–tcp17–tcp24
    mutant, 252
Teosinte branched1/Cincinnata/proliferating
    cell factor (TCP)
  domain, posttranslational modifications
    of, 145
  family. *See* Teosinte branched1/cycloidea/
    proliferating cell factor (TCP) family
  leaf phenotype of plants, 250
  proteins, 144, 381
    biochemical function of, 139
    DNA-binding activity, 381
    posttranslational modifications of, 148
  transcription factors
    activation and repression domains,
      147–148
    24 Arabidopsis TCP proteins, alignment
      of, 143
    CYC/TB1 genes from dicotyledonous
      species, phylogenetic tree of, 141
    domain, structure and function, 142–147
      proteins bind DNA as dimers, 145–147
      proteins form homo-and heterodimers,
        143–144
      proteins recognize specific DNA
        sequences, 144–145
      TCP domain basic region binds
        DNA, 144
    factors as intrinsically disordered
      proteins, 148
    maximum likelihood (ML) phylogenetic
      tree of Arabidopsis factor, 140
    overview of, 139
    posttranslational modifications of,
      148–149
    proteins, evolution of, 139–142
    sequences recognized by, 146
Teosinte branched1/cycloidea/proliferating
    cell factor (TCP) family, 50, 250
  leaf development, control of, 250
    auxin signaling, 255
    CK signaling, 255
    class I, 252
    compound leaf development, 253
    CUC/KNOX1 *gene pathway*, 253
    CYC/TB1 genes, 252
    gibberellin signaling, 255
    growth-regulating factors, 254
    jasmonate signaling, 256
    plants, 250

posttranslational regulation, 256
regulated networks, 253
simple leaves, 250
  CIN genes, 250
shoot branching. *See* Shoot branching, TCP
  genes
Teosinte glume architecture 1 (tga1), 285
Tetracycline (Tet) repressor, 17
TGA1 transcription factors, 377
Thioredoxins (TRX), 377
Three amino acid loop extension (TALE)
  class protein, 217
  HD genes functions, in plant lineages, 217
  homeobox genes, 216, 217
  superfamily homeobox genes, 216
TIR–NBS–LRR–WRKY protein, 176
Tobacco etch virus (TEV), 27
Tobacco heterograft, 362
Tobacco plants
  KNOX-overexpressing, 220
Tomato
  axillary meristem formation, 237
  GAI, encoding protein, 363
  LA activity, 255
  lateral suppressor *(Ls)* mutant, 301
  Lycopersicum esculentum, 344
  mutant phenotypes and expression
    patterns, 234
  SlBRC1b loss-of-function *(SlBRC1b–RNAi)*
    plants, 256
  Solanum lycopersicum, 253, 284, 290
    LA and *KNOX1* activity, 254
Tradescantia pallida, 260
Transactivation domains (TAD), 140
Transcriptional activation domain
  (TAD), 405
Transcriptional complexes, 7
  activation domain (AD), 25, 122
  characteristics of, 9
  initiation and elongation, 37–39
  termination and reinitiation, 39
Transcriptional regulators
  GRAS (gibberelic acid insensitive repressor
    of *ga1-3* scarecrow) family, 313
  schematic representation, 333
  ubiquitin-dependent proteolysis of, 401
Transcription cycle
  Arabidopsis TF databases, comparison of, 45
  head and tail modules, role of, 42
  Med components, 42
  overview in eukaryotes, 36–39
    core promoter elements and general
      transcription factors, 36, 37
    nucleosome-modifying enzymes and
      chromatin remodelers, role
      of, 36–37
    productive transcription initiation and
      elongation, 37–39
    transcription termination and
      reinitiation, 39
  RNAPII transcriptional PIC
    Arabidopsis core components, 40
  RNAPII transcription cycle, 38
  TFIID subunits model interact with CPEs
    to facilitate PIC formation, 41
Transcription elongation complex, 37

Transcription factor (TF), 35, 101, 128, 359
  action, regulation of, 7–9
  basic helix-loop-helix (bHLH) family, 331
  characteristics of, 14
  classification of, 7, 44
  coding region of, 14
  DNA
    binding domains, 5–6, 273
    binding properties of, 7
    recognition by, 4–5
  in eukaryotes, 3–4
  function, 1, 8
  genes, general patterns of, 90
  GROWTHREGULATING FACTOR
    family, 270
  methods to study, 13–14
  NAM/ATAF1/CUC2 (NAC)
    proteins, 199
  nuclear localization of, 21
  phloem-mobile mRNAs, 361
  plant-specific *GRF* family, 271
  plant transcription factors, 9
  proteins, 73
    DNA interactions *in vivo*, methods to
      study, 21–25
    protein interactions, 6–7
    analysis, 25–29
  role of, 36
  stability of, 7
  strategies to study function, 15
  structure of, 4, 5
  in vitro protein–DNA interactions methods,
    18–21
  in vivo functional studies, 14–18
    expression of dominant negative
      forms, 16
    expression of dysregulated forms, 16
    expression of fusions to activating/
      repressor domains, 17
    gain-of-function mutants, 16
    inducible systems, 17–18
    overexpression/ectopic expression
      of, 14
    transactivation systems for expression,
      14–16
Transcription factor II D (TFIID), 3
  role of, 39
Transcription start site (TSS), 36
Transcriptome analyses, 224
Transcriptome shotgun assembly (TSA)
  sequences, 230
Transgenic *Arabidopsis* plants, 379
Transgenic plants expressing fluorescent
  fusion proteins, 27
Transient assays
  to analyze protein–DNA interactions
    in vivo, 24–25
Transmembrane domain (TMD), 385
  motif, 50
Triticum aestivum, 353
Trp-Arg-Cys motif, 270
Tuberization signal, long-distance, 362
Tubular ray flower, 260
Turnip crinkle virus interacting protein
  (TIP), 204
Two-component systems, 14

U

Ubiquitin (U)
  nonproteolytic (1-2) and proteolytic (3-6) functions of, 400
  proteolytic function, 402
Ubiquitin carboxy-terminal hydrolases (UCH), 397
Ubiquitin-dependent turnover, of transcriptional repressors, 404
Ubiquitin gene translation, 396
Ubiquitin proteasome system, 396
  conjugation, 396
  enzymes, classes of, 397
  modifications and outcomes, 399
  MYC2 activity, 404
  26S proteasome, 399
  transcription factor function, regulation of, 400
    nonproteolytic ubiquitination, 400
    proteolytic ubiquitination, 401
    regulating repressor/inhibitor abundance, 403
    regulating transcription factor abundance, 401
    transcription-coupled proteolysis, 405
Ubiquitinspecific processing protease (UBP), 397
U-box-type E3, 398

V

Vacuum infiltration, 22
Vascular plants, shoot apical meristem (SAM), 216
VHIID motif, 157
Volvox, 88
VSLTLGL box, 105

W

Wheat
  high-yielding semidwarf crop, 316
W motifs (WNY), 232
WOX family genes, 223
  plant lineages, 216
  in plant lineages, 216
WOX4 promotes, 223
WOX protein structure, 222
wox2 stimpy-like (stpl/wox8) double-mutants, 241
WRC domain, 273
  DNA-binding activity of, 270
WRKY-containing protein, 68
WRKY domains, 49
  structural comparison, 203
WRKY motif (WRKYGQK), 202
WRKY proteins, 63
WRKY transcription factors, 49–50, 202
  domains, neighbor-joining phylogenetic tree derived MUSCLE alignment of, 166
  domain, structure of, 164
  evolution of, 178
  fungal and *Physcomitrella* group III proteins, 167
  GCM1 and FLYWCH domains, comparison of, 172
  genes, counterparts in flowering plants, 170
  genes, distribution of, 165
  genes, evolution of, 164–173
    distribution of, 164–165
    in mosses and spikemosse, 169–171
    in multicellular green algae, 163
    nonplant genes, 165–169
    relationship with FLYWCH, GCM1, and BED proteins, 171–173
    R protein-genes, 173–177
    in unicellular green algae, 169
  genes, from *K. flaccidum*, 171
  group I-like WRKY proteins from social amoebae and amoebozoa, 168
  introduction, 139
  NBS–LRR–WRKY genes, 174
  reevaluation of evolution, 177
  R protein-WRKY (RW) families
    distribution of, 173
    HMMER analyses of, 176
    phylogenetic analyses of, 175
WUSCHEL (WUS)
  expression domain, 223
  expression termination, floral stem cells, 225
  function
    in stem cell regulation, 222
  gene function, multiple feedback loop, 222
  homeobox transcription factor, 241
  homodimerizes, 222
  mutant phenotype, 223
  targets, 224
  transcript expression, 223
  wus-1 loss-of-function, 222
  wus mutants, 223
WUSCHEL homeobox *(WOX)* gene family, 106, 222

X

X-ray crystallography, 64
Xyloglucan endotransglusosylase, 323

Y

Yeast
  activator, 25
  genes
    HIS3, 25
    LEU2, 25
  one-hybrid assay
    identification of transcription factors, 24
  two-hybrid (Y2H) analysis, 24–27, 120, 143, 379
  screening, 27
Yellow fluorescent protein (YFP), 28

Z

Zea mays ssp. *parviglumis*, 256
Zinc finger homeodomain 1 (ZFHD1), 206
Zinc (Zn) finger motif, 49, 115, 185
  in DNA binding, 270
Zinc (Zn) finger proteins, 183
Zmhdz10, from maize, 352
Zygomorphy, 250

Printed in the United States
By Bookmasters